Lecture Notes in Computer Science 2976

Edited by G. Goos, J. Hartmanis, and J. van Leeuwen

Springer
Berlin
Heidelberg
New York
Hong Kong
London
Milan
Paris
Tokyo

Martin Farach-Colton (Ed.)

LATIN 2004:
Theoretical Informatics

6th Latin American Symposium
Buenos Aires, Argentina, April 5-8, 2004
Proceedings

 Springer

Series Editors

Gerhard Goos, Karlsruhe University, Germany
Juris Hartmanis, Cornell University, NY, USA
Jan van Leeuwen, Utrecht University, The Netherlands

Volume Editor

Martin Farach-Colton
Rutgers University
Department of Computer Science, Piscataway, NJ 08855, USA
E-mail: farach@cs.rutgers.edu

Cataloging-in-Publication Data applied for

A catalog record for this book is available from the Library of Congress.

Bibliographic information published by Die Deutsche Bibliothek
Die Deutsche Bibliothek lists this publication in the Deutsche Nationalbibliografie;
detailed bibliographic data is available in the Internet at <http://dnb.ddb.de>.

CR Subject Classification (1998): F.2, F.1, E.1, E.3, G.2, G.1, I.3.5, F.3, F.4

ISSN 0302-9743
ISBN 3-540-21258-2 Springer-Verlag Berlin Heidelberg New York

Springer-Verlag is a part of Springer Science+Business Media

springeronline.com

© Springer-Verlag Berlin Heidelberg 2004
Printed in Germany

Typesetting: Camera-ready by author, data conversion by PTP-Berlin, Protago-TeX-Production GmbH
Printed on acid-free paper   SPIN: 10989654   06/3142   5 4 3 2 1 0

Preface

This volume contains the proceedings of the Latin American Theoretical Informatics (LATIN) conference that was held in Buenos Aires, Argentina, April 5–8, 2004.

The LATIN series of symposia was launched in 1992 to foster interactions between the Latin American community and computer scientists around the world. This was the sixth event in the series, following São Paulo, Brazil (1992), Valparaiso, Chile (1995), Campinas, Brazil (1998), Punta del Este, Uruguay (2000), and Cancun, Mexico (2002). The proceedings of these conferences were also published by Springer-Verlag in the Lecture Notes in Computer Science series: Volumes 583, 911, 1380, 1776, and 2286, respectively. Also, as before, we published a selection of the papers in a special issue of a prestigious journal.

We received 178 submissions. Each paper was assigned to four program committee members, and 59 papers were selected. This was 80% more than the previous record for the number of submissions. We feel lucky to have been able to build on the solid foundation provided by the increasingly successful previous LATINs. And we are very grateful for the tireless work of Pablo Martínez López, the Local Arrangements Chair. Finally, we thank Springer-Verlag for publishing these proceedings in its LNCS series.

December 2003   Martin Farach-Colton

Invited Presentations

Cynthia Dwork, Microsoft Research, USA
Mike Paterson, University of Warwick, UK
Yoshiharu Kohayakawa, Universidade de São Paulo, Brazil
Jean-Eric Pin, CNRS/Université Paris VII, France
Dexter Kozen, Cornell University, NY, USA

Organization

Program Chair   Martin Farach-Colton, Rutgers University, USA

Local Arrangments Chair   Pablo Martínez López, Univ. Nacional de La Plata, Argentina

Steering Committee   Ricardo Baeza Yates, Univ. de Chile, Chile
Gaston Gonnet, ETH Zurich, Switzerland
Claudio Lucchesi, Univ. de Campinas, Brazil
Imre Simon, Univ. de São Paulo, Brazil

Program Committee

Michael Bender, SUNY Stony Brook, USA
Gerth Brodal, University of Aarhus, Denmark
Fabian Chudak, ETH, Switzerland
Mary Cryan, University of Leeds, UK
Pedro D'Argenio, UNC, Argentina
Martin Farach-Colton (Chair), Rutgers University, USA
David Fernández-Baca, Iowa State University, USA
Paolo Ferragina, Università di Pisa, Italy
Juan Garay, Bell Labs, USA
Claudio Gutiérrez, Universidad de Chile, Chile
John Iacono, Polytechnic University, USA
Bruce Kapron, University of Victoria, Canada
Valerie King, University of Victoria, Canada
Marcos Kiwi, Universidad de Chile, Chile
Sulamita Klein, Univ. Federal do Rio de Janeiro, Brazil
Stefan Langerman, Université Libre de Bruxelles, Belgium
Moshe Lewenstein, Bar Ilan University, Israel
Alex López-Ortiz, University of Waterloo, Canada
Eduardo Sany Laber, PUC-Rio, Brazil
Pablo E. Martínez López, UNLP, Argentina
S. Muthukrishnan, Rutgers Univ. and AT&T Labs, USA
Sergio Rajsbaum, Univ. Nacional Autónoma de México, Mexico
Andrea Richa, Arizona State University, USA
Gadiel Seroussi, HP Labs, USA
Alistair Sinclair, UC Berkeley, USA
Danny Sleator, Carnegie Mellon University, USA

Local Arrangements Committee

Eduardo Bonelli, Universidad Nacional de La Plata
Carlos "Greg" Diuk, Universidad de Buenos Aires
Santiago Figueira, Universidad de Buenos Aires
Carlos López Pombo, Universidad de Buenos Aires
Matías Menni, Universidad Nacional de La Plata
Pablo E. Martínez López (Chair), Univ. de La Plata
Alejandro Russo, Universidad de Buenos Aires
Marcos Urbaneja Sánchez, Universidad Nacional de La Plata
Hugo Zaccheo, Universidad Nacional de La Plata

Referees

Dimitris Achlioptas	Moses Charikar	Gudmund S. Frandsen
Ali Akhavi	Chandra Chekuri	Antonio Frangioni
David Aldous	Koen Claessen	Ari Freund
Jorge Almeida	Don Coppersmith	Daniel Fridlender
Greg Aloupis	Massimo Coppola	Alan Frieze
Andris Ambainis	Ricardo Corin	Fabio Gadducci
Eric Bach	Peter Csorba	Naveen Garg
Pablo Barcelo	Ricardo Dahab	Leszek Gasieniec
Alexander Barg	Ivan Damgaard	Vincenzo Gervasi
Elad Barkan	Gianna Del Corso	Jovan Golic
Paulo Barreto	Erik Demaine	Roberto Grossi
Tomas Barros	Vinay Deolalikar	Antonio Gulli
Cecilia Bastarrica	Vania Maria F. Dias	Hermann Haeusler
Gabriel Baum	Irit Dinur	Petr Hajek
Amir Ben-Amram	Shlomi Dolev	Angele Hamel
Julien Bernet	Dan Dougherty	Darrel Hankerson
Javier Blanco	Vida Dujmovic	Carmit Harel
Paulo Blauth	Dannie Durand	Amir Herzberg
Hans Bodlaender	Jerome Durand-Lose	Alejandro Hevia
Philip Bohannon	Nadav Efraty	Steve Homer
Eduardo Bonelli	John Ellis	Carlos Hurtado
Prosenjit Bose	Hazel Everett	Ferran Hurtado
Hervé Brönnimann	Luerbio Faria	Lucian Ilie
Veronique Bruyere	Sándor P. Fekete	Neil Immerman
John Brzozowski	Claudson Ferreira	Andre Inacio Reis
Ayelet Butman	Bornstein	Gabriel Infante Lopez
Ying Cai	Santiago Figueira	Achim Jung
Carlile Campos Lavor	Celina M. H.	Charanjit Jutla
Héctor Cancela	de Figueiredo	Mehmet Hakan Karaata
Jean Cardinal	Philippe Flajolet	Hakan Karaata
Olivier Carton	Paola Flocchini	Makino Kazuhisa

Carmel Kent
Claire Kenyon
Tien Kieu
Tomi Klein
Jon Kleinberg
Lars R. Knudsen
Cetin Koc
Ulrich Kohlenbach
Goran Konjevod
Peter Kornerup
Margarita Korovina
Guy Kortsarz
Natalio Krasnogor
Danny Krizanc
Marcos Kurban
Alair Lago
Leema Lallmamode
Orlando Lee
Noa Lewenstein
Yehuda Lindell
Ricardo Linden
Claudia Linhares
Marina Lipshtein
Errol Lloyd
Martin Loebl
John Longley
Fabrizio Luccio
Alejandro Maass
Guido Macchi
Phil MacKenzie
Nelson Maculan
Francesco Maffioli
Greg Malewicz
Arnaldo Mandel
Giovanni Manzini
Alvaro Martín
Demetrio Martin Vilela
Carlos Alberto
   Martinhon
Brian Mayoh
Robert W. McGrail
Candido F.X.
   de Mendonca
Matías Menni
Andrea Mennucci
Peter Merz

Fatma Mili
Ruy Milidiu
Peter Bro Miltersen
Manuela Montangero
Pat Morin
Rémi Morin
Sergio Muñoz
Seffi Naor
Gonzalo Navarro
Alantha Newman
Stefan Nickel
Peter Niebert
Rolf Niedermeier
Soohyun Oh
Alfredo Olivero
Nicolas Ollinger
Melih Onus
Erik Ordentlich
Friedrich Otto
Daniel Panario
Alessandro Panconesi
Luis Pardo
Rodrigo Paredes
Ojas Parekh
Michal Parnas
Mike Paterson
Boaz Patt-Shamir
David Peleg
Marco Pellegrini
David Pelta
Daniel Penazzi
Pino Persiano
Raúl Piaggio
Benny Pinkas
Nadia Pisanti
Ely Porat
Daniele Pretolani
Corrado Priami
Cristophe Prieur
Kirk Pruhs
Geppino Pucci
Claude-Guy Quimper
Rajmohan Rajaraman
Desh Ranjan
Matt Robshaw
Ricardo Rodríguez

Alexander Russell
Andrei Sabelfeld
Kai Salomaa
Louis Salvail
Luigi Santocanale
Eric Schost
Matthias Schröder
Marinella Sciortino
Michael Segal
Arun Sen
Rahul Shah
Jeff Shallit
Scott Shenker
David Shmoys
Amin Shokrollahi
Igor Shparlinski
Riccardo Silvestri
Guillermo Simari
Imre Simon
Bjarke Skjernaa
Dan Spielman
Jessica Staddon
Mike Steele
William Steiger
Bernd Sturmfels
Subhash Suri
Maxim Sviridenko
Wojciech Szpankowski
Shang-Hua Teng
Siegbert Tiga
Loana Tito Nogueira
Yaroslav Usenko
Santosh Vempala
Newton Vieira
Narayan Vikas
Jorge Villavicencio
Alfredo Viola
Elisa Viso
Marcelo Weinberger
Nicolas Wolovick
David Wood
Jinyun Yuan
Michal Ziv-Ukelson

Sponsoring Institutions

sponsored by

El Ojo del Huracán
[comunicación visual
+ ilustración]

Table of Contents

Invited Speakers

Contributions

Analysis of Scheduling Algorithms for Proportionate Fairness

Mike Paterson

Department of Computer Science
University of Warwick, Coventry, UK

Abstract. We consider a multiprocessor operating system in which each current job is guaranteed a given proportion over time of the total processor capacity. A scheduling algorithm allocates units of processor time to appropriate jobs at each time step. We measure the goodness of such a scheduler by the maximum amount by which the cumulative processor time for any job ever falls below the "fair" proportion guaranteed in the long term.

In particular we focus our attention on very simple schedulers which impose minimal computational overheads on the operating system. For several such schedulers we obtain upper and lower bounds on their deviations from fairness. The scheduling quality which is achieved depends quite considerably on the relative processor proportions required by each job.

We will outline the proofs of some of the upper and lower bounds, both for the unrestricted problem and for restricted versions where constraints are imposed on the processor proportions. Many problems remain to be investigated and we will give the results of some exploratory simulations. This is joint research with Micah Adler, Petra Berenbrink, Tom Friedetzky, Leslie Ann Goldberg and Paul Goldberg.

M. Farach-Colton (Ed.): LATIN 2004, LNCS 2976, p. 1, 2004.

Advances in the Regularity Method

Yoshiharu Kohayakawa[*]

Instituto de Matemática e Estatística, Universidade de São Paulo
Rua do Matão 1010, 05508–090 São Paulo, Brazil
yoshi@ime.usp.br

A beautiful result of Szemerédi on the asymptotic structure of graphs is his regularity lemma. Roughly speaking, his result tells us that *any* large graph may be written as a union of a bounded number of induced, random looking bipartite graphs (the so called *ε-regular pairs*). Many applications of the regularity lemma are based on the following fact, often referred to as the *counting lemma*: Let G be an s-partite graph with vertex partition $V(G) = \bigcup_{i=1}^{s} V_i$, where $|V_i| = m$ for all i and all pairs (V_i, V_j) are ε-regular of density d. Then G contains $(1 + f(\varepsilon))d^{\binom{s}{2}}m^s$ cliques K_s, where $f(\varepsilon) \to 0$ as $\varepsilon \to 0$. The combined application of the regularity lemma followed by the counting lemma is now often called the *regularity method*.

In recent years, considerable advances have occurred in the applications of the regularity method, of which we mention two: (i) the regularity lemma and the counting lemma have been generalized to the hypergraph setting and (ii) the case of sparse graphs is now much better understood.

In the sparse setting, that is, when n-vertex graphs with $o(n^2)$ edges are involved, most applications have so far dealt with random graphs. In this talk, we shall discuss a new approach that allows one to apply the regularity method in the sparse setting in purely *deterministic* contexts.

We cite an example. Random graphs are known to have several fault-tolerance properties. The following result was proved by Alon, Capalbo, Rödl, Ruciński, Szemerédi, and the author, making use of the regularity method, among others. *The random bipartite graph $G = G(n, n, p)$, with $p = cn^{-1/2k}(\log n)^{1/2k}$ and k a fixed positive integer, has the following fault-tolerance property with high probability: for any fixed $0 \leq \alpha < 1$, if c is large enough, even after the removal of any α-fraction of the edges of G, the resulting graph still contains all bipartite graphs with at most $a(\alpha)n$ vertices in each vertex class and maximum degree at most k, for some $a: [0, 1) \to (0, 1]$.*

Clearly, the above result implies that certain sparse fault-tolerant bipartite graphs exist. With the techniques discussed in this talk, one may prove that the celebrated norm-graphs of Kóllar, Rónyai, and Szabó, of suitably chosen density, are concrete examples.

This is joint work with V. Rödl and M. Schacht (Emory University, Atlanta).

[*] Partially supported by MCT/CNPq (ProNEx Project Proc. CNPq 664107/1997–4) and by CNPq (Proc. 300334/93–1 and 468516/2000–0)

M. Farach-Colton (Ed.): LATIN 2004, LNCS 2976, p. 2, 2004.

Fighting Spam: The Science

Cynthia Dwork

Microsoft Research, Silicon Valley Campus; 1065 La Avenida, Mountain View, CA 94043 USA; dwork@microsoft.com

Consider the following simple approach to fighting spam [5]:

> If I don't know you, and you want your e-mail to appear in my inbox, then you must attach to your message an easily verified "proof of computational effort", just for me and just for this message.

If the proof of effort requires 10 seconds to compute, then a single machine can send only 8,000 messages per day. The popular press estimates the daily volume of spam to be about 12-15 billion messages [4,6]. At the 10-second price, this rate of spamming would require at least 1,500,000 machines, working full time.

The proof of effort can be the output of an appropriately chosen *moderately hard* function of the message, the recipient's e-mail address, and the date and time. To send the same message to multiple recipients requires multiple computations, as the e-mail addresses vary. Similarly, to send the same (or different) messages, repeatedly, to a single recipient requires repeated computation, as the dates and times (or messages themselves) vary.

Initial proposals for the function [5,2] were CPU-intensive. To decrease disparities between machines, Burrows proposed replacing the original CPU-intensive pricing functions with memory-intensive functions, a suggestion first investigated in [1].

Although the architecture community has been discussing the so-called "memory wall" – the point at which the memory access speeds and CPU speeds have diverged so much that improving the processor speed will not decrease computation time – for almost a decade [7], there has been little theoretical study of the memory-access costs of computation. A rigorous investigation of memory-bound pricing functions appears in [3], where several candidate functions (including those in [1]) are analyzed, and a new function is proposed. An abstract version of the new function is proven to be secure against amortization by a spamming adversary.

References

1. M. Abadi, M. Burrows, M. Manasse, and T. Wobber, Moderately Hard, Memory-Bound Functions, *Proceedings of the 10th Annual Network and Distributed System Security Symposium*, 2003.

M. Farach-Colton (Ed.): LATIN 2004, LNCS 2976, pp. 3–4, 2004.

2. A. Back, Hashcash - A Denial of Servic Counter-Measure,
   http://www.cypherspace.org/hashcash/hashcash.pdf.
3. C. Dwork, A. Goldberg, and M. Naor, On Memory-Bound Functions for Fighting
   Spam, *Advances in Cryptology – CRYPTO 2003, LNCS 2729*, Springer, 2003, pp.
   426–444.
4. Rita Chang, "Could spam kill off e-mail?" PC World October 23, 2003. See
   http://www.nwfusion.com/news/2003/1023couldspam.html.
5. C. Dwork and M. Naor, Pricing via Processing, Or, Combatting Junk Mail,*Advances
   in Cryptology – CRYPTO'92, LNCS 740*, Springer, 1993, pp. 139–147.
6. Spam Filter Review, Spam Statistics 2004,
   http://www.spamfilterreview.com/spam-statistics.html.
7. Wm. A. Wulf and Sally A. McKee, Hitting the Memory Wall: Implications of the
   Obvious, *Computer Architecture News 23*(1), 1995, pp. 20–24.

The Consequences of Imre Simon's Work in the Theory of Automata, Languages, and Semigroups

Jean-Eric Pin

CNRS / Università Paris VII, France

Abstract. In this lecture, I will show how influential has been the work of Imre in the theory of automata, languages and semigroups. I will mainly focus on two celebrated problems, the restricted star-height problem (solved) and the decidability of the dot-depth hierarchy (still open). These two problems lead to surprising developments and are currently the topic of very active research. I will present the prominent results of Imre on both topics, and demonstrate how these results have been the motor nerve of the research in this area for the last thirty years.

M. Farach-Colton (Ed.): LATIN 2004, LNCS 2976, p. 5, 2004.

Querying Priced Information in Databases: The Conjunctive Case

Extended Abstract

Sany Laber[1]*, Renato Carmo[3,2]**, and Yoshiharu Kohayakawa[2]***

[1] Departamento de Informática da Pontifícia Universidade Católica do Rio de Janeiro
R. Marquês de São Vicente 225, Rio de Janeiro RJ, Brazil
laber@info.puc-rio.br
[2] Instituto de Matemática e Estatística da Universidade de São Paulo
Rua do Matão 1010, 05508–090 São Paulo SP, Brazil
{renato,yoshi}@ime.usp.br
[3] Departamento de Informática da Universidade Federal do Paraná
Centro Politécnico da UFPR, 81531–990, Curitiba PR, Brazil
renato@inf.ufpr.br

Abstract. Query optimization that involves *expensive predicates* have received considerable attention in the database community. Typically, the output to a database query is a set of tuples that satisfy certain conditions, and, with expensive predicates, these conditions may be computationally costly to verify. In the simplest case, when the query looks for the set of tuples that simultaneously satisfy k expensive predicates, the problem reduces to ordering the evaluation of the predicates so as to minimize the time to output the set of tuples comprising the answer to the query.

Here, we give a simple and fast deterministic k-approximation algorithm for this problem, and prove that k is the best possible approximation ratio for a deterministic algorithm, even if exponential time algorithms are allowed. We also propose a randomized, polynomial time algorithm with expected approximation ratio $1 + \sqrt{2}/2 \approx 1.707$ for $k = 2$, and prove that $3/2$ is the best possible expected approximation ratio for randomized algorithms.

1 Introduction

The main goal of *query optimization in databases* is to determine how a query over a database should be processed in order to minimize the user response time. A typical query extracts the tuples from a database relation that satisfy a set of conditions, or *predicates*, in database terminology. For example, consider

* Partially supported by FAPERJ (Proc. E-26/150.715/2003) and CNPq (Proc. 476817/2003-0)
** Partially supported by CAPES (PICDT) and CNPq (Proc. 476817/2003-0)
*** Partially supported by MCT/CNPq (ProNEx, Proc. CNPq 664107/1997-4) and CNPq (Proc. 300334/93–1, 468516/2000–0 and Proc. 476817/2003-0)

M. Farach-Colton (Ed.): LATIN 2004, LNCS 2976, pp. 6–15, 2004.

the set of tuples $D = \{(a_1, b_1), (a_1, b_2), (a_1, b_3), (a_2, b_1)\}$ (see Fig. 1(a)) and a conjunctive query that seeks to extract the subset of tuples (a_i, b_j) for which a_i satisfies predicate P_1 and b_j satisfies predicate P_2. Clearly, these predicates can be viewed together as a 0/1-valued function δ defined on the set of tuple elements $\{a_1, a_2, b_1, b_2, b_3\}$, with the convention that, $\delta(a_i) = 1$ if and only if $P_1(a_j)$ holds and $\delta(b_j) = 1$ if and only if $P_2(b_j)$ holds. The answer to the query is the set of pairs (a_i, b_j) with $\bar{\delta}(a_i, b_j) = \delta(a_i)\delta(b_j) = 1$. The query optimization problem that we consider is that of determining a strategy for evaluating $\bar{\delta}$ so as to compute this set of tuples by evaluating as few values of the function δ as possible (or, more generally, with the total cost for evaluating the function $\bar{\delta}$ minimal).

It is usually the case that the cost (measured as the computational time) needed to evaluate the predicates of a query can be assumed to be bounded by a constant so that the query can be answered by just scanning through all the tuples in D while evaluating the corresponding predicates.

In the case of computationally expensive predicates, however, e.g., when the database holds complex data as images and tables, this constant may happen to be so large as to render this strategy impractical. In such cases, the different costs involved in evaluating each predicate must also be taken into account in order to keep user response time within reasonable bounds.

Among several proposals to model and solve this problem (see, for example, [1,3,5]), we focus on the improvement of the approach proposed in [8] where, differently from the others, the query evaluation problem is reduced to an optimization problem on a hypergraph (see Fig. 1).

1.1   Problem Statement

A *hypergraph* is a pair $G = (V(G), E(G))$ where $V(G)$, the set of *vertices* of G, is a finite set and each *edge* $e \in E(G)$ is a non empty subset of $V(G)$.

The size of the largest edge in G is called the *rank* of G and is denoted $r(G)$. A hypergraph G is said to be *uniform* if each edge has size $r(G)$, and is said to be *k-partite* if there is a partition $\{V_1, \ldots, V_k\}$ of $V(G)$ such that no edge contains more than one vertex in the same partition class. A *matching* in a hypergraph G is a set $M \subseteq E(G)$ with no two edges in M sharing a common vertex. A hypergraph G is said to be a *matching* if $E(G)$ is a matching.

Given a hypergraph G and a function $\delta : V(G) \to \{0, 1\}$ we define an *evaluation* of (G, δ) as a set $\mathsf{E} \subseteq V(G)$ such that, knowing the value of $\delta(v)$ for each $v \in \mathsf{E}$, one may determine, for each $e \in E(G)$, the value of

$$\bar{\delta}(e) = \prod_{v \in e} \delta(v). \tag{1}$$

Given a hypergraph G and a function $\gamma : V(G) \to \mathbb{R}$ we define the *cost* of a set $X \subseteq V(G)$ by $\gamma(X) = \sum_{v \in X} \gamma(v)$.

An instance to the Dynamic Multipartite Ordering problem (DMO) is an $r(G)$-partite, uniform hypergraph G, together with functions δ and γ as above. The objective in DMO is to determine an evaluation of minimum cost for (G, δ, γ). Observe that while the value of $\gamma(v)$ is known in advance for each $v \in V(G)$, the function δ is 'unknown' to us at first. More precisely, the value of $\delta(v)$ becomes known only when $\delta(v)$ is actually evaluated, and this evaluation costs $\gamma(v)$. The restriction of DMO to instances in which $r(G) = 2$ deserves special attention and will be referred to as the Dynamic Bipartite Ordering problem (DBO).

Before we proceed, let us observe that DMO models our database problem as follows: the sets in the partition $\{V_1, \ldots, V_k\}$ of $V(G)$ correspond to the k different attributes of the relation that is being queried and each vertex of G corresponds to a distinct attribute value (tuple element). The edges correspond to tuples in the relation, $\gamma(v)$ is the time required to evaluate δ on v and $\delta(v)$ corresponds to the result of a predicate evaluated at the corresponding tuple element.

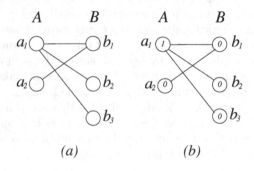

(a)   *(b)*

Fig. 1. The set of tuples $\{(a_1, b_1), (a_1, b_2), (a_1, b_3), (a_2, b_1)\}$ and an instance for DBO

Figure 1(b) shows an instance of DBO. The value of $\delta(v)$ is indicated inside each vertex v. Suppose that $\gamma(a_1) = 3$ and $\gamma(b_1) = \gamma(b_2) = \gamma(b_3) = 2$. In this case, any strategy that starts evaluating $\delta(a_1)$ will return the evaluation $\{a_1, b_1, b_2, b_3\}$, of cost 9. However, the evaluation of minimum cost for this instance is $\{b_1, b_2, b_3\}$, of cost 6. This example highlights the key point: the problem is to devise a strategy for dynamically choosing, based on the function γ and the values of δ already revealed, the next vertex v whose δ-value should be evaluated, so as to minimize the final, overall cost.

Let $\mathcal{A}$ be an algorithm for DMO and let $\mathcal{I} = (G, \delta, \gamma)$ be an instance to DMO. We will denote the evaluation computed by $\mathcal{A}$ on input $\mathcal{I}$ by $\mathcal{A}(\mathcal{I})$. Establishing a measure for the performance of a given algorithm $\mathcal{A}$ for DMO is somewhat delicate: for example, a worst case analysis of $\gamma(\mathcal{A}(\mathcal{I}))$ is not suitable since any correct algorithm should output an evaluation comprising all vertices in $V(G)$ when $\delta(v) = 1$ for every $v \in V(G)$ (if G has no isolated vertices). This remark motivates the following definition.

Given an instance $\mathcal{I} = (G, \delta, \gamma)$, let E be an evaluation for $\mathcal{I}$ and let $\gamma^*(\mathcal{I})$ denote the cost of a minimum cost evaluation for $\mathcal{I}$. We define the *deficiency of evaluation* E *(with respect to $\mathcal{I}$)* as the ratio $d(E, \mathcal{I}) = \gamma(E)/\gamma^*(\mathcal{I})$. Given an algorithm $\mathcal{A}$ for DMO, we define the *deficiency of $\mathcal{A}$* as the worst case deficiency of the evaluation $\mathcal{A}(\mathcal{I})$, where $\mathcal{I}$ ranges over all possible instances of the problem, that is, $d(\mathcal{A}) = \max_\mathcal{I} d(\mathcal{A}(\mathcal{I}), \mathcal{I})$.

If $\mathcal{A}$ is a randomized algorithm, $d(\mathcal{A}(\mathcal{I}), \mathcal{I})$ is a random variable, and the *expected deficiency* of $\mathcal{A}$ is then defined as the maximum over all instances of the mean of this random variable, that is,

$$d(\mathcal{A}) = \max_\mathcal{I} \mathbb{E}\left[d(\mathcal{A}(\mathcal{I}), \mathcal{I})\right] = \max_\mathcal{I} \mathbb{E}\left[\gamma(\mathcal{A}(\mathcal{I}))\right]/\gamma^*(\mathcal{I}).$$

Clearly, we wish to devise fast algorithms whose (expected) deficiency is as close to 1 as possible. In this paper, we are concerned with designing algorithms for DMO, analyzing them and establishing bounds for their deficiency.

1.2 Statement of Results

In Sect. 2 we start by giving lower bounds on the deficiency of deterministic and randomized algorithms for DMO (see Theorem 1). It is worth noting that these lower bounds apply even if we allow exponential time algorithms. We then present an optimal deterministic algorithm for DMO with time complexity $O(|E(G)| \log r(G))$, developed with the primal-dual approach. As an aside, we remark that this algorithm does not need to know the whole hypergraph in advance in order to solve the problem, since it scans the edges (tuples), evaluating each of them as soon as they become available. This is a most convenient feature for the database application that motivates this work. We also note that Feder et al. [4] independently obtained similar results.

In Sect. 3, for any given $0 \le \varepsilon \le 1 - \sqrt{2}/2$, we present a *randomized*, polynomial time algorithm $\mathcal{R}_\varepsilon$ for DBO whose expected deficiency is at most $2 - \varepsilon$. The best expected deficiency is achieved when $\varepsilon = 1 - \sqrt{2}/2$. However, the smaller the value of ε, the smaller is the probability that a particular execution of $\mathcal{R}_\varepsilon$ will return a truly poor result: we show that the probability that $d(\mathcal{R}_\varepsilon(\mathcal{I}), \mathcal{I}) \le 1 + 1/(1 - \varepsilon)$ holds is 1.

The deficiency of $\mathcal{R}_\varepsilon$ is not assured to be highly concentrated around the expectation. In Sect. 3.1, we show that this limitation is *inherent to the problem*, rather than a weakness of our approach: for any $0 \le \varepsilon \le 1$, no randomized algorithm can have deficiency smaller than $1 + \varepsilon$ with probability larger than $(1 + \varepsilon)/2$. The proof of this fact makes use of *Yao's Minimax Principle* [9].

The reader is referred to the full version of this extended abstract for the proofs of the results (or else [6]).

1.3 Related Work

The problem of optimizing queries with expensive predicates has gained some attention in the database community [1,3,5,7,8]. However, most of the proposed approaches [1,3,5] do not take into account the fact that an attribute value may appear in different tuples in order to decide how to execute the query. In this sense, they do not view the input relation as a general hypergraph, but as a set of tuples without any relation among them (i.e., as a matching hypergraph). The Predicate Migration algorithm proposed in [5], the main reference in this subject, may be viewed as an optimal algorithm for a variant of DMO, in which the input graph is always a matching, the probability p_i of a vertex from V_i (ith attribute) evaluating to true ($\delta(v) = 1$) is known, and the objective is to minimize the expected cost of the computed evaluation (we omit the details).

The idea of processing the hypergraph induced by the input relation appears first in [8], where a greedy algorithm is proposed with no theoretical analysis. The distributed case of DBO, in which there are two available processors, say P_A and P_B, responsible for evaluating δ on the nodes of the vertex classes A and B of the input bipartite graphs is studied in [7]. The following results are presented in [7]: a lower bound of 3/2 on the deficiency of any randomized algorithm, a randomized polynomial time algorithm of expected deficiency 8/3, and a linear time algorithm of deficiency 2 for the particular case of DBO with constant γ. We observe that the approach here allows one to improve some of these results.

In this extended abstract, we restrict our attention to *conjunctive* queries (in the sense of (1)). However, much more general queries could be considered. For example, $\bar{\delta} \colon E(G) \to \{0,1\}$ could be any formula in the first order propositional calculus involving the predicates represented by δ. In [2], Charikar et al. considered the problem of querying priced information. In particular, they considered the problem of evaluating a query that can be represented by an "AND/OR tree" over a set of variables, where the cost of probing each variable may be different. The framework for querying priced information proposed in that paper can be viewed as a restricted version of the problem described in this paragraph, where the input hypergraph has one single edge. It would be interesting to investigate DMO with such generalized queries.

1.4 Preliminaries

Let $\mathcal{I} = (G, \delta, \gamma)$ be an instance to DMO. The *neighbourhood* of $v \in V(G)$ is the set $\Gamma(v) = \{u \in V(G) - \{v\} \colon \{u, v\} \subseteq e \text{ for some } e \in E(G)\}$. For any $X \subseteq V(G)$, we let $V_0(X) = \{v \in X \colon \delta(v) = 0\}$, $V_1(X) = \{v \in X \colon \delta(v) = 1\}$, $\Gamma(X) = \bigcup_{v \in X} \Gamma(v)$ and $\Gamma_1(X) = \Gamma(V_1(X))$.

A *cover* for G is a set $C \subseteq V(G)$ such that every edge of G has at least one vertex in C. A *minimum cover* for (G, γ) is a cover C for G such that $\gamma(C)$ is minimal. Observe that any evaluation for $\mathcal{I}$ must contain a cover for G as a subset, otherwise the $\bar{\delta}$-value of at least one edge cannot be determined.

Let us now restrict our attention to DBO, the restricted case of DMO where G is a bipartite graph. Let $\mathcal{I} = (G, \delta, \gamma)$ be an instance to DBO. For a cover C for G, we call $\mathsf{E}(C) = C \cup \Gamma_1(C)$ the C-*evaluation for* $\mathcal{I}$. It is not difficult to see that a C-evaluation for $\mathcal{I}$ is indeed an evaluation for $\mathcal{I}$. Moreover, since any evaluation for (G, δ) must contain some cover for G and $\Gamma_1(V(G))$, it is not difficult to conclude that the deficiency of a C-evaluation for an instance to DBO has deficiency at most 2, whenever C is a minimum cover for (G, γ). This observation appears in [7] for the distributed version of DBO.

An optimal cover C for (G, γ), and as a consequence $\mathsf{E}(C)$, may be computed in polynomial time if G is a bipartite graph. We use COVER to denote an algorithm that outputs $\mathsf{E}(C)$ for some minimum cover C. Since 2 is a lower bound for the deficiency of any deterministic algorithm for DBO (see Sect. 2), we have that COVER is a polynomial time, optimal deterministic algorithm for DBO. This algorithm plays an important role in the design of the randomized algorithm proposed in Sect. 3.

2   An Optimal Polynomial Deterministic Algorithm

We start with some lower bounds for the deficiency of algorithms for DMO. It is worth noting that *these bounds apply even to algorithms of exponential time/space complexity.*

Theorem 1. (i) *For any given deterministic algorithm $\mathcal{A}$ for DMO and any hypergraph G with at least one edge, there exist functions γ and δ such that $d(\mathcal{A}(G, \delta, \gamma)) \geq r(G)$.*

(ii) *For any given randomized algorithm $\mathcal{B}$ for DMO and any hypergraph G with at least one edge, there exist functions γ and δ such that $d(\mathcal{B}(G, \delta, \gamma)) \geq (r(G) + 1)/2$.*

2.1   An Optimal Polynomial Deterministic Algorithm for DMO

We will now introduce a polynomial time, deterministic algorithm for DMO that has deficiency at most $r(G)$ on an instance $\mathcal{I} = (G, \delta, \gamma)$. In view of Theorem 1, this algorithm has the best possible deficiency for a deterministic algorithm.

Let (G, δ, γ) be a fixed instance to DMO, and let $E_i = \{e \in E(G): \bar{\delta}(e) = i\}$ and $W_i = \bigcup_{e \in E_i} e$ $(i \in \{0, 1\})$.

We let $G[E_i]$ be the hypergraph with vertex set W_i and edge set E_i. Let γ_0^* be the cost of a minimum cover for $(G[E_0], \gamma)$, among all covers for $(G[E_0], \gamma)$ that contain vertices in $V_0 = V_0(V(G)) = \{v \in V(G): \delta(v) = 0\}$ only. Then $\gamma^*(G, \delta, \gamma) = \gamma_0^* + \gamma(W_1)$.

Let us look at γ_0^* as the optimal solution of the following Integer Programming problem, which we will denote by $L_I(G, \delta, \gamma)$:

$$\min \left\{ \sum_{v \in V_0} \gamma(v) x_v : \sum_{v \in e \cap V_0} x_v \geq 1 \text{ for all } e \in E_0 \text{ and } x_v \in \{0, 1\} \text{ for all } v \in V_0 \right\}.$$

Let us denote by $L(G, \delta, \gamma)$ the linear relaxation of $L_I(G, \delta, \gamma)$, where the restrictions $x_v \in \{0, 1\}$ are replaced by $x_v \geq 0$ for all $v \in V_0$. The dual $L(G, \delta, \gamma)^D$ of $L(G, \delta, \gamma)$ is

$$\max \left\{ \sum_{e \in E_0} y_e : \sum_{e: \, v \in e} y_e \leq \gamma(v) \text{ for all } v \in V_0 \text{ and } y_e \geq 0 \text{ for all } e \in E_0 \right\}.$$

Lemma 2. *Let (G, δ, γ) be an instance to DMO and let $\bar{y} : E_0 \to \mathbb{R}$ be a feasible solution of $L(G, \delta, \gamma)^D$. Any evaluation E of (G, δ) satisfying*

$$\gamma(v) \leq \sum_{e: \, v \in e} \bar{y}_e \quad \text{for all } v \in \mathsf{E} - W_1, \tag{2}$$

has deficiency at most $r(G)$.

The algorithm presented below uses a primal-dual approach to construct a vector $y \colon E \to \mathbb{R}$ and an evaluation E such that both the restriction of y to E_0 and E satisfy the the conditions of Theorem 2.

Our algorithm maintains for each $e \in E(G)$ a value y_e and for every $v \in V(G)$ the value $r_v = \sum_{e \colon v \in e} y_e$. At each step, the algorithm selects an unevaluated edge e and increases the corresponding dual variable y_e until it "saturates" the next non-evaluated vertex v (r_v becomes equal to $\gamma(v)$). The values of r_u ($u \in e$) are updated and the vertex v is then evaluated. If $\delta(v) = 0$, then the edge e is added to E_0 along with all other edges that contain v, and the algorithm proceeds to the next edge. Otherwise the algorithm increases the value of the dual variable y_e until it "saturates" another unevaluated vertex in e and executes the same steps until either e is put into E_0 or there are no more unevaluated vertices in e, in which case e is put in E_1.

Algorithm $\mathcal{PD}(G, \delta, \gamma)$

1. Start with E_0, E_1 and E as empty sets, $r_v = 0$ for all $v \in V(G)$ and $y_e = 0$ for all $e \in E(G)$
2. While $E(G) \neq E_1 \cup E_0$
   a) Select an edge $e \in E(G) - (E_1 \cup E_0)$
   b) While $e \not\subseteq \mathsf{E}$ and $e \notin E_0$
      i.   select a vertex $v \in e - \mathsf{E}$ such that $\gamma(v) - r_v$ is minimum
      ii. add $\gamma(v) - r_v$ to y_e and to each r_u such that $u \in e$
      iii. insert v in E
      iv. If $\delta(v) = 0$, insert in E_0 every edge $e' \in E(G)$ such that $v \in e'$
   c) If $e \notin E_0$, insert e in E_1
3. Return E

Lemma 3. *Let (G, δ, γ) be an instance to* DMO. *At the end of the execution of $\mathcal{PD}(G, \delta, \gamma)$, the restriction of y to E_0 is a feasible solution to $L(G, \delta, \gamma)^D$ and E is an evaluation of (G, δ) satisfying* (2). *Algorithm $\mathcal{PD}(G, \delta, \gamma)$ runs in time $O(|E(G)| \log r(G))$.*

Theorem 4. *Algorithm $\mathcal{PD}$ is a polynomial time, optimal deterministic algorithm for* DMO.

3   The Bipartite Case and a Randomized Algorithm

Let $0 \leq \varepsilon \leq 1 - \sqrt{2}/2$. In this section, we present $\mathcal{R}_\varepsilon$, a polynomial time randomized algorithm for DBO with the following properties: for every instance $\mathcal{I}$, we have

$$\mathbb{E}\left[d(\mathcal{R}_\varepsilon(\mathcal{I}))\right] \leq 2 - \varepsilon \tag{3}$$

and

$$\mathbb{P}\left(d(\mathcal{R}_\varepsilon(\mathcal{I})) \leq 1 + \frac{1}{1 - \varepsilon}\right) = 1. \tag{4}$$

Thus, $\mathcal{R}_\varepsilon$ provides a trade-off between expected deficiency and worst case deficiency. At one extreme, when $\varepsilon = 1 - \sqrt{2}/2$, we have expected deficiency $1.707\ldots$ and worst case deficiency up to 2.41 for some particular execution. At the other extreme $(\varepsilon = 0)$, we have a deterministic algorithm with deficiency 2.

The key idea in $\mathcal{R}_\varepsilon$'s design is to try to understand under which conditions the COVER algorithm described in Sect. 1.4 does not perform well. More exactly, given an instance $\mathcal{I}$ to DBO, a minimum cover C for (G, δ), and $\varepsilon > 0$, we turn our attention to the instances $\mathcal{I}$ having $d(\mathsf{E}(C), \mathcal{I}) \geq 2 - \varepsilon$.

One family of such instances can be constructed as follows. Consider an instance (G, δ, γ) to DBO where G is a matching of n edges, the vertex classes of G are A and B, $\delta(v) = 1$ for every $v \in A$ and $\delta(v) = 0$ for every $v \in B$ and $\gamma(v) = 1$ for every $v \in V(G)$. Clearly, B is an optimum evaluation for $\mathcal{I}$, with cost n. On the other hand, note that the deficiency of the evaluation $\mathsf{E}(C)$ which is output by COVER depends on which of the 2^n minimum covers of G is chosen for C. In the particular case in which $C = A$, we have $d(\mathsf{E}(C), \mathcal{I}) = 2n/n = 2$.

This example suggests the following idea. If C is a minimum cover for (G, γ) and nonetheless $\mathsf{E}(C)$ is not a "good evaluation" for $\mathcal{I} = (G, \delta, \gamma)$, then there must be another cover C' of G whose intersection with C is "small" and still C' is not "far from being" a minimum cover for G. The following lemma formalizes this idea.

Lemma 5. *Let $\mathcal{I} = (G, \delta, \gamma)$ be an instance to DBO, let C be a minimum cover for (G, δ) and let $0 < \varepsilon < 1$. If $d(\mathsf{E}(C)) \geq 2 - \varepsilon$, then there is a vertex cover C_ε for G such that $\gamma(C_\varepsilon) \leq (\gamma(C - C_\varepsilon))/(1 - \varepsilon)$.*

Let $\mathcal{I} = (G, \delta, \gamma)$, C and ε be as in the statement of Lemma 5. Let C' be a minimum cover for $(G, \gamma_{C,\varepsilon})$, where $\gamma_{C,\varepsilon}$ is given by $\gamma_{C,\varepsilon}(v) = (1 - \varepsilon)\gamma(v)$ if $v \notin C$ and $\gamma_{C,\varepsilon}(v) = (2 - \varepsilon)\gamma(v)$ otherwise.

We can formulate the problem of finding a cover C_ε satisfying $\gamma(C_\varepsilon) \leq \gamma(C - C_\varepsilon)/(1 - \varepsilon)$ as a linear program in order to conclude that such a cover exists if and only if $\gamma_{C,\varepsilon}(C') \leq \gamma(C)$. Furthermore, if $\gamma_{C,\varepsilon}(C') \leq \gamma(C)$ then $\gamma(C') \leq \gamma(C - C')/(1 - \varepsilon)$.

This last remark, together with Lemma 5, provides an efficient way to verify whether or not a particular minimum cover C is going to give a good evaluation for (G, δ, γ).

The cover C' above can be computed in polynomial time in those cases where G is bipartite, we can devise the following randomized algorithm for DBO.

Algorithm $\mathcal{R}_\varepsilon(G, \delta, \gamma)$

1. $C \leftarrow$ a minimum cover for (G, γ)
2. $C' \leftarrow$ a minimum cover for $(G, \gamma_{C,\varepsilon})$
3. If $\gamma_{C,\varepsilon}(C') > \gamma(C)$, then return $\mathsf{E}(C)$
4. Let $p = (1 - 3\varepsilon + \varepsilon^2)/(1 - \varepsilon)$
5. Pick $x \in [0, 1]$ uniformly at random. Return $\mathsf{E}(C)$ if $x < p$ and $\mathsf{E}(C')$ otherwise

---

The correctness of algorithm $\mathcal{R}_\varepsilon$ follows from the fact that $\mathcal{R}_\varepsilon$ always outputs a cover evaluation (see Sect. 1.4). Properties (3) and (4) of the evaluation computed by $\mathcal{R}_\varepsilon$, claimed at the beginning of Sect. 3, are assured by the next result.

Theorem 6. *Let* $0 \le \varepsilon \le 1 - \sqrt{2}/2$. *For any instance* $\mathcal{I} = (G, \delta, \gamma)$ *we have* $\mathbb{E}\left[d(\mathcal{R}_\varepsilon(\mathcal{I}))\right] \le 2 - \varepsilon$ *and* $\mathbb{P}\left(d(\mathcal{R}_\varepsilon(\mathcal{I})) \le (2 - \varepsilon)/(1 - \varepsilon)\right) = 1$.

Theorem 6 is tight when $\varepsilon = 1 - \sqrt{2}/2$. Indeed, consider the instance $\mathcal{I} = (G, \delta, \gamma)$, where G is a complete bipartite graph with bipartition $\{A, B\}$, where $|B| = 1.41|A| \approx \sqrt{2}|A|$, $\delta(a) = 0$ for every $a \in A$, $\delta(b) = 1$ for every $b \in B$, and $\gamma(v) = 1$ for every $v \in V(G)$. Clearly, A is an evaluation of cost $|A|$ since it only checks the vertices in A. The set B, however, is a minimum cover for $(G, \gamma_{C,\varepsilon})$ and $\gamma_{C,\varepsilon}(B) \le \gamma(A)$. Hence, $\mathcal{R}_\varepsilon(\mathcal{I})$ returns $\mathsf{E}(A)$ with probability $1/2$ and $\mathsf{E}(B)$ with probability $1/2$, so that the expected deficiency is close to $1 + \sqrt{2}/2$.

3.1   Lower Bound for Randomized Algorithms

We have proved so far that algorithm $\mathcal{R}_\varepsilon$, for $\varepsilon = 1 - \sqrt{2}/2$, has expected deficiency $\le 1 + \sqrt{2}/2 = 1.707\ldots$. However, $\mathcal{R}_\varepsilon$ does not achieve this deficiency with high probability. For the instance described above, $\mathcal{R}_\varepsilon$ attains deficiency 2.41 with probability $1/2$ and deficiency 1 with probability $1/2$. One can speculate whether a more dynamic algorithm would not have smaller (closer to 1.5) deficiency with high probability. In this section, we prove that this is not possible, that is, no randomized algorithm for DBO can have deficiency smaller than μ for any given $1 \le \mu \le 2$ with probability close to 1 (see Theorem 8). *We shall prove this considering instances* $\mathcal{I} = (G, \delta, \gamma)$ *with* G *a balanced, complete bipartite graph on* n *vertices and with* $\gamma \equiv 1$ *only*. All instances in this section are assumed to be of this form.

Let $\mathcal{A}$ be a randomized algorithm for DBO and let $1/2 \le \lambda \le 1$. Given an instance $\mathcal{I} = (G, \delta, \gamma)$ where $|V(G)| = n$, let $P(\mathcal{A}, \mathcal{I}, \lambda n) = \mathbb{P}(\gamma(\mathcal{A}(\mathcal{I})) \ge \lambda n)$ and let $P(\mathcal{A}, \lambda n) = \max_{\mathcal{I}} P(\mathcal{A}, \mathcal{I}, \lambda n)$. Given a deterministic algorithm $\mathcal{B}$ and an instance $\mathcal{I}$ for DBO, we define the *payoff of* $\mathcal{B}$ *with respect to* $\mathcal{I}$ as $g(\mathcal{B}, \mathcal{I}) = 1$ if $\gamma(\mathcal{B}(\mathcal{I})) \ge \lambda n$ and $g(\mathcal{B}, \mathcal{I}) = 0$ otherwise.

One may deduce from Yao's minimax principle [9] that, for any randomized algorithm $\mathcal{A}$, we have $\max_{\mathcal{I}} \mathbb{E}\left[g(\mathcal{A}, \mathcal{I})\right] \ge \max_p \mathbb{E}\left[g(\mathsf{opt}, \mathcal{I}_p)\right]$, where opt is an optimal deterministic algorithm, in the average case sense, for the probability distribution p over the set of possible instances for DBO. (In the inequality above, the expectation is taken with respect to the coin flips of $\mathcal{A}$ on the left-hand side and with respect to p on the right-hand side; we write $\mathcal{I}_p$ for an instance generated according to p.)

Since a randomized algorithm can be viewed as a distribution probability over the set of deterministic algorithms, we have $\mathbb{E}\left[g(\mathcal{A}, \mathcal{I})\right] = P(\mathcal{A}, \mathcal{I}, \lambda n)$ and hence $\max_{\mathcal{I}} \mathbb{E}\left[g(\mathcal{A}, \mathcal{I})\right] = P(\mathcal{A}, \lambda n)$. Moreover, $\mathbb{E}\left[g(\mathsf{opt}, \mathcal{I}_p)\right]$ is the probability that the cost of the evaluation computed by the optimal algorithm for the distribution p is at least λn. Thus, if we are able to define a probability distribution p over the set of possible instances and analyze the optimal algorithm for such a distribution, we obtain a lower bound for $P(\mathcal{A}, \lambda n)$.

Let n be an even positive integer and let G be a complete bipartite graph with $V(G) = \{1, \ldots, n\}$. Let the vertex classes of G be $\{1, \ldots, n/2\}$ and $\{n/2 +$

$1, \ldots, n\}$. Let $\gamma(v) = 1$ for all $v \in V(G)$. For $1 \leq i \leq n$, define the function $\delta_i \colon V(G) \to \{0, 1\}$ putting $\delta_i(v) = 1$ if $i = v$ and $\delta_i(v) = 0$ otherwise. Consider the probability distribution p where the only instances with positive probability are $\mathcal{I}_i = (G, \delta_i, \gamma)$ $(1 \leq i \leq n)$ and all these instances are equiprobable, with probability $1/n$ each. A key property of these instances is that the cost of the optimum evaluation for all of them is $n/2$, since all the vertices of the vertex class of the graph that does not contain the vertex with δ-value 1 must be evaluated in order to determine the value of all edges. We have the following lemma.

Lemma 7. *Let* opt *be an optimal algorithm for the distribution probability* p. *Then* $\mathbb{E}\left[g(\mathsf{opt}, \mathcal{I}_p)\right] \geq 1 - \lambda$.

Since $\gamma^*(\mathcal{I}_j) = n/2$ for $1 \leq j \leq n$, we have the following result.

Theorem 8. *Let* $\mathcal{A}$ *be a randomized algorithm for* DBO *and let* $1 \leq \mu \leq 2$ *be a real number. Then there is an instance* $\mathcal{I}$ *for which* $\mathbb{P}\left(d(\mathcal{A}(\mathcal{I}), \mathcal{I}) \geq \mu\right) \geq 1 - \mu/2$.

References

1. L. Bouganim, F. Fabret, F. Porto, and P. Valduriez. Processing queries with expensive functions and large objects in distributed mediator systems. In *Proc. 17th Intl. Conf. on Data Engineering*, April 2-6, 2001, Heidelberg, Germany, pages 91–98, 2001.
2. M. Charikar, R. Fagin, V. Guruswami, J. Kleinberg, P. Raghavan, and A. Sahai. Query strategies for priced information (extended abstract). In *Proceedings of the 32nd ACM Symposium on Theory of Computing*, Portland, Oregon, May 21–23, 2000, pages 582–591, 2000.
3. S. Chaudhuri and K. Shim. Query optimization in the presence of foreign functions. In *Proc. 19th Intl. Conf. on VLDB*, August 24-27, 1993, Dublin, Ireland, pages 529–542, 1993.
4. T. Feder, R. Motwani, L. O'Callaghan, R. Panigrahy, and D. Thomas. Online distributed predicate evaluation. Preprint 2003.
5. J. M. Hellerstein. Optimization techniques for queries with expensive methods. *ACM Transactions on Database Systems*, 23(2):113–157, June 1998.
6. E. Laber, R. Carmo, and Y. Kohayakawa. Querying priced information in databases: The conjunctive case. Technical Report RT–MAC–2003–05, IME–USP, São Paulo, Brazil, July 2003.
7. E. S. Laber, O. Parekh, and R. Ravi. Randomized approximation algorithms for query optimization problems on two processors. In *Proceedings of ESA 2002*, pages 136–146, Rome, Italy, September 2002.
8. F. Porto. *Estratégias para a Execução Paralela de Consultas em Bases de Dados Científicos Distribuídos*. PhD thesis, Departamento de Informática, PUC-Rio, Apr. 2001.
9. A. C. Yao. Probabilistic computations: Toward a unified measure of complexity. In *18th Annual Symposium on Foundations of Computer Science*, pages 222–227, Long Beach, Ca., USA, Oct. 1977. IEEE Computer Society Press.

Sublinear Methods for Detecting Periodic Trends in Data Streams

Funda Ergun[1]*, S. Muthukrishnan[2]**, and S. Cenk Sahinalp[3]***

[1] Department of EECS, Case Western Reserve University. afe@eecs.cwru.edu
[2] Department of Computer Science, Rutgers University. muthu@cs.rutgers.edu
[3] Department of EECS, Case Western Reserve University. cenk@eecs.cwru.edu

Abstract. We present sublinear algorithms — algorithms that use significantly less resources than needed to store or process the entire input stream – for discovering representative trends in data streams in the form of periodicities. Our algorithms involve sampling $\tilde{O}(\sqrt{n})$ positions. and thus they scan not the entire data stream but merely a sublinear sample thereof. Alternately, our algorithms may be thought of as working on streaming inputs where each data item is seen once, but we store only a sublinear – $\tilde{O}(\sqrt{n})$ – size sample from which we can identify periodicities. In this work we present a variety of definitions of periodicities of a given stream, present sublinear sampling algorithms for discovering them, and prove that the algorithms meet our specifications and guarantees. No previously known results can provide such guarantees for finding any such periodic trends. We also investigate the relationships between these different definitions of periodicity.

1 Introduction

There is an abundance of time series data today collected by a varying and ever-increasing set of applications. For example, telecommunications companies collect traffic information–number of calls, number of dropped calls, number of bytes sent, number of connections etc. at each of their network links at small, say 5-minute, intervals. Such data is used for business decisions, forecasting, sizing, etc. based on trend analysis. Similarly time-series data is crucially used in decision support systems in many arenas including finance, weather prediction, etc.

There is a large body of work in time series data management, mainly on indexing, similarity searching, and mining of time series data to find various events and patterns. In this work, we are motivated by applications where the data is critically used for "trend analysis". We study a specific representative trend of time series, namely, *periodicity*. No real life time series is exactly periodic; i.e., repetition of a single pattern over and over again does not occur. For example,

* Supported in part by NSF CCR 0311548.
** Supported by NSF EIQ 0087022, NSF ITR 0220280 and NSF EIA 02-05116.
*** Supported in part by NSF CCR-0133791,IIS-0209145.

the number of bytes sent over an IP link in a network is almost surely not a perfect repeat of a daily, weekly or a monthly trend. However, many time series data are likely to be "approximately" periodic.

The main objective of this paper is to determine if a time series data stream is approximately periodic. The area of Signal Analysis in Applied Mathematics largely deals with finding various periodic components of a time series data stream. A significant body of work exists on stochastic or statistical time series trend analysis about predicting future values and outlier detection that grapples with the almost periodic properties of time series data.

In this paper we take a novel approach based on combinatorial pattern matching and random sampling to defining approximate periodicity and discovering approximate periodic behavior of time series data streams. The period of a data sequence is defined in terms of its self-similarity; this can be either in terms of the distance between the sequence and an appropriately shifted version of itself, or else in terms of the distance between different portions of the sequence. Motivated by these, our approach involves the following. We define several notions of self-distance for the input data streams for capturing the various combinatorial notions of approximate periodicity. Data streams with small self-distances are deemed to be approximately periodic; given time series data stream $S = S[1] \cdots S[n]$, we may define its self-distance (with respect to a candidate period p) as $\sum_{i \neq j} d(S[jp+1 : (j+1)p], S[ip+1 : (i+1)p])$, for some suitable distance function $d(., .)$ that captures the similarity between a pair of segments. We may now consider the time series data to be approximately periodic if the distance is below a certain threshold.

In this paper, we study algorithmic problems in discovering combinatorial periodic trends in time series data. Our main contributions are as follows.

1. We formulate different self-distances for defining approximately periodicity for time series data streams. Approximate periodicity in this sense will also indicate that only a small number of entries of the data set need to be changed to make it exactly periodic.

2. We present *sublinear* algorithms for determining if the input data stream is approximately periodic. In fact, our algorithms rely only on sampling a sublinear— $\tilde{O}(\sqrt{n})$—number of positions in the input.

A technical aspect of our approach is that we keep a small pool of random samples, even if we do not know in advance what the period might be. We show that there is always a subsample of this pool sufficient to compute the self-distance under any potential period. In this sense, we "recycle" the random samples for one approximate period to perform computations for other periods. For two notions of periodicity we define here, our methods are quite simple; for the third notion, the sampling (in Section 3.1) is more involved with two stages where the second stage depends on the first.

Related Work. Algorithmic literature on time series data analysis mostly focuses on indexing and searching problems, based on various distance measures amongst multiple time series data. Common distance measures are L_p norms, hierarchical distances motivated by wavelets, etc.[1]

Although most available papers do not consider the combinatorial periodicity notions we explore here, one relevant paper [6] aims to find "average period" of a given time series data in a combinatorial fashion. This paper describes $O(n \log n)$ space algorithms to estimate average periods by using *sketches*.

Our work here deviates from that in [6] in a number of ways. First, we present the first known $o(n)$, in fact, $O(\sqrt{n} \cdot \text{polylog } n)$ space algorithm for periodic trend analysis in contrast to the $\omega(n)$ space methods in [6]. We do not know of a way to employ sketches to design algorithms with our guarantees. Sampling seems to be ideal for us here: with a small number of samples we are able to perform computations for multiple period lengths. Second, we consider more general periodic trends than those in [6].

Sampling algorithms are known for computing Fourier coefficients with sublinear space [2]. However this algorithm is quite complex and expensive, using $(B \log n)^{O(1)}$ samples for finding B significant periodic components - the $O(1)$ factor is rather large. In general, there is a rich theory of sampling in time series data analysis [10,9]; our work is interesting in the way that it recycles random samples among multiple computations, and adds to this growing knowledge. Our methods are more akin to sublinear methods for property testing; see [4] for an overview. In particular, in parallel with this work and independent of it, authors in [1] present sublinear sampling methods for testing whether the edit distance between two strings is at least linear or at most n^α for $\alpha < 1$ by obtaining a directed sample set where the queries are at times evenly spaced within the strings.

2   Notions of Approximate Periodicity

Our definitions of approximate periodicity are based on the notion of *exact periodicity* from combinatorial pattern matching. We will first review that notion before presenting our main results.

Let S denote a time series data stream where each entry $S[i]$ is from a constant size alphabet σ. We denote by $S[i : j]$ the segment of S between the ith and the jth entries (inclusive). The exact periodicity of a data stream S with respect to a period of size p can be described in two alternative but equivalent ways as follows.

Definition 1. *We say that a data stream S of size n is exactly p-periodic if either*

a. *its size p suffix and size p prefix are identical; i.e., $S[1 : n - p] = S[p + 1 : n]$, or alternatively,*
b. *S consists of repetitions of the same block B of size p; i.e. $S = B^k B'$ where $B \in \sigma^p$, B' is a prefix of B and $k = \lfloor n/p \rfloor$.*

---

[1] A survey is in the tutorial offered at KDD 2000 [7]; see also [8].

When examining p-periodicity of a data stream S, we denote by b_i^p, the ith block of S of size p, that is, $S[(i-1)p+1 : (i-1)p]$. Notice that $S = b_1^p, b_2^p, \ldots b_k^p, b'$ where $k = \lfloor n/p \rfloor$ and b' is the length $n - kp$ suffix of S. When the choice of p is clear from the context, we drop it; i.e. we write $S = b_1, b_2, \ldots b_k, b'$. For simplicity, unless otherwise noted, we assume that the stream consists of a whole number of blocks, i.e., $n = kp$ for some $k > 0$, for any p under consideration. Any unfinished block at the end of the stream can be extended with *don't care* symbols until the desired format is obtained.

2.1 Self Distances and Approximate Periodicity

The above definitions of exact periodicity can be relaxed into a notion of approximate periodicity as follows. Intuitively, a data stream S can be considered approximately periodic if it can be made exactly periodic by changing a small number of its entries. To formally define approximate periodicity, we present the notion of a "self-distance" for a data stream. We will call a stream S approximately periodic if its self-distance is "small".

In what follows we introduce three self-distance measures (*shiftwise, blockwise* and *pairwise* distances, denoted respectively as D^p, E^p and F^p) each of which is defined with respect to a "base" distance between two streams. We will first focus on the Hamming distance $h(.,.)$ as the base distance for all three measures and subsequently discuss how to generalize our methods to other base distances.

Shiftwise Self Distance. We first relax Definition [a] of exact periodicity to obtain what we call the *shiftwise self-distance* of a data stream. As a preliminary step we define a simple notion of self-distance that we call the *single-shift self-distance* as follows.

Definition 2. *The single-shift self-distance of a data stream S with respect to period size p is $DS^p(S) = h(S[p+1 : n], S[1 : n - p])$.*

If one assumes for the sake of simplicity that $n = kp$, then it is possible to write $S = b_1^p b_2^p \ldots b_k^p$, and alternatively define the single-shift self-distance of S as $DS^p(S) = \sum_{i=1}^{k-1} h(b_i^p, b_{i+1}^p)$. Note that S is exactly p-periodic if and only if $DS^p(S) = 0$.

Unfortunately the single-shift self-distance of S fails to provide a satisfactory basis for approximate periodicity. A small $DS^p(S)$ does not necessarily imply that S can be made exactly p-periodic by changing a small number of its entries: Let $p = 1$ and $S = 00000000001111111111$. It is easy to see that $DS^1(S) = 1$. However, to make S periodic with $p = 1$ (in fact with *any* p) one needs to change a linear number of entries of S.

Even though S is "self similar" under $DS^1()$, it is clearly far from being exactly periodic as stipulated in Definition 1. Thus while Definition 1 (a) and (b) are equivalent in the context of exact periodicity, their simple relaxations for approximate periodicity can be quite different.

It is possible to generalize the notion of single-shift self-distance of S towards a more robust measure of self-similarity. Observe that if a data stream S is exactly p-periodic, it is also exactly $2p$-, $3p$-, ... periodic; i.e., if $DS^p(S) = 0$, then $DS^{2p}(S) = DS^{3p}(S) = \ldots = 0$. However, when $DS^p(S) = \ell > 0$ one can not say much about $DS^{2p}(S), DS^{3p}(S), \ldots$ in relation to ℓ. In fact, given S and p, $DS^{ip}(S)$ can grow linearly with i: observe in the example above that $DS^1(S) = 1$, $DS^2(S) = 2, \ldots DS^i(S) = i \ldots DS^{n/2}(S) = n/2$. A more robust notion of shiftwise self-distance can thus consider the self-distance of S w.r.t. all multiples of p as follows.

Definition 3. *The shiftwise self-distance of a given data stream S of length n with respect to p is defined as*

$$D^p(S) = \max_{j=1,\ldots n/p} h(S[jp+1:n], S[1:n-jp]).$$

In the subsequent sections we show that the shiftwise self-distance can be used to relax both definitions of exact periodicity up to a constant factor.

Blockwise Self Distance. Shiftwise self-distance is based on Definition [a] of exact periodicity. We now define a self-distance based on the alternative definition, which relates to the "average trend" of a data stream $S \in \sigma^n$ ([6]) defined in terms of a "representative" block b_i^p of S. More specifically, given block b_j^p of S, we consider the distance of the given string from one which consists only of repetitions of b_j^p. Define $E_j^p(S) = \sum_{\forall \ell} h(b_\ell^p, b_j^p)$. Based on this the notion of average trend, our alternative measure of self-distance for S (also used in [6]) is obtained as follows.

Definition 4. *The blockwise self-distance of a data stream S of length n w.r.t. p is defined as $E^p(S) = \min_i E_i^p(S)$.*

Blockwise self-distance is closely related to the shiftwise self-distance as will be shown in the following sections.

Pairwise Self-Distance. We finally present our third definition of self-distance, which, for a given p, is based on comparing all pairs of size p blocks. We call this distance the *pairwise self-distance* and define it as follows.

Definition 5. *Let S consist of k blocks $b_1^p, \ldots, b_k^p$, each of size p. The pairwise self-distance of S with respect to p and discrepancy δ is defined as*

$$F^p(S, \delta) = \frac{1}{k^2} |\{(b_i, b_j) \ : \ h(b_i, b_j) > \delta p\}|.$$

Observe that $F^p(S, \epsilon)$ is the ratio of "dissimilar" block pairs to all possible block pairs and thus is a natural measure of self-distance. A pairwise self-distance of ϵ reflects an accurate measure of the number of entries that need to be changed to make S exactly p-periodic up to an additive factor of $O((\epsilon + \delta)n)$ and thus is closely related to the other two self-distances.

3   Sublinear Algorithms for Measuring Self-Distances and Approximate Periodicity

A data stream S is thought of as being approximately p-periodic if its self-distance $(D^p(S), E^p(S)$ or $F^p(S, \delta))$ is below some threshold ϵ. Below, we present sublinear algorithms for testing whether a given data stream S is approximately periodic under each of the three self-distance measures. We also demonstrate that all the three definitions of approximate periodicity are closely related and can be used to estimate the minimum number of entries that must be changed to make a data stream exactly periodic.

We first define approximate periodicity under the three self-distance measures.

Definition 6. *A data stream $S \in \sigma^n$ is ϵ-approximately p-periodic with respect to D^p (resp. E^p and F^p) if $D^p(S) \leq \epsilon n$ (resp. $E^p(S) \leq \epsilon n$ and $F^p(S, \delta) \leq \epsilon n$) for some $p \leq n/2$.*

3.1   Checking Approximate Periodicity under D^p

We now show how to check whether S is ϵ-approximately p-periodic for a fixed $p \leq n/2$ under D^p. We generalize this to finding the smallest p for which S is ϵ-approximately p-periodic following the discussion on the other similarity measures.

We remind the reader that as typical of probabilistic tests, our method distinguishes self-distances of over ϵn from those below $\epsilon' n$. In our case, $\epsilon' = c\epsilon$ for some small constant $0 < c < 1$ which results from using probabilistic bounds.[2] The behavior of our method is not guaranteed when the self-distance is between ϵn and $\epsilon' n$.

We first observe that to estimate $DS^p(S)$ within a constant factor, it suffices to use a constant number of samples from S. More precisely, Given $S \in \sigma^n$ and $p \leq n/2$, one can determine whether $DS^p(S) \leq \epsilon n$ or $DS^p(S) \geq \epsilon' n$ with constant probability using $O(1)$ random samples from S. This is because, all we need to do is to estimate whether $h(S[p + 1 : n], S[1 : n - p])$ below $\epsilon' n$ or above ϵn. A simple application of Chernoff bounds shows us that comparing a constant number of sample pairs of the form $(S[i], S[i + p])$ is sufficient to obtain a correct answer with constant probability.

Recall that to test whether S is ϵ-approximately p-periodic, we need to compute each $DS^{ip}(S)$ separately for $ip \leq n/2$. When p is small, there are a linear number of such distances that we need to compute. If we choose to compute each

---

[2] Depending on ϵ, one has an amount of freedom in choosing c; for instance, $c = 1/2$ can be achieved through an application of Chernoff's or even Markov's inequality and the confidence obtained can be boosted through increasing the number of samples logarithmically in the confidence parameter. This will hold for the rest of this paper as well, and we will use ϵ and ϵ' without mentioning their exact relationship with this implicit understanding.

one separately, with different random samples (with the addition of a polylogarithmic factor for guaranteeing correctness for each period tested) this translates
into a superlinear number of samples. To economize on the number of samples
from S, we show how to "recycle" a sublinear pool of samples. This is viable as
our analysis does not require the samples to be determined independently.

Note that the definition of approximate periodicity w.r.t. D^p leads to the
following property analogous to that of exact periodicity.

Property 1. If S is ϵ-approximately p-periodic under D^p then it is ϵ-
approximately ip-periodic for all $i \leq n/2p$.

Our ultimate goal thus is to find the smallest p for which S is ϵ-approximately
p-periodic. We now explore how many samples are needed to estimate $DS^p(S)$
in the above sense for *all* $p = 1, 2, \cdots n/2$, which is sufficient for achieving our
goal.

In order to estimate $DS^p(S)$ for a specific p we need to compare $O(1)$ sample
pairs of the form $(S[i], S[i + p])$. We now would like to determine the number
of samples $S[i]$ required to guarantee that a sufficient number of sample pairs
$(S[i], S[i+p])$ will be available for each $n/2 \geq p \geq 1$. The following lemma states
that a pool of $O(\sqrt{n} \cdot \text{polylog } n)$ samples suffices.

Lemma 1. *A uniformly random sample pool of size $O(\sqrt{n} \cdot polylog\ n)$ from S
guarantees that $\Omega(1)$ sample pairs of the form $(S[i], S[i + p])$ are available for
every $1 \leq p \leq n/2$ with high probability.*

Proof. For any given p, one can use the birthday paradox using $O(\sqrt{n})$ samples
to show that availability of $\Omega(1)$ sample pairs of the form $(S[i], S[i + p])$ with
constant probability, say $1 - \rho$. For all possible values of p, the probability that
at least one of them will *not* provide enough samples is at most $1 - (1 - \rho)^{n/2}$.
Repeating the sampling $O(\text{polylog } n)$ times, this failure probability can be reduced to any desired $1/\text{poly } n$.   □

The lemma above demonstrates that by using $O(\sqrt{n} \cdot \text{polylog } n)$ independent
random samples from S one can test whether S is ϵ-approximately p-periodic
for any p. The below theorem follows.

Theorem 1. *It is possible to test whether a given $S \in \sigma^n$ is ϵ-approximately p-
periodic or is not ϵ'-approximately p-periodic under D^p by using $O(\sqrt{n} \cdot polylog\ n)$
samples and space with high probability.*

3.2   Checking Approximate Periodicity under E^p

Even though the blockwise self-distance $E^p(S)$ seems to be quite different from
shiftwise self-distance $D^p(S)$, we show that the two measures are closely related.
In fact we show that $D^p(S)$ and $E^p(S)$ are within a factor of 2 of each other:

Theorem 2. *Given $S \in \sigma^n$ and $p \leq n/2$, $E^p(S)/2 \leq D^p(S) \leq 2E^p(S)$.*

Proof. We first show the upper bound.

Let $b_i = b_i^p$ be the representative trend of S (of size p), that is, $i = argmin_{1 \leq j \leq k} \sum_{\ell=1}^{k} h(b_\ell, b_j)$. By definition, $D^p(S) = max_{1 \leq j \leq k} DS^{jp}(S) = max_j \sum_{\ell=1}^{k-j} h(b_\ell, b_{\ell+j})$.

By the triangular inequality, $D^p(S) \leq max_j[\sum_{\ell=1}^{k-j} h(b_\ell, b_i) + \sum_{\ell=1}^{k-j} h(b_i, b_{\ell+j})] \leq max_j[\sum_{\ell=1}^{k} h(b_\ell, b_i) + \sum_{\ell=1}^{k} h(b_i, b_\ell)]$. Since h is symmetric, this is at most $2\sum_{\ell=1}^{k} h(b_i, b_\ell)$, which is exactly $2 \cdot E^p(S)$.

For the lower bound, note that $E^p(S) \leq \frac{1}{k} \sum_{j=1}^{k} \sum_{\ell=1}^{k} h(b_\ell, b_j)$. But $D^p(S) \geq \frac{1}{k} \sum_{j=1}^{k} \sum_{\ell=1}^{k-j} h(b_{\ell+j}, b_j) \geq \frac{1}{2k} \sum_{j=1}^{k} \sum_{\ell=1}^{k} h(b_\ell, b_j) \geq E^p(S)/2$. □

As Theorem 2 implies, the two notions of self-distance (under Hamming measure) are equivalent up to a factor of 2. We have shown how to test whether the shiftwise self-distance of S, $D^p(S)$ is no more than some ϵn for any given p by using only a sublinear ($O(\sqrt{n} \cdot \text{polylog } n)$) number of samples from S and similar space. The above lemma implies that this is also doable for $E^p(S)$; i.e. one can test whether the blockwise self-distance of S is no more than some ϵn for any given p by using $O(\sqrt{n} \cdot \text{polylog } n)$ samples from S and similar space.

The method presented in [6] can also perform this test by first constructing from S a superlinear ($O(kn \log n)$) size pool of "sketches"; here k is the size of an individual sketch which depends on their confidence bound. Since this pool can be too large to fit in main memory, a scheme is developed to retrieve the pool from secondary memory in smaller chunks. In contrast, our overall memory requirement (and sample size) is sublinear; this comes at a price of some small loss of accuracy.

Due to the fact that $D^p()$ and $E^p()$ are within a factor 2 of each other, they can be estimated in the same manner. Thus, the theorem below follows from its counterpart for D^p, (Theorem 3), which states that approximate p-periodicity can be efficiently checked.

Theorem 3. *It is possible to test whether a given $S \in \sigma^n$ is ϵ-approximately p-periodic or is not ϵ'-approximately p-periodic under E^p by using $O(\sqrt{n} \cdot \text{polylog } n)$ samples and space with high probability.*

Here the "gap" between ϵ and ϵ' is within factor 4 of the gap for $D^p()$.

Non-Hamming Measures. We showed above how to test whether a data stream S of size n is ϵ-approximately p-periodic using self-distances $D^p()$ and $E^p()$. We assumed that the comparison of blocks was done in terms of the Hamming distance. We now show how to use other distances of interest.

First, consider the L_1 distance. Note that, since our alphabet σ is of constant size, the L_1 distance between two data streams is within a constant factor of their Hamming distance. More specifically, let $q = |\sigma|$. Then, for any $R, S \in \sigma^n$, $q \cdot h(R, S) \geq L_1(R, S)$. Thus, the method of estimating the Hamming distance will satisfy the requirements of our test for L_1 albeit with different constant factors. Let D' and E' be the self-distance measures which modify the Hamming distance based measures of D and E by the use of L_1 distance. Then, for any given p our estimate $D'^p(S)$ will still be within at most a constant factor of $E'^p(S)$.

Now consider the L_2 distance. Again, assuming that our alphabet σ is of size q one can observe that, if $h(R, S) = p$, then $\sqrt{p} \leq L_2(R, S) \leq q\sqrt{p}$. Thus, by making the necessary adjustments to the allowed distance, one can obtain a test with different constant factors as with the L_1 distance. In fact, a similar argument holds for any L_i distance.

Similar discussions apply for F^p as well and are hence omitted.

3.3   Checking Approximate Periodicity under F^p

Recall that F^p is a measure of the frequency of dissimilar blocks of size p in S. In this section, we show how to efficiently test whether S is ϵ-approximately p-periodic under F^p (for any p where p is not known a priori); we will later employ this technique to find all periods and the smallest period of S efficiently. In order to be able to estimate $F^p(S, \delta)$ for all p, we would like to compare pairs of blocks explicitly. This requires as many as polylogarithmic sample pairs *within each pair of blocks* (b_i, b_j) *of size* p that we compare. Unfortunately, our pool of samples from the previous section turns out to be too small to yield enough sample pairs of the above kind for all p – in fact, it can be seen easily that a sublinear uniform random sample pool will never achieve the desired sample distribution and the desired confidence bounds in this case. Instead, we present a more directed sampling scheme, which will collect a sublinear size sample pool and still have enough samples to perform the test for any period p.

A Two-Phase Scheme to Obtain The Sample Pool. To achieve a sublinear sample pool from S which will have enough per block samples, we obtain our samples in two phases.

In the first phase we obtain a uniform sample pool from S, as in the previous section, of size $O(\sqrt{n} \cdot \text{polylog } n)$; these samples are called *primary samples*.

In the second phase, we obtain, for each primary sample $S[i]$, a polylogarithmic set of *secondary samples* distributed identically around i (respecting the boundaries of S). To do this, we pick $O(\text{polylog } n)$ offsets relative to a generic location i as follows. We pick $O(\log n)$ neighborhoods of size 1, 2, 4, 8, ... n around i.[3] Neighborhood k refers to the interval $S[i - 2^{k-1} : i + 2^{k-1} - 1]$; e.g., neighborhood 3 (of size 8) of $S[i]$ is $S[i - 4 : i + 3]$. From each neighborhood we pick $O(\text{polylog } n)$ uniform random locations and note their positions relative to i. Note that the choosing of offsets is performed only once for a generic i; the same set of offsets will later be used for all primary samples.

To obtain the secondary samples for any primary sample $S[i]$, we sample the locations indicated by the offset set with respect to location i (as long as the sample location is within S).[4] Note that the secondary samples for any two

---

[3]   Since we are only choosing offsets, we allow neighborhoods to go past the boundaries of S. We handle invalid locations during the actual sampling. Also, for simplicity, we assume n to be a power of 2.

[4]   For easier use in the algorithm later, for each sample the size of the neighborhood from which it is picked is also noted.

primary samples $S[i]$ and $S[j]$ are located identically with around respective locations i and j.

Estimating F^p. We can now use standard techniques to decide whether $F^p(S, \delta)$ is large or small. We start by uniformly picking primary sample pairs $(S[i], S[j])$ such that $i - j$ is a multiple of p.[5] Call the size p blocks containing $S[i]$ and $S[j]$ b_k and b_l. We can now proceed to check whether $h(b_k, b_l)$ is large by comparing these two blocks at random locations. To obtain the necessary samples for this comparison, we use our sample pool and the neighborhoods used in creating it as follows. We consider the smallest neighborhood around $S[i]$ which contains b_k and use the secondary samples of $S[i]$ from this neighborhood that fall within b_k. We then pick samples from b_l in a similar way and compare the samples from b_k and b_l to check $h(b_k, b_l)$. We repeat the entire procedure for the next block pair until sufficient block pairs have been tested.

To show that this scheme works, we first show that we have sufficient primary samples for any given p to compare enough pairs of blocks. To do this, for any p, we need to pick $O(\text{polylog } n)$ pairs of size p blocks uniformly, which is possible given our sample set as the following simple lemma demonstrates.

Lemma 2. *Consider all sample pairs $(S[i], S[j])$ from a set of $O(\sqrt{n} \cdot \text{polylog } n)$ primary samples uniformly picked from a data stream S of length n. Given any $0 < p \leq n/2$, the following hold with high probability:*

(a) There are $\Omega(\text{polylog } n)$ pairs $(S[i], S[j])$ that one can obtain from the primary samples such that $i - j$ is a multiple of p.

(b) Consider block pair (b_i, b_j) containing a sample pair $(S[i], S[j])$ as described in (a). (b_i, b_j) is uniformly distributed in the space of all block pairs of of S of size p.[6]

Proof. (a) follows easily from Lemma 1.

To see (b), consider two block pairs (b_i, b_j) and (b_k, b_l). There are p sample pairs which will induce the picking of the former pair, and the same holds for the latter pair. Thus, any block pair will be picked with equal probability.

Thus, our technique allows us to have, for any p, a polylogarithmic size uniform sample of block pairs of size p. Now, consider the secondary samples within a block that we pick for comparing two blcoks as explained before. It is easy to see that these particular samples are uniformly distributed within their respective blocks, since secondary samples within any one neighborhood are uniformly distributed. Additionally, they are located at identical locations within their blocks. All we need is there to be a sufficient number of such samples, which we argue below.

---

[5] There are several simple ways of doing this without violating our space bounds which involve time/space tradeoffs that are not immediately relevant to this paper. Additionally, picking the pairs without replacement makes the final analysis more obvious but makes the selection process slightly more complicated.

[6] For simplicity we assume that p divides n; otherwise one needs to be a little careful during the sampling to take care of the boundaries.

Lemma 3. *Let $S[i]$ and $S[i + rp]$ be two primary samples. Let b_l and b_m be the blocks of size p that contain $S[i]$ and $S[i + rp]$ respectively. Then, with the sampling scheme desribed above we will have picked sufficient secondary samples to tell whether $h(b_l, b_m) \geq \delta p$ high probability.*

Proof. Consider t such that $2^{t-1} < p \leq 2^t$. The $t + 1$-neighborhood of $S[i]$ is of size at most $4p$, and contains b_l. Since b_l occupies at least $1/4$ of this neighborhood, it is expected to contain at least a quarter of the secondary samples of $S[i]$ from this neighborhood, which will be uniformly distributed in b_l. The case is the same for b_m and the samples it contains. As a result, we have $\Omega(\text{polylog } n)$ uniform random samples from both b_l and b_m, which, as we argued before, can be viewed as pairs of points located identically within their respective blocks. Then, one can test whether $h(b_l, b_m)$ with high probability by comparing the corresponding sample pairs from each block.   □

Combining the choice of blocks and the comparison of block pairs, we obtain the following theorem.

Theorem 4. *It is possible to test whether a given $S \in \sigma^n$ is ϵ-approximately p-periodic or is not ϵ'-approximately p-periodic under F^p by using $O(\sqrt{n} \cdot \text{polylog } n)$ samples and space with high probability.*

Since our algorithm does not require advance knowledge of p, to find all periods, the smallest period, etc. under this measure, it suffices to try the test with different values of p without increasing the sample size, as we argue in the next section.

3.4   Checking Periodicity for All Periods

So far we have focused on testing periodicity for one period. In general we not have access to a hypothetical period and may want to know whether a data stream S is ϵ-approximately periodic with any period, and/or what its smallest period p is. These can easily be determined once the particular similarity measure is evaluated for all possible p. Since D^p and E^p involve computing similarities for all p, for these two measures it is easy to extend the computation to all p. As for F^p, checking for approximate periodicity for a fixed p is easy, but the trivial technique of picking blocks and sampling will not extend to efficiently checking for all p. However, our technique as described in the previous section is specially designed so that its sample set will work with high probability for any and every valid p. Thus, checking periodicity for varying periods is now possible by using sublinear samples.

Theorem 5. *Given $S \in \sigma^n$, it is possible to perform any of the following tasks under D^p, E^p, and F^p by using $O(\sqrt{n} \cdot \text{polylog } n)$ independent random samples from S and similar space:*
a) to find out if S is ϵ-approximately p-periodic,
b) to find all periods p (and thus the smallest period) for which S is ϵ-approximately p-periodic.

Note that if the smallest approximate period of S is determined to be p then we guarantee that $D^p(S) \leq \epsilon' n$ and there exists no $j < p$ such that $D^j(S) < \epsilon n$. The same holds for E^p and F^p as well.

3.5 Relationships between Three Notions of Approximate Periodicity

We have already demonstrated that E^p and D^p are equivalent up to a factor of 2. We now demonstrate that F^p is closely related to both of these definitions of self-similarity in the sense that all three measures can be used to test whether a data stream S is *almost periodic* as follows.

Definition 7. *A data stream S of size n is called almost p-periodic w.r.t. γ if γn of its entries can be changed to make S exactly p-periodic.*

The next lemma relates the notion of approximate periodicity under E^p (and thus D^p) with that of almost periodicity.

Lemma 4. *If a data stream S is ϵ-approximate p-periodic under E^p then it is almost p-periodic w.r.t. γ for $\epsilon/2 \leq \gamma \leq \epsilon$.*

Proof. Let $B = argmin_{b \in \sigma^p} \sum_{\ell=1}^{k} h(b_\ell, b)$. Clearly S is almost p-periodic if $\sum_{\ell=1}^{k} h(b_\ell, B) \leq \gamma n$. Similarly let $b_i = b_i^p$ be the representative trend of S; i.e. $i = argmin_{1 \leq j \leq k} \sum_{\ell=1}^{k} h(b_\ell, b_j)$. However:

$$k\epsilon = k \sum_{\ell=1}^{k} h(b_\ell, b_i)$$

$$\leq \sum_{\forall \ell} \sum_{\forall j} h(b_\ell, b_j) \leq \sum_{\forall \ell} \sum_{\forall j} h(b_\ell, B) + h(B, b_j) \leq 2k \sum_{\ell=1}^{k} h(b_\ell, B) = 2k\gamma.$$

The second part of the inequality is trivial.   □

Finally one can easily verify the following relationship between approximate periodicity under F^p and almost periodicity.

Lemma 5. *If for a given data stream S, $F^p(S, \delta) \leq \epsilon$ then S can be made exactly periodic by changing $O((\delta + \epsilon)n)$ of its entries.*

4 Concluding Remarks

We introduced new notions of time series data streams being approximately periodic based on significance of combinatorial scores in terms of self-distances. We presented the first known sublinear–$\tilde{O}(\sqrt{n})$ space– algorithms for detecting such approximate periodicities in time series data streams based on sampling, and reusing these random samples for multiple potential period lengths. Besides such periodicities, there may be other representative trends in a data stream; it could be interesting to develop efficient, sublinear sampling algorithms for detecting such trends.

References

1. T. Batu, F. Ergun, J. Kilian, A. Magen, S. Raskhodnikova, R. Rubinfeld and R. Sami. *A sublinear algorithm for weakly approximating edit distance.* STOC 2003, 316–324.
2. A. Gilbert, S. Guha, P. Indyk, S. Muthukrishnan and M. Strauss. *Near-optimal sparse fourier representations via sampling.* Proc. STOC 2002. 152–161.
3. O Goldreich, S. Goldwasser and D. Ron. *Property testing and its connection to learning and approximation,* Journal of the ACM 45(4):653–750, 1998.
4. R. Rubinfeld. Talk on sublinear algorithms. http://external.nj.nec.com/homepages/ronitt/
5. R. Rubinfeld and M. Sudan, *Robust Characterization of Polynomials with Applications to Program Testing,* SIAM Journal of Computing 25(2):252–271, 1996.
6. P. Indyk and N. Koudas and S. Muthukrishnan *Identifying Representative Trends in Massive Time Series Data Sets Using Sketches.* Proc. VLDB 2000. 363–372.
7. G. Das and D. Gunopoulos. *Time Series Similarity Measures.* http://www.acm.org/sigs/sigkdd/kdd2000/Tutorial-Das.htm
8. G. Kollios. *Timeseries Indexing.* http://www.cs.bu.edu/faculty/gkollios/ada01/LectNotes/tsindexing.ppt
9. F. Olken and D. Rotem. *Random sampling from databases: A Survey. Bibliography at* http://pueblo.lbl.gov/ olken/mendel/sampling/bibliography.html
10. S. Chaudhuri, G. Das, M. Datar, R. Motwani and V. Narasayya. *Overcoming Limitations of Sampling for Aggregation Queries.* Proc. ICDE 2001.

An Improved Data Stream Summary: The Count-Min Sketch and Its Applications

Graham Cormode[1]* and S. Muthukrishnan[2]**

[1] Center for Discrete Mathematics and Computer Science (DIMACS) Rutgers University, Piscataway NJ. graham@dimacs.rutgers.edu

[2] Division of Computer and Information Systems, Rutgers University. muthu@cs.rutgers.edu

Abstract. We introduce a new *sublinear space* data structure—the *Count-Min Sketch*— for summarizing data streams. Our sketch allows fundamental queries in data stream summarization such as point, range, and inner product queries to be approximately answered very quickly; in addition, it can be applied to solve several important problems in data streams such as finding quantiles, frequent items, etc. The time and space bounds we show for using the CM sketch to solve these problems significantly improve those previously known — typically from $1/\varepsilon^2$ to $1/\varepsilon$ in factor.

1 Introduction

We consider a vector a, which is presented in an implicit, incremental fashion. This vector has dimension n, and its current state at time t is $a(t) = [a_1(t), \ldots a_i(t), \ldots, a_n(t)]$. Initially, a is the zero vector, $a_i(0) = 0$ for all i. Updates to individual entries of the vector are presented as a stream of pairs. The tth update is (i_t, c_t), meaning that $a_{i_t}(t) = a_{i_t}(t-1) + c_t$, and $a_{i'}(t) = a_{i'}(t-1)$ for all $i' \neq i_t$. At any time t, a *query* calls for computing certain functions of interest on $a(t)$.

This setup is the *data stream* scenario that has emerged recently. Algorithms for computing functions within the data stream context need to satisfy the following desiderata. First, the space used by the algorithm should be small, at most poly-logarithmic in n, the space required to represent a explicitly. Since the space is sublinear in data and input size, the data structures used by the algorithms to represent the input data stream is merely a summary—aka a *sketch* or synopsis [10]—of it; because of this compression, almost no function that one needs to compute on a can be done precisely, so some approximation is provably needed. Second, processing an update should be fast and simple; likewise, answering queries of a given type should be fast and have usable accuracy guarantees. Typically, accuracy guarantees will be made in terms of a pair of user specified parameters, ε and δ, meaning that the error in answering a query is within a factor of ε with probability δ. The space and update time will consequently depend on ε and δ; our goal will be limit this dependence as much as is possible.

Many applications that deal with massive data, such as Internet traffic analysis and monitoring contents of massive databases, motivate this one-pass data stream setup.

* Supported by NSF ITR 0220280 and NSF EIA 02-05116.
** Supported by NSF EIA 0087022, NSF ITR 0220280 and NSF EIA 02-05116.

There has been a frenzy of activity recently in the Algorithm, Database and Networking communities on such data stream problems, with multiple surveys, tutorials, workshops and research papers. See [7,3,16] for detailed description of the motivations driving this area.

In recent years, several different sketches have been proposed in the data stream context that allow a number of simple aggregation functions to be approximated. Quantities for which efficient sketches have been designed include the L_1 and L_2 norms of vectors [2], the number of distinct items in a sequence (ie number of non-zero entries in $a(t)$) [8], join and self-join sizes of relations (representable as inner-products of vectors $a(t), b(t)$) [2,1], item and range sum queries [12,4]. These sketches are of interest not simply because they can be used to directly approximate quantities of interest, but also because they have been used considerably as "black box" devices in order to compute more sophisticated aggregates and complex quantities: quantiles [13], wavelets [12], and histograms [11]. Sketches thus far designed are typically linear functions of their input, and can be represented as projections of an underlying vector representing the data with certain randomly chosen projection matrices. This means that it is easy to compute certain functions on data that is distributed over sites, by casting them as computations on their sketches. So, they are suited for distributed applications too.

While sketches have proved powerful, they have the following drawbacks.

- Although sketches use small space, the space used typically has a $\Omega(1/\varepsilon^2)$ multiplicative factor. This is discouraging because $\varepsilon = 0.1$ or 0.01 is quite reasonable and already, this factor proves expensive in space, and consequently, often, in per-update processing and function computation times as well.
- Many sketch constructions require time linear in the size of the sketch to process each update to the underlying data [2,13]. Sketches are typically a few kilobytes up to a megabyte or so, and processing this much data for every update severely limits the update speed.
- Sketches are typically constructed using hash functions with strong independence guarantees, such as p-wise independence [2], which can be complicated to evaluate, particularly for a hardware implementation. One of the fundamental questions is to what extent such sophisticated independence properties are needed.
- Many sketches described in the literature are good for one single, pre-specified aggregate computation. Given that in data stream applications one typically monitors multiple aggregates on the same stream, this calls for using many different types of sketches, which is a prohibitive overhead.
- Known analyses of sketches hide large multiplicative constants in big-Oh notation.

Given that the area of data streams is being motivated by extremely high performance monitoring applications—eg., see [7] for response time requirements for data stream algorithms that monitor IP packet streams—these drawbacks ultimately limit the use of many known data stream algorithms within suitable applications.

We will address all these issues by proposing a new sketch construction, which we call the *Count-Min*, or CM, sketch. This sketch has the advantages that: (1) space used is proportional to $1/\varepsilon$; (2) the update time is significantly sublinear in the size of the sketch; (3) it requires only pairwise independent hash functions that are simple to construct; (4) this sketch can be used for several different queries and multiple applications; and (5) all the constants are made explicit and are small. Thus, for the applications we discuss, our constructions strictly improve the space bounds of previous results from $1/\varepsilon^2$ to $1/\varepsilon$ and the time bounds from $1/\varepsilon^2$ to 1, which is significant.

Recently, a $\Omega(1/\varepsilon^2)$ space lower bound was shown for a number of data stream problems: approximating frequency moments $F_k(t) = \sum_k (a_i(t))^k$, estimating the number of distinct items, and computing the Hamming distance between two strings [17]. It is an interesting contrast that for a number of similar seeming problems (finding Heavy Hitters and Quantiles in the most general data stream model) we are able to give an $O(\frac{1}{\varepsilon})$ upper bound. Conceptually, CM Sketch also represents progress since it shows that pairwise independent hash functions suffice for many of the fundamental data stream applications. From a technical point of view, CM Sketch and its analyses are quite simple. We believe that this approach moves some of the fundamental data stream algorithms from the theoretical realm to the practical. Our results have some technical nuances: (1) The accuracy estimates for individual queries depend on the L_1 norm of $a(t)$ in contrast to the previous works that depend on the L_2 norm. (2) Most prior sketch constructions relied on embedding into small dimensions to estimate norms. Avoiding such embeddings allows our construction to avoid $\Omega(\frac{1}{\varepsilon^2})$ lower-bounds on these embeddings.

2   Preliminaries

We consider a vector a, which is presented in an implicit, incremental fashion. This vector has dimension n, and its current state at time t is $a(t) = [a_1(t), \ldots a_i(t), \ldots a_n(t)]$. For convenience, we shall usually drop t and refer only to the current state of the vector. Initially, a is the zero vector, 0, so $a_i(0)$ is 0 for all i. Updates to individual entries of the vector are presented as a stream of pairs. The tth update is (i_t, c_t), meaning that

$$a_{i_t}(t) = a_{i_t}(t-1) + c_t; \quad a_{i'}(t) = a_{i'}(t-1) \ i' \neq i_t$$

In some cases, c_ts will be strictly positive, meaning that entries only increase; in other cases, c_ts are allowed to be negative also. The former is known as the *cash register* case and the latter the *turnstile* case [16]. There are two important variations of the turnstile case to consider: whether a_is may become negative, or whether the application generating the updates guarantees that this will never be the case. We refer to the first of these as the *general* case, and the second as the *non-negative* case. Many applications that use sketches to compute queries of interest—such as monitoring database contents, analyzing IP traffic seen in a network link—guarantee that counts will never be negative. However, the general case occurs in important scenarios too, for example in distributed settings where one considers the subtraction of one vector from another, say.

At any time t, a *query* calls for computing certain functions of interest on $a(t)$. We focus on approximating answers to three types of query based on vectors a and b.

- A *point query*, denoted $\mathcal{Q}(i)$, is to return an approximation of a_i.
- A *range query* $\mathcal{Q}(l, r)$ is to return an approximation of $\sum_{i=l}^{r} a_i$.
- An *inner product query*, denoted $\mathcal{Q}(a, b)$ is to approximate $a \odot b = \sum_{i=1}^{n} a_i b_i$.

These queries are related: a range query is a sum of point queries; both point and range queries are specific inner product queries. However, in terms of approximations to these queries, results will vary. These are the queries that are fundamental to many applications in data stream algorithms, and have been extensively studied. In addition, they are of interest in non-data stream context. For example, in databases, the point and range queries are of interest in summarizing the data distribution approximately; and inner-product queries allow approximation of join size of relations. Fuller discussion of these aspects can be found in [9,16].

We will also study use of these queries to compute more complex functions on data streams. As examples, we will focus on the two following problems. Recall that $||a||_1 = \sum_{i=1}^{n} |a_i(t)|$; more generally, $||a||_p = (\sum_{i=1}^{n} |a_i(t)|^p)^{1/p}$.

- (ϕ-Quantiles) The ϕ-quantiles of the cardinality $||a||_1$ multiset of (integer) values each in the range $1 \ldots n$ consist of those items with rank $k\phi||a||_1$ for $k = 0 \ldots 1/\phi$ after sorting the values. Approximation comes by accepting any integer that is between the item with rank $(k\phi - \varepsilon)||a||_1$ and the one with rank $(k\phi + \varepsilon)||a||_1$ for some specified $\varepsilon < \phi$.
- (Heavy Hitters) The ϕ-heavy hitters of a multiset of $||a||_1$ (integer) values each in the range $1 \ldots n$, consist of those items whose multiplicity exceeds the fraction ϕ of the total cardinality, i.e., $a_i \geq \phi||a||_1$. There can be between 0 and $\frac{1}{\phi}$ heavy hitters in any given sequence of items. Approximation comes by accepting any i such that $a_i \geq (\phi - \epsilon)||a||_1$ for some specified $\varepsilon < \phi$.

Our goal is to solve the queries and the problems above using a sketch data structure, that is using space and time significantly sublinear—polylogarithmic—in input size n and $||a||_1$. All our algorithms will be approximate and probabilistic; they need two parameters, ε and δ, meaning that the error in answering a query is within a factor of ε with probability δ. Both these parameters will affect the space and time needed by our solutions. Each of these queries and problems has a rich history of work in the data stream area. We refer the readers to surveys [16,3], tutorials [9], as well as the general literature.

3 Count-Min Sketches

We now introduce our data structure, the Count-Min, or CM, sketch. It is named after the two basic operations used to answer point queries, counting first and computing the minimum next. We use e to denote the base of the natural logarithm function, ln.

Data Structure. A *Count-Min (CM) sketch* with parameters (ε, δ) is represented by a two-dimensional array counts with width w and depth d: $count[1, 1] \ldots count[d, w]$. Given parameters (ε, δ), set $w = \lceil \frac{e}{\varepsilon} \rceil$ and $d = \lceil \ln \frac{1}{\delta} \rceil$. Each entry of the array is initially zero. Additionally, d hash functions $h_1 \ldots h_d : \{1 \ldots n\} \rightarrow \{1 \ldots w\}$ are chosen uniformly at random from a pairwise-independent family.

Update Procedure. When an update (i_t, c_t) arrives, meaning that item a_{i_t} is updated by a quantity of c_t, then c_t is added to one count in each row; the counter is determined by h_j. Formally, set $\forall 1 \leq j \leq d : count[j, h_j(i_t)] \leftarrow count[j, h_j(i_t)] + c_t$ The space used by Count-Min sketches is the array of wd counts, which takes wd words, and d hash functions, each of which can be stored using 2 words when using the pairwise functions described in [15].

4 Approximate Query Answering Using CM Sketches

For each of the three queries introduced in Section 2: Point, Range, and Inner Product queries, we show how they can be answered using Count-Min sketches.

4.1 Point Query

We first show the analysis for point queries for the non-negative case.

Estimation Procedure. The answer to $Q(i)$ is given by $\hat{a}_i = \min_j count[j, h_j(i)]$.

Theorem 1. *The estimate $\hat{a}_i$ has the following guarantees: $a_i \leq \hat{a}_i$; and, with probability at least $1 - \delta$, $\hat{a}_i \leq a_i + \varepsilon ||a||_1$.*

Proof. We introduce indicator variables $I_{i,j,k}$, which are 1 if $(i \neq k) \wedge (h_j(i) = h_j(k))$, and 0 otherwise. By pairwise independence of the hash functions, then

$$E(I_{i,j,k}) = \Pr[h_j(i) = h_j(k)] \leq 1/\operatorname{range}(h_j) = \frac{\varepsilon}{e}.$$

Define the variable $X_{i,j}$ (random over the choices of h_i) to be $X_{i,j} = \sum_{k=1}^n I_{i,j,k} a_k$. Since all a_i are non-negative in this case, $X_{i,j}$ is a non-negative variable. By construction, $count[j, h_j(i)] = a_i + X_{i,j}$. So, clearly, $\min count[j, h_j(i)] \geq a_i$. For the other direction, observe that

$$E(X_{i,j}) = E\left(\sum_{k=1}^n I_{i,j,k} a_k\right) \leq \sum_{k=1}^n a_k E(I_{i,j,k}) \leq \frac{\varepsilon}{e}||a||_1$$

by pairwise independence of h_j, and linearity of expectation. By the Markov inequality,

$$\begin{aligned}
\Pr[\hat{a}_i > a_i + \varepsilon ||a||_1] &= \Pr[\forall j.\ count[j, h_j(i)] > a_i + \varepsilon ||a||_1] \\
&= \Pr[\forall j.\ a_i + X_{i,j} > a_i + \varepsilon ||a||_1] \\
&= \Pr[\forall j.\ X_{i,j} > eE(X_{i,j})] < e^{-d} \leq \delta \quad \blacksquare
\end{aligned}$$

The time to produce the estimate is $O(\ln \frac{1}{\delta})$ since finding the minimum count can be done in linear time; the same time bound holds for updates. The constant e is used here to minimize the space used: more generally, we can set $w = e/b$ and $d = \log_b \frac{1}{\delta}$ for any $b > 1$ to get the same accuracy guarantee. Choosing $b = e$ minimizes the space used, since this solves $\frac{d(wd)}{db} = 0$, giving a cost of $(2 + \frac{e}{\varepsilon}) \ln \frac{1}{\delta}$ words. For implementations, it may be preferable to use other (integer) values of b for simpler computations or faster updates.

The best known previous result using sketches was in [4]: there sketches were used to approximate point queries. Results were stated in terms of the frequencies of individual items. For arbitrary distributions, the space used is $O(\frac{1}{\varepsilon^2} \log \frac{1}{\delta})$, and the dependency on ε is $\frac{1}{\varepsilon^2}$ in every case considered.

In the full version of this paper[1] we describe how all existing sketch constructions can be viewed as variations of a common construction. This emphasizes the importance of our attempt to find the simplest sketch construction which has the best guarantees and smallest constants. A similar result holds when entries of the implicit vector a may be negative, which is the general case. Details of this appear in the full version of this paper.

---

[1] To appear in Journal of Algorithms

4.2   Inner Product Query

Estimation Procedure. Set $(\widehat{a \odot b})_j = \sum_{k=1}^{w} count_a[j,k] * count_b[j,k]$. Our estimation of $\mathcal{Q}(a,b)$ for non-negative vectors a and b is $\widehat{a \odot b} = \min_j (\widehat{a \odot b})_j$.

Theorem 2. $a \odot b \leq \widehat{a \odot b}$ and, with probability $1 - \delta$, $\widehat{a \odot b} \leq a \odot b + \varepsilon ||a||_1 ||b||_1$.

Proof.

$$(\widehat{a \odot b})_j = \sum_{i=1}^{n} a_i b_i + \sum_{p \neq q, h_j(p) = h_j(q)} a_p b_q$$

Clearly, $a \odot b \leq \widehat{a \odot b}_j$ for non-negative vectors. By pairwise independence of h,

$$E(\widehat{a \odot b}_j - a \odot b) = \sum_{p \neq q} \Pr[h_j(p) = h_j(q)] a_p b_q \leq \sum_{p \neq q} \frac{\varepsilon a_p b_q}{e} \leq \frac{\varepsilon ||a||_1 ||b||_1}{e}$$

So, by the Markov inequality, $\Pr[\widehat{a \odot b} - a \odot b > \varepsilon ||a||_1 ||b||_1] \leq \delta$, as required. ∎

The space and time to produce the estimate is $O(\frac{1}{\varepsilon} \log \frac{1}{\delta})$. Updates are performed in time $O(\log \frac{1}{\delta})$.

Join size estimation is important in database query planners in order to determine the best order in which to evaluate queries. The *join size* of two database relations on a particular attribute is the number of items in the cartesian product of the two relations which agree the value of that attribute. We assume without loss of generality that attribute values in the relation are integers in the range $1 \ldots n$. We represent the relations being joined as vectors a and b so that the values a_i represents the number of tuples which have value i in the first relation, and b_i similarly for the second relation. Then clearly $a \odot b$ is the join size of the two relations. Using sketches allows estimates to be made in the presence of items being inserted to and deleted from relations. The following corollary follows from the above theorem.

Corollary 1. *The Join size of two relations on a particular attribute can be approximated up to $\varepsilon ||a||_1 ||b||_1$ with probability $1 - \delta$, by keeping space $O(\frac{1}{\varepsilon} \log \frac{1}{\delta})$.*

Previous results have used the "tug-of-war" sketches [1]. However, here some care is needed in the comparison of the two methods: the prior work gives guarantees in terms of the L_2 norm of the underlying vectors, with additive error of $\varepsilon ||a||_2 ||b||_2$; here, the result is in terms of the L_1 norm. In some cases, the L_2 norm can be quadratically smaller than the L_1 norm. However, when the distribution of items is non-uniform, for example when certain items contribute a large amount to the join size, then the two norms are closer, and the guarantees of the CM sketch method is closer to the existing method. As before, the space cost of previous methods was $\Omega(\frac{1}{\varepsilon^2})$, so there is a significant space saving to be had with CM sketches.

4.3   Range Query

Estimation Procedure. We will adopt the use of *dyadic ranges* from [13]: a dyadic range is a range of the form $[x2^y + 1 \ldots (x+1)2^y]$ for parameters x and y. Keep $\log_2 n$ CM sketches, in order to answer range queries $\mathcal{Q}(l, r)$ approximately. Any range query can be reduced to at most $2 \log_2 n$ *dyadic range* queries, which in turn can each be reduced to a single point query. Each point in the range $[1 \ldots n]$ is a member of $\log_2 n$ dyadic ranges, one for each y in the range $0 \ldots \log_2(n) - 1$. A sketch is kept for each set of dyadic ranges of length 2^y, and update each of these for every update that arrives. Then, given a range query $\mathcal{Q}(l, r)$, compute the at most $2 \log_2 n$ dyadic ranges which canonically cover the range, and pose that many point queries to the sketches, returning the sum of the queries as the estimate. Let $a[l, r] = \sum_{i=l}^{r} a_i$ be the answer to the query $\mathcal{Q}(l, r)$ and let $\hat{a}[l, r]$ be the estimate using the procedure above.

Theorem 3. $a[l, r] \leq \hat{a}[l, r]$ *and with probability at least* $1 - \delta$,

$$\hat{a}[l, r] \leq a[l, r] + 2\varepsilon \log n \|a\|_1.$$

Proof. Applying the inequality of Theorem 1, then $a[l, r] \leq \hat{a}[l, r]$. Consider each estimator used to form $\hat{a}[l, r]$; the expectation of the additive error for any of these is $2 \log n \frac{\varepsilon}{e} \|a\|_1$, by linearity of expectation of the errors of each point estimate. Applying the same Markov inequality argument as before, the probability that this additive error is more than $2\varepsilon \log n \|a\|_1$ for any estimator is less than $\frac{1}{e}$; hence, for all of them the probability is at most δ. ∎

The time to compute the estimate or to make an update is $O(\log(n) \log \frac{1}{\delta})$. The space used is $O(\frac{\log(n)}{\varepsilon} \log \frac{1}{\delta})$.

The above theorem states the bound for the standard CM sketch size. The guarantee will be more useful when stated without terms of $\log n$ in the approximation bound. This can be changed by increasing the size of the sketch, which is equivalent to rescaling ε. In particular, if we want to estimate a range sum correct up to $\varepsilon' \|a\|_1$ with probability $1 - \delta$ then set $\varepsilon = \frac{\varepsilon'}{2 \log n}$. The space used is $O(\frac{\log^2(n)}{\varepsilon'} \log \frac{1}{\delta})$. An obvious improvement of this technique in practice is to keep exact counts for the first few levels of the hierarchy, where there are only a small number of dyadic ranges. This improves the space, time and accuracy of the algorithm in practice, although the asymptotic bounds are unaffected.

The best previous bounds for this problem in the turnstile model are given in [13], where range queries are answered by keeping $O(\log n)$ sketches, each of size $O(\frac{1}{\varepsilon'^2} \log(n) \log \frac{\log n}{\delta})$ to give approximations with additive error $\varepsilon \|a\|_1$ with probability $1 - \delta'$. Thus the space used there is $O(\frac{\log^2 n}{\varepsilon'^2} \log \frac{\log n}{\delta})$ and the time for updates is linear in the space used. The CM sketch improves the space and time bounds; it improves the constant factors as well as the asymptotic behavior. The time to process an update is significantly improved, since only a few entries in the sketch are modified, rather than a linear number.

5 Applications of Count-Min Sketches

By using CM sketches, we show how to improve best known time and space bounds for the two problems from Section 2.

5.1   Quantiles in the Turnstile Model

In [13] the authors showed that finding the approximate ϕ-quantiles of the data subject to insertions and deletions can be reduced to the problem of computing range sums. Put simply, the algorithm is to do binary searches for ranges $1 \ldots r$ whose range sum $a[1, r]$ is $k\phi\|a\|_1$ for $1 \leq k \leq \frac{1}{\phi} - 1$. The method of [13] uses *Random Subset Sums* to compute range sums. By replacing this structure with Count-Min sketches, the improved results follow immediately. By keeping $\log n$ sketches, one for each dyadic range and setting the accuracy parameter for each to be $\varepsilon/\log n$ and the probability guarantee to $\delta\phi/\log(n)$, the overall probability guarantee for all $1/\phi$ quantiles is achieved.

Theorem 4. ε-*approximate* ϕ-*quantiles can be found with probability at least* $1 - \delta$ *by keeping a data structure with space* $O(\frac{1}{\varepsilon} \log^2(n) \log(\frac{\log n}{\phi\delta}))$. *The time for each insert or delete operation is* $O(\log(n) \log(\frac{\log n}{\phi\delta}))$, *and the time to find each quantile on demand is* $O(\log(n) \log(\frac{\log n}{\phi\delta}))$.

Choosing CM sketches over Random Subset Sums improves both the query time and the update time from $O(\frac{1}{\varepsilon^2} \log^2(n) \log \frac{\log n}{\varepsilon\delta})$, by a factor of more than $\frac{34}{\varepsilon^2} \log n$. The space requirements are also improved by a factor of at least $\frac{34}{\varepsilon}$.

It is illustrative to contrast our bounds with those for the problem in the weaker Cash Register Model where items are only inserted (recall that in our stronger Turnstile model, items are deleted as well). The previously best known space bounds for finding approximate quantiles is $O(\frac{1}{\varepsilon}(\log^2 \frac{1}{\varepsilon} + \log^2 \log \frac{1}{\delta}))$ space for a randomized sampling and $O(\frac{1}{\varepsilon} \log(\varepsilon\|a\|_1))$ space for a deterministic solution [14]. These bounds are not completely comparable, but our result is the first on the more powerful Turnstile model to be comparable to the Cash Register model bounds in the leading $1/\varepsilon$ term.

5.2   Heavy Hitters in the Turnstile Model

We adopt the solution given in [5], which describes a divide and conquer procedure to find the heavy hitters. This keeps sketches for computing range sums: $\log n$ different sketches, one for each different dyadic range. When an update (i_t, c_t) arrives, then each of these is updated as before. In order to find all the heavy hitters, a parallel binary search is performed, descending one level of the hierarchy at each step. Nodes in the hierarchy (corresponding to dyadic ranges) whose estimated weight exceeds the threshold of $(\phi+\varepsilon)\|a\|_1$ are split into two ranges, and investigated recursively. All single items found in this way whose approximated count exceeds the threshold are output.

We instead must limit the number of items output whose true frequency is less than the fraction ϕ. This is achieved by setting the probability of failure for each sketch to be $\frac{\delta\phi}{2\log n}$. This is because, at each level there are at most $1/\phi$ items with frequency more than ϕ. At most twice this number of queries are made at each level, for all of the $\log n$ levels. By scaling δ like this and applying the union bound ensures that, over all the queries, the total probability that any one (or more) of them overestimated by more than a fraction ε is bounded by δ, and so the probability that every query succeeds is $1 - \delta$. It follows that

Theorem 5. *The algorithm uses space* $O(\frac{1}{\varepsilon} \log(n) \log\left(\frac{2\log(n)}{\delta\phi}\right))$, *and time* $O(\log(n) \log\left(\frac{2\log n}{\delta\phi}\right))$ *per update. Every item with frequency at least* $(\phi + \varepsilon)\|a\|_1$ *is output, and with probability* $1 - \delta$ *no item whose frequency is less than* $\phi\|a\|_1$ *is output.*

The previous best known bound appears in [5], where a non-adaptive group testing approach was described. Here, the space bounds agree asymptotically but have been improved in constant factors; a further improvement is in the nature of the guarantee: previous methods gave probabilistic guarantees about outputting the heavy hitters. Here, there is absolute certainty that this procedure will find and output every heavy hitter, because the CM sketches never underestimate counts, and strong guarantees are given that no non-heavy hitters will be output. This is often desirable.

In some situations in practice, it is vital that updates are as fast as possible, and here update time can be played off against search time: ranges based on powers of two can be replaced with an arbitrary branching factor k, which reduces the number of levels to $\log_k n$, at the expense of costlier queries and weaker guarantees on outputting non-heavy hitters.

6   Conclusions

We have introduced the Count-Min sketch, and shown how to estimate fundamental queries such as point, range or inner product queries as well as solve more sophisticated problems such as quantiles and heavy hitters. The space and/or time bounds of our solutions improve previously best known bounds for these problems. Typically the improvement is from $1/\varepsilon^2$ factor to $1/\varepsilon$ which is significant in real applications. Our CM sketch is quite simple, and is likely to find many applications, including in hardware solutions for these problems.

We have recently applied these ideas to the problem of change detection on data streams [6], and we also believe that it can be applied to improve the time and space bounds for constructing approximate wavelet and histogram representations of data streams [11]. Also, the CM Sketch can also be naturally extended to solve problems on streams that describe multidimensional arrays rather than the unidimensional array problems we have discussed so far.

Our CM sketch is not effective when one wants to compute the norms of data stream inputs. These have applications to computing correlations between data streams and tracking the number of distinct elements in streams, both of which are of great interest. It is an open problem to design extremely simple, practical sketches such as our CM Sketch for estimating such correlations and more complex data stream applications.

References

1. N. Alon, P. Gibbons, Y. Matias, and M. Szegedy. Tracking join and self-join sizes in limited storage. In *Proceedings of the Eighteenth ACM Symposium on Principles of Database Systems (PODS '99)*, pages 10–20, 1999.

2. N. Alon, Y. Matias, and M. Szegedy. The space complexity of approximating the frequency moments. In *Proceedings of the Twenty-Eighth Annual ACM Symposium on the Theory of Computing*, pages 20–29, 1996. Journal version in *Journal of Computer and System Sciences*, 58:137–147, 1999.
3. B. Babcock, S. Babu, M. Datar, R. Motwani, and J. Widom. Models and issues in data stream systems. In *Proceedings of Symposium on Principles of Database Systems (PODS)*, pages 1–16, 2002.
4. M. Charikar, K. Chen, and M. Farach-Colton. Finding frequent items in data streams. In *Proceedings of the International Colloquium on Automata, Languages and Programming (ICALP)*, pages 693–703, 2002.
5. G. Cormode and S. Muthukrishnan. What's hot and what's not: Tracking most frequent items dynamically. In *Proceedings of ACM Principles of Database Systems*, pages 296–306, 2003.
6. G. Cormode and S. Muthukrishnan. What's new: Finding significant differences in network data streams. In *Proceedings of IEEE Infocom*, 2004.
7. C. Estan and G. Varghese. Data streaming in computer networks. In *Proceedings of Workshop on Management and Processing of Data Streams*, http://www.research.att.com/conf/mpds2003/schedule/estanV.ps, 2003.
8. P. Flajolet and G. N. Martin. Probabilistic counting. In *24th Annual Symposium on Foundations of Computer Science*, pages 76–82, 1983. Journal version in *Journal of Computer and System Sciences*, 31:182–209, 1985.
9. M. Garofalakis, J. Gehrke, and R. Rastogi. Querying and mining data streams: You only get one look. In *Proceedings of the ACM SIGMOD International Conference on Management of Data*, 2002.
10. P. Gibbons and Y. Matias. Synopsis structures for massive data sets. *DIMACS Series in Discrete Mathematics and Theoretical Computer Science*, A, 1999.
11. A. Gilbert, S. Guha, P. Indyk, Y. Kotidis, S. Muthukrishnan, and M. Strauss. Fast, small-space algorithms for approximate histogram maintenance. In *Proceedings of the 34th ACM Symposium on Theory of Computing*, pages 389–398, 2002.
12. A. Gilbert, Y. Kotidis, S. Muthukrishnan, and M. Strauss. Surfing wavelets on streams: One-pass summaries for approximate aggregate queries. In *Proceedings of 27th International Conference on Very Large Data Bases*, pages 79–88, 2001. Journal version in *IEEE Transactions on Knowledge and Data Engineering*, 15(3):541–554, 2003.
13. A. C. Gilbert, Y. Kotidis, S. Muthukrishnan, and M. Strauss. How to summarize the universe: Dynamic maintenance of quantiles. In *Proceedings of 28th International Conference on Very Large Data Bases*, pages 454–465, 2002.
14. M. Greenwald and S. Khanna. Space-efficient online computation of quantile summaries. *SIGMOD Record (ACM Special Interest Group on Management of Data)*, 30(2):58–66, 2001.
15. R. Motwani and P. Raghavan. *Randomized Algorithms*. Cambridge University Press, 1995.
16. S. Muthukrishnan. Data streams: Algorithms and applications. In *ACM-SIAM Symposium on Discrete Algorithms*, http://athos.rutgers.edu/~muthu/stream-1-1.ps, 2003.
17. D. Woodruff. Optimal space lower bounds for all frequency moments. In *ACM-SIAM Symposium on Discrete Algorithms*, 2004.

Rotation and Lighting Invariant
Template Matching

Kimmo Fredriksson[1], Veli Mäkinen[2*], and Gonzalo Navarro[3**]

[1] Department of Computer Science, University of Joensuu.
kfredrik@cs.joensuu.fi
[2] Department of Computer Science, University of Helsinki.
vmakinen@cs.helsinki.fi
[3] Center for Web Research, Department of Computer Science, University of Chile.
gnavarro@dcc.uchile.cl

Abstract. We address the problem of searching for a two-dimensional pattern in a two-dimensional text (or image), such that the pattern can be found even if it appears rotated and brighter or darker than its occurrence. Furthermore, we consider approximate matching under several tolerance models. We obtain algorithms that are almost worst-case optimal. The complexities we obtain are very close to the best current results for the case where only rotations, but not lighting invariance, are supported. These are the first results for this problem under a combinatorial approach.

1 Introduction

We consider the problem of finding the occurrences of a two-dimensional *pattern* of size $m \times m$ cells in a two-dimensional *text* of size $n \times n$ cells, when all possible rotations of the pattern are allowed and also pattern and text may have differences in brightness. This stands for *rotation and lighting invariant template matching*. Text and pattern are seen as images formed by cells, each of which has a gray level value, also called a color.

Template matching has numerous important applications from science to multimedia, for example in image processing, content based information retrieval from image databases, geographic information systems, processing of aerial images, to name a few. In all these cases, we want to find a small subimage (the pattern) inside a large image (the text) permitting rotations (a small degree or any). Furthermore, pattern and text may have been photographed under different lighting conditions, so one may be brighter than the other.

The traditional approach to this problem [2] is to compute the cross correlation between each text location and each rotation of the pattern template. This

---

* A part of the work was done while visiting University of Chile under a researcher exchange grant from University of Helsinki.
** Funded by Millennium Nucleus Center for Web Research, Grant P01-029-F, Mideplan, Chile.

M. Farach-Colton (Ed.): LATIN 2004, LNCS 2976, pp. 39–48, 2004.

can be done reasonably efficiently using the Fast Fourier Transform (FFT), requiring time $O(Kn^2 \log n)$ where K is the number of rotations sampled. Typically K is $O(m)$ in the two-dimensional (2D) case, and $O(m^3)$ in the 3D case, which makes the FFT approach very slow in practice. In addition, lighting-invariant features may be defined in order to make the FFT insensitive to brightness. Also, in many applications, "close enough" matches of the pattern are also accepted. To this end, the user may specify, for example, a parameter κ such that matches that have at most κ differences with the pattern should be accepted, or a parameter δ such that gray levels differing by less than δ are considered equal. The definition of the matching conditions is called the "matching model".

Rotation invariant template matching was first considered from a combinatorial point of view in [8,9]. Since then, several fast filters have been developed for diverse matching models [10,7,6]. These represent large performance improvements over the FFT-based approach. The worst-case complexity of the problem was also studied [1,7]. However, lighting invariance has not been considered in this scenario.

On the other hand, *transposition invariant* string matching was considered in music retrieval [3,11]. The aim is to search for (one-dimensional) patterns in texts such that the pattern may match the text after all its characters (notes) are shifted by some value. The reason is that such an occurrence will sound like the pattern to a human, albeit in a different scale. In this context, efficient algorithms for several approximate matching functions were developed in [12].

We note that transposition invariance becomes lighting invariance when we replace musical notes by gray levels of cells in an image. Hence, the aim of this paper is to enrich the existing algorithms for rotation invariant template matching [7] with the techniques developed for transposition invariance [12] so as to obtain rotation and lighting invariant template matching. It turns out that lighting invariance can be added at very little extra cost. The key technique exploited is *incremental distance computation*; we show that several transposition invariant distances can be computed incrementally taking the computation done with the previous rotation into account in the next rotation angle.

Let us now determine which are the reasonable matching models. In [7], some of the models considered were useful only for binary images, a case where obviously we are not interested in this paper. We will address models that make sense for gray level images. We define three transposition-invariant distances: $d_{\mathrm{H}}^{t,\delta}$, which counts how many pattern and text cells differ by more than δ; $d_{\mathrm{MAD}}^{t,\kappa}$, which is the maximum color difference between pattern and text cells when up to κ outliers are permitted; and $d_{\mathrm{SAD}}^{t,\kappa}$, which is the sum of absolute color differences between pattern and text cells permitting up to κ outliers. Table 1 shows our complexities to compute these distances for every possible rotation of a pattern centered at a fixed text position. Variable σ is the number of different gray levels (assume $\sigma = \infty$ if the alphabet is not a finite discrete range). A lower bound to this problem is $O(m^3)$, achieved in [7] without lighting invariance.

We also define two search problems, consisting in finding all the transposition-invariant rotated occurrences of P in T such that: (1) there are at most κ cells of P differing by more than δ from their text cell (δ-matching); or (2) the sum

Table 1. Worst-case complexities to compute the different distances defined.

Distance	Complexity
$d_{\mathrm{H}}^{t,\delta}$	$\min(\log m, \sigma + (\delta+1))m^3$
$d_{\mathrm{MAD}}^{t,\kappa}$	$(\min(\kappa,\sigma) + \log\min(m,\sigma))m^3$
$d_{\mathrm{SAD}}^{t,\kappa}$	$(\min(\kappa,\sigma) + \log\min(m,\sigma))m^3$

of absolute difference between cells in P and T, except for κ outliers, does not exceed γ (γ-matching). Note that δ-matching can be solved by examining every text cell and reporting it if $d_{\mathrm{H}}^{t,\delta}(P,T) \leq \kappa$, or if $d_{\mathrm{MAD}}^{t,\kappa}(P,T) \leq \delta$ around the text cell. Hence any $O(f(m))$ algorithm for computing $d_{\mathrm{H}}^{t,\delta}$ or $d_{\mathrm{MAD}}^{t,\kappa}(P,T)$ yields an $O(f(m)n^2)$ algorithm for δ-matching. Similarly, γ-matching can be reduced to checking whether $d_{\mathrm{SAD}}^{t,\kappa}(P,T) \leq \gamma$ around each cell. Without transposition invariance all searching worst cases are $O(m^3n^2)$ [7].

We remark that we have developed algorithms that work on arbitrary alphabets, but we have also taken advantage of the case where the alphabet is a discrete range of integer values.

A full version of this paper [5] considers also (δ, γ)-matching and optimal average case search complexities.

2 Definitions

Let $T = T[1..n, 1..n]$ and $P = P[1..m, 1..m]$ be arrays of unit squares, called *cells*, in the (x, y)-plane. Each cell has a value in an alphabet called Σ, sometimes called its gray level or its color. A particular case of interest is that of Σ being a finite integer range of size σ. The corners of the cell for $T[i, j]$ are $(i-1, j-1), (i, j-1), (i-1, j)$ and (i, j). The center of the cell for $T[i, j]$ is $(i-\frac{1}{2}, j-\frac{1}{2})$. The array of cells for pattern P is defined similarly. The center of the whole pattern P is the center of the cell in the middle of P. Precisely, assuming for simplicity that m is odd, the center of P is the center of cell $P[\frac{m+1}{2}, \frac{m+1}{2}]$.

Assume now that P has been moved on top of T using a rigid motion (translation and rotation), such that the center of P coincides exactly with the center of some cell of T (*center-to-center assumption*). The location of P with respect to T can be uniquely given as $((i, j), \theta)$ where (i, j) is the cell of T that matches the center of P, and θ is the angle between the x-axis of T and the x-axis of P. The (approximate) occurrence between T and P at some location is defined by comparing the values of the cells of T and P that overlap. We will use the centers of the cells of T for selecting the comparison points. That is, for the pattern at location $((i, j), \theta)$, we look which cells of the pattern cover the centers of the cells of the text, and compare the corresponding values of those cells (Fig. 1).

More precisely, assume that P is at location $((i, j), \theta)$. For each cell $T[r, s]$ of T whose center belongs to the area covered by P, let $P[r', s']$ be the cell of P such that the center of $T[r, s]$ belongs to the area covered by $P[r', s']$. Then $M(T[r, s]) = P[r', s']$, that is, our algorithms compare the cell $T[r, s]$ of T against the cell $M(T[r, s])$ of P.

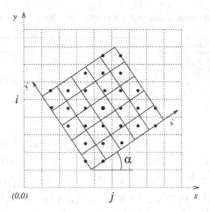

Fig. 1. Each text cell is matched against the pattern cell that covers the center of the text cell.

Hence the *matching function* M is a function from the cells of T to the cells of P. Now consider what happens to M when angle θ grows continuously, starting from $\theta = 0$. Function M changes only at the values of θ such that some cell center of T hits some cell boundary of P. It was shown in [8] that this happens $O(m^3)$ times, when P rotates full 2π radians. This result was shown to be also a lower bound in [1]. Hence there are $\Theta(m^3)$ relevant orientations of P to be checked. The set of angles for $0 \le \theta \le \pi/2$ is

$A = \{\beta, \pi/2 - \beta \mid \beta = \arcsin\frac{h+\frac{1}{2}}{\sqrt{i^2+j^2}} - \arcsin\frac{j}{\sqrt{i^2+j^2}}; \ i = 1, 2, \ldots, \lfloor m/2 \rfloor; j =$

$0, 1, \ldots, \lfloor m/2 \rfloor; h = 0, 1, \ldots, \lfloor \sqrt{i^2+j^2} \rfloor \}$. By symmetry, the set of possible angles θ, $0 \le \theta < 2\pi$, is $\mathcal{A} = A \ \cup \ A + \pi/2 \ \cup \ A + \pi \ \cup \ A + 3\pi/2$.

Furthermore, pattern P matches at location $((i, j), \theta)$ with lighting invariance if there is some integer transposition t such that $T[r, s] + t = P[r', s']$ for all $[r', s']$ in the area of P.

Once the position and rotation $((i, j), \theta)$ of P in T define the matching function, we can compute different kinds of *distances* between the pattern and the text. Lighting-invariance versions of the distances choose the transposition minimizing the basic distance. Interesting distances for gray level images follow.

Hamming Distance (H): The number of times $T[r, s] \ne P[r', s']$ occurs, over all the cells of P, that is, $d_{\mathrm{H}}(i, j, \theta, t) = \sum_{r',s'}[\text{if } T[r, s] + t \ne P[r', s'] \text{ then } 1 \text{ else } 0]$, and $d_{\mathrm{H}}^{\mathrm{t}}(i, j, \theta) = \min_t d_{\mathrm{H}}(i, j, \theta, t)$. This can be extended to distance d_{H}^{δ} and its transposition-invariant version $d_{\mathrm{H}}^{\mathrm{t},\delta}$, where colors must differ by more than δ in order to be considered different, that is, $T[r, s] + t \notin [P[r', s'] - \delta, P[r', s'] + \delta]$.

Maximum Absolute Differences (MAD): The maximum value of $|T[r, s] - P[r', s']|$ over all the cells of P, that is, $d_{\mathrm{MAD}}(i, j, \theta, t) = \max_{r',s'} |T[r, s] + t - P[r', s']|$, and $d_{\mathrm{MAD}}^{\mathrm{t}}(i, j, \theta) = \min_t d_{\mathrm{MAD}}(i, j, \theta, t)$. This can be extended to distance $d_{\mathrm{MAD}}^{\kappa}$ and its transposition-invariant version $d_{\mathrm{MAD}}^{\mathrm{t},\kappa}$, so that up to κ pattern cells are freed from matching the text. Then the problem is to compute the MAD distance with the best choice of κ outliers that are not included in the maximum.

Sum of Absolute Differences (SAD): The sum of the $|T[r,s]-P[r',s']|$ values over all the cells of P, that is, $d_{\mathrm{SAD}}(i,j,\theta,t) = \sum_{r',s'} |T[r,s]+t-P[r',s']|$, and $d_{\mathrm{SAD}}^t(i,j,\theta) = \min_t d_{\mathrm{SAD}}(i,j,\theta,t)$. Similarly, this distance can be extended to d_{SAD}^κ and its transposition-invariant version $d_{\mathrm{SAD}}^{t,\kappa}$, where up to κ pattern cells can be removed from the summation.

3 Efficient Algorithms

In [1] it was shown that for the problem of the two dimensional pattern matching allowing rotations the worst case lower bound is $\Omega(n^2 m^3)$. We have shown in [7] a simple way to achieve this lower bound for any of the distances under consideration (without lighting invariance). The idea is that we will check each possible text center, one by one. So we have to pay $O(m^3)$ per text center to achieve the desired complexity. What we do is to compute the distance we want for each possible rotation, by reusing most of the work done for the previous rotation. Once the distances are computed, it is easy to report the triples (i,j,θ) where these values are smaller than the given thresholds (δ and/or γ). Only distances d_{H} (with $\delta = 0$) and d_{SAD} (with $\kappa = 0$) were considered.

Assume that, when computing the set of angles $\mathcal{A} = (\beta_1, \beta_2, \ldots)$, we also sort the angles so that $\beta_i < \beta_{i+1}$, and associate with each angle β_i the set $\mathcal{C}_i$ containing the corresponding cell centers that must hit a cell boundary at β_i. This is done in a precomputation step that depends only on m, not on P or T. Hence we can evaluate the distance functions (such as d_{SAD}) incrementally for successive rotations of P. That is, assume that the distance has been evaluated for β_i, then to evaluate it for rotation β_{i+1} it suffices to re-evaluate the cells restricted to the set $\mathcal{C}_i$. This is repeated for each $\beta \in \mathcal{A}$. Therefore, the total time for evaluating the distance for P centered at some position in T, for all possible angles, is $O(\sum_i |\mathcal{C}_i|)$. This is $O(m^3)$ because each fixed cell center of T, covered by P, can belong to some $\mathcal{C}_i$ at most $O(m)$ times. To see this, note that when P is rotated the whole angle 2π, any cell of P traverses $O(m)$ cells of T.

If we want to add lighting invariance to the above scheme, a naive approach is to run the algorithm for every possible transposition, for a total cost of $O(n^2 m^3 \sigma)$. In case of a general alphabet there are $O(m^2)$ relevant transpositions at each rotation (that is, each pattern cell can be made to match its corresponding text cell). Hence the cost raises to $O(n^2 m^5)$.

In order to do better, we must be able to compute the optimal transposition for the initial angle and then maintaining it when some characters of the text change (because the pattern has been aligned over a different text cell). If we take $f(m)$ time to do this, then our lighting invariant algorithm becomes worst-case time $O(n^2 m^3 f(m))$. In the following we show how can we achieve this for each of the distances under consideration.

3.1 Distance $d_{\mathrm{H}}^{t,\delta}$

As proved in [12], the optimal transposition for Hamming distance is obtained as follows. Each cell $P[r', s']$, aligned to $T[r, s]$, *votes* for a range of transpositions $[P[r', s'] - T[r, s] - \delta, P[r', s'] - T[r, s] + \delta]$, for which it would match. If a transposition receives v votes, then its Hamming distance is $m^2 - v$. Hence, the transposition that receives most votes is the one yielding distance $d_{\mathrm{H}}^{t,\delta}$. Let us now separate the cases of integer and general alphabets.

Integer alphabet. The original algorithm [12] obtains $O(\sigma + |P|)$ time on integer alphabet, by bucket-sorting the range extremes and then traversing them linearly so as to find the most voted transposition (a counter is incremented when a range starts and decremented when it finishes).

In our case, we have to pay $O(\sigma + m^2)$ in order to find the optimal transposition for the first rotation angle. The problem is how to recompute the optimal transposition once some text cell $T[r, s]$ changes its value (due to a small change in rotation angle). The net effect is that the range of transpositions given by the old cell value loses a vote and a new range gains a vote.

We use the fact that the alphabet is an integer range, so there are $O(\sigma)$ possible transpositions. Each transposition can be classified according to the number of votes it has. There are $m^2 + 1$ lists L_i, $0 \le i \le m^2$, containing the transpositions that currently have i votes. Hence, when a range of transpositions loses/gains one vote, the $2\delta + 1$ transpositions are moved to the lower/upper list. An array pointing to the list node where each transposition appears is necessary to efficiently find each of those $2\delta + 1$ transpositions. We need to keep control of which is the highest-numbered non-empty list, which is easily done in constant time per operation because transpositions move only from one list to the next/previous. Initially we pay $O(\sigma + m^2)$ to initialize all the lists and put all the transpositions in list L_0, then $O((\delta + 1)m^2)$ to process the votes of all the cells, and then $O(\delta + 1)$ to process each cell that changes. Overall, when we consider all the $O(m^3)$ cell changes, the scheme is $O(\sigma + (\delta + 1)m^3)$. This is our complexity to compute distance $d_{\mathrm{H}}^{t,\delta}$ between a pattern and a text center, considering all possible rotations and transpositions.

General alphabet. Let us resort to a more general problem of *dynamic range voting*: In the static case we have a multiset $S = \{[\ell, r]\}$ of one-dimensional closed ranges, and we are interested in obtaining a point p that is included in most ranges, that is $\mathrm{maxvote}(S) = \max_p |\{[\ell, r] \in S \mid \ell \le p \le r\}|$. In the dynamic case a new range is added to or an old one is deleted from S, and we must be able to return $\mathrm{maxvote}(S)$ after each update.

Notice that our original problem of computing $d_{\mathrm{H}}^{t,\delta}$ from one rotation angle to another is a special case of dynamic range voting; multiset S is $\{[P[r', s'] - T[r, s] - \delta, P[r', s'] - T[r, s] + \delta] \mid M(T[r, s]) = P[r', s']\}$ in one rotation angle, and in the next one some $T[r, s]$ changes its value. That is, the old range is deleted and the new one is inserted, after which $\mathrm{maxvote}(S)$ is requested to compute the distance $d_{\mathrm{H}}^{t,\delta} = m^2 - \mathrm{maxvote}(S)$ in the new angle.

We show that dynamic range voting can be supported in $O(\log |S|)$ time, which immediately gives an $O(m^3 \log m)$ time algorithm for computing $d_H^{t,\delta}$ between a pattern and a text center, considering all rotations and transpositions.

First, notice that the point that gives maxvote(S) can always be chosen among the endpoints of ranges in S. We store each endpoint e in a balanced binary search tree with key e. Let us denote the leaf whose key is e simply by (leaf) e. With each endpoint e we associate a value vote(e) (stored in leaf e) that gives the number $|\{[\ell,r] \mid \ell \le e \le r, [\ell,r] \in S\}|$, where the set is considered as a multiset (same ranges can have multiple occurrences). In each internal node v, value maxvote(v) gives the maximum of the vote(e) values of the leaves e in its subtree. After all the endpoints e are added and the values vote(e) in the leaves and values maxvote(v) in the internal nodes are computed, the static case is solved by taking the value maxvote($root$) $=$ maxvote(S) in the root node of the tree.

A straightforward way of generalizing the above approach to the dynamic case would be to recompute all values vote(e) that are affected by the insertion/deletion of a range. This would, however, take $O(|S|)$ time in the worst case. To get a faster algorithm, we only store the changes of the votes in the roots of certain subtrees so that vote(e) for any leaf e can be computed by summing up the changes from the root to the leaf e.

For now on, we refer to vote(e) and maxvote(v) as virtual values, and replace them with counters diff(v) and values maxdiff(v). Counters diff(v) are defined implicitly so that for all leaves of the tree it holds

$$\text{vote}(e) = \sum_{v \in \text{path}(root,e)} \text{diff}(v), \tag{1}$$

where path($root,e$) is the set of nodes in the path from the root to a leaf e (including the leaf). We note that there are several possible ways to choose diff(v) values so that they satisfy the definition. Values maxdiff(v) are defined recursively as

$$\max(\text{maxdiff}(v.left) + \text{diff}(v.left), \text{maxdiff}(v.right) + \text{diff}(v.right)), \tag{2}$$

where $v.left$ and $v.right$ are the left and right child of v, respectively. In particular, maxdiff(e) $= 0$ for any leaf node e. One easily notices that

$$\text{maxvote}(v) = \text{maxdiff}(v) + \sum_{v' \in \text{path}(root,v)} \text{diff}(v'),$$

which also gives as a special case Equation (1) once we notice that maxvote(e) $=$ vote(e) for each leaf node e.

Our goal is to maintain diff() and maxdiff() values correctly during insertions and deletions. We have three different cases to consider: (i) How to compute the value diff(e) for a new endpoint of a range, (ii) how to update the values of diff() and maxdiff() when a range is inserted/deleted, and (iii) how to update the values during rotations to rebalance the tree.

Case (i) is handled by storing in each leaf an additional counter end(e). It gives the number of ranges whose rightmost endpoint is e. Assume that this value is computed for all existing leaves. When we insert a new endpoint e, we either find a leaf labeled e or otherwise there is a leaf e' after which e is inserted. In the first case vote(e) remains the same and in the latter case vote(e) = vote(e') − end(e'), because e is included in the same ranges as e' except those that end at e'. Notice also that vote(e) = 0 in the degenerate case when e is the leftmost leaf. The +1 vote induced by the new range whose endpoint e is, will be handled in case (ii). To make vote(e) = $\sum_{v' \in \text{path}(root,e)} \text{diff}(v')$, we fix diff($e$) so that vote($e$) = diff($e$) + $\sum_{v' \in \text{path}(root,v)} \text{diff}(v')$, where v is the parent of e. Once the maxdiff() values are updated in the path from e to the root, we can conclude that all the necessary updates are done in $O(\log |S|)$ time.

Let us then consider case (ii). Recall the one-dimensional range search on a balanced binary search tree (see e.g. [4], Section 5.1). We use the fact that one can find in $O(\log |S|)$ time the minimal set of nodes, say F, such that the range $[\ell, r]$ of S is *partitioned* by F; the subtrees starting at nodes of F contain all the points in $[\ell, r] \cap S$ and only them. It follows that when inserting (deleting) a range $[\ell, r]$, we can set diff(v) = diff(v) + 1 (diff(v) = diff(v) − 1) at each $v \in F$. This is because all the values vote(e) in these subtrees change by ±1 (including leaves ℓ and r). Note that some diff(v) values may go below zero, but this does not affect correctness. To keep also the maxdiff() values correctly updated, it is enough to recompute the values in the nodes in the paths from each $v \in F$ to the root using Equation (2); other values are not affected by the insertion/deletion of the range $[\ell, r]$. The overall number of nodes that need updating is $O(\log |S|)$.

Finally, let us consider case (iii). Counters diff(v) are affected by tree rotations, but in case a tree rotation involving e.g. subtrees $v.left$, $v.right.left$ and $v.right.right$ takes place, values diff(v) and diff($v.right$) can be "pushed" down to the roots of the affected subtrees, and hence they become zero. Then the tree rotation can be carried out, also maintaining subtree maxima easily.

Hence, each insertion/deletion takes $O(\log |S|)$ time, and maxvote(S) = maxdiff($root$) + diff($root$) is readily available in the root node.

3.2   Distance $d_{\text{MAD}}^{t,\kappa}$

Let us start with $\kappa = 0$. As proved in [12], the optimal transposition for distance d_{MAD}^{t} is obtained as follows. Each cell $P[r', s']$, aligned to $T[r, s]$, votes for transposition $P[r', s'] - T[r, s]$. Then, the optimal transposition is the average between the minimum and maximum vote, and d_{MAD}^{t} distance is the difference of maximum minus minimum, divided by two. An $O(|P|)$ algorithm follwed.

We need $O(m^2)$ to obtain the optimal transposition for the first angle, zero. Then, in order to address changes of text characters (because, due to angle changes, the pattern cell was aligned to a different text cell), we must be able to maintain minimum and maximum votes. Every time a text character changes, a vote disappears and a new vote appears. We can simply maintain balanced search trees with all the current votes so as to handle any insertion/deletion of votes in $O(\log(m^2)) = O(\log m)$ time, knowing the minimum and maximum at any time. If we have an integer alphabet of size σ, there are only $2\sigma + 1$ possible votes, so it is not hard to obtain $O(\log \sigma)$ complexity. Hence d_{MAD}^t distance between a pattern and a text center can be computed in $O(m^3 \log m)$ or $O(m^3 \log \min(m, \sigma))$ time, for all possible rotations and transpositions.

In order to account for up to κ outliers, it was already shown in [12] that it is optimal to choose them from the pairs that vote for maximum or minimum transpositions. That is, if all the votes are sorted into a list $v_1 \ldots v_{m^2}$, then distance $d_{\mathrm{MAD}}^{t,\kappa}$ is the minimum among distances d_{MAD}^t computed in sets $v_1 \ldots v_{m^2-\kappa}$, $v_2 \ldots v_{m^2-\kappa+1}$, and so on until $v_{\kappa+1} \ldots v_{m^2}$. Moreover, the optimum transposition of the i-th value of this list is simply the average of maximum and minimum, that is, $(v_{m^2-\kappa-1+i} + v_i)/2$.

So our algorithm for $d_{\mathrm{MAD}}^{t,\kappa}$ is as follows. We make our tree threaded (each node points to its predecessor and successor in the tree), so we can easily access the $\kappa + 1$ smallest and largest votes. After each change in the tree, we retraverse these $\kappa + 1$ pairs and recompute the minimum among the $v_{m^2-\kappa-1+i} - v_i$ differences. This takes $O(m^3(\kappa + \log m))$ time. In case of an integer alphabet, since there cannot be more than $O(\sigma)$ different votes, this can be done in time $O(m^3(\min(\kappa, \sigma) + \log \min(m, \sigma)))$.

3.3   Distance $d_{\mathrm{SAD}}^{t,\kappa}$

Let us first consider case $\kappa = 0$. This corresponds to the SAD model of [12], where it was shown that, if we collect votes $P[r', s'] - T[r, s]$, then the median vote (either one if $|P|$ is even) is the transposition that yields distance d_{SAD}^t. Then the actual distance can be obtained by using the formula for d_{SAD}, and an $O(|P|)$ time algorithm was immediate.

In this case we have to pay $O(m^2)$ to compute the distance for the first rotation, and then have to manage to maintain the median transposition and current distance when some text cells change their value due to small rotations.

We maintain a balanced and threaded binary search tree for the votes, plus a pointer to the median vote. Each time a vote changes because a pattern cell aligns to a new text cell, we must remove the old vote and insert the new one. When insertion and deletion occur at different halves of the sorted list of votes (that is, one is larger and the other smaller than the median), the median may move by one position. This is done in constant time since the tree is threaded.

The distance value itself can change. One change is due to the fact that one of the votes changed its value. Given a fixed transposition, it is trivial to remove the appropriate summand and introduce a new one in the formula for d_{SAD}. Another change is due to the fact that the median position can change from a value in the sorted list to the next or previous. It was shown in [12] how to modify in constant time distance d_{SAD}^t in this case: If we move from transposition v_j to v_{j+1}, then all the j smallest votes increase their value by $v_{j+1} - v_j$, and the $m - j$ largest votes decrease by $v_{j+1} - v_j$. Hence distance d_{SAD} at the new transposition is the value at the old transposition plus $(2j - m)(v_{j+1} - v_j)$.

Hence, we can traverse all the rotations in time $O(m^3 \log m)$. This can be reduced to $O(m^3 \log \min(m, \sigma))$ on finite integer alphabet, by noting that there cannot be more than $O(\sigma)$ different votes, and taking some care in handling repeated values inside single tree nodes.

To compute distance $d_{\text{SAD}}^{t,\kappa}$, we have again that the optimal values to free from matching are those voting for minimum or maximum transpositions. If we remove those values, then the median lies at positions $m - \lceil \kappa/2 \rceil \ldots m + \lceil \kappa/2 \rceil$ in the list of sorted votes, where m is the median position for the whole list.

Hence, instead of maintaining a pointer to the median, we maintain two pointers to the range of $\kappa+1$ medians that could be relevant. It is not hard to maintain left and right pointers when votes are inserted and deleted in the set. All the median values can be changed one by one, and we can choose the minimum distance among the $\kappa+1$ options. This gives us an $O(m^3(\kappa+\log m))$ time algorithm to compute $d_{\text{SAD}}^{t,\kappa}$. On integer alphabet, this is $O(m^3(\kappa + \log \min(m, \sigma)))$, which can be turned into $O(m^3(\min(\kappa, \sigma) + \log \min(m, \sigma)))$ by standard tricks using the fact that there are $O(\sigma)$ possible median votes that have different values.

References

1. A. Amir, A. Butman, M. Crochemore, G. Landau, and M. Schaps. Two-dimensional pattern matching with rotations. In *Proc. CPM'03*, LNCS 2676, pages 17–31, 2003.
2. L. G. Brown. A survey of image registration techniques. *ACM Computing Surveys*, 24(4):325–376, 1992.
3. T. Crawford, C. Iliopoulos, and R. Raman. String matching techniques for musical similarity and melodic recognition. *Computing in Musicology*, 11:71–100, 1998.
4. M. de Berg, M. van Kreveld, M. Overmars, and O. Schwarzkopf. *Computational Geometry: Algorithms and Applications*. Springer-Verlag, 2nd rev. edition, 2000.
5. K. Fredriksson, V. Mäkinen, and G. Navarro. Rotation and lighting invariant template matching. Technical Report TR/DCC-2003-3, Dept. of Comp.Sci., Univ. of Chile, 2003. ftp://ftp.dcc.uchile.cl/pub/users/gnavarro/lighting.ps.gz.
6. K. Fredriksson, G. Navarro, and E. Ukkonen. *Faster than FFT: Rotation Invariant Combinatorial Template Matching*, volume II, pages 75–112. Trans.Res.Net., 2002.
7. K. Fredriksson, G. Navarro, and E. Ukkonen. Optimal exact and fast approximate two dimensional pattern matching allowing rotations. In *Proc. CPM'02*, LNCS 2373, pages 235–248, 2002.
8. K. Fredriksson and E. Ukkonen. A rotation invariant filter for two-dimensional string matching. In *Proc. CPM'98*, LNCS 1448, pages 118–125, 1998.
9. K. Fredriksson and E. Ukkonen. Combinatorial methods for approximate image matching under translations and rotations. *Patt. Rec. Lett.*, 20(11–13):1249–1258, 1999.
10. G. Navarro K. Fredriksson and E. Ukkonen. An index for two dimensional string matching allowing rotations. In *Proc. IFIP TCS'00*, LNCS 1872, pages 59–75, 2000.
11. K. Lemström and J. Tarhio. Detecting monophonic patterns within polyphonic sources. In *Proc. RIAO'00*, pages 1261–1279, 2000.
12. V. Mäkinen, G. Navarro, and E. Ukkonen. Algorithms for transposition invariant string matching. In *Proc. STACS'03*, LNCS 2607, pages 191–202, 2003. Extended version as TR/DCC-2002-5, Dept. of Computer Science, Univ. of Chile.

Computation of the Bisection Width for Random d-Regular Graphs[*]

Josep Díaz[1], Maria J. Serna[1], and Nicholas C. Wormald[2]

[1] Dept. Llenguatges i Sistemes, Universitat Politecnica de Catalunya.
{diaz,mjserna}@lsi.upc.es
[2] Dept. Combinatorics and Optimization, University of Waterloo.
nwormald@uwaterloo.ca

Abstract. In this paper we provide an explicit way to compute asymptotically almost sure upper bounds on the bisection width of random d-regular graphs, for any value of d. We provide the bounds for $5 \leq d \leq 12$. The upper bounds are obtained from the analysis of the performance of a randomized greedy algorithm to find bisections of d-regular graphs. We also give empirical values of the size of bisection found by the algorithm for some small values of d and compare it with numerical approximations of our theoretical bounds. Our analysis also gives asymptotic lower bounds for the size of the maximum bisection.

1 Introduction

Given a graph $G = (V, E)$ with $|V| = n$ and n even, a *bisection* of V is a partition of V into two parts each of cardinality $n/2$, and its *size* is the number of edges crossing between the parts. A *minimum bisection* is a bisection of V with minimal size. The decision problem related to finding a minimum bisection is known to be NP-complete [10], even for 3-regular graphs [5]. (See for example [7] for further results an applications on graph bisection). The size of a minimum bisection is called the *bisection width* and the *min bisection problem* consists of finding a minimum bisection in a given G. In the present paper, we give a family of randomized algorithms which give asymptotic upper bounds as $n \to \infty$ on the bisection width of almost all d-regular graphs, where d is fixed.

Plenty of results are known on bisection width. With respect to lower bounds, in 1975 Fiedler gave a spectral lower bound of $\lambda_2 n/4$ applicable for any graph, where λ_2 is the second eigenvalue of the Laplacian of the graph [9]. In 1984, Bollobás provided a lower bound of $(\frac{d}{4} - \frac{\sqrt{d \ln 2}}{2})n$, for almost all d-regular graphs

---

[*] The work of the first and second authors was partially supported by the IST programme of the EU under contract IST-1999-14186 (ALCOM-FT). The first author is also supported by the Distinció per a la recerca of the Generalitat de Catalunya. The second author is also supported by the Spanish CICYT project TIC-2002-04498-C05-03. The third author is supported by the Canada Research Chairs Program and partially by the Australian Research Council when this author was at the University of Melbourne.

M. Farach-Colton (Ed.): LATIN 2004, LNCS 2976, pp. 49–58, 2004.
© Springer-Verlag Berlin Heidelberg 2004

[3]. Later Kostochka and Melnikov proved that almost all cubic graphs have bisection width greater than $0.101n$ [12]. Using spectral techniques, Bezrukov et al. gave lower bounds of $0.082n$ for the bisection width of cubic Ramanujan graphs, and of $0.176n$ for the case of 4-regular Ramanujan graphs [2].

Regarding upper bounds, Kostochka and Melnikov proved that asymptotically as $n \to \infty$, all d regular graphs have bisection width of at most $\frac{d-2}{4}n + O(d\sqrt{n}\log n)$ [12]. Later, Alon proved that for $n > 40d^9$, all d-regular graphs have bisection width at most $(\frac{d}{4} - \frac{3\sqrt{d}}{32\sqrt{2}})n$ [1]. More recently, Monien and Preis [14] gave upper bounds on the bisection width of $(\frac{1}{6} + \epsilon)n$ for 3-regular graphs and of $(0.4 + \epsilon)n$ for 4-regular graphs, for any ϵ, when n is larger than some function of the chosen ϵ. To the best of our knowledge, the most recent result on bisection width was given in [6], where it was proved that the bisection width of a random 4-regular graph on n vertices is asymptotically smaller than $(\frac{1}{3} + \epsilon)n$, with probability tending to 1 (a.a.s.). This result was proved by analysing a simple greedy algorithm, a variant of which only yielded bisections of width 0.174 for a random cubic graph on n vertices.

The problem of finding the maximum bisection size has also received considerable attention. This problem is again NP-hard even for planar graphs [11]. It is known to be solvable in polynomial time for graphs of bounded treewidth [11].

The maximum bisection size is a lower bound on the maximum size (number of edges) of a bipartite subgraph. Locke [13] showed that a d-regular graph which is not complete or a cycle has a bipartite subgraph with at least $(nd/4)d/(d-1)$ edges if d is even, and at least $(nd/4)((d+1)/d + 2/d^2)$ edges if d is odd. Shearer [15] improved this result to $(nd/4) + n\sqrt{d}/8\sqrt{2}$ for triangle-free graphs, a property which a positive fraction of random regular graphs have. Our lower bounds for maximum bisection in random d-regular graphs easily exceed these bounds.

In Section 2 we present a basic randomized algorithm to find a (small) bisection of a graph by 2-colouring its vertices in a greedy way. The next vertex to be coloured is chosen according to a prioritisation scheme. The priority depends on to the status of a vertex with respect to the number of neighbours it has of either colour. Many different priority schemes were considered, each specified by a list of the types of vertices (i.e. their possible status with respect to colours of neighbours).

This prioritisation scheme is significant both as a simplification and as a generalisation of the method in [6], where only 3-regular and 4-regular graphs were considered. It is a generalisation because, to each of the algorithms given there, there are corresponding algorithms of the general type considered in this paper which have equivalent asymptotic performance (although the algorithms do not give identical results). It is a simplification, because the method in [6] was to specify one or two main phases of the algorithm. In each main phase, two types of vertex were coloured, with one of the specified types having priority. Such detailed control of the algorithm is difficult to generalise to higher d because of the difficulty of knowing which types of vertices might be available when one phase ends and a new phase starts. The key idea in the present paper is to specify which types of vertices should have priority over which others, throughout the whole algorithm. The transitions between phases then become automatic. It is hard to substantiate this claim without looking at the algorithm in more detail, since even the definition of a phase becomes more delicate with this approach. A similar effect occurred in the analysis of greedy algorithms for finding independent sets in random regular graphs [17], but the situation there was considerably simpler. In that case, the prioritisation was merely according to the degree of a vertex during a deletion algorithm, whilst in the present case, the best prioritisation list is much harder to determine. Moreover, in the present case, the algorithm in some sense returns to phase which it visited earlier, and this did not happen in [17].

In Section 3, we *sketch* the analysis of the performance using the differential equation method. For any given d, we choose the appropriated priority list, set the equations and solve them numerically to find the asymptotic bisection width for the random d-regular graphs under consideration. In the same section, we produce empirical evidence indicating that there are two types of optimal priority lists of the vertices: one for even values of d and the other for odd values of d. In Section 4, we give empirical results comparing the values obtained numerically from the differential equations with the bisection width obtained by the randomized greedy algorithm. In Section 5, we discuss the maximum bisection results. It should be emphasised that the main contribution of this paper is to give better asymptotic bounds for the bisection width of d-regular graphs ($d > 4$), and the algorithm produced in Section 2 is only of methodological value.

2   The Priority-Greedy Algorithm

In this section, we describe a family of randomized greedy procedures to find a bisection for d-regular graphs. We also introduce some generic notation to be used later.

Given a graph, and given a partial assignment of colours red (R) and blue (B) to its vertices, we classify the uncoloured vertices according to the colours of their neighbours: An uncoloured vertex is of *Type* (r, b), if it has r neighbours coloured R and b neighbours coloured B.

For $r \leq b$, we say that a pair of uncoloured vertices is a *symmetric pair* if their types are (r, b) and (b, r) for some r and b. We then call this an (r, b)-*symmetric* pair, or a symmetric pair of *type* (r, b).

The greedy procedure works by colouring vertices chosen randomly in symmetric pairs, to maintain balancedness, and repeatedly uses the *majority operation* (Maj), that colours each vertex of an (r, b)-symmetric pair, $r < b$, with the majority colour among its neighbours, and, given an (r, b)-symmetric pair with $r = b$, randomly colours one vertex of the pair R and the other B .

We assume that the symmetric pair types have the priorities $0, 1, 2, \ldots$ associated with them (a larger number denotes higher priority). The priority-greedy algorithm for random d-regular graphs is given in Figure 1.

Initial step: input prio(r, b) for all $r \le b \le d$;
                      select two non-adjacent vertices u.a.r., colour one with R and
                      the other with B.

Main iteration: **while** there is at least one uncoloured symmetric pair **do**
                              let (r, b) denote the highest priority type of
                                  an uncoloured symmetric pair;
                              select u.a.r. an (r, b)-symmetric pair
                                  and perform Maj;

Clean up: colour any remaining uncoloured vertices, half of them R
                      and half B, in any manner, and output the bisection R, B.

Fig. 1. Algorithm priority-greedy for obtaining a bisection of a d-regular graph

This algorithm takes as input a predetermined priority list assigning a distinct priority, prio(r, b), to each symmetric pair type (r, b). We impose the conditions on all priority lists that prio($0, 0$) $= 0 <$ prio(r, b) whenever $(r, b) \ne (0, 0)$. Note that the priority of pairs (r, b) with $r + b = d$ is immaterial since colouring vertices in such pairs cannot affect the remainder of the algorithm. So for simplicity we assume that all such vertices have negative priority, and only those with $r + b < d$ need to be specified.

3   Analysis: The Differential Equation System

We follow the description in [6], extending it to the d-regular setting for arbitrary d. The algorithms considered there give equivalent results in special cases of the priority-greedy algorithm, for particular priority lists and for $d = 3$ and 4. (Notice, that the algorithms described in [6] for $d = 4$ and $d = 4$ are different than the *general* algorithm presented in this paper)

One method of analysing the performance of a randomized algorithm is to use a system of differential equations to express the expected changes in the variables describing the state of the algorithm during its execution. An exposition of this method can be found in [18], which includes various examples of graph-theoretic optimisation problems.

We use the pairing model to generate n-vertex d-regular graphs u.a.r. Briefly, to generate such a random graph, it is enough to begin with dn points in n cells, and choose a random perfect matching of the points, which we call a *pairing*. The corresponding pseudograph (possibly with loops or multiple edges) has the cells as vertices and the pairs as edges. Any property a.a.s. true of the random pseudograph is also a.a.s. true of the restriction to random graphs, with no loops or multiple edges, and this restricted probability space is uniform (see for example [4,19] for a full description).

We consider the priority-greedy algorithm applied directly to the random pairing. As discussed in [18], the random pairing can be generated pair by pair, and at each step a point p can be chosen by any rule whatsoever, as long as the other point in the pair is chosen u.a.r. from the remaining unused points. We call this step *exposing* the pair containing p. At each step of the priority-greedy algorithm in which a vertex is coloured, we expose all pairs containing points in that vertex.

We now give an analysis of the algorithm which is not rigorous but will presumably yield the same bounds rigorously by introducing technical arguments as in [6]. Informally speaking, in a typical part of the algorithm, there will be symmetric pairs of one particular type, (r_0, b_0), which are plentiful in the graph but are quite regularly chosen in the main iteration of the algorithm. Symmetric pairs of types with higher priority may also be regularly chosen, but will be rare and regularly be used up entirely (at which point another pair of type (r_0, b_0) will be used). In this situation, we say that (r_0, b_0) is the *basic* type. The algorithm will typically pass through *phases*, determined by points at which, roughly speaking, the basic type changes. A phase finishes when either symmetric pairs with higher priorities than the current basic type become plentiful, or those with the current basic type become very scarce. The boundaries of the phases are best defined precisely in terms of the solution of a set of differential equations which we now proceed to derive.

At each point in the algorithm, let Z_{rb} represent the number of uncoloured vertices of type (r, b), and let W denote the number of points not yet involved in exposed pairs. Then $W = \sum_{r+b<d}(d - r - b)Z_{rb}$.

From this point onwards, we assume the reader is thoroughly familiar with the argument in [6], and omit any justifications are identical to those appearing there. Let $d_{r,b}$ denote the expected contribution to ΔZ_{rb}, the increment of Z_{rb}, due to exposing the pair containing a point in a vertex u which has just been coloured red by the priority-greedy algorithm. Then the probability that the other point chosen in the pair is in a vertex v of type (i, j) is $(d - i - j)Z_{i,j}/(W - 1)$ (except for a correction of size $O(1/W)$ due to the change in status of u). Hence, ignoring terms of size $O(1/W)$, $d_{r,b} = (\alpha_{d+1,r+b}Z_{r-1,b} - \alpha_{d,r+b}Z_{r,b})/W$ for $r + b \leq d$, where $\alpha_{x,y}$ is $x - y$ when $x > y$, 0 otherwise.

In the following we continue to ignore terms of size $O(1/W)$. The equations due to the case that u is coloured blue are $d_{r,b} = (\alpha_{d+1,r+b}Z_{r,b-1} - \alpha_{d,r+b}Z_{r,b})/W$ for $r+b \leq d$. Let $\bar{d}_{r,b}$ the expected increment due to the colouring of a symmetric pair. Making the assumption of having *rb-symmetry* (for all i and j, $Z_{ij} = Z_{ji}$), and adding the effects from a point in the red vertex and a point in the blue, gives $\bar{d}_{r,b} = (\alpha_{d+1,r+b}(Z_{r,b-1} + Z_{r-1,b}) - 2\alpha_{d,r+b}Z_{r,b})/W$, for $r + b \leq d$.

Let $\phi_{r,b}$ denote the probability of processing an (r, b)-symmetric pair at some step in a given phase. This will be examined non-rigorously, since we to some extent ignore the history and current state of the process. (The equations which this heuristic argument culminates in can be used to define another process which can be analysed rigorously; see [6] for details.) Assume that at the beginning of a new phase, (r_0, b_0) is the basic type. Then these vertices are plentiful, and thus for a considerable part of the algorithm, no vertices of lower priority will be chosen for Maj. We calculate the ϕ's for that symmetric pair type and all others with higher priority.

Let B' denote the set of types of symmetric pairs with higher priority than the basic, (r_0, b_0), and let $B' = B \cup \{(r_0, b_0)\}$.

Given the assumption about the ϕ's, the expected number of points in a blue (or red) vertex when Maj is performed is $c = \sum_{(r',b')\in B}(d - r' - b')\phi_{r',b'}$. This leads to

$$\phi_{r,b} = c\,\bar{d}_{r,b}, (r,b) \in B', \text{ and } \sum_{(r,b)\in B} \phi_{r,b} = 1. \tag{1}$$

These three equations are easy to solve, and we find $c = (1 - \phi_{r_0,b_0})/S$, with $S = \sum_{(r',b')\in B'} \bar{d}_{r',b'}$, which yields $c = (d - r_0 - b_0)/T$, with $T = 1 + \sum_{(r',b')\in B'}(r' + b' - r_0 - b_0)\bar{d}_{r',b'}$. Now $\phi_{r,b}$ is determined for $(r,b) \in B'$, and a little computation produces $\phi_{r_0,b_0} = 1 + \sum_{(r',b')\in B'}(r' + b' - d)\bar{d}_{r',b'}$. Assuming rb-symmetry, and assuming validity of the equations for the ϕ's, the expected increments of the random variables $Z_{r,b}$ at each iteration is given by $\mathbf{E}[\Delta(Z_{r,b})] = c\bar{d}_{r,b} - (1 + \delta_{rb})\phi_{r,b}$ where δ_{rb} is the Kronecker delta (1 if $r = b$, 0 otherwise). The terms subtracted are due to the change in types of the symmetric pair of vertex being coloured; in the case $r = b$, two vertices of type (r,r) are lost.

As done in [6] for the case $d = 4$, we may express the above expected increments as a set of differential equations, where each $\mathbf{E}[\Delta(Z_{r,b})]$ is expressed as the differential $Z'_{r,b}$ (all as functions of the number t of iterations). We scale both time and the variables by dividing by n, and denote $Z_{r,b}/n$ by $z_{r,b}$ t/n by x and W/n by w. Then the equations are $z'_{r,b} = (\alpha_{d+1,r+b}(z_{r,b-1} + z_{r-1,b}) - 2\alpha_{d,r+b}z_{r,b})\frac{C}{w} - (1+\delta_{rb})\theta_{r,b}$, where $w = w(x) = \sum_{r+b\leq d} \alpha_{d,r+b}z_{r,b}$, $\theta_{r,b} = \theta_{r,b}(x)$ represents $\phi_{r,b}(t/n)$ and can be defined as before but with $Z_{r,b}$ replaced by $z_{r,b}(x)$ (and the same goes for $\bar{d}_{r,b}$), and $C = C(x) = \sum_{(r',b')\in B}(d - r' - b')\theta_{r',b'}$ (after manipulation of the equations above).

The increase in the size of the bisection due to a vertex of type (r,b) being coloured red is r, and the symmetric vertex being coloured blue, and of type (b,r), also increases the bisection by r. Thus, the expected increase per algorithm step is $2\sum_{(r,b)\in B} r\,\phi_{r,b}$. Letting z denote the bisection size (divided by n), this suggests the equation $z' = 2\sum_{(r,b)\in B} r\,\phi_{r,b}$.

These are the differential equations for a phase with (r_0, b_0) being the basic type. The phase will end when either $\theta_{r_0,b_0} = 0$ (in which case, the basic type will have priority $\texttt{prio}(r_0, b_0)+1$) or when z_{r_0,b_0} begins to go negative (in which case, the basic type (r,b) in the next phase will be whichever type has highest priority among those with $z_{r,b} > 0$). There is the possibility of a phase of zero length, if these criteria immediately apply to the new basic type; then the next basic type can be determined using the same rule. The variables' initial conditions at the start of a phase are just their values at the end of the previous phase. The whole calculation begins with basic type $(0,1)$ and with all variables equal to 0, except for $z_{0,0} = 1$. The size of the bisection is represented by the value of z (scaled up by a factor n) when the values of all the $z_{r,b}$ reach 0 simultaneously. The last few phases are those in which the basic type (r_0, b_0) has $r_0 + b_0 = d$, and in practice, these may be skipped if the appropriate quantity is added to z.

After trying many different priority lists, solving the resulting system of differential equations using a Runge-Kutta method of order 2, we focused on

Table 1. Results for lists A and B, rounded up

d	5	6	7	8	9	10	11	12
list A	0.5028	0.6675	0.8502	1.0391	1.2317	1.4278	1.624	1.823
list B	0.5247	0.6674	0.8590	1.0386	1.2318	1.4278	1.624	1.823
min(A,B)	0.5028	0.6674	0.8502	1.0386	1.2317	1.4278	1.624	1.823

two priority lists which appear to give the best results. The following determine the order amongst those (r, b) with $r + b < d$:

List A: $\mathtt{prio}(i, j) > \mathtt{prio}(k, l)$ iff $j - i < l - k$ or ($j - i = l - k$ and $i > k$).
List B: Same as List A but swapping $\mathtt{prio}(0,2)$ with $\mathtt{prio}(\lfloor d/2 \rfloor - 1, \lfloor d/2 \rfloor)$.

For example, with $d = 5$, List A places the types in the following order: $(0,0)$, $(1,1)$, $(2,2)$, $(0,1)$, $(1,2)$, $(0,2)$, $(1,3)$, $(0,3)$, $(0,4)$; List B puts $(0,2)$ before $(1,2)$ but retains all other relative rankings.

List A appears, from the results of the calculations, to perform better for d odd and List B performs better for d even. However, for larger d, this is not clearly demonstrated to the accuracy with which we can confidently quote the results (due to errors inherent in numerical solution of the differential equations). The bounds obtained with Lists A and B for $d \leq 12$ are given in Table 1. Machine power was too limiting to go much further than this with sufficient accuracy, but the further digits obtained, which we do not report here, suggested that the ranking of Lists A and B according to the parity of d continues, at least up to $d = 12$.

As mentioned above, this argument is not rigorous, and in particular the concept of the probability measured by $\phi_{r,b}$ was not well specified. In the remainder of this section we sketch how the results could be turned into rigorous upper bounds using the type of argument in [6].

A major complicating factor is that the rate of change of variables is not smooth: when a vertex of one type is coloured, the effects are different from that of another type being processed. The values of the ϕ's can be estimated as in [17] by breaking the process up into pieces called *clutches* in [8], separated by the steps in which pairs of basic type are processed. Alternatively, the number of pairs of each type in a clutch can be estimated and an expression found for $\phi_{r,d}$. The heuristic argument above can be justified in a more direct way by considering a different, deprioritized algorithm as in [20]. At the beginning of the deprioritized algorithm, the first ϵn steps each randomly choose a pair of type $(0, 0)$ and applies Maj. This produces a plentiful supply of vertices of all types. Then, during the algorithm proper, the type of symmetric pair to be coloured in a given step is chosen randomly with probability $\phi_{r,b}$ for type (r, b), where ϕ is calculated using equations (1). The expected changes in the values of the $Z_{i,j}$ in one step can then be calculated (asymptotically), given their values at the beginning of the step. The differential equations we have derived above apply, but with the slightly different initial conditions determined by ϵ. When the variable z_{r_0,b_0} corresponding to the basic type of symmetric pair reaches 0, or the corresponding θ_{r_0,b_0} reaches 0, the phase ends. Inductively, one may apply the differential equation method (see [19]) to show that the variables of the

Table 2. Size of the bisection obtained by the greedy algorithm for five graphs with $n = 10^5$ (e5-*) and two graphs with $n = 2 \times 10^5$ (2e5-*). The asymptotic almost sure upper bound from the differential equation analysis is given in the left column.

		e5-1	e5-2	e5-3	e5-4	e5-5	av:e5	2e5-1	2e5-2	av:2e5
d=5	avg	0.5046	0.5087	0.5042	0.5046	0.5038	0.5052	0.5036	0.5040	0.5038
	max	0.5060	0.5463	0.5060	0.5061	0.5060		0.5045	0.5052	
0.5028	min	0.5026	0.5026	0.5019	0.5035	0.5025		0.5032	0.5031	
d=6	avg	0.6692	0.6710	0.6695	0.6690	0.6689	0.6695	0.6687	0.6691	0.6689
	max	0.6719	0.6849	0.6720	0.6724	0.6711		0.6692	0.6703	
0.6674	min	0.6675	0.6682	0.6670	0.6672	0.6675		0.6677	0.6683	
d=7	avg	0.8517	0.8515	0.8517	0.8522	0.8516	0.8517	0.8511	0.8516	0.8513
	max	0.8536	0.8534	0.8541	0.8530	0.8545		0.8533	0.8524	
0.8502	min	0.8443	0.8482	0.8502	0.8511	0.8498		0.8496	0.8509	
d=8	avg	1.0410	1.0397	1.0403	1.0406	1.0404	1.0407	1.0407	1.0402	1.0404
	max	1.0438	1.0420	1.0426	1.0427	1.0422		1.0432	1.0416	
1.0386	min	1.0388	1.0378	1.0384	1.0379	1.0378		1.0392	1.0384	
d=9	avg	1.2340	1.2339	1.2336	1.2333	1.2335	1.2336	1.2325	1.2342	1.2333
	max	1.2374	1.2366	1.2353	1.2352	1.2357		1.2331	1.2364	
1.2317	min	1.2312	1.2306	1.2313	1.2308	1.2309		1.2316	1.2326	
d=10	avg	1.4292	1.4303	1.4289	1.4296	1.4300	1.4296	1.4291	1.4288	1.4289
	max	1.4326	1.4327	1.4299	1.4315	1.4333		1.4300	1.4301	
1.4278	min	1.4265	1.4293	1.4275	1.4280	1.4281		1.4283	1.4272	
d=11	avg	1.6257	1.6256	1.6256	1.6256	1.6262	1.6258	1.6251	1.6254	1.6254
	max	1.6276	1.6276	1.6286	1.6273	1.6288		1.6264	1.6270	
1.624	min	1.6227	1.6234	1.6236	1.6240	1.6239		1.6243	1.6240	
d=12	avg	1.8245	1.8248	1.8241	1.8241	1.8257	1.8246	1.8241	1.8244	1.8244
	max	1.8264	1.8273	1.8270	1.8260	1.8273		1.8248	1.8254	
1.823	min	1.8231	1.8214	1.8226	1.8226	1.8223		1.8237	1.8236	

process with probability $1 + o(1)$ follow the solution of the differential equations in each phase; that is, $Z_{r,b}(t) = n z_{r,b}(t/n) + o(n)$ for all relevant t. Finally, it can be shown that letting $\epsilon \to 0$, the differential equation solution trajectory for the deprioritized algorithm approaches arbitrarily close to the trajectory for the original algorithm. The desired result follows from this.

4   The Experimental Upper Bounds

We have also generated a set of d-regular graphs for each $d = 5$ to 12, following the method described in [16]. We repeated the algorithm 10 times on each of the graphs, with priorities given by List A for d odd and List B for d even. The results, for five graphs with 10^5 vertices and two graphs with 2×10^5 vertices, for each $d = 5, \ldots, 12$ are summarised in Table 2. The left column of the table includes the bound via the differential equations. The mean, max and min of the bisection values obtained using Algorithm 1 are given for each graph, and the means are also averaged over all graphs of each of the two sizes. For any reader interested in checking the experiment, the graphs generated can be found at: *http://www.lsi.upc.es/~mjserna/dregraphs.html*

5   Maximum Bisection

Let us consider the variation of the priority-greedy algorithm obtained replacing the majority operation (Maj) with the minority operation, that assigns to a vertex the colour minority among its coloured neighbours. Let us call this variation max priority-greedy

Define an edge to be *fully coloured* when both its ends are finally coloured. A fully coloured edge is *mono-coloured* if both ends have the same colour and *bicoloured* if both ends have different colour. So the edges mono-coloured by priority-greedy get bicoloured by max priority-greedy and vice versa, whenever the vertices of the graph are treated in the same order (which happens with the same probability, in both cases). That is, every edge that counts in the bisection for one algorithm does not count in the other and vice versa. So, taking into account that the total number of edges in a d-regular graph is $dn/2$, we have the following complementary bounds for the maximum bisection: the size of the maximum bisection in a random d-regular graph is a.a.s. at least $dn/2 - c_d n$, where c_d is the min value given in Table 1 in column d.

6 Conclusions and Open Problems

In this paper we have proposed a randomized greedy procedure which bounds the bisection width of any d-regular graph, and analyzed its typical performance on random d-regular graphs. The algorithm uses a predefined list of priorities. Furthermore, a related algorithm shows complementary bounds for the maximum bisection size. We sketch a proof that for any given list, and any $d \geq 3$, the values of the size of the bisection obtained by the algorithm are concentrated around the value determined by the solution of a set of differential equations.

Experimentally, we notice that a good list of priorities is given by List A for $d \geq 6$ even and List B for $d \geq 5$ odd. It remains an open problem to search for other possible lists of priorities that improve the outcome of the algorithm. In Table 1, we get the asymptotic bisection width as solution to the differential equations (the tables reflect the constant to be multiplied by n). We may compare with the asymptotic lower bound of Bollobás and the asymptotic value of the upper bound of Alon, which is a deterministic one. For instance, for $d = 5$, Bollobás' lower bound yields $0.31917n$, Alon's upper bound yields $1.15118n$, while our upper bound is $0.5028n$. Furthermore, the complementary lower bounds we get on max bisection are well above the known lower bounds for all d-regular graphs, triangle-free or not.

Moreover, as can bee seen from Table 2, even for rather small values of n, the size of the bisection obtained by the algorithm is close to the solution determined by the differential equations. As n grows, this phenomenon strengthens.

As mentioned above, in [6] there is an analytic expression for the bound obtained on the bisection width of a random 4-regular graph, obtained from differential equations corresponding to the priority-greedy algorithm. When run for $d = 4$, Lists A and B give the same theoretical result, because the types $(0, 2)$ and $(1, 2)$ never become basic: there is only one phase, with $(0, 1)$ basic (see [6], where the algorithm is expressed in a different way but gives the same differential equations).

Several open problems remain, the first being to improve the upper bound. One way to do this may be to find better priority lists. Another question is whether there is some simpler way to analyse these greedy algorithms rigorously.

References

1. N. Alon. On the edge-expansion of graphs. *Combinatorics, Probability and Computing*, 6:145–152, 1997.
2. S. Bezrukov, R. Elsasser, B. Monien, R. Preis, and J. Tillich. New spectral lower bounds on the bisection width of graphs. In U. Brandes and D. Wagner, editors, *Graph-Theoretic Concepts in Computer Science*, LNCS 1928, pages 23–34. Springer, 2000.
3. B. Bollobas. The isoperimetric number of random regular graphs. *European Journal of combinatorics*, 9:241–244, 1984.
4. B. Bollobas. *Random Graphs. Second Edition*. Cambridge University Press, Cambridge, 2001.
5. T. Bui, S. Chaudhuri, T. Leighton, and M. Sipser. Graph bisection algorithms with good average case behavior. *Combinatorica*, 7:171–191, 1987.
6. J. Díaz, N. Do, M. J. Serna, and N. C. Wormald. Bounds on the max and min bisection of random cubic and 4-regular graphs. *Theoretical Computer Science*, 307:531–547, 2003.
7. J. Díaz, J. Petit, and M. Serna. A survey on graph layout problems. *ACM Computing Surveys*, 34:313–356, 2002.
8. W. Duckworth and N. C. Wormald. Minimum independent dominating sets of random cubic graphs. *Random Structures and Algorithms*, 21:147–161, 2002.
9. M. Fiedler. A property of the eigenvectors of nonnegative symmetric matrices and its application to graph theory. *Czechoslovak Mathematical Journal*, 25:619–633, 1975.
10. M. R. Garey and D. S. Johnson. *Computers and Intractability: A Guide to the Theory of NP-Completeness*. Freeman, San Francisco, 1979.
11. K. Jansen, M. Karpinski, A. Lingas, and E. Seidel. Polynomial time approximation schemes for MAX-BISECTION on planar and geometric graphs. In A. Ferreira and H. Reichel, editors, *Symposium on Theoretical Aspects of Computer Science*, LNCS 2010, pages 365–375. Springer, 2001.
12. A. Kostochka and L. Melnikov. On bounds of the bisection width of cubic graphs. In J. Nesetril and M. Fiedler, editors, *Czechoslovakian Symposium on Combinatorics, Graphs and Complexity*, pages 151–154. Elsevier Science Publishers, 1992.
13. S. C. Locke. Maximum k-colorable subgraphs. *Journal of Graph Theory*, 6(2):123–132, 1982.
14. B. Monien and R. Preis. Upper bounds on the bisection width of 3- and 4-regular graphs. In A. P. J. Sgall and P. Kolman, editors, *Mathematical Foundations of Computer Science*, LNCS 2136, pages 524–536. Springer, 2001.
15. J. B. Shearer. A note on bipartite subgraphs of triangle-free graphs. *Random Structures Algorithms*, 3(2):223–226, 1992.
16. A. Steger and N. C. Wormald. Generating random regular graphs quickly. *Combinatorics, Probabability and Computing*, 8:377–396, 1999.
17. N. C. Wormald. Differential equations for random processes and random graphs. *Annals of Applied Probability*, 5:1217–1235, 1995.
18. N. C. Wormald. The differential equation method for random graph processes and greedy algorithms. In M. Karoński and H. Prömel, editors, *Lectures on Approximation and Randomized Algorithms*, pages 73–155. PWN, Warsaw, 1999.
19. N. C. Wormald. Models of random regular graphs. In *Surveys in Combinatorics*, pages 239–298. Cambridge University Press, 1999.
20. N. C. Wormald. Analysis of greedy algorithms on graphs with bounded degree. *Discrete Mathematics*, 2003.

Constrained Integer Partitions

Christian Borgs[1], Jennifer T. Chayes[1], Stephan Mertens[2], and Boris Pittel[3]

[1] Microsoft Research, Redmond, WA 98052
[2] Institut für Theoretische Physik, Otto-von-Guericke Universität, D-39016
Magdeburg, Germany
[3] Department of Mathematics, Ohio State University, Columbus, Ohio 43210

Abstract. We consider the problem of partitioning n integers into two
subsets of given cardinalities such that the discrepancy, the absolute
value of the difference of their sums, is minimized. The integers are
i.i.d. random variables chosen uniformly from the set $\{1, \ldots, M\}$. We
study how the typical behavior of the optimal partition depends on n, M
and the bias s, the difference between the cardinalities of the two subsets
in the partition. In particular, we rigorously establish this typical behav-
ior as a function of the two parameters $\kappa := n^{-1} \log_2 M$ and $b := |s|/n$
by proving the existence of three distinct "phases" in the κb-plane, characterized by the value of the discrepancy and the number of optimal
solutions: a "perfect phase" with exponentially many optimal solutions
with discrepancy 0 or 1; a "hard phase" with minimal discrepancy of
order $Me^{-\Theta(n)}$; and a "sorted phase" with an unique optimal partition
of order Mn, obtained by putting the $(s + n)/2$ smallest integers in one
subset.

1 Introduction

Phase transitions in random combinatorial problems have been the subject of
much recent attention. The random optimum partitioning problem is the only
NP-hard problem for which the existence of a sharp phase transition has been
rigorously established, as have many detailed properties of the transition ([2],
see [3] for a short overview). Here we study a constrained version of the random
optimum partitioning problem, and extend some of the results of [2] to that case.
Complete proofs of the results announced here will be given in [1].

The integer optimum partitioning problem is a classic problem of combinatorial optimization which has been studied in the theoretical computer science
([9], [4], [10], [11]), artificial intelligence ([8]), theoretical physics ([7], [5], [6],
[12], [13]) and mathematics ([3], [2]) communities. The problem is to partition a
given set of n integers into two subsets in order to minimize the absolute value of
the difference between the sum of the integers in the two subsets, the so-called
discrepancy. Notice that for any given set of integers, the discrepancies of all
partitions have the same parity, namely that of the sum of the n integers. We
call a partition *perfect* if its discrepancy is 0, when this sum is even, or 1, when
this sum is odd. The decision question is whether there exists a perfect partition.
In the uniformly random version, an instance is a given a set of n i.i.d. integers

M. Farach-Colton (Ed.): LATIN 2004, LNCS 2976, pp. 59–68, 2004.
© Springer-Verlag Berlin Heidelberg 2004

60   C. Borgs et al.

drawn uniformly at random from $\{1, 2, \ldots, M\}$. We will sometimes use the notation $m = \log_2 M$; notice that each of the random integers has m binary bits. Previous work had established a sharp transition as a function of the parameter $\kappa := m/n$, characterized by a dramatic change in the probability of a perfect partition. For M and n tending to infinity in the limiting ratio $\kappa = m/n$, the probability of a perfect partition tends to 1 for $\kappa < 1$, while the probability tends to 0 for $\kappa > 1$. This result was suggested by the work of one of the authors [12] and proved in a paper by the three other authors [2].

The location of the phase transition for the unconstrained problem immediately yields a one-dimensional phase diagram as a function of κ: For $\kappa \in (0, \kappa_c)$ with $\kappa_c = 1$, the system is in a "perfect phase" in which the probability of a perfect partition tends to 1 as M and n tend to infinity in the fixed function κ. For $\kappa \in (\kappa_c, \infty)$, the probability of a perfect partition tends to 0, and moreover, there is an unique optimal partition. We call this the "hard phase," since for $\kappa > \kappa_c$, it is presumably computationally difficult to find the optimal partition.

Here we consider a constrained variant of the problem in which we require that the two subsets have given cardinalities; we say that the difference of the two cardinalities is the *bias*, s, of the partition. We establish the two-dimensional phase diagram of the random constrained integer partitioning problem as a function of the parameters $\kappa := m/n$ and $b := |s|/n$. See Fig. 1. In addition to the extensions of the perfect and hard phases, we establish the existence of a new phase which we call the "sorted phase."

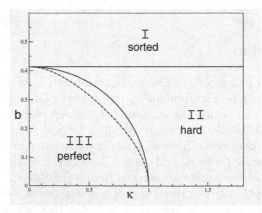

Fig. 1. Phase diagram of the constrained integer partitioning problem.

The sorted phase is easy to understand. One way to meet the bias constraint is to take the $(s + n)/2$ smallest integers and put them in one subset of the partition. We define the sorted phase as the subset of the κb-plane where the sorted partition is optimal. We prove that the sorted phase is given by the condition $b > b_c := \sqrt{2} - 1$; see region I in Fig. 1. Moreover, we show that the minimal discrepancy in this phase is of the order Mn.

Our analysis of the perfect and hard phases for $b < b_c$ is much more difficult. In this region, we use integral representations for the number of partitions with a given discrepancy and bias; these representations generalize those used in [2]. The asymptotic analysis of the resulting two-dimensional random integrals leads to saddle point equations for a saddle point described in terms two real parameters η and ζ. For discrepancies of order $o(M)$ (including, in particular, the case of perfect partitions), the saddle point equations determining ζ and η are:

$$\int_0^1 x \tanh(\zeta x + \eta)\, dx = 0, \quad \int_0^1 \tanh(\zeta x + \eta)\, dx = -b. \tag{1}$$

The solution (ζ, η) of these equations can be used to define the two convex curves in Fig. 1. To this end, let

$$L(\zeta, \eta) := b\eta + \int_0^1 \log(2\cosh(\zeta x + \eta))\, dx \tag{2}$$

$$\rho(\zeta, \eta) := 1 - \frac{\tanh(\zeta + \eta) - \tanh(\eta)}{2\zeta}. \tag{3}$$

For (ζ, η) a solution of (1), we then define

$$\kappa_-(b) := -\log_2 \rho(\zeta, \eta), \quad \kappa_c(b) := \frac{1}{\log 2} L(\zeta, \eta). \tag{4}$$

From bottom to top, the two convex curves joining $(0, b_c)$ and $(1, 0)$ in Fig. 1 are then given by $\kappa = \kappa_-(b)$ and $\kappa = \kappa_c(b)$.

Our results prove that, in the region $\kappa < \kappa_-(b)$, with probability tending to one as n tends to infinity (or, more succinctly, with high probability, w.h.p.) there exist perfect partitions; see region III in Fig. 1. Moreover the number of perfect partitions is about $2^{(\kappa_c - \kappa)n}$ in this "perfect phase." We also prove that w.h.p. there are no perfect partitions in the region $b < b_c$ and $\kappa > \kappa_c(b)$, which we call this the "hard phase." Our results leave open the question of what happens in the narrow region $\kappa_- < \kappa < \kappa_c$, and also whether the optimal partition is unique in the hard phase.

We also prove that these phase transitions correspond to qualitative changes in the solution space of the associated linear programming problem (LPP). In the actual optimum partitioning problem, each integer is put in one subset or the other. The relaxed version is defined by allowing any fraction of each integer to be put in either of the two partitions. Here we show the following. In the sorted phase, i.e. for $b > b_c = \sqrt{2} - 1$, w.h.p. the LPP has a unique solution given by the sorted partition itself. For $b < b_c$, i.e. in the perfect and hard phases, w.h.p. the relaxed minimum discrepancy is zero, and the total number of optimal basis solutions is exponentially large, of order $2^{k_c(b)n + O_p(n^{1/2})}$. Finally, in the perfect and hard phases, we consider the fraction of these basis solutions whose integer-valued components form an optimal integer partition of the subproblem with the corresponding subset of the weights. We show that this fraction is exponentially small. Moreover, except for the crescent-shaped region between $\kappa = \kappa_-(b)$ and $\kappa = \kappa_c(b)$, we show that the fraction is strictly exponentially smaller in the hard phase than in the perfect phase. This fraction thus represents some measure of the algorithmic difficulty of the problem.

In the next section, we define the problem in detail, and precisely state our main results. In Section 3, we introduce our integral representation and show how it leads to the relevant saddle point equations. We also give a brief heuristic derivation of some of the phase boundaries. The complete proofs are quite involved, and are presented in the full paper version [1] of this extended abstract.

2 Statement of Main Results

Let $X_1, \ldots, X_n$ be n i.i.d. random variables distributed uniformly on $\{1, \ldots, M\}$. We use $\mathbb{P}$ and $\mathbb{E}$ to denote the corresponding probability and expectation induced by $\mathbf{X} = (X_1, \ldots, X_n)$. We are interested in the case when M grows exponentially with n, and define κ as the exponential rate, i.e. $\kappa = n^{-1} \log_2 M$. To avoid trivial counterexamples, we assume that κ stay bounded away from 0 and ∞ as $n \to \infty$.

A partition of integers into two disjoint subsets is coded by an n-long binary sequence $\sigma = (\sigma_1, \ldots, \sigma_n)$, $\sigma_j \in \{-1, 1\}$; so the subsets are $\{j : \sigma_j = 1\}$ and $\{j : \sigma_j = -1\}$. Obviously σ and $-\sigma$ are the codes of the same partition. Given a partition σ, we define its *discrepancy*, $d(\mathbf{X}, \sigma) = |\sigma \cdot \mathbf{X}|$, and *bias*, $s(\sigma) = \sigma \cdot \mathbf{e} = |\{j : \sigma_j = 1\}| - |\{j : \sigma_j = -1\}|$. Here $\sigma \cdot \mathbf{X} = \sum_{j=1}^n \sigma_j X_j$ and $\mathbf{e}$ is the vector $(1, \ldots, 1)$. Clearly $s(\sigma)$ is an integer in $\{-n, \ldots, n\}$, so let $s \in \{-n, \ldots, n\}$ and define the bias density $b = |s|/n$ so that $b \in [0, 1]$. Note that by symmetry it suffices to consider $s(\sigma) \in \{0, \ldots, n\}$, so we will often take a non-negative integer $s \in \{0, \ldots, n\}$, in which case $s = bn$. We define an *optimum partition* as a partition σ that minimizes the discrepancy $d(\mathbf{X}, \sigma)$ among all the partitions with bias equal to s, and a *perfect partition* as a partition σ with $|d(\mathbf{X}, \sigma)| \leq 1$.

Theorems 2, 3 and 4 below describe our main results on the phases labelled I, II, and III in Fig. 1 in the introduction. In the statement of these theorems we will use the parameters $\zeta, \eta, \kappa_c(b)$ and $\kappa_-(b)$ defined in (1) – (4). Before getting to principal results, we begin with an existence statement.

Theorem 1. *Let $b < b_c$, where $b_c = \sqrt{2} - 1$. Then the saddle point equations (1) have a unique solution $(\zeta, \eta) = (\zeta(b), \eta(b))$.*

Let $Z_n(\ell, s) = Z_n(\ell, s; \mathbf{X})$ denote the random number of partitions σ with $\sigma \cdot \mathbf{X} = \ell$ and $\sigma \cdot \mathbf{e} = s$. Since $s(\sigma)$ has the same parity as n, and $\mathbf{X} \cdot \sigma$ has the same parity as $\sum_{j=1}^n X_j$, we will only consider values of s which have the same parity as n, and values of ℓ which have the same parity as $\sum_{j=1}^n X_j$. In the theorems below, we will not state these restrictions explicitly.

To formulate our results in a compact form, we use a shorthand $a_n < a$ ($a_n > a$, resp.) instead of $\limsup a_n < a$ ($\liminf a_n > a$, resp.), even when the n-dependence of a_n is only implicit, as in $\kappa = n^{-1} \log_2 M$ and $b = |s/n|$. We also use the notation $f_n = O_p(g_n)$ and $f_n = o_p(h_n)$ if f_n/g_n is bounded in probability and f_n/h_n goes to zero in probability, respectively. Also, as is customary, we will say that an event happens with high probability (w.h.p.) if the probability of this event approaches 1 as $n \to \infty$. In all our statements n, M, s and ℓ will be integers with $n \geq 1$, $M \geq 1$ and $s \geq 0$.

Our main results in the perfect phase are summarized in the next theorem.

Theorem 2. *Let $\ell = o(Mn^{1/2})$, $b < b_c$ and $\kappa < \kappa_-(b)$. Then w.h.p. $Z_n(\ell, s) \geq 1$ and*

$$Z_n(\ell, s) = 2^{[\kappa_c(b)-\kappa]n} e^{S_n n^{1/2}+o(n^{1/2})}, \tag{5}$$

where S_n converges in probability to a Gaussian with mean zero and variance $\sigma^2 = Var(\log(2\cosh(\zeta U + \eta)))$, with U uniformly distributed on $[0, 1]$. Consequently, w.h.p., there exist exponentially many perfect partitions, with $\ell = 0$ if $\sum_j X_j$ is even, and $|\ell| = 1$ if $\sum_j X_j$ is odd.

Our next theorem, which describes our main results on the hard phase, has two parts: The first shows that there are no perfect partitions above $\kappa = \kappa_c(b)$, and the second gives a bound on the number of optimum partition for $\kappa > \kappa_-$. To state the theorem, let $d_{opt} = d_{opt}(n; s)$ denote the discrepancy of the optimal partition, and let $Z_{opt} = Z_{opt}(n; s)$ denote the number of optimal partitions.

Theorem 3. *Let $b < b_c$.*

1. *If $\kappa > \kappa_c(b)$, then there exists a $\delta > 0$ such that with probability $1 - O(e^{-\delta \log^2 n})$ there are no perfect partitions, and moreover*

$$d_{opt} \geq 2^{n[\kappa - \kappa_c(b)] - O_p(n^{1/2})}. \tag{6}$$

2. *If $\kappa > \kappa_-(b)$ and $\varepsilon > 0$, then there exists a constant $\delta > 0$ such that*

$$d_{opt} \leq 2^{n[\kappa - \kappa_-(b)+\varepsilon]} \quad and \quad Z_{opt} \leq 2^{n[\kappa_c(b)-\kappa_-(b)+\varepsilon]}, \tag{7}$$

both with probability $1 - O(e^{-\delta \log^2 n})$.

Our main result on the sorted phase is the following theorem.

Theorem 4. *Let $b > b_c$. Then w.h.p. the optimal partition is uniquely obtained by putting $(s+n)/2$ smallest integers X_j in one part, and the remaining $(n-s)/2$ integers into another part. W.h.p., d_{opt} is asymptotic to $\frac{Mn}{4}[(1+b)^2 - 2]$, i.e., of order Mn.*

By this theorem, for b sufficiently large, the partition is determined by the decreasing order of weights X_j, but not by the actual values of X_j.

It is a rather common idea to approximate an optimization problem defined with integer-valued variables by its relaxed version, where the variables are now allowed to assume any value within the real intervals whose endpoints are the admissible values of the original integer variables. In our case, the relaxed version is a linear programming problem (LPP) which can be stated as follows. Find the minimum value d_{opt} of d, subject to linear constraints

$$-d \leq \sum_j \sigma_j X_j \leq d, \quad \sum_j \sigma_j = s, \quad and \quad -1 \leq \sigma_j \leq 1, \quad (1 \leq j \leq n). \tag{8}$$

As usual, the LPP has at least one basis solution, i.e. a solution (σ, d_{opt}), which is an extreme (vertex) point of the polyhedron defined by the constraints (8). Let $N(\sigma) := |\{j : \sigma_j \in (-1,1)\}|$ be the number of components of σ which are non-integer. It is easy for the reader to verify that $N(\sigma)$ is either 0 or 2 for all basis solutions σ.

Our next theorem shows that the LPP inherits the phase diagram of the optimum partition problem, and moreover provides a limited way to quantify the relative algorithmic difficult of the optimal partition problem in the three regions. For $b > b_c$ the solutions of the initial partition problem and of its LPP version coincide. For $b < b_c$ they are very far apart, in terms of the *ratio* of respective optimal discrepancies. To state this precisely, we define $F_n(\kappa, b)$ to be the fraction of basis solutions σ with the property that the deletion of the $N(\sigma)$ components of σ with values in $(-1,1)$ produces an optimal integer partition for the corresponding subproblem with weights X_i.

Theorem 5.   *1. If $b > b_c$, then w.h.p. the sorted partition is a unique solution of the LPP, and thus $d_{opt}^{LPP} = \Theta(Mn)$ and $F_n(\kappa, b) = 1$.*

2. If $b < b_c$, then w.h.p. $d_{opt}^{LPP} = 0$. In addition, w.h.p. there are $2^{[\kappa_c(b)+o(1)]n}$ basis solutions, each having either none or exactly two components $\sigma_i \neq \pm 1$.

3. If $b < b_c$, then w.h.p. $F_n(\kappa, b) = 2^{-[\kappa+o(1)]n}$ for $\kappa < \kappa_-(b)$, and $2^{-[\kappa_c(b)+o(1)]n} \leq F_n(\kappa, b) \leq 2^{-[\kappa_-(b)+o(1)]n}$ for $\kappa > \kappa_-(b)$.

Remark 1. (i) If one assume that the number of optimal partitions Z_{opt} in the hard phase grows subexponentially with probability at least $1 - o(n^{-2})$ (see [1] for a motivation of this assumption), our upper bound on the fraction $F_n(\kappa, b)$ in the hard phase can be improved to match the lower bound, yielding $F_n(\kappa, b) = 2^{-n[\kappa_c(b)+o(1)]}$ in the hard phase.

(ii) If, on the other hand, the asymptotics of Theorem 2 hold up to κ_c, more precisely, if one assumes that for $b < b_c$ and $\kappa < \kappa_c(b)$, the bound $Z_n(\ell, s) = 2^{n[\kappa_c(b)-\kappa+o(1))]}$ holds with probability least $1 - o(n^{-2})$, then a bound of the form $F_n(\kappa, b) = 2^{-n[\kappa+o(1)]}$ can be extended to all $\kappa < \kappa_c$.

3   Outline of Proof Strategy

3.1   Sorted Partitions

We first discuss our strategy to prove that in region I, the optimal partition is sorted and has discrepancy of order Mn. Let us first recall that M is assumed to grow exponentially with n, so that, in particular, $n^2 = o(M)$. As a consequence, w.h.p. no two weights are equal, and there is a unique reordering of the weights $X_1, \ldots, X_n$ such that their sizes are increasing, $X_{\pi(1)} < X_{\pi(2)} < \cdots < X_{\pi(n)}$, where $\pi(1), \ldots, \pi(n)$ is a suitable permutation of $1, \ldots, n$.

Given a bias $s > 0$, we need to find an optimum partition that puts $k = (s + n)/2$ integers in one part, and the remaining $n - k$ integers into another part. One such feasible partition is obtained if we select the k smallest integers for the first part; we call it the sorted partition. It is coded by the σ, with $\sigma_{\pi(i)} = 1$ for $i \leq k$ and $\sigma_{\pi(i)} = -1$ for $i > k$. If the total weight of $(n - k)$ largest weights is, at most, the total weight of k smallest weights, then it is intuitively clear that the sorted partition is optimal. More precisely: if $\delta_s(\boldsymbol{X}) = \sum_{j=1}^{k} X_{\pi(j)} - \sum_{j=k+1}^{n} X_{\pi(j)} \geq 0$ then the sorted partition is the unique, optimal partition and $d_{opt} = \delta_s(\boldsymbol{X})$.

To establish the boundary of the sorted phase, we thus have to determine the region of the phase diagram in which w.h.p. $\delta_s(\boldsymbol{X}) \geq 0$. Rather than considering a sorted partition with *fixed* bias $s = bn$, let us instead consider a random sorted partition with $\sigma_j = 1$ for $X_j \leq M_0$, and $\sigma_j = -1$ for $X_j > M_0$, where $M_0 \in \{1, \ldots, M\}$ is chosen so that the *expected* bias density is b, i.e. $M_0 = M(b+1)/2$. The expected discrepancy of such a partition is

$$\frac{n}{M} \sum_{j=1}^{M_0} m - \frac{n}{M} \sum_{j=M_0}^{M} m = \left[\left(\frac{b+1}{2} \right)^2 - \frac{1}{2} + O(M^{-1}) \right] Mn. \tag{9}$$

The right hand side of (9) is positive and of order Mn iff $(b+1)^2/4 - 1/2 > 0$, or equivalently $b > b_c = \sqrt{2} - 1$. In Sect. 6 of [1], we prove the required concentration, implying that the condition $b > b_c$ is both necessary and sufficient for $\delta_s(\mathbf{X})$ to be, w.h.p., positive and of order Mn.

3.2 Integral Representations

Let us now turn to the much more difficult region $b < b_c$. Guided by the results of [2], one might hope to prove that, as the parameter $\kappa = n^{-1} \log_2 M$ is varied, the model undergoes a phase transition between a region with exponentially many perfect partitions and a region with no perfect partitions.

A starting point in [2] was an integral (Fourier-inversion) type formula for $Z_n(\ell) = Z_n(\ell; \mathbf{X})$, the total number of σ's such that $\sigma \cdot \mathbf{X} = \ell$, namely

$$Z_n(\ell) = \frac{2^n}{\pi} \int_{x \in (-\pi/2, \pi/2]} \cos(\ell x) \prod_{j=1}^{n} \cos(x X_j) \, dx. \tag{10}$$

We need to derive a two-dimensional counterpart of this formula for $Z_n(\ell, s)$. To this end, let us first note that $s = 2|\{j : \sigma_j = 1\}| - n$, so that a generic value s of $s(\sigma)$ must meet the condition $n + s \equiv 0 \pmod 2$. In a similar way, we get that $\sigma \cdot \mathbf{X}$ has the same parity as the sum $\sum_j X_j$. Keeping this in mind, we have that on the event $\{\sum_j X_j \equiv \ell \pmod 2\}$, for $n + s \equiv 0 \pmod 2$,

$$\mathbb{I}(\sigma \cdot \mathbf{X} = \ell, \sigma \cdot \mathbf{e} = s) = \frac{1}{\pi^2} \iint_{x,y \in (-\pi/2, \pi/2]} e^{i(\sigma \cdot \mathbf{X} - \ell)x} e^{i(\sigma \cdot \mathbf{e} - s)y} \, dx dy, \tag{11}$$

thus extending (4.6) in [2]. Multiplying both sides of the identity by 2^n, and summing over all σ, we obtain

$$Z_n(\ell, s) = \frac{2^n}{\pi^2} \iint_{(-\frac{\pi}{2}, \frac{\pi}{2}]^2} e^{-i(\ell x + sy)} \prod_{j=1}^{n} \cos(x X_j + y) \, dx dy \tag{12}$$

$$= 2^n \, \mathbb{P}_{1/2} \left(\sigma \cdot \mathbf{X} = \ell, \sigma \cdot \mathbf{e} = s \big| \mathbf{X} \right),$$

where $\sigma = (\sigma_1, \ldots, \sigma_n)$ is a sequence of i.i.d. Bernoulli random variables with probability of $\sigma_i = \pm 1$ equal to $1/2$.

We would like to estimate the asymptotics of the integral in (12), which is equivalent to proving a local limit theorem for the conditional probability in (12). In general, to compute—via local limit theorems—the probability that some random variable A takes the value a, it must be the case that the corresponding expectation of A is near a. Thus the analogue of the representation (12) for the unconstrained problem was well adapted to the analysis of perfect partitions. Indeed, in that case, we wanted to estimate $\mathbb{P}_{1/2}(|\sigma \cdot \mathbf{X}| \le 1|\mathbf{X})$, and we had $\mathbb{E}_{1/2}(\sigma \cdot \mathbf{X}|\mathbf{X}) = 0$. However, in the constrained case, this strategy cannot be expected to work for $b > 0$, since $s = bn$ is very far from the expectation of $\sigma \cdot \mathbf{e}$, namely $\mathbb{E}_{1/2}(\sigma \cdot \mathbf{e}|\mathbf{X}) = 0$.

To resolve this difficulty, we introduce a *two-parameter* family of distributions for σ_j as follows: Given $\xi, \eta \in \mathbb{R}$, let $\sigma = (\sigma_1, \ldots, \sigma_n)$ be a sequence of random variables such that, conditioned on $\mathbf{X}$, $\sigma_1, \ldots, \sigma_n$ are mutually independent, and

$$\mathbb{P}(\sigma_j = 1|\mathbf{X}) = P(\xi X_j + \eta) \quad \text{with} \quad P(u) := \frac{e^{-u}}{2\cosh u}. \tag{13}$$

It is not hard to show [1] that, in terms of these random variables, $Z_n(\ell, s)$ can be rewritten as

$$Z_n(\ell, s) = e^{nL_n(\xi, \eta; \mathbf{X})} \, \mathbb{P}(\sigma \cdot \mathbf{X} = \ell, \, \sigma \cdot \mathbf{e} = s|\mathbf{X})$$

$$= e^{nL_n(\xi, \eta; \mathbf{X})} \iint_{x,y \in (-\pi/2, \pi/2]} \frac{1}{\pi^2} e^{-i(\ell x + sy)} \mathbb{E}\left(e^{i(x\sigma \cdot \mathbf{X} + y\sigma \cdot \mathbf{e})}|\mathbf{X}\right) dx\, dy, \tag{14}$$

where

$$L_n(\xi, \eta; \mathbf{X}) := \frac{\ell\xi}{n} + \frac{s\eta}{n} + \frac{1}{n}\sum_{j=1}^{n} \log(2\cosh(\xi X_j + \eta)). \tag{15}$$

3.3   Saddle Point Equations and Their Solution

Given ξ, η, we now face the problem of determining an asymptotic value of the *local* probability in (14). This will obviously be easier if the chosen parameters ℓ and s are among the more likely values of $\sigma \cdot \mathbf{X}$ and $\sigma \cdot \mathbf{e}$, respectively. A natural choice is to take ℓ and s equal to their expected values, that is $\mathbb{E}(\sigma \cdot \mathbf{X}|\mathbf{X}) = \ell$ and $\mathbb{E}(\sigma \cdot \mathbf{e}|\mathbf{X}) = s$, or explicitly,

$$\sum_{j=1}^{n} X_j \tanh(\xi X_j + \eta) = -\ell, \qquad \sum_{j=1}^{n} \tanh(\xi X_j + \eta) = -s. \tag{16}$$

Note that the equations (16) also arise naturally in an apparently different approach to estimate the integral in (12), the "method of steepest descent." In our context, this corresponds to a complex shift of the integration path, i.e., to changing the path of integration for x to the complex path from $-\pi/2 + i\xi$ to $-\pi/2 + i\xi$, and the path of integration for y to the complex path from $-\pi/2 + i\eta$ to $-\pi/2 + i\eta$, where ξ and η are determined by a suitable saddle point condition. For general ξ and η, this leads to (14), while the saddle point conditions turn out to be nothing but (16).

Both approaches raise the question of uniqueness and existence of a solution to the saddle point equations (16). While uniqueness follows from an abstract convexity argument which holds independent of the value of the parameters, existence turns out to be much more difficult and requires that s is smaller than some critical value $s_c = s_c(\mathbf{X})$. In the actual proof, we modify this approach a little since the solution $\xi = \xi(\mathbf{X})$, $\eta = \eta(\mathbf{X})$ does not lend itself to a rigorous analysis of $\mathbb{P}(\sigma \cdot \mathbf{X} = \ell, \sigma \cdot e = s | \mathbf{X})$. Instead, we will resort to "suboptimal" $\xi = \zeta/M$, η, where ζ, η are nonrandom constants, and (ζ, η) is a solution of nonrandom equations, obtained by replacing the (scaled) sums in (16) with their weak-law limits, and the parameters ℓ and s by their scaled version ℓ/Mn and $b = s/n$. Since we are mainly interested in $\ell = 0, \pm 1$ (corresponding to perfect partitions) with $\ell/Mn = o(1)$, this leads to the equations (1) given in the introduction. As shown in Sect. 4 of [1], these equations have a (unique) solution $\zeta = \zeta(b)$, $\eta = \eta(b)$ iff $b < b_c = \sqrt{2} - 1$, the same b_c that determines the sorted phase.

3.4 Asymptotic Behavior of $Z_n(\ell, s)$.

Assuming that $b < b_c$, let us now consider the right hand side of (14) with (ζ, η) taken to be the solution of (1). Then we can try to prove a local limit theorem for the *conditional* probability in (14), giving the approximation

$$\mathbb{P}(\sigma \cdot \mathbf{X} = \ell, \sigma \cdot \mathbf{e} = s | \mathbf{X}) \sim \frac{2}{\pi \sqrt{\det Q}}, \tag{17}$$

where

$$Q = \begin{pmatrix} \mathrm{Var}(\sigma \cdot X) & \mathrm{cov}(\sigma \cdot \mathbf{X}, \sigma \cdot \mathbf{e}) \\ \mathrm{cov}(\sigma \cdot \mathbf{X}, \sigma \cdot \mathbf{e}) & \mathrm{Var}(\sigma \cdot \mathbf{e}) \end{pmatrix}. \tag{18}$$

Here the (co)variances are conditioned on $\mathbf{X}$, so, e.g., $Q_{11} = \mathrm{Var}(\sigma \cdot \mathbf{X} | \mathbf{X})$. Next we appeal to the weak law of large numbers to further approximate the matrix elements of Q by $Q_{11} \sim n^2 M^2 R_{11}$, $Q_{12} \sim n^2 M R_{12}$, $Q_{21} \sim n^2 M R_{21}$ and $Q_{22} \sim n^2 R_{22}$, where R is a deterministic matrix depending on ζ, η. This gives

$$\mathbb{P}(\sigma \cdot \mathbf{X} = \ell, \sigma \cdot \mathbf{e} = s | \mathbf{X}) \sim \frac{1}{nM} \frac{2}{\pi \sqrt{\det R}}. \tag{19}$$

In a similar way we approximate the exponent $L_n(\xi(\mathbf{X}), \eta(\mathbf{X}); \mathbf{X})$ in (14) by its weak limit, which is just the function $L(\zeta, \eta)$ introduced in (2). For $|\ell| \leq 1$ and $M = 2^{\kappa n}$, we thus approximate $\log_2 Z_n(\ell, s)$ by $n(L(\zeta, \eta) - \kappa) = n(\kappa_c(b) - \kappa)$, suggesting that for $\kappa < \kappa_c(b)$ there are exponentially many perfect partitions, while for $\kappa > \kappa_c(b)$ there are none.

However, this informal argument is too naive, and neglects several important error terms. Indeed, the above approximation for $\log_2 Z_n(\ell, s)$ could not possibly hold for $\kappa > \kappa_c(b)$ since, $Z_n(\ell, s)$ is an integer, and thus cannot be asymptotically equivalent to an exponentially small, yet positive number. This means that a rigorous proof must be based on the condition $\kappa < \kappa_c(b)$. In fact, we will need the stronger condition $\kappa < \kappa_-(b)$; see [1] for the (quite painful) details.

References

1. C. Borgs, J.T. Chayes, S. Mertens, and B. Pittel. Phase diagram for the constrained integer partitioning problem. *Preprint*, 2003.
2. C. Borgs, J.T. Chayes, and B. Pittel. Phase transition and finite-size scaling for the integer partitioning problem. *Rand. Struc. Alg.*, 19:247–288, 2001.
3. C. Borgs, J.T. Chayes, and B. Pittel. Sharp threshold and scaling window for the integer partitioning problem. *Proc. 33rd ACM Symp. on Theor. of Comp.*, pages 330–336, 2001.
4. B.Yakir. The differencing algoritm LDM for partitioning; a proof of a conjecture of Karmakar and Karp. *Math. of Operations Res.*, 21:85–99, 1996.
5. F.F. Ferreira and J.F. Fontanari. Probabilistic analysis of the number partitioning problem. *J. Phys. A: Math. Gen.*, 31:3417–3428, 1998.
6. F.F. Ferreira and J.F. Fontanari. Statistical mechanics analysis of the continuous number partitioning problem. *Physica A*, 269:54–60, 1999.
7. Y. Fu. The use and abuse of statistical mechanics in computational complexity. In *Lectures in the Science of Complexity; Proceedings of the 1988 Complex Systems Summer School, Santa Fe, New Mexico, 1988, edited by D.L. Stein.* Addison-Wesley, Reading, MA, 1989.
8. I.P. Gent and T. Walsh. In *Proc. of the 12th European Conference on Artificial Intelligence, Budapest, Hungary, 1996, edited by W. Wahlster*, pages 170–174. John Wiley & Sons, New York, NY, 1996.
9. N. Karmarkar and R.M. Karp. The differencing method of set partitioning. Technical Report UCB/CSD 82/113, Computer Science Division (EECS), University of California, Berkeley, 1982.
10. N. Karmarkar, R.M. Karp, G.S. Lueker, and A.M. Odlyzko. Probabilistic analysis of optimum partitioning. *J. Appl. Prob.*, 23:626–645, 1986.
11. G.S. Lueker. Exponentially small bounds on the expected optimum of the partition and subset sum problem. *Rand. Struc. Alg.*, 12:51–62, 1998.
12. S. Mertens. Phase transition in the number partitioning problem. *Phys. Rev. Lett.*, 81:4281–4284, 1998.
13. S. Mertens. Random costs in combinatorial optimization. *Phys. Rev. Lett.*, 84:1347–1350, 2000.

Embracing the Giant Component

Abraham Flaxman[1], David Gamarnik[2], and Gregory B. Sorkin[2]

[1] Carnegie Mellon University, Department of Mathematical Sciences,
Pittsburgh PA 15213, USA.
abie@cmu.edu
[2] IBM T.J. Watson Research Center, Department of Mathematical Sciences,
Yorktown Heights NY 10598, USA.
daveg@us.ibm.com, sorkin@watson.ibm.com

Abstract. Consider a game in which edges of a graph are provided a
pair at a time, and the player selects one edge from each pair, attempting
to construct a graph with a component as large as possible. This game
is in the spirit of recent papers on *avoiding* a giant component, but here
we *embrace* it.

We analyze this game in the offline and online setting, for arbitrary
and random instances, which provides for interesting comparisons. For
arbitrary instances, we find a large lower bound on the competitive ratio.
For some random instances we find a similar lower bound holds with
high probability (**whp**). If the instance has $\frac{1}{4}(1+\epsilon)n$ random edge pairs,
when $0 < \epsilon \leq 0.003$ then any online algorithm generates a component
of size $O((\log n)^{3/2})$ **whp**, while the optimal offline solution contains a
component of size $\Omega(n)$ **whp**. For other random instances we find the
average-case competitive ratio is much better than the worst-case bound.
If the instance has $\frac{1}{2}(1 - \epsilon)n$ random edge pairs, with $0 < \epsilon \leq 0.015$, we
give an online algorithm which finds a component of size $\Omega(n)$ **whp**.

1   Introduction

A pair of recent papers [BF01,BFW02] analyze the "Achlioptas process", where
a collection of random edge pairs is given a pair at a time, and the object is
to select one edge from each pair to *avoid* having a (suitably defined) giant
component in the resulting graph. Without any intelligent selection process,
a giant component forms after about $\frac{1}{2}n$ edges; [BF01] shows that a strategy
exists which accepts at least $0.535n$ edges without forming a giant component;
[BFW02] shows (among other things) that no more than about 0.964 edges may
be accepted.

It is equally natural to ask the opposite question,

> What can you do to encourage a random graph to form
> a giant component, using fewer than $(1 + \epsilon)n/2$ edges?

In fact, it is so natural we learned that Bohman and Kravitz are studying it
independently [BK03].

M. Farach-Colton (Ed.): LATIN 2004, LNCS 2976, pp. 69–79, 2004.

We now define the problem of Embracing the Giant Component (EGC) more precisely. An instance I consists of a sequence of m pairs of edges on n vertices. (If you like, I may be regarded as an element of $\left[\binom{n}{2}\right]^{2m}$.) Edges, including those in a pair, may or may not be distinct. A solution is a choice of one edge from each pair, and its value is the order (number of vertices) in the largest component in the graph consisting of the chosen edges. $EGC(I)$ is the maximum value of a solution for instance I.

We focus on *online* versions of EGC, in which we see the pairs one at a time and must select our edge before seeing the next pair, but we also consider *offline* versions, in which we see all m pairs before making our choice. In either case, we consider edge pairs chosen *randomly* (defining an average-case behavior) or arbitrarily (chosen adversarially).

In addition to being a natural graph-game problem, EGC has two other sources of interest. First, imagine that you are a company trying to build up a network of some sort, each new link you build must be in response to a customer demand, and your budget allows you to spend at a rate which satisfies only half of all new requests. Presuming that a large connected component in the network is beneficial to your customers and to you, your goal is to solve an optimization problem very similar to EGC. Of course any real-world problem would be much more complicated, with different costs and benefits for different links, the ability to wait longer or shorter times to see more choices, and so forth, but it is conceivable that there are real-world problems whose mathematical core is captured by EGC.

The second motivation is that EGC provides an example of a problem for which the competitive ratio is awful in the worst case, but, for certain parameters, quite reasonable in an average case; a previous example was given by [SSS02]. For certain other parameters, EGC has a lower bound on average-case competitive ratio that is almost as awful as in the worst case.

1.1 Worst Case

We first observe that in the worst case, it is hard to solve offline EGC exactly (to select edges giving a component as large as possible), or even to approximate it to better than some fixed factor.

Theorem 1. *Offline EGC is MAX SNP-hard.*

In the online setting, it is natural to measure performance in terms of the *competitive ratio*, the ratio z_{opt}/z_{online} between the sizes of the components produced by the best possible offline and online algorithms. The next theorem shows that in the worst case, the competitive ratio is as bad as it conceivably could be.

Theorem 2. *The worst-case competitive ratio for EGC is $(m+1)/2$. Specifically, for every online algorithm, there is a sequence of m edge pairs for which the algorithm produces a collection of isolated edges, yet the optimal solution has a component on $m+1$ vertices.*

As we remark after the proof of this theorem, a competitive-ratio lower bound of $\Omega(n/\log n)$ holds even for randomized online algorithms against an oblivious adversary.

1.2 Average Case

We define $I_{n,m}$ to be a *random* instance of EGC in which each edge of each pair is chosen independently, uniformly at random from the edge set of the complete graph K_n.

Our main intention is to compare the average-case competitive ratio with the worst-case lower bound in Theorem 2. To do so, we need some idea of the optimal offline value of $\mathrm{EGC}(I_{n,m})$. We will see that these random instances exhibit a sharp threshold in objective value at $m = \frac{1}{4}n$, which we will prove by analyzing a greedy heuristic for offline EGC.

Throughout the paper, we will rely on a "component-identification algorithm". This algorithm, and our method of analysis, is quite standard in the random-graph literature; see for example the giant-component chapter of *Random Graphs* [JLR00, pp. 108–111].

Our component-identification algorithm, Algorithm $\mathcal{A}$, maintains two set of vertices, called *unborn*, U_i, and *alive*, A_i. Initially, a single vertex is alive, $A_1 = \{v_1\}$, and the remainder are unborn, $U_1 = [n] \setminus \{v_1\}$. At step i, we look at all the neighbors of some vertex $v_i \in A_i$. We kill v_i and give birth to all its unborn neighbors (formally, let $P_i = U_i \cap N(v_i)$ be the *progeny* of v_i, and set $A_{i+1} = A_i \setminus \{v_i\} \cup P_i$ and $U_{i+1} = U_i \setminus P_i$).

Our greedy heuristic is very similar to Algorithm $\mathcal{A}$. Roughly, we try starting at each vertex, and using the first edge we see from each pair. We will elaborate on this description in the proof of Theorem 3.

Theorem 3. *For any fixed $\epsilon > 0$, for $m = \frac{1}{4}(1 - \epsilon)n$, we have $\mathrm{EGC}(I_{n,m}) = O(\log n)$ while for $m = \frac{1}{4}(1 + \epsilon)n$, our greedy heuristic finds a solution showing $\mathrm{EGC}(I_{n,m}) = \Omega(n)$ **whp**.*

The below-the-threshold half of the theorem follows from well-known results in the theory of random graphs, since the union of all the edges in all the pairs is a random graph with $\frac{1}{2}(1 - \epsilon)n$ edges, which is below the threshold for a giant component (see, for example, [JLR00]).

It is interesting to note that below the threshold, the largest component in the union of the edges contains at most one edge from each pair **whp**, so for $m = \frac{1}{4}(1 - \epsilon)n$, we can solve $\mathrm{EGC}(I_{n,m})$ optimally **whp**.

The above-the-threshold half of the theorem is proved in Section 3, in a manner similar to the analysis of the giant component above the threshold in $G_{n,p}$.

Our next theorem shows that even in the average case, any online algorithm performs much worse than offline.

Theorem 4. *For $\epsilon \leq 0.003$, on instances $I_{n,m}$ with $m = \frac{1}{4}(1+\epsilon)n$, every online algorithm finds a component of size only $O((\log n)^{3/2})$ **whp**.*

Theorem 3 and 4 together give a lower bound on the *average-case competitive ratio* for EGC: the ratio of offline solution to online is $\Omega(n/(\log n)^{3/2})$ **whp**. This shows that the lower bound on competitive ratio for EGC is more robust than Theorem 2 alone indicates.

Theorem 4 is only true for some range of ϵ, however. For example, if $\epsilon > 1$, then taking the first edge from each pair yields a random graph above the giant component threshold, and so this trivial algorithm has a constant competitive ratio. We go slightly beyond the trivial bound in the next theorem.

Consider Algorithm $\mathcal{C}$, which does the following: for some γ to be determined later, for the first γn choices we take the first edge of each pair. For the remaining $m - \gamma n$ choices, we take the first edge unless it touches an isolated vertex, in which case we take the second edge.

Theorem 5. *For $\epsilon \leq 0.015$, on instances $I_{n,m}$ with $m = \frac{1}{2}(1-\epsilon)n$, Algorithm $\mathcal{C}$ yields a component of size $\Omega(n)$ **whp**.*

2   Proofs of Worst-Case Theorems

Proof of Theorem 1: To show the hardness of approximating EGC, we reduce from MAX 3SAT-5. MAX 3SAT-5 is a structured relative of MAX 3-SAT, introduced by Feige, where every variable appears in exactly 5 clauses and a variable does not appear in a clause more than once. Feige proves that there is some $\epsilon > 0$ for which it is NP-hard to distinguish a satisfiable instance from an instance with at most $(1 - \epsilon)m$ satisfiable clauses [Fei98].

Given a MAX 3SAT-5 instance, we make a EGC instance with $n+3m+1$ vertices by including a vertex for each literal, ℓ, 3 vertices for each clause C_1, C_2, C_3, and an additional "root" vertex, r. We model the assignment by n edge pairs which decide if each variable is true or false: let pair i be $(\{r, x_i\}, \{r, \overline{x_i}\})$. We include $3m$ additional pairs: for each clause C, let ℓ be the j-th literal in C (where $j \in \{1, 2, 3\}$), and include a pair of the form $(\{\ell, C_j\}, \{C_j, C_{j+1(\mod 3)}\})$. If the assignment is satisfiable, there is a way of selecting edges which yields a component of size $3m + n + 1$. On the other hand, any selection of edges from the first n pairs corresponds naturally to some assignment. If a literal is not selected, then since it appears in at most 5 clauses and is not connected to the root, it can be in a component of size at most 16. Since it is NP-hard to distinguish satisfiable 3SAT-5 instances from instances with at most $(1 - \epsilon)m$ clauses satisfiable, it is also NP-hard to distinguish instances of EGC with a component of size $\frac{8}{5}m + 1$ from those with a component of size at most $(\frac{8}{5} - \epsilon)m + 1$. $\square$

Proof of Theorem 2: We will present a sequence of edge pairs, depending on the previous choices of the algorithm. The edge pairs will all come from a complete binary tree, and the edges in each pair will be siblings, i.e., of the form $(\{x, y\}, \{x, y'\})$. Whatever the algorithm chooses at step i — and for a fixed deterministic algorithm this choice is predictable — we make it wish it chose otherwise. So, if the algorithm selects edge $\{x, y\}$, the next pair we give it is $(\{y', z\}, \{y', z'\})$.

Thus, the online algorithm obtains a graph with only isolated edges, while making the opposite choice at every step would yield a component with m edges.

$\square$

Of course, the same $(m+1)/2$ ratio also applies to *randomized* online algorithm, if the adversary is allowed to see the algorithm's choice before constructing the next pair. Even if the adversary is required to fix a sequence of pairs in advance, and even if she does not know what randomized algorithm is being used, there is an almost equally bad instance. It is given by a random path down the tree and the siblings of the path edges, each edge paired with its sibling, the pairs presented in order from root to leaf. At each step, the online algorithm has probability only $1/2$ of choosing the path edge rather than its sibling, and hence **whp** gets a largest component of size only $O(\log n)$.

3   Proofs of Average-Case Theorems

Proof of Theorem 3: We repeat more formally the greedy heuristic sketched in the introduction, in a form conducive to analysis. Algorithm $\mathcal{B}$ repeats the following n times, starting with each possible vertex for v_1. At each step, we maintain two sets of vertices, called *unborn*, U_i; and *alive*, A_i. Initially, a single vertex is alive, $A_1 = \{v_1\}$, and the remainder are unborn, $U_1 = [n] \setminus \{v_1\}$. At step i, we choose some vertex $v_i \in A_i$ and identify all previously unidentified pairs with an edge incident to v_i. For each such edge pair, we use the edge incident to v_i. We let $P_i = N(v_i) \cap U_i$ denote the set of newly discovered vertices, and we set $A_{i+1} = (A_i \setminus \{v_i\}) \cup P_i$ and $U_{i+1} = U_i \setminus P_i$.

For analysis, it is convenient to work with an instance resembling $G_{n,p}$. Let $I_{n,p}$ be a random instance formed by including each pair of edges independently with probability p. Thus, our probability space is $\{0,1\}^{\binom{n}{2}^2}$ with the product measure. We will show that the threshold value is $p = \frac{1}{n^3}$, which has expected $\frac{n}{4}$ pairs. We do so by analyzing the behavior of Algorithm $\mathcal{B}$ on $I_{n,(1+\epsilon)n^{-3}}$, which proceeds in two claims. We will then translate this result to random instances $I_{n,m}$ where $m = \frac{1}{4}(1+\epsilon)n$.

Let $p = (1+\epsilon)n^{-3}$, and let $\beta, \delta > 0$ be such that $(1+\epsilon)(1-\beta)^3 = 1+\delta$ and let $t_0 = 8(1+\delta)\delta^{-2} \log n$ and $t_1 = \beta n$.

Claim 1: Running Algorithm $\mathcal{B}$ on $I_{n,p}$ with any starting vertex v_1, either the algorithm halts before step t_0 or for all $t_0 \leq t \leq t_1$ we have $|A_t| \geq 1$ **whp**.

For this it is sufficient, for each t with $t_0 \leq t \leq t_1$, to identify t vertices of the component. Before we have identified a size βn component, there are at least $(1-\beta)n$ unborn vertices. So there are at least $2(n(1-\beta))\binom{(1-\beta)n}{2} \approx (n(1-\beta))^3$ candidate edge pairs which contribute a unique vertex to P_i. Thus we have

$$\mathbb{P}\left[\sum_{i=1}^{t} |P_i| \leq t\right] \leq \mathbb{P}\left[\sum_{i=1}^{t} \mathrm{B}((n(1-\beta))^3, p) \leq t\right].$$

We also have

$$\mathbb{E}\left[\sum_{i=1}^{t} B(n^3(1-\beta)^3, p)\right] = (1+\epsilon)(1-\beta)^3 t = (1+\delta)t,$$

so we use a standard Chernoff bound to show the probability that Algorithm $\mathcal{B}$, starting at any vertex, halts at time t for any $t_0 \leq t \leq t_1$ is at most

$$n\sum_{t=t_0}^{t_1} \mathbb{P}\left[\sum_{i=1}^{t} B\left(n^3(1-\beta)^3, p\right) \leq t\right] \leq n\sum_{t=t_0}^{t_1} \exp\left(\frac{-(\delta t)^2}{2(1+\delta)t}\right)$$

$$\leq nt_1 e^{-\delta^2 t_0/2(1+\delta)}$$

$$\leq n^{-2}.$$

Claim 2: There is some vertex v so that starting Algorithm $\mathcal{B}$ on v yields a component of size at least t_0 **whp**.

For this, we start Algorithm $\mathcal{B}$ on some vertex v, and if it fails to discover t_0 vertices, we start it on an unexplored vertex, v', and keep going. Each run, we expose at most t_0 vertices, so if we fail t_0 times, the number of edges exposed at each step dominates $B(n^3(1-\beta)^3, p)$. Now, for Algorithm $\mathcal{B}$ to fail in every run, we must have that the total number of vertices exposed is less than the number of steps. But

$$\mathbb{P}\left[\sum_{i=1}^{t_0^2} B(n^3(1-\beta)^3, p) \leq t_0^2\right] \leq e^{-\delta^2 t_0^2/2(1+\delta)} = o(n^{-2}).$$

Therefore, we have some vertex where Algorithm $\mathcal{B}$ runs for at least t_0 steps with probability $1 - o(n^{-2})$.

Claims 1 and 2 imply Algorithm $\mathcal{B}$ finds a component of size $t_1 = \beta n$ in $I_{n,p}$ **whp**.

To translate this result from $I_{n,p}$ to $I_{n,m}$, note that the probability $I_{n,p}$ has exactly $\frac{n^4}{4}p = \frac{1}{4}(1+\epsilon)n =: m$ edge pairs is

$$\mathbb{P}_{I_{n,p}}[\mathcal{M}] = \binom{\frac{n^4}{4}}{m}p^m(1-p)^{\frac{n^4}{4}-m} = O(n^{-1/2}),$$

where $\mathcal{M}$ denotes the event that I has m distinct edge pairs. Also note that the probability $I_{n,m}$ consists of m distinct edge pairs is

$$\mathbb{P}_{I_{n,m}}[\mathcal{M}] = \prod_{i=0}^{m-1}\left(1 - \frac{i}{\binom{n}{2}^2}\right) = 1 - O(n^{-2}).$$

And, by symmetry, for any particular instance I^* we have

$$\mathbb{P}_{I_{n,m}}[I_{n,m} = I^* \mid \mathcal{M}] = \mathbb{P}_{I_{n,p}}[I_{n,p} = I^* \mid \mathcal{M}].$$

So the probability of any event $\mathcal{E}$ in the $I_{n,m}$ model is related to the probability in the $I_{n,p}$ model by

$$
\begin{aligned}
\mathbb{P}_{I_{n,m}}[\mathcal{E}] &= \mathbb{P}_{I_{n,m}}[\mathcal{E} \mid \mathcal{M}]\mathbb{P}_{I_{n,m}}[\mathcal{M}] + \mathbb{P}_{I_{n,m}}[\mathcal{E} \mid \overline{\mathcal{M}}]\mathbb{P}_{I_{n,m}}[\overline{\mathcal{M}}] \\
&\leq \mathbb{P}_{I_{n,m}}[\mathcal{E} \mid \mathcal{M}] + O(n^{-2}) \\
&= \mathbb{P}_{I_{n,p}}[\mathcal{E} \mid \mathcal{M}] + O(n^{-2}) \\
&= O(n^{1/2})\mathbb{P}_{I_{n,p}}[\mathcal{M}]\mathbb{P}_{I_{n,p}}[\mathcal{E} \mid \mathcal{M}] + O(n^{-2}) \\
&\leq O(n^{1/2})\mathbb{P}_{I_{n,p}}[\mathcal{E}] + O(n^{-2}).
\end{aligned}
$$

Since the failure probability was $O(n^{-2})$ in the $I_{n,p}$ model, it is $O(n^{-3/2})$ for $I_{n,m}$. $\qquad\square$

Proof of Theorem 4: We will analyze a wider class of algorithms. Instead of requiring the algorithm to choose edges at each step of the process, we will generate the first γn pairs, and allow the process to keep any components in the union with at least 2 edges, and additionally to keep up to γn of the isolated edges. Then we will generously allow the process to keep all edges from an additional $\frac{1}{4}(1+\epsilon)n - \gamma n$ pairs. A nonrigorous intuition for our proof is that the first γn pairs are "pretty much" isolated edges, and so the graph resulting from this process "looks like" the union of $\gamma n + 2(\frac{1}{4}(1+\epsilon) - \gamma)n = \frac{1}{2}(1+\epsilon - 2\gamma)n$ random edges. For $\gamma > \frac{1}{2}\epsilon$ such a graph is below the threshold for the giant component.

This heuristic argument does not translate directly into a rigorous proof because the union of the edges in the γn pairs is not a collection of isolated edges. To work around this, we will bound the contribution of the components of 3 or more vertices in the union of the first γn pairs. Note that, by symmetry, it makes no difference which γn isolated edges the algorithm selects, so in this wider class of algorithms, the results of any selection process are the same. To prove the theorem, we decompose the graph into two parts. Let G' be the union of γn isolated edges and the components containing at least 3 vertices in the union of $2\gamma n$ random edges. Let G'' be the union of $2(\frac{1}{4}(1+\epsilon) - \gamma)n = \frac{1}{2}(1+\epsilon - 4\gamma)n$ edges. We show that **whp** $G' \cup G''$ contains no component of size exceeding

$$
t_1 = \delta^{-1}6(1-\gamma)^{-2}(\log n)^{3/2}.
$$

To simplify calculations, we make G'' a realization of $G_{n,p}$, with $p = (1 + \epsilon - 4\gamma)/n$. We will translate our results to $G_{n,m}$ at the end of the proof. Let $\epsilon, \gamma, \delta > 0$ so that

$$
(1 + \epsilon - 4\gamma)\left(e^{-4\gamma} + 4\gamma e^{-8\gamma} + 2\gamma + \frac{(4\gamma)^2 e^{3(1-4\gamma)}}{1 - 4\gamma e^{1-4\gamma}}\right) = 1 - 2\delta.
$$

Note that such ϵ, γ, δ exist, for example taking $\epsilon = 0.003, \gamma = 0.003$ and $\delta \approx 0.003177$.

Let T_k denote the number of components in G' with k vertices. Given the values of the T_k's, we use an exposure procedure similar to Algorithm $\mathcal{A}$ to prove

$G' \cup G''$ has no large components. We expose all the vertices adjacent to v_i in G'' and if we discover a vertex of a connected component of G', we add every vertex of this component to the set A_{i+1}. Thus, at each step, and conditioned on any history, the size of P_i is stochastically dominated by

$$\sum_{k=1}^{n} k \, \mathrm{B}(kT_k, p),$$

and the probability that we discover a component of size exceeding t_1 is bounded by

$$\mathbb{P}\left[\sum_{i=1}^{t_1}\sum_{k=1}^{n} |P_i| \leq t_1\right] \leq \mathbb{P}\left[\sum_{i=1}^{t_1}\sum_{k=1}^{n} k \, \mathrm{B}(kT_k, p) \geq t_1\right].$$

Let $\mathcal{E}_1$ denote the event that G' contains no component with more than $K = 6(1-\gamma)^{-2}\log n$ vertices. Standard arguments show $\mathbb{P}[\overline{\mathcal{E}_1}] \leq O(n^{-2})$ (see, for example, [JLR00]). If $\mathcal{E}_1$ holds, we need only consider the sum of weighted Binomial r.v.'s up to the K-th term. In other words, $\mathcal{E}_1$ implies

$$\sum_{k=1}^{n} k \, \mathrm{B}(kT_k, p) = \sum_{k=1}^{K} k \, \mathrm{B}(kT_k, p).$$

Let $Z = \sum_{k=1}^{K} k^2 T_k p$. Note that Z is the expectation of the sum above conditioned on the T_k's. Also note that the value of Z is dependent on G' only. We now obtain a bound on Z that holds **whp**.

We use a tree census results for sparse random graphs. It is known that in $G_{n,m=cn/2}$ we have the following: (see, for example, Pittel, [Pit90])

$$\mathbb{E}[T_k] \sim n(k^{k-2}c^{k-1}e^{-ck}/k!),$$

which, in G' applies to T_k with $k \geq 3$. We also have $T_2 \leq \gamma n$, and $\mathbb{E}[T_1] = e^{-4\gamma}n + 4\gamma e^{-8\gamma}n - 2T_2$. So we have

$$\mathbb{E}[Z] = \mathbb{E}\left[\sum_{k=1}^{K} k^2 T_k p\right]$$

$$\leq (1+\epsilon-4\gamma)\left(e^{-4\gamma} + 4\gamma e^{-8\gamma} + 2\frac{T_2}{n} + \sum_{k=3}^{K} k^k(4\gamma)^{k-1}e^{-4\gamma k}/k!\right)$$

$$\leq (1+\epsilon-4\gamma)\left(e^{-4\gamma} + 4\gamma e^{-8\gamma} + 2\gamma + (4\gamma)^{-1}\sum_{k=3}^{\infty}\left(4\gamma e^{1-4\gamma}\right)^k\right)$$

$$= (1+\epsilon-4\gamma)\left(e^{-4\gamma} + 4\gamma e^{-8\gamma} + 2\gamma + \frac{(4\gamma)^2 e^{3(1-4\gamma)}}{1-4\gamma e^{1-4\gamma}}\right)$$

$$= 1 - 2\delta.$$

Let $\mathcal{E}_2$ be the event that $Z \leq 1 - \delta$. We use a form of the Azuma-Hoeffding inequality due to McDiarmid (see [Hoe63,McD89]) to show $\mathcal{E}_2$ holds **whp**. Note that changing one edge of G' can create or destroy at most two components in G'. So this can change the value of Z by at most $2K^2p$. Therefore,

$$\mathbb{P}[\overline{\mathcal{E}_2}] = \mathbb{P}[Z \geq 1 - \delta]$$
$$\leq \mathbb{P}[Z \geq \mathbb{E}[Z] + \delta]$$
$$\leq \exp\left(-\frac{2\delta^2}{(2\gamma n)(2K^2p)^2}\right)$$
$$\leq \exp\left(-\frac{\delta^2 n}{4K^4}\right).$$

Conditioning on $\mathcal{E}_1$ and $\mathcal{E}_2$, we have

$$\sum_{i=1}^{t_1} \sum_{k=1}^{n} k\,\mathrm{B}(kT_k, p) = \sum_{i=1}^{t_1} \sum_{k=1}^{K} k\,\mathrm{B}(kT_k, p), \tag{1}$$

and

$$\mathbb{E}\left[\sum_{i=1}^{t_1} \sum_{k=1}^{K} k\,\mathrm{B}(kT_k, p)\right] = Zt_1 \leq (1 - \delta)t_1.$$

To bound the probability that sum (1) is larger than t_1, we use the following Chernoff bound, from [AS00, Theorem A.1.18].

Theorem 6. *Let X_i, $1 \leq i \leq n$ be independent random variables with each $\mathbb{E}[X_i] = 0$ and no two values of any X_i ever more than one apart. Set $S = X_1 + \cdots + X_n$. Then $\mathbb{P}[S > a] < \exp(-2a^2)$.*

Applying this to (1), we have

$$\mathbb{P}\left[\sum_{i=1}^{t_1} \sum_{k=1}^{n} k\,\mathrm{B}(kT_k, p) \geq t_1 \,\middle|\, \mathcal{E}_1, \mathcal{E}_2\right] \leq \mathbb{P}\left[\sum_{i=1}^{t_1} \sum_{k=1}^{K} k\,\mathrm{B}(kT_k, p) \geq Zt_1 + \delta t_1\right]$$
$$\leq \mathbb{P}\left[\sum_{i=1}^{t_1} \sum_{k=1}^{K} \sum_{j=1}^{kT_k} \frac{(\mathrm{Be}(p) - p)k}{K} \geq \delta t_1/K\right]$$
$$\leq \exp\left(-2\left(\delta t_1/K\right)^2\right)$$
$$= n^{-2}.$$

So the probability there exists any vertex on which we run for at least t_1 steps is at most n^{-1} by the union bound.

Since $\mathbb{P}[\overline{\mathcal{E}_1}] + \mathbb{P}[\overline{\mathcal{E}_2}] \leq o(n^{-1})$, we complete the theorem by observing that $G_{n,p}$ has at least $\frac{1}{2}(1 + \epsilon - 4\gamma)n$ edges with constant probability, and extra edges can only increase the size of the largest component, so our claim also holds in $G_{n,m}$. $\qquad\square$

Proof of Theorem 5: We bound the size of components formed by this process by exposing edges starting from a vertex v_1 and tracking the number of vertices unborn and alive.

Consider decomposing the graph into G', the edges selected before time γn, and G'', the edges selected after.

To simplify calculations, we take G' to be a realization of $G_{n,p}$, with $p = 2\gamma/n$. Also, we generate G'' by applying our selection rule to a realization of $I_{n,p'}$, with $p' = 2(1 - \epsilon - 2\gamma)n^{-3}$ (recall this is an instance where every pair of edges is included independently with probability p'). Thus the expected number of edges in $G' \cup G''$ is $\gamma n + \frac{1}{2}(1 - \epsilon - 2\gamma)n = \frac{1}{2}(1 - \epsilon)n$, as it should be.

Let $\alpha, \beta, \gamma, \delta, \epsilon, \eta, \theta > 0$ be such that $(1 - \beta)(2\gamma) = 2\gamma - \epsilon/2 + \delta$, and

$$(1 - \beta - (1 + \eta)e^{-2\gamma})(1 - \beta)^2/2 + (1 - \beta - (1 + \eta)e^{-2\gamma})(1 - \delta)e^{-2\gamma}(1 - \beta) = \alpha$$

and

$$\alpha 2(1 - \epsilon - 2\gamma) = 1 - 2\gamma - \epsilon/2 + \theta.$$

Note that such parameters exist, for example $\beta = \eta = 10^{-6}$, $\gamma = 0.4$, $\epsilon = 0.015$, yielding $\alpha \approx 0.521$, $\delta \approx 0.007$, and $\theta \approx 0.0002$. Let $t_0 = 8\max\{\delta^{-2}(1 - \beta)\gamma, \theta^{-2}(1 - \epsilon - 2\gamma)\}\log n$ and $t_1 = \beta n$. We wish to bound the probability Algorithm $\mathcal{C}$ halts with a component of size $t_0 \leq t \leq t_1$. For this, it is sufficient to bound the probability $\sum_{i=1}^{t} |P_i| \leq t$ for all $t_0 \leq t \leq t_1$. We decompose P_i into $P_i = P_i' \cup P_i''$, where P_i' are the progeny contributed by edges in G' and P_i'' are the progeny contributed by edges in G'' (and not by edges in G'). If $\sum_{i=1}^{t} |P_i| \leq t$, then either $\sum_{i=1}^{t} |P_i'| \leq (2\gamma - \epsilon/2)t$ or $\sum_{i=1}^{t} |P_i''| \leq (1 - 2\gamma + \epsilon/2)t$.

Now, at any step t, conditioned on any history that has not yet discovered βn vertices, we have $|P_i'|$ stochastically dominates $B((1 - \beta)n, p)$ and

$$\mathbb{P}\left[\sum_{i=1}^{t} |P_i'| \leq (2\gamma - \epsilon/2)t\right] \leq \mathbb{P}\left[\sum_{i=1}^{t} B((1 - \beta)n, p) \leq (2\gamma - \epsilon/2)t\right]$$
$$\leq e^{-\delta^2 t/4\gamma(1-\beta)}$$
$$\leq n^{-2}.$$

Let $\mathcal{E}_3$ denote the event that G' contains $(1 \pm \eta)e^{-2\gamma}n$ isolated vertices. We omit a simple calculation using Chebyschev's inequality to show $\mathcal{E}_3$ holds **whp**.

Conditioning on $\mathcal{E}_3$, we have that at any step t, conditioned on any history that has not yet discovered βn vertices, there are at least $(1 - \beta - (1 + \eta)e^{-2\gamma})n$ vertices that are not isolated in G' and are still in U_i. So there are at least $((1 - \beta - (1 + \eta)e^{-2\gamma})(1 - \beta)^2/2 + (1 - \beta - (1 + \eta)e^{-2\gamma})(1 - \delta)e^{-2\gamma}(1 - \beta))n^3 = \alpha n^3$ unexposed edge pairs which would cause our selection rule to place a vertex in P_i'. So

$$\mathbb{P}\left[\sum_{i=1}^{t} |P_i''| \leq (1 - 2\gamma - \epsilon/2)t\right] \leq \mathbb{P}\left[\sum_{i=1}^{t} B(\alpha n^3, p') \leq (1 - 2\gamma - \epsilon/2)t\right]$$
$$\leq e^{-\theta^2 t/4\alpha(1-\epsilon-2\gamma)} \leq n^{-4}.$$

Thus by the union bound, the probability that some component has size t for $t_0 \leq t \leq t_1$ is $O(n^{-2})$.

Now, as in the proof of Theorem 3, we argue that some component has size at least t_0 **whp**. The argument is identical to the earlier theorem, and the size of the progeny at each stage is bounded identically to the previous paragraph, so we omit further details.

Finally, we observe that the probability there is no giant component is $O(n^{-2})$, so we can convert to the original model as in Theorem 3.   □

References

[AS00]   Noga Alon and Joel H. Spencer, *The probabilistic method*, second ed., Wiley-Interscience Series in Discrete Mathematics and Optimization, Wiley-Interscience [John Wiley & Sons], New York, 2000, With an appendix on the life and work of Paul Erdős. MR 2003f:60003

[BF01]   Tom Bohman and Alan Frieze, *Avoiding a giant component*, Random Structures Algorithms **19** (2001), no. 1, 75–85. MR 2002g:05169

[BFW02]   Tom Bohman, Alan Frieze, and Nicholas C. Wormald, *Avoiding a giant component II*, manuscript, 2002.

[BK03]   Tom Bohman and David Kravitz, *Creating a giant component*, manuscript, 2003.

[Fei98]   Uriel Feige, *A threshold of* $\ln n$ *for approximating set cover*, J. ACM **45** (1998), no. 4, 634–652. MR 2000f:68049

[Hoe63]   Wassily Hoeffding, *Probability inequalities for sums of bounded random variables*, J. Amer. Statist. Assoc. **58** (1963), 13–30. MR 26 #1908

[JLR00]   Svante Janson, Tomasz Łuczak, and Andrzej Rucinski, *Random graphs*, Wiley-Interscience Series in Discrete Mathematics and Optimization, Wiley-Interscience, New York, 2000. MR 2001k:05180

[McD89]   Colin McDiarmid, *On the method of bounded differences*, Surveys in combinatorics, 1989 (Norwich, 1989), London Math. Soc. Lecture Note Ser., vol. 141, Cambridge Univ. Press, Cambridge, 1989, pp. 148–188. MR 91c:05077

[Pit90]   Boris Pittel, *On tree census and the giant component in sparse random graphs*, Random Structures Algorithms **1** (1990), no. 3, 311–342. MR 92f:05087

[SSS02]   Mark Scharbrodt, Thomas Schickinger, and Angelika Steger, *A new average case analysis for completion time scheduling*, Proceedings of the 34th Annual ACM Symposium on Theory of Computing (STOC), 2002, pp. 170–178.

Sampling Grid Colorings with Fewer Colors

Dimitris Achlioptas[1], Mike Molloy[2*], Cristopher Moore[3], and
Frank Van Bussel[4]

[1] Microsoft Research optas@microsoft.com
[2] Dept of Computer Science, University of Toronto, and Microsoft Research [‡]
molloy@cs.toronto.edu
[3] Computer Science Department, University of New Mexico [§] moore@santafe.edu
[4] Dept of Computer Science, University of Toronto fvb@cs.toronto.edu

Abstract. We provide an optimally mixing Markov chain for 6-colorings
of the square grid. Furthermore, this implies that the uniform distribution on the set of such colorings has strong spatial mixing. Four and five
are now the only remaining values of k for which it is not known whether
there exists a rapidly mixing Markov chain for k-colorings of the square
grid.

1 Introduction

Sampling and counting graph colorings is a fundamental problem in computer
science and discrete mathematics. Much focus has gone towards attacking this
problem using rapidly mixing Markov chains; see for example [3,12,14,18,22].

Sampling graph colorings is also of fundamental interest in statistical physics.
Graph colorings correspond to the zero-temperature case of the *antiferromagnetic Potts model*, a model of magnetism on which physicists have performed
extensive numerical experiments (see for instance [10,16,15]). Physicists wish to
estimate physical quantities such as spatial correlations and magnetization, and
to do this they attempt to sample random states using Markov chains.

Moreover, *optimal temporal mixing*, i.e. a mixing time of $O(n \log n)$, is deeply
related to the physical properties of the system [9]. In particular, it implies *spatial
mixing*, i.e. the exponential decay of correlations, and thus the existence of a finite
correlation length and the uniqueness of the Gibbs measure. Therefore, optimal
mixing of natural Markov chains for q-colorings of the grid is considered a major
open problem in physics (see e.g. [21]). Physicists have conjectured [10,21] that
the q-state Potts model has spatial mixing for $q \geq 4$. This has been established
rigorously for $q \geq 7$ by Bubley, Dyer and Greenhill [3] who showed that all
4-regular triangle-free graphs, such as the square grid, have optimal mixing.

---

[*] **Contact author:** M. Molloy, Dept of Computer Science, University of Toronto, 10
Kings College Rd, Toronto, ON M5S 3G1, Canada.
[‡] Some of this work was done while visiting the Fields Institute. Supported by NSERC
and a Sloan Research Fellowship
[§] Supported by NSF grant PHY-0200909

M. Farach-Colton (Ed.): LATIN 2004, LNCS 2976, pp. 80–89, 2004.
© Springer-Verlag Berlin Heidelberg 2004

Our main result is that the square grid has optimal mixing for $q = 6$. We prove this by considering the following Markov chain, often called the *block heat-bath dynamics*, which we call $M(i, j)$. At each step, we choose an $i \times j$ subgrid S of G, uniformly at random from amongst all such subgrids (i.e. its upper-left vertex is chosen uniformly from the vertices of G). Let C be the set of q-colorings of S which are consistent with the coloring of $G \setminus S$. We choose a uniformly random coloring $c \in C$ and recolor S with c. Our main theorem is:

Theorem 1. $M(2, 3)$ *on 6-colorings of the square grid mixes in* $O(n \log n)$ *time.*

We prove Theorem 1 for a variety of boundary conditions: on the torus, on finite rectangular regions with fixed colorings on their boundary, and on finite rectangular regions with free boundary conditions. Our method is similar to that of [3] in that it consists of a computer-assisted proof of the existence of a path coupling. At the same time, we exploit the specific geometry of the square grid to consider a greater variety of neighborhoods. Moreover, the calculations necessary to find a good coupling in our setting are far more complicated than those in [3] and require several new ideas to become computationally tractable.

Using the comparison method of Diaconis and Saloff-Coste [5,17], Theorem 1 implies that the Glauber and Kempe chain Markov chains also mix in polytime:

Theorem 2. *The Glauber dynamics and the Kempe chain dynamics on 6-colorings of the square grid mix in* $O(n^2 \log n)$ *time.*

Like Theorem 1, this result holds both on the torus and on finite rectangular regions with fixed or free boundary conditions.

Consider now a finite region V and two colorings C, C' of its boundary that differ at a single site v, and a subregion $U \subseteq V$ such that the distance from v to the nearest point $u \in U$ is ℓ. Let μ and μ' denote the probability distributions on colorings of U, given the uniform distribution on colorings of V conditioned on C and C' respectively. We say that q-colorings have *optimal spatial mixing* if there are constants $\alpha, \beta > 0$ such that $\|\mu - \mu'\| \le \beta |U| \exp(-\alpha \ell)$. In other words, conditioning on particular colors appearing on vertices far away from v has an exponentially small effect on the conditional distribution of the color of v in a uniformly random coloring of the grid. Physically, this means that correlations decay exponentially as a function of distance, and that the system has a unique Gibbs measure and no phase transition.

The following recent result of Dyer, Sinclair, Vigoda and Weitz [9] (see also the lecture notes by Martinelli [14]) relates optimal temporal mixing with spatial mixing: if the boundary constraints are *permissive*, i.e. a finite region can always be colored no matter how we color its boundary and the heat-bath dynamics on some finite block mixes in $O(n \log n)$ time, then the system has strong spatial mixing. As they point out, q-colorings are permissive for any $q \ge \Delta + 1$. Thus, the fact that $M(2, 3)$ has optimal temporal mixing implies a strong result about spatial correlations.

Corollary 3. *The uniform measure on the set of q-colorings of the square grid, or equivalently the zero-temperature antiferromagnetic q-state Potts model on the square lattice, has strong spatial mixing for $q \ge 6$.*

As mentioned above, physicists conjecture spatial mixing for $q \geq 4$. In the last section we discuss to what extent our techniques might be extended to $q = 4, 5$.

1.1   Markov Chains, Mixing Times, and Earlier Work

Given a Markov chain M, let π be its stationary distribution and P_x^t be the probability distribution after t steps starting with an initial point x. Then, for a given $\epsilon > 0$, the ϵ-*mixing time* of M is

$$\tau_\epsilon = \max_x \min \left\{ t : \; \left\| P_x^t - \pi \right\| < \epsilon \right\}$$

where $\left\| P_x^t - \pi \right\|$ denotes the total variation distance

$$\left\| P_x^t - \pi \right\| = \frac{1}{2} \sum_y \left| P_x^t(y) - \pi(y) \right| \; .$$

In this paper we will often adopt the common practice of suppressing the dependence on ϵ, which is typically logarithmic, and speak just of the mixing time τ for fixed small ϵ. Thus the mixing time becomes a function of n, the number of vertices, alone. We say that a Markov chain has *rapid mixing* if $\tau = \text{poly}(n)$, and *optimal (temporal) mixing* if $\tau = O(n \log n)$.

The most common Markov chain for this system is *Glauber dynamics*. There are several variants of this in the literature, but for colorings we fix the following definition. At each step, choose a random vertex $v \in G$. Let S be the set of colors, and let T be the set of colors taken by v's neighbors. Then choose a color c uniformly at random from $S \setminus T$, i.e. from among the colors consistent with the coloring of $G - \{v\}$, and recolor v with c. Independently, Jerrum [12] and Salas and Sokal [18] proved that for q-colorings on a graph of maximum degree Δ the Glauber dynamics has optimal mixing for $q > 2\Delta$, and Bubley and Dyer [2] showed that it mixes in $O(n^3)$ time when $q = 2\Delta$. Note that $M(1,1)$ is simply the Glauber dynamics.

Dyer and Greenhill [7] considered a "heat bath" Markov chain which updates both ends of a random edge simultaneously, and showed that it has optimal mixing for $q \geq 2\Delta$. By widening the updated region to include a site and all of its neighbors, Bubley, Dyer and Greenhill [3] showed optimal mixing for $q \geq 7$ for 4-regular triangle-free graphs, such as the square grid.

Another Markov chain commonly used by physicists is the Kempe chain algorithm, which they call the zero-temperature case of the Wang-Swendsen-Kotecký algorithm [23,24]. It works as follows: we choose a random vertex v and a color b which differs from v's current color a. We construct the largest connected subgraph containing v which is colored with a and b, and recolor this subgraph by switching a and b. In a major breakthrough, Vigoda [22] showed that a similar Markov chain has optimal mixing for $q > (11/6)\Delta$, and this implied that the Glauber dynamics and the Kempe chain algorithm both have rapid mixing for $q \geq (11/6)\Delta$. However, for the square grid this gives only $q \geq 8$.

For $q = 3$ on the square grid, Luby, Randall and Sinclair [13] showed that a Markov chain including "tower moves" has rapid mixing for any finite simply-connected region with fixed boundary conditions, and Randall and Tetali [17] showed that this implies rapid mixing for the Glauber dynamics as well. Recently Goldberg, Martin and Paterson [11] proved rapid mixing for the Glauber dynamics on rectangular regions with free boundary conditions, i.e., with no fixed coloring of the vertices on their boundary. However, the technique of [13, 11] relies on a bijection between 3-colorings and random surfaces through a "height representation" which does not hold for other values of q.

2 Coupling

We consider two parallel runs of our Markov chain, $M(2,3)$, with initial colorings X_0, Y_0. We will couple the steps of these chains in such a way that (i) each chain runs according to the correct distribution on its choices and (ii) with high probability, $X_t = Y_t$ for some $t = O(n \log n)$. A now standard fact in this area is that this implies that the chain mixes in time $O(n \log n)$, i.e. this implies Theorem 1; this fact was first proved by Aldous [1] (see also [8]).

Bubley and Dyer [2] introduced the very useful technique of Path Coupling, via which it suffices to do the following: Consider any two *not necessarily proper* colorings X, Y which differ on exactly one vertex, and carry out a single step of the chain on X and on Y, producing two new colorings X', Y'. We will prove that we can couple these two steps such that (i) each step is selected according to the correct distribution, and (ii) the expected number of vertices on which X', Y' differ is at most $1 - \epsilon/n$ for some constant $\epsilon > 0$. See, e.g., [8] for the formal (by now standard) details as to why this suffices to prove Theorem 1.

We perform the required coupling as follows. We pick a uniformly random 2×3 subgrid S, and let C_X and C_Y denote the set of permissible recolorings of S according to X, Y respectively. For each $c \in C_X$, we define a carefully chosen probability distribution p_c on the colorings of C_Y. We pick a uniformly random member $c_1 \in C_X$ and in X we recolor S with c_1 to produce X'. We then pick a random member $c_2 \in C_Y$ according to the distribution p_{c_1} and in Y we recolor S with c_2 to produce Y'. Trivially, the pair S, c_1 is chosen according to the correct distribution. In order to ensure that the same is true of S, c_2, we must have the following property for the set of distributions $\{p_c : c \in C_X\}$:

$$\text{for each } c_2 \in C_Y, \qquad \tfrac{1}{|C_X|} \sum_{c \in C_X} p_c(c_2) = \tfrac{1}{|C_Y|} \ . \tag{1}$$

Suppose that v is the vertex on which X, Y differ. If $v \in S$ then $C_X = C_Y$, so we can simply define $c_2 = c_1$ (i.e. $p_c(c) = 1$ for each c) and this ensures that $X' = Y'$. If S does not contain v or any neighbor of v, then again $C_X = C_Y$ and by defining $c_2 = c_1$ we ensure that X', Y' differ only on v. If v is not in S but is adjacent to a vertex in S, then $C_X \neq C_Y$ so our coupling becomes very complicated and it is quite possible that $c_2 \neq c_1$ and so X', Y' will differ on one or more vertices of S as well as on v.

For any pair C_X, C_Y, we let $H(C_X, C_Y)$ denote the expected number of vertices in S on which c_1, c_2 differ. For every possible pair C_X, C_Y we obtain a coupling satisfying (1) and:

$$H(C_X, C_Y) < 0.52 \ . \tag{2}$$

Proof of Theorem 1. As described above, it suffices to prove that for any choice of X, Y differing only at v, the expected number of vertices on which X', Y' differ is less than $1 - \epsilon/n$ for some $\epsilon > 0$. The probability that S contains v is $|S|/n = 6/n$; for any such choice of S, $X' = Y'$. The probability that S contains a neighbor of v but does not contain v is easily seen to be $10/n$; for any such choice of S, the expected number of vertices on which X', Y' differ is less than 1.52. Therefore, the overall expected number of vertices on which X', Y' differ is less than

$$1 \times (n - 16)/n + 0 \times 6/n + 1.52 \times 10/n = 1 - 0.8/n \ .$$

$\square$

Of course, we still need to prove that the desired couplings exist for each possible C_X, C_Y. These couplings were found with the aid of computer programs. In principle, for any pair C_X, C_Y, searching for the coupling that minimizes $H(C_X, C_Y)$ subject to (1) is simply a matter of solving a linear program and so can be done in polytime. However, the number of variables is $|C_X| \times |C_Y|$ which can be as high as roughly $(5^6)^2$. Furthermore, the number of possible pairs X, Y is roughly 6^{10}, and even after eliminating pairs which are redundant by symmetry, it is enormous. Thus, finding these couplings is computationally intensive. To help we designed a fast heuristic which, rather than finding the best coupling for a particular pair, just found a very good coupling; i.e. one that satisfied (2). The code used can be found at www.cs.toronto.edu/~fvb. We provide a more detailed description in the next section.

3   The Programs Used

Method of the computation: Let R denote the *rim* vertices, that is, those vertices which are adjacent to but outside of the subgrid S. We call a (not necessarily proper) coloring of the vertices of R a *rim coloring*. For each possible pair of rim colorings X, Y which differ only at a vertex $v \in R$, we need to find a coupling between the extensions C_X and C_Y of X, Y to S, so that the couplings satisfy (1) and (2). These couplings were found with a small suite of programs working in two phases. In the first phase, exhaustive lists of pairs of rim colorings (reduced by equivalence with respect to allowable grid colorings) were generated. In the second phase, for each pair X, Y, we generated C_X, C_Y separately; these were then coupled, satisfying (1), in a (close to) optimal way to obtain a bound on $H(C_X, C_Y)$ that satisfies (2).

Implementation: All programs take the following parameters: number of colors, grid dimensions, and an integer denoting the position of the distinguished vertex v with respect to the grid (0 if adjacent to the corner, $+i$ if adjacent to the i-th vertex along the top of the grid, and $-i$ if adjacent to the i-th vertex along the side of the grid). If one specifies a rim coloring X and the position of v, then this determines Y (up to equivalence via permutation of colors). For each coloring X we determined a good coupling for each non-equivalent position of v.

By default the programs generate rim colorings and couplings on the assumption that the subgrid is not on the boundary of the supergrid (i.e. all rim vertices potentially constrain the allowable subgrid colorings). Boundary cases, however, can easily be simulated by using values in the rim colorings that are outside the range determined by the number-of-colors parameter. Grids on the boundary were only checked when the analysis of the non-boundary grids yielded promising values; in all cases we found that the maximum cost for boundary subgrids was lower than for the associated non-boundary subgrids.

Generating rim colorings: Since the calculations required for phase 2 were much more time-consuming than those for phase 1, the rim coloring generation procedure was designed to minimize the number of colorings output rather than the time used generating them. A rim coloring is represented by a vector of the colors used on the rim, starting from the distinguished vertex v and going clockwise around the subgrid. Since we can assume by symmetry that the color used for v is 0 in X and 1 in Y, 0|1 is always the first element (0 is used in the actual output, 1 is understood). The following reductions were applied to avoid equivalent rim colorings: reduction by color isomorphism (colors 2 and above), by exchange of colors 0 and 1, by exchange of colors of vertices adjacent to the corners of the subgrid, and by application of flip symmetries where applicable.

Finding a coupling for particular rim colorings: Two programs were used for each rim coloring X, and position i of v. In each, the initial operation is the generation of all compatible grid colorings; this is done separately for $col(v) = 0$ and $col(v) = 1$ (i.e. for C_X, C_Y). The first program creates a set of linear programming constraints that is readable by the program lp-solve (by Michel Berkelaar of the Eindhoven University of Technology; it is available with some Linux distributions). As mentioned above, time and space requirements made use of this procedure feasible only for checking individual rim colorings, and even then the subgrid size had to be fairly modest. The second program calculates an upper bound on the optimal cost using a greedy algorithm to create a candidate coupling. Given sets of colorings C_X and C_Y (of size m_X and m_Y respectively), the algorithm starts by assigning "unused" probabilities $\frac{1}{m_Y}$ and $\frac{1}{m_X}$ to the individual colorings. Then, for each distance $d = 0, 1, ..., n$, for each coloring c_1 in C_X it traverses C_Y looking for a coloring c_2 which differs from c_1 on exactly d vertices. When such an c_2 is found it removes the coloring with the lower unused probability p from its list and reduces the unused probability p' of the other to $p' - p$; the distance $d \cdot p$ is added to the total distance so far. The order in which the lists of colorings C_X and C_Y is traversed does affect the solution, so an optional argument is available that allows the user to select one of several alternatives.

This heuristic does not guarantee an optimal solution, and with some grids and particular rim colorings the coupling it generates is far from the best. However, for the rim colorings we are most interested in (ones where $H(C_X, C_Y)$ is high for all couplings) it seems to consistently give results that are optimal or very close (within 2%). We cannot give a rigorous bound on the running time, but a cursory analysis and empirical evidence suggest that it runs in roughly $\mathcal{O}(m \log m)$ time, where m is the number of compatible grid colorings. Because the heuristic is so much faster than the LP solver, our general procedure was as follows: (1) Use the heuristic with the default traversal order to calculate bounds on the expected distance for all the rim colorings generated in phase one. (2) When feasible, use the LP solver on those rim colorings that had the highest value of $H(C_X, C_Y)$, to obtain an exact value for the maximum. (3) For larger grids / more colors than could be comfortably handled by the LP solver, use all available traversal orders on those rim colorings that had the maximum value of $H(C_X, C_Y)$ to obtain as tight a bound as possible within a feasible time.

Results of the computations: Computations were run on various grid sizes and numbers of colors in order to check the correctness of the programs, and as well collect data which could be used to estimate running times and maximum expected distance for larger subgrid dimensions. For 7 and 8 colorings our results corresponded well with previous work on the problem (eg [2]).

For 6 colorings, we checked $1 \times k$ subgrids for $k \leq 5$, as well as 2×2 and 2×3 subgrids; for all but the last of these the maximum expected distance we obtained was too large to give us rapid mixing. The 2×3 subgrid has 2 non-equivalent positions with respect to the rim, the corner (position 0, 8 rim vertices are adjacent to this position) and the side (position 1, 2 rim vertices adjacent). For each X, Y with v in position 0, we obtained a coupling satisfying:

$$H(C_X, C_Y) \leq 0.5118309760.$$

For each X, Y with v in position 1, we obtained a coupling satisfying:

$$H(C_X, C_Y) \leq 0.4837863092.$$

Thus, in each case we satisfy (2), as required.

A slightly stronger output: By examining the problem a bit more closely, we see that condition (2) is sufficient, but not necessary, for our purposes. Let H_i denote the maximum of $H(C_X, C_Y)$ over the couplings found for all pairs X, Y where v is in position i, and let $mult_i$ denote the number of rim vertices adjacent to position i. Then, being more careful about the calculation used in the proof of Theorem 1, and extending it to a general $a \times b$ subgrid, we see that the overall expected number of vertices on which X', Y' differ is at most

$$1 \times (n - 2(a + b))/n + 0 \times ab/n + \sum_i mult_i \times H_i/n,$$

a smaller value than that used in the proof of Theorem 1, where we (implicitly) used $(\max_i H_i) \times \sum_i mult_i$ rather than $\sum_i mult_i \times H_i$. Our programs actually compute this smaller value. Even so, we could not obtain suitable couplings for any grid size smaller than 2×3.

4   Rapid Mixing: Glauber and Kempe Chain Dynamics

In this section we prove Theorem 2, showing rapid mixing for the Glauber and Kempe chain dynamics, by following the techniques and presentation of Randall and Tetali [17].

Suppose Q is a Markov chain whose mixing time we would like to bound, and $\tilde{Q}$ is another Markov chain for which we already have a bound. Let E and $\tilde{E}$ and denote the edges of these Markov chains, i.e. the pairs (x, y) such that the transition probabilities $Q(x, y)$ and $\tilde{Q}(x, y)$, respectively, are positive. Now, for each edge of $\tilde{Q}$, i.e. each $(x, y) \in \tilde{E}$, choose a fixed path $\gamma_{x,y}$ using the edges of Q: that is, choose a series of states $x = x_1, x_2, \ldots, x_k = y$ such that $(x_i, x_{i+1}) \in E$ for $1 \le i < k$. Denote the length of such a path $|\gamma_{x,y}| = k$. Furthermore, for each $(z, w) \in E$, let $\Gamma(z, w) \subseteq \tilde{E}$ denote the set of pairs (x, y) such that $\gamma_{x,y}$ uses the edge (z, w). Finally, let

$$A = \max_{(z,w) \in E} \left[\frac{1}{Q(z,w)} \sum_{(x,y) \in \Gamma(z,w)} |\gamma_{x,y}| \, \tilde{Q}(x,y) \right] .$$

Note that A depends on our choice of paths.

By combining bounds on the mixing time in terms of the spectral gap [6,20, 19] with an upper bound on Q's spectral gap in terms of $\tilde{Q}$'s due to Diaconis and Saloff-Coste [5], we obtain the following upper bound on Q's mixing time:

Theorem 4. *Let Q and $\tilde{Q}$ be reversible Markov chains on q-colorings of a graph of n vertices whose unique stationary distribution is the uniform distribution. Let $\tilde{\lambda}_1$ be the largest eigenvalue of $\tilde{Q}$'s transition matrix smaller than 1, let τ_ϵ and $\tilde{\tau}_\epsilon$ denote the ϵ-mixing time of Q and $\tilde{Q}$ respectively, and define A as above. Then for any $\epsilon \le 1/4$,*

$$\tau_\epsilon \le \frac{4 \log q}{\tilde{\lambda}_1} A n \tilde{\tau}_\epsilon .$$

We omit the proof. The reason for the additional factor of n is the fact that the upper and lower bounds on mixing time in terms of the spectral gap are $\log \pi(x)^{-1}$ apart, where π is the uniform distribution. Since there are at most q^n colorings, we have $\log \pi(x)^{-1} \le n \log q$. On the grid, it is easy to see that there are an exponentially large number of q-colorings for $q \ge 3$, so removing this factor of n would require a different comparison technique.

Now suppose that $\tilde{Q}$ is the block dynamics and Q is the Glauber or Kempe chain dynamics. We wish to prove Theorem 2 by showing that $\tau_\epsilon = O(n \tilde{\tau}_\epsilon)$. By adding self-loops with probability greater than $1/2$ to the block dynamics, we can ensure that the eigenvalues of $\tilde{Q}$ are positive with only a constant increase in the mixing time. Therefore, it suffices to find a choice of paths for which A is constant. Since, for all three of these Markov chains, each move occurs with probability $\Theta(1/n)$, if $|\gamma_{x,y}|$ and $|\Gamma(z, w)|$ are constant then so is A.

In fact, for $q \ge \Delta + 2$, we can carry out a block move on any finite neighborhood with Glauber moves. We need to flip each site u in the block to its new color; however, u's flip is blocked by a neighbor v if v's current color equals u's new color. Therefore, we first prepare for u's flip by changing v to a color which differs from u's new color as well as that of v's Δ neighbors. If the neighborhood has m sites, this gives $|\gamma_{x,y}| \leq m(\Delta + 1)$, or $|\gamma_{x,y}| \leq 30$ for $M(2,3)$. (With a little work we can reduce this to 13.)

For the Kempe chain dynamics, recall that each move of the chain chooses a vertex v and a color b other than v's current color. If b is the color which the Glauber dynamics would assign to v, then none of v's neighbors are colored with b, and the Kempe chain move is identical to the Glauber move. Since this happens with probability $1/q$, the above argument applies to Kempe chain moves as well, and again we have $|\gamma_{x,y}| \leq m(\Delta+1)$. Moreover, we only need to consider moves that use Kempe chains of size 1.

Finally, since each site appears in only $m = 6$ blocks, the number of block moves that use a given Glauber move or a given Kempe chain move of size 1 is bounded above by m times the number of pairs of colorings of the block. Thus $|\Gamma(z,w)| \leq m(q^m)^2$, and we are done.

An interesting open question is whether we can prove *optimal* temporal mixing for the Glauber or Kempe chain dynamics. One possibility is to use log-Sobolev inequalities as in [4]. We leave this as a direction for further work.

5   Conclusion: Can We Reach Smaller q?

We have run our programs on 2×4 and 3×3 subgrids to see if we could achieve rapid mixing on 5 colors, but in both cases the largest values of $H(C_X, C_Y)$ were too high. It does seem likely that rapid mixing on 5 colors is possible by recoloring a 3×4 subgrid, based on the decrease of the ratio of $\max E[dist] \cdot |R|$ to $|S|$ as the dimensions increase; similar reasoning leads us to believe that rapid mixing using a $2 \times k$ subgrid is possible, but we would probably need a 2×10 grid or larger to achieve success. Unfortunately, doing the calculations for 3×4 is a daunting proposition. The problem is exponential in two directions at once (number of rim colorings, and number of grid colorings for each rim coloring), so we get huge increases in running time when we move up a level.

Acknowledgments. We are grateful to Leslie Goldberg, Dana Randall and Eric Vigoda for helpful discussions, and to an anonymous referee for pointing out a technical error. Recently, L. Goldberg, R. Martin and M. Paterson have reported obtaining the main result of this paper independently.

References

1. Aldous, D.: Random walks on finite groups and rapidly mixing Markov chains. Séminaire de Probabilités XVII 1981/82 (Dold, A. and Eckmann, B., eds.), Springer Lecture Notes in Mathematics **986** (1986) 243–297
2. Bubley, R., Dyer, M.: Path coupling: a technique for proving rapid mixing in Markov chains. Proc. 28th Ann. Symp. on Found. of Comp. Sci. (1997) 223–231

3. Bubley, R., Dyer, M., Greenhill, C.: Beating the 2Δ bound for approximately counting colourings: A computer-assisted proof of rapid mixing. Proc. 9th Ann. ACM-SIAM Symposium on Discrete Algorithms (1998) 355–363

4. Cesi, F.: Quasi-factorization of the entropy and logarithmic Sobolev inequalities for Gibbs random fields. Probability Theory and Related Fields **120** (2001) 569–584

5. Diaconis, P., Saloff-Coste, L.: Comparison theorems for reversible Markov chains. Annals of Applied Probability **6** (1996) 696–730

6. Diaconis, P., Stroock, D.: Geometric bounds for eigenvalues of Markov chains. Annals of Applied Probability **1** (1991) 36–61

7. Dyer, M., Greenhill, C.: A more rapidly mixing Markov chain for graph colorings. Random Structures and Algorithms **13** (1998) 285–317

8. Dyer, M., Greenhill, C.: Random walks on combinatorial objects. Surveys in Combinatorics, 1999 (Lamb, J. and Preece, D., eds.), Cambridge University Press, J., 1999, 101–136

9. Dyer, M., Sinclair, A., Vigoda, E., Weitz, D.: Mixing in time and space for lattice spin systems: a combinatorial view. Proc. RANDOM (2002) 149–163

10. Ferreira, S., Sokal, A.: Antiferromagnetic Potts models on the square lattice: a high-precision Monte Carlo study. J. Statistical Physics **96** (1999) 461–530

11. Goldberg, L., Martin, R., Paterson, M.: Random sampling of 3-colourings in $\mathbb{Z}^2$. Random Structures and Algorithms (to appear)

12. Jerrum, M.: A very simple algorithm for estimating the number of k-colorings of a low-degree graph. Random Structures and Algorithms **7** (1995), 157–165

13. Luby, M., Randall, D., Sinclair, A.: Markov chain algorithms for planar lattice structures. SIAM Computing **31** (2001) 167–192

14. Martinelli, F.: Lectures on Glauber dynamics for discrete spin models. Lectures on Probability Theory and Statistics, Saint-Flour 1997, Springer Lecture Notes in Mathematics **1717** (1999) 93-191

15. Moore, C., Newman, M.: Height representation, critical exponents, and ergodicity in the four-state triangular Potts antiferromagnet. J. Stat. Phys. **99** (2000) 661–690

16. Moore, C., Nordahl, M., Minar, N., Shalizi, C.: Vortex dynamics and entropic forces in antiferromagnets and antiferromagnetic Potts models. Physical Review E **60** (1999) 5344–5351

17. Randall, D., Tetali, P.: Analyzing Glauber dynamics by comparison of Markov chains. J. Mathematical Physics **41** (2000) 1598–1615

18. Salas, J., Sokal, A.: Absence of phase transition for antiferromagnetic Potts models via the Dobrushin uniqueness theorem. J. Statistical Physics **86** (1997) 551–579

19. Sinclair, A.: Algorithms for random generation and counting: a Markov chain approach. Birkhauser, Boston, 1993, pp. 47–48.

20. Sinclair, A., Jerrum, M.: Approximate counting, uniform generation, and rapidly mixing Markov chains. Information and Computation **82** (1989) 93–133

21. Sokal, A.: A personal list of unsolved problems concerning lattice gases and antiferromagnetic Potts models. Talk presented at the conference on Inhomogeneous Random Systems, Université de Cergy-Pontoise, January 2000. Markov Processes and Related Fields **7** (2001) 21–38

22. Vigoda, E.: Improved bounds for sampling colorings. J. Mathematical Physics **41** (2000) 1555–1569

23. Wang, J., Swendsen, R., Kotecký, R.: Physical Review Letters **63** (1989), 109–

24. Wang, J., Swendsen, R., Kotecký, R.: Physical Review B **42** (1990), 2465–

The Complexity of Finding
Top-Toda-Equivalence-Class Members*

Lane A. Hemaspaandra[1], Mitsunori Ogihara[1], Mohammed J. Zaki[2], and
Marius Zimand[3]

[1] Department of Computer Science, University of Rochester, Rochester, NY 14627
{lane,ogihara}@cs.rochester.edu,
[2] Department of Computer Science, Rensselaer Polytechnic Institute, Troy, NY 12180
zaki@cs.rpi.edu
[3] Department of Computer and Information Sciences, Towson Univ., Baltimore, MD 21252
mzimand@towson.edu

Abstract. We identify two properties that for P-selective sets are effectively computable. Namely we show that, for any P-selective set, finding a string that is in a given length's top Toda equivalence class (very informally put, a string from Σ^n that the set's P-selector function declares to be most likely to belong to the set) is $\mathrm{FP}^{\Sigma_2^p}$ computable, and we show that each P-selective set contains a weakly-$\mathrm{P}^{\Sigma_2^p}$-rankable subset.

1 Introduction

P-selectivity is a generalization of P. A set A is in P if there is a polynomial-time algorithm which, given any string x, determines whether x belongs to A. In contrast, a set A is P-selective if there is a polynomial-time algorithm (called a P-selector function) that, given any two strings x and y, outputs one of those strings, and such that the algorithm has the property that if at least one of x or y is in A, then the one the algorithm outputs belongs to A. Informally, it always places a bet on one of them being in the set, and it wins whenever winning such a bet is possible.

The book [18] provides a recent overview of the state of research regarding P-selectivity theory (see also the somewhat older article [2]). Nickelsen's thesis [24] and the recent survey article by Nickelsen and Tantau [25] are also very good starting points regarding the study of partial information classes such as the P-selective sets.

A key notion used in P-selectivity theory is the notion of a Toda equivalence class which, very loosely put, is a strongly connected component of the graph induced by a given P-selector function on the strings of a given length. This paper studies the complexity of finding a string from a given P-selector function's top Toda equivalence class—that is, a string from the unique strongly connected component that can be reached from no other strongly connected component.

P-selectivity theory has many features making it an interesting complexity-related research area. P-selectivity represents a natural generalization of feasible decidability, it

---

* Supported in part by NSF grants INT-9815095/DAAD-315-PPP-gü-ab, EIA-0080124, DUE-9980943, and EIA-0205061, and NIH grant P30-AG18254.

has a well-studied analog in computability theory, namely, the semi-recursive sets [21]; the exploration of P-selectivity had a strong, unexpected impact on the study of NP functions, solutions and ambiguity, namely, the nondeterministic version of selectivity is the central tool used to show that SAT has unique solutions only if the polynomial hierarchy collapses [12]; and, relatedly, selectivity has proven central in understanding more generally whether one can reduce the number of solutions of NP functions [12,26, 23,15].

Informally, P-selectivity captures the notion of sets for which there is a polynomial-time algorithm f telling which of any two given elements is "logically no less likely to be in the set" (see Definition 2.1). Such sets are called P-selective. P-selective sets can be arbitrarily complex: For every tally set A, there is a P-selective set that is Turing-equivalent to A [27,28]. In particular, some P-selective sets are uncomputable. Despite this, in the present paper we identify natural tasks that are computable for all P-selective sets. Indeed, these natural tasks are even computable within relatively low levels of the polynomial hierarchy.

The first such task is to produce, for an arbitrary P-selective set A (which, w.l.o.g., has at least one commutative P-selector function), at each length n, a string that is "most likely" to be in A. Let us explain a bit more what we mean by this. Each commutative polynomial-time P-selector function f for A will implicitly specify a structure of equivalence classes of strings (at a given length) that are equally likely (according to f) to be in A, and these classes can be ordered with respect to the order that one might informally call *no-less-likely-to-be-in-A* (we will explain how to define such equivalence classes, which themselves depend on only the P-selector function and not on A, in rigorous detail in Section 2 after introducing the tournament-graph model that is useful in their definition; in brief, two strings of the same length are said to be equivalent exactly if there are chains of applications of the P-selector function leading from each to the other). We will call those classes Toda equivalence classes, in light of Toda [33], and the related order will be called a Toda order. Informally put, a Toda equivalence class ζ, with respect to commutative P-selector function f, of length n strings has the property that for *every* set A for which f is a P-selector, either all the strings in ζ are in A or none of the strings in ζ are in A. And (restricting ourselves as we will globally do to just looking at strings all of the same length) each Toda equivalence class is a maximal set of strings for which this can be said.

The Toda-class approach's ordering implications play a central role in a wide range of results, ranging for example from the study of whether P-selective sets can be hard for standard complexity classes [33] to the study of associative P-selector functions [9, 10]. In this paper, we seek to better understand the Toda classes' own complexity. We show that finding an element in the top Toda class of a length (very informally and intuitively—and not quite correctly—put, a string that among the strings of that length is "most likely" to be in A) can be done with an $\mathrm{FP}^{\Sigma_2^p}$ computation.

The second task we study is that of weak-P-rankability. A function f *weakly ranks* a set A if, for any string x that is in A, $f(x)$ returns the rank of x in A; in other words, it says how many strings lexicographically less than or equal to x are in A. A set A is weakly-P-rankable if it has a function f that is computable in polynomial time and that weakly ranks A. The relation between P-selectivity and weak-P-rankability has been studied extensively by Hemaspaandra, Zaki, and Zimand in a paper [19] that focused on polynomial-time semi-rankable sets, i.e., sets that are simultaneously P-selective and weakly-P-rankable. It is shown there that there are P-selective sets that are not weakly-P-rankable. In partial contrast, in the present paper we show that any infinite P-selective set has an infinite subset that is weakly rankable by an $\mathrm{FP}^{\Sigma_2^p}$ function. We also obtain a result about the relationship between P-selectivity and weak-P-rankability. All the P-selective sets that have been considered in the literature are either standard left cuts or are structurally similar to a standard left cut (more precisely $\leq_m^p$-reducible to a standard left cut). We show that if a standard left cut is weakly-P-rankable, then it is in P. Regarding this section (Section 3), we particularly commend to the reader's attention the proof of Lemma 3.3 (omitted here; see the full version), which we feel to be a novel technique, and which is given prompt application in yielding the theorems that are stated immediately after it.

For space reasons, all proofs and much discussion/explanation is omitted here; please see the full version [16].

2 Instantiating the Top Toda Class

Let $\mathbb{N} = \{0, 1, 2, \ldots\}$ and let $\mathbb{N}^+ = \{1, 2, 3, \ldots\}$. Our alphabet will be $\Sigma = \{0, 1\}$. For any set A, $\|A\|$ denotes the cardinality of A. For any string $x \in \Sigma^*$, $|x|$ denotes the length of x. For any set A and any nonnegative integer n, $A^{=n}$ denotes $\{x \mid x \in A \wedge |x| = n\}$. For any set A and any string x, $A^{\leq x}$ denotes $\{y \mid y \in A \wedge y \leq_{lex} x\}$, where $\leq_{lex}$ denotes $\leq$ with respect to the standard lexicographical ordering. For the definitions of standard complexity classes such as P, NP, Σ_k^p, etc., we refer the reader to, for example, the handbook [14]. As is standard, FP denotes the class of all (total, single-valued) polynomial-time computable functions, $\Sigma_2^p = \mathrm{NP}^{\mathrm{NP}}$, $\Sigma_3^p = \mathrm{NP}^{\mathrm{NP}^{\mathrm{NP}}}$, $\mathrm{E} = \bigcup_{k \geq 0} \mathrm{DTIME}[2^{kn}]$, and $\mathrm{NE} = \bigcup_{k \geq 0} \mathrm{NTIME}[2^{kn}]$. Given classes $\mathcal{C}$ and $\mathcal{D}$, as is standard we say that $\mathcal{C}$ is $\mathcal{D}$-immune (equivalently, $\mathcal{C}$ is immune to $\mathcal{D}$) if there is an infinite set $A \in \mathcal{C}$ such that no infinite subset of A is a member of $\mathcal{D}$.

Definition 2.1. *[27,29] A set A is P-selective if there is a (total, single-valued) polynomial-time computable function f such that, for every x and y, it holds that*

1. $f(x, y) = x$ or $f(x, y) = y$, and

2. $\{x, y\} \cap A \neq \emptyset \Rightarrow [(x \in A$ and $f(x, y) = x)$ or $(y \in A$ and $f(x, y) = y)]$.

We use P-sel to denote the class of all sets that are P-selective.

The function f appearing in Definition 2.1 is called a *P-selector* or a *P-selector function*. A P-selector (function) f is *commutative* if it has the property that for all x and y in Σ^*, $f(x, y) = f(y, x)$. Each P-selective set A has a commutative P-selector function, because we can replace an arbitrary P-selector f for A with $f'(x, y) = f(\min(x, y), \max(x, y))$. It is easy to see that f' is a commutative P-selector for A. Since all P-selective sets have commutative P-selector functions, it is very common in the literature to focus on commutative P-selector functions, and we do so here.

A tournament graph $G = (V_G, E_G)$ is a complete oriented graph, i.e., a directed graph having the property that, for every two (possibly equal) nodes a and b, $||\{(a, b), (b, a)\} \cap E_G\}|| = 1$. A commutative P-selector induces a tournament graph G_f on Σ^*. The nodes are the strings in Σ^* and there is an edge (x, y) (also denoted $x \geq_f y$) exactly if $f(x, y) = x$. When speaking in plain text, we will use synonymously with this the terms x *beats* y, x *wins at* y, and y *loses to* x. (Note that each x both wins at itself and loses to itself, and our tournament graphs have self-loops at each node.) We say $x >_f y$ if $x \geq_f y$ and $x \neq y$.

Let $G_{f,n}$ be the subgraph of G_f induced by the nodes in Σ^n. Two nodes $x, y \in \Sigma^n$ are *Toda-equivalent*, notated $x \equiv_{Toda} y$, if $G_{f,n}$ contains a path from x to y and a path from y to x. The relation $\equiv_{Toda}$ is an equivalence relation. We will denote the equivalence class of $x \in \Sigma^n$ by $[x]_f$. For strings x and y of the same length, we order their equivalence classes as follows: $[x]_f \geq [y]_f$ holds exactly if $x \geq_f y$. For strings x and y of the same length, we say that $[x]_f > [y]_f$ exactly if $[x]_f \geq [y]_f$ and $[x]_f \neq [y]_f$.

(It is easy to see that these relations are consistent when applied only among strings all of the same length. Note that we neither define nor ever use $[w]_f > [z]_f$ or $[w]_f \geq [z]_f$ for the case where $|w| \neq |z|$. Our focus will always be on collections of same-length strings. However, if one wanted to compare different-length strings, for the purposes of this paper let us say that equivalence classes of different lengths are always, by definition, incomparable, and so viewed as being over all of Σ^* our classes form a partial rather than a total order. Since our focus is within a length and there our order is never undefined, we will for simplicity simply use the term "order.")

Of course, $[y]_f < [x]_f$ means the same as $[x]_f > [y]_f$, and $[y]_f \leq [x]_f$ means the same as $[x]_f \geq [y]_f$.

The following fact holds.

Fact 2.2.

1. *If A is a P-selective set having commutative function f as a P-selector then, for all $x \in \Sigma^n$, $A \cap [x]_f = [x]_f$ or $A \cap [x]_f = \emptyset$.*

2. *Let A be a P-selective set having commutative function f as a P-selector. If $A \cap [x]_f = [x]_f$ then for all y such that $|y| = |x|$ and $[y]_f \geq [x]_f$, it holds that $A \cap [y]_f = [y]_f$. (In fact, it even holds that if $A \cap [x]_f = [x]_f$ then for all y (regardless of $|y|$) such that $f(x, y) = y$ it holds that $A \cap [y]_f = [y]_f$.)*

These equivalence classes (related to strings of length n) form a partition of Σ^n. In particular, there are strings $x_1^n, x_2^n, \ldots, x_k^n \in \Sigma^n$ such that the equivalence classes $[x_1^n]_f, \ldots, [x_k^n]_f$ form a partition of Σ^n and $[x_1^n]_f > [x_2^n]_f > \ldots > [x_k^n]_f$. The set $[x_1^n]_f$ is called the *top Toda class* (at length n with respect to f).

Definition 2.3. *Let f be a commutative P-selector function. A function g is a BestAtLength function (for f) if, for each n, $g(1^n) \in [x_1^n]_f$, i.e., it outputs an element of the top Toda class at length n with respect to f.*

Each string that belongs to the top Toda class (at length n with respect to f) will be called a *top Toda element* (at length n with respect to f).

Note that Definition 2.3 does not mention what set f is a P-selector for, since each BestAtLength function will work equally well for each of the potentially uncountably many sets for which f is a P-selector.

Theorem 2.4. *Every commutative P-selector f has a BestAtLength function computable in* $\mathrm{FP}^{\Sigma_2^p}$.

It is natural to ask if there is a more efficient BestAtLength function. The next result shows that it is unlikely that the BestAtLength function of every commutative P-selector is in FP because this would imply $\mathrm{E} = \mathrm{NE}$. The proof is relativizable, so it follows that there is an oracle relative to which the BestAtLength function for some commutative P-selector is not in FP. The issue of whether the statement of Theorem 2.4 can be improved to $\mathrm{FP}^{\mathrm{NP}}$ remains open.

Theorem 2.5. *1. If* $\mathrm{P} = \mathrm{NP}$, *then every commutative P-selector f has a BestAtLength function in* FP.

   2. If every commutative P-selector f has a BestAtLength function in FP *then* $\mathrm{E} = \mathrm{NE}$.

Since there are relativized worlds in which E and NE differ, Theorem 2.5 (in its relativized version, which also holds) yields the following corollary.

Corollary 2.6. *There is an oracle relative to which there is a commutative P-selector f such that no BestAtLength function for f is in* FP.

Theorem 2.5 relativizes. That is, for each set B it holds that: If every commutative P^B-selector f has a BestAtLength function in FP^B then $\mathrm{E}^B = \mathrm{NE}^B$. Though neither $\mathrm{NP} \cap \mathrm{coNP}$ nor $\mathrm{coNE} \cap \mathrm{NE}$ is currently known to have complete sets (see [31,4,5] regarding $\mathrm{NP} \cap \mathrm{coNP}$), one nonetheless can (by a "set by set" argument—[11, Corollary 6] for example employs a set-by-set argument in the completely different setting of Karp–Lipton-type results), in light of the facts that $\mathrm{E}^{\mathrm{NP} \cap \mathrm{coNP}} = \mathrm{coNE} \cap \mathrm{NE}$ and $\mathrm{NE} = \mathrm{coNE} \cap \mathrm{NE} \iff \mathrm{NE} = \mathrm{coNE}$, as an application of this see that the following holds.

Corollary 2.7. *If every commutative* $\mathrm{FP}^{\mathrm{NP} \cap \mathrm{coNP}}$*-selector f has a BestAtLength function in* $\mathrm{FP}^{\mathrm{NP} \cap \mathrm{coNP}}$ *then* $\mathrm{coNE} = \mathrm{NE}$.

The reader may naturally wonder whether the $\mathrm{E} = \mathrm{NE}$ conclusion of part 2 of Theorem 2.5 can be strengthened to a $\mathrm{P} = \mathrm{NP}$ conclusion. We do not have a definitive answer; the fact that BestAtLength functions have tally inputs seems a difficult impediment to proving this. We do note that it is the only impediment. That is, if we make a new notion of BestAtLength function, let us call it BestBelowUsAtLength, that (for a fixed commutative P-selector function f) takes as its input $\langle 1^n, z \rangle$ and (i) when $|z| \neq n$ outputs "illegal z"; and (ii) when $|z| = n$ and $[z]_f$ is not in the bottom Toda equivalence class (with respect to f at length n) outputs a string w, $|w| = n$, such that $[w]_f$ is the topmost Toda equivalence class (with respect to f at length n) that is less than $[z]_f$ (i.e., such that $[z]_f > [w]_f$ yet, for each $v \in \Sigma^n$, it holds that $[w]_f < [v]_f \Rightarrow [v]_f \geq [z]_f$). (When $|z| = n$ and $[z]_f$ is in the bottom Toda equivalence class (with respect to f at length n), a BestBelowUsAtLength function can output whatever lie it likes.) Under this definition, it is easy to see that a $\mathrm{P} = \mathrm{NP}$ conclusion holds.

Proposition 2.8. *If every commutative P-selector f has a BestBelowUsAtLength function in* FP *then* $\mathrm{P} = \mathrm{NP}$.

An element x in the top Toda class at length n has the property that for every string $y \in \Sigma^n$ there is a path from x to y in the induced tournament graph. We next investigate a related notion. It is known (this was noted as early as the 1950s [22]) that in a tournament graph there is at least one node v such that for every other node w there is a path of length at most 2 from v to w. This property has played an important role in improving from quasilinear to linear the amount of nondeterminism used to accept the P-selective sets with optimal nonuniform advice ([13], see also [17]) and in understanding the complexity of the reachability problem in tournaments [32]. A node v with the above property is called a *king*. A function g is a Find-a-King function for a commutative P-selector f if for all n, on input 1^n, f outputs a king of the graph $G_{f,n}$.

Proposition 2.9. *Every commutative P-selector function f has a Find-A-King function computable in* $\mathrm{FP}^{\Sigma_3^p}$.

It is interesting to note that building a top Toda element seems to be easier than building a king (as indicated by the previous theorems), yet recognizing a top Toda element seems (given our current stage of knowledge) to be a more difficult task than recognizing a king. This holds because a string $x \in \Sigma^n$ is a top Toda element at length n exactly if for any string $y \in \Sigma^n$ there is a path from x to y; checking this condition is in PSPACE. One can observe that this problem is also in the advice class $\mathrm{PP}/\,\mathrm{linear}$ (that is it can be done in PP given an advice string of size $O(n)$). Indeed, for an arbitrary P-selector f, let us consider the Toda-equivalence classes for the strings in Σ^n sorted according to the order relation defined above: $[x_1]_f > [x_2]_f > \ldots > [x_k]_f$. Then any element x in $[x_1]_f$ beats at least the elements in $\{x\} \cup (\bigcup_{2 \leq i \leq k} [x_i]_f)$ and thus has outdegree at least $1 + \sum_{2 \leq i \leq k} ||[x_i]_f||$. Also note that any element (of Σ^n) that is not in $[x_1]_f$ cannot beat any element in $[x_1]_f$ and thus its outdegree is at most $\sum_{2 \leq i \leq k} ||[x_i]_f||$. Thus if we use as the advice string the binary encoding of $1 + \sum_{2 \leq i \leq k} ||[x_i]_f||$ and if we check versus a threshold the outdegree of a node via a PP computation, we can check whether a string is in the top Toda class.

On the other hand, a string x in $G_{n,f}$ is a king node if and only if $(\forall y_1 : |y_1| = |x|)(\exists y_2 : |y_2| = |x|)[(x \text{ beats } y_1) \vee ((x \text{ beats } y_2) \wedge (y_2 \text{ beats } y_1))]$. Thus checking whether a string x is a king in $G_{n,f}$ can be done with a Π_2^p computation. We note that this fact, and also Proposition 2.9, could be shown indirectly using Tantau's recent work on the complexity of succinct tournament reachability [32].

Though all P-selective sets are known to be in the class $\mathrm{NP}/\,\mathrm{linear}$ [17], which was an improvement from even earlier work [8] that implicitly placed them in $\mathrm{PP}/\,\mathrm{linear}$, it is an open question whether the $\mathrm{PP}/\,\mathrm{linear}$ result for top Toda element recognition given above can itself be strengthened to $\mathrm{NP}/\,\mathrm{linear}$. We conjecture that it cannot. Also, we note in passing that in the profoundly different model in which the tournament—far from being uniformly specified via a P-selector function—can be explored only via queries to a black box, and our input additionally includes the set of nodes inducing via that black box a tournament over which a king (or a certain sequence of kings) is sought, bounds on the necessary and sufficient numbers of queries to find such have been studied in, for example, [30].

3   P-Selectivity and Ranking

While P-selectivity is an extension of polynomial-time decidability, the notion of polynomial-time weak rankability goes in the opposite direction. It describes sets that are so simple that there exists a polynomial-time algorithm that, on inputs that are elements in the set, outputs the number of elements in the set up to that element. Weak-P-rankability can be an attribute of extremely complex sets. One further generalization is to allow the ranking functions to belong to some broader family of functions (e.g., having some complexity bound less stringent than polynomial time). There have been many papers studying the issue of which sets can be ranked [3,7,1,20].

Definition 3.1. *For any set B and any string x, define $rank_B(x) = ||B^{\leq x}||$.*

1. *[7] A set A is* strongly-P-rankable *if there is a polynomial-time computable function f such that $(\forall x \in \Sigma^*) [f(x) = rank_A(x)]$. We also use* strongly-P-rankable *to denote the class of all sets that are strongly-P-rankable.*

2. *[3] A set A is* P-rankable *if there is a polynomial-time computable function f such that (a) $(\forall x \in A) [f(x) = rank_A(x)]$ and (b) $(\forall x \notin A) [f(x) = $ "not in A"]. We also use* P-rankable *to denote the class of all sets that are P-rankable.*

3. *[7] A set A is* weakly-P-rankable *if there is a polynomial-time computable function f such that $(\forall x \in A) [f(x) = rank_A(x)]$. We also use* weakly-P-rankable *to denote the class of all sets that are weakly-P-rankable.*

4. *Let $\mathcal{F}$ be a family of functions mapping strings into natural numbers. A set A is* weakly-$\mathcal{F}$-rankable *if there is a function $f \in \mathcal{F}$ such that $(\forall x \in A) [f(x) = rank_A(x)]$. We also use* weakly-$\mathcal{F}$-rankable *to denote the class of all sets that are weakly-$\mathcal{F}$-rankable.*

Note that, immediately from the definitions, strongly-P-rankable $\subseteq$ P-rankable $\subseteq$ weakly-P-rankable. (The first inclusion is easy to see in light of the fact that, for each $x \neq \epsilon, x \in A \iff rank_A(x) > rank_A(x - 1)$, where $x - 1$ denotes the immediate lexicographic predecessor of x.) Also note that for $x \notin A$, the definition of weakly-P-rankable sets puts no constraint on the behavior of f on input x other than that f must run in polynomial time. This is a point of similarity with P-selectivity useful for the following refinement of P-selectivity, which has been introduced in [19]. The refinement adds the requirement that when at least one of the inputs belongs to the set, we output not merely an input that is in the set, but also output its correct ranking information.

Definition 3.2. *[19] A set A is* polynomial-time semi-rankable *if there is a (total, single-valued) function f such that, for every x and y,*

1. *$(\exists n) [f(x,y) = \langle x, n \rangle$ or $f(x,y) = \langle y, n \rangle]$, and*

2. *$\{x, y\} \cap A \neq \emptyset \Rightarrow [(x \in A$ and $f(x,y) = \langle x, rank_A(x) \rangle)$ or $(y \in A$ and $f(x,y) = \langle y, rank_A(y) \rangle)]$.*

In such a case, we say that f is a semi-ranking function *for A. We use* P-sr *to denote the class of sets that are polynomial-time semi-rankable.*

As noted in [19], P-sr = P-sel ∩ weakly-P-rankable. Among other results, it is shown in [19] that P-sr is a proper set of P-sel, i.e., there are P-selective sets that are not weakly-P-rankable (there are also sets that are weakly-P-rankable but not P-selective). In the full version of this paper, we provide a new and short argument showing this fact, and even that there are P-selective sets that are not weakly-rankable by any function in, for example, the arithmetical hierarchy.

It is shown in [19] that P-sr has structural properties different from those of P-sel. For example, unlike P-sel, P-sr is not closed under complementation, union with P sets, or join with P sets. It is natural to ponder whether P-sel can be separated from the class of weakly-P-rankable sets or from P-sr in a stronger sense, namely with immunity. We note first that the above approach is useless because $L(0) = \emptyset$ and any standard left cut $L(\gamma)$, $0 < \gamma < 1$, contains the subset $\{0^j \mid j \in \mathbb{N}\}$, which belongs to P-sr. Perhaps somewhat surprisingly it turns out, as we will show as Theorem 3.5, that the statement "P-sel is weakly-P-rankable immune" implies $P \neq \Sigma_2^p$ (equivalently, $P \neq NP$) and thus it seems beyond reach at this time.

A set S if P-printable [6] if there exists a polynomial-time algorithm such that, for each $n \in \mathbb{N}$, on input 0^n the algorithm outputs exactly the members of S having length at most n. Note that every P-printable set is sparse and belongs to P. The above definition can be relativized in the standard way.

Lemma 3.3. *Let Q be any set. If A is P-selective, S is P^Q-printable, and $A \cap S$ is infinite, then $A \cap S$ (and thus also S, and most particularly also A) has an infinite weakly-FP^Q-rankable subset.*

The following results immediately follow from Lemma 3.3.

Theorem 3.4. *P-sel is not bi-immune to the class of weakly-P-rankable sets.*

Theorem 3.5. *Any infinite P-selective set A has an infinite weakly-$FP^{\Sigma_2^p}$-rankable subset.*

Corollary 3.6. *If $P = NP$, then P-sel is not immune to the class of weakly-P-rankable sets.*

The reader may wish to compare Theorem 3.5 with the following result—neither of which seems to imply the other—from [9] (see also [10]) regarding printability [6]: Each infinite P-selective set B has an infinite $FP^{B \oplus \Sigma_2^p}$-printable subset.

An important subclass of P-sel is the class of sets that are standard left cuts, a class that we have already used. Recall that for each real $0 \leq \gamma < 1$ the standard left cut of γ is the set $L(\gamma) = \{\beta_1\beta_2 \cdots \beta_z \mid z \in \mathbb{N} \wedge (\forall j : 1 \leq j \leq z)[\beta_j \in \{0,1\}] \wedge \sum_{1 \leq i \leq z} \frac{\beta_i}{2^i} < \gamma\}$. All the P-selective sets that have been constructed in the literature are either standard left cuts or are $\leq_m^p$-equivalent to a standard left cut and, in fact, proving that there is a P-selective set that is not $\leq_m^p$-equivalent to a standard left cut is known to be as hard as showing $P \neq PP$ [8]. In contrast, we observe that standard left cuts that are weakly-rankable are always in P.

Theorem 3.7. *If A is a standard left cut, then A is weakly-P-rankable if and only if A is strongly-P-rankable.*

Since strongly-P-rankable $\subseteq$ P-rankable $=$ P $\cap$ weakly-P-rankable, we have the following immediate corollary.

Corollary 3.8. *Each weakly-*P*-rankable standard left cut belongs to* P.

Acknowledgments. We are grateful to A. Kaplan and B. Serog for helpful conversations, and to an anonymous LATIN '04 referee for helpful comments.

References

1. A. Bertoni, M. Goldwurm, and N. Sabadini. The complexity of computing the number of strings of given length in context-free languages. *Theoretical Computer Science*, 86(2):325–342, 1991.
2. D. Denny-Brown, Y. Han, L. Hemaspaandra, and L. Torenvliet. Semi-membership algorithms: Some recent advances. *SIGACT News*, 25(3):12–23, 1994.
3. A. Goldberg and M. Sipser. Compression and ranking. *SIAM Journal on Computing*, 20(3):524–536, 1991.
4. Y. Gurevich. Algebras of feasible functions. In *Proceedings of the 24th IEEE Symposium on Foundations of Computer Science*, pages 210–214. IEEE Computer Society Press, Nov. 1983.
5. J. Hartmanis and N. Immerman. On complete problems for NP $\cap$ coNP. In *Proceedings of the 12th International Colloquium on Automata, Languages, and Programming*, pages 250–259. Springer-Verlag *Lecture Notes in Computer Science #194*, July 1985.
6. J. Hartmanis and Y. Yesha. Computation times of NP sets of different densities. *Theoretical Computer Science*, 34(1–2):17–32, 1984.
7. L. Hemachandra and S. Rudich. On the complexity of ranking. *Journal of Computer and System Sciences*, 41(2):251–271, 1990.
8. E. Hemaspaandra, A. Naik, M. Ogihara, and A. Selman. P-selective sets and reducing search to decision vs. self-reducibility. *Journal of Computer and System Sciences*, 53(2):194–209, 1996.
9. L. Hemaspaandra, H. Hempel, and A. Nickelsen. Algebraic properties for deterministic selectivity. In *Proceedings of the 4th Annual International Computing and Combinatorics Conference*, pages 49–58. Springer-Verlag *Lecture Notes in Computer Science #2108*, Aug. 2001.
10. L. Hemaspaandra, H. Hempel, and A. Nickelsen. Algebraic properties for deterministic and nondeterministic selectivity. Technical Report TR-778, Department of Computer Science, University of Rochester, Rochester, NY, May 2002. Revised, May 2003.
11. L. Hemaspaandra, A. Hoene, A. Naik, M. Ogiwara, A. Selman, T. Thierauf, and J. Wang. Nondeterministically selective sets. *International Journal of Foundations of Computer Science*, 6(4):403–416, 1995.
12. L. Hemaspaandra, A. Naik, M. Ogihara, and A. Selman. Computing solutions uniquely collapses the polynomial hierarchy. *SIAM Journal on Computing*, 25(4):697–708, 1996.
13. L. Hemaspaandra, C. Nasipak, and K. Parkins. A note on linear-nondeterminism, linear-sized, Karp–Lipton advice for the P-selective sets. *Journal of Universal Computer Science*, 4(8):670–674, 1998.
14. L. Hemaspaandra and M. Ogihara. *The Complexity Theory Companion*. Springer-Verlag, 2002.
15. L. Hemaspaandra, M. Ogihara, and G. Wechsung. Reducing the number of solutions of NP functions. *Journal of Computer and System Sciences*, 64(2):311–328, 2002.

16. L. Hemaspaandra, M. Ogihara, M. Zaki, and M. Zimand. The complexity of finding top-Toda-equivalence-class members. Technical Report TR-808, Department of Computer Science, University of Rochester, Rochester, NY, Aug. 2003.
17. L. Hemaspaandra and L. Torenvliet. Optimal advice. *Theoretical Computer Science*, 154(2):367–377, 1996.
18. L. Hemaspaandra and L. Torenvliet. *Theory of Semi-Feasible Algorithms*. Springer-Verlag, 2003.
19. L. Hemaspaandra, M. Zaki, and M. Zimand. Polynomial-time semi-rankable sets. In *Journal of Computing and Information*, 2(1), Special Issue: *Proceedings of the 8th International Conference on Computing and Information*, pages 50–67, 1996. CD-ROM ISSN 1201-8511/V2/#1.
20. D. Huynh. The complexity of ranking simple languages. *Mathematical Systems Theory*, 23(1):1–20, 1990.
21. C. Jockusch. Semirecursive sets and positive reducibility. *Transactions of the AMS*, 131(2):420–436, 1968.
22. H. Landau. On dominance relations and the structure of animal societies, III: The condition for score structure. *Bulletin of Mathematical Biophysics*, 15(2):143–148, 1953.
23. A. Naik, J. Rogers, J. Royer, and A. Selman. A hierarchy based on output multiplicity. *Theoretical Computer Science*, 207(1):131–157, 1998.
24. A. Nickelsen. Polynomial-time partial information classes. Wissenschaft und Technik Verlag, 2001. Also Ph.D. thesis, Technische Universität Berlin, Berlin, Germany, 1999.
25. A. Nickelsen and T. Tantau. Partial information classes. *SIGACT News*, 34(1), 2003.
26. M. Ogihara. Functions computable with limited access to NP. *Information Processing Letters*, 58(1):35–38, 1996.
27. A. Selman. P-selective sets, tally languages, and the behavior of polynomial time reducibilities on NP. *Mathematical Systems Theory*, 13(1):55–65, 1979.
28. A. Selman. Analogues of semirecursive sets and effective reducibilities to the study of NP complexity. *Information and Control*, 52(1):36–51, 1982.
29. A. Selman. Reductions on NP and P-selective sets. *Theoretical Computer Science*, 19(3):287–304, 1982.
30. J. Shen, L. Sheng, and J. Wu. Searching for sorted sequences in tournaments. *SIAM Journal on Computing*, 32(5):1201–1209, 2003.
31. M. Sipser. On relativization and the existence of complete sets. In *Proceedings of the 9th International Colloquium on Automata, Languages, and Programming*, pages 523–531. Springer-Verlag *Lecture Notes in Computer Science #140*, July 1982.
32. T. Tantau. A note on the complexity of the reachability problem for tournaments. Technical Report TR01-092, Electronic Colloquium on Computational Complexity, http://www.eccc.uni-trier.de/eccc/, Nov. 2001.
33. S. Toda. On polynomial-time truth-table reducibilities of intractable sets to P-selective sets. *Mathematical Systems Theory*, 24(2):69–82, 1991.

List Partitions of Chordal Graphs

Tomás Feder[1], Pavol Hell[2], Sulamita Klein[3], Loana Tito Nogueira[4], and
Fábio Protti[5]

[1] 268 Waverley St., Palo Alto, CA 94301, USA
`tomas@theory.stanford.edu`
[2] School of Computing Science, Simon Fraser University
Burnaby, B.C., Canada V5A 1S6
`pavol@cs.sfu.ca`
[3] IM and COPPE-Sistemas, Universidade Federal do Rio de Janeiro
Caixa Postal 68511, 21945-970, Rio de Janeiro, RJ, Brasil
`sula@cos.ufrj.br`
[4] COPPE-Sistemas, Universidade Federal do Rio de Janeiro
Caixa Postal 68511, 21945-970, Rio de Janeiro, RJ, Brasil
`loana@cos.ufrj.br`
[5] IM and NCE, Universidade Federal do Rio de Janeiro
Caixa Postal 2324, 20001-970, Rio de Janeiro, RJ, Brasil
`fabiop@nce.ufrj.br`

Abstract. In an earlier paper we gave efficient algorithms for partitioning chordal graphs into k independent sets and ℓ cliques. This is a natural generalization of the problem of recognizing split graphs, and is NP-complete for graphs in general, unless $k \leq 2$ and $\ell \leq 2$. (Split graphs have $k = \ell = 1$.)
In this paper we expand our focus and consider general M-partitions, also known as trigraph homomorphisms, for the class of chordal graphs. For each symmetric matrix M over $0, 1, *$, the M-partition problem seeks a partition of the input graph into independent sets, cliques, or arbitrary sets, with some pairs of sets being required to have no edges, or to have all edges joining them, as encoded in the matrix M. Such partitions generalize graph colorings and homomorphisms, and arise frequently in the theory of graph perfection. We show that many M-partition problems that are NP-complete in general become solvable in polynomial time for chordal graphs, even in the presence of lists. On the other hand, we show that there are M-partition problems that remain NP-complete even for chordal graphs. We also discuss forbidden subgraph characterizations for the existence of M-partitions.

1   Introduction

The *M-partition problem* was introduced in [8]. Let M be a symmetric $m \times m$ matrix with entries $M_{i,j} \in \{0, 1, *\}$. An instance of the M-partition problem is a graph G. A solution for the instance is a partition of vertices in G into m *parts*, corresponding to the rows (and columns) of the matrix M, such that for distinct

M. Farach-Colton (Ed.): LATIN 2004, LNCS 2976, pp. 100–108, 2004.

vertices x and y of the graph G, placed in parts i and j (possibly with $i = j$) respectively, we have the following:

- if $M(i,j) = 0$, then xy is not an edge of G;
- if $M(i,j) = 1$, then xy is an edge of G.

(If $M(i,j) = *$, then xy may or may not be an edge in G.)

An instance of the *list M-partition problem* is a graph G, together with a collection of *lists* $L(x), x \in V(G)$, each list being a set of parts. A solution for the instance of list M-partition is a solution for the corresponding M-partition, such that each vertex x is placed in a part $i \in L(x)$.

List M-partitions generalize list colorings, retractions, and list homomorphisms [7], and are of interest in the theory of perfect graphs [3,4]. Many well-known problems seeking, say, clique cutsets, homogeneous sets, skew cutsets, joins, etc., can be formulated as list M-partition problems [8]. Moreover, the study of list M-partition problems can lead to efficient solutions of some of these problems [4].

In [8] we have given polynomial time algorithms for many list M-partition problems, and quasi-polynomial $(O(n^{\log n}))$ time algorithms for certain others. In [6] we have shown that all list M-partition problems are solvable in quasi-polynomial time, or are NP-complete. (We call such a result a *quasi-dichotomy.*) Many of our quasi-polynomial time algorithms from [8] were improved to polynomial time algorithms in [2,4], but it is not known whether all list M-partition problems are polynomial time solvable or NP-complete; this is known as the *Dichotomy Problem* for list M-partitions.

In this paper, we consider the restrictions of both the M-partition and the list M-partition problems to instances G that are chordal graphs. The two corresponding problems will be called the *chordal M-partition problem* and *chordal list M-partition problem*.

There are several classical examples to suggest that M-partitions of chordal graphs can be found in polynomial time. For instance, k-colorability of chordal graphs (M is the $k \times k$ matrix with 0 on the diagonal and $*$ everywhere else) can be decided efficiently using a perfect elimination ordering [9]; in fact, the algorithm either produces a k-coloring of the input graph or produces the unique forbidden subgraph K_{k+1}. A similar result is known about clique covering (M is the $\ell \times \ell$ matrix with 1 on the diagonal and $*$ elsewhere). In [10] we have shown more generally that there is a polynomial time recognition algorithm, and a forbidden subgraph characterization, of graphs that can be partitioned into k independent sets and ℓ cliques (M has k zeros and ℓ ones on the diagonal, $*$ everywhere else).

We further extend these results to the list M-partition problem. We also extend the class of matrices M for which we can give polynomial time algorithms, and forbidden subgraph characterizations. However, we also find M-partition problems that remain NP-complete for chordal graphs, even in the absence of lists. Certain dichotomy and quasi-dichotomy results will also be claimed.

2   Matrices M with $0, 1$ Diagonal

If the diagonal of the matrix M contains no $*$, we have several large classes of polynomially solvable list M-partition problems, including the list versions of the above problem of partitioning G into k independent sets and ℓ cliques.

Consider first the case where the $k \times k$ matrix M has zero diagonal.

Theorem 1. *If all diagonal entries of M are zero, then the chordal list M-partition problem can be solved in polynomial time.*

Proof. A chordal graph G which admits an M-partition with such a matrix M cannot have a clique with $k + 1$ vertices; hence it must have treewidth at most $k - 1$. The existence of a list M-partition of a graph of bounded treewidth can be tested by standard dynamic programming techniques [1,5,11]. Recall that a *tree decomposition* of a graph G is a pair (X, U) where U is a tree and $X = (X_i)_{i \in V(U)}$ is a collection of subsets of $V(G)$ whose union equals $V(G)$, such that each edge xy of G is included in some X_i, and such that for each vertex x of G, the set of all X_i containing x forms a subtree of G. The treewidth of a decomposition is the maximum value of $|X_i| - 1$, and the treewidth of a graph is the minimum treewidth of a decomposition.

A tree decomposition in which U has a fixed root r is called *nice* [1] if each node of the rooted tree U has at most two children, and the following conditions are satisfied: If i has two children, say j and h (a *join node*), then $X_i = X_j = X_h$; if i has one child j then X_i is obtained from X_j by adding (an *introduce node*) or deleting (a *forget node*) a single vertex of G, and if $|X_i| = 1$ for each leaf (*start node*) i of U. It is known that a nice tree decomposition of a chordal graph of bounded treewidth can be obtained in linear time [1].

Given a nice tree decomposition (X, U) of G with root r, we denote by G_i the subgraph of G induced by the union of X_i and all X_j where j is a descendant of i. Let $F(i)$ be the set of all pairs (Π, S), where Π is an assignment of the vertices in X_i to parts, obtained by restricting a list M-partition Σ of G_i, and S is the set of those parts in the partition Σ which contain vertices of $G_i - X_i$. Note that each $F(i)$ has at most $(2k)^k$ elements.

We can compute the set $F(i)$ for any node, once all its descendants j have had their values $F(j)$ calculated. This is not hard to see, considered separately the start, introduce, forget, and join nodes. For instance, suppose i is a forget node, with the unique child j, and $X_i = X_j - x$. For each $(\Pi, S) \in F(j)$ we add to $F(i)$ the pair (Π', S'), where Π' is Π restricted to X_i and S' equals either S, if the part a that x was assigned in Π was already present in S, or equals $S \cup a$. On the other hand, if i is an introduce node, with the unique child j and $X_j = X_i - x$, then for each $(\Pi, S) \in F(j)$ we consider all possible values x can take with the current assignment Π, because of the adjacencies of x in X_j, and also because of the non-adjacencies of x in $G_i - X_i$; it is for this purpose that we keep track of the set S.

The above proof yields an algorithm for the list M-partition problem restricted to graphs of treewidth at most $k - 1$ (and hence for all chordal graphs), of complexity $O(n(2k)^k)$; the complexity analysis is easily adapted from that of [5].

We next consider the case where the $\ell \times \ell$ matrix M has all diagonal entries one. Let G with lists $L(x)$ be an instance of the chordal list M-partition problem. A *rectangle* in G is a collection of sublists $L_x \subseteq L(x), x \in V(G)$, such that any choice of parts from L_x for each x constitutes a solution.

Theorem 2. *If all diagonal entries of M are one, then the chordal list M-partition problem can be solved in time polynomial in n^ℓ. The set of solutions to an instance is the union of at most $n^{2\ell}$ rectangles, and can be found in polynomial time.*

Proof. Consider a perfect elimination ordering of the graph. If there are ℓ parts, then choose ℓ pairs (x_i, y_i) of vertices in the input graph G, where x_i will be the first vertex in the perfect elimination ordering to go to part i, and y_i the last vertex in the perfect elimination ordering to go to part i. This involves $n^{2\ell}$ possible choices. For each choice, remove part i from the list of any vertex that occurs either before x_i or after y_i in the elimination ordering.

Remove from all lists of vertices z forbidden parts j given their adjacency or non-adjacency to the vertices x_i, y_i that go to part i according to what M requires. That is, vertex z cannot go to part j if there is an edge zx_i or an edge zy_i in G and the entry $M(i, j) = 0$. Similarly, vertex z cannot go to part j if there is no edge zx_i or no edge zy_i in G and the entry $M(i, j) = 1$.

Finally, assign parts to vertices from their resulting reduced lists L_x arbitrarily. Suppose z_i, z_j end up in parts i, j respectively and are adjacent, but $M(i, j) = 0$. Say z_i occurs before z_j in the perfect elimination ordering. Then z_i is adjacent to y_i, since $M(i, i) = 1$. Thus y_i and z_j are both neighbors of z_i, and both occur after z_i, so y_i is adjacent to z_j by the definition of a perfect elimination ordering. Since $M(i, j) = 0$, part j would have been removed from the list of z_j.

In the other case, suppose z_i, z_j end up in parts i, j respectively and are not adjacent, but $M(i, j) = 1$. Say z_i occurs before z_j. Then x_i is adjacent to z_i since $M(i, i) = 1$. Also x_i is adjacent to z_j since $M(i, j) = 1$. Thus x_i is adjacent to both z_i and z_j, and both occur after x_i, so z_i is adjacent to z_j by the definition of a perfect elimination ordering, a contradiction.

Thus we end up with $n^{2\ell}$ families of solutions, each family given only by restrictions on possible parts for each element, so that each family is a rectangle. $\square$

In the rest of the paper we often focus on $(k + \ell) \times (k + \ell)$ matrices M which consist of a $k \times k$ diagonal matrix A and an $\ell \times \ell$ diagonal matrix B, with an off-diagonal $k \times \ell$ matrix C (and its $\ell \times k$ transpose). We shall call such matrices A, B, C-*block* matrices.

Assume now that all diagonal entries of A are zero, and all diagonal entries of B are one. We shall also consider restrictions on C.

Feder, Hell, Klein, and Motwani [8] showed the following. Let $\mathcal{A}$ and $\mathcal{B}$ be two classes of graphs that are closed under taking induced subgraphs, and for which membership can be tested in polynomial time. Suppose further that there exists a constant c such that any graph both in $\mathcal{A}$ and $\mathcal{B}$ has at most c vertices. They consider the question of partitioning the vertices of a graph G into two sets S_A and S_B so that the subgraph G_A induced by S_A is in $\mathcal{A}$, and the subgraph G_B induced by S_B is in $\mathcal{B}$. They show that there are at most n^{2c} such partitions, and that all such partitions can be found in polynomial time.

In our application, we let $\mathcal{A}$ be the class of graphs without a clique with $k+1$ vertices, and $\mathcal{B}$ the class of graphs without an independent set of $\ell+1$ vertices. A chordal graph without a clique with $k+1$ vertices and without an independent set with $\ell+1$ vertices has at most $c = k\ell$ vertices, since it is k-colorable, and thus a union of k independent sets.

Given an instance G, let S_A be the set of vertices that are placed in the parts corresponding to the $k \times k$ matrix A, and let S_B be the set of vertices that go to the parts corresponding to the $\ell \times \ell$ matrix B. It follows that $S_A \in \mathcal{A}, S_B \in \mathcal{B}$.

Suppose first the k by ℓ matrix C is all $*$. Then for each of the n^{2c} valid partitions into two graphs G_A and G_B, we can restrict the lists for G_A to parts in A and solve the problem for matrix A on G_A with the algorithm of Theorem 1. Similarly, we can restrict the lists for G_B to parts in B and solve the problem for matrix B on G_B with the algorithm of Theorem 2.

More generally, we call a matrix C *horizontal* if all entries of C corresponding to a part i in A are the same, and *vertical* if all entries of C corresponding to a part j in B are the same. Finally, we call matrix C *crossed* if the entries of C are all 0 or $*$ (or all 1 or $*$) and every zero (respectively every one) belongs to either a row or a column of all zeros (respectively all ones).

Theorem 3. *Suppose M is an A, B, C-block matrix.*

If all diagonal entries of A are zero, all diagonal entries of B are one, and if C is either horizontal, vertical, or crossed, then the chordal list M-partition can be solved in time polynomial in $n^{k\ell}$.

Proof. For each choice of G_A and G_B, if all entries of C corresponding to a part i in A are zero (respectively one) then it suffices to remove the part i from the list of any vertex v of G_A that has a neighbor in G_B (respectively for which some vertex of G_B is not a neighbor). Similarly, if all entries of C corresponding to a part j in B are zero (respectively one) then it suffices to remove the part j from the list of any vertex v of G_B that has a neighbor in G_A (respectively for which some vertex of G_A is not a neighbor). Once the conditions given by C are met, we can replace C by an all $*$ matrix and solve the problem for G_A and G_B using Theorems 1 and 2.

If C is vertical, then the complexity can be improved to $n^{2l+O(1)}(2k)^k$.

We can generalize this result to matrices A which consist of diagonal blocks A_i with zero diagonals, and matrices B which consist of diagonal blocks B_i with all diagonal entries one, as long as all entries of A not in the diagonal blocks are one, all entries of B not in the diagonal blocks are zero, and all block matrices $C_{i,j}$ of C corresponding to A_i, B_j are either horizontal, vertical, or crossed.

3   NP-Complete Problems

Consider a fixed bipartite graph H. The *list H-coloring problem* is defined as follows: An instance is a bipartite graph G with lists (white vertices of G have lists consisting of white vertices of H and similarly for black vertices), and a solution is a mapping of vertices of G to vertices of H so that adjacency is preserved and each vertex of G is mapped to a member of its list. (Such a mapping is called a *list H-coloring* of G.) Feder, Hell and Huang [7] showed that the list H-coloring problem is polynomial time solvable if the bipartite graph H is the complement of a circular arc graph (*a cocircular graph*), and is NP-complete otherwise. Based on this result, it will be possible to find NP-complete chordal list M-partition problems.

Given a bipartite graph H with k white vertices (forming the set V_A) and ℓ black vertices (forming the set V_B), the *matrix corresponding to H* is the $k \times \ell$ matrix C with $C(i, j) = *$ if ij is an edge in H (with $i \in V_A, j \in V_B$), and with $C(i, j) = 0$ otherwise.

Theorem 4. *Let M be an A, B, C-block matrix.*

Suppose A does not contain any 1's, and B does not contain any 0's. If C is the matrix corresponding to a bipartite graph H that is not a cocircular graph, then the chordal list M-partition problem is NP-complete.

Proof. Consider an instance G of the list H-coloring problem, and define the graph G' to be obtained from G by adding all edges between pairs of black vertices. (The lists of G' remain the same as in G.) It is easy to see that G has a list H-coloring if and only if G' has a list M-partition. Since G' is a split graph (it can be partitioned into a clique and an independent set), it is also chordal [9].

The same result holds if C is obtained from the matrix corresponding to a bipartite graph H by replacing each 0 with a 1. (This follows by replacing the bipartite graph G' by the bipartite complement G'' of G.)

The proof implies that the list M-partition problems corresponding to graphs that are not cocircular are NP-complete even for split graphs. It is easy to see that, in the special case when A is an all zero matrix and B is an all one matrix, we in fact obtain the following dichotomy (again, valid also for matrices C obtained by replacing all zeros by ones):

Theorem 5. *Let M be an A, B, C-block matrix.*

If $A = 0$, $B = 1$, and C is the matrix corresponding to a bipartite graph H, then the chordal list M-partition problem is polynomial if H is a cocircular graph and is NP-complete otherwise.

A similar quasi-dichotomy result can be derived from the theorem of Feder and Hell [6], who showed that on general instances, all list M-partition problems are quasi-polynomial or NP-complete. In particular, if A and B are as above (all-zero and all-one matrices), and C is an arbitrary matrix (not necessarily corresponding to a graph H), then it can be shown that in fact the chordal list M-partition problem is quasi-polynomial or NP-complete.

Several generalizations of these dichotomy and quasi-dichotomy results can be proved: It is enough to assume, for instance, that B (instead of being an all one matrix) has ones on the diagonal and no zeros. The quasi-dichotomy also applies if B is only assumed to have ones on the diagonal, as long as A has zeros on the diagonal and no $*$'s. In this case, if additionally C has no zeros (or no ones), we have dichotomy. These results will be proved elsewhere.

We now focus on constructing NP-complete M-partition problems (without lists). Let H again be a bipartite graph. The H-*retraction* problem is the restriction of the list H-coloring problem to instances G containing H as a subgraph, and with lists either $L(g) = g$, if $g \in V(H)$, or $L(g) = V(H)$, otherwise. A list H-coloring of G is called an H-*retraction* of G, in this situation. Many bipartite graphs H are known to yield NP-complete H-retraction problems, although a complete classification of complexity is not known, and dichotomy has not been proved, for H-retractions. In particular, it is known that if H is an even cycle of length greater than four, the H-retraction problem is NP-complete [7].

Theorem 6. *For every bipartite graph H such that the H-retraction problem is NP-complete, there exists a matrix M_H such that the M_H-partition problem (without lists) is also NP-complete.*

Proof. Let H be a bipartite graph such that the H-retraction problem is NP-complete. We first extend the graph H to a larger bipartite graph H', by attaching to each white vertex of H a path of length five and to each black vertex of H a path of length four. Note that all the leaves (vertices of degree one) of H' are black.

We now introduce an auxiliary problem, which we shall call the *weak H'-retraction problem*. Suppose that the bipartite graph H' has k black vertices, forming the set V_B, and let L denote the set of all black leaves of H'. An instance of the weak H-retraction problem is a bipartite graph G with a specified set X of k black vertices, such that each vertex of G not in X has at most one neighbour in X. A solution to the instance is an edge-preserving and color-preserving mapping of the vertices of G to the vertices of H such that X is mapped bijectively to V_B. We now show that the H-retraction problem reduces to the weak H'-retraction problem.

Suppose G is an instance of the H-retraction problem, i.e., a bipartite graph containing H. We transform G to an instance G' (with a set X) of the weak H'-retraction problem as follows: Let X be another copy of the set V_B, disjoint from G. Consider the union of G and X, and identify each vertex of L in X with the corresponding vertex of L in G. Finally, add internally disjoint paths of length four joining all pairs of vertices of X which correspond to vertices in V_B of distance two or four in H'. Call the resulting graph G'. We now argue that G admits an H-retraction if and only if G' admits a weak H'-retraction.

On the one hand, suppose f is an H-retraction of G. Then f, extended by taking each vertex of $X - L$ to the corresponding vertex of V_B is a weak H'-

retraction of G'. For the other direction, we note that any bijection between X and V_B has to map vertices of L to vertices of L, since leaves in H' have exactly two vertices in H' at distance two or four, while black vertices of H' that are not leaves have at least three vertices in H' at distance two or four. Therefore, any weak H'-retraction of G' which maps the vertices of X bijectively to the vertices of V_B must map the copy of H' in G' isomorphically to H'. It follows that G admits an H'-retraction, which can easily be modified to an H-retraction by mapping all the added paths of H' into H.

Next, we define a matrix M_H such that the chordal M_H-partition problem (without lists) is NP-complete, as claimed in the theorem. The matrix M_H will be an A, B, C-block matrix in which the diagonal matrix A is an all zero matrix; the diagonal matrix B has all diagonal entries one and all other entries $*$; and finally, the matrix C will be the matrix corresponding to the bipartite graph H'.

We now reduce the weak H'-retraction problem to the M_H-partition problem. Given an instance G' for the weak H'-retraction problem, we construct an instance G'' of the M_H-partition problem as follows. We replace each white vertex a of G' by a set $I(a)$ of $k + 1$ independent vertices (where $k = |V_B|$), and each black vertex b of G' by a clique $K(b)$ of two vertices. Whenever a and b are adjacent in G', all vertices of I_a are adjacent to all vertices of K_b in G''. Finally, we add all edges between K_b and $K_{b'}$ unless both b and b' are in X. Note that each vertex every $I(a)$ is adjacent to at most one $K(b)$ with $b \in X$.

We claim that G' admits a weak H'-retraction if and only if G'' admits an M_H-partition. Indeed, if f is a weak H'-retraction of G', all vertices of a set $I(a)$ can be placed in the part $f(a)$ and all vertices of a set $K(b)$ can be placed in the part $f(b)$. Conversely, each M_H-partition of G'' must place at least one of the two vertices in any $K(b)$ to a part in B, since A is an all-zero matrix. Also, if b, b' are both in X, these vertices must be placed in distinct parts of B. By a similar argument, at least one vertex of each $I(a)$ must be placed in a part in A, since the vertices placed to parts in B are covered by k cliques. This way we deduce an H'-retraction of G'.

It remains to argue that the instance G'' is a chordal graph. We first note that each vertex of every $I(a)$ is only adjacent to vertices in $K(b)$ with $b \notin X$ expect possibly in one $K(b)$ with $b \in X$. According to the definition of G'', these vertices are all mutually adjacent, i.e., a clique. Thus we can repeatedly remove simplicial vertices (vertices whose neighbours form a clique) from the sets $I(a)$, until G'' is reduced to the union of the $K(b)$, which is clearly chordal.

4   Conclusions

We also have some forbidden subgraph charaterizations of M-partitionable chordal graphs. It is well-known, for example, that a chordal graph G is k-colorable if and only if it does not contain a K_{k+1}. In [8] we have extended this as follows: A chordal graph G can be partitioned into k cliques and ℓ independent sets if and only if it does not contain an induced subgraph isomorphic to $(\ell + 1)K_{k+1}$. For many other matrices M it is possible to characterize non-M-

partitionable chordal graphs G by a finite number of forbidden subgraphs. At the same time, it follows from our results that, unless P=NP, this is not the case for all matrices M.

Once again, we shall consider only A, B, C-block matrices M, with A having zero diagonal and B having a diagonal of ones. Moreover, we shall assume that all off-diagonal entries of A are the same, say a, all off-diagonal entries of B are the same, say b, and all entries of C are the same, say c. Note that we may assume $a \neq 0$ and $b \neq 1$, otherwise we may replace M by a matrix with $k = 1$ or $\ell = 1$ respectively. The result of [10] states that when $a = b = c = *$, a chordal graph is non-M-partitionable if and only if it contains in induced subgraph isomorphic to $(\ell + 1)K_{k+1}$.

We can show that if $c \neq *$, the non-M-partitionable chordal graphs can always be characterized by a finite number of forbidden subgraphs, all with at most $(k + 1)(\ell + 1)$ vertices.

This bound does not always apply: if $c = *$, we know that in the particular case of $k = 1$ and $b = 0$, the largest minimal forbidden subgraph has on the order of ℓ^2 vertices.

Nevertheless, even in the case $c = *$ we can prove that there is always only a finite number of obstructions. If $a = 1$, the best bounds we currently have for the size of minimal obstructions are t^k, where $t = O(\ell)$ if $b = *$, and $t = O(\ell^2)$ if $b = 0$. If the remaining case $a = *, b = 0$, we have the bound $2(k + 1)^{(k+2)\ell+1}$.

We will return to these results in a future paper.

References

1. H. L. Bodlaender and T. Kloks, Efficient and constructive algorithms for the pathwidth and treewidth of graphs, J. Algorithms 21 (1996) 358–402.
2. K. Cameron, E. M. Eschen, C. T. Hoang, and R. Sritharan, The list partition problem for graphs, SODA 2004.
3. M. Chudnovsky, N. Robertson, P. Seymour, and R. Thomas, The strong perfect graph theorem, manuscript, 2002.
4. C. M. H. de Figueiredo, S. Klein, Y. Kohayakawa, and B. A. Reed, Finding skew partitions efficiently, Journal of Algorithms 37 (2000) 505-521.
5. J. Diaz, M. Serna, and D. M. Thilikos, The complexity of parametrized H-colorings: a survey, Dimacs Series in Discrete Mathematics, 2003.
6. T. Feder and P. Hell, List constraint satisfaction and list partition, submitted to SIAM J. Comput. (2003).
7. T. Feder, P. Hell, and J. Huang, List homomorphisms and circular arc graphs, Combinatorica 19 (1999) 487–505.
8. T. Feder, P. Hell, S. Klein, and R. Motwani, Complexity of list partitions, Proc. 31st Ann. ACM Symp. on Theory of Computing (1999) 464–472. SIAM J. Comput., in press.
9. M. C. Golumbic, Algorithmic Graph Theory and Perfect Graphs, Academic Press, New York, 1980.
10. P. Hell, S. Klein, L.T. Nogueira, F. Protti, Partitioning chordal graphs into independent sets and cliques, Discrete Applied Math., to appear.
11. A. Proskurowski and S. Arnborg, Linear time algorithms for NP-hard problems restricted to partial k-trees, Discrete Applied Math. 1989 (23) 11–24.

Bidimensional Parameters and Local Treewidth[*]

Erik D. Demaine[1], Fedor V. Fomin[2], Mohammad Taghi Hajiaghayi[1], and
Dimitrios M. Thilikos[3]

[1] MIT Laboratory for Computer Science, 200 Technology Square, Cambridge,
Massachusetts 02139, USA, {edemaine,hajiagha}@mit.edu
[2] Department of Informatics, University of Bergen, N-5020 Bergen, Norway,
fomin@ii.uib.no
[3] Departament de Llenguatges i Sistemes Informàtics, Universitat Politècnica de
Catalunya, Campus Nord – Mòdul C5, c/Jordi Girona Salgado 1-3, E-08034,
Barcelona, Spain, sedthilk@lsi.upc.es

Abstract. For several graph theoretic parameters such as vertex cover
and dominating set, it is known that if their values are bounded by k
then the treewidth of the graph is bounded by some function of k. This
fact is used as the main tool for the design of several fixed-parameter
algorithms on minor-closed graph classes such as planar graphs, single-
crossing-minor-free graphs, and graphs of bounded genus. In this paper
we examine the question whether similar bounds can be obtained for
larger minor-closed graph classes, and for general families of parameters
including all the parameters where such a behavior has been reported so
far.
Given a graph parameter P, we say that a graph family $\mathcal{F}$ has the
parameter-treewidth property for P if there is a function $f(p)$ such that
every graph $G \in \mathcal{F}$ with parameter at most p has treewidth at most
$f(p)$. We prove as our main result that, for a large family of parameters called *contraction-bidimensional parameters*, a minor-closed graph
family $\mathcal{F}$ has the parameter-treewidth property if $\mathcal{F}$ has bounded local treewidth. We also show "if and only if" for some parameters, and
thus this result is in some sense tight. In addition we show that, for
a slightly smaller family of parameters called *minor-bidimensional parameters*, all minor-closed graph families $\mathcal{F}$ excluding some fixed graphs
have the parameter-treewidth property. The bidimensional parameters
include many domination and covering parameters such as vertex cover,
feedback vertex set, dominating set, edge-dominating set, q-dominating
set (for fixed q). We use these theorems to develop new fixed-parameter
algorithms in these contexts.

1   Introduction

The last ten years has witnessed the rapid development of a new branch of computational complexity, called parameterized complexity; see the book of Downey

---

[*] The last author was supported by EC contract IST-1999-14186: Project ALCOM-
FT (Algorithms and Complexity) – Future Technologies and by the Spanish CICYT
project TIC-2002-04498-C05-03 (TRACER)

M. Farach-Colton (Ed.): LATIN 2004, LNCS 2976, pp. 109–118, 2004.

& Fellows [14]. Roughly speaking, a parameterized problem with parameter k is *fixed-parameter tractable (FPT)* if it admits an algorithm with running time $f(k)|I|^{O(1)}$. (Here f is a function depending *only* on k and $|I|$ is the size of the instance.)

A celebrated example of a fixed-parameter tractable problem is VERTEX COVER, asking whether an input graph has at most k vertices that are incident to all its edges. When parameterized by k, the k-VERTEX COVER problem admits a solution as fast as $O(kn + 1.285^k)$ [7]. Moreover, if we restrict k-VERTEX COVER to planar graphs then it is possible to design FPT-algorithms where the contribution of k in the non-polynomial part of their complexity is subexponential. The first algorithm of this type was given by Alber et al. (see [2]). Recently, Fomin and Thilikos reported a $O(k^4 + 2^{4.5\sqrt{k}} + kn)$ algorithm for planar k-VERTEX COVER [19].

However, not all parameterized problems are fixed-parameter tractable. A typical example of such a problem is DOMINATING SET, asking whether an input graph has at most k vertices that are adjacent to the rest of the vertices. When parameterized by k, the k-DOMINATING SET Problem is known to be $W[2]$-complete and thus it is not expected to be fixed-parameter tractable. Interestingly, the fixed-parameter complexity of the same problem can be distinct for special graph classes. During the last five years, there has been substantial work on fixed-parameter algorithms for solving the k-DOMINATING SET on planar graphs and different generalizations of planar graphs. For planar graphs Downey and Fellows [14], suggested an algorithm with running time $O(11^d n)$. Later the running time was reduced to $O(8^d n)$ [2]. An algorithm with a sublinear exponent for the problem with running time $O(4^{6\sqrt{34d}} n)$ was given by Alber et al. [1]. Recently, Kanj & Perkovuć [23] improved the running time to $O(2^{27\sqrt{d}} n)$ and Fomin & Thilikos to $O(2^{15.13\sqrt{d}} d + n^3 + d^4)$ [18]. The fixed-parameter algorithms for extensions of planar graphs like bounded-genus graphs and graphs excluding single-crossing graphs as minors are introduced in [11,9,15].

In the majority of these results, the design of FPT algorithms for solving problems such as k-VERTEX COVER or k-DOMINATING SET in a sparse graph class $\mathcal{F}$ is based on the following lemma: every graph G in $\mathcal{F}$ where the value of the parameter is at most p has treewidth bounded by $f(p)$, where f is a function depending only on $\mathcal{F}$. With some work (sometimes very technical), a tree decomposition of width $O(f(p))$ is constructed and standard dynamic-programming techniques on graphs of bounded treewidth are implemented. Of course this method can not be applied for any graph class $\mathcal{F}$. For instance, the n-vertex complete graph K_n has a dominating set of size one and treewidth equal to $n - 1$. So the emerging question is: For which (larger) graph classes and for which parameters can the "bounding treewidth method" be applied? In this paper we give a *complete* characterization of minor-closed graph families for which the aforementioned "bounding treewidth method" can be applied for a wide family of graph parameters. For a given parameter P, we say that a graph family $\mathcal{F}$ has the *parameter-treewidth property* for P if there is a function $f(p)$ such for every graph $G \in \mathcal{F}$ where $P(G) \leq p$ implies that G has treewidth at most $f(p)$. Our main result is that for a large family of parameters called *contraction-bidimensional parameters*, a minor-closed graph family $\mathcal{F}$ has the parameter-treewidth property if $\mathcal{F}$ has bounded local treewidth. Moreover, we show that the inverse is also correct if some simple condition is satisfied by P. In addition we show that, for a slightly smaller family of parameters called *minor-bidimensional parameters*, every minor-closed graph family $\mathcal{F}$ excluding some fixed graph has the parameter-treewidth property. The bidimensional-parameter family includes many domination and covering parameters such as vertex cover, feedback vertex set, dominating set, edge-dominating set, and q-dominating set (for fixed q) (see also [11] for more examples).

The proof of the main result uses the characterization of Eppstein for minor-closed families of bounded local treewidth [16] and Diestel et al.'s modification of the Robertson & Seymour excluded-grid-minor theorem [13]. In addition, the proof is constructive and can be used for constructing fixed-parameter algorithms to decide bidimensional parameters on minor-closed families of bounded local treewidth. In this sense, we extend to fixed-parameter algorithms the result of Frick & Grohe [21] that, for each property ϕ definable in first-order logic, and for each class of minor-closed graphs of bounded local treewidth, there is a (non-fixed-parameter) $O(n^{1+\epsilon})$-time algorithm deciding whether a given graph has property ϕ.

A preliminary and special case of our result, concerning only the dominating set parameter, appeared in [20] with a different and more complicated proof. Also, another proof of the same result appeared in [10]. In this paper we present shorter and more elegant proofs of the combinatorial results of [20] and [10] while we extend their applicability to general families of parameters.

2   Definitions and Preliminary Results

Let G be a graph with vertex set $V(G)$ and edge set $E(G)$. We let n denote the number of vertices of a graph when it is clear from context. For every nonempty $W \subseteq V(G)$, the subgraph of G induced by W is denoted by $G[W]$. We define the *q-neighborhood* of a vertex $v \in V(G)$, denoted by $N_G^q[v]$, to be the set of vertices of G at distance at most q from v. Notice that $v \in N_G^q[v]$. We put $N_G[v] = N_G^1[v]$. We also often say that a vertex v *dominates* subset $S \subseteq V(G)$ if $N_G[v] \supseteq S$.

Given an edge $e = \{x, y\}$ of a graph G, the graph G/e is obtained from G by contracting the edge e; that is, to get G/e we identify the vertices x and y and remove all loops and duplicate edges. A graph H obtained by a sequence of edge contractions is said to be a *contraction* of G. A graph H is a *minor* of a graph G if H is the subgraph of a contraction of G. We use the notation $H \preceq G$ [resp. $H \preceq_c G$] for H a minor [a contraction] of G. A family (or class) of graphs $\mathcal{F}$ is *minor-closed* if $G \in \mathcal{F}$ implies that every minor of G is in $\mathcal{F}$. A minor-closed graph family $\mathcal{F}$ is *H-minor-free* if $H \notin \mathcal{F}$.

The $m \times m$ grid is the graph on $\{1, 2, \ldots, m^2\}$ vertices $\{(i, j) \colon 1 \le i, j \le m\}$ with the edge set

$$\{(i, j)(i', j') \colon |i - i'| + |j - j'| = 1\}.$$

For $i \in \{1, 2, \ldots, m\}$ the vertex set (i, j), $j \in \{1, 2, \ldots, m\}$, is referred as the *ith row* and the vertex set (j, i), $j \in \{1, 2, \ldots, m\}$, is referred to as the *ith column* of the $m \times m$ grid. The vertices (i, j) of the $m \times m$ grid with $i \in \{1, m\}$ or $j \in \{1, m\}$ are called *boundary* vertices and the rest of the vertices are called *non-boundary* vertices.

The notion of treewidth was introduced by Robertson and Seymour [25]. A *tree decomposition* of a graph G is a pair $(\{X_i \mid i \in I\}, T = (I, F))$, with $\{X_i \mid i \in I\}$ a family of subsets of $V(G)$ and T a tree, such that

1. $\bigcup_{i \in I} X_i = V(G)$;
2. for all $\{v, w\} \in E(G)$, there is an $i \in I$ with $v, w \in X_i$; and
3. for all $i_0, i_1, i_2 \in I$, if i_1 is on the path from i_0 to i_2 in T, then $X_{i_0} \cap X_{i_2} \subseteq X_{i_1}$.

The *width* of the tree decomposition $(\{X_i \mid i \in I\}, T = (I, F))$ is $\max_{i \in I} |X_i| - 1$. The treewidth $\mathbf{tw}(G)$ of a graph G is the minimum width of a tree decomposition of G.

We need the following facts about treewidth. The first fact is trivial.

– For any complete graph K_n on n vertices, $\mathbf{tw}(K_n) = n - 1$.

The second fact is well known but its proof is not trivial. (See e.g., [12].)

– The treewidth of the $m \times m$ grid is m.

The next fact we need is the improved version of the Robertson & Seymour theorem on excluded grid minors [26] due to Diestel et al. [13]. (See also the textbook [12].)

Theorem 1 ([13]). *Let r, m be integers, and let G be a graph of treewidth at least $m^{4r^2(m+2)}$. Then G contains either K_r or the $m \times m$ grid as a minor.*

A *parameter* P is any function mapping graphs to nonnegative integers. The *parameterized problem* associated with P asks, for a fixed k, whether $P(G) \leq k$ for a given graph G.

A parameter P is $g(r)$-*minor-bidimensional* if (i) contracting an edge, deleting an edge, or deleting a vertex in a graph G cannot increase $P(G)$, and (ii) there exists a function g such that, for the $r \times r$ grid R, $P(R) \geq g(r)$. Similarly, a parameter P is $g(r)$-*contraction-bidimensional* if (i) contracting an edge in a graph G cannot increase $P(G)$, and (ii) there exists a function g such that, for any $r \times r$ augmented grid R of constant span, $P(R) \geq g(r)^1$. Here an $r \times r$ *augmented grid of span s* is an $r \times r$ grid with some extra edges such that each vertex is attached to at most s non-boundary vertices of the grid. We assume that $g(r)$ is monotone and invertible for $r \geq 0$. We note that a $g(r)$-minor-bidimensional parameter is also a $g(r)$-contraction-bidimensional parameter. One can easily

---

[1] Closely related notions of bidimensional parameters are introduced by the authors in [9].

observe that many parameters such as minimum sizes of dominating set, q-dominating set (distance q-dominating set for a fixed q), vertex cover, feedback vertex set, and edge-dominating set (see exact definitions of the corresponding parameters in [11]) are $\Theta(r^2)$-minor- or $\Theta(r^2)$-contraction-bidimensional parameters. Another example of contraction-bidimensional parameter is the minimum length in TSP (Travelling salesman problem), i.e. the smallest number of edges in a walk containing all vertices of a graph.

Here, we present a theorem for minor-bidimensional parameters on general minor-closed classes of graphs excluding some fixed graphs, whose intuition plays an important role in the main result of this paper.

Theorem 2. *If a $g(r)$-minor-bidimensional parameter P on an H-minor-free graph G has value at most p, then $\mathbf{tw}(G) \leq 2^{O(g^{-1}(p)\log g^{-1}(p))}$. (The constant in the O notation depends on H.)*

Proof. By Theorem 1, since G is H-minor-free (and thus $K_{|V(H)|}$-minor-free), we know if m is the largest integer such that $\mathbf{tw}(G) \geq m^{4|V(H)|^2(m+2)}$, then G has an $m \times m$ grid as a minor. Since P is $g(r)$-minor-bidimensional, $p \geq g(m)$ and thus we obtain the desired bound.

Theorem 2 can be applied for minor-bidimensional parameters such as vertex cover or feedback vertex set.

The notion of local treewidth was introduced by Eppstein [16] (see also [22]). The *local treewidth* of a graph G is

$$\mathbf{ltw}(G, r) = \max\{\mathbf{tw}(G[N_G^r[v]]) \colon v \in V(G)\}.$$

For a function $f \colon N \to N$ we define the minor-closed class of graphs of bounded local treewidth

$$\mathcal{L}(f) - \{G \colon \vee H \preceq G \ \forall r \geq 0, \ \mathbf{ltw}(H, r) \leq f(r)\}.$$

Also we say that a minor-closed class of graphs $\mathcal{C}$ has bounded local treewidth if $\mathcal{C} \subseteq \mathcal{L}(f)$ for a function f.

Well-known examples of minor-closed classes of graphs of bounded local treewidth are graphs of bounded treewidth, planar graphs, graphs of bounded genus, and single-crossing-minor-free graphs.

Many difficult graph problems can be solved efficiently when the input is restricted to graphs of bounded treewidth (see e.g., Bodlaender's survey [5]). Eppstein [16] made a step forward by proving that some problems like subgraph isomorphism and induced subgraph isomorphism can be solved in linear time on minor-closed graphs of bounded local treewidth. Also the classic Baker's technique [4] for obtaining approximation schemes on planar graphs for different NP-hard problems can be generalized to minor-closed families of bounded local treewidth. (See [16] for a generalization of these techniques.)

An *apex graph* is a graph G such that, for some vertex v (the *apex*), $G - v$ is planar. The following result is due to Eppstein [16].

Theorem 3 ([16]). *Let $\mathcal{F}$ be a minor-closed family of graphs. Then $\mathcal{F}$ is of bounded local treewidth if and only if $\mathcal{F}$ does not contain all apex graphs.*

3   Main Theorem

Due to space restriction we omit the proofs of the following two combinatorial lemmas.

Lemma 1. *Suppose we have a $m \times m$ grid H and a subset S of vertices in the central $(m - 2k) \times (m - 2k)$ subgrid H', where $s = |S|$ and $k = \lfloor \sqrt[4]{s} \rfloor$. Then H has as a minor the $k \times k$ grid R such that each vertex in R is a contraction of at least one vertex in S and other vertices in H.*

Lemma 2. *Let $G \in \mathcal{L}(f)$ be a graph containing the $m \times m$ grid H as a subgraph, $m > 2k$, where $k = f(2) + 1$. Then the central $(m - 2k) \times (m - 2k)$ subgrid H' has the property that every vertex $v \in V(G)$ is adjacent to less than k^4 vertices in H'.*

Now we are ready to present the main result of this paper.

Theorem 4. *Let P be a $g(r)$-contraction-bidimensional parameter. Then for any function $f \colon \mathbb{N} \to \mathbb{N}$ and any graph $G \in \mathcal{L}(f)$ on which parameter P has value at most p, we have $\mathbf{tw}(G) \leq 2^{O(g^{-1}(p) \log g^{-1}(p))}$. (The constant in the O notation depends on $f(1)$ and $f(2)$.)*

Proof. Let $r = f(1) + 1$ and $k = f(2) + 1$. Let $G \in \mathcal{L}(f)$ be a graph on which the parameter P has value p. Let m be the largest integer such that $\mathbf{tw}(G) \geq m^{4r^2(m+2)}$. Without loss of generality, we assume G is connected, and $m > 2k$ (otherwise, $\mathbf{tw}(G)$ is a constant since both r and k are constants.) Then G has no complete graph K_r as a minor. By Theorem 1, G contains an $m \times m$ grid H as a minor. Thus there exists a sequence of edge contractions and edge/vertex deletions reducing G to H. We apply to G the edge contractions from this sequence, we ignore the edge deletions, and instead of deletion of a vertex v, we only contract v into one of its neighbors. Call the new graph G', which has the $m \times m$ grid H as a subgraph and in addition $V(G') = V(H)$. Since parameter P is contraction-bidimensional, its value on G' will not increase. By Lemma 2, we know that the central $(m - 2k) \times (m - 2k)$ subgrid H' of H has the property that every vertex $v \in V(G')$ is adjacent to less than k^4 vertices in H'.

Now, suppose in graph G', we further contract all $2k$ boundary rows and $2k$ boundary columns into two boundary rows and two boundary columns (one on each side) and call the new graph G''. Note that here G'' and H' have the same set of vertices. The degree of each vertex of G'' to the vertices that are not on the boundary is at most $(k + 1)^2 k^4$, which is a constant since k is a constant. Here the factor $(k + 1)^2$ is for the boundary vertices each of which is obtained by contraction of at most $(k + 1)^2$ vertices. Again because parameter P is contraction-bidimensional, its value on G'' does not increase and thus it is at most p. On the other hand, since the parameter is $g(r)$-contraction-bidimensional, its value on graph G'' is at least $g(m - 2k)$. Thus $g^{-1}(p) \geq m - 2k$, so $m = O(g^{-1}(p))$. Therefore, the treewidth of the original graph G is at most $2^{O(g^{-1}(p) \log g^{-1}(p))}$ as desired.

A direct corollary of Theorem 4 is the following.

Lemma 3. *Let P be a contraction-bidimensional parameter. A minor-closed graph class $\mathcal{F}$ has the parameter-treewidth property for P if $\mathcal{F}$ is of bounded local treewidth.*

The *apex graphs* A_i, $i = 1, 2, 3, \ldots$, are obtained from the $i \times i$ grid by adding a vertex v adjacent to all vertices of the grid. It is interesting to see that, for a wide range of parameters, the inverse of Lemma 3 also holds.

Lemma 4. *Let P be any contraction-bidimensional parameter where $P(A_i) = O(1)$ for any $i \geq 1$. A minor-closed graph class $\mathcal{F}$ has the parameter-treewidth property for P only if $\mathcal{F}$ is of bounded local treewidth.*

Proof. The proof follows from Theorem 3. The apex graph A_i, has diameter ≤ 2 and treewidth $\geq i$. So a minor-closed family of graphs with the parameter-treewidth property for P cannot contain all apex graphs and hence it is of bounded local treewidth.

Typical examples of parameters satisfying Lemmas 3 and 4 are dominating set and its generalization q-dominating set, for a fixed constant q (in which each vertex can dominate its q-neighborhood). These parameters are $\Theta(r^2)$-contraction-bidimensional and their value is 1 for any apex graph $A_i, i \geq 1$.

We can strengthen the "if and only if" result provided by Lemmas 3 and 4 with the following lemma. We just need to use the fact that if the value of P is less than the value of P' then the parameter-treewidth property for P implies the parameter-treewidth property for P' as well.

Lemma 5. *Let P be a parameter whose value is lower bounded by some contraction-bidimensional parameter and let $P(A_i) = O(1)$ for any $i \geq 1$. Then a minor-closed graph class $\mathcal{F}$ has the parameter-treewidth property for P if and only if $\mathcal{F}$ is of bounded local treewidth.*

Lemma 5 can apply for parameters that are not necessarily contraction-bidimensional. As an example we mention the *clique-transversal number* of a graph, i.e., the minimum number of vertices meeting all the maximal cliques of a graph.[2] It is easy to see that this parameter always exceeds the domination number (the size of a minimum dominating set) and that any graph in A_i has a clique-transversal set of size 1.

Another application is the *Π-domination number*, i.e., the minimum cardinality of a vertex set that is a dominating set of G and satisfies some property Π in G. If this property is satisfied for any one-element subset of $V(G)$ then we call it *regular*. Examples of known variants of the parameterized dominating set problem corresponding to the Π-domination number for some regular property

---

[2] The clique-transversal number is not contraction-bidimensional because an edge contraction may create a new maximal clique and the value of the clique-transversal number may increase.

Π are the following parameterized problems: the independent dominating set problem, the total dominating set problem, the perfect dominating set problem, and the perfect independent dominating set problem (see the exact definitions in [1]).

We summarize the previous observations with the following:

Corollary 1. *Let P be any of the following parameters: the minimum cardinality of a dominating set, the minimum cardinality of a q-dominating set (for any fixed q), the minimum cardinality of a clique-transversal set, or the minimum cardinality of a dominating set with some regular property Π. A minor-closed family of graphs $\mathcal{F}$ has the parameter-treewidth property for P if and only if $\mathcal{F}$ is of bounded local treewidth. The function $f(p)$ in the parameter-treewidth property is $2^{O(\sqrt{p}\log p)}$.*

4   Algorithmic Consequences and Concluding Remarks

Courcelle [6] proved a meta-theorem on graphs of bounded treewidth; he showed that, if ϕ is a property of graphs that is definable in monadic second-order logic, then ϕ can be decided in linear time on graphs of bounded treewidth. Frick and Grohe [21] extended this result to graphs of bounded local treewidth; they showed that, for each property ϕ that is definable in first-order logic and for each minor-closed class of graphs of bounded local treewidth, there is an $O(n^{1+\epsilon})$-time algorithm deciding whether a given graph has property ϕ. However Frick & Grohe's proof is not constructive. It uses a transformation of a first-order logic formula into a "local formula" according to Gaifman's theorem and even the complexity of this transformation is unknown.

Using Theorems 2 and 4, we can extend the result of Frick & Grohe for fixed-parameter algorithms and show that any minor-bidimensional property that is solvable in polynomial time on graphs of bounded treewidth is also fixed-parameter tractable on general minor-closed graph families excluding some fixed graphs, and similarly for any contraction-bidimensional property on minor-closed graph families of bounded local treewidth. In contrast to the work of Frick & Grohe, the running time of our algorithm is explicit.

Theorem 5. *Let P be a parameter such that, given a tree decomposition of width at most w for a graph G, the parameter can be decided in $h(w)n^{O(1)}$ time. Now, if P is a $g(r)$-minor-bidimensional parameter and G belongs to a minor-closed graph family excluding some fixed graphs, or P is a $g(r)$-contraction-bidimensional parameter and G belongs to a minor-closed family of graphs of bounded local treewidth, then we can decide P on G in $h(2^{O(g^{-1}(k)\log g^{-1}(k))})n^{O(1)} + 2^{2^{O(g^{-1}(k)\log g^{-1}(k))}}n^{3+\epsilon}$ time.*

Proof. The algorithm is as follows. First we check whether $\mathbf{tw}(G)$ is in $2^{O(g^{-1}(k)\log g^{-1}(k))}$. By Theorems 2 and 4, if it is not, parameter P has value more than k on graph G. This step can be performed by Amir's algorithm [3], which for a given graph G and integer ω, either reports that the treewidth of G

is at least ω, or produces a tree decomposition of width at most $(3 + \frac{2}{3})\omega$ in time $O(2^{3.698\omega}n^3\omega^3 \log^4 n)$. Thus by using Amir's algorithm we can either compute a tree decomposition of G of size $2^{O(g^{-1}(k) \log g^{-1}(k))}$ in time $2^{2^{O(g^{-1}(k) \log g^{-1}(k))}} n^{3+\epsilon}$, or conclude that the treewidth of G is not in $2^{O(g^{-1}(k) \log g^{-1}(k))}$.

Now if we find a tree decomposition of the aforementioned width, we can decide P on G in time $h(2^{O(g^{-1}(k) \log g^{-1}(k))})n^{O(1)}$ time. The running time of this algorithm is the one mentioned in the statement of the theorem.

For example, let G be a graph from a minor-closed family $\mathcal{F}$ of bounded local treewidth. Since the dominating set of a graph with a given tree decomposition of width at most ω can be computed in time $O(2^{2\omega}n)$ [1], Theorem 5 gives an algorithm which either computes a dominating set of size at most p, or concludes that there is no such a dominating set in $2^{2^{O(\sqrt{p} \log p)}} n^{O(1)}$ time. The same result holds also for computing the minimum size of a q-dominating set. Indeed, Theorem 5 can be applied because the q-dominating set of a graph with a given tree decomposition of width at most ω can be computed in time $O(q^{O(\omega)}n)$ [8]. Also, algorithms on graphs of bounded treewidth for clique-transversal set, and Π-domination set appeared in [24] and [1] respectively. Using these algorithms, and the fact that all these parameters are lower bounded by the domination number, the methodology of the proof of Theorem 5 can give algorithmic results for clique-transversal set and Π-domination set with the same running times as in the case of dominating set (i.e., $2^{2^{O(\sqrt{p} \log p)}} n^{O(1)}$).

Finally, we mention some open problems. For planar graphs and for some of their extensions, it is known that for any graph G from these classes with dominating set of size at most p, we have $\mathbf{tw}(G) = O(\sqrt{p})$. It is tempting to ask if such a much smaller bound holds for all minor-closed families of bounded local treewidth. This will provide subexponential fixed-parameter algorithms on graphs of bounded local treewidth for the dominating set problem.

References

1. J. ALBER, H. L. BODLAENDER, H. FERNAU, T. KLOKS, AND R. NIEDERMEIER, *Fixed parameter algorithms for dominating set and related problems on planar graphs*, Algorithmica, 33 (2002), pp. 461–493.
2. J. ALBER, H. FAN, M. FELLOWS, AND R. H. FERNAU, AND R. NIEDERMEIER, *Refined search tree technique for dominating set on planar graphs*, MFCS 2001, Springer, vol. 2136, Berlin, 2000, pp. 111–122.
3. E. AMIR, *Efficient approximation for triangulation of minimum treewidth*, in Uncertainty in Artificial Intelligence: Proceedings of the Seventeenth Conference (UAI-2001), San Francisco, CA, 2001, Morgan Kaufmann Publishers, pp. 7–15.
4. B. S. BAKER, *Approximation algorithms for NP-complete problems on planar graphs*, J. Assoc. Comput. Mach., 41 (1994), pp. 153–180.
5. H. L. BODLAENDER, *A tourist guide through treewidth*, Acta Cybernetica, 11 (1993), pp. 1–23.
6. B. COURCELLE, *Graph rewriting: an algebraic and logic approach*, In Handbook of theoretical computer science, Vol. B, pp. 193–242. Elsevier, Amsterdam, 1990.

7. J. CHEN, I. A. KANJ, AND W. JIA. *Vertex cover: further observations and further improvements*, Journal of Algorithms, 41(2):280–301, 2001.

8. E. D. DEMAINE, F. V. FOMIN, M. T. HAJIAGHAYI, AND D. M. THILIKOS, *Fixed-Parameter Algorithms for the (k, r)-Center in Planar Graphs and Map Graphs*, ICALP 2003, Springer, Lecture Notes in Computer Science, Berlin, vol. 2719, 2003, pp. 829-844.

9. E. D. DEMAINE, F. V. FOMIN, M. T. HAJIAGHAYI, AND D. M. THILIKOS, *Subexponential parameterized algorithms on graphs of bounded genus and H-minor-free graphs*, To appear in SODA 2004.

10. E. D. DEMAINE, M. T. HAJIAGHAYI, *Fixed Parameter Algorithms for Minor-Closed Graphs (of Locally Bounded Treewidth)*, to appear in SODA 2004.

11. E. D. DEMAINE, M. T. HAJIAGHAYI, AND D. M. THILIKOS, *Exponential speedup of fixed parameter algorithms on $K_{3,3}$-minor-free or K_5-minor-free graphs*, ISAAC 2002, Springer, Lecture Notes in Computer Science, Berlin, vol. 2518, 2002, pp. 262–273.

12. R. DIESTEL, *Graph theory*, vol. 173 of Graduate Texts in Mathematics, Springer-Verlag, New York, second ed., 2000.

13. R. DIESTEL, T. R. JENSEN, K. Y. GORBUNOV, AND C. THOMASSEN, *Highly connected sets and the excluded grid theorem*, J. Combin. Theory Ser. B, 75 (1999), pp. 61–73.

14. R. G. DOWNEY AND M. R. FELLOWS, *Parameterized complexity*, Springer-Verlag, New York, 1999.

15. J. ELLIS, H. FAN, AND M. FELLOWS, *The dominating set problem is fixed parameter tractable for graphs of bounded genus*, SWAT 2002, Springer, Lecture Notes in Computer Science, Berlin, vol. 2368, 2002, pp. 180–189.

16. D. EPPSTEIN, *Diameter and treewidth in minor-closed graph families*, Algorithmica, 27 (2000), pp. 275–291.

17. J. FLUM AND M. GROHE, *Fixed-parameter tractability, definability, and model-checking*, SIAM J. Comput. 31:113-145, 2001.

18. F. V FOMIN AND D. M. THILIKOS, *Dominating sets in planar graphs: branch-Width and exponential speed-up*, SODA 2003, pp. 168–177.

19. F. V FOMIN AND D. M. THILIKOS, *A Simple and Fast Approach for Solving Problems on Planar Graphs*, to appear in STACS 2004.

20. F. V FOMIN AND D. M. THILIKOS, *Dominating sets and local treewidth*, ESA 2003, Springer-Verlag Lecture Notes in Computer Science, vol. 2832, 2003, pp. 221-229.

21. M. FRICK AND M. GROHE, *Deciding first-order properties of locally tree-decomposable graphs*, J. ACM, 48 (2001), pp. 1184 – 1206.

22. M. GROHE, *Local tree-width, excluded minors, and approximation algorithms*. To appear in Combinatorica.

23. I. KANJ AND L. PERKOVIĆ, *Improved parameterized algorithms for planar dominating set*, MFCS 2002, Springer, Lecture Notes in Computer Science, Berlin, vol.2420, 2002, pp. 399–410.

24. M. S.CHANG, T. KLOKS, AND C. M. LEE, Maximum clique transversals, WG 2001, Lecture Notes in Computer Science, Berlin, vol. 2204, pp. 300–310, 2001.

25. N. ROBERTSON AND P. D. SEYMOUR, *Graph minors. II. Algorithmic aspects of tree-width*, J. Algorithms, 7 (1986), pp. 309–322.

26. N. ROBERTSON AND P. D. SEYMOUR, *Graph minors. V. Excluding a planar graph*, J. Comb. Theory Series B, 41 (1986), pp. 92–114.

Vertex Disjoint Paths on Clique-Width Bounded Graphs

Extended Abstract

Frank Gurski and Egon Wanke

Department of Computer Science, D-40225 Düsseldorf, Germany
{gurski,wanke}@cs.uni-dueseldorf.de

Abstract. We show that l vertex disjoint paths between l pairs of vertices can be found in linear time for co-graphs but is NP-complete for graphs of NLC-width at most 4 and clique-width at most 7. This is the first inartificial graph problem known to be NP-complete on graphs of bounded clique-width but solvable in linear time on co-graphs and graphs of bounded tree-width.

1 Introduction

The clique-width of a graph is defined by a composition mechanism for vertex-labeled graphs [CO00]. The operations are the vertex disjoint union, the addition of edges between vertices controlled by a label pair, and the relabeling of vertices. The clique-width of a graph G is the minimum number of labels needed to define it. The NLC-width of a graph is defined by a composition mechanism similar to that for clique-width [Wan94]. Every graph of clique-width at most k has NLC-width at most k and every graph of NLC-width at most k has clique-width at most $2k$ [Joh98]. The only essential difference between the composition mechanisms of clique-width bounded graphs and NLC-width bounded graphs is the addition of edges. In an NLC-width composition the addition of edges is combined with the union operation. This union operation applied to two graphs G and J is controlled by a set S of label pairs such that for every pair $(a, b) \in S$ all vertices of G labeled by a will be connected with all vertices of J labeled by b. We use both notations, because it is sometimes much more comfortable to use NLC-width expressions instead of clique-width expressions and vice versa, respectively.

Clique-width and NLC-width bounded graphs are particularly interesting from an algorithmic point of view. A lot of NP-complete graph problems can be solved in polynomial time for graphs of bounded clique-width if the clique-width or NLC-width expression for the graph is explicitly given. For example, all graph properties which are expressible in monadic second order logic with quantifications over vertices and vertex sets (MSO$_1$-logic) are decidable in linear time on clique-width bounded graphs [CMR00]. The MSO$_1$-logic has been extended by counting mechanisms which allow the expressibility of optimization problems concerning maximal or minimal vertex sets [CMR00]. All graph problems expressible in extended MSO$_1$-logic can be solved in polynomial time on

M. Farach-Colton (Ed.): LATIN 2004, LNCS 2976, pp. 119–128, 2004.
© Springer-Verlag Berlin Heidelberg 2004

clique-width bounded graphs. Furthermore, there are also a lot of NP-complete graph problems which are not expressible in extended MSO_1-logic like Hamiltonicity, partition problems, and bounded degree subgraph problems but which can also be solved in polynomial time on clique-width bounded graphs [EGW01, KR01,Tod03].

If a graph G has clique-width (NLC-width) at most k then the edge complement $\overline{G}$ has clique-width at most $2k$ (NLC-width at most k) [CO00,Wan94]. Distance hereditary graphs have clique-width at most 3 [GR99]. The set of all graphs of clique-width at most 2 or NLC-width 1 is the set of all labeled cographs. Brandstädt et al have analyzed the clique-width of graphs defined by forbidden one-vertex extensions of P_4 [BDLM02]. The clique-width and NLC-width of permutation graphs, interval graphs, grids and planar graphs is not bounded [GR99]. An arbitrary graph with n vertices has clique-width at most $n - r$, if $2^r < n - r$, and NLC-width at most $\lceil \frac{n}{2} \rceil$ [Joh98]. Every graph of tree-width[1] at most k has clique-width at most $3 \cdot 2^{k-1}$ [CR01]. In [GW00], it is shown that every graph of clique-width or NLC-width k which does not contain the complete bipartite graph $K_{n,n}$ for some $n > 1$ as a subgraph has tree-width at most $3k(n - 1) - 1$. The recognition problem for graphs of clique-width or NLC-width at most k is still open for $k \geq 4$ and $k \geq 3$, respectively. Clique-width of at most 3 is decidable in polynomial time [CHL+00]. NLC-width of at most 2 is decidable in polynomial time [Joh00]. Clique-width of at most 2 and NLC-width 1 is decidable in linear time [CPS85]. The clique-width of tree-width bounded graphs is also computable in linear time [EGW03].

In this paper, we analyze the problem of finding vertex disjoint paths. Given l pairs of vertices $(s_1, t_1), \ldots, (s_l, t_l)$ and l integers $r_1, \ldots, r_l$, we consider the problem of finding r_i paths between s_i and t_i for $i = 1, \ldots, l$ whose inner vertices are all distinct. The vertex disjoint paths problem can be solved in polynomial time if l is fixed, i.e., not part of the input, and $r_i = 1$ for $i = 1, \ldots, l$ [RS95]. It is NP-complete for $l = 2$ if r_1 and r_2 are unbounded [EIS76]. It is also NP-complete if the number l of vertex pairs is unbounded and $r_i = 1$ for $i = 1, \ldots, l$ [MP95].

In Section 3, we show that the vertex disjoint paths problem for co-graphs can be solved in linear time for $r_i = 1$, $1 \leq i \leq l$, and in polynomial time for unbounded r_is.

Nishizeki, Vygen, and Zhou have shown in [NVZ01] that the edge disjoint paths problem is NP-complete for graphs of tree-width at most 2. The edge disjoint paths problem for a graph G can be solved with an algorithm for the vertex disjoint paths problem on the line graph of G. The line graph of a graph G has a vertex for every edge of G and an edge between two vertices if the corresponding edges of G are adjacent. In Section 4, we show that the line graph of a graph of tree-width k has clique-width at most $2k+3$ and NLC-width at most $k + 2$. This implies our main result that the vertex disjoint paths problem where all $r_i = 1$ is NP-complete for graphs of clique-width at most 7 and NLC-width at most 4. This is the first inartificial NP-complete graph problem shown to be NP-complete for graphs of bounded clique-width but solvable in linear time for

---

[1] See Robertson and Seymour [RS86] for a definition of tree-width.

co-graphs and graphs of bounded tree-width [Sch94]. It is also the first problem that separates co-graphs and clique-width bounded graphs from a complexity point of view. By inartificial we mean that the problem is not exclusively defined for the purpose to be NP-complete for clique-width bounded graphs and solvable in polynomial time for co-graphs and tree-width bounded graphs.

2 Preliminaries

Let $[k] := \{1, \ldots, k\}$ be the set of all integers between 1 and k. We work with finite undirected labeled *graphs* $G = (V_G, E_G, \mathrm{lab}_G)$, where V_G is a finite set of *vertices* labeled by some mapping $\mathrm{lab}_G : V \to [k]$ and $E_G \subseteq \{\{u, v\} \mid u, v \in V_G, \ u \neq v\}$ is a finite set of *edges*. A labeled graph $J = (V_J, E_J, \mathrm{lab}_J)$ is a *subgraph* of G if $V_J \subseteq V_G$, $E_J \subseteq E_G$ and $\mathrm{lab}_J(u) = \mathrm{lab}_G(u)$ for all $u \in V_J$. J is an *induced subgraph* of G if additionally $E_J = \{\{u, v\} \in E_G \mid u, v \in V_J\}$. The labeled graph consisting of a single vertex labeled by $a \in [k]$ is denoted by $\bullet_a$.

Definition 1 (Clique-width, [CO00]). *Let k be some positive integer. The class CW_k of labeled graphs is recursively defined as follows.*

1. *The single vertex graph $\bullet_a$ for some $a \in [k]$ is in CW_k.*
2. *Let $G = (V_G, E_G, \mathrm{lab}_G) \in CW_k$ and $J = (V_J, E_J, \mathrm{lab}_J) \in CW_k$ be two vertex disjoint labeled graphs. Then $G \oplus J := (V', E', \mathrm{lab}')$ defined by $V' := V_G \cup V_J$, $E' := E_G \cup E_J$, and*

$$\mathrm{lab}'(u) := \begin{cases} \mathrm{lab}_G(u) \ \textit{if } u \in V_G \\ \mathrm{lab}_J(u) \ \textit{if } u \in V_J \end{cases}, \ \forall u \in V'$$

   is in CW_k.
3. *Let $a, b \in [k]$ be two distinct integers and $G = (V_G, E_G, \mathrm{lab}_G) \in CW_k$ be a labeled graph then*

   a) *$\rho_{a \to b}(G) := (V_G, E_G, \mathrm{lab}')$ defined by*

$$\mathrm{lab}'(u) := \begin{cases} \mathrm{lab}_G(u) \ \textit{if } \mathrm{lab}_G(u) \neq a \\ b \qquad\quad \textit{if } \mathrm{lab}_G(u) = a \end{cases}, \ \forall u \in V_G$$

   is in CW_k and

   b) *$\eta_{a,b}(G) := (V_G, E', \mathrm{lab}_G)$ defined by*

$$E' := E \cup \{\{u, v\} \mid u, v \in V_G, \ u \neq v, \ \mathrm{lab}(u) = a, \ \mathrm{lab}(v) = b\}$$

   is in CW_k.

Definition 2 (NLC-width, [Wan94]). *Let k be some positive integer. The class NLC_k of labeled graphs is recursively defined as follows.*

1. *The single vertex graph $\bullet_a$ for some $a \in [k]$ is in NLC_k.*

2. Let $G = (V_G, E_G, lab_G) \in NLC_k$ and $J = (V_J, E_J, lab_J) \in NLC_k$ be two vertex disjoint labeled graphs and $S \subseteq [k]^2$ be a relation, then $G \times_S J := (V', E', lab')$ defined by $V' := V_G \cup V_J$,

$$E' := E_G \cup E_J \cup \{\{u, v\} \mid u \in V_G,\ v \in V_J,\ (lab_G(u), lab_J(v)) \in S\},$$

and

$$lab'(u) := \begin{cases} lab_G(u) & \text{if } u \in V_G \\ lab_J(u) & \text{if } u \in V_J \end{cases},\ \forall u \in V'$$

is in NLC_k.

3. Let $G = (V_G, E_G, lab_G) \in NLC_k$ and $R : [k] \to [k]$ be a function, then $\circ_R(G) := (V_G, E_G, lab')$ defined by $lab'(u) := R(lab_G(u))$, $\forall u \in V_G$ is in NLC_k.

The *clique-width* (*NLC-width*) of a labeled graph G is the least integer k such that $G \in CW_k$ ($G \in NLC_k$, respectively). An expression X built with the operations $\bullet_a, \oplus, \rho_{a \to b}, \eta_{a,b}$ for integers $a, b \in [k]$ is called a *clique-width k-expression*. An expression X built with the operations $\bullet_a, \times_S, \circ_R$ for $a \in [k]$, $S \subseteq [k]^2$, and $R : [k] \to [k]$ is called an *NLC-width k-expression*. The graph defined by expression X is denoted by $\text{val}(X)$.

A *path* p of *length* $r - 1$ in a graph $G = (V_G, E_G, lab_G)$ is a sequence of r vertices $p = (u_1, \ldots, u_r)$ such that $\{u_i, u_{i+1}\} \in E_G$ for $i = 1, \ldots, r - 1$ is an edge of G. Two paths $p = (u_1, \ldots, u_r)$ and $q = (v_1, \ldots, v_{r'})$ are *vertex disjoint* if $\{u_2, \ldots, u_{r-1}\} \cap \{v_1, \ldots, v_{r'}\} = \emptyset$ and $\{u_1, \ldots, u_r\} \cap \{v_2, \ldots, v_{r'-1}\} = \emptyset$. That is, the inner vertices of p do not occur in q, and vice versa.

We analyze the following decision problem.

PROBLEM: **Vertex Disjoint Paths**
INSTANCE: Graph $G = (V_G, E_G, lab_G)$, l vertex pairs $(s_1, t_1), \ldots, (s_l, t_l)$, all $s_1, \ldots, s_l$ and $t_1, \ldots, t_l$ distinct, and l positive integers $r_1, \ldots, r_l$.
QUESTION: Are there r_i paths between s_i and t_i for $i = 1, \ldots, l$ such that all $\sum_{i=1}^l r_i$ paths are mutually vertex disjoint.

3   Polynomial Time Solutions

Let G be a co-graph defined by some NLC-width 1-expression X. Expression X can be found for a given graph G in linear time using any linear time recognition algorithm for co-graphs that computes the co-tree for G, see for example [CPS85].

The vertices of the vertex pairs $(s_1, t_1), \ldots, (s_l, t_l)$ are called *terminal vertices* and all other vertices are called *free vertices*. We assume that the terminal vertices are explicitly specified in the NLC-width 1-expression X. That is, we know which sub-expression $\bullet_1$ of X represents terminal vertex s_i or t_i for $1 \leq i \leq l$.

It is well known that a graph is a co-graph if and only if it has no induced P_4, i.e., it has no induced path of length 3. That is, to solve the vertex disjoint paths problem for co-graphs we only have to look for paths that consist of exactly two or three vertices, the start vertex s_i, at most one free vertex u_i, and the target vertex t_i, for $1 \leq i \leq l$.

Theorem 1. *The vertex disjoint paths problem, for $r_i = 1$, $i = 1, \ldots, l$, is decidable in linear time for co-graphs.*

Proof. If two terminal vertices s_i, t_i, $1 \leq i \leq l$, are adjacent, we can remove them from G by modifying X. This can be done in linear time. The remaining graph is still a co-graph. Hence, we can assume that all s_i, t_i, $1 \leq i \leq l$, are non-adjacent.

The paths (s_i, u_i, t_i) of length two, where s_i and t_i are not adjacent in G, can only be constructed in a composition step of the form $Y \times_{\{(1,1)\}} Z$, where either val(Y) or val(Z) contains both terminal vertices s_i, t_i and the other graph val(Z) or val(Y), respectively, contains the free vertex u_i. Since all vertices have the same label, label 1, every free vertex of val(Y) will be connected by operation $Y \times_{\{(1,1)\}} Z$ with every terminal vertex of val(Z), and vice versa.

Let G' be the induced subgraph of G represented by a subexpression X' of X. Then let

1. $R(X')$ be the number pairs s_i, t_i, $1 \leq i \leq l$, contained in G',
2. $F(X')$ be the number of free vertices in G', and
3. $M(X')$ be the maximal number of vertex disjoint paths (s_i, u_i, t_i) in subgraph G', for free vertices u_i, $1 \leq i \leq l$.

By definition, graph G defined by X has a solution for the vertex disjoint paths problem if and only if $R(X) = M(X)$. The main part now is to show that all $R(X')$, $F(X')$, and $M(X')$ are computable in linear time. This can be done recursively as follows:

- Let $X' = \bullet_1$.
  If the single vertex u of val(X') is a free vertex then
  $R(X') = 0$, $F(X') = 1$, and $M(X') = 0$
  otherwise, if u is a terminal vertex, then
  $R(X') = 0$, $F(X') = 0$, and $M(X') = 0$.
- Let $X' = Y \times_\emptyset Z$. (Operation $\times_\emptyset$ does not create any new edge.)
  Let $R_{Y,Z}$ be the number of pairs s_i, t_i, $1 \leq i \leq l$, such that val(Y) contains terminal vertex s_i or t_i and val(Z) contains the other terminal vertex t_i or s_i, respectively. Then
  $R(X') = R(Y) + R(Z) + R_{Y,Z}$,
  $F(X') = F(Y) + F(Z)$, and
  $M(X') = M(Y) + M(Z)$.
- Let $X' = Y \times_{\{(1,1)\}} Z$.
  Then
  $R(X') = R(Y) + R(Z) + R_{Y,Z}$,
  $F(X') = F(Y) + F(Z)$, and
  $M(X') = M(Y) + M(Z) + \min\{R(Y) - M(Y), F(Z) - M(Z)\}$
  $\qquad\qquad\qquad\qquad + \min\{R(Z) - M(Z), F(Y) - M(Y)\}$.
  $F(Z) - M(Z)$ free vertices of val(Z) can be used to realize some of the still missing $R(Y) - M(Y)$ paths in val(Y), and vice versa.

This computation can be done in linear time by traversing bottom-up the NLC-width 1-expression. The necessary values $R_{Y,Z}$ at the operations $\times_\emptyset$ and

$\times_{\{(1,1)\}}$ can be computed in a preprocessing. We just have to compute in the expression tree of X the nearest common ancestor for every pair (s_i, t_i). Such a preprocessing can be done for all pairs (s_i, t_i) together in linear time with respect to the size of the expression tree of X, see [AGKR02].

Theorem 2. *The vertex disjoint paths problem is decidable in polynomial time for co-graphs.*

Proof. In a first step, we decrement r_i by one if s_i and t_i are adjacent in G. Then we define a bipartite graph H as follows. H has a vertex $w_{i,j}$ for $1 \leq i \leq l$ and $1 \leq j \leq r_i$ and a vertex v_k for every free vertex u_k of G. Vertex v_k of H is connected with all vertices $w_{i,j}$, $1 \leq i \leq l$, $1 \leq j \leq r_i$, in H if and only if free vertex u_k of G is adjacent to s_i and t_i in G.

Graph H is bipartite and can be constructed in polynomial time. Graph G has r_i paths of length 2 between s_i and t_i, $1 \leq i \leq l$, if and only if H has a matching of size $\sum_{i=1,\ldots,l} r_i$. A maximum matching in bipartite graphs with n vertices and m edges can be found in time $O(\sqrt{n} \cdot m \cdot \log(\frac{n^2}{m})/\log(n))$ [FM91].

4   NP-Completeness

If we consider edge disjoint paths instead of vertex disjoint paths, then we get the *edge disjoint paths problem*. The edge disjoint paths problem for a graph G can be solved with an algorithm for the vertex disjoint paths problem applied to the *line graph* of G. The line graph of a graph $G = (V_G, E_G)$ has a vertex for every edge of G and an edge between two vertices if the corresponding edges in G have a common vertex. The edge disjoint paths problem is NP-complete even for series parallel graphs, i.e. for graphs of tree-width at most 2 [NVZ01] and $r_i = 1$, $i = 1, \ldots, l$.

We now show that the line graph of a graph of tree-width k has NLC-width at most $k + 2$. Graphs of tree-width at most k are also called *partial k-trees*. A partial k-tree is a subgraph of a *k-tree* which can recursively be defined as follows. The complete graph with k vertices is a k-tree. If G is a k-tree then the graph obtained by inserting a new vertex u and k edges between u and all vertices of a k vertex complete subgraph of G is again a k-tree.

Theorem 3. *The line graph of a partial k-tree has NLC-width at most $k + 2$.*

Proof. It suffices to show that the line graph of a k-tree G has NLC-width at most $k+2$, because the line graph of every subgraph of G is an induced subgraph of the line graph of G, and the set of all graphs of NLC-width at most $k + 2$ is closed under induced subgraphs.

Let $G = (V_G, E_G)$ be a k-tree with n vertices. Let $o = (u_1, \ldots, u_n)$ be an order of the n vertices of G. Let $N^-(G, o, i)$ and $N^+(G, o, i)$ for $i = 1, \ldots, n$ be the set of neighbors u_j of vertex u_i with $j < i$ and $j > i$, respectively. That is,

$$N^-(G, o, i) := \{u_j \mid \{u_i, u_j\} \in E_G \ \wedge \ j < i\} \text{ and}$$

$$N^+(G, o, i) := \{u_j \mid \{u_j, u_i\} \in E_G \ \wedge \ j > i\}.$$

A vertex order $(u_1, \ldots, u_n)$ for G is called a *perfect elimination order* (PEO) for G if the vertices of $N^+(G, o, i)$ for $i = 1, \ldots, n$ induce a complete subgraph of G.

There is always a PEO $o = (u_1, \ldots, u_n)$ for G such that the vertices of every $N^+(G, o, i)$ for $i = 1, \ldots, n - k$ induce a k vertex complete subgraph and the vertices of every $N^+(G, o, i)$ for $i = n - k + 1, \ldots, n$ induce an $n - i$ vertex complete subgraph of G. Here we can use, for example, the reverse order of the vertices from the recursive definition of the k-tree.

The structure of the k-tree G can be characterized by the tree $T(G, o) = (V_T, E_T)$ defined as follows. Let $o = (u_1, \ldots, u_n)$ be a perfect elimination order for G.

$$V_T := V_G \qquad E_T := \{\{u_i, u_j\} \in E_G \mid i < j \ \wedge \ \forall j', i < j' < j, \{u_i, u_{j'}\} \notin E_G\}.$$

Graph $T(G, o)$ is a tree, because every vertex u_i, $i < n$, of $T(G, o)$ is incident with exactly one edge $\{u_i, u_j\}$ for $j > i$.

Let col be a $k + 1$-coloring of G, i.e., col : $V_G \to [k + 1]$ is a mapping with $\operatorname{col}(u_i) \neq \operatorname{col}(u_j)$ for all $\{u_i, u_j\} \in E_G$. It is easy to see that each k-tree is $k + 1$ colorable if we assign to u_i any color not used by the vertices of $N^+(G, o, i)$.

We now recursively define for $i = 1, \ldots, n$ an NLC-width $k + 2$-expression X_i as follows. Let $N^-(T(G, o), o, i) = \{u_{j_1}, \ldots, u_{j_m}\}$ and $N^+(G, o, i) = \{u_{l_1}, \ldots, u_{l_r}\}$. Note that $\{u_{j_1}, \ldots, u_{j_m}\}$ is defined by tree $T(G, o)$ and $\{u_{l_1}, \ldots, u_{l_r}\}$ is defined by G.

1. If $m = 1$ then let $Y_i = X_{j_1}$. If $m > 1$ then let $Y_i = X_{j_1} \times_I \cdots \times_I X_{j_m}$, where $I = \{(s, s) \mid s \in [k + 1]\}$ is the identity between the labels $1, \ldots, k + 1$. Graph val(Y_i) is the disjoint union of m graphs val(X_{j_1}), $\ldots$, val(X_{j_m}), where equal labeled vertices from different graphs are joined by an edge. Note that relation I uses only the labels $1, \ldots, k + 1$. The label $k + 2$ is exclusively used for the vertices that will not be connected with other vertices in any further composition step.

2. If $r > 0$ then let Z_i be an NLC-width $k + 1$-expression that defines a complete graph with r vertices labeled by $\operatorname{col}(u_{l_1}), \ldots, \operatorname{col}(u_{l_r})$. Note that these $r \leq k$ labels are distinct and do not include the color of u_i.

3. Then we define

$$X_i = \begin{cases} \circ_R(Y_i \times_S Z_i) & \text{if } m > 0 \text{ and } r > 0 \\ Z_i & \text{if } m = 0 \text{ and } r > 0 \\ \circ_R(Y_i) & \text{if } m > 0 \text{ and } r = 0 \end{cases}$$

where $S = \{(s, s) \mid s \in [k + 1]\} \cup \{(\operatorname{col}(u_i), s) \mid s \in [k + 1]\}$ and

$$R(s) = \begin{cases} s & \text{if } s \neq \operatorname{col}(u_i) \\ k + 2 & \text{if } s = \operatorname{col}(u_i) \end{cases}.$$

It remains to show that the NLC-width $k + 2$-expression X_n defines the line graph of k-tree G.

By the definition of Z_i for $i = 1, \ldots, n$ there is a one to one correspondence between the edges of G and the vertices of the graph defined by X_i. We say, the vertex of the graph defined by Z_i which is labeled by s represents the edge between u_i and the unique vertex of $N^+(G, o, i)$ colored by s. In this way, there is also a one to one correspondence between the edges of G and the vertices of the graph defined by X_n, although all labels are finally changed into label $k + 2$. The vertices in the graph defined by X_i which are labeled by some label of $[k+1]$ will represent exactly those edges of G with one end vertex from $\{u_1, \ldots, u_i\}$ and one end vertex from $\{u_{i+1}, \ldots, u_n\}$.

Let us describe more precisely the graphs that the expressions X_i define for $i = 1, \ldots, n$. Let G_i, $1 \le i \le n$, be the subgraph of G induced by the vertices $\{u_1, \ldots, u_i\}$. Let G_i' be the connected component of G_i to which u_i belongs, and let $\widetilde{G}_i$ be the graph G_i' extended by all the edges (and their end vertices) of G that have exactly one end vertex in G_i'. A simple induction on i will show that X_i defines the line graph of $\widetilde{G}_i$.

Basis: Let $i = 1$.
Graph $\widetilde{G}_1$ consists of $1 + |N^+(G, o, 1)|$ vertices and $|N^+(G, o, 1)|$ edges between u_1 and the vertices from $N^+(G, o, 1)$. In this case, graph $\mathrm{val}(X_1)$ defines a complete graph with $|N^+(G, o, 1)|$ vertices labeled by the colors of the vertices of $N^+(G, o, 1)$. The graph defined by X_1 obviously represents the line graph of $\widetilde{G}_1$.

Induction: Let $i > 1$.
Let $N^-(T(G, o), o, i) = \{u_{j_1}, \ldots, v_{j_m}\}$ and $N^+(G, o, i) = \{u_{l_1}, \ldots, u_{l_r}\}$. If $m = 0$, then as in the case where $i = 1$, $\widetilde{G}_i$ consists of $1 + |N^+(G, o, i)|$ vertices and $|N^+(G, o, i)|$ edges between u_i and the vertices from $N^+(G, o, i)$. Here also, expression X_i defines a complete graph with $|N^+(G, o, i)|$ vertices labeled by the colors of the vertices of $N^+(G, o, i)$.

If $m > 0$, then we first define an expression Y_i for the union of all the graphs defined by the expressions $X_{j_1}, \ldots, X_{j_m}$ in which equal labeled vertices from different graphs are connected by edges. These equal labeled vertices from different graphs have to be connected by edges because they will represent edges with one end vertex from $\{u_1, \ldots, u_{i-1}\}$ and the same end vertex from $\{u_i, \ldots, u_n\}$.

By the inductive hypothesis and the additionally inserted edges, expression $Y_i \times_S Z_i$ now defines a graph that represents the line graph of G_i. Relation S connects a vertex u of the graph defined by Y_i and a vertex v of the graph defined by Z_i if the following hold true.

1. Both vertices have the same label from $[k + 1]$.
   Then u and v represent two adjacent edges $\{u_{i'}, u_{j'}\}$ and $\{u_i, u_{j'}\}$ of G, respectively, where $i' < i < j'$.
2. The label of u is the color $\mathrm{col}(u_i)$ of u_i in G.
   Then u represents an edge $\{u_{i'}, u_i\}$ of G where $i' < i$. These edges are all adjacent with the edges represented by the vertices of the graph defined by Z_i.

The final relabeling $\circ_R$ changes label $\mathrm{col}(u_i)$ into label $k + 2$, because the vertices labeled by $\mathrm{col}(u_i)$ now represent only edges e of G such that for all edges adjacent to e both end vertices are already contained in the graph defined by X_i.

Since $\widetilde{G}_n$ is the graph G, we have defined an NLC-width $k+2$-expression for the line graph of G.

For a given graph $G = (V_G, E_G)$, l vertex pairs $(u_1, v_1), \dots, (u_l, v_l)$, and l integers $r_1, \dots, r_l$ let $G' = (V_{G'}, E_{G'})$ be the graph G with $2 \cdot l$ additional vertices $u'_1, \dots, u'_l, v'_1, \dots, v'_l$ and $2 \cdot l$ additional edges $\{u'_1, u_1\}, \dots, \{u'_l, u_l\}, \{v'_1, v_1\}, \dots,$ $\{v'_l, v_l, \}$. Let H be the line graph of G' and s_i and t_i be the vertices of H that represents the edge $\{u'_i, u_i\}$ and $\{v'_i, v_i\}$, $1 \leq i \leq l$, of G'. Then there are r_i mutually vertex disjoint paths in H between s_i and t_i for $i = 1, \dots, l$ if and only if there are r_i mutually edge disjoint paths in G between u_i and v_i for $i = 1, \dots, l$. If G is a partial 2-tree than G' is a partial 2-tree. Since the problem of finding edge disjoint paths is NP-complete for partial 2-tree, see [NVZ01], we have shown the following theorem.

Theorem 4. *The vertex disjoint paths problem is NP-complete for graphs of NLC-width at most 4.*

A simple modification of the proof of Theorem 3 shows that the line graph of every partial k-tree has clique-width at most $2k + 3$. This implies that the vertex disjoint paths problem is NP-complete for graphs of clique-width at most 7. Theorem 3 also implies that the chromatic index of a graph of bounded tree-width can be solved in polynomial time, because the chromatic number problem for NLC-width and clique-width bounded graphs can be solved in polynomial time [EGW01]. This re-proves a result by Bodlaender [Bod90]. Note also that the proof of Theorem 3 is constructive, i.e., an NLC-width expression and clique-width expression can simply be constructed from a given partial k-tree.

References

[AGKR02] S. Alstrup, C. Gavoille, H. Kaplan, and T. Rauhe. Nearest common ancestors: A survey and a new distributed algorithm. In *Proceedings of the Annual ACM Symposium on Parallel Algorithms and Architectures*, pages 258–264. ACM, 2002.

[BDLM02] A. Brandstädt, F. F. Dragan, H.-O. Le, and R. Mosca. New graph classes of bounded clique width. In *Proceedings of Graph-Theoretical Concepts in Computer Science*, volume 2573 of *LNCS*, pages 57–67. Springer-Verlag, 2002.

[Bod90] H.L. Bodlaender. Polynomial algorithms for chromatic index and graph isomorphism on partial k-trees. *Journal of Algorithms*, 11(4):631–643, 1990.

[CHL+00] D.G. Corneil, M. Habib, J.M. Lanlignel, B. Reed, and U. Rotics. Polynomial time recognition of clique-width at most three graphs. In *Proceedings of Latin American Symposium on Theoretical Informatics (LATIN '2000)*, volume 1776 of *LNCS*. Springer-Verlag, 2000.

[CMR00] B. Courcelle, J.A. Makowsky, and U. Rotics. Linear time solvable optimization problems on graphs of bounded clique-width. *Theory of Computing Systems*, 33(2):125–150, 2000.

[CO00] B. Courcelle and S. Olariu. Upper bounds to the clique width of graphs. *Discrete Applied Mathematics*, 101:77–114, 2000.

[CPS85]   D.G. Corneil, Y. Perl, and L.K. Stewart. A linear recognition algorithm for cographs. *SIAM Journal on Computing*, 14(4):926–934, 1985.

[CR01]   D.G. Corneil and U. Rotics. On the relationship between clique-width and treewidth. In *Proceedings of Graph-Theoretical Concepts in Computer Science*, volume 2204 of *LNCS*, pages 78–90. Springer-Verlag, 2001.

[EGW01]   W. Espelage, F. Gurski, and E. Wanke. How to solve NP-hard graph problems on clique-width bounded graphs in polynomial time. In *Proceedings of Graph-Theoretical Concepts in Computer Science*, volume 2204 of *LNCS*, pages 117–128. Springer-Verlag, 2001.

[EGW03]   W. Espelage, F. Gurski, and E. Wanke. Deciding clique-width for graphs of bounded tree-width. *Journal of Graph Algorithms and Applications - Special Issue of JGAA on WADS 2001*, 7(2):141–180, 2003.

[EIS76]   S. Even, A. Itai, and A. Shamir. On the complexity of timetable and multicommodity flow problems. *SIAM Journal on Computing*, 5:691–703, 1976.

[FM91]   T. Feder and R. Motwani. Clique partitions, graph compression and speeding up algorithms. In *Proceedings of the Annual ACM Symposium on Theory of Computing*, pages 123–133. ACM, 1991.

[GR99]   M.C. Golumbic and U. Rotics. On the clique-width of some perfect graph classes. In *Proceedings of Graph-Theoretical Concepts in Computer Science*, volume 1665 of *LNCS*, pages 135–147. Springer-Verlag, 1999.

[GW00]   F. Gurski and E. Wanke. The tree-width of clique-width bounded graphs without $K_{n,n}$. In *Proceedings of Graph-Theoretical Concepts in Computer Science*, volume 1938 of *LNCS*, pages 196–205. Springer-Verlag, 2000.

[Joh98]   Ö. Johansson. Clique-decomposition, NLC-decomposition, and modular decomposition - relationships and results for random graphs. *Congressus Numerantium*, 132:39–60, 1998.

[Joh00]   Ö. Johansson. NLC_2-decomposition in polynomial time. *International Journal of Foundations of Computer Science*, 11(3):373–395, 2000.

[KR01]   D. Kobler and U. Rotics. Polynomial algorithms for partitioning problems on graphs with fixed clique-width. In *Proceedings of the ACM-SIAM Symposium on Discrete Algorithms*, pages 468–476. ACM-SIAM, 2001.

[MP95]   M. Middendorf and F. Pfeiffer. On the complexity of the disjoint paths problems. *Combinatorica*, 35(1):97–107, 1995.

[NVZ01]   T. Nishizeki, J. Vygen, and X. Zhou. The edge-disjoint paths problem is NP-complete for series-parallel graphs. *Discrete Applied Mathematics*, 115:177–186, 2001.

[RS86]   N. Robertson and P.D. Seymour. Graph minors II. Algorithmic aspects of tree width. *Journal of Algorithms*, 7:309–322, 1986.

[RS95]   N. Robertson and P.D. Seymour. Graph minors XIII. The disjoint paths problem. *Journal of Combinatorial Theory, Series B*, 63(1):65–110, 1995.

[Sch94]   P. Scheffler. A practical linear time algorithm for vertex disjoint paths in graphs with bounded treewidth. Technical Report 396, Dept. of Mathematics, Technische Universität Berlin, 1994.

[Tod03]   I. Todinca. Coloring powers of graphs of bounded clique-width. In *Proceedings of Graph-Theoretical Concepts in Computer Science*, volume 2880 of *LNCS*, pages 370–382. Springer-Verlag, 2003.

[Wan94]   E. Wanke. k-NLC graphs and polynomial algorithms. *Discr. Applied Mathematics*, 54:251–266, 1994. revised version: "http://www.cs.uni-duesseldorf.de/~wanke".

On Partitioning Interval and Circular-Arc Graphs into Proper Interval Subgraphs with Applications

Frédéric Gardi*

Laboratoire d'Informatique Fondamentale,
Parc Scientifique et Technologique de Luminy,
Case 901 – 163, Avenue de Luminy,
13288 Marseille Cedex 9, France
Frederic.Gardi@lif.univ-mrs.fr

Abstract. In this note, we establish that any interval or circular-arc graph with n vertices admits a partition into $O(\log n)$ proper interval subgraphs. This bound is shown to be asymptotically sharp for an infinite family of interval graphs. Moreover, the constructive proof yields a linear-time and space algorithm to compute such a partition. The second part of the paper is devoted to an application of this result, which has actually inspired this research: the design of an efficient approximation algorithm for a $\mathcal{NP}$-hard problem of planning working schedules.

1 Introduction

An undirected graph G=(V,E) is an *interval graph* if to each vertex $v \in V$ can be associated an open (resp. closed) interval I_v of the real line, such that any pair of distinct vertices u, v are connected by an edge of E if and only if $I_u \cap I_v \neq \emptyset$. The family $\{I_v\}_{v \in V}$ is an *interval representation* of G; the left and right endpoints of I_v are respectively denoted by $le(I_v)$ and $re(I_v)$. The edges of the complement graph $\overline{G}$ are transitively orientable by setting $u \to v$ if $r_u < l_v$; the orientation of the edges induces a partial order called interval order (we shall write $I_u \prec I_v$ if $r_u < l_v$). In the same way, the intersection graph of collections of arcs on a circle is called *circular-arc graph*. A circular-arc representation of an undirected graph G which fails to cover some point p on the circle will be topologically the same as an interval representation of G. In effect, we can cut the circle at p and straighten it out a line, the arcs becoming intervals. It is easy to notice therefore, that every interval graph is a circular-arc graph.

An interval graph G is called *proper interval graph* if there is an interval representation of G such that no interval contains properly another. A nice result of Roberts (1969, *cf.* [13,6]) establishes that proper interval graphs coincide with *unit interval graphs*, the interval graphs having an interval representation such that all intervals have the same size, and $K_{1,3}$-*free interval graphs*, the interval graphs without induced copy of a tree composed of one central vertex and three leaves.

---

* The author is a Ph.D. student in Computer Science and Mathematics.

M. Farach-Colton (Ed.): LATIN 2004, LNCS 2976, pp. 129–140, 2004.
© Springer-Verlag Berlin Heidelberg 2004

The main result. Interval and circular-arc graphs have been intensively studied for several decades by both discrete mathematicians and theoretical computer scientists. These two classes of graphs are particulary known for providing numerous models in diverse areas like scheduling, genetics, psychology, sociology, archæology and others. For surveys on all results and applications concerning interval and circular-arc graphs, the interested reader is referred to [13,6,8].

In this note, the problem of *partitioning interval or circular-arc graphs into proper interval subgraphs* is investigated. Two questions can be raised concerning this problem. The first, rather asked by the mathematician is: could you find good lower and upper bounds on the size of a minimum partition of an interval or circular-arc graph into proper interval subgraphs ? The second, rather asked by the computer scientist is: could you find an efficient algorithm to compute such a minimum partition ? An answer to the first question is given in this paper, through the following theorem. Although the result provides some advances on the second question (discussed in Conclusion), this one remains open at our knowledge.

Theorem 1. *Any interval graph or circular-arc graph with n vertices admits a partition into $O(\log n)$ proper interval subgraphs. Moreover, this bound is asymptotically sharp for an infinite family of interval graphs.*

The constructive proof of the result (described Section 2) yields a linear-time and space algorithm to compute such a partition. Thereby, this result could find applications in the design of approximation algorithms for hard problems on interval or circular-arc graphs, since many untractable problems for these graphs become easier for proper interval graphs. In the second part of the paper, we present such a kind of application in the area of working schedules planning, which has actually inspired this research.

Applications. The problem of *planning working schedules* holds an important place in operations research and business administration. In a schematic way, the problem consists in the *assignment of fixed tasks to employees in the form of shifts*. The tasks of the shift allocated to an employee, which induce his working schedules, must be pairwise *disjoint* (non-intersecting). Here a problem derived from schedules planning problems solved by the firm PROLOGIA - Groupe Air Liquide [12] is considered. This fundamental problem, denoted WSP, is defined as follows. Let $\{T_i\}_{i=1,...,n}$ be a set of tasks having respective starting and ending dates (l_i, r_i). The *regulation* imposes that any employee cannot execute more than k tasks. Given that the tasks allocated to an employee must not overlap, build an optimal planning according to the following objectives: on a *first level*, reduce the number of shifts or employees (*productivity*) and then on a *second level*, balance the planning (*social*) and prevent as well as possible the future modifications of the planning (*robustness*).

Since the tasks are simply some intervals of the real line, the WSP problem can be reformulated in graph-theoretic terms as the problem of *coloring an interval graph such that each color marks at most k vertices*. When the planning is *cyclic*, we obtain the same coloring problem with *circular-arc graphs*. In this model, the optimization criteria become respectively: to minimize the number of colors (P), balance the number of vertices in each color class (S) and maximize the smallest gap existing between two consecutive intervals or arcs having the same color (R). In fact, the criterion R prevents overlappings when some intervals or arcs are delayed or put forward. Hence, a solution to WSP is called (P)-*optimal* (resp. (S, R)-*optimal*) if it is optimal according to criterion P (resp. criteria S and R). Then, a $(P|S, R)$-*optimal* solution is defined to be one which is (S, R)-optimal *among* all (P)-optimal solutions.

The complexity of WSP for interval graphs was recently investigated with the single optimization criterion P. Bodlaender and Jansen [2] have shown that this is a $\mathcal{NP}$-hard problem even for fixed $k \geq 4$; the problem for $k = 3$ remains an open question at our knowledge. For $k = 2$, this is solved in linear time and space by matching techniques [1,5]. Unless $\mathcal{P} = \mathcal{NP}$, the inherent hardness of the problem condemns us to design efficient heuristics for finding "good" solutions. In this way, linear-time approximations are presented for the WSP in the second part of the paper (Section 3). A classical algorithm is briefly described which achieves a constant worst-case ratio for the single criterion P. Unfortunately, such an algorithm offers no guarantee on the satisfiability of criteria S and R. Surprisingly, the WSP problem for proper interval graphs is proved to be solvable in a $(P|R, S)$-optimal way by a greedy algorithm. Thus, an idea is to partition the input interval graph into proper interval subgraphs and solve optimally the problem on each subgraph using the greedy. Obviously, the quality of such a *local optimization* depends strongly on how the input interval graph is partitionned. Hence, the theorem previously cited enables us to design a new algorithm which achieves a logarithmic worst-case ratio for criterion P, but moreover guarantees that $(P|R, S)$-*optima* are reached in a logarithmic number of subproblems. Finally, we remark that in real-life situations, *ie.* under certain conditions, the logarithmic worst-case ratio becomes *constant*.

Preliminaries. Before giving the first results, some useful notations and definitions are detailed. All the graph-theoretic terms not defined here can be found in [13,6]. Let $G = (V, E)$ be an undirected graph. For simplicity, n and m denote respectively the number of vertices and edges of G throughout the paper. A *complete set* or *clique* is a set of pairwise connected vertices. The *clique number* $\omega(G)$ is the cardinality of the largest clique in G. On the opposite, an *independent set* or *stable* is a set of pairwise non-connected vertices. A *coloring* of G associates to each vertex one color in such a way that two connected vertices have different colors. In fact, a coloring of G corresponds to a partition of G into stables. The *chromatic number* $\chi(G)$ is the cardinality of a partition of G into the least number of stables. In the same way, $\chi(G, k)$ is defined to be the size of a minimum partition of G into stables of size at most k. The quality of our approximation algorithms in relation to the criterion P is measured by their *worst-case ratio* defined as $\sup_G \{|\mathcal{S}|/\chi(G, k)\}$ where $\mathcal{S}$ is any partition of G into stables of size at most k output by the algorithm.

2   The Proof of Theorem 1

Although offering only a linear upper bound, the following lemma is crucial in the proof of the theorem.

Lemma 1. *Let $G = (V, E)$ be a $K_{1,t}$-free interval graph with $t \geq 3$. Then G admits a partition into $\lfloor t/2 \rfloor$ proper interval subgraphs. Moreover, this partition is computed in linear time and space.*

Proof. An algorithm is proposed for computing such a partition. Synthetically, the algorithm extracts and colors greedily some cliques of G with the set of colors $\{1, \dots, \lfloor t/2 \rfloor\}$; the output is the partition of G induced by these $\lfloor t/2 \rfloor$ colors.

> **Algorithm ColorCliques**
> **input:** a $K_{1,t}$-free interval graph $G = (V, E)$ with $t \geq 3$;
> **output:** a partition of G into $\lfloor t/2 \rfloor$ proper interval subgraphs;
> **begin**
>   compute an interval representation $I_1, \dots, I_n$ of G;
>   order $I_1, \dots, I_n$ according to the left endpoints;
> $C^1 \leftarrow \cdots \leftarrow C^{\lfloor t/2 \rfloor} \leftarrow \emptyset, i \leftarrow 1, j \leftarrow 1$;
>   **while** $i \leq n$ **do**
> $C_j \leftarrow \{I_i\}, I_{left} \leftarrow I_i, i \leftarrow i + 1$;
>   **while** $i \leq n$ and $I_{left} \cap I_i \neq \emptyset$ **do**
> $C_j \leftarrow C_j \cup \{I_i\}$;
>   **if** $re(I_i) < re(I_{left})$ **then** $I_{left} \leftarrow I_i$;
> $i \leftarrow i + 1$;
> $c \leftarrow (j - 1) \bmod \lfloor t/2 \rfloor + 1, C^c \leftarrow C^c \cup \{C_j\}, j \leftarrow j + 1$;
>   **return** $C^1, \dots, C^{\lfloor t/2 \rfloor}$;
> **end;**

Since computing an ordered interval representation is done in $O(n + m)$ time and space [4,9], the algorithm runs in linear time and space. This correctness is established by showing that the color class $\mathcal{C}^c$ induces a proper interval graph for any $c \in \{1, \dots, \lfloor t/2 \rfloor\}$. Let $\mathcal{C}^c = \{C_1^c, \dots, C_q^c\}$ be the set of cliques assigned to $\mathcal{C}^c$ by the algorithm (in the order of their extraction). If $q \leq 2$ then $\mathcal{C}^c$ is trivially $K_{1,3}$-free. Otherwise, suppose that $\mathcal{C}^c$ contains an induced subgraph $K_{1,3}$ with I_a its central vertex and $I_b \prec I_c \prec I_d$ its three leaves. Clearly, the leaves belong to disjoint cliques: set $I_b \in C_u^c$, $I_c \in C_v^c$ and $I_d \in C_w^c$ with $u < v < w \in \{1, \dots, q\}$. According to the algorithm, I_a belongs necessarily to C_u^c. Now, from every clique C_j colored by the algorithm between C_u^c and C_w^c, select the interval having the smallest right endpoint in C_j and add it to the set S initially empty. We claim that S induces a stable of size at least $2\lfloor t/2 \rfloor + 1$. If two intervals of S are intersecting, then they belong to the same colored clique, a contradiction. At least $\lfloor t/2 \rfloor$ cliques are colored by the algorithm from C_u^c to C_v^c exclusive and still at least $\lfloor t/2 \rfloor$ from C_v^c to C_w^c exclusive. Thus, S contains at least $2\lfloor t/2 \rfloor + 1$ elements, which proves the claim. Since $I_a \in C_u^c$ and $I_a \cap I_d \neq \emptyset$, I_a intersects every interval in S except maybe the one most to right which belongs to C_w^c. This last interval is replaced in S by the interval I_d; in effect, I_d cannot intersect the last but one interval of S (otherwise $I_d \notin C_w^c$, a contradiction). Finally, since $2\lfloor t/2 \rfloor + 1 \geq t$ for all $t \geq 3$, we obtain that at least t disjoint intervals are overlapped by I_a, which is in contradiction with the fact that G is $K_{1,t}$-free. Therefore, the color class C^c induces well a $K_{1,3}$-free interval graph, ie. a proper interval graph by Roberts theorem (cf. [13,6]), and the whole correctness of the algorithm is established.   □

Remark. In Algorithm ColorCliques, the assignment of colors is done according to the basic ordering $\{1, \ldots, \lfloor t/2 \rfloor\}$. The correctness holds by using any permutation of the set $\{1, \ldots, \lfloor t/2 \rfloor, 1, \ldots, \lfloor t/2 \rfloor\}$, repeated as many time as necessary to complete the assignment (the proof remains the same). Notably, this implies that there exists at least $(2t)!/2^t t!$ non-isomorphic partitions of a $K_{1,t}$-free interval graph into proper interval graphs. Note that determining the minimum value t for which G is $K_{1,t}$-free can be done in $O(n^2)$ time by computing the largest stable [7] contained in each interval of its representation $I_1, \ldots, I_n$.

Lemma 2. *Any interval graph $G = (V, E)$ admits a partition into less than $\lceil \log_3((n + 1)/2) \rceil$ $K_{1,5}$-free interval subgraphs. Moreover, this partition is computed in linear time and space.*

Before giving the proof of the lemma, we need to establish this useful claim.

Claim. Any interval graph $G = (V, E)$ admits an open (resp. closed) interval representation such that every interval has positive integer endpoints lower than n (resp. $2n$). Moreover, this representation is computed in linear time and space.

Proof. Let $A = (a_{ij})$ be the maximal cliques-versus-vertices incidence matrix of G. A $(0, 1)$-matrix has the consecutive 1's property for columns if its rows can be permuted in such a way that the 1's in each column occur consecutively. A well-known characterization of interval graphs is that the matrix A has the consecutive 1's property for columns and no more than n rows (Fulkerson-Gross 1965, cf. [6]). Thereby, consider a representation of A with the 1's consecutive in each column and for each $v \in V$, set $le(I_v) = \min\{i \mid a_{iv} = 1\}$ and $re(I_v) = \max\{i \mid a_{iv} = 1\}$. Clearly, the open interval representation $\{I_v\}_{v \in V}$ is such that every endpoint is in $\{1, \ldots, n\}$. This interval representation is correct because two intervals are intersecting if and only if their two corresponding vertices are connected. Computing the matrix A with consecutive 1's is done in $O(n + m)$ time and space [9]. Therefore, the complexity of the previous construction is linear. Finally, a closed interval representation is obtained from the previous open interval representation. Sort all the endpoints (left and right mixed) in the ascendant order. For $i = 1, \ldots, 2n$, assign to the ith endpoint the value i and then redefine the n intervals as closed with their new endpoints in $\{1, \ldots, 2n\}$. Since the order on the endpoints is unchanged, the interval graph remains the same. Moreover, sorting $2n$ integers in $\{1, \ldots, n\}$ is done in $O(n)$ time using $O(n)$ space, which concludes the proof.   □

Proof (of Lemma 2). According to the Claim, compute in linear time and space an open interval representation $I_1, \ldots, I_n$ of G with endpoints in $\{1, \ldots, n\}$ and denote by ℓ the maximum length of an interval ($\ell \leq n - 1$). Then, partition the intervals according to their length into $\lceil \log_3((\ell + 2)/2) \rceil$ subsets as follows: $\mathcal{I}_1$ contains the intervals of length $\{1, 2, 3, 4\}$, $\mathcal{I}_2$ the intervals of length $\{5, 6, \ldots, 16\}, \ldots, \mathcal{I}_i$ the intervals of length $\{2.3^{i-1} - 1, \ldots, 2.3^i - 2\}$. We affirm that each subset $\mathcal{I}_i$ induces a $K_{1,5}$-free interval graph. Indeed, the contrary implies that one interval of $\mathcal{I}_i$ contains properly three disjoint intervals whose sum of lengths is lower than $2.3^i - 4$, which is a contradiction (the minimum sum of three intervals is $3(2.3^{i-1} - 1) = 2.3^i - 3$). Note that the proof remains correct by starting with a closed interval representation with endpoints in $\{1, \ldots, 2n\}$ and partitioning such that each set $\mathcal{I}_i$ contains the intervals of length $\{4.3^{i-1} - 3, \ldots, 4.3^i - 4\}$ for $i = 1, \ldots, \lceil \log_3((\ell + 4)/4) \rceil$ (here $\ell \leq 2n - 1$). $\quad\square$

Remark. In fact, we can prove more generally that any interval graph $G = (V, E)$ admits a partition into $O(\log_t n)$ $K_{1,t+2}$-free interval subgraphs for any integer $t \geq 3$.

Proposition 1. *Any interval graph (resp. circular-arc graph) $G = (V, E)$ admits a partition into less than $2\lceil \log_3((n + 1)/2) \rceil$ (resp. $2\lceil \log_3((n + 1)/2) \rceil + 1$) proper interval subgraphs. Moreover, this partition is computed in linear time and space.*

Proof. The proof of the bound for interval graphs follows immediately the combination of Lemmas 2 and 1 (with $t = 5$). For circular-arc graphs, compute first a circular-arc representation of G in linear time and space [10]. Now, choose one point p on the circle and compute the set of vertices V^* corresponding to the arcs which contain p. By observing that V^* forms a clique and the subgraph induced by $V \setminus V^*$ is an interval graph, we obtain the desired bound for circular-arc graphs (any clique induces trivially a proper interval graph). $\quad\square$

The first half of Theorem 1 is established through the previous proposition, while the second is established via the next proposition.

Proposition 2. *For infinitely many r, the complete r-partite graph $H_r = (S_1 \cup \cdots \cup S_r, E)$ with $|S_1| = 1, \ldots, |S_r| = 3^{r-1}$ admits no partition into less than $\log_3(2n + 1)$ proper interval subgraphs.*

Proof. An interval representation of the graph H_r is built by defining recursively the r stables $S_1, \ldots, S_r$ as follows. The stable S_1 consists of one open interval of length 3^{r-1}. For all $i = 2, \ldots, r$, the stable S_i is obtained by copying the stable S_{i-1} and subdivising each interval of this one into three open intervals of equal length. The resulting stables $S_1, \ldots, S_r$ induce well a complete r-partite graph. Note that the number of vertices of H_r is given by $(*)$ $n = \sum_{i=1}^{r} 3^{i-1} = (3^r - 1)/2$.

Since any stable induces trivially a proper interval graph, H_r admits a partition into r proper interval subgraphs. Now, using induction, we show that any minimum partition of H_r into proper interval subgraphs has the cardinality $p(H_r) = r$. First, one can easily verify that $p(H_1) = 1$ or $p(H_2) = 2$; then, the induction basis is $p(H_{i-1}) = i - 1$ for $i > 2$. Now, suppose that $p(H_i) < i$ and consider a partition of H_i into $i-1$ sets $\mathcal{I}_1, \ldots, \mathcal{I}_{i-1}$ of proper intervals. Without loss of generality, the single interval $I^* \in S_1$ belongs to $\mathcal{I}_1$. We claim that the intervals of $\mathcal{I}_1 \setminus I^*$ induce at most two disjoint cliques. In effect, the contrary implies the existence of an induced subgraph $K_{1,3}$ in $\mathcal{I}_1$ (with I^* as central vertex and one interval in each disjoint clique as leaves). According to this claim, at least one interval of S_2 and all the intervals stemming from its subdivision in $S_3, \ldots, S_i$ do not belong to $\mathcal{I}_1$. Clearly, such a set of intervals induces the graph H_{i-1} and by induction hypothesis, needs $i - 1$ sets to be partitionned into proper interval subgraphs. However, only the $i - 2$ sets $\mathcal{I}_2, \ldots, \mathcal{I}_{i-1}$ are available to realize that, which leads to a contradiction. This completes the induction by obtaining that $p(H_i) = i$ for $i > 2$. The equality $(*)$ is finally used to obtain $p(H_r) = \log_3(2n + 1)$. $\qquad\square$

Corollary 1. *For every* $t \geq 3$, *a* $K_{1,t}$*-free interval graph with at most* $\lfloor (3t - 4)/2 \rfloor$ *vertices exists which admits no partition into less than* $\lfloor \log_3(t - 1) \rfloor + 1$ *proper interval subgraphs.*

Proof. The graph H_r defined in Proposition 2 is clearly $K_{1,t}$-free for $t \in \{3^{r-1} + 1, \ldots, 3^r\}$. By simple calculation, we deduce that H_r has at most $\lfloor (3t - 4)/2 \rfloor$ vertices and admits no partition into less than $\lfloor \log_3(t - 1) \rfloor + 1$ proper interval subgraphs for $t \in \{3^{r-1} + 1, \ldots, 3^r\}$. $\qquad\square$

3   Applications to Working Schedules Planning

A classical approximation. In this subsection, a classical algorithm is presented to approximate WSP with interval graphs. Here are two propositions, partially established in [5], which are behind its proof.

Proposition 3. *A minimum coloring of an interval graph $G = (V, E)$ such that the number $s(G)$ of stables consisting of only one vertex is as small as possible is computed in linear time and space.*

Proposition 4. *Let $G = (V, E)$ be an interval graph and k an integer. If G is colored such that each color is used at least k times, then G admits an optimal partition into $\lceil n/k \rceil$ stables of size at most k. Moreover, this partition is computed in linear time and space given the coloring in input.*

Algorithm 2-ApproxWSP
input: an interval graph $G = (V, E)$, an integer k;
output: a solution $\mathcal{S}$ to the WSP problem for G;
begin
    compute a minimum coloring $\mathcal{C} = \{S_1, \ldots, S_{\chi(G)}\}$ of G with $s(G)$ minimum;
    $\mathcal{S} \leftarrow \emptyset$;
    for each $S_i \in \mathcal{C}$ **do**
        if $|S_i| < k$ **then** $\mathcal{C} \leftarrow \mathcal{C} \setminus \{S_i\}$, $\mathcal{S} \leftarrow \mathcal{S} \cup \{S_i\}$;
    compute an optimal partition $\mathcal{S}_k$ of $\mathcal{C}$ into stables of size at most k;

$\mathcal{S} \leftarrow \mathcal{S} \cup \mathcal{S}_k;$
    return $\mathcal{S}$;
end;

Theorem 2. *Algorithm 2-ApproxWSP achieves in linear time and space the asymptotic worst-case ratio $2(k-1)/k$ for the criterion P. Moreover, this worst-case ratio is tight.*

Proof. Omitted here.   □

Remark. A similar algorithm can be designed to approximate WSP for circular-arc graphs with worst-case ratio 3: first determine in linear time a coloring using less than $2\,\omega(G)$ colors and then use Proposition 4, which remains correct for circular-arc graphs, to find a solution to WSP.

A greedy for proper interval graphs. Here a greedy algorithm is presented which solves the WSP problem for proper interval graphs.

> **Algorithm GreedyProperWSP**
> **input:** a proper interval graph $G = (V, E)$, an integer k;
> **output:** a solution $\mathcal{S}$ to the WSP problem for G;
> **begin**
>   compute a proper interval representation $I_1, \ldots, I_n$ of G;
>   order $I_1, \ldots, I_n$ according to the left endpoints;
>   compute $\omega(G)$ and $\chi(G, k) \leftarrow \max\{\omega(G), \lceil n/k \rceil\}$;
> $S_1 \leftarrow \cdots \leftarrow S_{\chi(G,k)} \leftarrow \emptyset$;
>   **for** i **from** 1 **to** n **do**
> $j \leftarrow (i-1) \bmod \chi(G, k) + 1,\ S_j \leftarrow S_j \cup \{I_i\}$;
> $\mathcal{S} \leftarrow \{S_1, \ldots, S_{\chi(G,k)}\}$;
>   **return** $\mathcal{S}$;
> **end**;

Computing an ordered proper interval representation of G is done in $O(n+m)$ time and space [3] and $\omega(G)$ is computed in $O(n)$ time [7]. Consequently, the algorithm runs in linear time and space.

Lemma 3. *The output solution $\mathcal{S}$ is $(P|S)$-optimal.*

Proof. First, we claim that the output stables $S_1, \ldots, S_{\chi(G,k)}$ have a size at most k. According to the algorithm, the stables have the same size (to within one unity if n is not a multiple of k). Then, the existence of one stable of size strictly larger than k implies that $n > k\chi(G, k)$, a contradiction. Additionally, this establishes the (S)-optimality of $\mathcal{S}$. Now, suppose that two intervals I_u, I_v with $u < v$ are intersecting in the stable S_j for any $j \in \{1, \ldots, \chi(G, k)\}$. By the algorithm, we have $u = j + \alpha\chi(G, k)$ and $v = j + \beta\chi(G, k)$ with $\alpha < \beta$. When the intervals are proper, the right endpoints have the same order as the left endpoints. Then, the intervals $I_u, I_{u+1}, \ldots, I_{v-1}, I_v$ include the portion $[l_v, r_u]$ of the real line and also induce a clique of size $v - u + 1 = (\beta - \alpha)\chi(G, k) + 1 \geq \chi(G, k) + 1$. Such a clique implies that $\omega(G) > \chi(G, k)$, which is a contradiction and the correctness of the solution S is entirely proved. To conclude, S is $(P|S)$-optimal because $\max\{\omega(G), \lceil n/k \rceil\}$ is a lower bound for $\chi(G, k)$. □

Lemma 4. *The output solution S is $(P|R)$-optimal.*

Proof (Sketch). The (P)-optimality of S is established by Lemma 2. Now, suppose that the set $S_1, \ldots, S_{\chi(G,k)}$ is not $(P|R)$-optimal. Define $S_1^*, \ldots, S_{\chi(G,k)}^*$ to be a $(P|R)$-optimal solution and g^* the minimum gap between two consecutive intervals of this solution. Remind that the intervals $I_1, \ldots, I_n$ are ordered according to the left endpoints and $I_{v,t}$ denotes the interval of rank t in the stable S_v^*. We claim that for all $i = 1, \ldots, n$, the interval $I_i \in S_u^*$ can be moved at the rank $t = \lfloor (i-1)/\chi(G,k) \rfloor + 1$ of the stable set S_v^* with $v = (i-1) \bmod \chi(G, k) + 1$, without decreasing g^*. After such an operation, the resulting set $S_1^*, \ldots, S_{\chi(G,k)}^*$ coincide exactly with the solution $S_1, \ldots, S_{\chi(G,k)}$ of the greedy, which establishes its $(P|R)$-optimality. The claim is proved by an inductive process whose initial step is done as follows. If $I_1 \in S_u^*$ with $u \neq 1$, exchange the entire set of intervals of S_u^* with the one of S_1^*. Clearly, g^* is not deteriored (no gap is modified) and I_1 is correctly placed. Now, the inductive step is proved; the intervals $I_1, \ldots, I_{i-1}$ are considered to be correctly placed. The interval $I_i \in S_u^*$ shall be moved to the stable S_v^* if $u \neq v$. Then, two cases are distinguished.

Case $u < v$:
$S_u^* = \{I_{u,1}, \ldots, I_{u,t}, I_i, \ldots, I_{u,j}, \ldots\}$ and $S_v^* = \{I_{v,1}, \ldots, I_{v,t-1}, I_{v,t}, \ldots, I_{v,j}, \ldots\}$. By induction hypothesis, we get $re(I_{v,t-1}) \leq re(I_{u,t})$ and $le(I_i) \leq le(I_{v,t})$. Since $re(I_{u,t}) < le(I_i)$, we obtain the inequalities (i) $re(I_{v,t-1}) \leq re(I_{u,t}) < lc(I_i) \leq le(I_{v,t})$ which allow us to redefine $S_u^* = \{I_{u,1}, \ldots, I_{u,t}, I_{v,t}, \ldots, I_{v,j}, \ldots\}$ and $S_v^* = \{I_{v,1}, \ldots, I_{v,t-1}, I_i, \ldots, I_{u,j}, \ldots\}$. Two gaps are changed: $le(I_i) - re(I_{u,t})$ in S_u^* becomes $le(I_{v,t}) - re(I_{u,t})$ and $le(I_{v,t}) - re(I_{v,t-1})$ in S_v^* becomes $le(I_i) - re(I_{v,t-1})$. According to (i), the new gaps are larger than the minimum of the two old ones.

Case $u > v$:
$S_u^* = \{I_{u,1}, \ldots, I_{u,t-1}, I_i, \ldots, I_{u,j}, \ldots\}$ and $S_v^* = \{I_{v,1}, \ldots, I_{v,t-1}, I_{v,t}, \ldots, I_{v,j}, \ldots\}$. Here induction hypothesis provide the inequalities (ii) $re(I_{v,t-1}) \leq re(I_{u,t-1}) < le(I_i) \leq le(I_{v,t})$ and we redefine $S_u^* = \{I_{u,1}, \ldots, I_{u,t-1}, I_{v,t}, \ldots, I_{v,j}, \ldots\}$ and $S_v^* = \{I_{v,1}, \ldots, I_{v,t-1}, I_i, \ldots, I_{u,j}, \ldots\}$. According to (ii), the two new gaps in S_u^* and S_v^* are still larger than the minimum of the two old ones.

The analysis of these two cases shows the correctness of the inductive step and completes the proof of the claim. □

Theorem 3. *Algorithm GreedyProperWSP determines in linear time and space $(P|S, R)$-optimal solutions to the problem WSP for proper interval graphs.*

The logarithmic approximation with sub-optima. According to the previous discussions, a new approximation algorithm is designed for WSP with interval graphs.

> **Algorithm log-ApproxWSP**
> **input:** an interval graph $G = (V, E)$, an integer k;
> **output:** a solution S to the WSP problem for G;
> **begin**
> $\quad S \leftarrow \emptyset$;
> $\quad$ **if** G is a proper interval graph **then** $S \leftarrow$ GreedyProperWSP(G, k);
> $\quad$ **else**
> $\quad\quad$ partition G into $B(n)$ proper interval subgraphs $G_1, \ldots, G_{B(n)}$;
> $\quad\quad$ **for** each subgraph G_i **do** $S \leftarrow S \cup$ GreedyProperWSP(G, k);
> $\quad$ **return** S;
> **end;**

Theorem 4. *Algorithm* log-*ApproxWSP achieves in linear time and space the absolute worst-case ratio* $\min\{k, B(n)\}$ *with* $B(n) = 2\lceil \log_3((n+1)/2)\rceil$ *for the criterion* P *and guarantees that* $(P|S, R)$-*optima are reached in* $B(n)$ *subproblems. Moreover, the worst-case ratio is asymptotically tight.*

Proof. Correctness and complexity follow from Theorems 1 and 3, plus the fact that recognizing a proper interval graph is done in linear time and space [3]. To complete the proof, the worst-case ratio is established. If G is a proper interval graph then S is optimal. Otherwise, we have $|S| = \sum_{i=1}^{B(n)} \chi(G_i, k)$. By using the inequalities $\sum_{i=1}^{B(n)} \chi(G_i, k) \leq n \leq k \cdot \chi(G, k)$ and $\sum_{i=1}^{B(n)} \chi(G_i, k) \leq \sum_{i=1}^{B(n)} \chi(G, k) \leq B(n) \cdot \chi(G, k)$, we obtain the result.

Finally, an interval graph G is given which tights asymptotically the ratio $\min\{k, B(n)\}$ with $B(n) = 2\lfloor \log_3((n+1)/2)\rfloor$ and $k = B(n)$. The complete proof is not detailed here; without loss of generality, we assume that n is a multiple of $B(n)$ and set $N(n) = n/B(n) - 1$. The interval graph is modeled by the following set of open intervals. For $i = 1, \ldots, B(n)/2$, take one interval $(1, 2.3^i - 1)$, one interval $(1, 2.3^{i-1})$, $N(n)$ intervals $(2.3^{i-1}, 4.3^{i-1} - 1)$ and $N(n)$ intervals $(4.3^{i-1} - 1, 2.3^i - 2)$ (see Fig. 1 above for an example of construction). Note that the endpoints are well in $\{1, \ldots, n\}$ and G is not a proper interval graph. In this case, one can verify that the approximation ratio of Algorithm log-ApproxCIG$_k$ is

$$\frac{|S|}{\chi(G, k)} = \frac{(B(n)/2)(2N(n)+1)}{2(B(n)/2) + N(n) - 1} = B(n) \cdot \frac{n - B(n)/2}{n + B^2(n) - 2B(n)} \xrightarrow[n \to \infty]{} B(n) = k.$$

$\square$

Remark. Algorithm log-ApproxCIG$_k$ produces $(P|S, R)$-optimal solutions when G is a proper interval graph. Besides, in real-life situations [12], the minimum

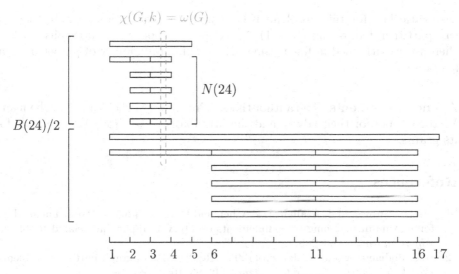

Fig. 1. An example of construction which tights the worst-case ratio with $n = 24$ ($k = B(24) = 4$, $N(24) = 5$): $\chi(G, k) = 8$ and $|\mathcal{S}| = 22$.

value t for which G is $K_{1,t}$-free is generally *small* (≤ 9). This allows direct partitionings into proper interval subgraphs by Algorithm ColorCliques and also the obtaining of *constant* worst-case ratios (≤ 4) for the criterion P. For example, for tasks of $1, 2, 3$ or 4 hours, we can obtain a 2-approximation and for tasks of $1, 2, \ldots, 8$ hours, a 3-approximation. Moreover, Algorithm log-ApproxWSP can be easily adapted for circular-arc graphs. In this case, its "real-life" worst-case ratio is nearly the same than the one obtained by the classical approach.

4   Conclusion

As a conclusion, we discuss some projections on the complexity of determining a minimum partition of a interval graph into proper interval subgraphs. In effect, answering to the mathematician has provided some hints for answering to the computer scientist.

First, we know now that a minimum partition of a $K_{1,5}$-free interval graph G into proper interval subgraphs is computed in linear time and space: if G is not a proper interval graph, then we can use Lemma 1 to partition G into 2 proper interval graphs (recognizing proper interval graph is done in linear time and space [3]). For $K_{1,6}$-free interval graphs (and also for arbitrary interval graphs), we conjecture that the problem is $\mathcal{NP}$-complete.

Finding a polynomial-time approximation algorithm with constant worst-case ratio for the problem seems to be difficult too. However, combining the previous remark with Lemma 2 enables us to design a linear-time approximation algorithm, similar to Algorithm log-ApproxWSP, which achieves the worst-case ratio $\ln n$ for this problem: if G is not a proper interval graph, then we can partition it into $\lceil \log_3((n+1)/2) \rceil$ $K_{1,5}$-free proper interval graphs (each of then are partitionned in linear time into a minimum number of proper interval subgraphs).

Acknowledgements. The author thanks Professors Michel Van Caneghem and Victor Chepoi for their advice and the firm PROLOGIA - Groupe Air Liquide for its grants.

References

1. M.G. Andrews, M.J. Atallah, D.Z. Chen and D.T. Lee (2000). Parallel algorithms for maximum matching in complements of interval graphs and related problems. *Algorithmica* 26, 263–289.
2. H.L. Bodlaender and K. Jansen (1995). Restrictions of graph partition problems. Part I. *Theoretical Computer Science* 148, 93–109.
3. D.G. Corneil, H. Kim, S. Natarajan, S. Olariu and A. Sprague (1995). Simple linear time recognition of unit interval graphs. *Information Processing Letters* 55, 99–104.
4. D.G. Corneil, S. Olariu and L. Stewart (1998). The ultimate interval graph recognition algorithm ? In *Proc. 9th Annual ACM-SIAM Symposium on Discrete Algorithms*, pages 175–180, ACM Publications, New-York, NY.
5. F. Gardi (2003). Efficient algorithms for disjoint matchings among intervals and related problems. In *Proc. 4th International Conference on Discrete Mathematics and Theoretical Computer Science, LNCS* 2731, 168–180.
6. M.C. Golumbic (1980). *Algorithmic Graph Theory and Perfect Graphs*. Computer Science and Applied Mathematics Series, Academic Press, New-York, NY.
7. U.I. Gupta, D.T. Lee and J.Y.-T. Leung (1982). Efficient algorithms for interval graphs and circular-arc graphs. *Networks* 12, 459–467.
8. P.C. Fishburn (1985) *Interval Orders and Interval Graphs*. John Wiley & Sons, New-York, NY.
9. M. Habib, R. McConnel, C. Paul and L. Viennot (2000). Lex-BSF and partition refinement, with applications to transitive orientation, interval graph recognition and consecutive ones testing. *Theoretical Computer Science* 234, 59–84.
10. R. McConnel (2001). Linear time recognition of circular-arc graphs. In *Proc. 42nd Annual IEEE Symposium on Foundations of Computer Science*, pages 386–394, IEEE Computer Society Publications, Los Alamitos, CA.
11. S. Olariu (1991). An optimal greedy heuristic to color interval graphs. *Information Processing Letters* 37, 21–25.
12. BAMBOO - Planification by PROLOGIA - Groupe Air Liquide. http://prologianet.univ-mrs.fr/bamboo/bamboo_planification.html
13. F.S. Roberts (1978). *Graph Theory and its Application to the Problems of Society*. SIAM Publications, Philadelphia, PA.

Collective Tree Exploration

Pierre Fraigniaud[1]*, Leszek Gąsieniec[2]**, Dariusz R.Kowalski[3,4]***, and
Andrzej Pelc[5]†

[1] CNRS-LRI, Université Paris-Sud, 91405 Orsay, France, http://www.lri.fr/~pierre.
[2] Department of Computer Science, The University of Liverpool, Liverpool L69 7ZF, UK,
leszek@csc.liv.ac.uk.
[3] Max-Planck-Institut für Informatik, Stuhlsatzenhausweg 85, Saarbrücken, 66123 Germany.
[4] Instytut Informatyki, Uniwersytet Warszawski, Banacha 2, 02-097 Warszawa, Poland,
darek@mimuw.edu.pl.
[5] Département d'informatique, Université du Québec en Outaouais, Hull, J8X 3X7, Canada,
pelc@uqo.ca.

Abstract. An n-node tree has to be explored by k mobile agents (robots), starting
in its root. Every edge of the tree must be traversed by at least one robot, and
exploration must be completed as fast as possible. Even when the tree is known
in advance, scheduling optimal collective exploration turns out to be NP-hard. We
investigate the problem of distributed collective exploration of unknown trees. Not
surprisingly, communication between robots influences the time of exploration.
Our main communication scenario is the following: robots can communicate by
writing at the currently visited node previously acquired information, and reading
information available at this node. We construct an exploration algorithm whose
running time for any tree is only $O(k/\log k)$ larger than optimal exploration
time with full knowledge of the tree. (We say that the algorithm has *overhead*
$O(k/\log k)$). On the other hand we show that, in order to get overhead sublinear
in the number of robots, some communication is necessary. Indeed, we prove that
if robots cannot communicate at all, then every distributed exploration algorithm
works in time $\Omega(k)$ larger than optimal exploration time with full knowledge, for
some trees.

1 Introduction

A collection of robots (mobile agents), initially located at one node of an undirected
connected graph, have to explore this graph and return to the starting point. The graph is
explored if every edge is traversed by at least one robot. Every robot traverses any edge in
unit time, and the time of collective exploration is the maximum time used by any robot

* Research supported by the Actions Spécifiques CNRS *Dynamo* and *Algorithmique des grands
graphes*, and by the project *PairAPair* of the ACI *Masse de données*.
** Research partially supported by the EPSRC grant GR/R84917/01.
*** This work was done in part during this author's stay at the Research Chair in Distributed
Computing of the Université du Québec en Outaouais, as a postdoctoral fellow. Research
supported in part by KBN grant 4T11C04425.
† Research supported in part by NSERC grant OGP 0008136 and by the Research Chair in
Distributed Computing of the Université du Québec en Outaouais.

M. Farach-Colton (Ed.): LATIN 2004, LNCS 2976, pp. 141–151, 2004.
© Springer-Verlag Berlin Heidelberg 2004

from the group. It turns out that scheduling optimal collective exploration is NP-hard, even in the simplest case, when the explored graph is a tree and when it is known in advance. However, most often, exploration problems are studied in the case of unknown graphs (cf. [1,6,12,14,15,16,17,21]). This is also the approach adopted in the present paper. We restrict attention to *trees* and, unlike in the above quoted papers, we consider exploration by *many* robots. The goal is to collectively explore the tree in the shortest possible time. Since the explored tree is not known in advance, a collective exploration algorithm can have different performance in different trees. In order to measure the quality of such an algorithm, we compare its performance to the performance of the optimal exploration algorithm which knows the tree in advance (recall that designing such an optimal exploration is NP-hard). A collective exploration algorithm $\mathcal{A}$ for k robots (working in unknown trees) is said to have *overhead* Q, if Q is the supremum of ratios $\mathcal{A}(k, T, r)/opt(k, T, r)$, where $\mathcal{A}(k, T, r)$ is the exploration time of tree T by algorithm $\mathcal{A}$, when robots start at node r, and $opt(k, T, r)$ is the optimal exploration time of T by k robots starting at r, assuming that T and r are known. The supremum is taken over all trees T and starting nodes r. Hence overhead is a measure of performance similar to competitive ratio for on-line algorithms. We seek collective exploration algorithms with low overhead. If the explored tree was known in advance, any exploration algorithm could be viewed as centralized, since it could assume knowledge of global history by any robot at any step. However, in our case, when the topology of the tree is unknown, distributed control of robots implies that their knowledge at any step of the exploration depends on communication between them. Below we specify communication scenarios.

1.1 The Model

We consider k robots initially located at the root r of an unknown tree T. Robots have distinct identifiers. Apart from that, they are identical. Each robot knows its own identifier and follows the same exploration algorithm which has the identifier as a parameter. The network is anonymous, i.e., nodes are not labeled, and ports at each node have only local labels which are distinct integers between 1 and the degree of the node. The robots move as follows. At every exploration step, every robot either traverses an edge incident to its current position, or remains in the current position. A robot traversing an edge knows local port numbers at both ends of the edge.

Our main communication scenario, called *exploration with write-read communication*, is the following. In every step of the algorithm every robot performs the following three actions: it moves to an adjacent node, writes some information in it, and then reads all information available at this node, including its degree. Alternatively, a robot can remain in the current node, in which case it skips the writing action. Actions are assumed to be synchronous: if A is the set of robots that enter v in a given step, then first all robots from A enter v, then all robots from A write and then all robots currently located at v (those from A and those that have not moved from v in the current step) read.

We also consider two extreme communication scenarios. In one, called *exploration without communication*, all robots are oblivious of each other. I.e., at each step, every robot knows only the route it traversed until this point (which is the sequence of exit and entry port numbers), and degrees of all nodes it visited. In the other, called *exploration with complete communication*, all robots can instantly communicate at each step.

In all scenarios, a robot, currently located at a node, does not know the other endpoints of yet unexplored incident edges. If the robot decides to traverse such a new edge, the choice of the actual edge belongs to the adversary, as we are interested in the worst-case performance.

1.2   Our Results

As a preliminary result, we show that the problem of finding optimal collective exploration, if the tree and the starting node are known in advance, is NP-hard. Our main result concerns collective distributed exploration of unknown trees by k robots, under the write-read communication scenario. We construct an exploration algorithm with overhead $O(k/\log k)$. Indeed, our algorithm explores any n-node tree of diameter D in time $O(D + n/\log k)$. We first describe our algorithm for the stronger scenario, exploration with complete communication, and then we show how to simulate this algorithm in the write-read model, without changing time complexity. We also prove that any algorithm must have overhead at least $2 - 1/k$ under the complete communication scenario. (This lower bound obviously carries over to the write-read communication scenario.) On the other hand we show that, in order to get overhead sublinear in the number of robots, some communication is necessary. Indeed, we prove that, under the scenario without communication, every distributed collective exploration algorithm must have overhead $\Omega(k)$. Since this is the overhead of an algorithm using only one out of k robots, our lower bound shows that exploration without communication does not allow any effective splitting of the task among robots. Comparing the upper bound on time for the scenario with write-read communication with the lower bound for the scenario without communication, shows that this difference of communication capability influences the *order of magnitude* of time of collective exploration. Even limited communication permitted by our write-read model allows robots to effectively collaborate in executing the exploration task.

1.3   Related Work

Exploration and navigation problems for robots in an unknown environment have been thoroughly investigated in recent literature (cf. the survey [23]). There are two types of models for these problems. In one of them a particular geometric setting is assumed, e.g., unknown terrain with convex obstacles [11], or room with polygonal [13] or rectangular [7] obstacles. Another approach is to model the environment as a graph, assuming that the robot may only move along its edges. The graph setting can be further specified in two different ways. In [1,8,9,14] the robot explores strongly connected directed graphs and it can move only in the direction from head to tail of an edge, not vice-versa. In [6, 12,15,16,17,21] the explored graph is undirected and the robot can traverse edges in both directions. In some papers, additional restrictions on the moves of the robot are imposed. It is assumed that the robot has either a restricted tank [6,12], forcing it to periodically return to the base for refueling, or that it is tethered, i.e., attached to the base by a rope or cable of restricted length [17]. It is proved in [17] that exploration can be done in time $O(e)$ under both scenarios, where e is the number of edges in the graph.

   Exploration of anonymous graphs presents a different type of challenges. In this case it is impossible to explore arbitrary graphs if no marking of nodes is allowed. Hence the scenario adopted in [8,9] was to allow *pebbles* which the robot can drop on nodes to recognize already visited ones, and then remove them and drop in other places. In [9] the authors compared exploration power of one robot to that of two cooperating robots with a constant number of pebbles. In [8] it was shown that one pebble is enough if the robot knows an upper bound on the size of the graph, and $\Theta(\log \log n)$ pebbles are necessary and sufficient otherwise.

In all the above papers, except [9], exploration was performed by a single robot. Exploration by many robots was investigated mostly in the context of graphs known in advance. In [18], approximation algorithms were given for the collective exploration problem in arbitrary graphs. In [4,5] the authors constructed approximation algorithms for the collective exploration problem in weighted trees. It was also observed in [4] that scheduling optimal collective exploration in weighted trees is NP-hard even for two robots. However, the argument from [4] does not work if all weights of edges are equal to 1, which we assume. It should also be noted that, while in [4,5] exploration was centralized, the main focus of this paper is a distributed approach to collective tree exploration.

Another interesting study of collective exploration in unknown environments can be found, e.g., in [24,20], in the context of a *search problem* in geometric trees and simple polygons. Finally, collective exploration is also related to the *freeze-tag* problem [2,3] in which a set of "asleep" robots must be awaken, starting with only one "awake" robot.

2 Exploration with Complete Communication

It is possible to prove that the problem of scheduling optimal collective exploration, if the tree and the starting node are known in advance (i.e., the problem of finding an exploration scheme working in time $opt(k, T, r)$), is NP-hard. However, due to the space constraints the proof of this fact is omitted here.

In this section we describe and analyze an exploration algorithm for k robots, with overhead $O(k/\log k)$, under a communication model stronger than write-read communication, namely exploration with complete communication. At every step of exploration all robots exchange messages containing all information acquired so far.

We will use the following terminology. We denote by T_u the subtree of the explored tree T, rooted at node u. T_u is *explored*, if every edge of T_u has been traversed by some robot. Otherwise, it is called *unexplored*. T_u is *finished*, if it is explored and either there are no robots in it, or all robots in it are in u. Otherwise, it is called *unfinished*. T_u is *inhabited*, if there is at least one robot in it.

Algorithm Collective Exploration
Fix a step i of the algorithm and a node v in which some robots are currently located. There are three possible (exclusive) cases.
Case 1. *Subtree T_v is finished.*
Action: if $v \neq r$, all robots from v go to the parent of v, else all robots from v stop.
Case 2. *There exists a child u of v such that T_u is unfinished.*
Let $u_1,..., u_j$ be children of v for which the corresponding trees are unfinished, ordered in increasing order of corresponding local port numbers at v. Let x_l be the number of robots currently located in T_{u_l}. Partition all robots from v into sets $A_1,...,A_j$ of sizes

$y_1,...,y_j$, respectively, so that integers $x_l + y_l$ differ by at most 1. The partition is done in such a way that indices l for which integers $x_l + y_l$ are larger by one than for some others, form an initial segment $[1, ..., z]$ in $1, ..., j$. (We will show in the proof of Lemma 1 that such a partition can be constructed). Moreover, sets A_l are formed one-by-one, by inserting robots from v in order of increasing identifiers. (Thus, the partition into sets $A_1,...,A_j$ can be done distributedly by robots from v, using knowledge that they currently have).

Action: all robots from set A_l go to u_l, for $l = 1, ..., j$.

Case 3. *For all children u of v, trees T_u are finished, but at least one T_u is inhabited.*

Action: all robots from v remain in v.

The following lemmas will be used in the analysis of this algorithm.

Lemma 1. *Let v be any node of tree T and let i be a fixed step of Algorithm Collective Exploration. Then numbers of robots in unfinished subtrees T_u, for all children u of v, differ by at most 1.*

Lemma 2. *Let T_v be a subtree of tree T, and let i be the first step in which a robot enters v in the execution of Algorithm Collective Exploration. If T_v has m edges then T_v is finished by step $i + 2m$.*

Lemma 3. *Algorithm Collective Exploration works in time $O(D + n/\log k)$ for all n-node trees of diameter D.*

Proof. Consider Algorithm Collective Exploration, working on a tree T of diameter D, rooted at r. Define a path $S = (a_0, a_1, ...)$ in T as follows. $a_0 = r$. Suppose that a_j is already defined. Among all children of a_j consider those nodes v for which T_v was finished last (there can be several such children). Define a_{j+1} to be such a child with smallest port label. The length $|S|$ of S is at most D. Intuitively, the path S leads to one of the leaves explored very late.

For any positive integer i and for any $j = 0, \dots, \log k$, denote by $p_i(j)$ the largest index of a node v on path S such that there are at least 2^j robots in T_v after step i. We will say that $p_i(j)$ *corresponds* to the node with this index. Define nodes $w_i(l)$, for $|S| \geq l \geq 1$, as follows. Let $w_i(l)$ denote the lth node on S which has at least two children u_1 and u_2, such that T_{u_1} and T_{u_2} are inhabited after step i. Let $d_i(l)$, for $l \geq 1$, denote the number of such children of node $w_i(l)$.

Define i_0 to be the last step of the algorithm satisfying the following condition: for all $i \leq i_0$, $p_i(0)$ is smaller than the length of S. We first consider only steps of the algorithm until step i_0. We define two types of such steps. A step $i \leq i_0$ of the algorithm is of type

A. if $\sum_l d_i(l) \geq \frac{1}{2}\log k$;
B. if $|\{j : p_{i+1}(j) \neq p_i(j)\}| \geq \frac{1}{4}(\log k + 1)$.

We now show that all steps of the algorithm are of one of the above types. The proof of this fact is split into the following three claims.

Claim 1. *Fix a step $i \leq i_0$ of the algorithm, and consider a node $w_i(l)$, for some $l \geq 1$. Then $|\{j : p_i(j) \text{ corresponds to node } w_i(l)\}| \leq d_i(l) + 1$.*

Let v denote the successor of $w_i(l)$ on path S (v exists by definition of i_0). Let j_0 be the smallest element in the set $\{j : p_i(j)$ corresponds to node $w_i(l)\}$. The number of robots in T_v, after step i, is $x < 2^{j_0}$. By the definition of $d_i(l)$ and by Lemma 1, the number of robots in $T_{w_i(l)}$ is less than $(x+1) \cdot d_i(l)$. We have

$$
\begin{aligned}
(x+1) \cdot d_i(l) &\leq x \cdot d_i(l) + d_i(l) \\
&\leq x \cdot 2^{d_i(l)} + 2^{d_i(l)} \\
&\leq x \cdot 2^{d_i(l)} + x \cdot 2^{d_i(l)} \\
&= x \cdot 2^{d_i(l)+1} \\
&< 2^{j_0 + d_i(l)+1} .
\end{aligned}
$$

Hence, if $p_i(j)$ corresponds to $w_i(l)$ then $j < j_0 + d_i(l) + 1$. This proves Claim 1.

Claim 2. *Fix a step $i \leq i_0$ of the algorithm. If $p_i(j)$ does not correspond to any $w_i(l)$, for $l \geq 1$, then $p_{i+1}(j) \neq p_i(j)$.*

Consider $p_i(j)$ satisfying the assumption of Claim 2. Let v denote the corresponding node on path S, and let v' denote the successor of v on S. The number of robots in T_v is equal to the number of robots in v plus the number of robots in $T_{v'}$, in view of the fact that $p_i(j)$ does not correspond to any $w_i(l)$ and of the definition of $w_i(l)$. In step $i+1$, all robots from v move to v', and all robots located in $T_{v'}$ remain in $T_{v'}$. (Indeed, since $i \leq i_0$, $T_{v'}$ has not yet been explored, hence it has not been finished, and all subtrees rooted at siblings of v' are finished and not inhabited, by the assumption that $p_i(j)$ does not correspond to any $w_i(l)$.) Hence $p_{i+1}(j)$ corresponds to v'. This proves Claim 2.

Claim 3. *All steps $i \leq i_0$ of the algorithm are either of type A or of type B.*

Fix a step $i \leq i_0$ of the algorithm and suppose that it is not of type A. Hence $\sum_l d_i(l) < \frac{1}{2} \log k$. Since $d_i(l) \geq 2$, for all $l \geq 1$, the number of indices l for which $d_i(l)$ are defined, is less than $\frac{1}{4} \log k$. It follows that $\sum_l (d_i(l)+1) < \frac{1}{2} \log k + \frac{1}{4} \log k = \frac{3}{4} \log k$. By Claim 1, the number of integers j, such that $p_i(j)$ does not correspond to any $w_i(l)$, for $l \geq 1$, is larger than $\log k + 1 - \frac{3}{4} \log k > \frac{1}{4}(\log k + 1)$. By Claim 2, the number of integers j, such that $p_{i+1}(j) \neq p_i(j)$, is also larger than $\frac{1}{4}(\log k + 1)$. Hence step i is of type B. This proves Claim 3.

We now estimate the number of steps of type A. Consider all subtrees T_u rooted at nodes u outside of S. Let x_u denote the number of edges of T_u. We have $\sum_u (x_u+1) \leq n$. Let t_u denote the number of steps during which T_u is inhabited. By Lemma 2, $\sum_u t_u \leq 2n$. In every step i of type A, at least $\sum_l (d_i(l) - 1)$ trees T_u are inhabited (subtrees T_u are rooted at nodes u outside of S, hence summands are $d_i(l) - 1$). Since $d_i(l) \geq 2$, we have $\sum_l (d_i(l) - 1) \geq (\sum_l d_i(l))/2 \geq \frac{1}{4} \log k$. Hence the number of steps of type A is at most $\frac{2n}{\frac{1}{4} \log k} = \frac{8n}{\log k}$.

Next, we estimate the number of steps of type B. We have

$$
\sum_{i \leq i_0} |\{j : p_{i+1}(j) \neq p_i(j)\}| = |\bigcup_{i \leq i_0} \bigcup_{j=0}^{\log k} \{(i,j) : p_{i+1}(j) \neq p_i(j)\}| =
$$

$$
|\bigcup_{j=0}^{\log k} \bigcup_{i \leq i_0} \{(i,j) : p_{i+1}(j) \neq p_i(j)\}| = \sum_{j=0}^{\log k} |\{i : p_{i+1}(j) \neq p_i(j)\}| \leq (\log k + 1) \cdot |S|,
$$

the last inequality following from the fact that before step i_0 all moves of robots on S are down the path S, and hence, for a given j, the size of the set $\{(i,j) : p_{i+1}(j) \neq p_i(j)\}$ is bounded by the length of S. For every step i of type B, we have $|\{j : p_{i+1}(j) \neq p_i(j)\}| \geq \frac{1}{4}(\log k + 1)$, hence the number of steps of type B is at most $\frac{|S|(\log k+1)}{\frac{1}{4}(\log k+1)} = 4|S|$. Hence, by Claim 3, we have $i_0 \leq \frac{8n}{\log k} + 4|S|$.

We finally show that the algorithm completes exploration by step $i_0 + 1 + |S|$. Let $i_1 = i_0 + 1$. Let X be the set of robots that are in the last node b of S after step i_1. In step $i_1 + 1$, all robots from X go to the parent of b, because b is a leaf. By definition of S, when a set of robots containing X moves from a node v' on S to its parent v, then $T_{v'}$ is finished and not inhabited, and consequently, by the construction of v', T_v is also finished. It follows that in the next step, all robots from v move to the parent of v. Hence the number of steps after i_1, needed to terminate the algorithm, is $|S|$. This implies that the algorithm terminates by step $i_1 + |S| = i_0 + 1 + |S|$. Hence the running time of the algorithm is at most $\frac{8n}{\log k} + 5|S| + 1 \in O(D + n/\log k)$.

Theorem 1. *Algorithm Collective Exploration has overhead $O(k/\log k)$.*

Proof. Consider any n-node tree T rooted at node r. If the diameter of T is at most $\frac{n\log k}{k}$ then the theorem follows from Lemma 3, because $opt(k,T,r) \geq 2(n-1)/k$. If the diameter of T is larger than $\frac{n\log k}{k}$ then $opt(k,T,r) \in \Omega(\frac{n\log k}{k})$, because at least one robot has to visit the leaf farthest from r. By Lemma 2, Algorithm Collective Exploration uses time $\leq 2n$, hence the overhead is $O(k/\log k)$ in this case as well.

We conclude this section by stating a lower bound on the overhead of any collective exploration under the complete communication scenario. Clearly, this lower bound also holds under the write-read communication scenario. The proof is omitted.

Theorem 2. *Any collective exploration algorithm for k robots has overhead $\geq 2 - 1/k$.*

3   Exploration with Write-Read Communication

In this section we show how Algorithm Collective Exploration can be simulated in our write-read model, without changing time complexity. Fix any node v of the tree. Let i denote the step number, and let p denote the port number at v corresponding to the parent of v; in the case $v = r$, we define $p = *$. We define the following sets:

- $\mathcal{P}_i$ is the set of ports at v corresponding to children which are roots of unfinished subtrees,
- $\mathcal{P}_i' \subseteq \mathcal{P}_i$ is the set of ports at v corresponding to children in whose subtrees there is one robot more than in subtrees of all other children. In the special case when all subtrees of children are inhabited by q robots, we define $\mathcal{P}_i' = \mathcal{P}_i$, if $q > 0$, and $\mathcal{P}_i' = \emptyset$, if $q = 0$.
- $\mathcal{R}_i$ is the set of identifiers of robots that are in v after step $i - 1$.

Let $\mathcal{K}_i = \{p, \mathcal{P}_i, \mathcal{P}'_i, \mathcal{R}_i\}$, if node v has been visited by step $i - 1$ of Algorithm Collective Exploration. Otherwise $\mathcal{K}_i$ is undefined. We refer to $\mathcal{K}_i$ as the knowledge at node v after step $i - 1$ of Algorithm Collective Exploration. The action performed by every robot located at v after step $i - 1$ depends only on $\mathcal{K}_i$ and on the identifier of the robot. Hence Algorithm Collective Exploration defines the following *action function H*. For any step i and any robot R located at v after step $i - 1$, the value of $H(\mathcal{K}_i, R)$ is one of the following:

- the port number α by which R leaves v in step i,
- 0, if R remains at v in step i,
- $*$, if R stops.

We construct a simulation of Algorithm Collective Exploration in the write-read communication model. The new algorithm is called Algorithm Write-Read. It operates in *rounds* logically corresponding to steps of Algorithm Collective Exploration. Each round $i > 0$ consists of three steps, $3i, 3i + 1, 3i + 2$, and round 0 consists of two steps, 1 and 2. Each step is in turn divided into three stages: in Stage 1 robots move, in Stage 2 they write information in their location, and in Stage 3 they read information previously written in their location.

Recall that, in the write-read model, any robot R entering node v can write some information in this node. In the Algorithm Write-Read, a robot R entering node v in step i using port α, writes the triplet (i, R, α) at node v. Denote by $\mathcal{I}_i$ the set consisting of the degree of v and of all triplets written at node v until step $i - 1$ of Algorithm Write-Read.

We now define the knowledge $\hat{\mathcal{K}}_i$ at v after round $i - 1$ of Algorithm Write-Read. If no triplets are written at node v then $\hat{\mathcal{K}}_i$ is not defined. Otherwise, we define $\hat{\mathcal{K}}_i = \{p, \hat{\mathcal{P}}_i, \hat{\mathcal{P}}'_i, \hat{\mathcal{R}}_i\}$, where $\hat{\mathcal{P}}_i, \hat{\mathcal{P}}'_i, \hat{\mathcal{R}}_i$ are defined with respect to Algorithm Write-Read (after round $i - 1$) in the same way as $\mathcal{P}_i, \mathcal{P}'_i, \mathcal{R}_i$ were defined with respect to Algorithm Collective Exploration (after step $i - 1$). We will show that, at the beginning of each round i of Algorithm Write-Read, any robot located at v knows $\hat{\mathcal{K}}_i$. Moreover, we will show that, for any v and any i, $\hat{\mathcal{K}}_i = \mathcal{K}_i$, and that $\hat{\mathcal{K}}_i$ is defined for exactly the same nodes as $\mathcal{K}_i$.

Knowledge $\hat{\mathcal{K}}_i$ is obtained from input $\mathcal{I}_{3i}$ by the following recursive procedure.

Procedure Knowledge Construction
Assume that knowledge $\hat{\mathcal{K}}_1$ is undefined at nodes other than r and that it equals to $\{*, \hat{\mathcal{P}}_1, \hat{\mathcal{P}}'_1, \hat{\mathcal{R}}_1\}$ at the root r, where $\hat{\mathcal{P}}_1$ is the set of all ports of r, $\hat{\mathcal{P}}'_1 = \emptyset$, and $\hat{\mathcal{R}}_1$ is the set of all robots. Suppose that we can compute $\hat{\mathcal{K}}_i$ from input $\mathcal{I}_{3i}$, at all nodes v. We show how to compute $\hat{\mathcal{K}}_{i+1}$ from $\mathcal{I}_{3i+3}$, at node v.
(1) If there are no triplets written at node v for steps smaller than $3i$ (i.e., $\hat{\mathcal{K}}_i$ is undefined) but there is some triplet $(3i, R, \alpha) \in \mathcal{I}_{3i+3}$ then we put: $p = \alpha$ (there is exactly one such α in this case), $\hat{\mathcal{P}}_{i+1}$ is the set of all ports at v other than α, $\hat{\mathcal{P}}'_{i+1} = \emptyset$, $\hat{\mathcal{R}}_{i+1}$ is the set of all robots R, such that a triplet $(3i, R, \alpha) \in \mathcal{I}_{3i+3}$ is written in v.
(2) Otherwise, we first put $\hat{\mathcal{K}}_{i+1} = \hat{\mathcal{K}}_i$, and then modify $\hat{\mathcal{K}}_{i+1}$, s.t.: p remains unchanged, $\hat{\mathcal{P}}_{i+1}$ is the set of all ports from $\hat{\mathcal{P}}_i$, except those ports α, for which there is a triplet $(3i + 1, R, \alpha) \in \mathcal{I}_{3i+3}$ at v (we discard those ports by which a robot entered confirming that the corresponding subtree is finished), $\hat{\mathcal{P}}'_{i+1}$ contains z initial ports from $\hat{\mathcal{P}}_{i+1}$, where z is the integer defined in step i of Algorithm Collective Exploration, $\hat{\mathcal{R}}_{i+1} := \hat{\mathcal{R}}_i \cup X \setminus Y$,

where X is the set of robots R for which $(3i, R, \alpha) \in \mathcal{I}_{3i+3}$, and Y is the set of robots R' for which $H(\mathcal{K}_i, R') = \alpha \neq 0$ (we add robots that entered v in step $3i$ and delete those that left v in this step).

Algorithm Write-Read

Round 0 - This is a special round used to distinguish the root r.

- - STEP I: - -

stage 1: do nothing.

stage 2: every robot R writes $(1, R, *)$ at node r.

stage 3: every robot R reads $\mathcal{I}_1$ at node r.

- - STEP 2: - - (do nothing).

Round $i > 0$ - Execution of each round is based on two assumptions

The assumptions after round $i - 1$ are: assumption A_i - $\hat{\mathcal{K}}_i$ is correctly computed by Procedure Knowledge Construction, using $\mathcal{I}_{3i}$; and assumption B_i - $\hat{\mathcal{K}}_i = \mathcal{K}_i$, at any node, and $\hat{\mathcal{K}}_i$ is defined for exactly the same nodes as $\mathcal{K}_i$.

The three steps of each round i have the following purpose. Step $3i$ is used to make the actual move of a robot to its new location, according to the simulated Algorithm Collective Exploration. Step $3i + 1$ is used to temporarily move robots from a node whose subtree is finished, to its parent w, in order to update information held at w, concerning children with finished subtrees. Step $3i + 2$ is used to move back robots that temporarily moved in Step $3i + 1$.

- - STEP $3i$: - -

stage 1: If R is at node r at the end of round $i - 1$, and $H(\mathcal{K}_i, R) = *$ for node r, then R stops. If R is at node v at the end of round $i - 1$, and $H(\mathcal{K}_i, R) = \alpha \notin \{0, *\}$, for node v, then R leaves v through port α.

stage 2: Every robot R that entered v through port α in Stage 1 of Step $3i$, writes $(3i, R, \alpha)$ in node v.

stage 3: Every robot R located at v reads $\mathcal{I}_{3i+1}$ (this is information held at v after Stage 2 of Step $3i$.)

- - STEP $3i + 1$: - -

stage 1: If $\mathcal{P}_i = \emptyset$ then every robot R located at v at the end of Step $3i$ leaves v through port p.

stage 2: Every robot R that entered v through port α in Stage 1 of Step $3i + 1$, writes $(3i + 1, R, \alpha)$ at node v.

stage 3: Every robot R located at v reads $\mathcal{I}_{3i+2}$.

- - STEP $3i + 2$: - -

stage 1: Every robot R that entered v through port α in Step $3i + 1$, leaves v through port α.

stage 2: Every robot R that entered v through port α in Stage 1 of Step $3i + 2$, writes $(3i + 2, R, \alpha)$ in node v.

stage 3: Every robot R located at v reads $\mathcal{I}_{3i+3}$.

Remark. The return moves of robots in stage 1 of step $3i + 2$ could be avoided. They are introduced to simplify analysis of knowledge update, and do not influence exploration complexity.

Lemma 4. *Assumptions A_i & B_i from Algorithm Write-Read are satisfied for all $i > 0$.*

Theorem 3. *Algorithm Write-Read works in time* $O(D + \frac{n}{\log k})$ *for all* n-*node trees of diameter* D.

Proof. By Lemma 3, it is enough to show that, for every tree T rooted at r, the number of rounds used by Algorithm Write-Read is not larger than the number of steps used by Algorithm Collective Exploration. Let i_0 denote the latter number. By Lemma 4, assumptions A_{i_0} and B_{i_0} are satisfied. By assumption B_{i_0}, all robots are at the root r after round $i_0 - 1$, because they are all at the root after step $i_0 - 1$ of Algorithm Collective Exploration. In Step $3i_0$ of Algorithm Write-Read, every robot R performs action $H(\hat{\mathcal{K}}_{i_0}, R)$, by assumption A_{i_0}. This action is equal $H(\mathcal{K}_{i_0}, R)$, by assumption B_{i_0}. By the definition of i_0 this action is stop. Hence all robots stop after round i_0 of Algorithm Write-Read.

Corollary 1. *Algorithm Write-Read has overhead* $O(k/\log k)$.

4   Conclusion

It can be proved (see full version of this paper) that, in the absence of communication between the robots, the overhead of any exploration algorithm is $\Omega(k)$, i.e., of the same order of magnitude as if only one out of k robots were used to explore the tree. While we showed that collective tree exploration can be done faster, if robots have some communication capabilities. This result should be considered a first step in the study of the impact of communication between robots on the efficiency of collective network exploration. Several related problems remain open, including: (1) find a tree exploration algorithm with constant overhead in the complete communication scenario; (2) find a good lower bound on the overhead of tree exploration for the write-read model; (3) generalize our results to exploration of arbitrary networks; and (4) consider other communication models in the context of collective network exploration.

References

1. S. Albers and M. R. Henzinger, Exploring unknown environments, *SIAM Journal on Computing*, **29** (2000), 1164-1188.
2. E. Arkin, M. Bender, S. Fekete, J. Mitchell, and M. Skutella, The freeze-tag problem: How to wake up a swarm of robots, In *13th ACM-SIAM Symp. on Disc. Alg.* (SODA'02), 568-577.
3. E. Arkin, M. Bender, D. Ge, S. He, and J. Mitchell. Improved approximation algorithms for the freeze-tag problem. In *15th ACM Symp. on Par. in Alg. and Arch.* (SPAA'03), 295-303.
4. I. Averbakh and O. Berman, A heuristic with worst-case analysis for minimax routing of two traveling salesmen on a tree, *Discr. Appl. Mathematics*, **68** (1996), 17-32.
5. I. Averbakh and O. Berman, $(p-1)/(p+1)$-approximate algorithms for p-traveling salesmen problems on a tree with minmax objective, *Discr. Appl. Mathematics*, **75** (1997), 201-216.
6. B. Awerbuch, M. Betke, R. Rivest and M. Singh, Piecemeal graph learning by a mobile robot, In *8th Conf. on Comput. Learning Theory* (COLT'95), 321-328.
7. E. Bar-Eli, P. Berman, A. Fiat and R. Yan, On-line navigation in a room, *Journal of Algorithms*, **17** (1994), 319-341.

8. M.A. Bender, A. Fernandez, D. Ron, A. Sahai and S. Vadhan, The power of a pebble: Exploring and mapping directed graphs, In *30th Ann. Symp. on Theory of Comp.* (STOC'98), 269-278.

9. M.A. Bender and D. Slonim, The power of team exploration: Two robots can learn unlabeled directed graphs, In *35th Ann. Symp. on Foundations of Comp. Science* (FOCS'96), 75-85.

10. P. Berman, A. Blum, A. Fiat, H. Karloff, A. Rosen and M. Saks, Randomized robot navigation algorithms, In *7th ACM-SIAM Symp. on Discrete Algorithms* (SODA'96), 74-84.

11. A. Blum, P. Raghavan and B. Schieber, Navigating in unfamiliar geometric terrain, *SIAM Journal on Computing*, **26** (1997), 110-137.

12. M. Betke, R. Rivest and M. Singh, Piecemeal learning of an unknown environment, *Machine Learning*, **18** (1995), 231-254.

13. X. Deng, T. Kameda and C. H. Papadimitriou, How to learn an unknown environment I: the rectilinear case, *Journal of the ACM*, **45** (1998), 215-245.

14. X. Deng and C. H. Papadimitriou, Exploring an unknown graph, *J. of Graph Th.*, **32** (1999), 265-297.

15. A. Dessmark and A. Pelc, Optimal graph exploration without good maps, In *10th European Symposium on Algorithms* (ESA'02), 374-386.

16. K. Diks, P. Fraigniaud, E. Kranakis and A. Pelc, Tree exploration with little memory, In *13th Ann. ACM-SIAM Symposium on Discrete Algorithms* (SODA'02), 588-597.

17. C.A. Duncan, S.G. Kobourov and V.S.A. Kumar, Optimal constrained graph exploration, In *12th Ann. ACM-SIAM Symp. on Discrete Algorithms* (SODA'01), 807-814.

18. G. N. Frederickson, M. S. Hecht and C. E. Kim, Approximation algorithms for some routing problems, *SIAM J. on Computing*, **7** (1978), 178-193.

19. M. Garey, and D. Johnson, Computers and Intractability, W.H. Freeman and Company, New York, 1979.

20. A. Lopez-Ortiz and S. Schuierer, On-line Parallel Heuristics and Robot Searching under the Competitive Framework. In *8th Scandinavian Work. on Alg. Theory* (SWAT'02), 260-269.

21. P. Panaite and A. Pelc, Exploring unknown undirected graphs, *J. of Algorithms*, **33** (1999), 281-295.

22. C. H. Papadimitriou and M. Yannakakis, Shortest paths without a map, *Theoretical Computer Science*, **84** (1991), 127-150.

23. N. S. V. Rao, S. Hareti, W. Shi and S.S. Iyengar, Robot navigation in unknown terrains: Introductory survey of non-heuristic algorithms, *Tech. Rep. ORNL/TM-12410*, Oak Ridge National Laboratory, July 1993.

24. S. Schuierer, On-line searching in geometric trees, In *Sensor Based Intelligent Robots*, LNAI 1724, 1999, 220-239.

Off-Centers: A New Type of Steiner Points for Computing Size-Optimal Quality-Guaranteed Delaunay Triangulations*

Alper Üngör

Department of Computer Science
Duke University, Durham, NC 27708, USA
ungor@cs.duke.edu

Abstract. We introduce a new type of Steiner points, called off-centers, as an alternative to circumcenters, to improve the quality of Delaunay triangulations. We propose a new Delaunay refinement algorithm based on iterative insertion of off-centers. We show that this new algorithm has the same quality and size optimality guarantees of the best known refinement algorithms. In practice, however, the new algorithm inserts about 40% fewer Steiner points (hence runs faster) and generates triangulations that have about 30% fewer elements compared with the best previous algorithms.

Keywords. Delaunay refinement, computational geometry, triangulations

1 Introduction

Meshes are heavily used in many applications including engineering simulations, computer-aided design, solid modeling, computer graphics, and scientific visualization. Most of these applications require that the shape of the mesh elements are of good quality and that the size of the mesh is small. An element is said to be good if its aspect ratio (circumradius over inradius) is bounded from above or its smallest angle is bounded from below. Mesh element quality is critical in determining interpolation error in the applications and hence is an important factor in the accuracy of simulations as well as the convergence speed. Mesh size, meaning the number of elements, is also a big factor in the running time of the applications algorithm. Between two meshes with the same quality bound, the one with fewer elements is preferred almost exclusively.

Among several types of domain discretizations, unstructured meshes, in particular Delaunay triangulations, are quite popular due to their theoretical guarantees as well as their practical performance. Earliest algorithms that provide both size optimality and quality guarantee used balanced quadtrees to generate first a nicely spread point set and then the Delaunay triangulation of these

---

* Research of the author is supported by NSF under grant CCR-00-86013.

M. Farach-Colton (Ed.): LATIN 2004, LNCS 2976, pp. 152–161, 2004.

points [1]. Subsequently, Delaunay refinement techniques are developed based on an incremental point insertion strategy and provide the same theoretical guarantees [9]. Over the last decade, Delaunay refinement has become much more popular than the quadtree-based algorithms mostly due to its superior performance in generating smaller meshes. Many versions of the Delaunay refinement is suggested in the literature [2,6,7,8,9,10,11]. We attribute the large amount of research on Delaunay refinement to its impact on a wide range of applications. It is important to generalize the input domains that the Delaunay refinement works, as well as to improve the performance of the algorithm. Even a small but reasonable reduction in mesh size translates to important savings in the running-time of the subsequent application algorithm.

The first step of a Delaunay refinement algorithm is the construction of a constrained or conforming Delaunay triangulation of the input domain. This initial Delaunay triangulation is likely to have bad elements. Delaunay refinement then iteratively adds new points to the domain to improve the quality of the mesh and to ensure that the mesh conforms to the boundary of the input domain. The points inserted by the Delaunay refinement are called *Steiner points*. A sequential Delaunay refinement algorithm typically adds one new vertex on each iteration. Each new vertex is chosen from a set of candidates — the circumcenters of bad triangles (to improve mesh quality) and the mid-points of input segments (to conform to the domain boundary). Ruppert [9] was the first to show that proper application of Delaunay refinement produces well-shaped meshes in two dimensions whose size is within a constant factor of the best possible. There are efficient implementations [10] as well as three-dimensional extensions of Delaunay refinement [4,10].

In this paper, we introduce a new type of Steiner points, called *off-centers*, as an alternative to circumcenters and propose a new Delaunay refinement algorithm. We show that this new algorithm has the same theoretical guarantees as the Ruppert's algorithm, and hence, generates quality-guaranteed size-optimal meshes. Moreover, experimental study indicates that our Delaunay refinement algorithm with off-centers inserts about 40% fewer Steiner points than the circumcenter insertion algorithms and results in meshes about 30% smaller in the number of elements. This implies substantial reduction not only in mesh generation time, but also in the running time of the application algorithm. For instance a quadratic-time application algorithm, if ran on the new meshes, would take about half the time it takes on the old meshes.

2   Preliminaries

In two dimensions, the input domain Ω is represented as a *planar straight line graph* (PSLG) — a proper planar drawing in which each edge is mapped to a straight line segment between its two endpoints [9]. The segments express the *boundaries* of Ω and the endpoints are the *vertices* of Ω. The vertices and boundary segments of Ω will be referred to as the input *features*. A vertex is incident to a segment if it is one of the endpoints of the segment. Two segments are incident if they share a common vertex. In general, if the domain is given as a collection of vertices only, then the boundary of its convex hull is taken to be the boundary of the input.

The *diametral circle* of a segment is the circle whose diameter is the segment. A point is said to *encroach* a segment if it is inside the segment's diametral circle.

Given a domain Ω embedded in $\mathbb{R}^2$, the *local feature size* of each point $x \in \mathbb{R}^2$, denoted by $\mathrm{lfs}_\Omega(x)$, is the radius of the smallest disk centered at x that touches two non-incident input features. This function is proven [9] to have the so-called *Lipschitz property*, i.e., $\mathrm{lfs}_\Omega(x) \leq \mathrm{lfs}_\Omega(y) + |xy|$, for any two points $x, y \in \mathbb{R}^2$.

Let P be a point set in $\mathbb{R}^d$. A simplex τ formed by a subset of P points is a *Delaunay simplex* if there exists a circumsphere of τ whose interior does not contain any points in P. This empty sphere property is often referred to as the *Delaunay property*. The Delaunay triangulation of P, denoted $Del(P)$, is a collection of all Delaunay simplices. If the points are in general position, that is, if no $d+2$ points in P are co-spherical, then $Del(P)$ is a simplicial complex. The Delaunay triangulation of a point set of size n can be constructed in $O(n \log n)$ time in two dimensions [5].

In the design and analysis of the Delaunay refinement algorithms, a common assumption made for the input PSLG is that the input segments do not meet at junctions with small angles. Ruppert [9] assumed, for instance, that the smallest angle between any two incident input segment is at least $90°$. A typical Delaunay refinement algorithm may start with the *constrained Delaunay triangulation* [3] of the input vertices and segments or the Delaunay triangulation of the input vertices. In the latter case, the algorithm first splits the segments that are encroached by the other input features. Alternatively, for simplicity, we can assume that no input segment is encroached by other input features. A preprocessing algorithm, which is also parallelizable, to achieve this assumption is given in [11].

Radius-edge ratio of a triangle is the ratio of its circumradius to the length of its shortest side. A triangle is considered *bad* if its radius-edge ratio is larger than a pre-specified constant $\beta \geq \sqrt{2}$. This quality measure is equivalent to other well-known quality measures, such as smallest angle and aspect ratio, in two dimensions [9].

3   Delaunay Refinement with Off-Centers

3.1   Off-Centers

The line that goes through the midpoint of an edge of a triangle and its circumcenter is called the *bisector* of the edge. Given a bad triangle pqr, suppose that its shortest edge is pq. Let c denote the circumcenter of pqr. We define the *off-center* to be the circumcenter of pqr if the radius-edge-ratio of pqc is smaller than or equal to β (Figure 1 (a)). Otherwise, the *off-center* is the point on the bisector (and inside the circumcircle), which makes the radius-edge ratio of the triangle based on p, q and the off-center itself exactly β (Figure 1 (b)). The circle that is centered at the off-center and goes through the endpoints of the

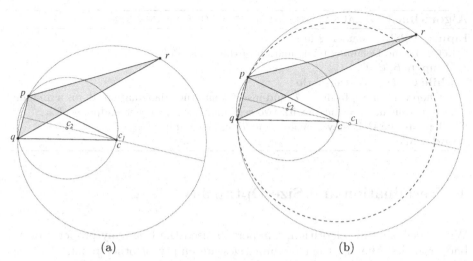

Fig. 1. The off-center and the circumcenter of triangle pqr is labeled c and c_1 respectively. The circumcenter of pqc is labeled as c_2. If $|cc_2| \le \beta|pq|$ then $c = c_1$ (a). Otherwise, $c \ne c_1$ and by construction $|cc_2| = \beta|pq|$ (b). The off-circle of pqr is same as the circumcircle in (a) and shown as dashed circle in (b).

shortest edge is called the *off-circle*. In the first case, off-circle is same as the circumcircle of the triangle. A bad triangle can have two shortest edges. In such cases, the off-center is defined once we arbitrarily choose one of the two edges as the shortest.

Notice in Figure 1 (b) that, if we were to insert the circumcenter c_1, the triangle pqc_1 would still be bad and require another circumcenter insertion. We instead suggest to insert just the off-center c. This, of course, is a simplified picture and the actual behavior of Delaunay refinement is more complicated. Nevertheless, this very observation is the main intuition behind the expectation of smaller size meshes. In other words, around a small feature we create a good element with the longest possible new features.

3.2   Algorithm

At each iteration, we choose a new point for insertion from a set of candidate points. There are two kinds of candidate points: (1) the off-centers of bad triangles, and (2) the midpoints of segments. Let $\dot{\mathcal{C}}$ denote the set of all candidate off-centers that do not encroach any segment. Let $\mathcal{C}$ denote their corresponding off-circles. Similarly, let $\dot{\mathcal{B}}$ denote the set of all candidate off-centers that do encroach some segment. Candidate off-centers of this second type are *rejected* from insertion. Let $\mathcal{B}$ denote their corresponding off-circles. The midpoint of a boundary segment is a *candidate* for insertion if it is encroached by an off-center in $\dot{\mathcal{B}}$. Let $\dot{\mathcal{D}}$ be all midpoint candidates. Then we suggest the following algorithm to incrementally insert the candidate points.

Algorithm 1 DELAUNAY REFINEMENT WITH OFF-CENTERS

Input: A PSLG domain Ω in $\mathbb{R}^2$
 Let T be the Delaunay triangulation of the vertices of Ω.
 Compute $\dot{\mathcal{B}}, \dot{\mathcal{C}}, \dot{\mathcal{D}}$;
 while $\dot{\mathcal{C}} \cup \dot{\mathcal{D}}$ is not empty **do**
   Choose a point q from $\dot{\mathcal{C}} \cup \dot{\mathcal{D}}$ and insert q into the triangulation. If q is a midpoint
   of a segment s, replace s with two segments from q to each endpoint of s;
   Update the Delaunay triangulation T and recompute $\dot{\mathcal{B}}, \dot{\mathcal{C}}, \dot{\mathcal{D}}$.
 end while

---

4   Termination and Size Optimality

When analyzing his algorithm, Ruppert [9] used the Delaunay property on the bad triangles, that is, their circumcircles are empty of other points. Unfortunately, the off-circles are not necessarily empty of other points. There is a small crescent-shape possibly non-empty region of each off-circle outside the corresponding circumcircle. This raises a challenge in our analysis. One easy way around this is to use a special insertion order among the off-centers. For instance, it is relatively easy to prove that the off-circle of the bad triangle that has the shortest edge is empty of all other points. Alternatively, an ordering that favors the bad triangles with the smallest circumradius serves for the same purpose. We could use one of these ordering strategies and apply the same arguments given in [9]. However, for the sake of a generic result, we opt for an arbitrary order in the analysis of our off-center insertion algorithm.

We prove that the meshes generated by the off-center insertion algorithm is size optimal using the same machinery as Ruppert [9]. Moreover, we adapt the terminology introduced in [10] which includes a clearer rewrite of Ruppert's results. We first prove that the edge length function is within a constant factor of the local feature size. Then, we conclude that the output mesh is size-optimal within a constant.

Let *insertion length* of a vertex u, denoted r_u, be the length of the shortest edge incident to u right after u is inserted (or were to be inserted if u is encroaching). If u is an input vertex its insertion length is the shortest edge incident to u in the initial Delaunay triangulation of the input. Also, for each Steiner vertex u, we define a *parent vertex*, denoted $\hat{u}$, as the most recently inserted endpoint of the shortest edge of the bad triangle responsible of the insertion of u. This definition applies also for vertices that are considered but not actually inserted due to encroachment.

Lemma 1. *Let pqr be a bad triangle with off-center u. Then, $r_u \geq C_0 |u\hat{u}|$, for some constant C_0. Moreover, $r_u \geq \beta r_{\hat{u}}$.*

Proof. Without loss of generality, let pq be the shortest edge of pqr and $\hat{u} = p$. Consider the following two cases:

– u *is the circumcenter of pqr:* By the Delaunay property, $r_u \geq |u\hat{u}|$, that is $C_0 = 1$. Moreover, since the triangle pqr is bad, $|u\hat{u}|/|pq| \geq \beta$. The distance from p to q is at least $r_p = r_{\hat{u}}$. Hence, $r_u \geq \beta r_{\hat{u}}$.

– u *is not the circumcenter of pqr:* Let m be the midpoint of the segment pq and c_2 be the circumcenter of pqu. See Figure 1. The intersection of the off-circle and the circumcircle is empty by the Delaunay property. So, as a conservative bound, r_u is at least $|um|$. By construction, $\angle pum = \arcsin(\frac{1}{2\beta})/2$. Also, on the right triangle pum, $\cos(\angle pum) = |um|/|u\hat{u}|$. Since $\beta \geq \sqrt{2}$, $|um| \geq |u\hat{u}| \cos(\arcsin(\frac{1}{2\sqrt{2}})/2)$. So, $C_0 = \cos(\arcsin(\frac{1}{2\sqrt{2}})/2) \approx 0.98$. Moreover,

$$r_u \geq |um| \geq |uc_2| \qquad \text{(because } \angle pc_2q < 90°)$$
$$= \beta|pq| \qquad\qquad \text{(by construction)}$$
$$\geq \beta r_{\hat{u}} \qquad\qquad\qquad \square$$

Lemma 2. *For each vertex u, either $r_u \geq \mathrm{lfs}_\Omega(u)$ or $r_u \geq C_1 r_{\hat{u}}$, for some constant C_1.*

Proof. We consider the following cases:

– u *is not a Steiner vertex:* Then, its nearest neighbor in the initial triangulation is at most $\mathrm{lfs}_\Omega(u)$ away, hence $r_u \geq \mathrm{lfs}_\Omega(u)$.

– u *is an off-center Steiner vertex:* Then, by Lemma 1 we know that $r_u \geq \beta r_{\hat{u}}$, that is $C_1 = 1$.

– u *is midpoint of an encroached subsegment s:* If $\hat{u}$ is an input vertex, or is a Steiner vertex on a segment then $r_u \geq \mathrm{lfs}_\Omega(u)$. Otherwise, $\hat{u}$ is an encroaching rejected circumcenter. Let v be the nearest endpoint of s from $\hat{u}$. By definition, $r_{\hat{u}}$ is at most $|\hat{u}v|$. Moreover, since $\hat{u}$ is inside the diametral circle of s, $|\hat{u}v| \leq \sqrt{2}r_u$. Therefore, $r_u \geq r_{\hat{u}}/\sqrt{2}$, that is $C_1 = 1/\sqrt{2}$. $\square$

Theorem 1. *The* DELAUNAY REFINEMENT WITH OFF-CENTERS *terminates.*

Proof. Let $\underline{\mathrm{lfs}}$ be the smallest distance between two non-incident features of the input PSLG. We prove, by contradiction that there are no edges shorter than $\underline{\mathrm{lfs}}$ introduced during the refinement. Suppose e is the first edge that is shorter than $\underline{\mathrm{lfs}}$. Then, at least one end-point of e is a Steiner vertex. Let v be the most recently inserted endpoint of e. Let $\hat{v}$ be the grandparent of v.

– If v is the off-center of a bad triangle, then by Lemma 2, $r_v \geq \beta r_{\hat{v}}$.

– If v is the midpoint of an encroached segment then there are two sub-cases. If $\hat{v}$ is the off-center of a bad triangle, then by Lemma 2, $r_v \geq r_{\hat{v}}/\sqrt{2} \geq \beta r_{\hat{v}}/\sqrt{2} \geq r_{\hat{v}}$. Otherwise, $\hat{v}$ is on a non-incident segment because of the PSLG input assumption. Then, clearly $r_v \geq \underline{\mathrm{lfs}}$.

In all cases, $r_v \geq r_u$ for some ancestor u of v. If $r_v < \underline{\mathrm{lfs}}$, then $r_u < \underline{\mathrm{lfs}}$, contradicting the assumption that e was the first such edge. Hence, the termination of the algorithm follows. This also implies that there are no bad triangles in the output mesh. $\square$

For each vertex u, let D_u be the ratio of $\mathrm{lfs}_\Omega(u)$ over r_u.

Lemma 3. *If $r_u \geq r_{\hat{u}}/C_2$ for some constant C_2, then $D_u \leq 1/C_0 + C_2 D_{\hat{u}}$.*

Proof.
$$
\begin{aligned}
D_u = \mathrm{lfs}_\Omega(u)/r_u &\leq (\mathrm{lfs}_\Omega(\hat{u}) + |u\hat{u}|)/r_u && \text{(By Lipschitz property)} \\
&\leq (D_{\hat{u}} r_{\hat{u}} + r_u/C_0)/r_u && \text{(By definition and Lemma 1)} \\
&\leq (D_{\hat{u}} C_2 r_u + r_u/C_0) r_u \\
&= C_2 D_{\hat{u}} + 1/C_0
\end{aligned}
$$
□

Lemma 4. *There exist fixed constants $C_T \geq 1$ and $C_S \geq 1$ such that, for each vertex u, $D_u \leq C_T$ if u is a Steiner or rejected off-center vertex and $D_u \leq C_S$ if u is a midpoint Steiner vertex.*

Proof. We prove the lemma by induction.

Basis: If $\hat{u}$ is an input vertex or on a segment, then $D_{\hat{u}} = \mathrm{lfs}_\Omega(\hat{u})/r_{\hat{u}} \leq 1$.

Induction hypothesis: Lemma holds for vertex $\hat{u}$. So, $D_{\hat{u}} \leq \max\{C_T, C_S\}$.

Induction: Now we make a case analysis:

- If u is an off-center of a bad triangle, then by Lemma 3 (where $C_2 = 1/\beta$ by Lemma 1) and the induction hypothesis, $D_u \leq \frac{1}{C_0} + \max\{C_T, C_S\}/\beta$. This implies that $D_u \leq C_T$ if

$$ C_T \geq \frac{1}{C_0} + \max\{C_T, C_S\}/\beta \tag{1} $$

- Otherwise, u is a midpoint of a subsegment s. If parent is an input vertex or on another segment, lemma holds by the basis of the induction. If $\hat{u}$ is a rejected off-center of a bad triangle, then by Lemma 2, $r_u \geq r_{\hat{u}}/\sqrt{2}$. So, by Lemma 3 (where $C_2 = \sqrt{2}$) and the induction hypothesis, $D_u \leq \frac{1}{C_0} + \sqrt{2} C_T$. This implies that $D_u \leq C_S$ if

$$ C_S \geq \frac{1}{C_0} + \sqrt{2} C_T \tag{2} $$

We choose $C_S = \frac{\beta(\sqrt{2}+1)}{C_0(\beta-\sqrt{2})}$ and $C_T = \frac{\beta+1}{C_0(\beta-\sqrt{2})}$, to satisfy both Inequalities (1) and (2). Hence the lemma holds.   □

Lemma 5. *For each vertex u of the output mesh, its nearest neighbor vertex v is at a distance at least $C_3 \mathrm{lfs}_\Omega(u)$ for some constant C_3.*

Proof. By Lemma 4, $\mathrm{lfs}_\Omega(u)/r_u \leq C_S$, for any vertex u. If u was inserted after v, then $|uv|$ is at least r_u. Hence, $|uv| \geq r_u \geq \mathrm{lfs}_\Omega(u)/C_S$, and the lemma holds. If v was inserted after u, then by Lemma 4 $|uv| \geq r_v \geq \mathrm{lfs}_\Omega(v)/C_S$. By Lipschitz property, $|uv| \geq (\mathrm{lfs}_\Omega(u) - |uv|)/C_S$. Hence, $|uv| \geq \mathrm{lfs}_\Omega(u)/(C_S+1)$, that is, $C_3 = 1/(C_S+1)$.   □

Local feature size for an output mesh M (which is a PSLG) is well-defined and denoted by $\mathrm{lfs}_M()$. Previous lemma essentially states that $\mathrm{lfs}_M(x) \geq \mathrm{lfs}_\Omega(x)$, $\forall x \in M$. We next state a theorem proven by Ruppert [9], which together with Lemma 5 leads to Theorem 3, the main result of this section.

Theorem 2 ([9]). *Suppose a triangulation M with radius-edge ratio bound β has the property that there is some constant C_4 such that $\mathrm{lfs}_M(p) \geq \mathrm{lfs}_\Omega(p)/C_4$, $\forall p \in \mathbb{R}^2$. Then, the size of T is less than C_5 times the size of any triangulation of the input Ω with bounded radius-edge ratio β, where $C_5 = O(C_4^2\beta)$.*

Theorem 3. *The* DELAUNAY REFINEMENT WITH OFF-CENTERS *algorithm generates a size-optimal mesh.*

5   Experiments

Implementing the Delaunay refinement with off-centers is as simple as replacing the circumcenter procedure in classical Delaunay refinement implementations with a new off-center procedure. Computing off-centers and circumcenters are very similar and take roughly the same time. Hence, savings in the number of Steiner points reported below also reflects the amount mesh generation time.

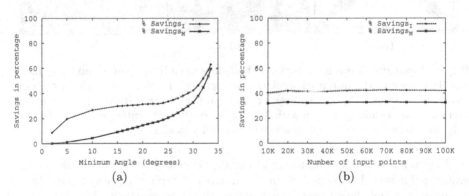

(a)   (b)

Fig. 2. (a) Percentage savings when the number of input points is 10K and the minimum angle threshold samples the interval [2°-34°]. (b) Percentage savings when the minimum angle threshold is 30° and the number of input points samples [10K-100K].

Earlier experiments with circumcenter insertion method indicates that the insertion order has an impact on the output mesh size. For instance, inserting the circumcenter of worst triangles first tends to result in smaller meshes. In this study, for fairness of comparison, we chose the ordering strategy that performs the best for the circumcenter insertion and use the same for the off-center insertion. For Delaunay refinement with circumcenters we used the CMU software `triangle`[1] [10], which is reported to have over thousand users.

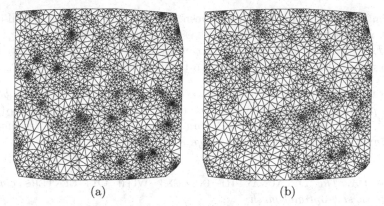

(a)   (b)

Fig. 3. Input consists of 500 points. Smallest angle in both meshes is 29°. Circumcenter insertion adds 2399 Steiner points resulting a mesh with 4579 triangles (a). Off-center insertion adds 1843 Steiner points resulting a mesh with 3479 triangles (b).

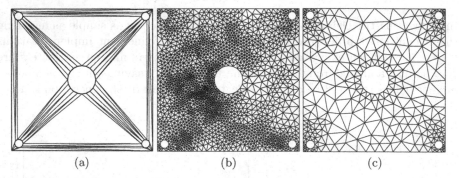

(a)   (b)   (c)

Fig. 4. Input PSLG is a plate with five holes described by 64 points and 64 segments. Smallest angle in the initial triangulation (a) is about 1°. Smallest angle in both output triangulations is 34°. Circumcenter insertion (**triangle** software) introduces 1984 Steiner points resulting a mesh with 3910 triangles (b). Off-center insertion introduces only 305 Steiner points resulting a mesh with 601 triangles (c).

Figure 2 illustrates a summary of our experimental results on randomly generated point sets. Let S_c and S_o be the number of Steiner points inserted by the circumcenter and the off-center insertion methods, respectively. Also, let M_c and M_o be the number of elements generated by the circumcenter and the off-center insertion methods, respectively. We report the following two measures:

$$Savings_I = \frac{S_c - S_o}{S_c}, \qquad Savings_M = \frac{M_c - M_o}{M_c}$$

Percentage savings both in the number of Steiner points and in the mesh size increases as the user specified minimum angle threshold gets higher (Figure 2 (a)). We also observed that for a given threshold angle, the savings remain consistent as we change the input size (Figure 2 (b)). For a visual comparison of

---

[1] Available at http://www-2.cs.cmu.edu/~quake/triangle.html

the off-center and the circumcenter insertion algorithms see Figure 3 where the input is a randomly generated point set.

6 Discussions

By definition, the off-center of some triangles is same as their circumcenter. The off-center and the circumcenter insertion algorithms are likely to generate very similar (sometimes the same) meshes when the initial triangulation is reasonably good to begin with. In most applications, however, tiny angles are ubiquitous in the initial Delaunay triangulation. Figure 4 demonstrates the output of the two algorithms in one such case. In this example, our off-center insertion algorithm gives a mesh that is a factor six smaller than the output of `triangle`. We also observed many other examples, where the off-center insertion algorithm terminates (computing a quality-bounded mesh) and `triangle` does not.

This new insertion scheme also leads to a parallel Delaunay refinement algorithm that takes only $O(\log(L/h))$ iterations to generate quality-guaranteed size-optimal meshes, where L is the diameter of the domain, and h is the smallest edge length in the initial triangulation. This is an improvement over the previously best known equivalent algorithm that runs $O(\log^2(L/h))$ iterations [11]. Due to space limitations, we do not include the description and the analysis of the new parallel off-center insertion algorithm in this publication. Furthermore, we plan to extend the off-center algorithm to three dimensions and explore its benefits both in theoretical and practical fronts.

References

1. M. Bern, D. Eppstein, and J. Gilbert. Provably good mesh generation. *J. Comp. System Sciences* 48:384–409, 1994.
2. L.P. Chew. Guaranteed-quality triangular meshes. TR-89-983, Cornell Univ., 1989.
3. L.P. Chew. Constrained Delaunay triangulations. *Algorithmica* 4:97–108, 1989.
4. T.K. Dey, C. L. Bajaj, and K. Sugihara. On good triangulations in three dimensions. *Int. J. Computational Geometry & Applications* 2(1):75–95, 1992.
5. H. Edelsbrunner. *Geometry and Topology for Mesh Generation.* Cambridge Univ. Press, 2001.
6. H. Edelsbrunner and D. Guoy. Sink insertion for mesh improvement. *Proc. 17th ACM Symp. Comp. Geometry*, 115–123, 2001.
7. G.L. Miller. A time efficient Delaunay refinement algorithm. *Proc. ACM-SIAM Symp. on Disc. Algorithms*, (to appear), 2004.
8. G.L. Miller, S. Pav, and N. Walkington. When and why Ruppert's algorithm works. *Proc. 12th Int. Meshing Roundtable*, 91–102, 2003.
9. J. Ruppert. A new and simple algorithm for quality 2-dimensional mesh generation. *Proc. 4th ACM-SIAM Symp. on Disc. Algorithms*, 83–92, 1993.
10. J.R. Shewchuk. *Delaunay Refinement Mesh Generation.* Ph.D. thesis, Carnegie Mellon University, 1997.
11. D.A. Spielman, S.-H. Teng, and A. Üngör. Parallel Delaunay refinement: Algorithms and analyses. *Proc. 11th Int. Meshing Roundtable*, 205–217, 2002.

Space-Efficient Algorithms for Computing the Convex Hull of a Simple Polygonal Line in Linear Time

Hervé Brönnimann[1]* and Timothy M. Chan[2]**

[1] Computer and Information Science, Polytechnic University, Six Metrotech Center,
Brooklyn, NY 11201, USA.
[2] School of Computer Science, University of Waterloo, Waterloo, ON N2L 3G1,
Canada.

Abstract. We present space-efficient algorithms for computing the convex hull of a simple polygonal line in-place, in linear time. It turns out that the problem is as hard as stable partition, i.e., if there were a truly simple solution then stable partition would also have a truly simple solution, and vice versa. Nevertheless, we present a simple self-contained solution that uses $O(\log n)$ space, and indicate how to improve it to $O(1)$ space with the same techniques used for stable partition. If the points inside the convex hull can be discarded, then there is a truly simple solution that uses a single call to stable partition, and even that call can be spared if only extreme points are desired (and not their order). If the polygonal line is closed, then the problem admits a very simple solution which does not call for stable partitioning at all.

1   Introduction

An algorithm is *space-efficient*, if its implementation requires little or no extra memory beyond that which is needed to store the input. In-place algorithms are tricky to devise due to the limited memory considerations. For the classical sorting problem, both quicksort and heapsort are in-place algorithms (it is well-known that the first can be implemented with logarithmic expected amount of extra memory, and the second with a constant amount [4]). It turns out that devising in-place merge and mergesort is a challenge [6,8,9]. Many other classical problems have been considered when space is dear.

Recently, several classical problems of computational geometry have been revisited with space requirements in mind. Two-dimensional convex hull of points is one of them that has been solved in almost every respect in the past twenty years: there are optimal, output-sensitive solutions which compute the smallest convex polygon enclosing a set of points in the plane. In [3], Brönnimann et al. give optimal in-place algorithms for computing two-dimensional convex hulls. For this problem, the points on the convex hull can be reordered at the beginning

* Research of this author has been supported by NSF CAREER Grant CCR-0133599.
** Research of this author has been supported in part by an NSERC Research Grant.

M. Farach-Colton (Ed.): LATIN 2004, LNCS 2976, pp. 162–171, 2004.
© Springer-Verlag Berlin Heidelberg 2004

of the array, so that the output merely consists of a permutation of the input (encoded in the input array itself), and the number of points on the hull.

Space-efficient algorithms have many advantages over their classical counterparts. Mostly, they necessitate little memory beyond the input itself, so they typically avoid virtual memory / paging and external I/O bottlenecks (unless the input itself is too large to fit in primary memory, in which case I/O-efficient algorithms can be used).

Convex hull of a simple polygonal line. Computing the convex hull of a simple polygonal (either open or closed) is another of the classical problems of computational geometry, and was long suspected to be solvable in linear time. Unfortunately, correct algorithms are outnumbered by the number of algorithms proposed in the literature that have turned out to be flawed [1]. Two algorithms that are correct are Lee's algorithm [10] (a variant of Graham's scan for closed polygonal chains) and Melkman's [11] (works for open polygonal chains as well, in an online fashion).

The problem is two-fold: the polygonal line can be either closed or open. For closed chains, we can use Lee's stack-based algorithm and implement the stack implicitly in the array, using the fact that the vertices on the convex hull are sorted. This leads to a linear-time constant-space solution presented in Section 2.1.

For open chains, the problem is complicated by the fact that either endpoint may lie inside the convex hull. The only solutions known use a deque and we show how to encode a deque implicitly with logarithmic extra storage in Section 3.1. We then improve the storage required to constant size using known techniques. Another solution can be obtained by using the relationship mentioned in [3] between convex hulls and stable partition and by reusing space from points that have been discovered to be non-extreme.

Other related work. There seems to be little point for in-place algorithms when the output requires linear space simply to write down, so one may assume that the output is a permutation of the input or can otherwise be represented in a small amount of memory (e.g., the answer to many geometric optimization problems typically consist of identifying a constant-sized subset of the input). Recently, Chen and Chan [5] proposed another model in which the output is written to an output stream and never read again; only a limited of extra memory is available for workspace. They gave a solution for the problem of counting or enumerating the intersections among a set of n line segments in the plane, with $O(\log^2 n)$ extra memory. There could be up to $\Theta(n^2)$ intersections, but they are written to the output stream and never needed again beyond what is stored in the logarithmic working memory. A similar model holds for other classical problems such as Voronoi diagrams and 3D convex hulls.

Equivalence with stable partition. There are linear-time algorithms for performing stable partition in-place (i.e., how to sort an array of 0's and 1's in-place, respecting the orders of the 0's and 1's); see papers by Munro, Raman, and

Salowe [12] and Katajainen and Pasanen [7]. These algorithms are not simple, however. Nevertheless, efficient implementations are provided as a routine in modern algorithm libraries, e.g. the C++ STL. A truly practical implementation may use available extra storage to speed up the computation, and only resort to the more involved algorithms mentioned above if no extra storage can be spared. Hence it makes sense to try and obtain simple algorithm that use stable partition as subroutine.

The partitioning problem itself is linear-time reducible to convex hull in the following way: Given an array A of 0's and 1's, compute the convex hull of the polygonal line defined by $B[i] = \left(i, i(n-i)(A[i] - 0.5)\right)$. These points lie on two parabolas $y = \pm\frac{1}{2}x(n-x)$, and therefore all appear on the boundary of the convex hull, first the points for which $A[i] = 0$ in order on the lower parabola and those for which $A[i] = 1$ in reverse order on the upper parabola. Thus the stable partition can be constructed in linear time by reversing the 1's in the final array. It thus appears difficult to have a truly simple linear-time algorithm for computing the convex hull of a simple polygonal line, given that no truly simple linear-time algorithm exists for stable partition.

It turns out that by using stable partition as an oracle, we obtain a very simple algorithm. If we are not interested in the order of the points on the convex hull, then we may even forego stable partition altogether, and therefore obtain a truly simple algorithm given in Section 3.3.

2   Closed Chains

For closed simple polygons, the solution turns out to be very simple: the input is represented in an array $A[1] \ldots A[n]$ of vertices, which can be cyclically permuted at will. One may therefore assume that $A[1]$ is a vertex of the convex hull (e.g. the vertex of minimum abscissa). There is an algorithm due to Lee which closely resembles Graham's scan and uses only one stack [10]. We give a brief outline of the algorithm first, then show how to implement it in a space-efficient manner.

2.1   Overview of Lee's Algorithm

In this description, we assume that the polygon is oriented counterclockwise and that $A[1]$ is minimal in some direction. Fortunately, the orientation of the polygon is given by the order type of $(A[n], A[1], A[2])$ and can be computed in $O(1)$ time. Should it turn otherwise, the whole array can be reversed. The invariant is the vertices on the stack form a convex polygon (when viewed as a circular sequence). That is, the vertices in the stack form a convex polygonal line, and the point at the bottom of the stack (i.e. $A[1]$) is always to the left of the line joining two consecutive points in the stack.

Lee's algorithm starts by pushing the first two vertices on the stack, and maintains the line L that connects the top two vertices on the stack, as well as the line L' joining the bottom and top vertices of the stack. When a point is processed, it may fall into several regions as depicted in figure 1(left). There are several cases:

- If current vertex is not to the left of L, restore the invariant by backtracking/deleting vertices from the top of the stack until a convex turn is encountered, or there is only one vertex left on the stack. Then push the current vertex on the stack and recompute L and L'.
- If current vertex is to the left of L but not to the left of the line L' (i.e. falls into the pink or yellow region), then ignore it and process next vertex.
- If current vertex is to the left of both L and L', then push the current vertex on the stack and recompute L and L'.

Note that in the third case, both lines L and L' always rotate counterclockwise, but in the first case, after popping vertices from the stack they may rotate either way. In particular, some previously processed vertices may end up on the left side of L'. The algorithm therefore does *not* maintain the invariant that the stack contains the convex hull of the vertices seen so far. It *does* however maintain the invariant that the stack contains the prefix of the convex hull ending at the vertex on top of the stack, and that the subsequent points (from the top of the stack to the current point) are to the right of L'. In particular, if the last point has maximal abscissa, the stack contains the lower convex hull of the points, a fact that will be useful later. And more importantly, Lee proved that if the chain is closed, then after all the points have been processed the stack contains the entire convex hull.

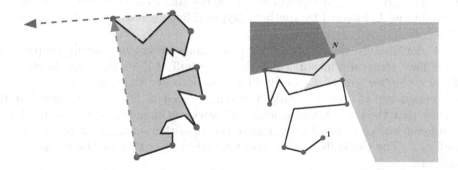

Fig. 1. (left) The two types of regions for processing a point in Lee's algorithm. The points on the stack are in blue, the blue line is L and the red line is L'. (right) The four types of regions for processing a point in Melkman's algorithm. (Both diagrams courtesy of Greg Aloupis).

2.2   A Space-Efficient Implementation

Implementing the previous algorithm in place is trivial since the stack can be stored in the prefix of the array. The minimal-abscissa vertex can be found and the entire array permuted, both in linear time and with a constant amount of extra memory. Moreover, the points inside the convex hull can be kept in the array, if we are careful to perform swaps instead of assignments when popping from and pushing into the stack. The algorithm then produces a permutation of the input and an index h such that $A[1]..A[h]$ form a convex polygon and the points in $A[h+1]..A[n]$ are inside the convex hull. Note that this algorithm is not much different than the in-place Graham-Andrew scan proposed by Brönnimann et al. [3], when the points have already been sorted by abscissa; the only modification is the use of the line L', and the fact that the points haven't been sorted but instead that they lie on a simple polygonal line. The runtime of the algorithm is clearly linear, and it uses $O(1)$ extra memory.

2.3 Open Chains: A Special Case

Although this algorithm only works for closed chains (see the next section for open chains), it also works for open chains in the special case where both endpoints are extreme vertices. For simplicity of the discussion, we assume that $A[1]$ has minimal abscissa, and $A[n]$ is maximal. This lets us use the fact that the convex hull is a concatenation of the lower hull (from $A[1]$ to $A[n]$) and the reverse of the upper hull (from $A[n]$ to $A[1]$).

As observed above, running the one-stack algorithm of from $A[1]$ to $A[n]$ would produce the lower hull, and from $A[n]$ back to $A[1]$ the upper hull. Unfortunately, we cannot run both, but we can separate the points above and below the line L joining $A[1]$ to $A[n]$ by using stable partition, and reversing the second half. This gives us a polygonal line $\mathcal{L}_1$ joining $A[1]$ to $A[n]$ that stays below the straight line L, followed by another polygonal line $\mathcal{L}_2$ starting at $A[n]$ that stays above L.

Unfortunately, the resulting polygonal lines are not necessarily simple, but they have structure: in particular the lower hull of $\mathcal{L}_1$ is the same as that for $\mathcal{L}$, and the vertices of $\mathcal{L}_1$ occur in the same order they occur on $\mathcal{L}$ (this is a consequence of Jordan's curve theorem, proved in [2]). This is sufficient to ensure that the one-stack algorithm still works as intended on $\mathcal{L}_1$ and produces the lower hull of $\mathcal{L}$. Similarly, running the algorithm on $\mathcal{L}_2$ produces the upper hull of $\mathcal{L}$. The two hulls can be concatenated in place to form the whole convex hull of $\mathcal{L}$.

3   Open Chains

While the former algorithm works easily for closed chains, it does not work for open polygonal chains, due to the fact that some vertices might be removed from the stack but appear to the left of the line L' and therefore contribute to the convex hull. Melkman [11] showed how to use a deque instead of a stack to cope with this problem. We give a brief outline of his algorithm first, then show how to adapt the implementation to make it space-efficient.

3.1   Overview of Melkman's Algorithm

The points are processed in the order in which they appear in the input array. Melkman's algorithm maintains their CH as a deque (insertion and deletion from both front and back), which is initially the triangle formed by the first three points. For simplicity, we describe a version that uses a deque and one extra point, stored in a special register, which contains the last point added to the convex hull.

When a point is processed, it can fall into four types of regions, depicted in figure 1(right); note that it cannot fall into any other region without violating the simplicity of the polygonal line. These regions are determined solely by the point in the special register and the front and back vertices of the queue. The invariant of the algorithm is that the special point and the vertices in the deque form a convex polygon, when viewed as a circular sequence . The simplicity of the polygonal line, together with Jordan's curve theorem, imply that when a point comes out of the yellow region, it always does so through one of the two edges of this polygon that join the special point to the top or the bottom vertex of the deque.

- If yellow, ignore this and all following vertices until one emerges into the other regions.
- If red, push the point in the special register onto the front of the deque, then insert the current point into the special register. To restore the invariant, backtrack/delete vertices from the front of the deque until a convex turn is encountered.
- If green, push the point in the special register onto the back of the deque, then insert the current point into the special register. Restore the invariant by backtracking/deleting vertices from the back of the deque until a convex turn is encountered.
- If blue, simply replace the point in the special register by the current point, and restore the invariants as in both cases red and green.

This process is repeated for every point in turn. Note that the algorithm is completely symmetric and therefore does not assume the any orientation of the polygonal line. In fact, the first point of the array does *not* need to appear on the final convex hull, nor does the chain need to be closed. The algorithm is online, meaning that points can be added in a streaming fashion.

3.2   A Space-Efficient Implementation Using Implicit Pointers

The main problem is how to implement a deque of n elements in place, i.e., using only the first n cells of the array when n points have been processed. This is a non-trivial task, at least as hard as stable partitioning. We show that techniques developed for stable partitioning can actually be adapted to solve our problem.

If we represent the deque as a doubly linked list, then each deque operation can trivially be accomplished in constant time. The problem with this approach is of course the extra space needed for the pointers. The key idea is that pointers need not be stored explicitly but can be encoded implicitly via permutations of the input elements: since the points in the deque form a convex polygon, they are sorted by, e.g., angular order. One way to do that is to fix the origin inside the convex hull (e.g., the barycenter of the first three points) and pick a direction (e.g., the horizontal). In the sequel, when we say that a is less than b, we mean that its principal polar angle is less than that of b.

In more details, we store the first few and last few elements in two separate small deques (the *front deque* and *back deque*). The rest of the elements are stored inside the given array, which is divided into blocks of size $s = 4\lceil \log_2 n \rceil$. The blocks are linked together, in order, by the following scheme: Within each block, we encode $2\lceil \log_2 n \rceil$ bits by pairing consecutive elements and permuting each pair (a, b) so that having a less than b means the corresponding bit is a 0 and vice versa. These bits form the two pointer fields (the successor and predecessor) of the doubly linked list.

Insertions/deletions to the front/back are done directly within the two small deques, whose sizes are kept between 0 and $2s$. When the size of the front/back deque reaches $2s$ (a *full event*), we extract s elements from it, form a new block, and update two pointers from and to the new block. When the size of the front/back deque reaches 0 (an *empty event*), we take out the first/last block of the linked list, and insert its s elements into the small deque; furthermore, to ensure that the used blocks occupy a prefix of the array, we swap the deleted block with a block at the end of the array and readjust a constant number of pointers. After a full event, the corresponding small deque has exactly s elements and hence the next event will not occur for another s operation. Each such event processing requires $O(s)$ time, but two events are separated by at least s insertions/deletions, so the amortized cost per insertion/deletion in the deque is $O(1)$.

The extra space used is $O(\log n)$, for the two small deques. By more theoretical tricks, the space complexity can be made even smaller. One option is to handle the small deques recursively, by dividing their elements into tinier blocks in the same manner. Similar to Munro, Raman, and Salowe's stable partition method [12], this should result in an $O(\log^* n)$ space bound. Another option is to recurse for just two levels until the deque size is small enough ($O(\log \log n)$) so that all pointers can be packed into a single word, and pointer manipulations can be done in $O(1)$ RAM operations by table lookup. This is analogous to Katajainen and Pasanen's stable partition method [7] and should yield $O(1)$ space. Since either option is probably too complicated for actual implementation, we will not elaborate further on these refinements.

At the end, to produce the convex hull vertices in order in a prefix of the array, we can simply perform repeated deletions from one end of the deque. Although consecutive pairs have been permuted by the above process, we can permute the pairs back, knowing that they should form a convex polygon. As before, by being careful, we can ensure that points not on the hull boundary remain in a suffix of the array.

3.3   A Simpler, "Destructive" Implementation

If points not on the hull boundary need not be in the final array and can be destroyed, we can give a simple algorithm that directly reduces the problem to stable partitioning. In fact, if the convex hull vertices need not be ordered in the final array (i.e., we just want to identify the set of extreme points), we can avoid the stable partitioning subroutine altogether and thus obtain a truly simple algorithm.

The problem is again how to implement the deque in-place. The key idea is this: if there are no deletions (i.e., all the points are on the boundary of the convex hull), then nothing needs to be done: all the vertices are extreme; in fact, in this case simply stably partitioning the points with respect to the line joining the first and the last point and reversing the second portion produces the convex hull. But if there are many deletions, cells of deleted elements can be used to store other information (pointers, for example).

We describe one approach based on this idea. For simplicity's sake, we assume that each cell can hold two extra bits for marking purposes (*live* or *dead*, and − or +); later we will discuss how this assumption can be removed. The deque has to be stored within the first n cells of the array, where n is the current number of insertions (*not* the number of elements currently in the deque).

Basically, the deque is decomposed into two stacks: elements of sign − in the array form the front part reversed, and elements of sign + form the back part. Insertion is straightforward: just put the new element at the end of the array, and mark it − or + depending on whether we are inserting to the front or back of the deque.

An element is deleted by marking it as dead. To speed up computation, we use dead cells to store pointers as follows: Consider the elements of the array of one sign, in left-to-right order. They form a sequence of alternating blocks of live elements and dead elements. The invariant is that the rightmost element of each dead block should hold a pointer (i.e., the index) to the rightmost element of the preceding live block. See figure 2 for an illustration, where the O's represent live elements and the X's represent dead elements.

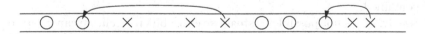

Fig. 2. Representing a deque (or two stacks) in a single array. Live and dead elements of one sign are shown.

It is not difficult to maintain this invariant after a deletion: just imagine when the rightmost O is changed to an X in figure 2; several cases may arise, but only a constant number of pointers need to be updated. To demonstrate the simplicity of the approach, we provide complete pseudocode of the insertion and deletion procedure below. Here, ℓ_σ points to the rightmost live element of sign $\sigma \in \{-, +\}$, and d_σ points to the rightmost dead element of sign σ.

Insert$_\sigma(x)$:
  1. $\ell_\sigma = k = k + 1$, $A[k] = x$
  2. mark $A[k]$ live and of sign σ

Delete$_\sigma()$:
  1. mark $A[\ell_\sigma]$ dead
  2. if $\ell_\sigma > d_\sigma$ then $d_\sigma = \ell_\sigma$
  3. $i =$ predecessor of ℓ_σ among elements of sign σ
  4. if predecessor exists then
  5.   if $A[i]$ is live then $\ell_\sigma = A[d_\sigma] = i$ else $\ell_\sigma = A[d_\sigma] = A[i]$
  6. else {
  7.   compress array by keeping only live elements
  8. $k =$ size of compressed array
  9.   reverse first half of array and switch the sign of these elements
  10. }

Searching for the predecessor of a live element (line 3) is a non-constant-time operation and requires time proportional to the distance between the element and its predecessor. However, this search is done at most once for each element (when its status changes from live to dead), so the total time is still linear in the size of the array.

One scenario has not yet been addressed: what if we run out of elements of one sign? This can be fixed by a standard amortization trick (used in the well-known two-stack simulation of a deque): we just re-divide the deque in the middle and start a new phase, as described in lines 7–10. (Notice that lines 7 and 9 can be done in-place easily.) If the i-th phase initially has k_i elements and ends after m_i insertion/deletion operations, then the phase requires $O(k_i + m_i)$ total time. Because the above strategy ensures that $m_i \geq k_i/2$, the running time of the i-th phase is $O(m_i)$ and the overall running time is $O(n)$, i.e., the amortized cost per update remains $O(1)$.

At the end, we can compress the array to remove all dead elements and thus have the convex hull vertices stored in a prefix of the array. If the vertices are required to be ordered, we can invoke a stable partition subroutine to put all $-$'s before all $+$'s and reverse the $-$ elements; otherwise, our algorithm is completely self-contained.

Finally, if it is not possible to steal two extra bits per cell, we can insert/delete to the deque only when we have gathered a pair of elements. We can permute the pair (a, b) so that having a left of b means the sign is $-$ and vice versa. A dead cell can be signaled by a pair (a, b) with either a or b a point at infinity.

4   Conclusion

The problem of computing the convex hull of a simple polygonal line is well-known to be solvable in linear time. In this paper, we have shown that it can be solved in linear time in-place, and that the in-place problem is as hard as stable partition in the following sense: any linear-time algorithm for one implies a not too convoluted linear-time algorithm for the other. Given that the algorithms for stable partition are rather involved, we do not expect an easy solution for this problem either. Nevertheless, we have given a simple $O(\log n)$-space solution which can be extended to an $O(1)$-space solution at the expense of the complexity of the implementation. If the chain is closed, the problem admits of a very simple in-place linear-time solution, which does not call for stable partitioning at all. If the chain is open but both endpoints are extreme, then a single call to stable partition and two calls to the same very simple in-place linear-time algorithm solve the problem.

References

1. G. Aloupis. A History of Linear-time Convex Hull Algorithms for Simple Polygons. http://cgm.cs.mcgill.ca/~athens/cs601/
2. J.-D. Boissonnat and M. Yvinec. *Algorithmic Geometry*. Cambridge University Press, 1998.
3. H. Brönnimann, J. Iacono, J. Katajainen, P. Morin, J. Morrison, and G.T. Toussaint. Space-efficient planar convex hull algorithms. To appear in *Theoretical Computer Science*. Special issue of selected papers from Latin American Theoretical INformatics (LATIN 2002).
4. T. Cormen, C. Leiserson, R. Rivest, and C. Stein. *Introduction to Algorithms*. 2nd edition, MIT Press, 2001.
5. E. Y. Chen and T. M. Chan. A Space-Efficient Algorithm for Segment Intersection. In *Proc. 15th Canadian Conference on Computational Geometry*, pp. 68-71, 2003.
6. V. Geffert, J. Katajainen, and T. Pasanen. Asymptotically efficient in-place merging. *Theoretical Computer Science* 237:159-181, 2000.
7. J. Katajainen and T. Pasanen. Stable minimum space partitioning in linear time. *BIT* 32:580-585, 1992.
8. J. Katajainen and T. Pasanen. Sorting multiset stably in minimum space. *Acta Informatica* 31:410-421, 1994.
9. J. Katajainen, T. Pasanen and J. Teuhola. Practical in-place mergesort. *Nordic Journal of Computing* 3:27-40, 1996.
10. D.T. Lee. On finding the convex hull of a simple polygon. *International Journal of Computing & Information Sciences* 12(2):87-98, 1983.
11. A. Melkman. On-line construction of the convex hull of a simple polygon. *Information Processing Letters* 25:11-12, 1987.
12. J. I. Munro, V. Raman, and J. S. Salowe. Stable in Situ Sorting and Minimum Data Movement. *BIT* 30(2): 220-234, 1990.

A Geometric Approach to the Bisection Method

Claudio Gutierrez[1], Flavio Gutierrez[2], and Maria-Cecilia Rivara[1]

[1] Department of Computer Science, Universidad de Chile
Blanco Encalada 2120, Santiago, Chile
{cgutierr,mcrivara}@dcc.uchile.cl
[2] Universidad de Valparaíso
Valparaíso, Chile

Abstract. The *bisection method* is the consecutive bisection of a triangle by the median of the longest side. This paper introduces a taxonomy of triangles that precisely captures the behavior of the bisection method. Our main result is an asymptotic upper bound for the number of similarity classes of triangles generated on a mesh obtained by iterative bisection, which previously was known only to be finite. We also prove that the number of directions on the plane given by the sides of the triangles generated is finite. Additionally, we give purely geometric and intuitive proofs of classical results for the bisection method.

1 Introduction

Longest-side bisection algorithms for the refinement of 2-dimensional triangulations were developed to fill a gap in the design of adaptive software for finite element applications to analyze physical problems described by partial differential equations, where the availability of algorithms able to produce automatic and local refinement of the mesh is crucial. A discussion of the algorithms and some generalizations can be found in [4,5]. These algorithms were designed to take advantage of the non-degeneracy properties of the iterative longest-side bisection (bisection method) of triangles, which essentially guarantee that consecutive bisections of the triangles nested in any triangle t_0 of smallest angle σ_0 produce triangles t (of minimum angle σ_t) such that $\sigma_t \geq \sigma_0/2$, and where the number of non-similar triangles generated is finite.

The systematic study of the bisection method began in a series of papers [2, 7,8,9,1] around two decades ago. First, Rosenberg and Stenger [7] proved that the method does not degenerate the smallest angle of the triangles generated by showing that it does not decrease beyond $\sigma/2$, where σ is the smallest angle from the triangle we started.

Then Kearfott [2] proved a bound on the behavior of the *diameter* (the length of the longest side of any triangle obtained). In [8] a better bound was presented for certain triangles. This bound was improved independently by Stynes [9] and Adler [1] for all triangles. From their proofs they also deduced that the number of classes of similarity of triangles generated is finite, although they give no bound.

There is very little research so far on complexity aspects of the bisection method. Although it is known that different types of triangles behave radically

M. Farach-Colton (Ed.): LATIN 2004, LNCS 2976, pp. 172–180, 2004.
© Springer-Verlag Berlin Heidelberg 2004

different under iterative bisection ("good" and "bad" triangles), no systematic classification of them is known.

This paper attempts to fill these gaps in the analysis of the bisection method. We present a precise taxonomy that captures the behavior of the bisection method for different types of triangles. We introduce as main parameter the smallest angle and prove that in the plane it predicts faithfully the behavior of the bisection method. We use this framework to prove new results and to give intuitive proofs of classical results.

The contributions of this paper are as follows:

- A taxonomy of triangles reflecting the behavior of the bisection method. We consider six classes of triangles, and two main groups.
- An asymptotic bound on the number of non-similar triangles generated. We prove a super-polynomial upper bound, identify the instances where this bound is polynomial, and describe worst case instances.
- An analysis of lower bounds on the smallest angle of triangles in the mesh obtained using the bisection method for each class of triangles defined.
- A proof that there is a finite number of directions in the plane generated by the corresponding segments (sides) of the triangles generated, and asymptotic bounds on this number.

Additionally, we present a unified view of the main known results for the bisection method from an elementary geometry point of view. This approach allows intuitive proofs and has the advantage of presenting the geometry inherent to the method.

2   Notation and Preliminaries

Capital letters denote points on the plane. In order to simplify we will avoid extra symbols and sometimes overload some notations. AB denotes a segment as well as the length of this segment usually denoted by $\overline{AB}$. An angle $\angle ACB$ denotes the actual instance as well as the value (measure) of it. A circumference of center A and radius r is denoted by $C(A, r)$.

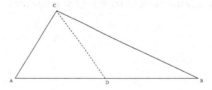

Fig. 1. Triangle ABC with $AB \geq BC \geq CA$. D is the midpoint of AB.

A *bisection*, by the median of the longest side, of triangle ABC with $AB \geq BC \geq CA$, is the figure obtained by tracing the segment CD, where D is the midpoint of the longest segment AB. See Figure 1. We will study the properties obtained by successively bisecting the triangles so obtained.

For a given triangle PQR, denote by σ_{PQR} (respectively γ_{PQR}) the value of the smallest (respectively greatest) angle in triangle PQR, and by β_{PQR} the remaining angle.

We will need a simple and useful technical lemma:

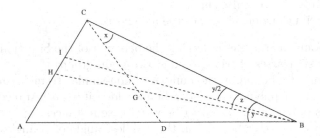

Fig. 2. BI is bisectriz, BH and CD are medians, G is center of gravity.

Lemma 1. For $\triangle ABC$ with $AB \geq BC \geq CA$, it holds $\angle BCD \geq \frac{1}{2}\angle DBC$.

Proof. (See Figure 2.) Let be ABC a triangle with $AB \geq BC \geq CA$, let BI the bisectriz of $\angle ABC$), let BH and CD be medians, and let G be its center of gravity. From $AB \geq BC \geq CA$ and elementary geometry it follows that $BG \geq GC$, hence $x \geq z \geq y/2$. Note that $x = y/2$ if only if $AB = AC$.

To simplify the study of the bisection method, it is convenient to group two or three consecutive bisections in triangle ABC, in what we will call a *step*, as follows. For this discussion refer to Figure 3. Let E be the middle-point of segment CB. Note that if $CD \geq CE$, then CD, DE and EF are consecutive bisections by the median of the longest side, and after these bisections we get exactly three non-similar triangles: ADC, CDE and CDB (all others are similar to one of these, see left side of Figure 3). We call these three consecutive bisections a *step of type A*. Note that $\triangle ADC$ is the only triangle that possibly generates new triangles non-similar to already generated ones.

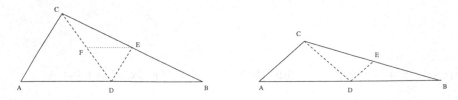

Fig. 3. Steps: Of type A on the left when $CD \geq CE$, and of type B on the right when $CD \leq CE$. Vertices D, E and F are midpoints of the corresponding segments.

On the other hand, if $CD \leq CE$, the longest side in triangle CDE is now CE. Hence we bisect only twice (CD and DE) and get two new triangles, namely ADC and CDE (see right side of Figure 3). We call these two consecutive bisections a *step of type B*. Note that for type B bisections, triangles ADC and CDE are the only triangles that could generate new triangles non-similar to already generated ones.

3   A Classification of Triangles

The behavior of the bisection method depends on the type of triangle to be bisected. We will partition the set of all triangles in classes that reflect this behavior by considering some elementary geometrical properties. The starting point will be a triangle ABC as in Figure 1.

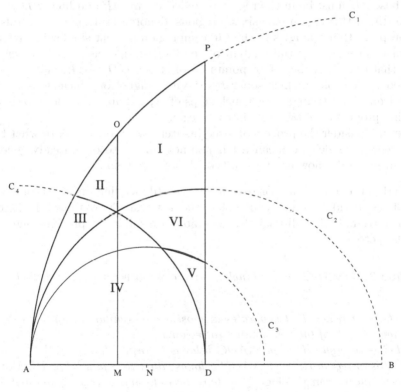

Fig. 4. Regions

Region	Defining properties	Other properties	step type
I	$AD \leq CD \leq AC$	$\gamma \leq \pi/2$	A
II	$AD \leq AC \leq CD$	$\gamma \leq \pi/2$	A
III	$AC \leq AD \leq CD$	$\gamma \leq \pi/2$	A
IV	$AC, CD \leq AD$	$\gamma \geq \pi/2$	A/B
V	$AD \leq AC;\ CD \leq CE$	$\gamma > \pi/2$	B
VI	$CD \leq AD \leq AC;\ CD \geq CE$	$\gamma \geq \pi/2$	A

The analysis is based on the geometrical places where vertex C of triangle ABC lies, assuming $AB \geq CB \geq CA$. For this discussion, we refer to Figure 4, where AB represents the longest side of the hypothetical triangle, D the midpoint of AB, M is the midpoint of AD, N is such that $AN = AB/3$, $MO \perp AB$ and $DP \perp AB$. The arc C_1 belongs to a circumference $C(B, \overline{AB})$, arc C_2 to $C(D, \overline{AD})$, arc C_3 to $C(N, \overline{AN})$ and finally arc C_4 to $C(A, \overline{AD})$.

From the condition $AB \geq BC \geq CA$, it follows that vertex C of a triangle with base AB must be in the region bounded by arc AP and lines PD and AD. We partition this region into six subregions, denoted by Roman numerals, with the property that triangles in the same subregion present similar behavior with regard to bisection by the median of the longest side, as stated in Lemma 2. Note that arc C_3 is the set of points C for which $CD = CE$, and is precisely the geometrical place which separates those triangles for which steps of type A apply from those triangles for which steps of type B apply. Table in page 5 lists defining properties of triangles in each region.

Let us consider the process of bisecting iteratively a triangle. In what follows by a "new triangle" we mean a triangle not similar to one already generated. We will proceed following steps of type A or B, as follows:

1. Perform a step of the corresponding type (depending on the triangle);
2. Choose nondeterministically one of the new triangles obtained. If there is no such triangle (i.e. all triangles generated are similar to previous ones), stop; else goto 1.

Lemma 2. *Let ABC be a triangle. For the iterative process described above it holds:*

1. *If C is in region I, it generates at most 4 non-similar triangles as shown in Figure 5, all of them belonging to region I.*
2. *If C is in region II, new $\triangle ADC$ belongs to region I.*
3. *If C is in region III, new $\triangle ADC$ belongs either to regions II or III. Moreover, in no more than $\lceil 5.7 \log(\frac{\pi}{6\sigma}) \rceil$ steps the only new triangle generated belongs to region II.*
4. *If C is in region IV or V, after no more than $\lceil (\gamma - \pi/2)/\sigma \rceil$ steps, the only new triangle has $\gamma \leq \pi/2$ (i.e. belongs to region I, II or III.)*
5. *If C is in region VI, new $\triangle ADC$ belongs to region I.*

Proof. 1. Follows from the analysis of the relations among sides of the triangles generated. See definition of region I and Figure 5.

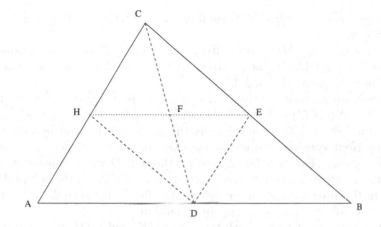

Fig. 5. After bisections in a triangle in Region I

2. Consider the triangle ABC' in region I, where C' is the reflex of C on the line MO. We know that $\triangle ADC'$ is in region I. Now observe that triangle $\triangle ADC$ is congruent to $\triangle ADC'$.

3. First, observe that $\triangle ADC$ has $\gamma \leq \pi/2$, and $\sigma_{ADC} \geq \frac{3}{2}\sigma_{ABC}$ (because $\sigma_{ADC} = \angle ADC$ and Lemma 1). Now, because at each step σ is increased by $3/2$, it is enough to find the smallest k such that $(\frac{3}{2})^k \sigma \geq \pi/6$, that is, $k \geq \log(\frac{\pi}{6\sigma})/\log(3/2)$. The solution, denoted by $k(\sigma)$, is $k(\sigma) = \lceil 5.7\log(\frac{\pi}{6\sigma}) \rceil$.

4. After one step, the only new triangles generated, $\triangle ADC$ and $\triangle CDE$, decrease their greatest angle by σ_{ABC}. Hence it is enough to find the smallest k such that $\gamma - k\sigma \leq \pi/2$. The solution depends on two parameters and is $\lceil (\gamma - \pi/2)/\sigma \rceil$.

5. Just observe that $\gamma_{ADC} \leq \pi/2$ and σ_{ADC} is the same as $\angle CAB$ of $\triangle ABC$.

4   Number of Similarity Classes of Triangles

We are ready to prove the main theorem:

Theorem 1. *Let ABC a triangle and σ its smallest angle.*

1. *The number of steps to be executed by the bisection method until no more non-similar triangles are generated is $\mathcal{O}(\sigma^{-1})$*
2. *If C is above arc C_3, then the number of non similar triangles generated by the bisection method is $\mathcal{O}(\log(\sigma^{-1}))$*
3. *The number of non similar triangles generated by the bisection method is $\mathcal{O}(\sigma^{\log \sigma})$.*

Proof. 1. Let us calculate the maximum number of steps to be executed before arriving to region I in the worst case. This occurs for triangles in regions IV or V. A rough upper bound in the number of steps is given by the sum $2 +$

$\lceil 5.7 \log(\frac{\pi}{6\sigma}) \rceil + \lceil (\gamma - \pi/2)/\sigma \rceil$ This number is asymptotically linear in σ^{-1} because $\pi/3 \leq \gamma < \pi$.

2. For a triangle ABC above arc C_3, the number $N(ABC)$ of non-similar triangles is $1 + N(ADC)$ (the 1 corresponds to $\triangle DBC$). The statement follows from Lemma 2, items 1, 2 and 3.

3. The complex case is region IV. (The analysis for region V is similar.) Here $N(ABC) = N(ADC) + N(CDE)$. First let us prove that $\sigma_{ADC} \geq \frac{3}{2}\sigma_{ABC}$. If C is to the left of MO, then σ_{ADC} is the angle $\angle ADC$ and by Lemma 1 we are done. Next consider the geometric place of the set of points C such that $\beta_{ABC} = \frac{3}{2}\sigma_{ABC}$. This is a line L passing through D with negative slope. If C lies to the right of L, then $\triangle ADC$ will be in region IV to the left of MO and we are in the previous case in one step. If C lies in between L and MO, then $\sigma_{ADC} = \angle CAD = \beta_{ABC} \geq \frac{3}{2}\sigma_{ABC}$ by definition.

Now, using the fact that both triangles ADC and CDE have γ diminished by σ, the fact already proven that $\sigma_{ADC} \geq \frac{3}{2}\sigma_{ABC}$, and observing that $\sigma_{DBC} \geq \sigma_{ABC}$, we have the following recurrence equation for the number $N(\gamma, \sigma)$ of non-similar triangles generated:

$$N(\gamma, \sigma) = N(\gamma - \sigma, \frac{3}{2}\sigma) + N(\gamma - \sigma, \sigma),$$

and Lemma 2.4 gives a bound to the number of necessary steps to take. It is not difficult to see that this recurrence essentially reduces to one of the type $f(n) = f(n/2) + f(n-1)$. This recurrence has no polynomial solution, and $\mathcal{O}(n^{\log n})$ is an upper bound, from where we get the bound $\mathcal{O}((\sigma^{-1})^{\log(\sigma^{-1})})$.

It it interesting to note that not only the number of non-similar triangles generated by the bisection method is finite, but a stronger result can be proved:

Proposition 1. *The bisection method generates a finite number of different directions in the plane. Moreover, in the worst case this number is $\mathcal{O}(\sigma^\sigma)$.*

Proof. Using Theorem 1, it is enough to show that in each step only finitely many new directions are added, and similar triangles generated use already generated directions. But we already know these facts from the analysis of the regions: at each step only one new direction is added except in regions IV and V where the number of directions is (possibly) doubled. Hence, a gross upper bound for the worst case is given by $\mathcal{O}(\sigma^\sigma)$.

5   Classical Results Revisited

Using only elementary geometric methods it is possible to re-prove classical results about the smallest angle and parallel iterative bisection in the bisection method.

Theorem 2. *1. The bisection method gives $\mu_{ABC} \geq \frac{1}{2}\sigma_{ABC}$, where μ_{ABC} is the smallest angle in the mesh obtained by iteratively bisecting triangle ABC. For triangles below arc C_2 it holds that $\mu_{ABC} = \sigma_{ABC}$.*

2. *For each triangle, no more than 5 bisections (2 steps) are necessary in order to diminish the longest side (called diameter) by one half. For simultaneous parallel bisections of all triangles in the mesh, it holds $d_j \leq c2^{-j/2}d_0$ for a small constant c depending on the regions and d_j the diameter after j (parallel) bisections.*

Proof. 1. First, checking case by case it follows that for triangles in regions below C_2 always holds $\sigma_{ADC} > \sigma_{ABC}$ and $\sigma_{DBC} > \sigma_{ABC}$. Second, for triangles in region III, the new triangle ADC has $\sigma_{ADC} \geq \frac{3}{2}\sigma_{ABC}$ (because $\sigma_{ADC} = \angle ADC$ and Lemma 1), and clearly $\sigma_{DBC} > \sigma_{ABC}$. For triangles ABC in region II, observe that $\sigma_{ABC} \leq \pi/6$ and $\sigma_{ADC} = \angle ACD > \sigma_{ABC}$. Finally, once a triangle is in region I, we have Figure 5, being the worst case when $C = P$.

2. The first sentence is an easy observation, the worst case being triangles in region I.

As for the diameter bound, using formula the area of a triangle $A = \frac{1}{2}bh$ and the fact that the area decreases exactly by half after a bisection, one gets immediately $b_j = (\frac{h_0}{h_j})\frac{b_0}{2^j}$, where the sub-indexes indicate sides corresponding to a triangle in the j-th (parallel) bisection.

Now the key point is to observe that: (i) for triangles whose vertex C is below arcs C_2 or C_4 the diameter decreases by half after two parallel bisections, i.e. $d_2 \leq d_0/2$; and (ii) the fact we already know that, as bisection progresses, triangles go "up" the level of arcs C_4 and C_2. Hence, h_j can be bound (in terms of b_j) because from the fact mentioned above that we can deduce that σ_j is no smaller than say $\pi/7$. Similarly, h_0 has a fixed bound in terms of b_0 (the worst case being $\sqrt{3}b_0/2$). Using these formulas we get $b_j^2 \leq c^2b_0^22^{-j}$, for some constant $c \leq \sqrt{3}$ (cf. also [1]). From here, taking square root we get the statement of the theorem.

6   Conclusion

We presented a taxonomy of triangles in the plane which captures the behavior of the bisection method. Besides allowing us to prove complexity results for the bisection method, this classification is useful to refine bounds for each class of triangles, and to determine more precisely lower bounds on the smallest angle μ_{ABC} in the mesh, as well as the number of non-similar triangles generated. The analysis could be further refined considering regions we did not separate, e.g. below arc C_2, above arc C_3 and to the left of MO in Figure 4. Further work includes use of this theoretical analysis to refine algorithms of bisection (4-edge partition, simple bisection, etc.) according to the type of triangle found in each iteration.

References

1. A. Adler, *On the Bisection Method for Triangles*, Mathematics of Computation, Vol. 40, Number 162, April 1983, pp. 571-574.

2. B. Kearfott, *A Proof of Convergence and an Error Bound for the Method of Bisection in R^n*, Mathematics of Computation, Vol. 32, Number 144, October 1978, pp. 1147-1153.
3. J. O'Rourke, *Computational Geometry Column 23*, International Journal of Computational Geometry & Applications, Vol. 4, No. 2 (1994), pp. 239-242.
4. M.C. Rivara, *Algorithms for refining triangular grids suitable for adaptive and multigrid techniques*, International journal for numerical methods in Engineering, vol.20, pp. 745-756, 1984.
5. M-C. Rivara, G. Irribarren, *The 4-Triangles Longest-side Partition of Triangles and Linear Refinement Algorithms*, Mathematics of Computation, Vol. 65, Number 216, October 1996, pp. 1485-1502.
6. M.C. Rivara, C. Levin, *A 3d Refinement Algorithm for adaptive and multigrid Techniques*, Communications in Applied Numerical Methods, vol. 8, pp. 281-290, 1992.
7. I.G. Rosenberg, F. Stenger, *A Lower Bound on the Angles of Triangles Constructed by Bisecting the Longest Side*, Mathematics of Computation, Vol. 29, Number 130, April 1975, pp. 390-395.
8. M. Stynes, *On Faster Convergence of the Bisection Method for certain Triangles*, Mathematics of Computation, Vol. 33, 1979, pp. 1195-1202.
9. M. Stynes, *On Faster Convergence of the Bisection Method for all Triangles*, Mathematics of Computation, Vol. 35, Number 152, October 1980, pp. 1195-1202.

Improved Linear Expected-Time Algorithms for Computing Maxima

H.K. Dai and X.W. Zhang

Computer Science Department, Oklahoma State University
Stillwater, Oklahoma 74078, U. S. A.
{dai, zxiwang}@cs.okstate.edu

Abstract. The problem of finding the maxima of a point set plays a fundamental role in computational geometry. Based on the idea of the certificates of exclusion, two algorithms are presented to solve the maxima problem under the assumption that N points are chosen from a d-dimensional hypercube uniformly and each component of a point is independent of all other components. The first algorithm runs in $O(N)$ expected time and finds the maxima using $dN + d \ln N + d^2 N^{1-1/d} (\ln N)^{1/d} + O(d N^{1-1/d})$ expected scalar comparisons. The experiments show the second algorithm has a better expected running time than the first algorithm while a tight upper bound of the expected running time is not obtained. A third maxima-finding algorithm is presented for N points with a d-dimensional component independence distribution, which runs in $O(N)$ expected time and uses $2dN + O(\ln N(\ln(\ln N))) + d^2 N^{1-1/d} (\ln N)^{1/d} + O(d N^{1-1/d})$ expected scalar comparisons. The substantial reduction of the expected running time of all three algorithms, compared with some known linear expected-time algorithms, has been attributed to the fact that a better certificate of exclusion has been chosen and more non-maximal points have been identified and discarded.

1 Preliminaries

The problem of finding all maxima of a point set plays a fundamental role in computational geometry since the maxima represent one of the characteristics of the point set, and this problem is closely related to convex hull problem. The problem occurs in many applications of diverse disciplines such as statistics, graphics, data analysis, economics, etc. Basically, a maximum is a point that is not dominated by any other point in the same point set. Domination is defined as follows by Preparata and Shamos [PS85]: Given a set S of N points, all points belong to the d-dimensional Euclidean space E^d with coordinates x_1, x_2, ..., x_d. A point p_1 dominates a point p_2 if $x_i(p_2) \le x_i(p_1)$ for $i = 1, 2, ..., d$. We refer that the point p_1 dominates the point p_2 as $p_2 \le p_1$. A point p in S is a maximal element of S if there does not exist any point q in S such that $p \le q$ and $p \ne q$. The maxima problem is to find all maximal elements (maxima) of a set S with respect to dominance.

M. Farach-Colton (Ed.): LATIN 2004, LNCS 2976, pp. 181–192, 2004.

Kung *et al.* [KLP75] showed that any algorithm that solves the maxima problem in two and three dimensions requires $\Omega(N \log N)$ time in the comparison-tree model. By using a divide-and-conquer approach, Kung *et al.* [KLP75] presented an algorithm to find all maxima for a set of N points in E^d, whose worst-case running time is $O(N(\log N)^{d-2}) + O(N \log N)$. This algorithm is referred to as the KLP algorithm here. Therefore, in the cases of 2- and 3-dimensional spaces, the running time of this algorithm will be bounded by $O(N \log N)$. Clearly, the KLP algorithm is optimal for the 2- and 3-dimensional spaces. By applying a multi-dimensional divide-and-conquer scheme, Bentley [Ben80] gave a simpler description of the KLP algorithm.

When the expected value of any variable, such as running time, is considered, the probability distributions of the coordinates of all input points must be specified. Under the hypothesis that the d components in each point are independent from continuous distributions, Bentley *et al.* [BKST78] showed that the expected number of maxima of a set of N points in E^d is $O((\log N)^{d-1})$. So the expected maxima are only a small part of the whole point set. The set satisfying this hypothesis is referred to having a component independence (CI) distribution by Bentley. Note that points uniform over any rectilinearly oriented hyperrectangle exhibit the CI property. Based on this hypothesis, Bentley *et al.* [BKST78] presented an algorithm to compute the maxima set with a linear expected running time $O(N)$, which is referred to as the BKST algorithm later.

Let us image what the set of maximal points of a set S looks like from a geometrical viewpoint. In the case of the 2-dimensional space, the set of the maximal points of S forms a structure, which monotonically decreases in the y-coordinates as the x-coordinates of the points increase. This kind of structure is called staircase structure. For a static point set S in the 2-dimensional space, it has been shown that the staircase structure can be computed in $O(N \log N)$. Regarding a dynamic point set S, Overmars and Van Leeuwen [OL81] designed a data structure which requires splitting and merging balanced trees when points are inserted and deleted. For each insertion and deletion, the required time is in $O(N \log^2 N)$. Both Fredrickson and Rodger [FR90] and Janardan [Jan91] developed a scheme which maintains the staircase structure of a set of maxima and allows the insertion in $O(\log N)$, and deletion in time $O(\log^2 N)$. In 1994, Kapoor [Kap94] designed an improved data structure, which maintains the staircase structure in $O(\log N)$ time.

Because the expected number of maxima of a set S of N points is $O((\log N)^{d-1})$ for a CI distribution, most of the points in S are not maxima when N is large. This is also demonstrated by the staircase structure of maxima in the 2-dimensional space. So it is possible for us to pick up an appropriate point and use this point to rule out all points dominated by it, which are not maxima of the set S. In 1990, this insight has been pointed out by Bentley *et al.* [BCL90] as follows: "A certificate of exclusion typically can quickly demonstrate that most of the N input points are not in the final output". They studied the case that N input points (with CI property) are distributed uniformly within a d-dimensional ($d \geq 2$) hypercube and presented an algorithm to find the maxima of these N points in $O(N)$ expected-time, which is referred to as the BCL hypercube algorithm here. Bentley et $al.$ [BCL90] also designed an algorithm for finding the maxima of N points chosen from a d-dimensional CI distribution in $O(N)$ linear expected-time, which is referred to as the BCL algorithm from now on.

2   Improved BCL Hypercube Algorithm-1

```
1: ImprovedBCLHypercube-1(S, d)
2: A₁ = ∅, B₁ = ∅, C₁ = ∅, p₁ =the point of (1 − (lnN/N)^{1/d}, ..., 1 − (lnN/N)^{1/d});
3: //Initialize three sets A₁, B₁, and C₁ to be empty and choose a certificate point p₁.
4: for each point q in the set S, do
5:         if q ≤ p₁ and q ≠ p₁, A₁ = A₁ ∪ {q};
6:         //The set A₁ contains all points different from p₁ that are dominated by p₁.
7:         else if p₁ ≤ q, C₁ = C₁ ∪ {q};
8:         //The set C₁ contains all points that dominate p₁.
9:         else B₁ = B₁ ∪ {q};
10:        //The set B₁ contains all points which are incomparable (with respect to "≤") to p₁.
11: if C₁ = ∅, compute the maxima of the set S by the KLP algorithm and return;
12: else do
13:        find the point p that has the maximum of x₁(p) · x₂(p) · x₃(p) · · · · · x_d(p) among all
           points in the set C₁;
14:        A = ∅, B = ∅, C = ∅;
15:        for each point q in the set B₁, do
16:                if q ≤ p and q ≠ p, A = A ∪ {q};
17:                //The set A contains all points different from p that are dominated by p.
18:                else if p ≤ q, C = C ∪ {q};
19:                //The set C contains all points that dominate p.
20:                else B = B ∪ {q};
21:                //The set B contains all points which are incomparable (with respect to "≤")
                   //to p.
22:        compute the maxima of the set B by the BKST algorithm and return.
```

As shown above, improved BCL hypercube algorithm-1 is based on the BCL hypercube algorithm and developed under the assumption that the points of the set S (with CI property) are distributed uniformly within a d-dimensional ($d \geq 2$) hypercube. The basic idea underlying improved BCL hypercube algorithm-1 is same as the BCL hypercube algorithm except applying another new certificate of exclusion to identify the non-maximal points that are discarded.

This basic idea can be illustrated in Figure 1 for a 2-dimensional unit square. First, a point $p_1 = (1 − (lnN/N)^{1/2}, 1 − (lnN/N)^{1/2})$ is chosen as the first certificate of exclusion for the set S as the BCL hypercube algorithm does. As Figure 1(a) shows, by using the point p_1, the whole unit hypercube can be partitioned into three sets: A_1, B_1 and C_1, which have the following properties respectively. All points in A_1 are dominated by the point p_1 while all points in C_1 dominate the point p_1. The point p_1 is incomparable (with respect to "≤") to each point in the set B_1. If the set C_1 is empty, the KLP algorithm will be applied to compute the maxima of the set S. In case that the set C_1 is non-empty, because the point p_1 dominates all points in the set A_1, none of these points will be a maximal element of the set S and therefore, they can be discarded. So, only the points in the set $B_1 \cup C_1$ are left for the continuing computation. The underlying idea in improved BCL hypercube algorithm-1 is that a better point p chosen from these points in the set C_1 can be used as the second certificate of exclusion to identify the non-maximal points in the set $B_1 \cup C_1$. As improved BCL hypercube algorithm-1 indicates, this point p is chosen from the set C_1 and has the maximum value of $x_1(p) \cdot x_2(p) \cdot x_3(p) \cdots \cdots x_d(p)$ among all points in the set C_1. After the point p is used to make a partition on the set $B_1 \cup C_1$, three subsets A, B and C will be obtained again, as shown in Figure 1(b). Clearly, all points in the set A are dominated by the point p and the set C contains only one point p. The set B contains all points incomparable to the point p. So all points in the set A can be discarded again and only the points in the set $B \cup C$ need to be computed. Finally, the BKST algorithm is applied to the set $B \cup C$ to compute the maxima set of the set S. The effectiveness of improved BCL hypercube algorithm-1 is determined by the number of points remained after the second partition because these points are kept for the future computation of maxima by applying the BKST algorithm, although extra cost needs to be paid for such a partition.

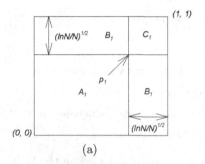
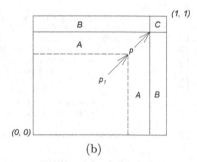

(a)   (b)

Fig. 1. An illustration of partitioning in improved BCL hypercube algorithm-1 for a 2-dimensional unit square: (a) the first partition; (b) the second partition.

The analysis of the expected run time and expected number of comparisons of improved BCL algorithm-1 is done as follows. At the beginning, Nd scalar comparisons are needed to partition the set S into three parts: A_1, B_1 and C_1. Then two cases need to be considered. As indicated by Bentley et $al.$ [BCL90], first, the case that the set C_1 is empty occurs with a probability at most $1/N$ on the average and the expected run time under this case to compute the maxima of the set S by applying the KLP algorithm is $O((logN)^{d-2}) + O(logN)$. Second, the case that the set C_1 is non-empty occurs with a probability at least $1 - 1/N$. Because the set C_1 is represented as a hypercube with volume of $\ln N/N$, the expected number of points in the set C_1 is $\ln N$. So, on the average, $(d-1)\ln N$ multiplications are needed to compute the quantity $x_1(q) \cdot x_2(q) \cdot x_3(q) \cdots \cdots x_d(q)$ for each point q in the set C_1. To find the maximum of this quantity, $\ln N - 1$ expected scalar comparisons are required. Then, the point p is used to partition the set $B_1 \cup C_1$ into three subsets: A, B and C and the BKST algorithm is applied to the set $B \cup C$. Because the expected number of points in the set $B_1 \cup C_1$ is bounded by $1 + dN^{1-1/d}(\ln N)^{1/d}$ under the case that C_1 is non-empty, the total extra cost of improved BCL hypercube algorithm-1 will be bounded by $d \ln N - 1 + d + d^2 N^{1-1/d}(\ln N)^{1/d}$. The expected run time of applying the BKST algorithm to the set $B \cup C$ is determined by its size. Under the assumption that the N points in the set S are distributed uniformly within the hypercube, the size of the set $B \cup C$ is bounded above by $N - N(1 - (1/N)^{1/d})^d$. Let $\alpha = (1/N)^{1/d}$ and the bound can be reduced as follows:

$$N - N(1 - (1/N)^{1/d})^d = N - N(1 - \alpha)^d \leq N(1 - (1 - d\alpha)), \;\; by \; Bernoulli's \; inequality$$
$$= Nd\alpha = dN^{1-1/d}.$$

According to Bentley et $al.$ [BKST78], the expected number of comparisons will be bounded by $O(dN^{1-1/d})$ if the BKST algorithm, which has a linear expected run time, is used to find the maxima of the subset $B \cup C$. Finally, the expected number of comparisons by applying improved BCL hypercube algorithm-1 to compute all maxima of a point set S with N points under the stated assumption at the beginning will be bounded by $dN + d \ln N + d^2 N^{1-1/d}(\ln N)^{1/d} + O(dN^{1-1/d})$ and the expected run time is $O(N)$. Compared with the expected number of comparisons $dN + O(dN^{1-1/d}(lnN)^{1/d})$ used by the BCL hypercube algorithm claimed by Bentley et $al.$ [BCL90], improved BCL hypercube algorithm-1 has a better performance on the average if the hidden coefficient in the asymptotic notation of the BKST algorithm is large enough.

3   Improved BCL Hypercube Algorithm-2

1: ImprovedBCLHypercube(S, d)
2: compute the maxima of the set returned by PARTITION(S, d) by the BKST algorithm and return.

3: PARTITION(S, d)
4: if the size of the set $S \leq 1$, return S.
5: for each point q in the set S, do
6:       compute $x_1(q) \cdot x_2(q) \cdot x_3(q) \cdot \cdots \cdot x_d(q)$;
7: find the point p that has the maximum product of $x_1(p) \cdot x_2(p) \cdot x_3(p) \cdot \cdots \cdot x_d(p)$ among all points in the set S;
8: $A = \emptyset, B_1 = \emptyset, B_2 = \emptyset, ..., B_d = \emptyset$;
9: for each point q in the set S, do
10:       if $q \leq p$ and $q \neq p$, $A = A \cup \{q\}$;
11:       else
12:             if $x_i(p) < x_i(q)$, $B_i = B_i \cup \{q\}$;
13: return $\{p\} \cup$PARTITION(B_1, d)$\cup$PARTITION(B_1, d) $\cup \cdots \cup$ PARTITION(B_d, d).

As shown above, improved BCL hypercube algorithm-2 is extended from the previously presented algorithm and developed under the same assumption. The basic idea upon which improved BCL hypercube algorithm-2 is based is as same as improved BCL hypercube algorithm-1 except the approach to identify the non-maximal points. Without using the certificate of exclusion only twice, this extended algorithm searches the non-maximal points by a recursive procedure PARTITION(S, d). This procedure, first, computes the product of $x_1(q) \cdot x_2(q) \cdot$

$x_3(q) \cdots \cdots x_d(q)$ for each point q in the set S and picks up the point p, which has the maximum product $x_1(p) \cdot x_2(p) \cdot x_3(p) \cdots \cdots x_d(p)$, as the certificate of exclusion for the set S. Then, it uses the point p to partition the set S into the subsets $\{p\}, A, B_1, B_2, ..., B_d$. Obviously, all points in the set A are dominated by the point p and, therefore, can be discarded. After that, the PARTITION(S, d) procedure partitions the sets $B_1, B_2, ..., B_d$ recursively to identify more non-maximal points and returned with the combined subsets. Finally, improved BCL algorithm-2 computes the maxima set of the input set S by applying the BKST algorithm to the returned set from the PARTITION(S, d) procedure.

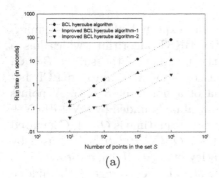 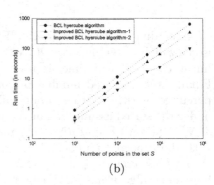

(a)   (b)

Fig. 2. The run times of the BCL hypercube algorithm and its improved versions versus the number of points in the set S: (a) in the 4-dimensional hypercube; (b) in the 6-dimensional hypercube.

4   Comparisons between Hypercube Algorithms

Experiments have been performed for the BCL hypercube algorithm, improved BCL hypercube algorithm-1 and algorithm-2 respectively. The input point set S has a random uniform distribution in a range from 0 to 1 on each coordinate that is independent with each other. The random number generator used in the implementation is recommended by Press *et al.* [PTVF93] and L'Ecuyer *et al.* [L'E88]. It combines two random generators to achieve this one. Both of them use the same algorithm with the Bays-Durham shuffle and added safeguards but with different parameters. The period of the combined generator is 2.3×10^{18} that makes period exhaustion practically impossible. Figures 2(a) and 2(b) show the curves of the run times versus the set size N for all three algorithms for the 4- and 6-dimensional hypercubes respectively. These figures indicate that, for both of 4- and 6-dimensional hypercubes, improved BCL hypercube algorithm-1 has a better performance than the BCL hypercube algorithm and improved BCL hypercube algorithm-2 has the best performance among all three algorithms for a given input point set S, whose size increases step by step.

Table 1. The number of points in the sets A, $B \cup C$, C and PARTITION(S, d) after the partitions in the BCL hypercube algorithm and its improved versions for the 4-dimensional hypercube.

	BCL hypercube algorithm			improved BCL hypercube algorithm-1		improved BCL hypercube algorithm-2
$N = \lvert S \rvert$	A	$B \cup C$	C	A	$B \cup C$	PARTITION(S, d)
1000	267	733	7	277	456	117
5000	2038	2962	7	1843	1119	295
10000	4635	5365	12	3560	1805	297
100000	64460	35540	12	24422	11118	636
500000	371579	128421	13	117582	10839	847

To explain the difference among the performances of improved BCL hypercube algorithm-1, algorithm-2 and the BCL hypercube algorithm, the numbers of points in the sets A, $B \cup C$, C and PARTITION(S, d), which are obtained after the certificates of exclusion are applied in all three algorithms for different input point sets for the 4-dimensional hypercube, are presented in Table 1.

For improved BCL hypercube algorithm-1 and the BCL hypercube algorithm, the application of the certificates of exclusion partitions the whole set S into the subsets and discards the identified non-maximal points. Finally, only the set $B \cup C$ is computed to obtain the maxima of the set S by applying the BKST algorithm. Therefore, compared with the BCL hypercube algorithm, the effectiveness of improved BCL hypercube algorithm-1 will be determined by the answer to the question whether the size of the final input set $B \cup C$ for the BKST algorithm is reduced to such an extent that the non-maximal points identified by it outweighs the extra cost paid for the second partition. The data in Table 1 presents a positive answer to this question. For instance, for the point set S with 5000000 points in the 4-dimensional hypercube, the set $B \cup C$ in the BCL hypercube algorithm has 128421 points after the first partition while the set $B \cup C$ in improved BCL hypercube algorithm-1 has 10839 points after the second partition. The same scenario happens for the point set S with the other sizes. This fact supports the basic idea in improved BCL hypercube algorithm-1 that more points can be discarded if a better certificate of exclusion in the set C is applied after the first partition. Figure 3(a) shows the ratio between the size of the set $B \cup C$ in improved BCL hypercube algorithm-1 and that in the BCL hypercube algorithm based on the data in Table 1. As this figure indicates, the ratio decreases as the number of points in the set S increases. It is understandable due to the fact that, on the average, the size of the set $B \cup C$ in improved BCL hypercube algorithm-1 is bounded by $dN^{1-1/d}$ and that of the set $B \cup C$ in the BCL hypercube algorithm is $dN^{1-1/d}(\ln N)^{1/d}$. So their ratio is bounded by $(\ln N)^{-1/d}$. Obviously, on the average, when N increases, this ratio will decrease. This is consistent with the trend in Figure 3(a).

Regarding improved BCL hypercube algorithm-2, the application of the certificates of exclusion recursively partitions the whole set S into the subsets and discards the identified non-maximal points. Therefore, only the set resulted from the PARTITION(S, d) procedure is computed by the BKST algorithm and the performance of improved BCL hypercube algorithm-2 is also determined by the size of this set plus the extra cost for the recursive partition. From Table 1, it can be seen that the PARTITION(S, d) procedure identified most of the non-maximal points. For example, for the point set S with 5000000 points, the BKST algorithm just needs to compute 847 points returned by the PARTITION(S, d) procedure while it needs to compute 128421 points in the BCL hypercube algorithm. As Figure 2(a) shows, the much smaller run time of improved BCL hypercube algorithm-2 than that of the BCL hypercube algorithm also shows the extra cost of recursive partition is much less significant than the gain resulted from the recursive partition. Figure 3(b) shows the ratio between the sizes of the final input set for the BKST algorithm in improved BCL algorithm-2 and that in the BCL hypercube algorithm based on the data in Table 1. As this figure indicates, the ratio also decreases as the number of points in the set S increases. Although a tight upper bound of the size of this set can not be obtained at this stage, a rough estimation can be achieved because this size is also bounded by $dN^{1-1/d}$, which can be drawn from the previous analysis. Therefore, the ratio is bounded by $(\ln N)^{-1/d}$, which indicates the trend in Figure 3(b) although this bound is very loose.

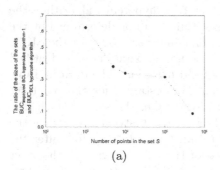
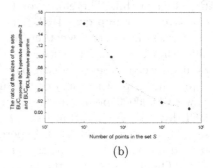

(a)   (b)

Fig. 3. The ratio between the size of $B \cup C$ in improved BCL hypercube algorithm and that in the BCL hypercube algorithm for the 4-dimensional unit hypercube: (a) improved BCL hypercube algorithm-1; (b) improved BCL hypercube algorithm-2.

5   Improved BCL Algorithm

```
1: ImprovedBCL(S, d)
2:   A₁ = ∅, B₁ = ∅, C₁ = ∅, p₁ =the point of (x₁, x₂, ..., x_d), where x_i is N(ln N/N)^{1/d}th largest
     element on the ith dimension;
3:   //Initialize three sets A₁, B₁, and C₁ to be empty and choose a certificate point p₁.
4:   for each point q in the set S, do
5:       if q ≤ p₁ and q ≠ p₁, A₁ = A₁ ∪ {q};
6:       //The set A₁ contains all points different from p₁ that are dominated by p₁.
7:       else if p₁ ≤ q, C₁ = C₁ ∪ {q};
8:       //The set C₁ contains all points that dominate p₁.
```

```
9:              else B₁ = B₁ ∪ {q};
10:             //The set B₁ contains all points which are incomparable (with respect to "≤") to p₁.
11:    if C₁ = ∅, compute the maxima of the set S by the KLP algorithm and return;
12:    else do
13:             find   the   point   p   that   has   the   maximum   of   OrderStatistic(x₁(p))   ·
       OrderStatistic(x₂(p))·
                OrderStatistic(x₃(p)) · · · · · OrderStatistic(x_d(p));
14:             //OrderStatistic(xᵢ(p)) is defined as n if xᵢ(p) is the nth smallest element on the ith
                //dimension for all points in the set C₁.
15:             A = ∅, B = ∅, C = ∅;
16:             for each point q in the set B₁ ∪ C₁, do
17:                     if q ≤ p and q ≠ p, A = A ∪ {q};
18:                     //The set A contains all points different from p that are dominated by p.
19:                     else if p ≤ q, C = C ∪ {q};
20:                     //The set C contains all points that dominate p.
21:                     else B = B ∪ {q};
22:                     //The set B contains all points which are incomparable (with respect to "≤")
                        //to p;
23:             compute the maxima of the set B ∪ C by applying the BKST algorithm and return.
```

Based on the BCL algorithm, improved BCL algorithm presented above is developed under the assumption that the input point set S has the component independent (CI) distribution. The basic idea underlying improved BCL algorithm is same as improved BCL hypercube algorithm-1 except that the certificates of exclusion are chosen based on order statistics. First, as the BCL algorithm presents, a point p_1, whose each coordinate is the $N(\ln N/N)^{1/d}$th largest element on each dimension, is chosen as the first certificate of exclusion for the set S. The order statistic is computed by using Floyd and Rivest's selection algorithm et al. [FR75], which selects the Mth largest element in a set of N elements with $N + min(M, N - M) + O(N^{1/2})$ expected comparisons. So this step requires $dN + O(N^{1/2})$ scalar comparisons. By using the point p_1, the whole set can be partitioned into three subsets: A_1, B_1 and C_1 and each has their own characteristics that are as same as these stated in the description of improved BCL hypercube algorithm-1. This partition procedure uses dN scalar comparisons. Because the probability that any point in the set S dominates the point p_1 is $\ln N/N$ due to CI property, two cases come up. First, the case that the set C_1 is empty occurs with a probability at most $1/N$ on the average and the expected run time under this case to compute the maxima of the set S by applying the KLP algorithm is $O((logN)^{d-2}) + O(logN)$ indicated by Bentley et al. [BCL90]. Second, the case that the set C_1 is non-empty occurs with a probability at least $1 - 1/N$. In this case, all points in the set A_1 are discarded and only the points in the set $B_1 \cup C_1$ are considered. Because of the CI property, the number of point in the set C_1 is also $\ln N$ on the average. Next, the second certificate of exclusion will be chosen based on order statistics. The order statistic of $x_i(q)$ for a point q in the set C_1 is defined as n if $x_i(q)$ is the nth smallest element on the ith dimension for all points in the set C_1. This second certificate of exclusion will be the point p that has the maximum of $OrderStatistic(x_1(p)) \cdot OrderStatistic(x_2(p)) \cdot OrderStatistic(x_3(p)) \cdot \cdots \cdot$ $OrderStatistic(x_d(p))$ among all points in the set C_1. Because this process uses a Quicksort to get the required point p_1, its expected run time is bounded by $O(\ln N(\ln(\ln N)))$. After that, the set $B_1 \cup C_1$ is partitioned into three subsets A, B and C by applying the point p and this step takes $d \ln N$ scalar comparisons on the average case. Then, only the points in $B \cup C$ are computed by the BKST algorithm. Due to the CI property, the average case analysis of improved BCL algorithm is same as improved BCL hypercube algorithm-1. Thus, we have the following conclusion. Improved BCL algorithm finds the maxima of a set S with N points with a d-dimensional CI distribution in $O(N)$ expected time and uses $2dN + O(\ln N(\ln(\ln N))) + d^2 N^{1-1/d}(lnN)^{1/d} + O(dN^{1-1/d})$ expected scalar comparisons. It has a better performance on the average than the BCL algorithm with expected scalar comparisons $2dN + O(dN^{1-1/d}(lnN)^{1/d})$, which is not $dN + O(dN^{1-1/d}(lnN)^{1/d})$ stated by Bentley *et al.* [BCL90].

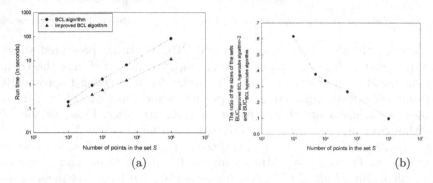

(a)   (b)

Fig. 4. (a) The run times of improved BCL algorithm and the BCL algorithm versus the number of points in the set S in the 4-dimensional space; (b) the ratio between the size of $B \cup C$ in improved BCL algorithm and that in the BCL algorithm for the 4-dimensional space.

Experiments on point sets with CI distribution (independent and uniformly distributed coordinates) have been performed for improved BCL algorithm and the BCL algorithm respectively. Figure 4(a), which shows the curves of the run times versus the set size N for both algorithms for the 4-dimensional space, indicates that improved BCL algorithm has a better performance than the BCL algorithm.To understand the difference between their performances, the numbers of points in the three sets A, B and C, which are obtained after the certificates of exclusion are applied in both algorithms for different input point sets for the 4-dimensional space, are studied. Like BCL hypercube algorithm-1, the effectiveness of improved BCL algorithm is also determined by whether the second partition identifies enough non-maximal points that are worth the extra partition cost. Figure 4(b) shows the ratio between the size of $B \cup C$ in improved BCL algorithm and that in the BCL algorithm. As it indicates, the ratio decreases as the number of points in the set S increases. Due to CI property, this happens because, on the average, the size of the set $B \cup C$ in improved BCL algorithm is bounded by $dN^{1-1/d}$ and that of the set $B \cup C$ in the BCL algorithm is $dN^{1-1/d}(\ln N)^{1/d}$. So their ratio is bounded by $(\ln N)^{-1/d}$. Clearly, on the average, when N increases, this ratio will decrease.

6   Conclusion

All three presented algorithms which find the maxima of a point set are based on the idea of the certificates of exclusion indicated by Bentley *et al.* [BCL90]. Improved BCL hypercube algorithm-1 and improved BCL hypercube algorithm-2 are presented to solve this maxima problem under the assumption that all points are chosen from a d-dimensional hypercube uniformly (with CI property). Improved BCL hypercube algorithm-1 runs at a $O(N)$ expected time and finds the maxima using $dN + d \ln N + d^2 N^{1-1/d}(\ln N)^{1/d} + O(dN^{1-1/d})$ expected scalar comparisons. The experiments[1] show improved BCL hypercube algorithm-2 has a better expected running time than improved BCL hypercube algorithm-1 and the BCL hypercube algorithm. Our current efforts are to obtain a tight upper bound on the expected number of points finally computed by the BKST algorithm in improved BCL hypercube algorithm-2. Improved BCL algorithm is presented for a CI distribution and has a expected running time $O(N)$ and uses $2dN + O(\ln N(\ln(\ln N))) + d^2 N^{1-1/d}(lnN)^{1/d} + O(dN^{1-1/d})$ expected scalar comparisons. As Table 1, Figures 3(a), 3(b), and 4(b) show, the substantial reduction of the expected-time of all three algorithms has been attributed to the fact that better certificates of exclusion are chosen and more non-maximal points have been identified and discarded for the computation. We conclude this paper with the possibility of extending the basic idea, on which improved BCL hypercube algorithm-1 and algorithm-2, and improved BCL algorithms are based, to similar certificate-based convex-hull algorithms [BCL90].

References

[BCL90]   J. L. Bentley, K. L. Clarkson, and D. B. Levine. Fast linear expected-time algorithms for computing maxima and convex hulls. In *Proceedings of the First Annual ACM-SIAM Symposium on Discrete Algorithms*, pages 179–187, 1990.

[Ben80]   J. L. Bentley. Multidimensional divide-and-conquer. *Communications of the Association for Computing Machinery*, 23(4):214–229, 1980.

[BKST78]   J. L. Bentley, H. T. Kung, M. Schkolnick, and C. D. Thompson. On the average number of maxima in a set of vectors and applications. *Journal of the Association for Computing Machinery*, 25:536–543, 1978.

[FR75]   R. W. Floyd and R. L. Rivest. Expected time bounds for selection. *Communications of the ACM*, 18:165–172, 1975.

[FR90]   G. N. Fredrickson and S. Rodger. A new approach to the dynamic maintenance of maximal points in the plane. *Discrete and Computational Geometry*, 5:365–374, 1990.

[Jan91]   R. Janardan. On the dynamic maintenance of maximal points in the plane. *Information Processing Letters*, 40(2):59–64, 1991.

[Kap94]   S. Kapoor. Dynamic maintenance of maxima of $2 - d$ point sets. In *Proceedings of the Tenth Computational Geometry*, pages 140–149, 1994.

[KLP75]   H. T. Kung, F. Luccio, and F. P. Preparata. On finding the maxima of a set of vectors. *Journal of the ACM*, 22(4):469–476, 1975.

---

[1] Complete implementations are available from the authors

[L'E88]   P. L'Ecuyer. Efficient and portable combined random number generators. *Communications of the ACM*, 31(6):742–751, 1988.

[OL81]   M. H. Overmars and J. L. Van Leeuwen. Maintenance of configuration in the plane. *Journal of Computer and System Sciences*, 23:253–257, 1981.

[PS85]   F. P. Preparata and M. I. Shamos. *Computational Geometry: An Introduction*. Springer-Verlag, New York, 1985.

[PTVF93]   W. H. Press, S. A. Teukolsky, W. T. Vetterling, and B. P. Flannery. *Numerical Recipes in C : the Art of Scientific Computing*. Cambridge University Press, 1993.

A Constant Approximation Algorithm for Sorting Buffers

Jens S. Kohrt[1]* and Kirk Pruhs[2]**

[1] Department of Mathematics and Computer Science
University of Southern Denmark, Odense, Denmark
svalle@imada.sdu.dk
[2] Computer Science Department
University of Pittsburgh
kirk@cs.pitt.edu

Abstract. We consider an algorithmic problem that arises in manu-facturing applications. The input is a sequence of objects of various types. The scheduler is fed the objects in the sequence one by one, and is equipped with a finite buffer. The goal of the scheduler/sorter is to maximally reduce the number of type transitions. We give the first polynomial-time constant approximation algorithm for this problem. We prove several lemmas about the combinatorial structure of optimal solu-tions that may be useful in future research, and we show that the unified algorithm based on the local ratio lemma performs well for a slightly larger class of problems than was apparently previously known.

1 Introduction

We consider an algorithmic problem that arises in some manufacturing applica-tions. The input is a sequence of objects of various types. The scheduler is fed the objects in the sequence one by one, and is equipped with a buffer that can hold up to k objects. When the scheduler receives a new object, the object is initially placed in the sorting buffer. Then the scheduler may eject any objects in the sorting buffer in arbitrary order. The sorting buffer may never hold more than k objects, and must end up empty. Thus the output from the sorting buffer is a permutation of the input objects. Informally, the goal of the scheduler is to minimize the number of transitions between objects of different type in the output sequence.

An example situation where this problem arises is the Daimler-Benz car plant in Sindelfingen, Germany [2]. Here the objects are cars, and the types are

---

* Supported in part by the Danish Natural Science Research Council (SNF) and in part by the Future and Emerging Technologies programme of the EU under contract number IST-1999-14186 (ALCOM-FT). Most of this work was done while visiting the Computer Science Department at the University of Pittsburgh.
** Supported in part by NSF grant CCR-0098752, NSF grant ANI-0123705, and NSF grant ANI-0325353, and and by a grant from the US Air Force.

M. Farach-Colton (Ed.): LATIN 2004, LNCS 2976, pp. 193–202, 2004.
© Springer-Verlag Berlin Heidelberg 2004

the final color that that particular car should be painted. Particular cars must be colored particular colors because of custom orders from customers/dealers. For example, a customer can go to http://mbusa.com/ and order a G55 AMG Mercedes-Benz SUV with any number of possible options, including the exterior paint color. It is reported in [2] that the performance of the final layer painting yield mainly depends on the batch size of cars that have to be painted the same color, and as a consequence the performance of these sorting buffers can have a great impact on the overall performance of the manufacturing.

For concreteness we will adopt terminology appropriate for the Daimler-Benz car plant example, and consider the types of the objects to be colors. The most obvious objective function would be to minimize the number of transitions between objects of different color in the output sequence. The corresponding maximization objective function would be to maximize the number of color transitions removed from the sequence. While it may not be completely obvious, it is not too difficult to see that in an optimal solution there are no color changes introduced into the output sequence that were not in the input sequence. Hence, one can then see that a solution is optimal for the minimization problem if and only if it is optimal for the maximization problem. Of course, the equivalence of the maximization and minimization problems does not hold in the context of approximation. Whether a minimization or a maximization approximation algorithm is most appropriate depends on the input.

As an example of the problem, and the notation that we adopt, consider the following example. The initial sequence is $r_1 g_1 r_2 g_2 r_3 g_3 b_1 r_4 b_2 r_5 b_3 r_6$ that contains 12 objects with 11 color changes. The letters denote colors and the subscripts denote different objects of the same color. If $k \geq 4$, then an optimal output solution is $g_1 g_2 g_3 r_1 r_2 r_3 r_4 r_5 r_6 b_1 b_2 b_3$. It can be achieved by storing r_1, r_2, and r_3 in the buffer until the buffer contains these red objects and b_1. Then r_1, r_2, r_3 can be output, and the buffer can store the blue objects until the end. This gives a value of 2 color changes for the minimization objective function, and a value of 9 color savings for the maximization objection function.

It is not known if this sorting buffers problem is NP-hard. It is not hard to see that there is an $O(n^{k+1})$-time dynamic programming algorithm for this problem. It is also not hard to see that there is an $O(n^{c+1})$-time dynamic programming algorithm for this problem, where c is the number of different colors. So if k or c is $O(1)$, then the problem can be solved exactly in polynomial time. It seems that there may be real-world applications where the number of colors is not too large. The best approximation result known for the minimization approximation, an approximation ratio of $O(\log^2 k)$, is obtained by the polynomial-time online algorithm Bounded Waste introduced in [6].

A related problem, Paint Blocking, is studied in [7]. In this case the input sequence is reordered twice using two different buffers of the same size. After the first reordering the number of transitions between different types of objects is counted, and the second reordering has to return the sequence to its original order. For the minimization problem, a 5-approximation algorithm is given in [7].

Our main algorithmic result is a polynomial-time $\frac{1}{20}$-approximation algorithm for the maximization problem. Thus, this is the first constant approximation for either the maximization or minimization version of the sorting buffers problem. In order to obtain this result we have to prove several combinatorial lemmas about the structure of optimal, and near optimal, solutions. We expect that lemmas will be useful in further investigations into this problem. In the process of developing our algorithm, we showed that the analysis of the unified algorithm in [1] can be generalized to a slightly larger class of problems than indicated in [1]. Essentially our formulation allows arbitrary pairwise restrictions on membership in the solution, while in [1] only transitive restrictions are allowed. It is certainly plausible that this generalization might be useful elsewhere.

2   Observations about the Optimal Schedule

For any algorithm ALG and any input sequence σ, we let $\mathrm{ALG}(\sigma)$ denote both the resulting schedule and the color savings created by this schedule. Let $\mathrm{OPT}(\sigma)$ denote both an optimal schedule on input σ and the color savings in this solution. We say that the algorithm ALG is a c-approximation algorithm, if for any input sequence σ, $\mathrm{ALG}(\sigma) \geq c \cdot \mathrm{OPT}(\sigma)$. A schedule is *lazy* if a color change is never created in the output sequence if there is another legal move that doesn't create a color change. As noted in [6], one can always change any algorithm into a lazy algorithm without any loss of performance. The following observations are then almost immediate consequences of this observation.

Lemma 1. *Consider an arbitrary input sequence. If two objects of the same color are adjacent in the input sequence, then there is an optimal schedule where these two objects are adjacent in the output sequence.*

Lemma 2. *For any optimal algorithm and any input sequence, we may assume that for any color, any two objects of this color have the same order in the input sequence and the output sequence.*

One consequence of Lemma 1 is that no color changes are created in the output sequence. Thus it makes sense to talk purely about the reduction of color savings in the optimal without needing to consider the possibility of color changes created in the optimal. This allows us to formally define our problem in a slightly non-intuitive manner, in which the input is a sequence of groups of objects with the same color. The fact that this problem statement is equivalent to the problem statement in the introduction essentially follows from the fact that we can restrict ourselves to lazy schedules.

Definition 1. *The Sorting Buffers Maximization Problem (SBMP) is defined as follows:*

- *The* input sequence σ *consists of a sequence of* groups. *Each group has a color and a size, i.e., the number of items it contains. No two consecutive groups have the same color.*

- *The* output sequence *is a permutation of the groups in the input sequence.*
- *The* buffer *B* *can contain a collection of groups with size at most* $k - 1$. *Initially the buffer is empty. Note that the* $k - 1$ *bound insures that there is one space in the buffer left to move objects through.*
- *Any algorithm for SBMP essentially repeats the following two steps until the entire input sequence has been read, and the buffer is empty.*
  1. *The next group in the input sequence is either appended to the output sequence or is put in the buffer (if space permits).*
  2. *Optionally one or more groups from the buffer are moved to the output sequence.*
- *When an algorithm ALG is run on an input sequence* σ *it gains a color savings of one each time the last group in the output sequence and the next group to be put there are the same color.*

In SBMP, a lazy schedule is now defined as follows: if the last group in the output sequence is of a color c, then first all groups of objects of color c in the buffer are output in first-in-first-out order, and then if the next group of objects in the input sequence has color c then this group is immediately output without being put into the buffer. As before, one can always change any algorithm/schedule into a lazy algorithm/schedule without loss of performance, and thus, we only consider lazy schedules. We now switch notation and use r_i to denote the ith group with color r.

We classify color savings between two groups, say r_i and r_{i+1} in a schedule in one of the following three ways:

- *Move out-saving (MOS):* In this case, r_i is placed in the output sequence before r_{i+1} is reached, and further all the groups, that are between r_i and r_{i+1} in the input sequence, appear after r_{i+1} in the output sequence. Hence, all the items between r_i and r_{i+1} in the input sequence were in the cache when r_{i+1} was output. The *out groups* for this MOS, are defined to be the groups between r_i and r_{i+1} in the input. An example of a MOS is (here the block label A is some arbitrary sequence of groups):

$$r_i \boxed{\quad A \quad} r_{i+1} \;\rightarrow\; r_i r_{i+1} \boxed{\qquad A \qquad}$$

Note that for a MOS it is not necessary that r_i be put in the cache, and thus we will assume that it is not in the optimal solution.

- *Move backward-saving (MBS):* In this case r_i is put into the buffer, and is not expelled before group r_{i+1} is reached. An example of a MBS is:

$$r_1 \boxed{\quad A \quad} r_2 \;\rightarrow\; \boxed{\quad A \quad} r_1 r_2$$

- *Move backward and out-saving (MBOS):* In this case, r_i placed in the output sequence before r_{i+1} is reached, and not all the groups, that are between r_i and r_{i+1} in the input sequence, appear after r_{i+1} in the output sequence. An example of a MBOS is:

$$r_i \boxed{\quad A \quad | \quad B \quad} r_{i+1} \;\rightarrow\; \boxed{\quad A \quad} r_i r_{i+1} \boxed{\quad B \quad}$$

This is a combination of the previous two. At first, the r_i is put in the buffer, and *moved backward*. Before r_{i+1} is reached, r_i is dropped in the output sequence, and the groups in B are *moved out*.

At the time that r_i is placed in the output, let g_j be the next group in the input sequence. Then the *drop point* is defined to be the point immediately before g_j in the input sequence. The *out groups* for this MBOS are defined to be the groups between the drop point of r_i and r_{i+1} in the input.

In all three cases, we say that r_i is the *first group* of the savings and r_{i+1} is the *last group*. We now give an illustrative example of these definitions in the following figure:

$$r_1\boxed{A}r_2\boxed{B}r_3\boxed{C\mid D}r_4\boxed{E}r_5\boxed{F}r_6 \rightarrow \boxed{A\mid B\mid C}r_1r_2r_3r_4r_5r_6\boxed{D\mid E\mid F}$$

The first two color savings, which correspond to the pairs $r_1 - r_2$ and $r_2 - r_3$, are MBS's. This is because r_1 is moved backward to r_2, then both of them are moved backward to r_3. After this the $r_1 - r_2 - r_3$ group is dropped in the output sequence. The groups in D are the out groups for the $r_3 - r_4$ MBOS. Thus, the color savings corresponding to the $r_3 - r_4$ pair is a MBOS. The drop point for r_3 is between the last group in C and the first group in D. The last two color changes are MOS's.

Lemma 3. *Let r_i and r_{i+1} be any two groups between which there is a MOS, and let A be the groups between r_i and r_{i+1} in the input sequence. Then:*

- *No group in A is part of a MOS nor is it the last group in a MBOS.*
- *No group before r_{i+1} in the input sequence is the first group in a MBOS with drop point between r_i and r_{i+1}.*

Proof. Each of these possibilities involve placing a group in the output sequence between r_i and r_{i+1}, which contradicts $r_i - r_{i+1}$ being a MOS.   □

Lemma 4. *Let r_i and r_{i+1} be any two groups between which there is a MBOS. Let A be the groups between r_i and the drop point for r_i in the input sequence. Let B be the groups between the drop point for r_i and r_{i+1} in the input sequence. Then:*

- *No group in B is part of a MOS, nor is it the last group in a MBOS.*
- *No group before r_{i+1} in the input sequence is the first group in a MBOS with drop point between the drop point of r_i and r_{i+1}.*

Proof. Similar to Lemma 3.   □

Lemma 5. *For any two distinct MOS's or MBOS's, the out groups do not overlap.*

Proof. For ease of reference, we denote the point in the input sequence succeeding the first group of a MOS the drop point of this savings. Let $r_i - r_{i+1}$ and $b_j - b_{j+1}$ be any two distinct pairs of groups between which there are a MOS or MBOS. Without loss of generality we assume that the drop point of $r_i - r_{i+1}$ occurs before the drop point of $b_j - b_{j+1}$. Then, by Lemmas 3 and 4 the drop point of $b_j - b_{j+1}$ is at the earliest after r_{i+1} in the input sequence. Thus, the out groups of the two savings do not overlap.

3   Reduction to Two Problems

We show that either an $\Omega(1)$ fraction of the savings in the optimal solution are MOS, or an $\Omega(1)$ fraction of the savings in the optimal solution are MBS. We then show how to efficiently construct a solution that is constant competitive with respect to the number of MOS color savings, and show how to efficiently construct a solution that is constant competitive with respect to the number of MBS color savings. By taking the better of these two solutions, we get a constant approximation algorithm.

We first note that we can disregard the MBOS.

Lemma 6. *For any input sequence σ, there exists a solution for which the total number of MBS's and MOS's equals at least half of the profit of an optimal solution.*

Proof. Let σ be any fixed input sequence, and let $\mathrm{OPT}(\sigma)$ be any fixed optimal schedule for σ. We gradually transform this schedule into a new schedule with the desired properties.

We consider all MBOS of $\mathrm{OPT}(\sigma)$ one by one in order of their first group starting from the end of input sequence and continuing to the beginning (in the opposite of the usual order). During this sweep we maintain the invariant that any group further towards the end of the input sequence which is the first group in a MBOS in the original optimal schedule either has been turned into the first group of a MBS or it has a unique associated MBS in the schedule. Furthermore, we do not change any MBS or MOS in the original schedule. As shown below we also ensure that no two MBOS's share the same associated MBS. Consequently the resulting schedule has at least half the profit of the optimal solution.

Let $r_i - r_{i+1}$ be a MBOS under consideration. Let A be the groups in σ between r_i and r_i's drop point, and B the groups in σ between r_i's drop point and r_{i+1}. That is this part of the input looks like, $\cdots r_i A B r_{i+1} \cdots$. Note that by Lemma 4 and 5, no group in B participates in a MOS, neither are they the last group of a MBOS.

First, if any group in B is the first group of a MBS, then we associate one of the corresponding MBS's with the r_i-r_{i+1} MBOS. Note, as a direct result of Lemma 5 no other MBOS is associated with this MBS.

If this is not the case, we instead transform the MBOS into a MBS. The transformation is by induction on the number of groups left in B. The base case is when no groups in B are left. In this case we have turned this MBOS into a MBS, since r_i is kept in the buffer until r_{i+1} is reached in the input sequence. For the induction step, let g_j be the group in B furthest to the beginning of the input sequence. If g_j is not the first group in a savings, then the same profit is gained, if g_j is before r_i in the output sequence. Consequently instead of placing r_i in the output sequence, just before g_j is met in the input sequence, we keep r_i in the buffer, output g_j directly, and only then we place r_i in the output sequence. Otherwise, if g_j is the first group of a savings, it must be a MBOS. In this case, we place g_j into the output sequence as soon as g_j is encountered, and then immediately afterwards place r_i in the output buffer. This may reduce the total profit gained by the solution by one, as the $g_j - g_{j+1}$ MBOS is deleted. But due to the invariant, this savings already has an associated MBS in the solution which pays for the deletion of this MBOS.   □

Definition 2. *Let the Reduced Sorting Buffers Maximization Problem (SBMP-R) be defined as the Sorting Buffers Maximization Problem (SBMP) (Definition 1), except that no group can participate in more than one color savings, and that profit is only gained for savings of type MBS or MOS.*

Note that in SBMP each group can participate in up to two color savings, one in front and one in back. As an example of this, look at the following input and output sequence:

$$r_1 \boxed{\quad A \quad} r_2 \boxed{\quad B \quad} r_3 \boxed{\quad C \quad} r_4 \rightarrow \boxed{\quad A \quad \mid \quad B \quad \mid \quad C \quad} r_1 r_2 r_3 r_4$$

A total of four groups are involved in the color savings. For SBMP, a total color savings of three is gained, whereas for SBMP-R, only a total colors savings of two is gained (the two blocks are $r_1 - r_2$ and $r_3 - r_4$). For SBMP-R, another solution gives rise to the same profit:

$$r_1 \boxed{\quad A \quad} r_2 \boxed{\quad B \quad} r_3 \boxed{\quad C \quad} r_4 \rightarrow \boxed{\quad A \quad} r_1 r_2 \boxed{\quad B \quad \mid \quad C \quad} r_3 r_4$$

By only looking at SBMP-R instead of SBMP, we loose an approximation factor of $\frac{1}{4}$.

Lemma 7. *Let σ be any input sequence for SBMP and SBMP-R. Let $OPT(\sigma)$ be an optimal solution for SBMP, and let $OPT_R(\sigma)$ be an optimal algorithm for SBMP-R. Then $OPT_R(\sigma) \geq \frac{1}{4} OPT(\sigma)$.*

Proof. By Lemma 6 there exists a schedule for which the total number of MOS's and MBS's is at least half of $OPT(\sigma)$. Let S be any such schedule, and divide the resulting output sequence of S into maximal runs of groups of the same color between which there is a MOS or MBS. So runs are either broken by a color change or a MBOS. For any such run with $i \geq 2$ groups, S has a profit of $i - 1$. This run can be divided into $\lfloor i/2 \rfloor$ disjoint pairs of color savings. Thus, for OPT_R the profit is at least $\lfloor i/2 \rfloor \geq \frac{i-1}{2}$, that is, at least one half of the profit gained by S on MOS's and MBS's. By Lemma 6 this is at least $OPT(\sigma)/4$.   □

Let SBMP-R-MOS be the problem of maximizing the number of color savings of type MOS where each group participates in at most one savings. Similarly, let SBMP-R-MBS be the problem of maximizing the number of color savings of type MBS where each group participates in at most one savings. In section 4, we give a polynomial time algorithm solving SBMP-R-MOS. In section 5, we give a polynomial time algorithm with an approximation factor of at least $\frac{1}{4}$ for SBMP-R-MBS. Either the optimal solution for SBMP-R-MOS is a $\frac{1}{5}$ approximation of the optimal solution for SBMP-R, or the optimal solution for SBMP-R-MBS is a $\frac{4}{5}$ approximation of the optimal solution for SBMP-R. Hence, the better of our solutions in section 4 and section 5 is a $\frac{1}{5}$ approximation of OPT_R. Consequently we have an $\frac{1}{20}$ approximation of SBMP.

Theorem 1. *There is an algorithm with an approximation factor of at least $\frac{1}{20}$ for SBMP running in polynomial time.*

4   A SBMP-R-MOS Algorithm

The next piece of the puzzle is a greedy algorithm solving SBMP-R-MOS exactly. Recall this problem is to maximize the number of MOS.

As before, we may assume that the ordering for groups of the same color does not change. Consequently it is sufficient to consider only MOS between two groups r_i and r_{i+1} of the same color, where no group of this color occurs between the two in the input sequence. Further, the total size of the groups in between r_i and r_{i+1} has to be at most $k-1$, or else they cannot be in the buffer. Again one can show, as in Lemma 3, that we may assume that no pair of MOS occur inside one another.

Instead of solving SBMP-R-MOS directly, we transform it into the problem of finding a maximum independent set in an interval graph. That is, the input is a sequence of intervals over the real line, and the desired output is a maximal cardinality disjoint set of intervals. If one orders the intervals by increasing right endpoint, and greedily selects intervals, it is a standard exercise to show that this greedy algorithm exactly solves the problem [3,4].

We now construct our reduction. For each possible MOS in the input sequence, we create a corresponding interval starting at the first group and ending at the last group of the MOS (inclusive). Then two intervals overlap, if and only if they cannot occur together. This is the case if they either share a group or if they would occur inside each other, if they were both used. Then the maximum MOS corresponds exactly to the maximal independent set in the resulting interval graph.

5   A SBMP-R-MBS Approximation Algorithm

As part of our approximation algorithm for SBMP-R-MBS we need a generalization of the unified algorithm introduced in [1]. First, we explain the generalization, and then we apply this result on our own problem by using a reduction from SBMP-R-MBS to The Resource Allocation Maximization Problem.

Definition 3 (RAMP). *The Resource Allocation Maximization Problem is defined as follows:*

- *The input consists of a number of instances I, each requiring the utilization of some limited resource. The amount of resource available is fixed over time, and its size is normalized to one.*
- *Instances I are each defined by the following four properties:*
    - *A half-open time interval $[s(I), e(I))$ during which the instance is to be executed. $s(I)$ and $e(I)$ are the start-time and end-time of the instance.*
    - *The amount of resource necessary or the width of the instance, $w(I)$, $0 \leq w(I) \leq 1$.*

- *The* profit $p(I) \geq 0$ *gained by using this instance*
- *The* activity $A(I)$ *which is a set of instances (always including I) which cannot occur at the same time as I. For any two instances I and J, then $I \in A(J)$ if and only if $J \in A(I)$.*
- *The output or a feasible schedule is a set of instances X, such that for each instance I in X the set $A(I) \cap X$ only contains I, and for each time instance t the total width of all instances in X which contains t is at most one. The total profit of X is the sum of the profit of the individual instances in X.*
- *The goal of RAMP is to find a feasible schedule with maximal profit.*

Note that the only way our problem differs from the original problem [1] is the way activities are defined. In our setting each instance I has its own set of other instances $A(I)$ which cannot occur at the same time as I. As an example, in our problem it is possible to have three instances, I_1, I_2, and I_3, such that $I_1 \in A(I_2)$ and $I_2 \in A(I_3)$, but $I_1 \notin A(I_3)$. In [1] this cannot be the case, since activities are transitive, i.e., $I_1 \in A(I_2)$ and $I_2 \in A(I_3)$ implies $I_1 \in A(I_3)$.

By following the analysis in [1], one can verify that it still holds in this more general setting. The reader is referred to [5] for a more detailed analysis. The main result about the obtained approximation ratio is the same as in [1]:

Lemma 8 (Lemma 3.2 from [1]). *Let $w_{\min}$ ($w_{\max}$) be the minimum (maximum) width of any instance in the input, and let α be some value larger than zero. The approximation factor of the unified algorithm is at least*

$$\frac{\min\{1, \alpha \cdot \max\{w_{\min}, 1 - w_{\max}\}\}}{1 + \alpha}$$

As noted in [1] the unified algorithm can easily be implemented in polynomial time. This result also applies for our slightly modified unified algorithm.

We now use this result to make an approximation algorithm for our original problem, SBMP-R-MBS, by reducing it to the generalized problem.

First note, that we may not, as in SBMP, assume that the ordering for groups of the same color does not change. This can be seen in the following example:

$$r_1 \boxed{A} \, r_2 \boxed{B} \, r_3 b_1 \boxed{C} \, b_2 r_4 \quad \rightarrow \quad \boxed{A} \boxed{B} \, r_2 r_3 \boxed{C} \, b_1 b_2 r_1 r_4$$

where r_1 contains one item, and all other groups each contain $k - 2$ items. If we want to move r_3 to r_4, then we cannot move b_1 to b_2. Thus, the only feasible solution with a profit of three is the one given above.

Further note, that as only MBS's are allowed, we may assume that the first group of a MBS is moved no farther backward than to the last group: moving the two groups farther back in the sequence does not give rise to more profit.

Also note that we can see the input sequence as a time line. Then a MBS starts/ends at the time corresponding to its first/last group (inclusive).

With the above in mind, we now construct the reduction. For each color r and for each pair of groups of this color r_i and r_j where r_i occurs before r_j in the input sequence and where the size of r_i is at most $k - 1$, we create an instance

$I_{r_i r_j}$ corresponding to the possible $r_i - r_j$ MBS. The width of $I_{r_i r_j}$ is the number of items in the first group normalized by $k - 1$, i.e., $w(I_{r_i r_j}) = \text{size}(r_i)/(k - 1)$. As all MBS's have a profit of one, $p(I_{r_i r_j}) = 1$. Further, the activity $A(I_{r_i r_j})$ contains $I_{r_i r_j}$ as well as those instances that use either r_i or r_j, i.e., exactly the instances which cannot occur at the same time as $I_{r_i r_j}$.

Lemma 9. *There is an algorithm running in polynomial time with an approximation factor of at least $\frac{1}{4}$ for SBMP-R-MBS.*

Proof. Suppose all instances have a width of at most $\frac{1}{2}$, i.e., $w_{max} \leq \frac{1}{2}$. In this case $\alpha = 2$ maximizes Lemma 8 with a performance guarantee of $\frac{1}{3}$.

Next, suppose all instances have a width of at least $\frac{1}{2}$, i.e., $w_{min} > \frac{1}{2}$. In this case no pair of intersecting instances may both be used. Consequently this problem is equivalent to the problem of finding the maximal independent set in an interval graph. Similar to Section 4, this can be solved exactly by a simple greedy algorithm.

In the general case we solve the problem separately for the instances of width at most a half and for the instances of width more than a half. Either the optimal solution for the former case is at least $\frac{3}{4}$ of the optimum, or the optimal solution for the latter case is at least $\frac{1}{4}$ of the optimum. Hence, the better of the two is at least a $\frac{1}{4}$ approximation of SBMP-R-MBS. $\square$

References

1. Amotz Bar-Noy, Reuven Bar-Yehuda, Ari Freund, Joseph (Seffi) Naor, and Baruch Schieber. A unified approach to approximating resource allocation and scheduling. *Journal of the ACM*, 48(5):1069–1090, 2001.
2. Sven Brückner, Jo Wyns, Patrick Peeters, and Martin Kollingbaum. Designing agent for manufacturing control. In *Proceedings of the 2nd AI & Manufacturing Research Planning Workshop*, pages 40–46, 1998.
3. Thomas H. Cormen, Charles E. Leiserson, Ronald L. Rivest, and Clifford Stein. *Introduction to Algorithms*, chapter 16. MIT Press and McGraw-Hill Book Company, 2nd edition, 2001.
4. U. I. Gupta, D. T. Lee, and Joseph Y.-T. Leung. An Optimal Solution for the Channel-Assignment Problem. *IEEE Transactions on Computers*, 28(11):807–810, 1979.
5. Jens S. Kohrt and Kirk Pruhs. A constant approximation algorithm for sorting buffers. Technical Report PP-2003-21, Department of Mathematics and Computer Science, University of Southern Denmark, Odense, 2003.
6. Harald Räcke, Christian Sohler, and Matthias Westermann. Online scheduling for sorting buffers. In Rolf H. Möhring and Rajeev Raman, editors, *Proceedings of 10th Annual European Symposium (ESA)*, volume 2461 of *Lecture Notes in Computer Science*, page 820 ff. Springer, September 2002.
7. Joel Scott Sokol. *Optimizing Paint Blocking in an Automobile Assembly Line: An Application of Specialized TSP's*. PhD thesis, Department of Electrical Engineering and Computer Science, MIT, June 1999.

Approximation Schemes for a Class of Subset Selection Problems

Kirk Pruhs[1][*] and Gerhard J. Woeginger[2]

[1] Department of Computer Science, University of Pittsburgh, USA
kirk@cs.pitt.edu
http://www.cs.pitt.edu/~kirk
[2] Department of Mathematics, University of Twente, The Netherlands
g.j.woeginger@math.utwente.nl
http://wwwhome.cs.utwente.nl/~woegingergj

Abstract. In paper we develop an easily applicable algorithmic technique/tool for developing approximation schemes for certain types of combinatorial optimization problems. Special cases that are covered by our result show up in many places in the literature. For every such special case, a particular rounding trick has been implemented in a slightly different way, with slightly different arguments, and with slightly different worst case estimations. Usually, the rounding procedure depended on certain upper or lower bounds on the optimal objective value that have to be justified in a separate argument. Our easily applied result unifies many of these results, and sometimes it even leads to a simpler proof.
We demonstrate how our result can be easily applied to a broad family of combinatorial optimization problems. As a special case, we derive the existence of an FPTAS for the scheduling problem of minimizing the weighted number of late jobs under release dates and preemption on a single machine. The approximability status of this problem has been open for some time.

1 Introduction

One of the commonly stated goals of algorithmic research is the development of a modestly-sized toolkit of widely applicable algorithmic techniques. The vision is that future researchers, particularly those without specialized training in algorithmics, could use these tools to quickly develop/analyze algorithms for new problems. In this paper, we develop an easily and widely applicable algorithmic technique/tool for developing approximation schemes for certain types of combinatorial optimization problems. This tool should save algorithmic researchers time, and is simple enough be used by researchers without specialized algorithmics training.

Over the years, there have evolved a number of standard approaches for designing approximation schemes; see for instance Horowitz & Sahni [8,9], Ibarra

[*] Supported in part by NSF grant CCR-0098752, NSF grant ANI-0123705, and NSF grant ANI-0325353.

& Kim [10], Sahni [18], and Woeginger [20]. (A review of basic definitions related to approximation schemes can be found in section 2.) We will investigate one of these standard approaches and demonstrate that it applies to a broad family of combinatorial optimization problems. The standard approach under investigation is the technique of *rounding the input*; this technique goes back to the 1970s and possibly has first been used in the paper by Horowitz & Sahni [8]. The family of combinatorial optimization problems under investigation is defined as follows.

Definition 1. *(Subset selection problems)*
A subset selection problem $\mathcal{P}$ is a combinatorial optimization problem whose instances $I = (X, w, S)$ consist of

- *a ground set X with $|X| = n$ elements;*
- *a positive integer weight $w(x)$ for every $x \in X$;*
- *a structure S that is described by $\ell(S)$ bits;*

The structure S specifies for every subset $Y \subseteq X$ whether Y is feasible or infeasible; this can be done within a time complexity polynomially bounded in n and $\ell(S)$.

If $\mathcal{P}$ is a minimization problem, then the goal is to find a feasible subset $Y \subseteq X$ that minimizes $w(Y) \doteq \sum_{y \in Y} w(y)$, and if $\mathcal{P}$ is a maximization problem, then the goal is to find a feasible subset $Y \subseteq X$ that maximizes $w(Y)$. $\square$

The class of subset selection problems described in Definition 1 is very general, and it contains many problems with very bad approximability behavior. For instance, the *weighted independent set* problem ("Given a graph with vertex weights, find the maximum weight subset of pairwise non-adjacent vertices") belongs to this class. It is known that weighted independent set does not possess *any* ρ-approximation algorithm with a fixed $\rho \geq 1$, unless P=NP (Håstad [7]). If we additionally impose condition (C) as in the following theorem, then the approximability behavior of subset selection problems improves considerably.

Theorem 1. *Let $\mathcal{P}$ be a subset selection problem with instances $I = (X, w, S)$ that satisfies the following condition:*

(C) There exists an algorithm that solves $\mathcal{P}$ to optimality whose running time is polynomially bounded in n, in $W := \sum_{x \in X} w(x)$, and in $\ell(S)$.

Then problem $\mathcal{P}$ has an FPTAS.

Theorem 1 is proved in Section 3. The proof is quite straightforward, and it mainly uses the folklore rounding tricks from the literature. The main contribution of this paper is to identify the neat and simple condition (C) that automatically implies the existence of an FPTAS. Special cases that are covered by Theorem 1 show up at many places in the literature. For every such special case, the rounding trick has been implemented in a slightly different way, with slightly different arguments, and with slightly different worst case estimations.

Usually, the rounding procedure depends on certain upper or lower bounds on the optimal objective value that have to be justified in a separate argument. Theorem 1 unifies many of these results, and sometimes it even leads to simpler proofs.

Sections 4 and 5 contain a number of optimization problems (from scheduling theory and from graph theory) that fit into the framework of Definition 1. These examples illustrate the wide applicability and ease of use of our result. As one special case, we prove in Theorem 2 that the scheduling problem $1 \mid pmtn, r_j \mid \sum w_j U_j$ (the problem of minimizing the weighted number of late jobs under release dates and preemption on a single machine) has a PTAS. The approximability status of this problem has been open for some time, and the question was considered to be difficult. In particular, the problem does not fit into the framework for FPTAS's established by Woeginger [20].

2 Basic Definitions

An algorithm that returns near-optimal solutions is called an *approximation algorithm*; if it does this in polynomial time, then it is called a *polynomial time* approximation algorithm. An approximation algorithm is called a *ρ-approximation algorithm*, if it always returns a near-optimal solution with cost at most a factor ρ above the optimal cost (for minimization problems) respectively at most a factor ρ below the optimal cost (for maximization problems). The value $\rho \geq 1$ is called the *worst-case performance guarantee* of this algorithm. A family of $(1 + \varepsilon)$-approximation algorithms over all real $\varepsilon > 0$ with polynomial running times is called a *polynomial time approximation scheme* or PTAS, for short. If the time complexity of a PTAS is also polynomially bounded in $1/\varepsilon$, then it is called a *fully* polynomial time approximation scheme or FPTAS, for short. With respect to relative performance guarantees, an FPTAS is essentially the strongest possible polynomial time approximation result that we can derive for an NP-hard problem (unless P=NP holds).

3 Proof of the Main Result

In this section we will prove the main result of the paper. The proof method is essentially due to Horowitz & Sahni [8]. The arguments for minimization problems and for maximization problems are slightly different. We start with the discussion of minimization problems.

Let $\varepsilon > 0$ be a small real number. Let $I = (X, w, S)$ be some instance of a minimization problem $\mathcal{P}$ that belongs to the class of subset selection problems as defined in Definition 1. Let $x_1, \ldots, x_n$ be an enumeration of the elements of the ground set X such that

$$w(x_1) \leq w(x_2) \leq \cdots \leq w(x_n). \tag{1}$$

For $k = 1, \ldots, n$, we define a so-called scaling parameter

$$Z^{(k)} \doteq \varepsilon \cdot \frac{1}{n} \cdot w(x_k). \tag{2}$$

We introduce a number of new instances $I^{(1)}, \ldots, I^{(n)}$ of problem $\mathcal{P}$. Every new instance $I^{(k)}$ has the same structure S and the same ground set X as I, but it has a different set of weights $w^{(k)}$. As a consequence, all instances $I^{(k)}$ have the same feasible solutions as the original instance I. The weights $w^{(k)}$ are defined as follows:

- For $i = 1, \ldots, k$, we set $w^{(k)}(x_i) = \lceil w(x_i)/Z^{(k)} \rceil$.
- For $i = k+1, \ldots, n$, we set $w^{(k)}(x_i) = n\lceil n/\varepsilon \rceil$.

The definition of the parameter $Z^{(k)}$ in (2) yields that $w^{(k)}(x_i) \leq \lceil n/\varepsilon \rceil$ for $1 \leq i \leq k$. Therefore, the overall weight $W^{(k)}$ of all elements in instance $I^{(k)}$ can be bounded as

$$W^{(k)} \leq \sum_{i=1}^{k} \lceil n/\varepsilon \rceil + \sum_{i=k+1}^{n} n\lceil n/\varepsilon \rceil \leq n^2 \lceil n/\varepsilon \rceil. \tag{3}$$

Hence, $W^{(k)}$ is polynomially bounded in n and in $1/\varepsilon$. If we feed instance $I^{(k)}$ to the exact algorithm in condition (C) in Theorem 1, then the running time is polynomially bounded in n, in $\ell(S)$, and in $1/\varepsilon$. That's precisely the type of time complexity that we need for an FPTAS. Hence we get Lemma 1.

Lemma 1. *Every instance $I^{(k)}$ can be solved to optimality within a time complexity polynomially bounded in n, $\ell(S)$, and $1/\varepsilon$.*   □

Next, let Y^* denote the optimal solution for the original instance I, and let OPT denote its optimal objective value $w(Y^*)$. Let $Y^{(k)} \subseteq X$ denote the optimal solution for instance $I^{(k)}$ for $k = 1, \ldots, n$. Let j denote the maximal index with $x_j \in Y^*$. Obviously,

$$\text{OPT} = w(Y^*) \geq w(x_j). \tag{4}$$

Furthermore, we claim that

$$Y^{(j)} \subseteq \{x_1, x_2, \ldots, x_j\}. \tag{5}$$

This statement is vacuously true for $j = n$. For $j \leq n-1$, we use that Y^* is *some* feasible solution for instance $I^{(j)}$, whereas $Y^{(j)}$ is the *optimal* feasible solution for instance $I^{(j)}$. Since $|Y^*| \leq j \leq n - 1$, this yields

$$w^{(j)}(Y^{(j)}) \leq w^{(j)}(Y^*) \leq |Y^*| \cdot w^{(j)}(x_j) \leq (n-1) \cdot \lceil n/\varepsilon \rceil. \tag{6}$$

By Inequality (6), the set $Y^{(j)}$ can not contain any of the expensive elements $x_{j+1}, \ldots, x_n$ that all have weight $n\lceil n/\varepsilon \rceil$. This proves (5).

We now analyze the quality of the feasible solution $Y^{(j)}$ for the original instance I. In the following chain of inequalities, the first inequality holds since

(5) implies $w(y) \leq Z^{(j)} \cdot w^{(j)}(y)$ for all $y \in Y^{(j)}$. The second inequality holds, since $Y^{(j)}$ is the optimal solution for weights $w^{(j)}(\cdot)$. The equation in the third line follows from (5). The inequality in the fourth line follows from $\lceil \alpha \rceil \leq \alpha + 1$. The inequality in the sixth line follows from $|Y^*| \leq n$ and from (2). The final inequality follows from (4).

$$\sum \{w(y) \mid y \in Y^{(j)}\} \leq Z^{(j)} \cdot \sum \{ w^{(j)}(y) \mid y \in Y^{(j)} \}$$

$$\leq Z^{(j)} \cdot \sum \{ w^{(j)}(y) \mid y \in Y^* \}$$

$$= Z^{(j)} \cdot \sum \{ \lceil w(y)/Z^{(j)} \rceil \mid y \in Y^* \}$$

$$\leq Z^{(j)} \cdot \sum \{ w(y)/Z^{(j)} + 1 \mid y \in Y^* \}$$

$$= \sum \{ w(y) \mid y \in Y^* \} + |Y^*| \cdot Z^{(j)}$$

$$\leq \text{OPT} + n \cdot \varepsilon \cdot \frac{1}{n} \cdot w(x_j)$$

$$\leq (1 + \varepsilon) \cdot \text{OPT}.$$

With this, it is clear how to get the FPTAS: We compute the optimal solutions $Y^{(k)} \subseteq X$ for the instances $I^{(k)}$ with $k = 1, \ldots, n$. By Lemma 1, this can be done with time complexity polynomially bounded in the length of the encoding of I and in $1/\varepsilon$. Then we compute the costs of $Y^{(k)}$ with respect to instance I, and we determine the best solution. By the above chain of inequalities, this best solution has objective value at most $(1 + \varepsilon)\text{OPT}$. This completes the proof of Theorem 1 for the case where $\mathcal{P}$ is a minimization problem.

Now let us discuss the case where $\mathcal{P}$ is a maximization problem. Consider an instance $I = (X, w, S)$ of $\mathcal{P}$, and enumerate the elements of the ground set X as in (1). In a preprocessing phase, we determine for every element x_k ($k = 1, \ldots, n$) whether there exists a feasible solution that contains x_k. This can be done as follows: We create a new instance $I^{(k)}$ from I by setting the weight of element x_k to n and by setting the weights of the remaining $n-1$ elements to 1. Clearly, x_k shows up in some feasible solution if and only if the optimal objective value of $I^{(k)}$ is greater or equal to n. Since the overall weight in instance $I^{(k)}$ is $2n - 1$, the algorithm from condition (C) can be used to solve it in time polynomially bounded in n and $\ell(S)$.

The main part of our algorithm is built around the maximal index j for which element x_j occurs in some feasible solution. This implies

$$\text{OPT} \geq w(x_j). \tag{7}$$

We introduce a scaling parameter $Z^\# \doteq \varepsilon \cdot \frac{1}{n} \cdot w(x_j)$. We define a new instance $I^\#$ from I that has new weights $w^\#$. For $i = 1, \ldots, j$ we set $w^\#(x_i) = \lfloor w(x_i)/Z^\# \rfloor$, and for $i = j+1, \ldots, n$ we set $w^{\#}(x_i) = 1$. Similarly as in the minimization case, instance $I^{\#}$ can be solved to optimality within a time complexity polynomially bounded in n, $\ell(S)$, and $1/\varepsilon$.

The optimal solution $Y^{\#}$ of $I^{\#}$ satisfies the following inequalities. These inequalities run in parallel to the inequalities for the minimization case. In the third line, we use (7) to bound $w(x_j)$.

$$
\sum \{ w(y) \mid y \in Y^{\#} \} \geq Z^{\#} \cdot \sum \{ w^{\#}(y) \mid y \in Y^{\#} \}
$$

$$
\geq Z^{\#} \cdot \sum \{ w^{\#}(y) \mid y \in Y^{*} \}
$$

$$
= Z^{\#} \cdot \sum \{ \lfloor w(y)/Z^{\#} \rfloor \mid y \in Y^{*} \}
$$

$$
\geq Z^{\#} \cdot \sum \{ w(y)/Z^{\#} - 1 \mid y \in Y^{*} \}
$$

$$
= \sum \{ w(y) \mid y \in Y^{*} \} - |Y^{*}| \cdot Z^{\#}
$$

$$
\geq \text{OPT} - n \cdot \varepsilon \cdot \frac{1}{n} \cdot w(x_j)
$$

$$
\geq (1 - \varepsilon) \cdot \text{OPT}.
$$

Hence, also maximization problems have an FPTAS. The proof of Theorem 1 is complete.

4   Example: Scheduling to Minimize the Weighted Number of Late Jobs

In this section, we will use the standard three-field scheduling notation (see e.g. Graham, Lawler, Lenstra & Rinnooy Kan [5] and Lawler, Lenstra, Rinnooy Kan & Shmoys [14]).

In the scheduling problem $1 \mid\mid \sum w_j U_j$, the input consists of n jobs J_j with positive integer processing times p_j, weights w_j, and due dates d_j $(j = 1, \ldots, n)$. All jobs are available for processing at time 0. In some schedule a job is *on-time* if its processing is completed by its deadline, and otherwise it is *late*. The goal is to schedule the jobs without interruption on a single machine such that the total weight of the late jobs is minimized. The problem $1 \mid\mid \sum w_j U_j$ is known to be NP-hard in the ordinary sense (Karp [11]).

Problem $1 \mid\mid \sum w_j U_j$ belongs to the class of subset selection problems as described in Definition 1. The ground set X consists of the n jobs with weights w_k and total weight $W = \sum_{i=1}^{n} w_k$. The structure S consists of the processing times p_j and the due dates d_j $(j = 1, \ldots, n)$. A subset Y of the jobs is feasible, if the remaining jobs in $X - Y$ can all be scheduled on-time on a single machine; clearly, this information is specified by the structure S. Lawler & Moore [15] give a dynamic programming formulation that solves $1 \mid\mid \sum w_j U_j$ in $O(nW)$ time.

Then our main result in Theorem 1 implies the following well-known result of Gens & Levner [4].

Corollary 1. *(Gens & Levner [4], 1981)*
There exists an FPTAS for minimizing the weighted number of late jobs in the scheduling problem $1 \mid\mid \sum w_j U_j$. ☐

A closely related problem is to maximize the total weight of the *on-time* jobs. Clearly, the algorithm of Lawler & Moore [15] also solves this maximization problem in $O(nW)$ time. We get the following result.

Corollary 2. *(Sahni [18], 1976)*
There exists an FPTAS for maximizing the weighted number of on-time jobs in the scheduling problem $1 \mid\mid \sum w_j U_j$. ☐

In the *0/1-knapsack problem*, the input consists of n pairs of positive integers (w_k, b_k) and a positive integer b: The weight w_k denotes the profit of the kth item, and b_k denotes the space occupied by this item. The goal is to select a subset Y that has the maximum profit subject to the condition that it does not occupy more than b space. The *0/1-knapsack problem* is NP-hard (Karp [11]), and it can be solved in $O(nW)$ time (see for instance Bellman & Dreyfus [1] or Martello & Toth [17]).

It is easy to see that the 0/1-knapsack problem belongs to the class of subset selection problems of Definition 1. In fact, it is a special case of the maximization version of the scheduling problem $1 \mid\mid \sum w_j U_j$ as described above: Essentially, the kth item corresponds to a job with processing time b_k, weight w_k, and (universal) due date b.

Corollary 3. *(Ibarra & Kim [10], 1975)*
The 0/1-knapsack problem possesses an FPTAS. ☐

Another closely related problem is $1 \mid pmtn, r_j \mid \sum w_j U_j$: There are n jobs J_j $(j = 1, \ldots, n)$ with processing times p_j, weights w_j, due dates d_j, and release dates r_j. In this variant, job J_j cannot be started before its release date r_j, but it may be preempted. Lawler [13] designs a (very complicated) dynamic program that solves $1 \mid pmtn, r_j \mid \sum w_j U_j$ in $O(n^3 W^2)$ time. We get the following (new) result.

Theorem 2. *There exists an FPTAS for minimizing the weighted number of late jobs in the scheduling problem* $1 \mid pmtn, r_j \mid \sum w_j U_j$.

5   Example: The Restricted Shortest Path Problem

An instance of the restricted shortest path problem (RSP, for short) consists of a directed graph $G = (V, A)$ and an integer bound T. Every arc $a \in A$ has a positive integer cost w_a and a positive integer transition time t_a. For a directed path Y in G, the cost $w(Y)$ and the transition time $t(Y)$ are defined as the sum of the costs and transition times, respectively, of the edges in the path Y. The goal is to find a path Y with $t(Y) \leq T$ from a specified source vertex to a specified target vertex, that minimizes the cost. RSP is NP-complete in the ordinary sense (Garey & Johnson [3]). Furthermore, RSP is solvable in $O(|A| \cdot W)$ time by dynamic programming; see for instance Warburton [19] or Hassin [6]. One way of doing this is to compute for every vertex $v \in V$ and for every cost $c \in \{0, \ldots, W\}$, the smallest possible transition time of a path from the source vertex to v with cost c.

Problem RSP belongs to the class of subset selection problems as described in Definition 1. The ground set X consists of the arcs $a \in A$ with costs w_a. The structure S consists of the graph G, of the transition times t_j, of the bound T, and of the source and sink vertices. A subset Y of the arcs is feasible, if it forms a path from source to sink with transition time $t(Y) \leq T$. Obviously, this feasibility information is encoded by the structure S. We get the following result.

Corollary 4. *(Hassin [6], 1992)*
The restricted shortest path problem RSP possesses an FPTAS.   □

The result in Corollary 4 has been established in 1987 by Warburton [19] for acyclic directed graphs and then in 1992 by Hassin [6] for arbitrary directed graphs. Lorenz & Raz [16] and Ergun, Sinha & Zhang [2] improve the time complexities of these approximation schemes.

References

1. R.E. BELLMAN AND S.E. DREYFUS (1962). *Applied Dynamic Programming.* Princeton University Press.
2. F. ERGUN, R. SINHA, AND L. ZHANG (2002). An improved FPTAS for restricted shortest path. *Information Processing Letters 83*, 287–291.
3. M.R. GAREY AND D.S. JOHNSON (1979). *Computers and Intractability.* W.H. Freeman and Co., New York.
4. G.V. GENS AND E.V. LEVNER (1981). Fast approximation algorithms for job sequencing with deadlines. *Discrete Applied Mathematics 3*, 313–318.
5. R.L. GRAHAM, E.L. LAWLER, J.K. LENSTRA, AND A.H.G. RINNOOY KAN (1979). Optimization and approximation in deterministic sequencing and scheduling: A survey. *Annals of Discrete Mathematics 5*, 287–326.
6. R. HASSIN (1992). Approximation schemes for the restricted shortest path problem. *Mathematics of Operations Research 17*, 36–42.
7. J. HÅSTAD (1999). Clique is hard to approximate within $n^{1-\epsilon}$. *Acta Mathematica 182*, 105–142.
8. E. HOROWITZ AND S. SAHNI (1974). Computing partitions with applications to the knapsack problem. *Journal of the ACM 21*, 277–292.
9. E. HOROWITZ AND S. SAHNI (1976). Exact and approximate algorithms for scheduling nonidentical processors. *Journal of the ACM 23*, 317–327.
10. O. IBARRA AND C.E. KIM (1975). Fast approximation algorithms for the knapsack and sum of subset problems. *Journal of the ACM 22*, 463–468.
11. R.M. KARP (1972). Reducibility among combinatorial problems. In: R.E. Miller and J.W. Thatcher, editors, *Complexity of Computer Computations*, Plenum Press, New York, 85–104.

12. E.L. LAWLER (1979). Fast approximation schemes for knapsack problems. *Mathematics of Operations Research 4*, 339–356.
13. E.L. LAWLER (1990). A dynamic programming algorithm for preemptive scheduling of a single machine to minimize the number of late jobs. *Annals of Operations Research 26*, 125–133.
14. E.L. LAWLER, J.K. LENSTRA, A.H.G. RINNOOY KAN, AND D.B. SHMOYS (1993). Sequencing and scheduling: Algorithms and complexity. In: S.C. Graves, A.H.G. Rinnooy Kan, and P.H. Zipkin (eds.) *Logistics of Production and Inventory*, Handbooks in Operations Research and Management Science 4, North-Holland, Amsterdam, 445–522.
15. E.L. LAWLER AND J.M. MOORE (1969). A functional equation and its application to resource allocation and sequencing problems. *Management Science 16*, 77–84.
16. D.H. LORENZ AND D. RAZ (2001). A simple efficient approximation scheme for the restricted shortest path problem. *Operations Research Letters 28*, 213–219.
17. S. MARTELLO AND P. TOTH [1990]. Knapsack problems: Algorithms and computer implementations. John Wiley & Sons, England.
18. S. SAHNI (1976). Algorithms for scheduling independent tasks. *Journal of the ACM 23*, 116–127.
19. A. WARBURTON (1987). Approximation of pareto optima in multiple-objective shortest path problems. *Operations Research 35*, 70–79.
20. G.J. WOEGINGER (2000). When does a dynamic programming formulation guarantee the existence of a fully polynomial time approximation scheme (FPTAS)? *INFORMS Journal on Computing 12*, 57–75.

Finding k-Connected Subgraphs with Minimum Average Weight

Prabhakar Gubbala and Balaji Raghavachari

Computer Science Department, University of Texas at Dallas, Richardson, TX 75080
{prabha,rbk}@utdallas.edu

Abstract. We consider the problems of finding k-connected spanning subgraphs with minimum *average* weight. We show that the problems are NP-hard for $k > 1$. Approximation algorithms are given for four versions of the minimum average edge weight problem:

1. 3-approximation for k-edge-connectivity,
2. $O(log k)$ approximation for k-node-connectivity
3. $2 + \epsilon$ approximation for k-node-connectivity in Euclidian graphs, for any constant $\epsilon > 0$,
4. 5.8-approximation for k-node-connectivity in graphs satisfying the triangle inequality.

1 Introduction

Given a graph $G = (V, E)$ that satisfies a specified property $\mathcal{P}$, and a cost function $c : E \mapsto \Re^+$, consider the problem of finding a subgraph $G = (V, E')$ that also satisfies $\mathcal{P}$, and has minimum average weight, i.e.,

$$\min \frac{\sum_{e \in E'} c(e)}{|E'|}, \text{ where } E' \text{ satisfies } \mathcal{P}.$$

In this paper, we consider the properties of k-edge-connectivity and k-node-connectivity. Our algorithm for k-node-connectivity also extends to any monotone property on graphs. $\mathcal{P}$ is a monotone property if G continues to satisfy $\mathcal{P}$ after the addition of arbitrary additional edges to G (connecting existing vertices). Connectivity and non-planarity are monotone properties, while acyclicity and planarity are not. We refer to the minimum-average-weight k-edge-connectivity problem as AVG-kEC, and the corresponding k-vertex-connectivity problem as AVG-kVC. Depending on the cost function on edges, the graph can be a general graph or a graph in a metric space (i.e., its edges satisfy the triangle inequality). We show that all the above versions of this problem are NP-hard and provide approximation algorithms for them.

1.1 Previous Work

For the minimization problem $\frac{\sum_{e \in E'} c(e)}{|E'|}$, lot of research has been done for the minimization of the numerator — that of finding minimum-cost k-vertex or k-edge connected subgraphs. The algorithm given by Frederickson and JáJá [4]

M. Farach-Colton (Ed.): LATIN 2004, LNCS 2976, pp. 212–221, 2004.

achieves a 3-approximation algorithm for finding a minimum-cost biconnected subgraph. Khuller and Vishkin [5] gave a 2-approximation algorithm for the minimum-cost k-edge-connectivity problem. Czumaj and Lingas [6,7] proposed a polynomial-time approximation scheme for minimum cost k-connectivity problem in Euclidian graphs. Cheriyan, Vempala and Vetta [2] gave an $O(log k)$-approximation algorithm for minimum cost k-vertex-connected subgraph problem in the general graphs if the the number of vertices is at least $6k^2$. The minimum mean cycle problem, another related problem, is to find a cycle in a graph with minimum average edge weight. Karp [8] gave a $O(nm)$-time algorithm for this in graphs with n vertices and m edges. Ahuja and Orlin [9] gave an $O(\sqrt{n}m \log nC)$-time scaling algorithm.

1.2   An Illustrative Example

Figure 1 illustrates an example that shows the difference between the edge connectivity problems of minimizing total weight versus minimizing average weight. Let us consider the Euclidian version of AVG-2EC. All vertices are in a straight line. The distance between vertices 1 and 2 is n units. The vertices from 2 to n are all uniformly spaced (one unit apart). Figure 1(a) shows an optimal min-cost 2-edge-connected subgraph whose total weight is $4n - 4$. To reduce its average edge weight, we can augment this solution by adding short edges, thus obtaining the solution in Figure 1(b), whose average edge weight is slightly less than 3. Figure 1(c) is an optimal solution of AVG-2EC, with an average cost of about 2.5 per edge. Note that if we start with the optimal min-cost 2-edge-connected subgraph shown, no matter how many small edges we add, we will never get an optimal solution to AVG-2EC.

2   Structure of AVG-kEC and AVG-kVC

We first observe a useful lemma that is satisfied by optimal solutions to AVG-kEC and AVG-kVC. We show that all edges of a graph whose cost is smaller than the optimal value must be in any optimal solution for the problems. Also, edges whose cost is more than the average must be critical, whose removal makes the solution infeasible.

Lemma 1. *Let* $G = (V, E)$ *be a given graph. Consider an optimal solution* $G^* = (V, E^*)$ *to either* AVG-kEC *or* AVG-kVC *on* G. *Let its value (average edge weight) be* c^*. *Then the following conditions are satisfied:*

1. *If* $c(e) < c^*$ *for any edge* $e \in E$, *then* $e \in E^*$.
2. *If* $c(e) > c^*$ *for any edge in* $e \in E^*$, *then* e *is a critical edge in* G^*, *i.e.,* $E^* - \{e\}$ *is not k connected.*

Proof. G^* is already k-connected (edge or vertex). It remains k-connected if we add more edges to it. If there is an edge e with $c(e) < c^*$ and $e \notin E^*$, then adding e to E^* is a feasible solution whose value is smaller than c^*, contradicting its

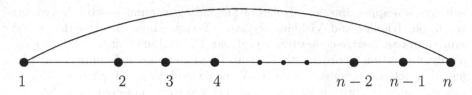

(a) A minimum weight 2-edge-connected subgraph.

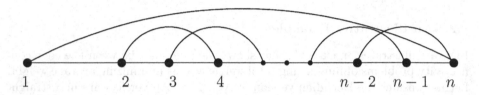

(b) Subgraph after "small" edges are added to above subgraph.

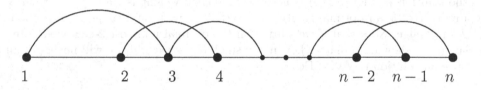

(c) An optimal AVG-2EC subgraph with average edge cost ≈ 2.5.

Fig. 1. An illustrative example

optimality. Therefore $e \in E^*$. Similarly, if an edge e with $c(e) > c^*$ is included in E^* and E^* is not critical, then $E^* - \{e\}$ is a feasible solution whose value is smaller than c^*, also a contradiction. Therefore, in an optimal solution to AVG-kEC and AVG-kVC, edges smaller than the optimal value must be included, and edges bigger than the optimal must be critical.

3   NP Hardness

Theorem 1. *Consider a k-edge-connected, undirected, edge-weighted graph $G = (V, E)$, for some integer $k \geq 2$. The problem of finding an optimal solution to AVG-kEC on G is NP-hard.*

Proof. We reduce the minimum-cardinality k-edge connected subgraph problem of an undirected graph to AVG-kEC problem. Let $G = (V, E)$ be a given graph with n nodes. We assign a cost of 1 to all edges of G. Select an arbitrary node $u \in V$. Add k new nodes $X = \{x_1, x_2, \ldots, x_k\}$, with edges of zero cost between every pair of nodes in $\{u\} \cup X$. For every k-connected subgraph of the new graph, there is a k-connected subgraph of the original graph of the same cost, obtained by deleting X and their incident edges. Consider an optimal AVG-kEC on the new graph. All edges of cost zero must be in any k-connected subgraph, since without them nodes in X will have degree less than k. This makes the average cost of an edge in an optimal solution to be strictly less than 1. Therefore, by Lemma 1, all its edges of cost 1 must be critical. Hence, the number of edges of cost 1 in it must be a minimum, implying that these edges correspond to a minimum-cardinality k-edge-connected subgraph of G.

4 Definitions

We use the term "k-connectivity" to mean both k-edge-connectivity and k-vertex-connectivity. It will be clear from the context which type of connectivity is used. Let $G = (V, E)$ be the given k connected graph with $|V| = n$, and a cost function $c(e)$ defined on its edges. Let $G^* = (V, E^*)$ be an optimal solution of either AVG-kEC or AVG-kVC. Let c^* be its value (i.e., the average edge weight of G^*). That means $c^* = \frac{\sum_{e \in E^*} c(e)}{|E^*|}$ is minimum of all spanning subgraphs which are also k connected. For any set of edges X and cost p, we define $B_p(X) = \{x \mid x \in X \text{ and } c(x) \geq p\}$, and $S_p(X) = \{x \mid x \in X \text{ and } c(x) < p\}$. We drop the subscript and write them as $B(X)$ and $S(X)$ when $p = 3c^*$. In other words, $B(X)$ are those edges of X that cost more than 3 times the optimal value (c^*) and $S(X)$ are edges of X that cost less than $3c^*$.

Let $\mathcal{P}$ be a graph property. $\mathcal{P}$ is defined to be a *monotone* property if whenever a graph $G = (V, E)$ satisfies $\mathcal{P}$, then so does $G' = (V, E \cup S)$ for any set of edges S. In other words, if a graph G satisfies $\mathcal{P}$, then so does any graph on the same set of nodes that contains G as a subgraph.

5 Vertex Connectivity Problems

In this section we consider finding a k-vertex-connected-subgraph with minimum average weight (AVG-kVC). The input is an integer k and a weighted undirected graph $G = (V, E)$ with vertex-connectivity at least k. We provide an algorithm for AVG-kVC with an approximation factor $\beta(\alpha/(\beta - 1) + 1)$, where $\beta > 1$ is a user-chosen parameter, and α is the approximation ratio of the algorithm used for finding a minimum-cost k-node-connected spanning subgraph. So here the property we consider is k-vertex connectivity. The following algorithm works for any monotone property on graphs.

5.1 Algorithm

This algorithm uses an approximation algorithm for minimum cost k-node connected subgraph as a subroutine. Let α the approximation ratio of the algorithm used for finding minimum-cost k-node connected subgraph. We start with the solution output by it as our intial solution. Calculate the average-weight of the present subgraph. Add a least cost edge which is not in the solution. If the inclusion of that edge decreases the average, we keep that edge. We try to add edges as long as it decreases the average edge cost of the current solution. We stop when the inclusion of the least cost edge which is not in the solution does not decrease the average weight of the edges.

Algorithm AVG-KVC:

1. Find an α-approximation of a minimum-cost k-vertex connected subgraph on the given graph $G = (V, E)$. Let $G^\alpha = (V, E^\alpha)$ be the solution.
2. $E' \leftarrow E^\alpha$
3. Repeat
   - Calculate the average weight of the edges in E':

$$avg_{app} = \frac{\sum_{e \in E'} c(e)}{|E'|}.$$

   - Let x be a least cost edge in $E - E'$, $E^{inc} = E' \cup \{x\}$. Calculate

$$avg_{inc} = \frac{\sum_{e \in E^{inc}} c(e)}{|E^{inc}|}.$$

   - If $avg_{inc} < avg_{app}$ then $E' = E' \cup \{x\}$.
   Until $avg_{inc} \geq avg_{app}$ or $E = E'$.

Theorem 2. *Given a k-node-connected, undirected, edge-weighted graph $G = (V, E)$, if there is a α-approximation algorithm for minimum-cost k-node connected subgraph, $\beta > 1$ is a constant, there is a polynomial-time algorithm that returns a feasible solution of AVG-KVC for which the average weight of the edges is within $\beta(\alpha/(\beta - 1) + 1)$ of c^*.*

Proof. Let E^* be an optimal AVG-KVC with value c^*, where

$$c^* = \frac{\sum_{e \in E^*} c(e)}{|E^*|}.$$

Observe that the average edge weight of the solution maintained by our algorithm for AVG-KVC decreases monotonically, since edges are added in nondecreasing order of weight, and only when their addition decreases the average weight of the solution. Consider some $\beta > 1$ such that at some point in time, the algorithm has added all edges that cost at most βc^* to the solution. We show that the average edge cost of the solution at that time (E') satisfies the theorem. Since the solution output by the algorithm is smaller than this value, the theorem will follow. Let γ be the number of such edges we added to E^α, the solution returned by the α-approximation algorithm for the k-vertex connectivity problem. In the following discussion, recall that $S_c(X)$ is the set of all edges of X whose cost is less than c.

$$
\begin{aligned}
avg_{app} &= \frac{\displaystyle\sum_{e \in E'} c(e)}{|E'|} \\
&\leq \frac{\displaystyle\sum_{e \in E^\alpha \cup S_{\beta c^*}(E')} c(e)}{|E'|} \\
&\leq \frac{\displaystyle\sum_{e \in E^\alpha} c(e) + \sum_{e \in S_{\beta c^*}(E')} c(e)}{|E'|}
\end{aligned}
$$

The cost of an α-approximate solution is less than α times the optimal solution of minimum-cost k-node connected subgraph. So, it is also less than α times any feasible solution, in particular, $\sum_{e \in E^\alpha} c(e) \leq \alpha \sum_{e \in E^*} c(e)$. It follows

$$
avg_{app} \leq \frac{\alpha \displaystyle\sum_{e \in E^*} c(e) + \gamma \, \beta \, c^*}{|E^\alpha| + \gamma} \tag{1}
$$

The number of edges in an optimal solution that costs more than $\beta \cdot c^*$ is not more than $(1/\beta)$ times total number of edges in it. Since our solution includes all edges that cost less than βc^*, the number of edges in our solution is at least $|E^*|(\beta - 1)/\beta$. Also $|E^\alpha| + \gamma \geq |E^*|(\beta - 1)/\beta$. Therefore

$$
\begin{aligned}
avg_{app} &\leq \frac{\alpha \displaystyle\sum_{e \in E^*} c(e)}{|E^*|(\beta - 1)/\beta} + \frac{\gamma \beta c^*}{|E^\alpha| + \gamma} \\
&\leq \alpha c^* \frac{\beta}{\beta - 1} + \beta c^*
\end{aligned}
$$

We can choose β for a given instance as the value that minimizes the above approximation ratio. For fixed α, we can use Newton's method and show that the ratio is minimum when $\beta = 1 + \sqrt{\alpha}$.

We now show how the result applies to different vertex connectivity problems. For Euclidian graphs, Czumaj and Lingas [6,7] gave a polynomial-time approximation scheme, i.e., $\alpha = 1 + \epsilon$, yielding a $2 + \epsilon$-approximation algorithm for AVG-KVC problem on Euclidean graphs. For general graphs, Cheriyan, Vempala and Vetta [2] gave a $O(\log k)$ approximation algorithm if the number of vertices is at least $6k^2$. So, $\alpha = O(\log k)$; we also get an $O(\log k)$ approximation algorithm. For metric graphs, $\alpha = 2$, and we choose $\beta = 1 + \sqrt{2}$, getting a 5.8-approximation.

6   Edge Connectivity Problems

In this section we consider finding a k-edge-connected-subgraph with minimum average weight (AVG-KEC). The input is an integer k and a weighted undirected graph $G = (V, E)$ with edge-connectivity at least k. We provide a 3-approximation algorithm for AVG-KEC.

We are able to do better for AVG-KEC because, there exists a 2-approximation algorithm for the k-edge-connectivity augmentation problem, whereas currently there is no constant factor approximation algorithm for the k-vertex-connectivity augmentation problem. Even for AVG-KVC on Euclidean graphs in the previous section, if we add some zero cost edges like we do in this algorithm, it ceases to be an Euclidean graph and there is no known constant factor approximation algorithm for finding a k-node connected subgraph.

6.1   Algorithm

From Lemma 1, it follows that the average of any feasible solution can be improved by adding any edge in $S(E)$ (edges whose cost is smaller than average). Also, c^* is at least as much as a smallest edge in the graph, but not greater than the biggest edge in the graph. We want to know the range of the c^* in such detail that we can define $S(E)$ uniquely. There are at most $\frac{n(n-1)}{2} - 1$ such different ranges possible for c^* which can define $S(E)$ uniquely, explained as follows. There are at most $\frac{n(n-1)}{2}$ distinct edge weights possible in a graph. We sort them and eliminate duplicates. If c^* lies between the ith smallest element and $(i + 1)^{th}$ smallest element, $S(E)$ is the set of i small elements. Since c^* lies between the smallest and largest edge costs, there could be at most $\frac{n(n-1)}{2} - 1$ such ranges possible. Since we are seeking a 3-approximation algorithm, we can add edges in $S(E)$.

We start with a graph with no edges. We guess $3c^*$ range as explained above. There are only $O(n^2)$ such ranges possible. Once we guess $3c^*$, we set to zero the cost of all edges in $S(E)$. The cost of the other edges remain the same. We now find a minimum weight spanning subgraph that is k-edge connected. The solution includes all the zero cost edges and other edges chosen by the approximation algorithm. Repeat the above scheme for all different ranges for $3c^*$ and take the solution which has minimum average.
    Algorithm AVG-KEC

1. Sort all the edges according to their weights. Put all edges of equal weight in to the same class. There are at most $\frac{n(n-1)}{2}$ different classess. Let L be the number of classes.

2. min-avg = ∞
3. for $i = 1$ to L
   - Take the given graph and set to zero the cost of all edges belonging to the first i classes.
   - Find a k-edge connected subgraph in the changed graph.
   - Let $G' = (V, E')$ be the graph returned by the algorithm.
     current-avg $= \dfrac{\sum_{e \in E'} c(e)}{|E'|}$
   - min-avg = min(min-avg, current-avg);

   Let min-avg be the average weight of the edges in the approximate solution and let app be the corresponding subgraph.

Theorem 3. *Given a k-edge-connected, undirected, edge-weighted graph $G = (V, E)$, there is a polynomial-time algorithm that returns a feasible solution of AVG-KEC on G for which the average weight of the edges is at most $3c^*$.*

Proof. Let $G' = (V, E')$ be the solution returned by Algorithm AVG-KEC above, and let $G^* = (V, E^*)$ be an optimum solution,

$$avg_{app} = \frac{\sum\limits_{e \in E'} c(e)}{|E'|} \; ; \; c^* = \frac{\sum\limits_{e \in E^*} c(e)}{|E^*|} \tag{2}$$

we want to prove

$$\frac{avg_{app}}{c^*} \leq 3 \tag{3}$$

E^* can be divided into two sets, consisting of all edges below $3c^*$ and the other consists of all edges above or equal to $3c^*$:

$$E^* = B(E^*) \cup S(E^*) \tag{4}$$

Similarly,

$$E' = B(E') \cup S(E') \tag{5}$$

E^* forms a k-edge connected subgraph. We can write it as $B(E^*) \cup S(E^*)$. Since E' contains all edges of the original graph whose cost is smaller than c^*, $S(E^*) \subseteq S(E')$. Therefore, $B(E^*) \cup S(E')$ is a k-edge connected graph. This is a feasible solution for the changed graph whose cost is $c(B(E^*))$. We use a 2-approximation algorithm to find a min-cost k-edge connnected subgraph in the changed graph [5]. The set of edges which incur cost in the above solution is $B(E')$, which is no more than 2 times the optimal solution for this problem. So it is not more that 2 times to any feasible solution.

$$\sum_{e \in B(E')} c(e) \leq 2 \sum_{e \in B(E^*)} c(e) \tag{6}$$

From equations (2) and (4), we can write

$$avg_{app} = \frac{\displaystyle\sum_{e \in B(E')} c(e) + \sum_{e \in S(E')} c(e)}{|B(E')| + |S(E')|} \tag{7}$$

If we remove some edges smaller than $3c^*$, the average of the edges increases. So,

$$\leq \frac{\displaystyle\sum_{e \in B(E')} c(e) + \sum_{e \in S(E' \cap E^*)} c(e)}{|B(E')| + |S(E') \cap E^*|} \tag{8}$$

$S(E')$ consists of all edges in the graph which are less than $3c^*$. So $S(E' \cap E^*) = S(E^*)$. Also, $S(E') \cap E^* = S(E^*)$. So,

$$= \frac{\displaystyle\sum_{e \in B(E')} c(e) + \sum_{e \in S(E^*)} c(e)}{|B(E')| + |S(E^*)|} \tag{9}$$

Observe that $|S(E^*)| \geq 2/3|E^*|$ because no more than one-third of the edges of E^* can be greater than three times its average weight. From (6) we have $\displaystyle\sum_{e \in B(E')} c(e) \leq 2 \sum_{e \in B(E^*)} c(e)$.

From the above two observations,

$$avg_{app} \leq \frac{2 \displaystyle\sum_{e \in B(E^*)} c(e) + \sum_{e \in S(E^*)} c(e)}{|B(E')| + 2/3|E^*|}$$

$$\leq \frac{2 \displaystyle\sum_{e \in E^*} c(e)}{2/3|E^*|}$$

$$\leq 3c^*$$

This completes the proof of Theorem 3.

Acknowledgement. We would like to thank Si Qing Zheng for posing AVG-KEC. This research was supported in part by the National Science Foundation under grant CCR-9820902.

References

1. J. Cheriyan and R. Thurimella, *Approximating minimum-size k-connected spanning subgraphs via Matching*, SIAM J. Comput., **30**, pp. 528-560, 2000.
2. J. Cheriyan, S. Vempala and A. Vetta, *Approximation algorithms for minimum-cost k-vertex connected subgraphs.*, STOC 2002: 306-312
3. C. G. Fernandes, *A better approximation for the minimum k-edge-connected spanning subgraph problem*, J. Algorithms, **28**, pp. 105-124, 1998.
4. G. N. Frederickson and J. JáJá, *Approximation algorithms for several graph augmentation problems*, SIAM J. Comput., **5**, pp. 25-53, 1982.
5. S. Khuller and U. Vishkin, *Biconnectivity approximations and graph carvings*, J. Assoc. Comput. Mach., **41**, pp. 214-235, 1994.
6. Artur Czumaj and Andrzej Lingas, *A Polynomial Time Approximation Scheme for Euclidean Minimum Cost k-Connectivity*, ICALP 1998, pp 682-694, 1998.
7. Artur Czumaj and Andrzej Lingas, *On Approximability of the Minimum-Cost k-Connected Spanning Subgraph Problem*, Proc. 10th Annual ACM-SIAM Symp. on Discrete. Algoithms (SODA), pp. 281-290, 1999.
8. R. M. Karp. *A characterization of the minimum cycle mean in a digraph*, Discrete Math, **23**, pp 309-311, 1978.
9. R. K. Ahuja and J. B. Orlin, *New scaling algorithms for assignment and minimum cycle mean problems*, Mathematical Programming, **54**, pp. 41-56, 1992.

On the (Im)possibility of Non-interactive Correlation Distillation

Ke Yang

Computer Science Department, Carnegie Mellon University,
5000 Forbes Ave. Pittsburgh, PA 15213, USA;
yangke@cs.cmu.edu

Abstract. We study the problem of non-interactive correlation distillation (NICD). Suppose Alice and Bob each has a string, denoted by $A = a_0a_1 \cdots a_{n-1}$ and $B = b_0b_1 \cdots b_{n-1}$, respectively. Furthermore, for every $k = 0, 1, ..., n - 1$, (a_k, b_k) is independently drawn from a distribution $\mathcal{N}$, known as the "noise mode". Alice and Bob wish to "distill" the correlation non-interactively, i.e., they wish to each apply a function to their strings, and output one bit, denoted by X and Y, such that Prob $[X = Y]$ can be made as close to 1 as possible. The problem is, for what noise model can they succeed? This problem is related to various topics in computer science, including information reconciliation and random beacons. In fact, if NICD is indeed possible for some general class of noise models, then some of these topics would, in some sense, become straightforward corollaries.

We prove two negative results on NICD for various noise models. We prove that for these models, it is impossible to distill the correlation to be arbitrarily close to 1. We also give an example where Alice and Bob can increase their correlation with one bit of communication. This example, which may be of its own interest, demonstrates that even the smallest amount of communication is provably more powerful than no communication.

1 Introduction

1.1 Non-interactive Correlation Distillation

Consider the following scenario. Let $\mathcal{N}$ be a distribution over $\Sigma \times \Sigma$, where Σ is an alphabet. We call $\mathcal{N}$ a "noise model." Suppose Alice and Bob each receives a string $A = a_0a_1 \cdots a_{n-1}$ and $B = b_0b_1 \cdots b_{n-1}$, respectively, as their local inputs. For every $k = 0, 1..., n - 1$, (a_k, b_k) is independently drawn from $\mathcal{N}$. Now, Alice and Bob wish to engage in a protocol to "distill" their correlation. An the end of the protocol, they wish to each output a bit, denoted by X and Y, respectively, such that both X and Y are "random enough", while Prob $[X = Y]$ can be made as close to 1 as possible, possibly by increasing n. We call such a protocol a *correlation distillation protocol*. Furthermore, if Alice and Bob wish to do so *non-interactively*, i.e., without communication, we call this "non-interactive correlation distillation" (NICD). Notice that in NICD, the most general thing

M. Farach-Colton (Ed.): LATIN 2004, LNCS 2976, pp. 222–231, 2004.
© Springer-Verlag Berlin Heidelberg 2004

for Alice and Bob to do is to each apply a function to their local inputs and outputs one bit. The problem of NICD is, for what noise model can Alice and Bob achieve this goal?

We note that NICD is indeed possible for many noise models. For example, if a noise model $\mathcal{N}$ is in fact "noiseless," i.e. Prob $_{(a,b)\in\mathcal{N}}[a = b] = 1$, then NICD is possible. However, we are interested in the "noisy" noise models, for example, the *binary symmetric model*, where Alice and Bob each has an unbiased bit as input, which agree with probability $1 - p$, and the *binary erasure model*, where Alice's input is an unbiased bit x, and Bob's input is x with probability $1-p$, and a special symbol $\perp$ with probability p. These models are extensively studied in the context of error correcting codes [3,9], where Alice encodes her information before sending it through a "noisy channel". It is known that there exists efficient encoding schemes that withstand these noise models and allow Alice and Bob to achieve almost perfect correlation. However, in the case of NICD, the "raw data" are already noisy. Can the techniques in error correcting codes be used here, and is NICD possible for these noise models?

1.2   Motivations and Related Work

Besides the obvious relation to error correcting codes, the study of NICD is naturally motivated by several other topics. We review these topics and discuss some of the related work.

Information Reconciliation. Information reconciliation is an extensively studied topic [4,13,6,7,8] with applications in quantum cryptography and information-theoretical cryptography. In this setting, Alice and Bob each receives a sequence of random bits drawn from a noise model, while Eve, the eavesdropper, also possesses some information about the their bits. Alice and Bob wish to "reconcile" their information via an "information reconciliation protocol", where they exchange information in a noiseless, public channel in order to agree on a random string U with very high probability. Therefore, information reconciliation protocols are somewhat like correlation distillation protocols. However, the primary concern for information reconciliation is *privacy*, i.e., that Eve gains almost no information about U. Notice that Eve can see the conversation between Alice and Bob, and thus maximum privacy would be achieved if information reconciliation can be performed without communication.

Random Beacons. A random beacon is an entity that broadcasts uncorrelated, unbiased random bits. The concept of random beacons were first introduced in 1983 by Rabin [15], who showed how they can be used to solve various problems in cryptography. From then on, random beacons have found many applications in security and cryptography [5,12,2,10]. There are many proposals to construct a *publicly verifiable* random beacon, among them are the ones that use the signals from a cosmic source [14]. In these proposals, Alice (as the beacon owner) and Bob (as the verifier) both point a telescope to an extraterrestrial object, e.g. a pulsar, and then measure the signals from it. Presumably these signals contain enough amount of randomness. Then Alice converts her measurement results into a sequence of random bits, and publishes them as beacon bits. Bob can then verify the bits by performing his own measurement and conversion. However, it is inevitable that there would be discrepancies in the results of Alice and Bob, due to measurement errors (described by a noise model). These discrepancies may cause the beacon bits published by Alice to disagree with the ones computed by Bob. One of the major concerns in the study on random beacons is to prevent *cheating* in the presence of measurement error. In other words, one needs to design a mechanism to prevent Alice from maliciously modifying her measurement data in order to affect the beacon bits, while pretending that the modification comes from the measurement error. Notice that in general, there is no communication between Alice and Bob. We note that if NICD is possible, then the cheating problem would be solved, since NICD protocols can be used to distill almost perfectly correlated bits. Then with very high probability, the bits output by Alice and Bob should agree, and this essentially removes the measurement error.

Related Work. As we have discussed, the problem of NICD lays, in some sense, at the foundations of both the studies of information reconciliation and random beacons. In fact, Researchers from both ares have, to some extent, considered the problem of NICD. In particular, a basic version of the problem concerning only the binary symmetric noise model was discovered and proven independently by several researchers since as early as 1991, including Alon, Maurer, and Wigderson [1] and Mossel and O'Donnell [14]. They proved that NICD is impossible over the binary symmetric noise model. Mossel and O'Donnell studied *multi-party* version of this problem, where $k > 2$ parties wish to agree on some random bits. They also only considered the binary symmetric noise model. In fact, we are not aware of any prior work that studies NICD beyond the binary symmetric noise model.

We stress the importance of understanding the problem of NICD for general noise models. As we have mentioned, this problem is important to both the studies of information reconciliation and random beacons. In both studies, there is no reason to assume that the binary symmetric noise model is the only reasonable one. As an example, the measurement of the signals from extraterrestrial objects is not unique, and different measurements may yield different noise models. If one of these noise models admits NICD, then the problems of information reconciliation and random beacon could, in some sense, be solved. Therefore, a better understanding of NICD over more general class of noise models would be very helpful.

1.3   Our Contribution

We study NICD beyond the binary symmetric noise model. First, we prove an impossibility result for NICD over a class of so-called "regular" noise models in Section 3. Intuitively, a noise model $\mathcal{N}$ is regular if it satisfies the following three requirements: that it is *symmetric*, i.e., $\mathcal{N}(a,b) = \mathcal{N}(b,a)$ for every $a,b \in \Sigma$;

that it is *locally uniform*, i.e., both the distributions of the local inputs of Alice and Bob are uniform; that it is *connected*, i.e., Σ cannot be partitioned into Σ_0 and Σ_1 such that $\mathcal{N}(a,b) = \mathcal{N}(b,a) = 0$ for all $a \in \Sigma_0$ and $b \in \Sigma_1$. Notice that if a noise model is not connected, that NICD is indeed possible for such a model. Suppose Σ is partitioned into Σ_0 and Σ_1. If Alice and Bob interpret symbols in Σ_0 as a "0" and symbols in Σ_1 as a "1", then they essentially have a noiseless binary noise model, which admits NICD.

In section 4, we move over to the binary erasure noise model. It is the simplest noise model that is not symmetric, and thus is not regular. The binary erasure model is also a realistic one. Consider as example the situation where Alice and Bob receive their inputs by observing a pulsar. It is quite likely that the noise of the measurements by Alice and Bob are of the "erasure-type", i.e., the corruption of information can be detected. Furthermore, it is also possible that Alice and Bob have different measurement apparatus and different levels of accuracy. In the random beacon problem, Alice (as the beacon owner) might own a more sophisticated (and more expensive) measuring device with higher accuracy, while Bob (as the verifier) has a more noisy measurement device. An extreme case would be that Alice has perfect accuracy in her measurement, but Bob's measurement is noisy. Such a situation can be described by the binary erasure noise model. We prove that NICD is impossible for this noise model as well.

The impossibility results we prove suggest that for many noise models, communication is essential for correlation distillation. Thus it is interesting to ask how much communication is essential, and in particular, if a single bit of communication helps. In Section 5, we answer this question in positive by presenting a protocol that non-trivially distills correlation from the binary symmetric noise model with one bit of communication. This result shows that even the minimal amount of communication is provably more powerful than no communications at all. The protocol itself may also be of its own interest.

Due to space limitation, some of the proofs are omitted and the readers are referred to the full version of this paper [16].

2   Preliminaries and Notations

We use $[n]$ to denote the set $\{0, 1, ..., n-1\}$. We often work with symbols from a particular *alphabet*, which is a finite set of cardinality q and is normally denoted by Σ. We often identify Σ with $[q]$.

All vectors are column vectors by default. A *string* is a sequence of symbols from an alphabet. We identify a string with a vector and use them interchangeably. For a string x of length n, we use $x[j]$ to denote its j-th entry, for $j = 0, 1, ..., n-1$. We use $\mathbf{1}_n$ to denote the all-one vector (whose each entry is 1) of dimension n. When the dimension is clear from the context, it is often omitted.

We identify a function with its truth table, which is written as a vector. For example, we view a function over $\{0,1\}^n$ also as a 2^n-dimensional vector. We assume a canonical ordering of n-bit strings.

We will work with tensor products. Let A and B both be vectors or both be matrices. We use $A \otimes B$ to denote the tensor product of A and B, and $A^{\otimes n}$ to denote the n-th tensor power of A, which is the tensor product of n copies of A.

Definition 1 (Noise Model). *A noise model over an alphabet Σ, often denoted by $\mathcal{N}$, is a probabilistic distribution over $\Sigma \times \Sigma$. The n-th tensor power of a noise model $\mathcal{N}$ is the distribution of a pair of length-n strings (A, B), where $A = a_0 a_1 \cdots a_{n-1}$ and $B = b_0 b_1 \cdots b_{n-1}$, and (a_k, b_k) is independently drawn from $\mathcal{N}$ for $k = 0, 1, ..., n-1$.*

In this paper we study *randomized, non-interactive* protocols. For the impossibility results in Section 3 and Section 4, we assume that Alice and Bob each outputs a single bit, since it suffices to prove a negative result on the "minimally useful" protocols. We shall consider protocols that outputs multiple bits in Section 5.

Since Alice and Bob do not communicate, the most general thing they can do is to apply a (randomized) function to their private inputs and outputs a bit.

Definition 2 (Protocols). *A protocol $\mathcal{P}$ over a noise model $\mathcal{N}$ is a family of function pairs (ϕ_n^A, ϕ_n^B) for $n > 0$, where $\phi_n^A, \phi_n^B : \Sigma^n \mapsto [-1, 1]$ are called the characteristic functions. The output of protocol $\mathcal{P}$ over noise model $\mathcal{N}$, denoted by $\mathcal{P}(\mathcal{N})$, is a sequence of distributions $\{\mathcal{D}_1, \mathcal{D}_2, ...\}$, where the n-th distribution $\mathcal{D}_n$ is of the bit pair (X_n, Y_n), defined as follows.*

$$(a, b) \leftarrow \mathcal{N}^{\otimes n}; x \leftarrow \phi_n^A(a), y \leftarrow \phi_n^B(b); \; X_n \leftarrow \mathsf{B}_{(1+x)/2}, Y_n \leftarrow \mathsf{B}_{(1+y)/2} : (X_n, Y_n)$$

Where B_p is the Bernoulli Distribution *of parameter p, defined as $\mathsf{B}_p(0) = 1 - p$ and $\mathsf{B}_p(1) = p$.*

Definition 3 (Statistical Distance). *The statistical distance between two probabilistic distributions A and B, denoted as $\mathsf{SD}(A, B)$, is defined to be $\mathsf{SD}(A, B) = \frac{1}{2} \sum_x |A(x) - B(x)|$ where the summation is taken over the support of A and B. If $\mathsf{SD}(A, B) \leq \epsilon$, we say A is ϵ-close to B.*

Definition 4 (δ-Locally Uniform Protocols). *A protocol $\mathcal{P}$ is δ-locally uniform over a noise model $\mathcal{N}$, if for every $n > 0$, both X_n and Y_n are δ-close to the uniform distribution over $\{0, 1\}$, where (X_n, Y_n) is the n-th distribution of $\mathcal{P}(\mathcal{N})$. A protocol is locally uniform if it is 0-locally uniform.*

Definition 5 (Correlation of Protocols). *The correlation of a protocol $\mathcal{P}$ over a noise model $\mathcal{N}$, denoted by $\mathsf{Cor}_{\mathcal{N}}[\mathcal{P}]$, is defined to be*

$$\mathsf{Cor}_{\mathcal{N}}[\mathcal{P}] = \liminf_n \{2 \cdot \mathsf{Prob}\,[X_n = Y_n] - 1\} \tag{1}$$

where (X_n, Y_n) is the n-th distribution of $\mathcal{P}(\mathcal{N})$.

3   An Impossibility Result for Regular Noise Models

We prove a general impossibility result for NICD over the regular noise models.

Definition 6 (Distribution Matrix). *Let $\mathcal{N}$ be a noise model over Σ, where $|\Sigma| = q$. We say a $q \times q$ matrix M is the* distribution matrix *for $\mathcal{N}$, if $M_{x,y} = \mathcal{N}(x, y)$ for all $x, y \in \Sigma$.[1] We write the distribution matrix of $\mathcal{N}$ by $M_{\mathcal{N}}$.*

Definition 7 (Regular Noise Model). *A $q \times q$ matrix M is* regular *if it is symmetric, and $\mathbf{1}_q$ is the unique eigenvector with the largest absolute eigenvalue. let ϵ be the difference between M's largest absolute eigenvalue and the second largest. We call $q \cdot \epsilon$ the* scaled eigenvalue gap *of M. A noise model $\mathcal{N}$ is* regular *if its distribution matrix is regular.*

Theorem 1. *If $\mathcal{N}$ is a regular noise model over Σ with scaled eigenvalue gap ϵ, then the correlation of any δ-locally uniform protocol over the $\mathcal{N}$ is at most $1 - \epsilon(1 - 4\delta^2)$.*

Notice that a distribution matrix M is non-negative (that every entry is non-negative). By the Perron-Frobenius Theorem [11], if M is symmetric, irreducible, and has $\mathbf{1}_q$ as an eigenvector, then $\mathbf{1}_q$ is the unique eigenvector with the largest eigenvalue, and thus M is regular.

Proof. Consider a protocol $\mathcal{P}$ over the noise model $\mathcal{N}$. We define $q = |\Sigma|$ and identify Σ with $[q]$ for the rest of the proof. We use M to denote the distribution matrix of $\mathcal{N}$ and denote the eigenvector of M by $v_0, v_1, ..., v_{q-1}$ with corresponding eigenvalues $\lambda_0, ..., \lambda_{q-1}$. We assume that $|\lambda_0| > |\lambda_1| \geq \cdots \geq |\lambda_{q-1}|$. Since M is regular, λ_0 is the unique largest eigenvalue that corresponds to eigenvector $\mathbf{1}_q$.

Since M is the distribution matrix, we know that the sum of all its entries is 1. Thus we have

$$1 = \mathbf{1}_q^T \cdot M \cdot \mathbf{1}_q = \lambda_0 \cdot \mathbf{1}_q^T \cdot \mathbf{1}_q = \lambda_0 \cdot q,$$

or $\lambda_0 = 1/q$. Since the scaled eigenvalue gap of M is ϵ, we know that $|\lambda_1| = (1 - \epsilon)/q$.

Consider the characteristic functions ϕ_n^A and ϕ_n^B. It is easy to see that

$$\text{Prob}\,[X_n = 1] = \frac{1}{2} \cdot \left[1 + \sum_{a \in \Sigma^n} \sum_{b \in \Sigma^n} \mathcal{N}^{\otimes n}(a, b) \cdot \phi^A(a) \right] \tag{2}$$

Clearly, $M^{\otimes n}$ is the distribution matrix for $\mathcal{N}^{\otimes n}$. We will be using a result about the eigenvalues and eigenvectors of $M^{\otimes}$, stated in Lemma 1.

Since $\mathcal{P}$ is δ-locally uniform, we have

$$\left| \sum_{a \in \Sigma^n} \sum_{b \in \Sigma^n} \mathcal{N}^{\otimes n}(a, b) \cdot \phi^A(a) \right| \leq 2\delta \tag{3}$$

---

[1] Here we identify Σ with $[q]$.

or $|(\phi^A)^T \cdot M^{\otimes n} \cdot \mathbf{1}_{q^n}| \leq 2\delta$, as we identify ϕ^A with the q^n-dimensional vector represented by its truth table. Since $\mathbf{1}_q$ is an eigenvector of M with eigenvalue $1/q$, $\mathbf{1}_{q^n}$ is an eigenvector of $M^{\otimes n}$ with eigenvalue $1/q^n$ (see Lemma 1). Since M is symmetric, so is $M^{\otimes n}$. Thus we have

$$|\mathbf{1}_{q^n}^T \cdot \phi^A| \leq 2\delta \cdot q^n. \tag{4}$$

Similarly we have

$$|\mathbf{1}_{q^n}^T \cdot \phi^B| \leq 2\delta \cdot q^n. \tag{5}$$

Now, we consider the correlation of $\mathcal{P}$. Let (X_n, Y_n) be the outputs of Alice and Bob. Then we have

$$2 \cdot \mathsf{Prob}\,[X_n = Y_n] - 1 = \sum_{A \in \Sigma^n} \sum_{B \in \Sigma^n} \mathcal{N}^{\otimes n}(A, B) \cdot \phi^A(A) \cdot \phi^B(B) \tag{6}$$

In other words, we have

$$2 \cdot \mathsf{Prob}\,[X_n = Y_n] - 1 = (\phi^A)^T \cdot M^{\otimes n} \cdot \phi^B \tag{7}$$

We diagonalize the matrix $M^{\otimes n}$. First we define a natural notion of inner product: $\langle A, B \rangle = \frac{1}{q^n} \sum_{x \in \Sigma^n} A[x] B[x]$. It is obvious that under this inner product, both ϕ_n^A and ϕ_n^B have norm at most 1. Since $M^{\otimes n}$ is symmetric, it has a set of eigenvectors that form an orthonormal basis. We denote the eigenvectors of $M^{\otimes n}$ by u_t with corresponding eigenvalues μ_t, where $t \in [q^n]$. We assume that $|\mu_0| \geq |\mu_1| \geq \cdots \geq |\mu_{q^n-1}|$. By Lemma 1, the eigenvalues μ_t are of the form $\prod_{i=1}^n \lambda_{k_i}$, where $k_i \in [q]$. Therefore $M^{\otimes n}$ has a unique maximum eigenvalue $\mu_0 = \lambda_0^n = 1/q^n$, which corresponds to the eigenvector $\mathbf{1}_q^{\otimes n} = \mathbf{1}_{q^n}$. The second largest absolute eigenvalue of $M^{\otimes n}$ is $|\mu_1| = \lambda_0^{n-1} \cdot |\lambda_1| = (1 - \epsilon)/q^n$.

Now we perform a Fourier analysis to vectors ϕ^A and ϕ^B. We write $\phi^A = \sum_{t \in [q^n]} \alpha_t \cdot u_t$ and $\phi^B = \sum_{t \in [q^n]} \beta_t \cdot u_t$. Then by Parseval, we have $\sum_t \alpha_t^2 \leq 1$, $\sum_t \beta_t^2 \leq 1$. Furthermore, from (4) and (5), we have $|\alpha_0| \leq 2\delta$ and $|\beta_0| \leq 2\delta$.

Putting things together, we have

$$\mathsf{Cor}_{\mathcal{N}^{\otimes n}}[\mathcal{P}] = (\phi^A)^T \cdot M^{\otimes n} \cdot \phi^B$$

$$= q^n \cdot \sum_{t \in [q^n]} \alpha_t \cdot \beta_t \cdot \mu_t$$

$$\leq \epsilon \cdot |\alpha_0 \beta_0| + (1 - \epsilon) \sum_{t \in [q^n]} |\alpha_t \cdot \beta_t| \qquad \text{(eigenvalue gap)}$$

$$\leq \epsilon \cdot 4\delta^2 + (1 - \epsilon) \left(\sum_t \alpha_t^2 \right) \left(\sum_t \beta_t^2 \right) \qquad \text{(Cauchy-Schwartz)}$$

$$\leq 1 - \epsilon(1 - 4\delta^2).$$

Lemma 1. *Let A be an $a \times a$ matrix of eigenvectors $v_0, ..., v_{a-1}$, with corresponding eigenvalues $\lambda_0, ..., \lambda_{a-1}$. Let B be a $b \times b$ matrix of eigenvectors $u_0, ..., u_{b-1}$, with corresponding eigenvalues $\mu_0, ..., \mu_{b-1}$. Then the eigenvalues of the matrix $A \otimes B$ are $v_i \otimes u_j$ with corresponding eigenvalues $\lambda_i \cdot \mu_j$, for $i \in [a]$ and $j \in [b]$.*

Definition 8 (Binary Symmetric Noise Model). *The binary symmetric noise model is a distribution over alphabet* $\{0,1\}$, *denoted by* $\mathcal{S}_p$ *and is defined as* $\mathcal{S}(0,0) = \mathcal{S}(1,1) = (1-p)/2$ *and* $\mathcal{S}(0,1) = \mathcal{S}(1,0) = p/2$.

Corollary 1. *The correlation of any locally uniform protocol over the binary symmetric noise model* $\mathcal{S}_p$ *is at most* $1 - 2p$.

It is easy to see that this bound is tight, since the naïve protocol where both Alice and Bob outputs their first bits is locally uniform with correlation $1 - 2p$.

Proof. Notice that $\mathcal{S}_p$ is regular with scaled eigenvalue gap $2p$.

This corollary was independently discovered by various researchers, including Alon, Maurer, and Wigderson [1], and Mossel and O'Donnell [14], and the latter attributing it as a "folklore".

4   The Binary Erasure Noise Model

We prove a similar impossibility result for another noise model, namely the binary erasure noise model. Intuitively, this model describes the situation where Alice sends an unbiased bit to Bob, which is erased (and replaced by a special symbol $\perp$) with probability p.

Definition 9 (Binary Erasure Noise Model). *The binary erasure noise model is a distribution over alphabet* $\{0,1,\perp\}$, *denoted by* $\mathcal{E}_p$ *and defined as* $\mathcal{E}(0,0) = \mathcal{E}(1,1) = (1-p)/2$, $\mathcal{E}(0,\perp) = \mathcal{E}(1,\perp) = p/2$.

Notice that in this model, Alice's input is the uniform distribution over $\{0,1\}$, and Bob's input is 0 and 1 with probability $(1-p)/2$ each, and $\perp$ with probability p. A naïve protocol under this model only uses the first pair of the inputs. Alice outputs her bit, and Bob outputs his bit if his input is 0 or 1, and outputs a random bit if his input is $\perp$. This is a locally uniform protocol with correlation $1 - p$. The next theorem shows that no protocol can do much better than the naïve protocol.

Theorem 2. *The correlation of any locally uniform protocol over the noise model* $\mathcal{E}_p$ *is at most* $\sqrt{1 - p(1 - 4\delta^2)}$.

We suspect that it is not a tight bound, but it is sufficient to show that it is bounded away from 1 and is independent from n. Therefore, even with perfect accuracy in Alice's measurement, NICD is impossible if Bob's measurement is noisy.

5   A One-Bit Communication Protocol

We present a protocol that non-trivially distills correlation over the binary symmetric noise model with one bit of communication. Recall that over no non-interactive, locally uniform protocols can have a correlation more than $1 - 2p$.

Now, we consider protocols with one bit of communication. Suppose Alice sends one bit to Bob, which Bob receives with perfect accuracy. With one bit of communication, Alice can generate an unbiased bit x and send it to Bob, and then Alice and Bob both output x. This protocol has perfect correlation. Thus, to make the problem non-trivial, we require that Alice and Bob must output two bits each. Suppose Alice outputs (X_1, X_2) and Bob outputs (Y_1, Y_2). We define the correlation of a protocol to be $2 \cdot \min_{i=1,2} \{\text{Prob } [X_i = Y_i]\} - 1$. In this situation, we say a protocol is *locally uniform*, if both (X_1, X_2) and (Y_1, Y_2) are uniformly distributed.

Now we describe a locally uniform protocol of correlation about $1 - 3p/2$. The protocol is called the "AND" protocol. Both Alice and Bob only take the first two bits as their inputs. Alice directly output her bits, and sends the AND of her bits to Bob. Then, intuitively, Bob "guesses" Alice's bits using the Bayes rule and outputs them. A technical issue is that Bob has to "balance" his output so that the protocol is still locally uniform. The detailed description is presented below.

STEP I Alice computes $r := a_1 \wedge a_2$, sends r to Bob, and outputs (a_1, a_2).
STEP II Bob, upon receiving r from Alice:
  IF $r = 1$ THEN output $(1, 1)$.
  ELSE IF $b_1 = b_2 = 1$ THEN output
  - $(0, 0)$ with probability $p/(2 - p)$;
  - $(0, 1)$ with probability $(1 - p)/(2 - p)$;
  - $(1, 0)$ with probability $(1 - p)/(2 - p)$;
  ELSE output (b_1, b_2).

We can easily verify (by a straightforward computation) the following result.

Theorem 3. *The AND protocol is a locally uniform protocol with correlation* $1 - \frac{3p}{2} + \frac{p^2}{4 - 2p}$. $\qquad\square$

This is a constant-factor improvement over the non-interactive case.

This result may seem a little surprising. It appears that Alice does not fully utilize the one-bit communication, since she sends an AND of two bits, whose entropy is less than 1. It is tempting to speculate that by having Alice send the XOR of the two bits, Alice and Bob can obtain better result, since Bob gets more information. Nevertheless, the XOR does not work, in some sense due to its "symmetry". Consider the case Alice sends the XOR of her bits to Bob. Bob can compute the XOR of his bits, and if the two XOR's agree, Bob knows that with high probability, both his bits agrees with Alice's. However, if the two XOR's don't agree, Bob knows one of his bits is "corrupted," but he has no information about which one. Furthermore, however Bob guesses, he will be wrong with probability 1/2. On the other hand, in the AND protocol, if Bob receives a "1" as the AND of the bits from Alice, he knows for sure that Alice has

$(1, 1)$ and thus he simply outputs $(1, 1)$; if $r = 0$ and $b_1 = b_2 = 1$, he knows that his input is "corrupted", and he "guesses" Alice's bit according to the Bayes rule of posterior probabilities. If Bob receives a "0" as the AND and $(b_1, b_2) \neq (1, 1)$, then the data looks "consistent" and Bob just outputs his bits. In this way, $1/4$ of the time (when Bob receives a 1), Bob knows Alice's bits for sure and can achieve perfect correlation; otherwise Alice and Bob behave almost like in the non-interactive case, which gives $1 - 2p$ correlation. So the overall correlation is about $1/4 \cdot 1 + (3/4) \cdot (1 - 2p) = 1 - 3p/2$.

Acknowledgment. The author thanks Luis von Ahn, Manuel Blum, Ning Hu, Joe Kilian, John Langford, and Steven Rudich for helpful discussions, and for Ueli Maurer and Avi Wigderson for providing the references.

References

1. N. Alon, U. Maurer, and A. Wigderson. private communication.
2. Y. Aumann and M.O. Rabin. Information theoretically secure communication in the limited storage space model. in *Crypto 99*:65-79, 1999.
3. R. E. Blahut, Theory and practice of error control codes. *Addison-Wesley*, 1983.
4. C. H. Bennett, F. Bessette, G. Brassard, L. Salvail, and J. Smolin. Experimental quantum cryptography. In *Journal of Cryptology*, vol. 5, no. 1, pp. 3–28, 1992.
5. C. H. Bennett, D. P. DiVincenzo, and R, Linsker. Digital recording system with time-bracketed authentication by on-line challenges and method for authenticating recordings. *US patent* 5764769 (1998).
6. G. Brassard and L. Salvail. Secret-key reconciliation by public discussion. In *Advances in Cryptology — EUROCRYPT '93*, LNCS 765, pp. 410–423, 1994.
7. C. Cachin and U. Maurer. Linking information reconciliation and privacy amplification. In *Journal of Cryptology*, vol. 10, no. 2, pp. 97-110, 1997.
8. C. Cachin and U.Maurer. Unconditional security against memory-bounded adversaries. In *Advances in Cryptology - CRYPTO '97*, LNCS 1294, pp. 292–306, 1997.
9. T. M. Cover and J. A. Thomas. Elements of information theory. *John Wiley and Sons*, 1991.
10. Y. Z. Ding. Oblivious transfer in the bounded storage model. In *Advances in Cryptology — CRYPTO 2001*, LNCS 2139, pages 155 – 177, 2001.
11. P. Lancaster and M. Tismenetsky. The theory of matrices, second edition, with applications. Academic Press, 1985.
12. U. M. Maurer. Conditionally-perfect secrecy and a provably secure randomized cipher. In *Journal of Cryptology*, 5:53-66, 1992.
13. U. M. Maurer. Secret key agreement by public discussion from common information. In *IEEE Transactions on Information Theory*, vol 39, pp. 733–742, May 1993.
14. E. Mossel and R. O'Donnell. Coin Flipping from a Cosmic Source: On Error Correction of Truly Random Bits. *manuscript*.
15. M. Rabin. Transaction Protection by Beacons. In *Journal of Computer and System Sciences*, 27(2):256-267, October 1983.
16. K. Yang. On the (Im)possibility of Non-interactive Correlation Distillation (full version). To appear in *ECCC*.

Pure Future Local Temporal Logics Are Expressively Complete for Mazurkiewicz Traces*

Volker Diekert[1] and Paul Gastin[2]

[1] FMI, Universität Stuttgart, Universitätsstrasse 38, D-70569 Stuttgart
diekert@fmi.uni-stuttgart.de
[2] LIAFA, Université Paris 7, 2, place Jussieu, F-75251 Paris Cedex 05
Paul.Gastin@liafa.jussieu.fr

Abstract. The paper settles a long standing problem for Mazurkiewicz traces: the pure future local temporal logic defined with the basic modalities *exists-next* and *until* is expressively complete. The analogous result with a global interpretation was solved some years ago by Thiagarajan and Walukiewicz (1997) and in its final form without any reference to past tense constants by Diekert and Gastin (2000). Each, the (previously known) global or the (new) local result generalizes Kamp's Theorem for words, because for sequences local and global viewpoints coincide. But traces are labelled partial orders and then the difference between an interpretation globally over cuts (configurations) or locally at points (events) is significant. For global temporal logics the satisfiability problem is non-elementary (Walukiewicz 1998), whereas for local temporal logics both the satisfiability problem and the model checking problem are solvable in PSPACE (Gastin and Kuske 2003) as in the case of words. This makes local temporal logics much more attractive.

Keywords: Temporal logics, Mazurkiewicz traces, concurrency.

1 Introduction

In various applications, the behaviour of a concurrent process is not represented by a string, but more accurately by some labelled partial order. This led Mazurkiewicz to the formulation of trace theory [14] which became a popular setting to study concurrency, see [7].

One advantage is that formal specifications of concurrent systems by temporal logic formulae have a direct (either global or local) interpretation for Mazurkiewicz traces. It is therefore no surprise that temporal logics for traces have received quite an attention, see [8,2,15,16,17,18,19]. In [20] (resp. finally in [4, 6]) it was shown that the basic global temporal logic with future tense operators and with (resp. without) past tense constants is expressively complete with respect to the first order theory. However the satisfiability problem for these global

---

* Work done while the second author stayed in Stuttgart within the MERCATOR program of the German Research Foundation DFG. Partial support of ACI Sécurité Informatique 2003-22 (VERSYDIS) is gratefully acknowledged.

M. Farach-Colton (Ed.): LATIN 2004, LNCS 2976, pp. 232–241, 2004.

logics is non-elementary [21]. The main reason for this high complexity is that the interpretation of a formula is defined with respect to a global configuration of the system, i.e., a finite prefix of the trace – and the prefix structure of traces is much more complex than in the case of linear orders (words). On the contrary, a local logic formula is evaluated at a local event of the system, i.e., at some vertex of the trace. The main advantage is that all local temporal logics over traces whose modalities are definable in monadic second order logic are decidable in PSPACE [10]. This is optimal since the PSPACE-hardness occurs already for words.

The better complexity makes local temporal logic much more attractive; and several attempts were made to prove expressive completeness. In [5] expressive completeness for the basic pure future local temporal logic is established, if the underlying dependence alphabet is a cograph. Moreover, one can hope to go beyond cographs, only if each trace is equipped with some bottom element or if we allow past tense modalities. This second approach is used in [11,12] to obtain expressive completeness for all dependence alphabet. In [11], the full power of *exists-previous* and *since* modalities equipped with filters is used. The result is improved in [12] where only past constants are necessary. Another temporal logic which is not local and based on more involved modalities (including both past tense and future tense) was shown to be expressively complete and decidable in PSPACE [1]. However, the most basic question remained open: whether expressive completeness holds for a pure future local temporal logic based upon *exists-next* and *until*, only. The present paper gives a positive answer to this question.

Note that the focus of this paper is only to obtain the simplest possible pure future and expressively complete local temporal logic. In order to express easily properties of systems one should instead introduce all convenient MSO modalities since the satisfiability and the model checking problem remains decidable in PSPACE regardless of the fixed set of modalities used [10].

For lack of space we give only main ideas and skip several proofs including interesting new techniques in Section 5. They can be found in the full version.

2   Preliminaries

A *dependence alphabet* is a pair (Σ, D) where the alphabet Σ is a finite set of actions and the *dependence relation* $D \subseteq \Sigma \times \Sigma$ is reflexive and symmetric. The *independence relation* I is the complement of D. For $a \in \Sigma$, the set of letters dependent of a is denoted by $D(a) = \{b \in \Sigma \mid (a, b) \in D\}$.

A *Mazurkiewicz trace* is an equivalence class of a labelled partial order $t = [V, \leq, \lambda]$ where V is a set of vertices labelled by $\lambda : V \to \Sigma$ and $\leq$ is a partial order over V satisfying the following conditions: For all $x \in V$, the downward set $\downarrow x = \{y \in V \mid y \leq x\}$ is finite, and for all $x, y \in V$, $(\lambda(x), \lambda(y)) \in D$ implies $x \leq y$ or $y \leq x$, and $x \lessdot y$ implies $(\lambda(x), \lambda(y)) \in D$, where $\lessdot = < \setminus <^2$ is the immediate successor relation in t. For $x \in V$, we also define $\Downarrow x = \{y \in V \mid y < x\}$, $\uparrow x = \{y \in V \mid x \leq y\}$, and $\Uparrow x = \{y \in V \mid x < y\}$.

The trace t is finite if V is finite and we denote by $\mathbb{M}(\Sigma, D)$ (or simply $\mathbb{M}$) the set of finite traces. By $\mathbb{R}(\Sigma, D)$ (or simply $\mathbb{R}$), we denote the set of finite or infinite traces (also called *real traces*). Let alph$(t) = \lambda(V)$ be the alphabet of t and alphinf$(t) = \{a \in \Sigma \mid \lambda^{-1}(a) \text{ is infinite}\}$ be the alphabet at infinity of t. For $A \subseteq \Sigma$, we let $\mathbb{R}_A = \{t \in \mathbb{R} \mid \text{alph}(t) \subseteq A\}$ and $\mathbb{M}_A = \{t \in \mathbb{M} \mid \text{alph}(t) \subseteq A\}$.

Let $t_1 = [V_1, \leq_1, \lambda_1]$ and $t_2 = [V_2, \leq_2, \lambda_2]$ be a pair of traces such that alphinf$(t_1) \times$ alph$(t_2) \subseteq I$. We then define the concatenation of t_1 and t_2 to be $t_1 \cdot t_2 = [V, \leq, \lambda]$ where $V = V_1 \cup V_2$ (assuming w.l.o.g. that $V_1 \cap V_2 = \emptyset$), $\lambda = \lambda_1 \cup \lambda_2$ and $\leq$ is the transitive closure of the relation $\leq_1 \cup \leq_2 \cup (V_1 \times V_2 \cap \lambda^{-1}(D))$. The set $\mathbb{M}$ of finite traces is then a monoid with the empty trace $1 = (\emptyset, \emptyset, \emptyset)$ as unit. The concatenation of two trace languages $K, L \in \mathbb{R}$ is $K \cdot L = \{r \cdot s \mid r \in K, s \in L \text{ and alphinf}(r) \times \text{alph}(s) \subseteq I\}$. We also use the infinite product $t = \prod_{i>0} t_i$ where $(t_i)_{i>0} \subseteq \mathbb{R}$ is a sequence of real traces such that alphinf$(t_i) \times$ alph$(t_j) \subseteq I$ for all $i < j$.

We denote by min(t) the set of minimal vertices of t. We let $\mathbb{R}^1 = \{t \in \mathbb{R} \mid |\min(t)| = 1\}$ be the set of traces with exactly one minimal vertex. To simplify the notation, we also use min(t) for the set $\lambda(\min(t))$ of labels of the minimal vertices of t. What we actually mean is always clear from the context.

If $U \subseteq V$ is an interval ($\uparrow x \cap \downarrow y \subseteq U$ for all $x, y \in U$) then $[U, \leq, \lambda]$ is a factor of t. We often identify U with $[U, \leq, \lambda]$. In particular, if $x \in V$ then $\downarrow x$ and $\Downarrow x$ are prefixes of t, and $\uparrow x$ and $\Uparrow x$ are suffixes of t. For $A \subseteq \Sigma$, the maximal prefix of t using actions from A only is $\mu_A(t) = \{x \in V \mid \lambda(\downarrow x) \subseteq A\}$.

3   Local Temporal Logics

The basic syntax of linear temporal logic $\text{LTL}_\Sigma = \text{LocTL}_\Sigma[\text{EX}, \text{U}]$ is given by

$$\varphi ::= a, \ (a \in \Sigma) \mid \neg\varphi \mid \varphi \vee \varphi \mid \text{EX}\,\varphi \mid \varphi \,\text{U}\, \varphi.$$

We give a standard locally defined semantics. Let $t \in \mathbb{R}$ be a real trace and $x \in t$ be a vertex in t. We have:

$$
\begin{aligned}
t, x &\models a && \text{if } \lambda(x) = a \\
t, x &\models \neg\varphi && \text{if } t \not\models \varphi \\
t, x &\models \varphi \vee \psi && \text{if } t \models \varphi \text{ or } t \models \psi \\
t, x &\models \text{EX}\,\varphi && \text{if } \exists y \ (x \lessdot y \text{ and } t, y \models \varphi) \\
t, x &\models \varphi \,\text{U}\, \psi && \text{if } \exists z \ (x \leq z \text{ and } t, z \models \psi \text{ and } \forall y \ (x \leq y < z) \Rightarrow t, y \models \varphi).
\end{aligned}
$$

We define some abbreviations. We write $\top$ for true, $\bot$ for false and $\mathsf{F}\,\varphi = \top\,\text{U}\,\varphi$ means that φ holds in the future. For $A \subseteq \Sigma$, we let $A = \bigvee_{a \in A} a$.

For $x \in t$ and $C \subseteq \Sigma$ with $C \times C \subseteq D$, we denote by x_C the unique minimal vertex of $\Uparrow x \cap \lambda^{-1}(C)$ if it exists, i.e., when $\Uparrow x \cap \lambda^{-1}(C) \neq \emptyset$. Note that $x < x_C$ if x_C exists. If $C = \{c\}$ is a singleton, then we simply write x_c instead of $x_{\{c\}}$. We write $x_a \parallel x_b$, if both x_a and x_b exist, but neither $x_a \leq x_b$ nor $x_a \geq x_b$.

Let $\mathcal{C} \subseteq 2^\Sigma \setminus \{\emptyset\}$ be a *covering of Σ by (dependence-)cliques*, this means that $C \times C \subseteq D$ for all $C \in \mathcal{C}$, and for all $a \in \Sigma$, we have $a \in C$ for some $C \in \mathcal{C}$. We consider the local temporal logic $\text{LocTL}(\mathcal{C}) = \text{LocTL}_\Sigma[(\mathsf{X}_a \leq \mathsf{X}_b), \mathsf{X}_C, \text{U}_C]$ whose syntax is given by

$$\varphi ::= a \mid (\mathsf{X}_a \leq \mathsf{X}_b) \mid \neg\varphi \mid \varphi \vee \varphi \mid \mathsf{X}_C\,\varphi \mid \varphi\,\text{U}_C\,\varphi$$

where C ranges over $\mathcal{C}$ and a, b range over Σ with $\{a, b\} \not\subseteq C$ for all $C \in \mathcal{C}$. If $C = \{a\}$ is a singleton, then we simply write X_a and U_a for $\mathsf{X}_{\{a\}}$ and $\mathsf{U}_{\{a\}}$. The semantics of $\mathrm{LocTL}(\mathcal{C})$ is defined as follows. First, $\varphi \, \mathsf{U}_C \, \psi = (\varphi \vee \neg C) \, \mathsf{U} \, (C \wedge \psi)$. Then, $t, x \models \mathsf{X}_C \, \varphi$ if x_C exists and $t, x_C \models \varphi$. Finally, $t, x \models (\mathsf{X}_a \leq \mathsf{X}_b)$ if x_a, x_b exist and $x_a \leq x_b$. If $a, b \in C$ then we have $(\mathsf{X}_a \leq \mathsf{X}_b) = \mathsf{X}_b \top \wedge \mathsf{X}_C(\neg b \, \mathsf{U}_C \, a)$. Hence, we can freely use in $\mathrm{LocTL}(\mathcal{C})$ all constants $(\mathsf{X}_a \leq \mathsf{X}_b)$ with $a, b \in \Sigma$.

We show that $\mathrm{LocTL}(\mathcal{C})$ is a fragment of $\mathrm{LocTL}_\Sigma[\mathsf{EX}, \mathsf{U}]$. First, we have $(\mathsf{X}_a \leq \mathsf{X}_b) = \bigvee_{c \in \Sigma}(\mathsf{X}_c \leq \mathsf{X}_a) \wedge (\mathsf{X}_c \leq \mathsf{X}_b) \wedge \mathsf{EX}(c \wedge \neg(\neg a \, \mathsf{U} \, b))$. Thus, it is enough to consider a conjunction $(\mathsf{X}_c \leq \mathsf{X}_a) \wedge \mathsf{EX}\, c$. For $\lambda(x) = a$, this is $\mathsf{EX}(c \wedge \mathsf{F}\, a)$. For $\lambda(x) \neq a$, this is $\mathsf{EX}(c \wedge \mathsf{F}\, a) \wedge \neg(\neg c \, \mathsf{U} \, a)$.

For the modality X_C, we have $\mathsf{X}_C \, \varphi = \bigvee_{c \in C}(\mathsf{X}_c \, \varphi \wedge \bigwedge_{d \in C \setminus \{c\}} \neg(\mathsf{X}_d \leq \mathsf{X}_c))$ and $\mathsf{X}_c \, \varphi = (\neg c \wedge (\bot \, \mathsf{U}_c \, \varphi)) \vee (c \wedge \mathsf{EX}(\bot \, \mathsf{U}_c \, \varphi))$.

We may also use $\varphi \, \mathsf{XU}_C \, \psi = \mathsf{X}_C(\varphi \, \mathsf{U}_C \, \psi)$. Note that the modalities X_a and U_a can be expressed in all logics $\mathrm{LocTL}(\mathcal{C})$: let $C \in \mathcal{C}$ such that $a \in C$, we have $\mathsf{X}_a \, \varphi = \neg a \, \mathsf{XU}_C \, (a \wedge \varphi)$ and $\varphi \, \mathsf{U}_a \, \psi = (\neg a \vee \varphi) \, \mathsf{U}_C \, (a \wedge \psi)$.

When $b, c \in \Sigma$ are such that $\Uparrow x \cap \lambda^{-1}(b) \neq \emptyset$ and $\Uparrow x_b \cap \lambda^{-1}(c) \neq \emptyset$, we let $x_{bc} = (x_b)_c$ be the minimal vertex of $\Uparrow x_b \cap \lambda^{-1}(c)$. We now define constants $(\mathsf{X}_{ac} = \mathsf{X}_{bc})$ for all $a, b, c \in \Sigma$ with $a \neq c \neq b$ by: $t, x \models (\mathsf{X}_{ac} = \mathsf{X}_{bc})$, if x_{ac}, x_{bc} exist and $x_{ac} = x_{bc}$. It is far from being obvious that the new constants $(\mathsf{X}_{ac} = \mathsf{X}_{bc})$ can be expressed in $\mathrm{LocTL}_\Sigma[\mathsf{EX}, \mathsf{U}]$. We will devote Section 7 to the proof of the next lemma.

Lemma 1. *The constants* $(\mathsf{X}_{ac} = \mathsf{X}_{bc})$ *can be expressed in* $\mathrm{LocTL}(\mathcal{C})$ *for all* $a, b, c \in \Sigma$ *with* $a \neq c \neq b$ *and all coverings of* Σ *by cliques* $\mathcal{C}$.

4   Lifting Lemma

In this section A denotes a subset of Σ and we let $\overline{A} = \Sigma \setminus A$ be its complement. For $x \in l \in \mathbb{R}$ we define $\mu_A(x, t)$ to be the prefix of $\uparrow x$ which is given by the set of vertices $\{z \in t \mid x \leq z$ and $\forall x < y \leq z, \lambda(y) \in A\}$.

Lemma 2. *Let* $x \in t \in \mathbb{R}$ *and* $a \in \Sigma$. *Then,* x_a *exists and* $x_a \in \mu_A(x, t)$ *if and only if* $t, x \models \mathsf{X}_a \top \wedge \bigwedge_{c \in \overline{A}} \neg(\mathsf{X}_c \leq \mathsf{X}_a)$.

Lemma 2 is easy to show. The aim of this section is to establish the following.

Theorem 3 (Lifting Lemma). *Let* $\varphi \in \mathrm{LocTL}_\Sigma[(\mathsf{X}_a \leq \mathsf{X}_b), \mathsf{X}_a, \mathsf{U}_a]$ *and* $A \subseteq \Sigma$. *Then we effectively find a formula* $\overline{\varphi}^A \in \mathrm{LocTL}_\Sigma[(\mathsf{X}_a \leq \mathsf{X}_b), \mathsf{X}_a, \mathsf{U}_a]$ *such that for all* $x \in t \in \mathbb{R}$ *we have:* $\mu_A(x, t), x \models \varphi$ *if and only if* $t, x \models \overline{\varphi}^A$.

The proof is done by structural induction on φ. We start with the following observations: $\overline{a}^A = a$ for all $a \in \Sigma$, $\overline{\varphi \wedge \psi}^A = \overline{\varphi}^A \wedge \overline{\psi}^A$, and $\overline{\neg \varphi}^A = \neg \overline{\varphi}^A$.

Now, $\mu_A(x, t), x \models (\mathsf{X}_a \leq \mathsf{X}_b)$ if and only if both $t, x \models (\mathsf{X}_a \leq \mathsf{X}_b)$ and $x_b \in \mu_A(x, t)$. However, $x_b \in \mu_A(x, t)$ can be expressed using Lemma 2.

Define $\varphi \, \mathsf{XU}_a \, \psi$ as $\mathsf{X}_a(\varphi \, \mathsf{U}_a \, \psi)$. Then, both X_a and U_a can be expressed in XU_a. Indeed, $\mathsf{X}_a \, \varphi = \bot \, \mathsf{XU}_a \, \varphi$ and $\varphi \, \mathsf{U}_a \, \psi = (a \wedge \psi) \vee ((\neg a \vee \varphi) \wedge \varphi \, \mathsf{XU}_a \, \psi)$. Thus it is enough to define $\overline{\varphi \, \mathsf{XU}_a \, \psi}^A$. This is the difficult part for which we establish first some auxiliary results.

Lemma 4. *Let $x \in t \in \mathbb{R}$ and $a \in \Sigma$ such that x_a exists and $x_a \in \mu_A(x, t)$. Define $B = \{a\} \cup \{b \in A \setminus \{a\} \mid t, x \models \bigwedge_{c \in \overline{A}} \neg(X_{ab} = X_{cb})\}$. Then we have $a \in B \subseteq A$ and $\mu_A(x, t) \cap \uparrow x_a = \mu_B(x_a, t)$.*

As a consequence of Lemmata 4 and 2 we obtain the following proposition.

Proposition 5. *Let $a \in B \subseteq A$. Then, there exists a formula $\mathrm{Switch}_{A,B,a} \in \mathrm{LocTL}_\Sigma[(X_d \le X_e), (X_{df} = X_{ef})]$ such that the following two assertions hold.*

1. *If $t, x \models \mathrm{Switch}_{A,B,a}$ then $x_a \in \mu_A(x, t)$ exists and $\mu_A(x, t) \cap \uparrow x_a = \mu_B(x_a, t)$.*
2. *If $x_a \in \mu_A(x, t)$ exists, then we have $t, x \models \mathrm{Switch}_{A,B,a}$ for some $a \in B \subseteq A$.*

Using an induction on the size of A, the remaining case of the proof of Theorem 3 follows easily from the following.

Lemma 6. *We have $\overline{\varphi \, \mathsf{XU}_a \, \psi}^A = \sigma_1 \vee \sigma_2$ where*

$$\sigma_1 = \bigvee_{B \subsetneq A} \mathrm{Switch}_{A,B,a} \wedge X_a \, \overline{\varphi \, \mathsf{U}_a \, \psi}^B,$$

$$\sigma_2 = \mathrm{Switch}_{A,A,a} \wedge \left((\mathrm{Switch}_{A,A,a} \wedge \overline{\varphi}^A) \, \mathsf{XU}_a \left(\overline{\psi}^A \vee (\overline{\varphi}^A \wedge \sigma_1) \right) \right).$$

5   Expressive Completeness of LocTL($\mathcal{C}$)

In the following, if a real trace $t \in \mathbb{R}$ has a unique minimal vertex x, then by $t \models \varphi$ we mean $t, x \models \varphi$. Hence, if $t \in \mathbb{R}$ and $x \in t$ is any vertex, then $t, x \models \varphi$ has the same meaning as $\uparrow x \models \varphi$ (if the reference to t is clear).

Now, we want to define *initial satisfiability*, i.e., when does a trace $t \in \mathbb{R}$ satisfies a local temporal logic formula φ. Since a trace t does not necessarily have a unique minimal position, there is no canonical way to choose an initial position in t. Our approach uses *rooted traces* as in e.g. [3]. Let $\# \notin \Sigma$ and $t = [V, \le, \lambda] \in \mathbb{R}(\Sigma, D)$. The rooted trace associated with t is

$$\#t = [V \cup \{\#\}, \le \cup (\{\#\} \times (V \cup \{\#\})), \lambda \cup (\# \mapsto \#)].$$

It is a trace over the alphabet $\Sigma' = \Sigma \cup \{\#\}$ and the dependence relation $D' = D \cup (\{\#\} \times \Sigma) \cup (\Sigma \times \{\#\})$. Then, for a formula $\varphi \in \mathrm{LocTL}(\mathcal{C})$, we define $\mathcal{L}_\Sigma(\varphi) = \{t \in \mathbb{R}(\Sigma, D) \mid \#t \models \varphi\}$. We simply write $\mathcal{L}(\varphi)$ when there is no ambiguity on the alphabet.

Alphabetic conditions can be easily expressed in LocTL($\mathcal{C}$). Therefore, for $A \subseteq \Sigma$, the languages $\mathbb{M}_A$, $\mathbb{R}_A$, (alphinf = A) = $\{t \in \mathbb{R} \mid \mathrm{alphinf}(t) = A\}$ and (min $\subseteq A$) = $\{t \in \mathbb{R} \mid \min(t) \subseteq A\}$ are definable in LocTL(Σ).

The first order theory of traces $\mathrm{FO}_\Sigma(<)$ is given by the syntax:

$$\varphi ::= P_a(x) \mid x < y \mid \neg\varphi \mid \varphi \vee \varphi \mid \exists x \varphi,$$

where $a \in \Sigma$ and $x, y \in \mathrm{Var}$ are first order variables. Given a trace $t = [V, \le, \lambda]$, we interpret each predicate P_a by the set $\{x \in V \mid \lambda(x) = a\}$ and the relation $<$ as the strict partial order relation of t. The semantics then lifts to all formulas as usual. For closed formulae we can define as usual the language $\mathcal{L}(\varphi) = \{t \in \mathbb{R} \mid t \models \varphi\}$. We say that a trace language $L \subseteq \mathbb{R}$ is expressible in $\mathrm{FO}_\Sigma(<)$ if there exists a sentence $\varphi \in \mathrm{FO}_\Sigma(<)$ such that $L = \mathcal{L}(\varphi)$.

Theorem 7. *Let $\mathcal{C}$ be a covering of Σ by cliques of (Σ, D). A real trace language over $\mathbb{R}(\Sigma, D)$ is expressible in $\mathrm{FO}_\Sigma(<)$ if and only if it is expressible in $\mathrm{LocTL}(\mathcal{C})$.*

Corollary 8. *The local temporal logic $\mathrm{LocTL}_\Sigma[\mathsf{EX}, \mathsf{U}]$ based on the modalities EX and U is expressively complete.*

The logic $\mathrm{LocTL}(\mathcal{C})$ is a fragment of $\mathrm{LocTL}_\Sigma[\mathsf{EX}, \mathsf{U}]$ and by its semantics it is clear that each real trace language expressible in $\mathrm{LocTL}_\Sigma[\mathsf{EX}, \mathsf{U}]$ is also expressible in $\mathrm{FO}_\Sigma(<)$. Therefore it is enough to prove the other direction of Theorem 7.

We use the algebraic notion of recognizability. Let $h : \mathbb{M} \to M$ be a morphism to a finite monoid M. For $s, t \in \mathbb{R}$, we say that s and t are h-similar, denoted $s \sim_h t$, if we can write $s = \prod_{i>0} s_i$ and $t = \prod_{i>0} t_i$ with $s_i, t_i \in \mathbb{M}$ and $h(s_i) = h(t_i)$ for all $i > 0$. The transitive closure $\approx_h$ of $\sim_h$ is an equivalence relation. For $t \in \mathbb{R}$, we denote by $[t]_{\approx_h}$ the equivalence class of t. When there is no ambiguity, we simply write $\sim$, $\approx$ and $[t]$. Since M is finite, the equivalence relation $\approx$ is of finite index with at most $|M|^2 + |M|$ equivalence classes. A trace language $L \subseteq \mathbb{R}$ is *recognized* by h if it is saturated by $\approx$ (or equivalently by $\sim$), i.e., $t \in L$ implies $[t] \subseteq L$ for all $t \in \mathbb{R}$.

A finite monoid M is *aperiodic* if there is an $n \geq 0$ such that $u^n = u^{n+1}$ for all $u \in M$. A trace language $L \subseteq \mathbb{R}$ is *aperiodic* if it is recognized by some morphism to a finite and aperiodic monoid.

Theorem 9 ([8,9]). *A language $L \subseteq \mathbb{R}(\Sigma, D)$ is expressible in $\mathrm{FO}_\Sigma(<)$ if and only if it is aperiodic.*

To prove that aperiodic trace languages are expressible in $\mathrm{LocTL}(\mathcal{C})$, we use an induction on Σ. If $\Sigma = \emptyset$ then there are only two trace languages: $\emptyset$ and $\mathbb{R} = \{1\}$ which are respectively defined by $\bot$ and $\top$. Assume now that $\Sigma \neq \emptyset$ and fix a covering $\mathcal{C}$ of Σ by cliques of (Σ, D). By induction, each aperiodic language $L \subseteq \mathbb{R}_A$ with $A \subsetneq \Sigma$ is expressible in $\mathrm{LocTL}(\mathcal{C}_{|A})$, where $\mathcal{C}_{|A} = \{C \cap A \mid C \in \mathcal{C}\} \setminus \{\emptyset\}$. In the following, we fix some $C \in \mathcal{C}$ and we let $A = \Sigma \setminus C \subsetneq \Sigma$. We use the unambiguous decomposition $\mathbb{R} = \mathbb{R}_A(\min \subseteq C)$.

Lemma 10. *Let $L \subseteq \mathbb{R}$ be a trace language recognized by h. Then, L is a finite union of languages of the form $(L_1 \cap \mathbb{R}_A)(L_2 \cap (\min \subseteq C))$, where the languages $L_1, L_2 \subseteq \mathbb{R}$ are recognized by h.*

Let $T = \{[t] \mid t \in \mathbb{R}^1 \cap C\mathbb{R}_A\}$. We consider T as a finite alphabet. Each trace $t \in (\min \subseteq C)$ has a unique C-*factorization* $t = \prod_{i<n} t_i$ with $n \in \mathbb{N} \cup \{\omega\}$ and $t_i \in \mathbb{R}^1 \cap C\mathbb{R}_A$ for all $i < n$. Hence, we can define a mapping $\sigma : (\min \subseteq C) \to T^\infty$ by $\sigma(t) = \prod_{i<n} [t_i]$ where $\prod_{i<n} t_i$ is the C-factorization of t.

Lemma 11. *Let $L \subseteq \mathbb{R}$ be recognized by h. Then, $L \cap (\min \subseteq C) = \sigma^{-1}(K)$ for some aperiodic word language $K \subseteq T^\infty$.*

The next lemma uses a classical result that aperiodic word languages $K \subseteq T^\infty$ are expressible in $\mathrm{LTL}_T[\mathsf{XU}]$. This result is based on Kamp's Thm. [13].

Lemma 12. *Suppose that each aperiodic trace language over A is expressible in $\mathrm{LocTL}(\mathcal{C}_{|A})$. Let $K \subseteq T^{\infty}$ be an aperiodic word language. There exists $\varphi \in \mathrm{LocTL}(\mathcal{C})$ such that for all $t \in (\min \subseteq C) \setminus \{1\}$, we have $\sigma(t) \in K$ if and only if $t \models \varphi$.*

We now turn to the proof of Theorem 7. Using Lemma 10, we have to show that if L_1 and L_2 are recognized by h then $L = (L_1 \cap \mathbb{R}_A)(L_2 \cap (\min \subseteq C))$ is expressible in $\mathrm{LocTL}(\mathcal{C})$.

Since $L_1 \cap \mathbb{R}_A$ is aperiodic, using the induction on the alphabet, we find $\varphi_1 \in \mathrm{LocTL}(\mathcal{C}_{|A})$ such that for all $t_1 \in \mathbb{R}_A$, $t_1 \in L_1$ iff $\#t_1 \models \varphi_1$. Let $\overline{\varphi_1}^A \in \mathrm{LocTL}(\mathcal{C})$ be the formula given by the Lifting Lemma (Theorem 3). For all $t \in \mathbb{R}$, we have $\#t \models \overline{\varphi_1}^A$ iff $\#\mu_A(t) \models \varphi_1$ iff $\mu_A(t) \in L_1$.

Since L_2 is recognized by h, using Lemma 11 we have $L_2 \cap (\min \subseteq C) = \sigma^{-1}(K)$ for some aperiodic word language $K \subseteq T^{\infty}$. By Lemma 12, we find $\varphi_2 \in \mathrm{LocTL}(\mathcal{C})$ such that for all $t_2 \in (\min \subseteq C) \setminus \{1\}$ we have $t_2 \models \varphi_2$ iff $\sigma(t_2) \in K$ iff $t_2 \in L_2$. Let $\widetilde{\varphi_2} = \neg \mathsf{X}_C \top \vee \mathsf{X}_C \varphi_2$ if $1 \in L_2$ and $\widetilde{\varphi_2} = \mathsf{X}_C \varphi_2$ otherwise.

We claim that $L = \mathcal{L}_\Sigma(\varphi)$ where $\varphi = \overline{\varphi_1}^A \wedge \widetilde{\varphi_2}$.

6   Process Based Logics Are Expressively Complete

We show that we can deal also with process-based logics as introduced in [19]. In this framework, we start with a finite set of processes $\mathcal{P} = \{1, \ldots, n\}$ and a mapping $p : \Sigma \to 2^{\mathcal{P}} \setminus \{\emptyset\}$. The execution of an action $a \in \Sigma$ requires the participation of all processes in the nonempty set $p(a)$. If $p(a) = \{i\}$ is a singleton then the action a is local to process i. Otherwise, the execution of a requires the synchronization of all processes in $p(a)$. The dependence relation is $D = \{(a,b) \in \Sigma^2 \mid p(a) \cap p(b) \neq \emptyset\}$. Hence the set $\mathcal{C} = \{p^{-1}(i) \mid i \in \mathcal{P}\}$ is a covering of Σ by cliques of (Σ, D).

Thanks to this more concrete view of the dependence alphabet based on processes, we can define temporal modalities that involve locations of actions. In [19], the formula $\mathcal{O}_i \varphi$ means that φ holds at the first event of process i that is not in the past of the current vertex. Clearly, this is not a future modality. Here, we use a future variant $\mathsf{X}_i \varphi$ meaning that φ holds at the first event of process i which is strictly above the current vertex. More formally, we define $\mathsf{X}_i \varphi := \mathsf{X}_{p^{-1}(i)} \varphi$. The until modality introduced in [19] is also not pure future. Here we use a future variant $\varphi \, \mathsf{U}_i \, \psi$ which means that on the sequence of events located on process i and above the current vertex we observe φ until ψ. More formally, we define $\varphi \, \mathsf{U}_i \, \psi := \varphi \, \mathsf{U}_{p^{-1}(i)} \, \psi$.

Since the set $\mathcal{C} = \{p^{-1}(i) \mid i \in \mathcal{P}\}$ is a covering of Σ by cliques of (Σ, D), a reformulation of Theorem 7 yields

Theorem 13. *Let $\mathcal{P}$ be a finite set of processes and $p : \Sigma \to 2^{\mathcal{P}} \setminus \{\emptyset\}$ be a location map. The process-based local logic $\mathrm{LocTL}[(\mathsf{X}_a \leq \mathsf{X}_b), \mathsf{X}_i, \mathsf{U}_i]$ based on the modalities X_i and U_i for $i \in \mathcal{P}$ and using only constants $(\mathsf{X}_a \leq \mathsf{X}_b)$ with $p(a) \cap p(b) = \emptyset$ is expressively complete.*

7 Removing Constants: Proof of Lemma 1

We prove Lemma 1 by showing how to express the constants $(X_{ac} = X_{bc})$ in terms of a Boolean combination of formulae of type $(X_d \leq X_e)$, X_d, and U_d for various $d, e \in \Sigma$. Note that the constants $(X_d < X_e)$ and $(X_d \| X_e)$ with obvious semantics may be used, too. The overall strategy is to proceed in $\mathcal{O}(n^3)$ rounds where $n = |\Sigma|$. In each round we introduce new formulae which are approximations of $(X_{ac} = X_{bc})$. At the end these approximations are getting so weak that we can replace them by *false*. In each round, when we replace an approximation we obtain a new formula of size $\mathcal{O}(n^2)$. Thus, overall $(X_{ac} = X_{bc})$ is replaced by a complex formula of exponential size in $|\Sigma|$.

Lemma 14. *1. Let z be a vertex such that $\lambda(z) = a$ and z_c exists. There exist letters $\{a_1, \ldots, a_{k-1}\} \subseteq \Sigma \setminus \{a, c\}$ such that $z < z_{a_1} < \cdots < z_{a_{k-1}} < z_c$ and $a = a_0 \,\text{—}\, a_1 \,\text{—}\, \cdots \,\text{—}\, a_{k-1} \,\text{—}\, a_k = c$ in (Σ, D).*

2. Let x be a vertex and $\{a_1, \ldots, a_{k-1}\} \subseteq \Sigma \setminus \{a, c\}$ such that $x_a < x_{aa_1} < \cdots < x_{aa_{k-1}} < x_{ac}$ and $a = a_0 \,\text{—}\, a_1 \,\text{—}\, \cdots \,\text{—}\, a_{k-1} \,\text{—}\, a_k = c$ in (Σ, D). If $x_a \| x_c$, then $x_{aa_i} = x_{ca_i}$ for some $1 \leq i < k$.

Let $a, c \in \Sigma$, $a \neq c$, and let $t \in \mathbb{R}$, $x \in t$ such that x_{ac} exists. Define $\delta_x(a, c)$ as the smallest integer $k \geq 1$ such that there exist letters $a_1, \cdots, a_{k-1}$ such that $x_a < x_{aa_1} < \cdots < x_{aa_{k-1}} < x_{ac}$ and $a = a_0 \,\text{—}\, a_1 \,\text{—}\, \cdots \,\text{—}\, a_{k-1} \,\text{—}\, a_k = c$ in (Σ, D). Note that such an integer k exists by Lemma 14 and $\delta_x(a, c) \leq |\Sigma| - 1$.

We also introduce the set $F_x(a, c)$ which consists of all pairs (d, e), $d \neq e$, such that either x_{de} does not exist or $x_{ac} < x_{de}$. Note that $|F_x(a, c)| \leq |\Sigma|^2 - |\Sigma|$. Throughout we use the following fact:

$$\text{if } x \leq y \text{ and } y_{fg} \leq x_{ac}, \text{ then } F_x(a, c) \subseteq F_y(f, g). \qquad (*)$$

Proposition 15. *Let $a, b, c \in \Sigma$ with $a \neq c \neq b$. For each triple (m, ℓ, r) with $0 \leq m \leq |\Sigma|^2 - |\Sigma|$, $0 \leq \ell \leq 2|\Sigma| - 2$, and $r \in \{0, 1\}$ we can define a formula $(X_{ac} = X_{bc}, m, \ell, r)$ in terms of $(X_d < X_e)$, X_d and U_d with $d, e \in \Sigma$ such that for all $x \in t \in \mathbb{R}$ the following assertions I and II are satisfied.*

I: *If $t, x \models (X_{ac} = X_{bc}, m, \ell, r)$, then $t, x \models (X_{ac} = X_{bc})$.*

II: *If the following four conditions $C_1, \ldots, C_4$ are simultaneously satisfied, then it holds: $t, x \models (X_{ac} = X_{bc}, m, \ell, r)$.*

C_1: *$t, x \models (X_{ac} = X_{bc})$.*
C_2: *$|F_x(a, c)| = |F_x(b, c)| \geq m$.*
C_3: *$\delta_x(a, c) + \delta_x(b, c) \leq \ell$.*
C_4: *$r = 1$ or $t, x \models (X_a \| X_b) \wedge \neg[(X_c < X_a) \wedge (X_c < X_b)]$.*

Corollary 16. *The formulae $(X_{ac} = X_{bc})$ and $(X_{ac} = X_{bc}, 0, 2|\Sigma| - 2, 1)$ are equivalent.*

Proof of Prop. 15. For $a = b$ we define $(\mathsf{X}_{ac} = \mathsf{X}_{bc}, m, \ell, r)$ by the formula $\mathsf{X}_a \mathsf{X}_c \top$ which simply states that x_{ac} exists. Obviously, I and II are both satisfied for $a = b$. Hence in the following we may assume $|\{a, b, c\}| = 3$. Consider a triple (m, ℓ, r). If now either $m > |\Sigma|^2 - |\Sigma| - 2$ or $\ell \leq 1$, then we define $(\mathsf{X}_{ac} = \mathsf{X}_{bc}, m, \ell, r)$ by *false*. Then, I and II hold.

In the following we may assume by induction that formulae are defined satisfying both I and II for all triples (m', ℓ', r') where either $m' > m$ or $m' = m$, $\ell' < \ell$ or $m' = m$, $\ell' = \ell$, and $r' < r$.

Case $r = 1$: We define $(\mathsf{X}_{ac} = \mathsf{X}_{bc}, m, \ell, 1)$ by $\varphi_0 \vee \varphi_1 \vee \varphi_2 \vee \varphi_3$ where:

$$\varphi_0 = (\mathsf{X}_a < \mathsf{X}_b) \wedge \mathsf{X}_a(\mathsf{X}_b < \mathsf{X}_c), \quad \varphi_1 = (\mathsf{X}_b < \mathsf{X}_a) \wedge \mathsf{X}_b(\mathsf{X}_a < \mathsf{X}_c),$$
$$\varphi_2 = (\mathsf{X}_{ac} = \mathsf{X}_{bc}, m, \ell, 0), \quad \varphi_3 = (\mathsf{X}_a \,\|\, \mathsf{X}_b) \wedge \psi_1 \wedge \psi_2,$$
$$\psi_1 = (\mathsf{X}_c < \mathsf{X}_a) \wedge (\mathsf{X}_c < \mathsf{X}_b), \quad \psi_2 = \psi_1 \,\mathsf{U}_c\, ((\mathsf{X}_{ac} = \mathsf{X}_{bc}, m, \ell, 0) \wedge \neg\psi_1).$$

Case $r = 0$: We define $(\mathsf{X}_{ac} = \mathsf{X}_{bc}, m, \ell, 0)$ by $\tau_0 \vee \tau_1 \vee \tau_2 \vee \tau_3$ where:

$$\tau_0 = (\mathsf{X}_a < \mathsf{X}_c) \wedge (\mathsf{X}_b < \mathsf{X}_c),$$
$$\tau_1 = (\mathsf{X}_c < \mathsf{X}_a) \wedge \bigvee_{b \neq b' \neq c} \tau(b, b') \wedge \mathsf{X}_c(\mathsf{X}_{ac} = \mathsf{X}_{b'c}, m, \ell - 1, 1),$$
$$\tau_2 = (\mathsf{X}_c < \mathsf{X}_b) \wedge \bigvee_{a \neq a' \neq c} \tau(a, a') \wedge \mathsf{X}_c(\mathsf{X}_{a'c} = \mathsf{X}_{bc}, m, \ell - 1, 1),$$
$$\tau_3 = \bigvee_{\substack{a \neq a' \neq c \\ b \neq b' \neq c}} \tau(a, a') \wedge \tau(b, b') \wedge \mathsf{X}_c(\mathsf{X}_{a'c} = \mathsf{X}_{b'c}, m, \ell - 2, 1),$$
$$\tau(d, d') = (\mathsf{X}_{dd'} = \mathsf{X}_{cd'}, m + 2, 2|\Sigma| - 2, 1) \wedge \mathsf{X}_d(\mathsf{X}_{d'} < \mathsf{X}_c).$$

We only give the proof of assertion II. Consider $x \in t \in \mathbb{R}$ such that $C_1, \ldots, C_4$ are all satisfied. In particular, x_{ac}, x_{bc} exist and we have $x_{ac} = x_{bc}$. If $x_a < x_c$ then $x_c = x_{ac} = x_{bc}$ hence also $x_b < x_c$ and $t, x \models \tau_0$. Similarly, if $x_b < x_c$ then $t, x \models \tau_0$. Hence in the following we assume that neither $x_a < x_c$ nor $x_b < x_c$.

There are three cases: 1) $x_c < x_a$, 2) $x_c < x_b$, and 3) neither $x_c < x_a$ nor $x_c < x_b$. These cases correspond to τ_1, τ_2, and τ_3, respectively. Since $r = 0$, C_4 implies $x_a \,\|\, x_b$ and $\neg(x_c < x_a \wedge x_c < x_b)$. Hence, in case 1, using $\neg(x_b < x_c)$ and $b \neq c$, we get $x_b \,\|\, x_c$. Similarly, in case 2 we have $x_a \,\|\, x_c$ and in case 3 we have both $x_a \,\|\, x_c$ and $x_b \,\|\, x_c$. One can prove the following.

Claim. If $x_a \,\|\, x_c$ then we find $a' \in \Sigma \setminus \{a, c\}$ such that both $\delta_y(a', c) < \delta_x(a, c)$ and $t, x \models \tau(a, a')$.

We come back to the proof of the three cases. We start with case 2). We have $x_c < x_b$ and $x_a \,\|\, x_c$. Let a' be given by the claim stated above, and let $y = x_c$. We can show that $C_1, \ldots, C_4$ hold for y, (a', b, c) and $(m, \ell - 1, 1)$. By induction, we get $t, y \models (\mathsf{X}_{a'c} = \mathsf{X}_{bc}, m, \ell - 1, 1)$ and therefore, $t, x \models \tau_2$.

Case 1) is symmetrical. For case 3), we apply twice the claim in order to get a' and b'. We can show that $C_1, \ldots, C_4$ hold for $y = x_c$, (a', b', c) and $(m, \ell - 2, 1)$. By induction, we get $t, y \models (\mathsf{X}_{a'c} = \mathsf{X}_{b'c}, m, \ell - 2, 1)$ and therefore, $t, x \models \tau_3$. $\square$

References

1. B. Adsul and M. Sohoni. Complete and tractable local linear time temporal logics over traces. In *Proc. of ICALP'02*, number 2380 in LNCS, pages 926–937. Springer Verlag, 2002.
2. R. Alur, D. Peled, and W. Penczek. Model-checking of causality properties. In *Proc. of LICS'95*, pages 90–100, 1995.
3. V. Diekert. A pure future local temporal logic beyond cograph-monoids. In M. Ito, editor, *Proc. of the RIMS Symposium on Algebraic Systems, Formal Languages and Conventional and Unconventional Computation Theory, Kyoto, Japan 2002*, 2002.
4. V. Diekert and P. Gastin. LTL is expressively complete for Mazurkiewicz traces. In *Proc. of ICALP'00*, number 1853 in LNCS, pages 211–222. Springer Verlag, 2000.
5. V. Diekert and P. Gastin. Local temporal logic is expressively complete for cograph dependence alphabets. In *Proc. of LPAR'01*, number 2250 in LNAI, pages 55–69. Springer Verlag, 2001.
6. V. Diekert and P. Gastin. LTL is expressively complete for Mazurkiewicz traces. *Journal of Computer and System Sciences*, 64:396–418, 2002.
7. V. Diekert and G. Rozenberg, editors. *The Book of Traces*. World Scientific, Singapore, 1995.
8. W. Ebinger. *Charakterisierung von Sprachklassen unendlicher Spuren durch Logiken*. Dissertation, Institut für Informatik, Universität Stuttgart, 1994.
9. W. Ebinger and A. Muscholl. Logical definability on infinite traces. *Theoretical Computer Science*, 154:67–84, 1996.
10. P. Gastin and D. Kuske. Satisfiability and model checking for MSO-definable temporal logics are in PSPACE. In *Proc. of CONCUR'03*, number 2761 in LNCS, pages 222–236. Springer Verlag, 2003.
11. P. Gastin and M. Mukund. An elementary expressively complete temporal logic for Mazurkiewicz traces. In *Proc. of ICALP'02*, number 2380 in LNCS, pages 938–949. Springer Verlag, 2002.
12. P. Gastin, M. Mukund, and K. Narayan Kumar. Local LTL with past constants is expressively complete for Mazurkiewicz traces. In *Proc. of MFCS'03*, number 2747 in LNCS, pages 429–438. Springer Verlag, 2003.
13. J.A.W. Kamp. *Tense Logic and the Theory of Linear Order*. PhD thesis, University of California, Los Angeles, California, 1968.
14. A. Mazurkiewicz. Concurrent program schemes and their interpretations. DAIMI Rep. PB 78, Aarhus University, Aarhus, 1977.
15. M. Mukund and P.S. Thiagarajan. Linear time temporal logics over Mazurkiewicz traces. In *Proc. of MFCS'96*, number 1113 in LNCS, pages 62–92. Springer Verlag, 1996.
16. P. Niebert. A ν-calculus with local views for sequential agents. In *Proc. of MFCS'95*, number 969 in LNCS, pages 563–573. Springer Verlag, 1995.
17. W. Penczek. Temporal logics for trace systems: On automated verification. *International Journal of Foundations of Computer Science*, 4:31–67, 1993.
18. R. Ramanujam. Locally linear time temporal logic. In *Proc. of LICS'96*, pages 118–128, 1996.
19. P.S. Thiagarajan. A trace based extension of linear time temporal logic. In *Proc. of LICS'94*, pages 438–447. IEEE Computer Society Press, 1994.
20. P.S. Thiagarajan and I. Walukiewicz. An expressively complete linear time temporal logic for Mazurkiewicz traces. In *Proc. of LICS'97*, pages 183–194, 1997.
21. I. Walukiewicz. Difficult configurations – on the complexity of LTrL. In *Proc. of ICALP'98*, number 1443 in LNCS, pages 140–151. Springer Verlag, 1998.

How Expressions Can Code for Automata

Sylvain Lombardy[1] and Jacques Sakarovitch[2]

[1] LIAFA, Université Paris 7,
2 place Jussieu, 75251 Paris cedex 05, France
lombardy@liafa.jussieu.fr
[2] LTCI, CNRS / ENST,
46 rue Barrault, 75634 Paris Cedex 13, France
sakarovitch@enst.fr

DEDICATED TO IMRE SIMON ON THE OCCASION OF HIS 60TH BIRTHDAY

Abstract. In this paper we investigate how it is possible to recover an automaton from a rational expression that has been computed from that automaton. The notion of derived term of an expression, introduced by Antimirov, appears to be instrumental in this problem. The second important ingredient is the co-minimization of an automaton, a dual and generalized Moore algorithm on non-deterministic automata. If an automaton is then sufficiently "decorated", the combination of these two algorithms gives the desired result. Reducing the amount of "decoration" is still the object of ongoing investigation.

1 A Natural Question

Kleene's theorem states the *equality of two families of languages*: the family of languages described by rational (*i.e.* regular) expressions coincides with the family of languages accepted (or recognized) by finite automata — equality which is often written as: $\mathrm{Reg}\,A^* = \mathrm{Rec}\,A^*$. Its proof amounts to showing the two inclusions:

$$\mathrm{Rec}\,A^* \subseteq \mathrm{Reg}\,A^* \qquad (1a) \qquad \text{and} \qquad \mathrm{Reg}\,A^* \subseteq \mathrm{Rec}\,A^* \qquad (1b)$$

and is *constructive*. In the earlier proofs, inclusion (1a) is established by an algorithm, say Φ, that takes an automaton $\mathcal{A}$ and produces a rational expression E — we thus can write $\mathsf{E} = \Phi(\mathcal{A})$ — such that the language denoted by E is equal to the language accepted by $\mathcal{A}$. And conversely, inclusion (1b) is obtained by showing that $\mathrm{Rec}\,A^*$ is (effectively) closed under union, product and star.

This closure proof is easily turned into an algorithm, say Ψ, that takes an expression E and computes an automaton $\mathcal{A} = \Psi(\mathsf{E})$ with the property that the language accepted by $\mathcal{A}$ is equal to the language denoted by E. It was not long before it was understood that these algorithms and their properties are as interesting in themselves as to be a piece of the proof of Kleene's theorem.

The problem we address here is to find Φ-type and Ψ-type algorithms which would be *inverse* of each other, that is which are going forth and back between

M. Farach-Colton (Ed.): LATIN 2004, LNCS 2976, pp. 242–251, 2004.
© Springer-Verlag Berlin Heidelberg 2004

expressions and automata not only at the level of the families but at the level of the individual objects. In order to understand the challenge of this problem, we have to say more about the Φ-type and Ψ-type algorithms.

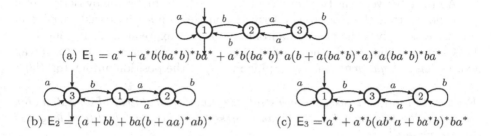

(a) $E_1 = a^* + a^*b(ba^*b)^*ba^* + a^*b(ba^*b)^*a(b + a(ba^*b)^*a)^*a(ba^*b)^*ba^*$

(b) $E_2 = (a + bb + ba(b + aa)^*ab)^*$

(c) $E_3 = a^* + a^*b(ab^*a + ba^*b)^*ba^*$

Fig. 1. The state elimination method on $\mathcal{P}_1$, the "divisor by 3"

The two better known algorithms of the Φ-type (*i.e.* from automata to expressions) are the so-called "McNaughton-Yamada" algorithm ([13]) and "state elimination method" (cf. [16,17] for instance)[1]. Although the computations involved in these algorithms are somewhat different (above all they are organized in a different way), they produce roughly the same results. Both algorithms depend on *an ordering* of the states of the automaton. Figure 1 shows the results of the state elimination method on a same automaton, with three different orderings of the states. The result, and in particular its size, may considerably vary with the ordering that is used. But one cannot avoid a combinatorial explosion in the general case[2]:

Fact 1 *The size of a rational expression* E *computed from a finite automaton* $\mathcal{A}$ *by the state elimination method may be exponential in the number of states of* $\mathcal{A}$.

There is a larger variety of algorithms turning a rational expression into a finite automaton — that is Ψ-type algorithms — both in results and in methods, than those of Φ-type. They fall roughly into two families.

The first class of algorithms yields what is often called the *Glushkov*, or the *position, automaton* of an expression ([10]). It is a non deterministic automaton with $n + 1$ states for an expression of *literal length* n. The Thompson construction ([15]) produces an automaton with ε-moves which is transformed into the position automaton when the ε-moves are eliminated in the adequate way. Let us denote by Ψ_p an algorithm that produces the position automaton.

The algorithms of the second class are based on the definition of the *derivation of an expression*. First introduced by Brzozowski ([5]), the definition of derivation has been slightly, but smartly, modified by Antimirov ([1]) and yields a non deterministic automaton which we propose to call the *derived term automaton* of the expression and which is smaller than or equal to the position

---

[1] A third one ([9]) gives rise to elegant proofs but is not useful for actual computations.

[2] *e.g.* an automaton whose underlying graph is the complete graph on the set of states.

automaton. The automaton of derived expressions computed in [5] is the determinized automaton of the derived term automaton. Champarnaud and Ziadi ([7]) have given an efficient method to compute the derived term automaton of an expression.

A bridge between the two families of algorithms was first given by Berry–Sethi who showed that the Brzozowski derivation applied on a "linearized" version of an expression gives the position automaton of that expression ([3,4]), and then by Champarnaud–Ziadi who showed that the derived term automaton of an expression is a *quotient* (*i.e.* a morphic image) of the position automaton ([8]).

Fact 2 *In the worst case, the minimal size (number of states) of an automaton accepting the language denoted by an expression is linear in the literal length of the expression*[3].

The juxtaposition of Facts 1 and 2 shows that there is no hope to find algorithms which are inverse of each other if we stay in these general families.

In [6], Caron and Ziadi describe an algorithm, say Θ, which decides whether or not an automaton $\mathcal{A}$ is *the* position automaton of a rational expression E; and if the answer is positive, Θ moreover computes E. Even if Θ is not properly a Φ-type algorithm since it does not compute an expression for any automaton, it holds:

$$\text{For any rational expression E,} \qquad \Theta(\Psi_\mathsf{p}(\mathsf{E})) = \mathsf{E} \ . \qquad (1)$$

Our purpose here is to describe a (slight) modification of Φ into Φ' and an algorithm Ω which given a rational expression E computes an equivalent automaton and such that, if E is obtained from an automaton $\mathcal{A}$ by a Φ'-type algorithm, then the result of Ω is precisely $\mathcal{A}$:

$$\text{For any automaton } \mathcal{A}, \qquad \Omega(\Phi'(\mathcal{A})) = \mathcal{A} \ . \qquad (2)$$

In the next section, we present the two main constructions on which such an Ω is built: the — barely modified — *(Antimirov) derivation of expressions* and the *co-minimization* of an automaton. We then (section 3) observe that these constructions yield the core of an algorithm and describe the *partial linearization* which makes the algorithm work in every case (Theorem 2). In conclusion, we mention several directions of investigation in order to minimize the linearization.

Space limitation has forced us to reduce to a very sketchy state the definitions and the description of the reduction of the linearization. In contrast, we have kept full length development for the examples together with their figures for we think they are the best introduction to the subject. A paper under the same title is to be published in a special issue of *TIA-RAIRO* dedicated to Imre Simon and gives a much more detailed presentation ([12]).

---

[3] *e.g.* $\mathsf{E} = f^*$ where f is a word.

2   The Ingredients of a Solution

In the sequel, A is an alphabet, *i.e.* a finite set of letters and A^* the free monoid generated by A. Rational expressions over A^* are the well-formed formulae with 0, 1, $a \in A$ as atomic formulae, $*$ as unary operator and $+$ and $\cdot$ as binary operators. The *constant term* of an expression E, denoted by $\mathsf{c}(\mathsf{E})$, is (the boolean) 1 or 0 according to whether the empty word belongs or not to the language denoted by E. It can easily be computed on the rational expression (*cf.* [1,11] for instance).

2.1   The Automaton of Derived Terms

Definition 1 ([1]). *Let* E *be a rational expression on* A *and let* a *be a letter in* A. *The* $\mathbb{B}$-*derivative*[4] *of* E *with respect to* a, *denoted* $\frac{\partial}{\partial a}\mathsf{E}$, *is a set of rational expressions on* A, *recursively defined by:*

$$\frac{\partial}{\partial a}0 = \frac{\partial}{\partial a}1 = \emptyset, \qquad \forall a, b \in A \qquad \frac{\partial}{\partial a}b = \begin{cases} \{1\} & \text{if} & b = a \\ \emptyset & \text{otherwise} \end{cases}$$

$$\frac{\partial}{\partial a}(\mathsf{E} + \mathsf{F}) = \frac{\partial}{\partial a}\mathsf{E} \cup \frac{\partial}{\partial a}\mathsf{F} \tag{3}$$

$$\frac{\partial}{\partial a}(\mathsf{E} \cdot \mathsf{F}) = \left[\frac{\partial}{\partial a}\mathsf{E}\right] \cdot \mathsf{F} \cup \mathsf{c}(\mathsf{E})\frac{\partial}{\partial a}\mathsf{F} \tag{4}$$

$$\frac{\partial}{\partial a}(\mathsf{E}^*) = \left[\frac{\partial}{\partial a}\mathsf{E}\right] \cdot \mathsf{E}^* \tag{5}$$

The induction implied by (3 – 5) should be interpreted by distributing derivation and product over union:

$$\frac{\partial}{\partial a}\left[\bigcup_{i \in I} \mathsf{E}_i\right] = \bigcup_{i \in I} \frac{\partial}{\partial a}\mathsf{E}_i, \qquad\qquad \left[\bigcup_{i \in I} \mathsf{E}_i\right] \cdot \mathsf{F} = \bigcup_{i \in I}(\mathsf{E}_i \cdot \mathsf{F}).$$

Definition 2. *Let* E *be a rational expression on* A *and* g *a non empty word of* A^*, *i.e.* $g = fa$ *with* a *in* A. *The* $\mathbb{B}$-*derivative of* E *with respect to* g, *denoted* $\frac{\partial}{\partial g}\mathsf{E}$, *is a set of rational expressions over* A, *recursively defined by the formulae (3 – 5) and by:*

$$\forall f \in A^*, \ \forall a \in A \qquad \frac{\partial}{\partial fa}\mathsf{E} = \frac{\partial}{\partial a}\left(\frac{\partial}{\partial f}\mathsf{E}\right). \tag{6}$$

We shall call derived term *of* E *any rational expression which belongs to a set* $\frac{\partial}{\partial g}\mathsf{E}$ *for some* g *in* A^*.

---

[4] We call it "$\mathbb{B}$-derivative" and not simply "derivative" for two reasons. First in order to avoid confusion with the derivation defined by Brzozowski, and second because this derivation is better understood — as we have explained in [11] — if the expressions are considered as expressions *with multiplicity*: "classical" expressions are expressions with multiplicity in the Boolean semiring $\mathbb{B}$.

Theorem 1 (Antimirov [1]). *The number of derived terms of a rational expression* E *is finite and smaller than or equal to the literal length of* E *plus 1.*

Example 1. Let $E_2 = (a + bb + ba(b + aa)^*ab)^*$ be the expression computed in Figure 1 (b). The computation of the derived terms of E_2 goes as follow [for sake of conciseness we put $H_1 = (b + aa)^*ab$]:

$$\frac{\partial}{\partial a} E_2 = \{E_2\}, \quad \frac{\partial}{\partial b} E_2 = \{bE_2, aH_1 E_2\}, \quad \frac{\partial}{\partial a} bE_2 = \emptyset, \quad \frac{\partial}{\partial b} bE_2 = \{E_2\},$$

$$\frac{\partial}{\partial a} [aH_1 E_2] = \{H_1 E_2\}, \quad \frac{\partial}{\partial b} [aH_1 E_2] = \emptyset,$$

$$\frac{\partial}{\partial a} [H_1 E_2] = \{bE_2, aH_1 E_2\}, \quad \frac{\partial}{\partial b} [H_1 E_2] = \{H_1 E_2\}.$$

Thus E_2 has 4 derived terms: E_2 itself, bE_2, $aH_1 E_2$ and $H_1 E_2$.

The above definition is the one given by Antimirov, which we have kept for accurate reference. We now slightly modify the definition of derived terms in order to reach our goal. For that purpose, we first define[5] a new operation on rational expressions which, roughly speaking, consists in decomposing an expression into a set of expressions whose first factor is not a sum.

Definition 3. *i) The set of* initial derived terms *of an expression* E *is a set* $d(E)$ *of expressions inductively defined by:*

$$d(0) = \{0\}, \qquad d(1) = \{1\}, \qquad d(a) = \{a\}, \quad \forall a \in A$$
$$d(E + F) = d(E) \cup d(F), \qquad d(E \cdot F) = [d(E)] \cdot F, \qquad d(E^*) = \{E^*\}.$$

ii) The set of derived terms *of* E *is redefined as the smallest set that contains the initial derived terms of* E *and that is closed under derivation (in the sense of Definition 1).*

In [1], Antimirov has defined an automaton by means of the derived terms and we use here the same construction *mutatis mutandis.*

Definition 4. *The* derived term automaton *of an expression* E *is the automaton* $\mathcal{A}_E$ *whose states are the derived terms of* E *and whose transitions are defined by:*
   i)   *the initial states are the initial derived terms of* E*;*
   ii) *a state* K *is final if and only if* $c(K) = 1$*;*
   iii) (K, a, K') *is a transition of* $\mathcal{A}_E$ *if and only if* K' *belongs to* $\frac{\partial}{\partial a} K$*.*

The automaton $\mathcal{A}_E$ recognizes the language denoted by E (the proof goes as in [1]). In the sequel, we denote by Δ the function that maps a rational expression onto its derived term automaton: $\Delta(E) = \mathcal{A}_E$ (and Δ is a Ψ-type algorithm).

---

[5] As we have explained in [11], this operation can be considered as a derivation with respect to the empty word.

Example 2 (Ex. 1 cont.). As $d(E_2) = E_2$, the new definition does not change the derived terms of E_2 and Figure 2 shows the derived term automaton of E_2.

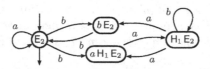

Fig. 2. The derived term automaton of $E_2 = (a + bb + ba(b + aa)^*ab)^*$

Example 3. Let us consider the automaton $\mathcal{A}_1$ of Figure 3 (a) and let E_4 be the expression obtained by the state elimination method on $\mathcal{A}_1$ using the order 1-2-3. It holds:

$$E_4 = a^* + a^*b(ba^*b)^*ba^* + \left(a^*a + a^*b(ba^*b)^*(a + ba^*a)\right) F_1 , \qquad \text{where}$$

$$F_1 = \left(b + ba^*a + (a + ba^*b)(ba^*b)^*(a + ba^*a)\right)^* \left(ba^* + (a + ba^*b)(ba^*b)^*ba^*\right) .$$

Then, $d(E_4) = \{a^*, a^*b(ba^*b)^*ba^*, a^*aF_1, a^*b(ba^*b)^*(a + ba^*a)F_1\}$. And the derived terms of E_4 are read on the automaton $\Delta(E_4)$ itself (Figure 3 (b)).

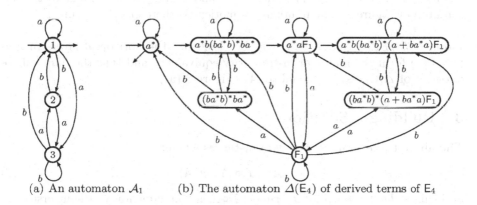

(a) An automaton $\mathcal{A}_1$ (b) The automaton $\Delta(E_4)$ of derived terms of E_4

Fig. 3. Anticipation of the algorithm

2.2   Minimal Co-quotient of an Automaton

The classical process of minimization of a deterministic automaton amounts to comparing transitions *going out* states of the automaton. It is less classical, but not new by far, to consider the same kind of process on automata that are not necessarily deterministic (*cf.* for instance the definition of *simulation* among *transition systems* [2]). We are interested here in the *dual* of such process; it could be defined on the *transposed* automata, but we prefer to give the direct definition using the *incoming transitions*.

Definition 5. *Let $\mathcal{A}$ be an automaton and Q its set of states.*

  i) *An equivalence $\sim$ on Q is an In-similarity equivalence if $p \sim p'$ implies :*

      a) *p is initial if and only if p' is initial;*

      b) *if there exists a transition (q, a, p), there exists a state q' such that:*
$$q' \sim q \text{ and } (q', a, p') \text{ is a transition of } \mathcal{A}.$$

  ii) *An automaton $\mathcal{B}$ is a co-quotient of $\mathcal{A}$ if there exist an In-similarity equivalence $\sim$ on Q and a bijection φ between the states of $\mathcal{B}$ and the classes of $\sim$ such that:*

      a) *a state r of $\mathcal{B}$ is initial iff the states of $\varphi(r)$ are initial;*

      b) *a state r of $\mathcal{B}$ is final iff at least one state of $\varphi(r)$ is final in $\mathcal{A}$;*

      c) *(s, a, r) is a transition of $\mathcal{B}$ iff, for every $p \in \varphi(r)$, there exists $q \in \varphi(s)$ such that (q, a, p) is a transition of $\mathcal{A}$.*

  iii) *If $\sim$ is the coarsest In-similarity equivalence on Q, $\mathcal{B}$ is the minimal co-quotient of $\mathcal{A}$ and $\mathcal{B}$ is said to be a co-minimal automaton.*

    The minimal co-quotient can be computed by a Moore algorithm that consists in refining the trivial partition by splitting the classes that are in contradiction with the In-similarity property. The algorithm stops as soon as the partition is an In-similarity equivalence. We denote this co-minimization algorithm by Υ. Let us note that $\Upsilon(\mathcal{A})$ is canonically attached to $\mathcal{A}$ and not to the language accepted by $\mathcal{A}$.

Example 4 (Ex. 1 cont.). The states $b\,\mathsf{E}_2$ and $a\,\mathsf{H}_1\,\mathsf{E}_2$ are In-similar in the automaton of Figure 2. The minimal co-quotient is therefore $\mathcal{P}_1$ (Figure 1).

Example 5 (Ex. 3 cont.). It should be clear that the horizontal layers in Figure 3 (b) form the maximal In-similarity equivalence and that the minimal co-quotient of $\Delta(\mathsf{E}_4)$ is thus equal to $\mathcal{A}_1$ (Figure 3 (a)).

3   Building a Solution

The above two examples show two instances where:

$$\mathcal{A} = \Upsilon \circ \Delta \circ \Phi(\mathcal{A}) \tag{7}$$

and this is *the main idea of the paper*: $\Upsilon \circ \Delta$ is "fundamentally" the inverse of Φ. Observe that this would not hold (*e.g.* in Example 3) if we had not modified the definition of the derivation. The same equality (and observation) hold if $\Upsilon \circ \Delta$ is applied to E_1 or to E_3 (Figure 1 (a) and (c)).

    However, it is clear that (7) cannot hold in full generality: if $\mathcal{A}$ is not co-minimal for instance, certainly (7) does not hold. But the situation is even more tricky and it may happen that (7) does not hold even for co-minimal automata, as shown by the following example.

Example 6. The automaton $\mathcal{P}_1'$ (Figure 4 (a)) is co-minimal. After two steps of the computation of Φ (following the indicated ordering), the configuration is the same as the one obtained after one step on the automaton of Figure 1 (b). Thus $\Phi(\mathcal{P}_1') = \mathsf{E}_2$ and $\Upsilon(\Delta(\mathsf{E}_2))$ is equal to $\mathcal{P}_1$ and not to $\mathcal{P}_1'$.

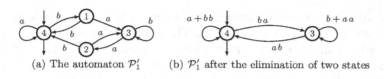

(a) The automaton $\mathcal{P}'_1$ (b) $\mathcal{P}'_1$ after the elimination of two states

Fig. 4. The state elimination method on the automaton $\mathcal{P}'_1$

A way of escaping the above mentioned difficulties is to "decorate" some labels of the automaton in order to indicate in the expression that some occurences of letters in the expression come from different transitions. We call this operation a *partial linearization* and we denoted it by Λ. The delinearization is a projection that we denote by Π. The aim is obviously to keep the linearization as small as possible. However, and as far as now, we can only prove the correctness of the algorithm if the linearization makes the automaton not only co-minimal but also *co-deterministic* (that is reverse deterministic). This will give sufficient conditions that are certainly not necessary. We come back to this question in the conclusion.

Theorem 2. *Let $\mathcal{A}$ be an automaton, Λ a partial linearization that makes $\Lambda(\mathcal{A})$ a minimal co-deterministic automaton and Π the corresponding delinearization. It then holds:*

$$\mathcal{A} = \Pi \circ \Upsilon \circ \Delta \circ \Phi \circ \Lambda(\mathcal{A}) \ . \tag{8}$$

If we come back to the notation of the introduction, Theorem 2 gives the Φ' and Ω we are looking for: $\Phi' = \Phi \circ \Lambda$ and $\Omega = \Pi \circ \Upsilon \circ \Delta$.

Example 7 (Ex. 6 cont.). $\mathcal{P}'_1$ is linearized into $\mathcal{P}''_1$ as shown on Figure 5 (a); the result of Φ is $\mathsf{E}'_2 = (a \mid \bar{b}\bar{b} + b\bar{a}(b + aa)^*ab)^*$, and $\Delta(\mathsf{E}'_2)$ is an automaton whose minimal co-quotient is $\mathcal{P}''_1$.

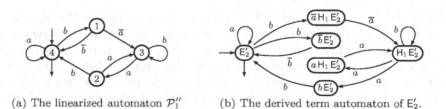

(a) The linearized automaton $\mathcal{P}''_1$ (b) The derived term automaton of E'_2.

Fig. 5. The complete algorithm on automaton $\mathcal{P}'_1$

Theorem 2 is a direct consequence of the following:

Theorem 3. *Let $\mathcal{A}$ be a co-deterministic automaton and $\mathsf{E} = \Phi(\mathcal{A})$ a rational expression computed from $\mathcal{A}$ by the state elimination method. Then, the derived term automaton $\Delta(\mathsf{E})$ of E is co-deterministic.*

Idea of the proof. Every occurence of a letter in E (and in any derived term of E) comes from a certain transition. The study of the way this transition is eliminated during Φ (whether it is a loop or whether or not the origin of the transition is smaller than the end state with respect to the ordering ω that was used for the elimination) allows to describe the form of the derived terms and to prove, by using the assumption of co-determinism that, for every letter a, there is at most one transition labelled by a that arrives in every derived term in the automaton $\Delta(E)$.   □

Proof of Theorem 2. The minimal co-quotient of any co-deterministic automaton is the co-minimal automaton of the language. Therefore, if $\mathcal{A}$ is the co-minimal automaton of the language, $\Upsilon(\Delta(E)) = \mathcal{A}$.   □

Remark 1. If we reverse our construction[6], Theorem 3 states that if $\mathcal{A}$ is deterministic, a (right) derived term automaton of $\Phi(\mathcal{A})$ can be built which is directly *a deterministic automaton*, which has a linear number of states (in the size of $\Phi(\mathcal{A})$) — and this without running any determinization.

4   Discussion

As we said, the conditions put on Λ in Theorem 2 are sufficient but far from being necessary. For instance, the automaton $\mathcal{A}_1$ in Example 3 is not co-deterministic and $\mathcal{A}_1 = \Upsilon(\Delta(\Phi(\mathcal{A}_1)))$ holds though. This example and other computations have led us to consider several ways that can help in distinguishing derived terms and thus reducing the role of Λ. As far as now, they can serve as heuristics and it is our current work to turn these ideas into precise statements. Let us quote here three of these directions of research.

Choosing a smart ordering. The automaton $\mathcal{P}'_1$ of Example 6 may give an illustration of this idea. The reader can check that if the elimination on $\mathcal{P}'_1$ is performed in the ordering: 3-2-4-1, the resulting expression is $E''_2 = \Phi(\mathcal{P}'_1) = (a + b(b + ab^*a(ab^*a)b)^*$ and $\mathcal{P}'_1 = \Upsilon(\Delta(\Phi(\mathcal{P}'_1)))$: no linearization at all is necessary. Thus, the search for a smart ordering may be — to some extend — an alternative to the linearization of $\mathcal{A}$.

Using the structure of $\Delta(\Phi(\mathcal{A}))$. Figure 3 (b) shows clearly the structure of the derived term automaton of an expression $E = \Phi(\mathcal{A})$ that is computed on a *strongly connected* automaton $\mathcal{A}$: the last p of $\mathcal{A}$ to be eliminated corresponds to a term that is a *cutvertex* in $\Delta(E)$ and this property holds inductively on subautomata. This observation is another way to distinguish states that are otherwise labelled by a same derived term. It thus lead to an improved version of Δ which may again depend on the ordering of the elimination.

Taking multiplicity into account. A fundamental property of the algorithms Φ and Δ is the fact that they respect the multiplicity of paths. The minimal co-

---

[6] We have defined a derivation that is applied to the left side of the expression; similar operation can be defined on the right side, and we would call the result the *right* derived terms.

quotient of $\Delta(\Phi(\mathcal{A}))$ when computed as an automaton with multiplicity (in $\mathbb{N}$) may contain transitions with coefficients larger than 1 (and this is clearly not what is wanted). Therefore, Υ has to be modified into an algorithm Υ' which computes a $\mathbb{B}$-automaton that is a *co-covering* of $\Delta(\Phi(\mathcal{A}))$ (as defined in [14]). But contrary to the minimal co-quotient, the minimal co-covering of an automaton is not necessary unique.

These examples and remarks give strong evidences that the computation of the derived terms of an expression is not only an algorithm that builds an equivalent automaton but also a way to retrieve the "track" of the states of an automaton when the expression has been computed from that automaton. How far these tracks are faithful, and how to read them efficiently are questions that are still under investigation.

References

1. V. ANTIMIROV, Partial derivatives of regular expressions and finite automaton constructions, *Theoret. Computer Sci.* **155** (1996), 291–319.
2. A. ARNOLD, *Systèmes de transitions finis et sémantique des processus communiquants*, Masson, 1992. English translation: *Finite transitions systems*, Prentice-Hall, 1994.
3. G. BERRY AND R. SETHI, From regular expressions to deterministic automata, *Theoret. Computer Sci.* **48** (1986), 117–126.
4. J. BERSTEL AND J.-E. PIN, Local languages and the Berry-Sethi algorithm, *Theoret. Computer Sci.* **155** (1996), 439–446.
5. J. A. BRZOZOWSKI, Derivatives of regular expressions. *J. Assoc. Comput. Mach.* **11** (1964), 481–494.
6. P. CARON AND D. ZIADI, Characterization of Glushkov automata, *Theoret. Computer Sci.* **233** (2000), 75–90.
7. J.-M. CHAMPARNAUD AND D. ZIADI, New finite automaton constructions based on canonical derivatives, *Pre-Proceedings of CIAA'00*, M. Daley, M. Eramian and S. Yu, eds, Univ. of Western Ontario, (2000), 36–43.
8. J.-M. CHAMPARNAUD AND D. ZIADI, Canonical derivatives, partial derivatives and finite automaton constructions, *Theoret. Computer Sci.* **289** (2002), 137–163.
9. J. H. CONWAY, *Regular algebra and finite machines*, Chapman and Hall, 1971.
10. V. GLUSHKOV, The abstract theory of automata, *Russian Mathematical Surveys* **16** (1961), 1–53.
11. S. LOMBARDY AND J. SAKAROVITCH, Derivatives of rational expressions with multiplicity, in *MFCS 02*, LNCS 2420 (2002), 471–482.
12. S. LOMBARDY AND J. SAKAROVITCH, How expressions code for automata, *TIARAIRO*, to appear.
13. R. MCNAUGHTON AND H. YAMADA, Regular expressions and state graphs for automata. *IRE Trans. on Electronic Computers* **9** (1960), 39–47.
14. J. SAKAROVITCH, A construction on automata that has remained hidden, *Theoret. Computer Sci.* **204** (1998), 205–231.
15. K. THOMPSON, Regular expression search algorithm, *Comm. Assoc. Comput. Mach.* **11** (1968), 419–422.
16. D. WOOD, *Theory of Computation*, Wiley, 1987.
17. S. YU, Regular languages, in *Handbook of Formal Languages*, G. Rozenberg and A. Salomaa (esd.), vol. 1, pp. 41–111, Elsevier, 1997.

Automata for Arithmetic Meyer Sets*

Shigeki Akiyama[1], Frédérique Bassino[2], and Christiane Frougny[3]

[1] Department of Mathematics, Faculty of Sciences, Niigata University, Ikarashi-2,
8050, Niigata 950-2181, Japan, akiyama@math.sc.niigata-u.ac.jp
[2] Institut Gaspard Monge, Université de Marne-la-Vallée, 5 Boulevard Descartes,
Champs-sur-Marne 77454 Marne-la-Vallée Cedex 2, France, bassino@univ-mlv.fr
[3] LIAFA, UMR 7089, 2 Place Jussieu, 75251 Paris Cedex 05, France, and
Université Paris 8, Christiane.Frougny@liafa.jussieu.fr

Abstract. The set $\mathbb{Z}_\beta$ of β-integers is a Meyer set when β is a Pisot number, and thus there exists a finite set F such that $\mathbb{Z}_\beta - \mathbb{Z}_\beta \subset \mathbb{Z}_\beta + F$. We give finite automata describing the expansions of the elements of $\mathbb{Z}_\beta$ and of $\mathbb{Z}_\beta - \mathbb{Z}_\beta$. We present a construction of such a finite set F, and a method to minimize the size of F. We obtain in this way a finite transducer that performs the decomposition of the elements of $\mathbb{Z}_\beta - \mathbb{Z}_\beta$ as a sum belonging to $\mathbb{Z}_\beta + F$.

1 Introduction

The so-called Meyer sets have been introduced by Meyer [11,12] under the name of "quasicrystals" in order to formalize the quasicrystals discovered by the physicists in the eighties. A set X is a *Delaunay set* if it is uniformly discrete and relatively dense. A set X is a *Meyer set* if it is a Delaunay set and there exists a finite set F such that $X - X \subset X + F$. There exist strong relations between Meyer sets and some algebraic integers. Recall that a *Pisot number* (or a Pisot-Vijayaraghavan number) is an algebraic integer > 1 such that all its algebraic conjugates have modulus strictly less than one. A *Salem number* is an algebraic integer such that every conjugate has modulus smaller than or equal to 1, and at least one of them has modulus 1. The following result from Meyer makes the connection between Meyer sets and those algebraic integers. If $X \subset \mathbb{R}^n$ is a Meyer set and if $\beta > 1$ is a real number such that $\beta X \subset X$ then β is a Pisot or a Salem number. Conversely for each n and for each Pisot or Salem number β, there exists a Meyer set $X \subset \mathbb{R}^n$ such that $\beta X \subset X$.

Note that all the quasicrystals encountered in the real world are linked to quadratic Pisot numbers, namely $\frac{1+\sqrt{5}}{2}$, $1 + \sqrt{2}$ and $2 + \sqrt{3}$.

In this paper we study Meyer sets $\mathbb{Z}_\beta$ associated with β-expansions, β being a Pisot number, and give a construction of a minimal finite set F such that $\mathbb{Z}_\beta - \mathbb{Z}_\beta \subset \mathbb{Z}_\beta + F$.

Lagarias [8] gave a general construction of a finite set F satisfying $X - X \subset X + F$ for a Delaunay set X such that $X - X$ is also a Delaunay set. But the

---

* Work supported by the CNRS/JSPS contract number 13569

M. Farach-Colton (Ed.): LATIN 2004, LNCS 2976, pp. 252–261, 2004.

sets obtained are huge and no method of minimization of these sets is known. Minimal sets F are given in [3] for $\mathbb{Z}_\beta$ when β is a quadratic Pisot unit. When β is a quadratic Pisot number, a possible set F for $\mathbb{Z}_\beta$ is exhibited in [6].

We first give finite automata describing the formal addition and substraction of beta-integers. We characterize the cases when the formal addition gives a system of finite type when the original system $\mathbb{Z}_\beta$ is of finite type.

We then give a construction of a family of finite sets F such that $\mathbb{Z}_\beta - \mathbb{Z}_\beta \subset \mathbb{Z}_\beta + F$, and a method to minimize the size of the sets F we built. We obtain in this way a finite transducer that performs the decomposition of the result of the formal substraction $\mathbb{Z}_\beta - \mathbb{Z}_\beta$ into a sum belonging to $\mathbb{Z}_\beta + F$.

2   Preliminaries

Let A be a finite alphabet. A concatenation of letters of A is called a *word*. The set A^* of all finite words equipped with the empty word ε and the operation of concatenation is a free monoid. We denote by a^k the word obtained by concatenating k letters a. The length of a word $w = w_0 w_1 \cdots w_{n-1}$ is denoted by $|w| = n$. One considers also infinite words $v = v_0 v_1 v_2 \cdots$. The set of infinite words on A is denoted by $A^{\mathbb{N}}$. An infinite word v is said to be *eventually periodic* if it is of the form $v = wz^\omega$, where w and z are in A^* and $z^\omega = zzz\cdots$. A *factor* of a finite or infinite word w is a finite word v such that $w = uvz$; if $u = \varepsilon$, the word v is a *prefix* of w. A prefix of w is *strict* if it is not equal to w.

Definitions and results on numeration systems can be found in [10, Chapter 7]. Let $\beta > 1$ be a real number. Any positive real number x can be represented in base β by the following greedy algorithm [14]. Denote by $\lfloor . \rfloor$ and by $\{ . \}$ the integral part and the fractional part of a number. There exists $k \in \mathbb{Z}$ such that $\beta^k \leq x < \beta^{k+1}$. Let $x_k = \lfloor x/\beta^k \rfloor$ and $r_k = \{x/\beta^k\}$. For $i < k$, put $x_i = \lfloor \beta r_{i+1} \rfloor$, and $r_i = \{\beta r_{i+1}\}$. Then $x = x_k \beta^k + x_{k-1}\beta^{k-1} + \cdots$. If $x < 1$, we get $k < 0$ and we put $x_0 = x_{-1} = \cdots = x_{k+1} = 0$. The sequence $(x_i)_{k \geq i \geq -\infty}$ is called the β-*expansion* of x, and is denoted by

$$\langle x \rangle_\beta = x_k x_{k-1} \cdots x_1 x_0 \cdot x_{-1} x_{-2} \cdots$$

most significant digit first. The part $x_{-1} x_{-2} \cdots$ after the "decimal" point is called the β-*fractional part* of x.

The digits x_i are elements of the *canonical* alphabet $A_\beta = \{0, \dots, \lfloor \beta \rfloor\}$ if $\beta \notin \mathbb{N}$ and $A_\beta = \{0, \dots, \beta - 1\}$ otherwise. When a β-expansion ends in infinitely many zeroes, it is said to be *finite*, and the 0's are omitted.

A finite or infinite word w on A_β which is the β-expansion of some number x is said to be *admissible*. Leading 0's are allowed.

The set $\mathbb{Z}_\beta$ of β-*integers* is the set of real numbers x such that the β-fractional part of $|x|$ is equal to 0,

$$\mathbb{Z}_\beta = \{x \in \mathbb{R} \mid \langle |x| \rangle_\beta = x_k \cdots x_0\} = \mathbb{Z}_\beta^+ \cup \mathbb{Z}_\beta^-$$

where $\mathbb{Z}_\beta^+$ is the set of non-negative beta-integers, and $\mathbb{Z}_\beta^- = -\mathbb{Z}_\beta^+$.

Denote by D_β the set of β-expansions of numbers of $[0,1)$ and the shift by σ. Then D_β is shift-invariant. Let S_β be its closure in $A_\beta^{\mathbb{N}}$. The set S_β is a symbolic dynamical system, called the β-*shift*. The set $\mathbb{Z}_\beta^+$ is equal to the set of finite factors of S_β.

There is a peculiar representation of the number 1 which plays an important role in the theory. It is denoted by $d_\beta(1)$, and computed by the following process [14]. Let the β-*transform* be defined on $[0,1]$ by $T_\beta(x) = \beta x \bmod 1$. Then $d_\beta(1) = (t_i)_{i \geq 1}$, where $t_i = \lfloor \beta T_\beta^{i-1}(1) \rfloor$. Note that $\lfloor \beta \rfloor = t_1$. We recall a result of Parry [13]: a sequence s of natural integers is an element of D_β if and only if for every $p \geq 1$, $\sigma^p(s)$ is strictly less in the lexicographic order than $d_\beta(1)$ if $d_\beta(1)$ is infinite, or less than $d_\beta^*(1) = (t_1 \cdots t_{m-1}(t_m - 1))^\omega$ if $d_\beta(1) = t_1 \cdots t_m$ is finite.

A word $w_1 \cdots w_n$ of A_β^* is said to be a *minimal forbidden* word for S_β if it is not a factor of S_β and if $w_1 \cdots w_{n-1}$ and $w_2 \cdots w_n$ are factors of S_β. Recall that a symbolic dynamical system is said to be *of finite type* if the set of its minimal forbidden words is finite. More generally it is said to be *sofic* if the set of its finite factors is recognized by a finite automaton. The β-shift is sofic if and only if $d_\beta(1)$ is eventually periodic, and it is of finite type if and only if $d_\beta(1)$ is finite. By abuse we say that the set $\mathbb{Z}_\beta$ of β-integers is of finite type (resp. sofic) if $d_\beta(1)$ is finite (resp. infinite eventually periodic). Recall that if β is a Pisot number, then $d_\beta(1)$ is finite or eventually periodic [2,15].

A set $X \subset \mathbb{R}^n$ is *uniformly discrete* if there exists a positive real r such that for any $x \in \mathbb{R}^n$, the open ball of center x and radius r contains at most one point of X. If $Y \subset X$ and X is uniformly discrete, then Y is uniformly discrete. A set $X \subset \mathbb{R}^n$ is *relatively dense* if there exists a positive real R such that for any $x \in \mathbb{R}^n$, the open ball of center x and radius R contains at least one point of X. If $X \subset Y$ and X is relatively dense, then Y is relatively dense. A set X is a *Delaunay set* if it is uniformly discrete and relatively dense. A set X is a *Meyer set* if it is a Delaunay set and there exists a finite set F such that $X - X \subset X + F$. Lagarias proved [8] that a set X is a Meyer set if and only if both X and $X - X$ are Delaunay sets. Note that when X is a Delaunay set, then $X - X$ is relatively dense, but not necessarily uniformly discrete. For example $X = \{n + \frac{1}{|n|+2}\}$ is a Delaunay set and $X - X$ has 1 as point of accumulation.

Proposition 1. [3] *If β is a Pisot number, then the set $\mathbb{Z}_\beta$ of β-integers is a Meyer set.*

3   Automata for Formal Addition and Substraction

In this section we construct automata that symbolically describe the elements of $\mathbb{Z}_\beta - \mathbb{Z}_\beta$ when β is a Pisot number. Note that

$$\mathbb{Z}_\beta - \mathbb{Z}_\beta = (\mathbb{Z}_\beta^+ - \mathbb{Z}_\beta^+) \cup (\mathbb{Z}_\beta^+ + \mathbb{Z}_\beta^+) \cup -(\mathbb{Z}_\beta^+ + \mathbb{Z}_\beta^+). \tag{1}$$

The reader is referred to [4] and [16] for definitions and results in automata theory. We introduce some notations. Denote by $L_\beta^+ \subset A_\beta^*$ the set of β-expansions of elements of $\mathbb{Z}_\beta^+$ with possible leading 0's. Set $\bar{k} = -k$, where k is an integer, and let $\overline{A_\beta} = \{\lfloor\beta\rfloor, \dots, \bar{1}, 0\}$. We denote by $L_\beta^- \subset \overline{A_\beta}^*$ the set $\{\overline{w} = \overline{w_N} \cdots \overline{w_0} \mid w = w_N \cdots w_0 = \langle -x\rangle_\beta, \ x \in \mathbb{Z}_\beta^- \}$.

When $d_\beta(1)$ is finite or eventually periodic, the set L_β^+ is recognizable by a finite automaton [5], of which we recall the construction. If $d_\beta(1) = t_1 \cdots t_m$ is finite, the automaton $\mathcal{A}_{\mathbb{Z}_\beta^+}$ recognizing L_β^+ has m states $q_1, \dots, q_m$. For each $1 \leq i \leq m-1$ there is an edge between q_i and q_{i+1} labelled by t_i. For each $1 \leq i \leq m$ there are t_i edges between q_i and q_1 labelled by $0, \dots, t_i - 1$. The initial state is q_1; every state is terminal.

If $d_\beta(1) = t_1 \cdots t_m (t_{m+1} \cdots t_{m+p})^\omega$ is infinite eventually periodic, the automaton $\mathcal{A}_{\mathbb{Z}_\beta^+}$ recognizing L_β^+ has $m + p$ states $q_1, \dots, q_{m+p}$. For each $1 \leq i < m + p - 1$ there is an edge between q_i and q_{i+1} labelled by t_i. For each $1 \leq i \leq m + p$ there are t_i edges between q_i and q_1 labelled by $0, \dots, t_i - 1$. There is an edge from q_{m+p} to q_{m+1} labelled by t_{m+p}. The initial state is q_1; every state is terminal.

Clearly the set L_β^- is recognizable by the same automaton as L_β^+, but with negative labels on edges. Then the automaton for $\mathbb{Z}_\beta$ is $\mathcal{A}_{\mathbb{Z}_\beta} = \mathcal{A}_{\mathbb{Z}_\beta^+} \cup \mathcal{A}_{\mathbb{Z}_\beta^-}$.

By a general construction one can compute the "sum" of two automata. Let $\mathcal{A}$ and $\mathcal{B}$ be two finite automata with labels in an alphabet of integers. One constructs a finite automaton $\mathcal{S}$ as follows :

- the set of states of $\mathcal{S}$ is the cartesian product $Q_\mathcal{S} = Q_\mathcal{A} \times Q_\mathcal{B}$
- there is an edge in $\mathcal{S}$ from (p, q) to (p', q') labelled by $a + b$ if and only if there is an edge from p to p' labelled by a in $\mathcal{A}$ and an edge from q to q' labelled by b in $\mathcal{B}$.
- the set of initial (resp. terminal) states is the cartesian product of the sets of initial (resp. terminal) states of $\mathcal{A}$ and $\mathcal{B}$.

Clearly the automaton $\mathcal{S}$ recognizes the set $\{s_N \cdots s_0 \mid N \geq 0, \ s_i = a_i + b_i, \ 0 \leq i \leq N, \ a_N \cdots a_0$ is recognized by $\mathcal{A}$ and $b_N \cdots b_0$ is recognized by $\mathcal{B}\}$.

The *formal addition* of elements of $\mathbb{Z}_\beta^+$ consists in adding elements without carry. More precisely,

$$L_\beta^+ + L_\beta^+ = \{(a_N + b_N) \cdots (a_0 + b_0) \mid a_N \cdots a_0, \ b_N \cdots b_0 \in \mathbb{Z}_\beta^+\} \subset \{0, \cdots, 2\lfloor\beta\rfloor\}^*.$$

Similarly the *formal subtraction* of elements of $\mathbb{Z}_\beta^+$ is defined by

$$L_\beta^+ - L_\beta^+ = \{(a_N - b_N) \cdots (a_0 - b_0) \mid a_N \cdots a_0, \ b_N \cdots b_0 \in \mathbb{Z}_\beta^+\} \subset \{-\lfloor\beta\rfloor, \cdots, \lfloor\beta\rfloor\}^*.$$

From the construction of the sum automaton follows

Proposition 2. *If $d_\beta(1)$ is finite or eventually periodic, the set $L_\beta^+ + L_\beta^+$ corresponding to the formal addition $\mathbb{Z}_\beta^+ + \mathbb{Z}_\beta^+$ and the set $L_\beta^+ - L_\beta^+$ corresponding to the formal subtraction $\mathbb{Z}_\beta^+ - \mathbb{Z}_\beta^+$ are recognizable by a finite automaton.*

By Equation (1), $\mathcal{A}_{\mathbb{Z}_\beta - \mathbb{Z}_\beta} = \mathcal{A}_{\mathbb{Z}_\beta^+ + \mathbb{Z}_\beta^+} \cup \mathcal{A}_{\mathbb{Z}_\beta^+ - \mathbb{Z}_\beta^+} \cup \mathcal{A}_{-(\mathbb{Z}_\beta^+ + \mathbb{Z}_\beta^+)}$. The automata given by this construction are generally not minimal.

Example 1. In the case where $\beta = \frac{1+\sqrt{5}}{2}$, $d_\beta(1) = 11$ and $d_\beta^*(1) = (10)^\omega$. We give below the minimal automata $\mathcal{A}_{\mathbb{Z}_\beta^+}$, $\mathcal{A}_{\mathbb{Z}_\beta^-}$, $\mathcal{A}_{\mathbb{Z}_\beta^+ + \mathbb{Z}_\beta^+}$, and $\mathcal{A}_{\mathbb{Z}_\beta^+ - \mathbb{Z}_\beta^+}$. Initial states are indicated by an incoming arrow, and every state is terminal.

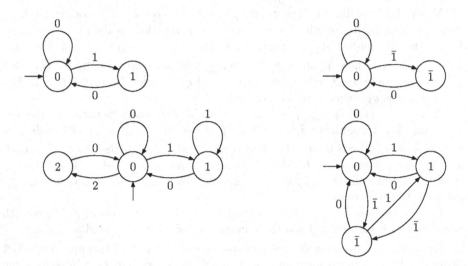

It is an interesting question to see what is the result of formal addition or subtraction when the system $\mathbb{Z}_\beta$ is of finite type. First recall that, from the result of Parry cited in Sect. 2, if $d_\beta(1) = t_1 \cdots t_m$, the set of minimal forbidden words for $\mathbb{Z}_\beta^+$ is the set $I_\beta = \{t_1 \cdots t_m\} \cup \{t_1 t_2 \cdots t_{p-1} x_p \mid t_p < x_p \leq t_1, \ 2 \leq p \leq m, \ x_p \in A_\beta\}$.

Proposition 3. *If $d_\beta(1) = t_1 \cdots t_m$ is finite, the formal subtraction $\mathbb{Z}_\beta^+ - \mathbb{Z}_\beta^+$ defines a system of finite type.*

Proof. Recall that if a word is admissible, any word with smaller nonnegative digits is admissible as well. Thus the set of forbidden words for the formal subtraction $\mathbb{Z}_\beta^+ - \mathbb{Z}_\beta^+$ is equal to $\{w, \overline{w} \mid w \in I_\beta\}$, which is finite.   □

The result for formal addition is quite different.

Proposition 4. *If $d_\beta(1) = t_1 \cdots t_m$ is finite, the formal addition $\mathbb{Z}_\beta^+ + \mathbb{Z}_\beta^+$ defines a system of finite type if and only if $t_m = t_1$ and, for each $2 \leq i \leq m - 1$, $t_i = t_1$ or $t_i = 0$.*

Corollary 1. *If $\beta < 2$ and $d_\beta(1)$ is finite then the formal addition $\mathbb{Z}_\beta^+ + \mathbb{Z}_\beta^+$ defines a system of finite type.*

The proof of Proposition 4 follows from several technical results.

Lemma 1. *Suppose that $d_\beta(1) = t_1 \cdots t_m$, and that there exists $2 \leq j \leq m$ with $0 < t_j < t_1$ (so $t_1 \geq 2$), and $t_i = 0$ or $t_i = t_1$ for $2 \leq i \leq j - 1$. Then the set of minimal forbidden words in the formal addition is infinite.*

Proof. For any $k \geq 1$ consider the word $u^{(k)} = [(t_1 + t_2)(t_2 + t_3) \cdots (t_{j-2} + t_{j-1})(t_{j-1} + t_j - 1)(t_j - 1 + t_1)]^k (t_1 + t_2)(t_2 + t_3) \cdots (t_{m-1} + t_m)$.
Let $w^{(k)} = (2t_1 - 1)u^{(k)}$. First we show that $w^{(k)}$ is forbidden in the formal addition system. This comes from the fact that $w^{(k)}$ is necessarily the digit-sum of the two words $x^{(k)} = (t_1 - 1)[t_1 \cdots t_{j-1}(t_j - 1)]^k t_1 \cdots t_{m-1}$ and $y^{(k)} = t_1[t_2 \cdots t_{j-1}(t_j - 1)t_1]^k t_2 \cdots t_m$. Clearly $y^{(k)}$ is not admissible for $\mathbb{Z}_\beta^+$ because it ends in the forbidden word $t_1 \cdots t_m$, and $x^{(k)}$ is admissible for $\mathbb{Z}_\beta^+$ and maximal in the sense that adding 1 to one of its digits makes the word not admissible.

Note that all strict prefixes of $y^{(k)}$ are admissible for $\mathbb{Z}_\beta^+$, so all strict prefixes of $w^{(k)}$ are also admissible.

Now we show that the word $u^{(k)}$ is admissible in the formal addition system. By hypothesis the digits $(t_i + t_{i+1})$ for $1 \leq i \leq j - 2$ are equal to $2t_1$, t_1 or 0. So $u^{(k)}$ can be obtained as the digit-sum of $v^{(k)}$ and $z^{(k)}$ with the following method: a digit $2t_1$ of $u^{(k)}$ gives a digit t_1 in $v^{(k)}$ and a digit t_1 in $z^{(k)}$; a digit t_1 of $u^{(k)}$ gives a digit $t_1 - 1$ in $v^{(k)}$ and a digit 1 in $z^{(k)}$; a digit 0 of $u^{(k)}$ gives a digit 0 in $v^{(k)}$ and in $z^{(k)}$. Since $0 < t_j < t_1$, the digits $t_{j-1} + t_j - 1$ and $t_j - 1 + t_1$ are $\leq 2t_1 - 2$, which is the sum of $t_1 - 1$ and $t_1 - 1$. The suffix $(t_j - 1 + t_1)(t_1 + t_2)(t_2 + t_3) \cdots (t_{m-1} + t_m)$ of $u^{(k)}$ is thus the digit-sum of $at_1 t_2 \cdots t_{m-1}$, with $a \leq t_1 - 1$, and of $bt_2 t_3 \cdots t_m$, with $b \leq t_1 - 1$. Hence $u^{(k)}$ is the digit-sum of $v^{(k)}$ and $z^{(k)}$, which are both admissible for $\mathbb{Z}_\beta^+$. $\square$

Lemma 2. *If $d_\beta(1) = t_1 \cdots t_m$ is finite and if $t_m = t_1$ and, for each $2 \leq i \leq m - 1$, $t_i = t_1$ or $t_i = 0$ then the formal addition is a system of finite type.*

Proof. As in Lemma 1 we consider the word $u^{(k)}$, with $t_j = t_1$ for a fixed j, $2 \leq j \leq m$. The difference with Lemma 1 is that now the suffix $s = (t_j - 1 + t_1)(t_1 + t_2)(t_2 + t_3) \cdots (t_{m-1} + t_m)$ is not admissible. Since $t_j = t_1$, s can be the the digit-sum of $(t_1 - 1)t_1 \cdots t_{m-1}$ and $t_1 t_2 \cdots t_m$, or of $(t_1 - 1)t_1 \cdots (t_{\ell-1})(t_\ell + 1)t_{\ell+1} \cdots t_{m-1}$ and $t_1 t_2 \cdots t_{\ell-1}(t_\ell - 1)t_{\ell+1} \cdots t_m$ if $t_\ell \neq 0$, for $2 \leq \ell \leq m - 1$. But none of the factors $t_1 \cdots t_{\ell-1}(t_\ell + 1)$ is admissible for $\mathbb{Z}_\beta^+$. By considering all the positions $2 \leq j \leq m$ in $u^{(k)}$, we see that it is not possible to construct an infinite family of minimal forbidden words of type $w^{(k)}$. $\square$

4   A Family of Finite Sets F

When β is a Pisot number, the set of beta-integers $\mathbb{Z}_\beta$ is a Meyer set so there exists a finite set F such that $\mathbb{Z}_\beta - \mathbb{Z}_\beta \subset \mathbb{Z}_\beta + F$. Our goal is to construct sets F as small as possible for $\mathbb{Z}_\beta$.

Remark 1. Note that there exist several sets F with minimal cardinality. For example when $\beta = (1 + \sqrt{5})/2$ then $\mathbb{Z}_\beta - \mathbb{Z}_\beta \subset \mathbb{Z}_\beta + F$, with $F = \{0, \beta - 1, -\beta + 1\}$, or $F = \{0, \beta - 2, -\beta + 2\}$ or $F = \{0, \beta - 1, -\beta + 2\}$.

We first define finite sets from which can be extracted the finite sets F.

Lemma 3. *Let β be a Pisot number of degree d, let $I \subset \mathbb{R}$ be an interval of length 1 and let U be the following set*

$$U = \left\{ x \in \mathbb{Z}[\beta] \mid x \in I \text{ and } \forall 2 \le j \le d, \, |x^{(j)}| < \frac{3\lfloor\beta\rfloor}{1 - |\beta^{(j)}|} \right\},$$

where $x^{(2)}, \ldots, x^{(d)}$ are the algebraic conjugates of x. Then U is finite, and there exists a subset F of U such that $\mathbb{Z}_\beta - \mathbb{Z}_\beta \subset \mathbb{Z}_\beta + F$.

Proof. As the maximal distance between two consecutive points of $\mathbb{Z}_\beta$ is equal to 1, one can find a set F such that $\mathbb{Z}_\beta - \mathbb{Z}_\beta \subset \mathbb{Z}_\beta + F$ in any interval I of length 1.

Fix an interval I of length 1 and $F \subset I$ as small as possible such that $\mathbb{Z}_\beta - \mathbb{Z}_\beta \subset \mathbb{Z}_\beta + F$. Let $x \in F$, then $x \in (\mathbb{Z}_\beta - \mathbb{Z}_\beta) - \mathbb{Z}_\beta$ and can be written as

$$x = \sum_{i=0}^{N} (a_i - b_i)\beta^i - \sum_{i=0}^{N} c_i \beta^i \quad \text{with } |a_i|, |b_i|, |c_i| \le \lfloor\beta\rfloor.$$

so

$$\forall 2 \le j \le d \quad x^{(j)} = \sum_{i=0}^{N} (a_i - b_i - c_i)(\beta^{(j)})^i \quad \text{with } |a_i - b_i - c_i| \le 3\lfloor\beta\rfloor.$$

As β is a Pisot number, for all $j \ge 2$, $|\beta^{(j)}| < 1$ and $|\sum_{i=0}^{N}(\beta^{(j)})^i| < (1 - |\beta^{(j)}|)^{-1}$. We obtain in this way the announced bound on the moduli of the conjugates of x and $x \in U$. So F is a subset of U.

As it contains only points of $\mathbb{Z}[\beta]$ with bounded modulus and whose all conjugates have bounded modulus, the set U is finite. Thus F is a finite set. $\square$

The choice of any interval $I \subset \,]-1, 1[$ of length 1 allows us to reduce the cardinality of the set containing a set F.

Lemma 4. *Let β be a Pisot number of degree d, let $I \subset \,]-1, 1[$ be an interval of length 1 and let U' be the following finite set*

$$U' = \left\{ x \in \mathbb{Z}[\beta] \mid x \in I \text{ and } \forall 2 \le j \le d, \, |x^{(j)}| < \frac{2\lfloor\beta\rfloor}{1 - |\beta^{(j)}|} \right\}.$$

Then there exists a subset F of U' such that $\mathbb{Z}_\beta - \mathbb{Z}_\beta \subset \mathbb{Z}_\beta + F$.

Proof. We choose here $I \subset \,]-1, 1[$ of length 1 and improve the bound on the moduli of the conjugates of x given in Lemma 3 by considering the decomposition

$$\mathbb{Z}_\beta - \mathbb{Z}_\beta = (\mathbb{Z}_\beta^+ - \mathbb{Z}_\beta^+) \cup (\mathbb{Z}_\beta^+ + \mathbb{Z}_\beta^+) \cup -(\mathbb{Z}_\beta^+ + \mathbb{Z}_\beta^+).$$

More precisely let $x \in F \subset I$, then $x \in (\mathbb{Z}_\beta - \mathbb{Z}_\beta) - \mathbb{Z}_\beta$ and can be written as

$$x = \sum_{i=0}^{N} (a_i - b_i)\beta^i - \sum_{i=0}^{N} c_i \beta^i.$$

We study $|a_i - b_i - c_i|$ according to the signs of a_i, b_i and c_i. In $\mathbb{Z}_\beta^+ - \mathbb{Z}_\beta^+$, the coefficients satisfy $|a_i - b_i| \leq \lfloor \beta \rfloor$. Moreover when $F \subset]-1, 1[$, $\mathbb{Z}_\beta^+ + \mathbb{Z}_\beta^+ \subset \mathbb{Z}_\beta^+ + F$ and $-\left(\mathbb{Z}_\beta^+ + \mathbb{Z}_\beta^+ \right) \subset \mathbb{Z}_\beta^- + F$, then we have $|a_i - c_i| \leq \lfloor \beta \rfloor$. So when $F \subset]-1, 1[$, we get in all cases $|a_i - b_i - c_i| \leq 2\lfloor \beta \rfloor$. Thus

$$\forall 2 \leq j \leq d \quad x^{(j)} = \sum_{i=0}^{N} (a_i - b_i - c_i)(\beta^{(j)})^i \quad \text{with } |a_i - b_i - c_i| \leq 2\lfloor \beta \rfloor,$$

and the announced bound on the moduli of the conjugates of x holds true.   □

Example 2. Let β be a quadratic Pisot unit, then the set U' contains 5 points.

5   A First Reduction of the Cardinality of the Sets Containing F

In order to reduce the size of the sets containing F we study the properties of the elements of F.

Lemma 5. *Let β be a Pisot number and let $F \subset (\mathbb{Z}_\beta - \mathbb{Z}_\beta) - \mathbb{Z}_\beta$. If $f \in F$ there exist a nonnegative integer N, and two finite words $b_N \cdots b_0$ and $a_N \cdots a_0$ respectively admissible for $\mathbb{Z}_\beta - \mathbb{Z}_\beta$ and $\mathbb{Z}_\beta$ such that*

$$f_0 = f, \quad \forall 0 \leq i \leq N \quad f_{i+1} = \frac{f_i - (b_i - a_i)}{\beta} \quad and \quad f_{N+1} = 0.$$

Proof. An element f in F can be written as $f = \sum_{i=0}^{N}(b_i - a_i)\beta^i$ with $x = \sum_{i=0}^{N} a_i \beta^i \in \mathbb{Z}_\beta$, $a_N \cdots a_0$ being admissible for $\mathbb{Z}_\beta$, and $y = \sum_{i=0}^{N} b_i \beta^i \in \mathbb{Z}_\beta - \mathbb{Z}_\beta$, $b_N \cdots b_0$ being admissible for $\mathbb{Z}_\beta - \mathbb{Z}_\beta$. Note that leading 0's are allowed.

With these notations we get for all $0 \leq i \leq N$, $f_i = \sum_{j=0}^{N-i}(b_{j+i} - a_{j+i})\beta^j$ and $f_{N+1} = 0$.   □

Let $V = \left\{ x \in \mathbb{Z}[\beta] \mid |x| < \frac{2\lfloor \beta \rfloor}{\beta - 1}, \text{ and } \forall 2 \leq j \leq d, |x^{(j)}| < \frac{2\lfloor \beta \rfloor}{1 - |\beta^{(j)}|} \right\}$. It is a finite set, with the following property that for all $f \in ((\mathbb{Z}_\beta - \mathbb{Z}_\beta) - \mathbb{Z}_\beta) \cap U'$, the elements $f_0, \ldots, f_N$ of any sequence associated with f according to Lemma 5 belong to V. Indeed, from Lemmas 4 and 5, when $F \subset U'$, for all i, $|b_i - a_i| \leq 2\lfloor \beta \rfloor$. So for $0 \leq i \leq N$ and $2 \leq j \leq d$, the conjugates $f_i^{(j)}$ of f_i satisfy $|f_i^{(j)}| \leq 2\lfloor \beta \rfloor/(1 - |\beta^{(j)}|)$. Moreover the smallest C such that $|x| < C$ implies $|(x - (b - a))/\beta| < C$ is $C = 2\lfloor \beta \rfloor/(\beta - 1)$.

Following [7], we define a directed graph G whose set of vertices is the set V and having an edge $x \xrightarrow{(b,a)} y$ labelled by (b, a) if $y = (x - (b - a))/\beta$.

Lemma 6. *Let $F \subset U'$ be a minimal set satisfying $\mathbb{Z}_\beta - \mathbb{Z}_\beta \subset \mathbb{Z}_\beta + F$. Let V_0 be the subset of V of vertices connected to 0 in G. Then $F \subset V_0$.*

From each vertex f of G which is in U' we look for a path from f to 0 in G which is successful in $\mathcal{A}_{\mathbb{Z}_\beta - \mathbb{Z}_\beta} \times \mathcal{A}_{\mathbb{Z}_\beta}$. Note that in G words are processed least significant digit first, contrarily to the automata for $\mathbb{Z}_\beta$ and $\mathbb{Z}_\beta - \mathbb{Z}_\beta$, where words are processed most significant digit first (*i.e.* from left to right). So we first define an automaton $\mathcal{G}_f$ having as underlying transition graph G with reversed edges, 0 as initial state and f as terminal state. We then compute the intersection automaton $\mathcal{I}_f = (\mathcal{A}_{\mathbb{Z}_\beta - \mathbb{Z}_\beta} \times \mathcal{A}_{\mathbb{Z}_\beta}) \cap \mathcal{G}_f$. The following result then holds true.

Proposition 5. *An element f of U' is in V_0 if and only if the language recognized by $\mathcal{I}_f$ is nonempty.*

Remark 2. The number of states of the automaton $\mathcal{I}_f$ constructed above is $\mathcal{O}\left(K^3 \times |V|\right)$ where K is the number of states of $\mathcal{A}_{\mathbb{Z}_\beta^+}$ and $|V|$ is the number of vertices of G.

6 Minimization of the Cardinality of the Set F

The finite sets $U' \cap V_0$ obtained by the previous construction are not minimal. An element $y \in \mathbb{Z}_\beta - \mathbb{Z}_\beta$ can be close to two different points of $\mathbb{Z}_\beta$, for example such that $x < y < x'$ with $x, x' \in \mathbb{Z}_\beta$ and $y = x + f = x' + f'$ with $f, f' \in U' \cap V_0$.

Theorem 1. *A minimal set $F \subset U' \cap V_0$ can be computed by an algorithm exponential in time and space. It consists in building a transducer which rewrites a representation of an element of $\mathbb{Z}_\beta - \mathbb{Z}_\beta$ into its representation in $\mathbb{Z}_\beta + F$.*

Proof. To find a minimal set $F \subset U' \cap V_0$ we proceed in two steps.

First for each $f \in U' \cap V_0$, we define a deterministic automaton $\mathcal{A}_f$ that recognizes the set of admissible words for $\mathbb{Z}_\beta - \mathbb{Z}_\beta$ that appear as the first component of the labels of the successful paths in $\mathcal{I}_f$. The automaton $\mathcal{A}_f$ is obtained by erasing the second component of the labels (that belongs to $\mathbb{Z}_\beta$) of the edges of $\mathcal{I}_f$ and determinizing the automaton defined in this way. The determinization of automata is based on the so-called subset construction (see [4]), which is exponential in space, and the automaton $\mathcal{A}_f$ has $\mathcal{O}(2^{Q_{\mathcal{I}_f}})$ states.

Next we look amongst all subsets of $U' \cap V_0$ for the smallest set F such that the language recognized by $\cup_{f \in F} \mathcal{A}_f$ contains an admissible representation of each element of $\mathbb{Z}_\beta - \mathbb{Z}_\beta$. To test the inclusion, we compute the complement $\mathcal{C}_F$ of $\cup_{f \in F} \mathcal{A}_f$. Then the language recognized by $\cup_{f \in F} \mathcal{A}_f$ contains an admissible representation of each element of $\mathbb{Z}_\beta - \mathbb{Z}_\beta$ if and only if the intersection of $\mathcal{C}_F$ and $\mathcal{A}_{\mathbb{Z}_\beta - \mathbb{Z}_\beta}$ is empty. Note that the complexity of the search amongst all subsets of $U' \cap V_0$ is exponential in time.

From the set F obtained above, we define a transducer that provides, given $y = \sum_{i=0}^{N} b_i \beta^i \in \mathbb{Z}_\beta - \mathbb{Z}_\beta$ where $b_N \ldots b_0$ is admissible for $\mathbb{Z}_\beta - \mathbb{Z}_\beta$, a decomposition $(a_N \ldots a_0, f)$ where $a_N \ldots a_0$ is admissible for $\mathbb{Z}_\beta$, $f \in F$ and $y = \sum_{i=0}^{N} a_i \beta^i + f$.

Consider the intersection automaton $\mathcal{I}_F = (\mathcal{A}_{\mathbb{Z}_\beta - \mathbb{Z}_\beta} \times \mathcal{A}_{\mathbb{Z}_\beta}) \cap \mathcal{G}_F$ (F is the set of terminal states of $\mathcal{G}_F$). For any element y admissible for $\mathbb{Z}_\beta - \mathbb{Z}_\beta$ there exists $f \in F$ such that y is the first component of the label of a successful path w ending in (s, f) where s is any state of $(\mathcal{A}_{\mathbb{Z}_\beta - \mathbb{Z}_\beta}) \times \mathcal{A}_{\mathbb{Z}_\beta}$ (by construction all states are terminal). Consequently we get $y = x + f$ where x is the second component of the label of the same path w and so is admissible for $\mathbb{Z}_\beta$.

More generally the first component of the labels of the edges in $\mathcal{I}_F$ can be interpreted as the inputs admissible for $\mathbb{Z}_\beta - \mathbb{Z}_\beta$ of the transducer, the second component as the corresponding outputs admissible for $\mathbb{Z}_\beta$. The associated element of F is given by the first component of the label of the state where the path ends.   □

To conclude, the method used here for determining minimal sets F could be generalized to more general Meyer sets related with integral matrices having β as spectral radius.

References

1. S. Akiyama, Self affine tiling and Pisot numeration system, in *Number theory and its applications*, K. Győry and S. Kanemitsu editors, Kluwer (1999) 7–17.
2. A. Bertrand, Développements en base de Pisot et répartition modulo 1, *C. R. Acad. Sc. Paris*, Série A, **285** (1977) 419–421.
3. Č. Burdik, Ch. Frougny, J.-P. Gazeau, R. Krejcar, Beta-integers as natural counting systems for quasicrystal, *J. of Physics A: Math. Gen.* **31** (1998) 6449–6472.
4. S. Eilenberg, *Automata, Languages and Machines*, Vol. A, Academic Press (1974).
5. Ch. Frougny and B. Solomyak, On representation of integers in linear numeration systems, in *Ergodic theory of $\mathbb{Z}^d$ actions* (Warwick, 1993–1994), London Math. Soc. Lecture Note Ser. **228**, Cambridge University Press (1996) 345–368.
6. L. S. Guimond, Z. Masáková, E. Pelantová, Arithmetics on beta-expansions, *Acta Arithmetica*, to appear.
7. K. H. Indlekofer, I. Katai, P. Racsko, Number systems and fractal geometry, in *Probability theory and applications*, Math. appl. **60**, Kluwer Acad. Publ. (1992) 319–334.
8. J. C. Lagarias, Meyer's concept of quasicrystal and quasiregular sets, *Commun. Math. Phys.* **179** (1996) 365–376.
9. J. C. Lagarias, Geometric models for quasicrystals : I. Delone sets of finite type, *Discrete Comput. Geom.* **21** (1999) 161–191.
10. M. Lothaire, *Algebraic combinatorics on words*, Cambridge University Press (2002).
11. Y. Meyer, *Algebraic numbers and harmonic analysis*, North-Holland (1972).
12. Y. Meyer, Quasicrystals, Diophantine approximation and algebraic numbers, in *Beyond Quasicrystals*, F. Axel, D. Gratias (Eds), Les Editions de Physique, Springer (1995).
13. W. Parry, On the β-expansions of real numbers, *Acta Math. Acad. Sci. Hungar.* **11** (1960) 401–416.
14. A. Rényi, Representations for real numbers and their ergodic properties, *Acta Math. Acad. Sci. Hung.* **8** (1957) 477–493.
15. K. Schmidt, On periodic expansions of Pisot numbers and Salem numbers, *Bull. London Math. Soc.* **12** (1980) 269–278.
16. J. Sakarovitch, *Eléments de théorie des automates*, Vuibert (2003).
17. W.P. Thurston, *Groups, tilings, and finite state automata*, Geometry supercomputer project research report GCG1, University of Minnesota (1989).

Efficiently Computing the Density
of Regular Languages*

Manuel Bodirsky[1], Tobias Gärtner[2], Timo von Oertzen[2], and
Jan Schwinghammer[3]

[1] Humboldt-Universität zu Berlin, Germany
[2] Universität des Saarlandes, Germany
[3] University of Sussex, UK

Abstract. A regular language L is called *dense* if the fraction f_m of
words of length m over some fixed signature that are contained in L tends
to one if m tends to infinity. We present an algorithm that computes the
number of accumulation points of (f_m) in polynomial time, if the regular
language L is given by a finite deterministic automaton, and can then
also efficiently check whether L is dense. Deciding whether the least accumulation point of (f_m) is greater than a given rational number, however,
is coNP-complete. If the regular language is given by a *non-deterministic*
automaton, checking whether L is dense becomes PSPACE-hard. We will
formulate these problems as convergence problems of partially observable
Markov chains, and reduce them to combinatorial problems for periodic
sequences of rational numbers.

1 Introduction

In computational logics, the complexity of *almost-sure validity* became a fundamental question to logical formalisms, besides e.g. the complexity of membership
test and validity. Grandjean [9] showed that almost-sure validity for first-order
logic in the finite is PSPACE-complete, whereas validity in the finite is undecidable, by Trakhtenbrot's theorem.

If we are considering formal languages, the corresponding concept to almost-
sure validity is the limit behavior of the density of a language. The density of
a language L over the alphabet Σ is the sequence $(f_m)_{m=0}^{\infty}$ of the fractions of
words of length m in the language, $f_m =_{\mathsf{df}} \frac{|L \cap \Sigma^m|}{|\Sigma|^m}$. The density of regular lan-
guages has already been studied in [2], and the methodology to analyze it using
formal power series is standard by now [11]. It is known that $(f_m)_m$ has finitely
many rational accumulation points [2]. However, to the best of our knowledge
the algorithmic complexity of e.g. computing the number of accumulation points
of (f_m) has not yet been discussed. We show that the computation of $\liminf f_m$
is coNP-hard, whereas $\lim f_m$ can be computed in time $O(n^3)$, if n is the size of

* Supported by the Deutsche Forschungsgemeinschaft (DFG) within the European
  graduate program 'Combinatorics, Geometry, and Computation' (No. GRK 588/2)

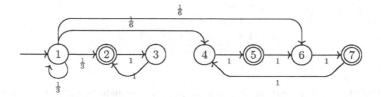

Fig. 1. Periodic and reducible Markov chain where the probability that the system is in the set $\{2, 5, 7\}$ converges to $\frac{1}{2}$.

the deterministic automaton accepting L. If the language is given by a nondeterministic automaton, the problem to decide whether the accepted language is dense becomes PSPACE-hard.

Partially observable Markov chains. The density problem for a deterministic automaton can be translated into a convergence problem for Markov chains. We view f_m as the probability that a word of length m, chosen uniformly at random, leads to an accepting state in the given finite deterministic automaton. The automaton can then be considered as the state space of a finite Markov chain, having transitions with probability $\frac{1}{|\Sigma|}$ for every labeled edge in the finite deterministic automaton. The accepting states of the automaton are the so-called set of observable states in the Markov chain. We are interested in the probability that the system is in the far future in this *observable* set of states.

The specification and verification of long-run average properties of probabilistic systems was recently studied by de Alfaro [6]. De Alfaro also presents efficient algorithms for model checking these long-time average properties using stable-state distributions of Markov chains. However, in general the Markov chain for the automaton is not *aperiodic*, and we are not interested in the *average* behaviour, but rather in the probability of a property of the system at some specific time point in the far future. There might be a limit probability, even though there is no stable-state distribution (see Fig. 1).

We show that computing the minimal or maximal accumulation point of (f_m) is coNP-complete. This means that we cannot expect to find an efficient algorithm that computes the minimum probability that a system has a certain property after a long run. However, it is possible to compute all the accumulation points in time polynomial in their number. In particular, if a property has a limit probability, i.e. if there is only one accumulation point, we will present an efficient algorithm to determine its value. Another problem will be the computation of the *number* of accumulation points.

Periodic Sequences of Rational Numbers. We will reduce the probabilistic problems above to equivalent combinatorial problems for periodic sequences. A *periodic sequence* over some field X is an infinite sequence $(\alpha[m])_{m=0}^{\infty}$ of elements in X such that there is an integer $p > 0$ so that $\alpha[m] = \alpha[m+p]$ for all $m \geq 0$. The least such integer is called the *period* $|\alpha|$ of α. If we add two sequences α and β

$$
\begin{array}{r}
1\ 1\ 3\ 0\ 2\ 2\ .\ .\ . \\
+\ 2\ 4\ 4\ 6\ 6\ 3\ 3\ 5\ 5\ 7\ .\ .\ . \\
+\ 6\ 4\ 2\ 3\ 1\ 4\ 5\ 3\ 1\ 2\ 5\ 3\ 4\ 2\ 0\ .\ .\ . \\
\hline
=\ 9\ 9\ 9\ 9\ 9\ 9\ 9\ 9\ 9\ 9\ 9\ 9\ 9\ 9\ 9\ .\ .\ .
\end{array}
$$

Fig. 2. A visualization of a sum of periodic sequences with period one.

componentwise, the result is obviously again a periodic sequence, and the period is at most $\mathsf{lcm}(|\alpha|, |\beta|)$. But sometimes a set of periodic sequences adds up to a sequence with a shorter period. Consider for example the sequences in Figure 2. The largest possible length of the sum of the sequences is $\mathsf{lcm}(6, 10, 15) = 30$, but in fact their sum has period one. We will investigate how to compute the sum of a set of periodic sequences without evaluating a possibly exponential number of entries in the sum.

2   Preliminaries

The long run behavior of the density of a deterministic finite automaton can be seen as a probabilistic process. If the regular language is given by a nondeterministic automaton, we first have to determinize the automaton, which might lead to an exponential blow-up of the size of the automaton. In fact, in this case the problem whether the language has a limit density becomes hard for PSPACE.

Proposition 1. *The problem to decide for a given nondeterministic finite automaton whether it accepts a dense language is PSPACE-hard.*

Proof. (sketch) We can adapt the classical proof showing that the non-universality problem for nondeterministic finite automata is PSPACE-hard [1]: Let M be a Turing Machine M accepting a language in polynomial space and let x an input. We can encode machine configurations (i.e., tape contents, state and head position) as words w, and computations of M as words $\#w_1\# \cdots \#w_k\#$. We can construct a (nondeterministic) automaton of size polynomial in M and x that rejects exactly the words $\#w_1\# \cdots \#w_k\#v$ where $w_1, \ldots, w_k$ represents an accepting computation of M on x and v is any word.

Clearly if there is such an accepting computation of M on x and n is the length of its representation, then the density of the language accepted by the automaton is at most $1 - 1/n < 1$. Conversely, this language is dense (in fact, universal) if x is not accepted by M.   □

To deal with the deterministic case, we recall in this section some notions common in the Markov chain literature.

Definition 1. *A* partially-observable Markov chain (POM) *can be described by a tuple* (V, A, s_0, O) *consisting of a finite set V of states and a function $A : V^2 \to$*

$[0, 1]$ *specifying* transition probabilities[1], *i.e. we have that* $\sum_{u \in V} A(u, v) = 1$ *for all* $v \in V$. *The* $|V|$-*dimensional vector* s_0 *is called the* initial distribution. *The set* $O \subseteq V$ *denotes the set of* observable states.

If we identify V with $\{1, \ldots, |V|\}$, the transition function A can be seen as a $|V| \times |V|$-matrix of rational numbers. This matrix $A = (a_{ij})_{i,j \in V}$ determines a directed weighted quasi graph, the *transition graph*, where there is an edge from v to u if $a_{uv} \neq 0$. We will freely use graph theoretic notions, and call a POM *strongly connected* (or *irreducible*), if its transition graph is. Strongly connected components of the transition graph with no outgoing edges are called *terminal components*. For simplicity we will identify a POM with its transition matrix or its transition graph, if initial distribution and labeling are clear from the context.

The *periodicity* of a strongly connected component in a POM is the greatest common divisor of the length of all the cycles in the underlying transition graph. The periodicity of a POM is the least common multiple of the periodicities of its terminal components. A POM is called *aperiodic* if its periodicity is one. An aperiodic and irreducible POM is called *ergodic*. We can draw POMs as graphs like in Figure 1, where we can see a transition graph of periodicity four.

A *distribution* s is a $|V|$-dimensional vector of numbers from $[0, 1]$. We denote the i-th component of this vector by $(s)_i$. A *run* of a POM is an (infinite) sequence $(s_m)_{m=0}^{\infty}$ of distributions, where s_0 is the initial distribution, and s_i is defined to be As_{i-1}, for $i \geq 1$. A *stable-state* distribution s is a distribution such that $As = s$.

For POMs we are not interested in stable-state distributions, but in the long-run behaviour of the sequence of probabilities $(f_m)_{m=0}^{\infty}$ that the system is in the set of observable states, where $f_m := \sum_{v \in O}(s_m)_v$. These are the problems for POMs we are investigating:

1. Check whether (f_m) converges.
2. Determine the *minimal* accumulation point of (f_m).
3. Determine the *number* of accumulation points of (f_m).
4. Determine the accumulation points of (f_m).

Convergence of aperiodic and irreducible Markov chains reduces to finding a stable state distribution of the Markov chain (see e.g. [3]):

Theorem 1 (Basic Limit Theorem). *Let A be an aperiodic and irreducible POM. Then $\lim_{m \to \infty} A^m s$ exists for all initial distributions s, and is independent of s.*

Moreover, we can efficiently find this limit distribution by finding the eigenvector to the eigenvalue 1 of A, i.e. solving a linear equation system. Since the POMs

---

[1] Note that we assume that the transition probabilities are real numbers. When analyzing the running time of algorithms dealing with POMs we will only count the number of additions and multiplications of field elements that we have to perform. For the application to densities of regular languages it suffices to represent the probabilities by rational numbers, and thus we will separately mention how to deal with rational numbers.

```
COMPUTE-LIMIT(B, s)
1   Compute terminal components B₁, ..., Bq.
2   for v : v ∉ B₁ ∪ ··· ∪ Bq
3   do reduce self loops at v
4       reduce edges to v.
5   for i = 0, ..., q
6   do λᵢ ← ∑ᵥ∈V, u∈Bᵢ (s)ᵥ Aᵥᵤ
7       bᵢ ← Eigenvector to eigenvalue 1 of
8          the transition matrix of Bᵢ.
9   return the vector ∑q_{i=1} λᵢb̄ᵢ .
```

Fig. 3. Computing the density of an aperiodic POM. This procedure is used by the algorithm in Figure 4. The sub-procedures *reduce self loops* and *reduce edges*, and correctness proofs can be found at [4].

```
COMPUTE-PERIOD(A, s)
1   Compute subchains A₁, ..., Ap
2   induced by the terminal components
3   of periodicity l₁, ..., lp.
4   for i = 0, ..., p;   j = 0, ..., lᵢ − 1
5   do αᵢ[j] ← ∑ᵥ∈O (COMPUTE-LIMIT(A^{lᵢ}, A^j s))ᵥ.
6   return α₁, ..., αp.
```

Fig. 4. The reduction of the convergence problems to period problems, calling the procedure COMPUTE-LIMIT of Figure 3. Periodicities of directed graphs are easy to compute (see, e.g., [10]).

considered in this paper are in general neither aperiodic nor strongly connected, we cannot apply this theorem directly.

3   Reducing the Problem

In this section show how to reduce the convergence problems mentioned in Section 2 to combinatorial problems of periodic sequences. The facts of this section are all essentially known [7], but we state them to emphasize their algorithmic aspects.

Let (V, A, s, O) be a POM, and suppose we are interested in the sequence of probabilities f_m that the system at time point m is in the set of observed states O. Only states in terminal components can contribute to the value of the accumulation points of f_m (see [7]), since the probability that the POM is in any other state converges to zero.

The main idea is to analyze each terminal component separately, computing the periodic contribution of every terminal component to the probabilities f_m. To this end we introduce the notion of a *subchain* of a POM: Given a set of states S, we replace all the outgoing edges of states that do not have a path into S by a self-loop.

Definition 2. *Let (V, A, s, O) be a POM, and $S \subseteq V$ be a set of states. Then we will define the* subchain (V, B, s, O) *of A induced by S. The transition function of B is defined for all $u, v \in V$ by*

$$
B(u, v) := \begin{cases} A(u, v) & \text{if there is a path in } A \text{ from } u \text{ into } S \\ 1 & \text{if } u = v, \text{ and there is no path in } A \text{ from } u \text{ into } S \\ 0 & \text{otherwise} \end{cases}
$$

Obviously, an induced subchain is a POMas well. Let $l_1, \ldots, l_p$ be the periodicity of the subchains $A_1, \ldots, A_p$ induced by the terminal components. Thus the periodicity of the POM is $l := \mathsf{lcm}(l_1, \ldots, l_p)$.

Now for each of these subchains A_i and for $0 \le j < l_i$, we define

$$
\alpha_i[j] := \lim_{m \to \infty} \sum_{v \in O} (A_i^{ml_i + j} s)_v \tag{1}
$$

As we will see in the next proposition, these limits exist and can be computed. The proposition states that the global accumulation points can be computed using the periodic contributions of the subchains induced by the terminal components:

Proposition 2. *Let A be a POM, and $A_1, \ldots, A_p$ the subchains induced by the terminal components $S_1, \ldots, S_p$. Let l_i denote the periodicity of A_i, and $l := \mathsf{lcm}(l_1, \ldots, l_p)$ the periodicity of A. Then for every $v \in S_i$ and every initial distribution s the following limits exists and can be computed, and we have:*

$$
\lim_{m \to \infty} (A^{lm} s)_v = \lim_{m \to \infty} (A_i^{l_i m} s)_v \tag{2}
$$

In particular, $\lim_{m \to \infty} f_{ml+j} = \sum_{1 \le i \le p} \alpha_i[j]$. A proof and an algorithm for computing the α_i can be found in the full version of the paper at [4].

4   Periodic Sequences of Field Elements

In the previous sections we saw how to compute certain characteristic periodic sequences of rational numbers that describe the long run behaviour of a POM with respect to the sequence of probabilities (f_m) that the system is in an observed state. If we want to know whether this probability converges, it suffices to check whether all accumulation points are equal. In this section we present a polynomial time algorithm that avoids to check in a brute-force way exponentially many different entries in the periodic sequence of the sum.

A sequence $(\alpha[i])_{i=0}^{\infty}$ of elements over some set X is *periodic* if there is an integer $p > 0$ so that $\alpha[m] = \alpha[m + p]$ for all $m \ge 0$. The least such integer is called the *period* $|\alpha|$ of α. Here we are interested in algorithmic problems for periodic sequences over real (or rational) numbers, and thus we assume that X

is a field. In Section 2 we asked several questions concerning the convergence of POMs. By the reduction of the previous section they correspond to the following problems for periodic sequences of integers $\alpha_1, \ldots, \alpha_k$, where a periodic sequence α is given by a finite sequence $\alpha[0], \ldots, \alpha[|\alpha| - 1]$ of elements in X:

Let β be the sequence defined by $\beta[j] := \sum_{i=1}^{k} \alpha_i[j]$;

1. Check whether $|\beta| = 1$.
2. Determine the minimal element $\min\{\beta[i] \mid 0 \leq i < |\beta|\}$ of the periodic sequence β.
3. Determine the length $|\beta|$ of the sequence β.
4. Determine the entries of the periodic sequence β.

These problems are in fact polynomial time equivalent to the corresponding problems for POMs: Assume we are given a set of periodic integer sequences. It is then easy to specify a POM and an observation set such that the respective convergence problem leads to the corresponding period sum problem.

For Problem 4, we have to measure the complexity of an algorithm in both n as above and $m := |\beta|$, because in this case the size of the output β itself might be exponential. The second problem turns out to be hard:

Proposition 3. *The problem to determine whether the minimal element in the sum of given periodic sequences is greater or equal than a given value k is coNP-complete.*

This can be proven by reduction of the complement of the NP-complete problem *simultaneous inequalities* [8], which stays hard even if the numbers of the instance are represented in unary (by inspection of the NP-hardness proof given in [12]). A proof can be found in the full version of the paper available at [4]. In the next section we will show that there is an efficient algorithm for Problem 1, 3 and 4.

5   An Efficient Algorithm for Periods over the Rationals

The main idea of the algorithms for the period sum problems presented here is to represent a periodic sequence α as the power series $p_\alpha(X)$ defined by $\sum_{i=1}^{\infty} \alpha[i - 1]X^{-i}$. Let $l := |\alpha|$; it is easy to verify the following closed representation of this power series:

$$p_\alpha(X) = \frac{\sum_{i=0}^{l-1} \alpha[i] X^{l-i-1}}{X^l - 1}.$$

Given any fraction of polynomials such that the denominator divides $X^l - 1$, this fraction can be expanded to such a representation of a periodic sequence. The sequence can then be determined by dividing the numerator by the denominator with a polynomial division; this can easily be verified for a denominator $X^l - 1$ and must therefore hold for any divisor of $X^l - 1$, because the result of a polynomial division is invariant under expansion and cancelation of the fraction.

For the period problems we are given the sequences $\alpha_1, \ldots, \alpha_k$, and want to analyze their sum. Adding up the fractions $p_{\alpha_1}(X), \ldots, p_{\alpha_k}(X)$ yields a fraction representation $\frac{u}{v}$ of the sum of the sequences. We would like to compute the potentially exponential period of $\frac{u}{v}$ without actually computing the entries of the periodic sequence. For the denominator v, we first compute the least common multiple of the denominators of all summands; since these are of the form $X^l - 1$, their zeroes are exactly all l-th roots of unity. We can therefore represent v by the list of its roots of unity. The entries of the list are stored as fractions $\frac{p_j}{q_j}$ such that $z_j = \exp(\frac{p_j}{q_j} 2\pi i)$ is the j-th root of unity in the list. For each z_j in the list, we test whether u also evaluates to 0 at z_j. If so, z_j can be canceled, and we eliminate it from the list.

We would like to compute the period of the resulting representation, i.e. the minimal m such that the fraction has denominator $X^m - 1$. Since every remaining z_j in the list is an l-th root for every multiple l of q_j, it suffices to find the minimal m that is a multiple of every q_j. Thus the period m is the smallest common multiple of the q_j.

If we are given periodic sequences of rational numbers, testing whether the numerator is zero at a root of unity can be done numerically. We first calculate the greatest common divisor of the numerator and the denominator, which is guaranteed to have only roots of unity as zeros. The minimal distance between two roots of unity is 2π times the distance of their representing fraction, which is limited by the inverse of the input size, and we can therefore test whether the polynomial contains a certain root of unity with a linear number of bits of precision. To approximate the values at the roots of unity up to n bits we need $O(n^2 \log n)$ time.

To actually compute the entries of the sequence sum (Problem 4) we perform the division of $\frac{u}{v}$ step by step, and stop after m steps.

Proposition 4. *Let* $n := |\alpha_1| + \cdots + |\alpha_k|$ *be the size of a set of periodic integer sequences, and* $m := |\alpha_1| \cdots + \alpha_k|$ *the length of their sum. Then the problem to calculate* m *is in* $O(n^2 \log n \log \log n)$. *Computing the entries of the sum takes* $O(n^2 \log n \log \log n + m)$ *operations. In particular, we can check in* $O(n^2 \log n \log \log n)$ *whether the sum has period one.*

Proof. The dominating step with respect to the input size n is the reduction of the at most n fractions to higher terms before adding them: Assuming an $n \log n \log \log n$ multiplication algorithm (see e.g. [5]), the algorithm runs within $O(n^2 \log n \log \log n)$. For large m, the performance of the division is the bottleneck, requiring m operations. $\qquad\square$

6   Conclusion

We reduced the problem to determine the limit behaviour of regular languages to convergence problems for partially observable Markov chains. In this more general setting we reduced the problem to combinatorial period problems over fields that can be solved efficiently using power series representations. It is possible to efficiently compute the potentially exponential number of accumulation points of the density of a regular language given by a deterministic automaton.

Moreover, we presented an algorithm that computes the accumulation points in time polynomial in the input and output. The overall running time of the algorithms for the tractable cases is dominated by the reduction to period problems over rational numbers, which involves the solution of a linear equation system.

If the language is given by a nondeterministic algorithm, we proved that the density problem is PSPACE-hard; we do not know whether it is PSPACE-complete. We would also like to know the computational complexity of checking whether a context free language, given by a generating grammar, is dense, i.e., its density converges to one.

References

1. A. V. Aho, J. E. Hopcroft, and J. D. Ullman. *The Design and analysis of algorithms*. Addison-Wesley, 1974.
2. J. Berstel. Sur la densité asymptotique de langages formels. *ICALP*, pages 345–358, 1972.
3. R. N. Bhattacharya and E. C. Waymire. *Stochastic processes with applications*. Wiley Series in Probability and Mathematical Statistics, New York, 1990.
4. M. Bodirsky, T. Gärtner, T. von Oertzen, and J. Schwinghammer. Efficiently computing the density of regular languages. Full version, available under http://www.informatik.hu-berlin.de/~bodirsky/publications.
5. P. Bürgisser, M. Clausen, and M. Shokrollahi. *Algebraic Complexity Theory*. Springer Verlag, 1997.
6. L. de Alfaro. How to specify and verify the long-run average behavior of probabilistic systems. In *Proc. 13th IEEE Symp. on Logic in Computer Science*, IEEE Computer Society Press, 1998.
7. F. R. Gantmacher. *The Theory of Matrices*. Chelsea Pub. Co., 1977.
8. M. Garey and D. Johnson. *A Guide to NP-completeness*. CSLI Press, 1978.
9. E. Grandjean. Complexity of the first-order theory of almost all finite structures. *Information and Control*, 57:180–204, 1983.
10. K. Mehlhorn and S. Näher. *LEDA. A platform for combinatorial and geometric computing*. Cambridge University Press, Cambridge, 1999.
11. A. Salomaa and M. Soittola. *Automata-Theoretic Aspects of Formal Power Series*. Springer-Verlag, 1978.
12. L. J. Stockmeyer and A. Meyer. Word problems requiring exponential time. In *Proc. 5th Ann. ACM Sypm. on Theory of Computing*, number 1–9 in Association of Computing Machinery, 1972.

Longest Repeats with a Block of Don't Cares

Maxime Crochemore[1,2]*, Costas S. Iliopoulos[2]**, Manal Mohamed[2]***, and
Marie-France Sagot[3]†

[1] Institut Gaspard-Monge, University of Marne-la-Vallée,
77454 Marne-la-Vallée CEDEX 2, France
maxime.crochemore@univ-mlv.fr
[2] Department of Computer Science, King's College London
London WC2R 2LS, England
mac,csi,manal@dcs.kcl.ac.uk
[3] Inria Rhône-Alpes, Laboratoire de Biométrie et Biologie Évolutive,
Université Claude Bernard, 69622 Villeurbanne cedex, France
Marie-France.Sagot@inria.fr

Abstract. We introduce an algorithm for extracting all longest repeats
with k don't cares from a given sequence. Such repeats are composed of
two parts separated by a block of k don't care symbols. The algorithm
uses suffix trees to fulfill this task and relies on the ability to answer the
lowest common ancestor queries in constant time. It requires $O(n \log n)$
time in the worst-case.

Keywords: Combinatorial Problems, String, Repeat Extraction, Don't
Care, Suffix Tree, Lowest Common Ancestor, Efficient Merging.

1 Introduction

In recent years, many combinatorial problems that originate in bioinformatics
have been studied. Here we consider a combinatorial problem on motifs. The
term *motif* [5] is often used in biology to describe similar functional components
that several biological sequences have in common. It can also be used to describe
any collection of similar substrings of a longer sequence. In nature, many motifs
are *composite*, *i.e.* they are composed of conserved parts separated by random
regions of variable lengths.

In this paper we explore a sub-problem that is important in the approach
to the combinatorics and the complexity of the original biologically motivated
topic. Thus, we concentrate on finding all *longest repeats with a block of k don't
cares*. Such repeats consist of two exact parts separated by a gap of fixed length
k. Hence, our aim is to find all such repeats and their positions in the string.

* Partially supported by CNRS, Wellcome Foundation, and Nato grants.
** Partially supported by a Marie Curie fellowship, Wellcome Foundation, Nato and
Royal Society grants.
*** Supported by an EPSRC studentship
† Partially supported by French Programme BioInformatique Inter EPST, Wellcome
Foundation, Royal Society and Nato grants.

M. Farach-Colton (Ed.): LATIN 2004, LNCS 2976, pp. 271–278, 2004.
© Springer-Verlag Berlin Heidelberg 2004

A closely related problem was studied by Brodal *et al.* [2]. They developed algorithms for finding all "maximal pairs with bounded gap". This notion refers to a non extendable substring having two occurrences within a bounded distance of each other. A restricted version of the same problem was considered by Kolpakov and Kucherov [7]. They proposed an algorithm for a fixed gap. The problem of finding longest repeats with no don't cares is a mere application of suffix trees [5].

In our method we use two suffix trees intensively, one for the original string and the other for its reverse. The use of a generalized suffix tree (for both the string and its reverse) would be possible but is not necessary because we do not need all the information it contains. We have not yet explored the possibility of using an affix tree [9] but there are some doubt that it will lead to a significant improvement on the asymptotic time complexity.

The paper is organized as follows: in Section 2, we state the preliminaries used throughout the paper. In Section 3, we define the longest repeat with k don't cares and describe in general how to find them using two suffix trees. In Section 4, we detail our algorithm. Finally in Section 5, we analyze the running time of the algorithm.

2   Preliminaries

Throughout the paper x denotes a *string* of length n defined on a finite alphabet Σ. We use $x[i]$, for $i = 1, 2, \ldots, n$, to denote the i-th letter of x, and $x[i..j]$ as a notation for the *substring* $x[i]x[i+1]\cdots x[j]$ of x. The string $\overleftarrow{x}$ denotes the *reverse* of x, such that $\overleftarrow{x}[1] = x[n], \ldots, \overleftarrow{x}[n] = x[1]$.

The *length* of a string w is denoted by $|w|$. If $w = uv$ then w is said to be the *concatenation* of the two strings u and v. The string w^k is the k-th power of w.

A symbol '$\diamond$' $\notin \Sigma$ is called a "don't care"; any other symbol is called *solid*. A don't care *matches* any other symbol, that is, $\diamond = \sigma$ for each $\sigma \in \Sigma \cup \{\diamond\}$. A pattern y over $\Sigma \cup \{\diamond\}$ is said to *occur* in x at position i if $y[j] = x[i + j - 1]$, for $1 \leq j \leq |y|$. A *motif* w denotes a pattern that occurs at least twice in x. We restrict the motifs to have a solid symbol at both ends, i.e., $w[1] \neq \diamond$ and $w[|w|] \neq \diamond$. The set $\mathcal{L}_w$ is the set of occurrence positions of a given motif w, where $\mathcal{L}_w = \{x[i..i + |w| - 1], 1 \leq i \leq n - |w| + 1\}$. Observe that $|\mathcal{L}_w| \geq 2$.

For a given string x and an integer k, a motif w of the form $L \diamond^k R$ is called *repeat with k don't cares*. The substrings L and R, respectively, are the left and right parts of w. The length of the longest such repeat in x is denoted by $lr_k(x)$. Later on, we use the following notion: a motif w is called *left maximal* (resp. *right maximal*) if w can not be extended to the left (resp. right) without losing one of its occurrences.

Here we present a method for finding all longest repeats with k contiguous don't cares and their positions. This method uses the suffix tree of x as a fundamental data structure. A complete description of suffix trees is beyond the scope of this paper, and can be found in [5] or [4]. However, for the sake of completeness, we will briefly review the notion.

Definition 1 (Suffix tree). *The suffix tree* $\mathcal{T}(x)$ *of the string* x *is the compacted trie of all suffixes of* $x\$$, *where* $\$ \notin \Sigma$. *Each leaf in* $\mathcal{T}(x)$ *represents a suffix* $x[i..n]$ *of* x *and is labelled with the index* i. *We refer to the set of indices stored at the leaves of the subtree rooted at node* v *as the leaf-list of* v; *it is denoted by* $LL(v)$. *Each edge in* $\mathcal{T}(x)$ *is labelled with a nonempty substring of* x *such that the path from the root to the leaf labelled with index* i *spells the suffix* $x[i..n]$. *We refer to the substring of* x *spelled by the path from the root to a node* v *as the label of* v, *and denote it by* ℓ_v. *The length of such a substring is the depth of* v *and we denote it by* d_v.

Several algorithms construct the suffix tree $\mathcal{T}(x)$ in $O(n)$ time, assuming an alphabet of fixed size (see for example [4] [5]). All the internal nodes in $\mathcal{T}(x)$ have an out-degree between 2 and $|\Sigma|$. Therefore, we can transform the suffix tree into a binary suffix tree $\mathcal{B}(x)$ by replacing every node v in $\mathcal{T}(x)$ with out-degree $d > 2$ by a binary tree with $d-1$ internal nodes and $d-2$ internal edges, where the d leaves are the d children of v. Since $\mathcal{T}(x)$ has n leaves, constructing the binary suffix tree $\mathcal{B}(x)$ requires adding at most $n-2$ new nodes. Each new node can be added in constant time. This implies that the binary suffix tree $\mathcal{B}(x)$ can be constructed in $O(n)$ time.

Our method makes use of the Schieber and Vishkin [8] *Lowest Common Ancestor* algorithm. For a given rooted tree T, the *lowest common ancestor* of two nodes u and v, $lca(u, v)$, is the deepest node in T that is ancestor of both u and v. After a linear-time preprocessing of a rooted tree, the lowest common ancestor of any two nodes can be found in constant time.

3   Longest Repeats with k Don't Cares

The *longest repeats with k don't cares* problem requires finding all longest repeats of the form $L \diamond^k R$, that appear in a given string x. In the notation, L and R are both over Σ and represent the left and the right parts, respectively, of the repeat. The parameter k is a given positive integer smaller than n. For example, if

$$x = BBAZYABAAAXBBAXZABAZAHIABAA$$

then the only longest repeat with 2 don't cares is $w = BBA \diamond \diamond ABA$ and its occurrence list is $\mathcal{L}_w\{1, 12\}$. Thus, $lr_2(x) = 8$. An obvious approach to solve this problem is as follows:

1. generate all possible repeated substrings in x;
2. for each pair of repeated substrings u and v, check whether there exist at least two pairs of occurrence positions i_1 and i_2 of u and j_1 and j_2 of v such that $j_l = i_l + |u| + k$;
3. calculate the length of the repeat with k don't cares $u \diamond^k v$;
4. report all longest ones.

This straightforward approach can be improved by dynamic programming yielding an $O(n^2)$ time algorithm.

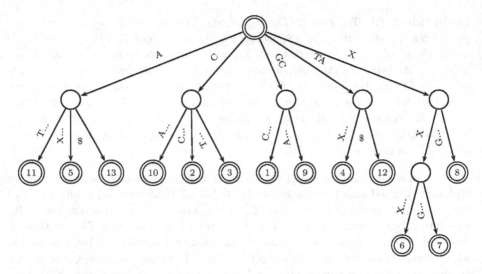

Fig. 1. The suffix tree of GCCTAXXXGCATA.

Our approach proceeds differently and results in an $O(n \log n)$ time algorithm. It starts by constructing the two suffix trees $\mathcal{T}(x)$ and $\mathcal{T}(\overleftarrow{x})$. The first suffix tree is used to generate the right part of the repeat, while the second suffix tree generates the left part. Observe that the label ℓ_u of each internal node $u \in \mathcal{T}(x)$ represents a right-maximal repeated substring of x which occurs at $LL(u)$. Similarly, the label ℓ_v for each internal node $v \in \mathcal{T}(\overleftarrow{x})$ represents the reverse of a left-maximal repeated substring of x ending at positions $\{j \mid j = n + 1 - i, i \in LL(v)\}$.

For simplicity, we replace each index i in $\mathcal{T}(\overleftarrow{x})$ by $n+1-i+(k+1)$. Our goal now, is to traverse both trees efficiently to find all pairs of nodes u and v where $u \in \mathcal{T}(x)$, $v \in \mathcal{T}(\overleftarrow{x})$, $|LL(u) \cap LL(v)| \geq 2$, and $d_u + d_v$ is maximum. For each pair u and v, the concatenation of the reverse of the label of v, k don't cares, and the label of u gives a longest repeat with k don't cares, i.e., $w = \overleftarrow{\ell_v} \diamond^k \ell_u$. Observe that, $lr_k = d_v + k + d_u$.

For example, if $x = $ GCCTAXXXGCATA and $k = 1$, then Fig. 1 and Fig. 2 represent the suffix trees of, respectively, x and $\overleftarrow{x}$. Note that, each index i in $\mathcal{T}(\overleftarrow{x})$ has been replaced by $16 - i$. The node in $\mathcal{T}(x)$ labelled by TA and node in $\mathcal{T}(\overleftarrow{x})$ labelled by CG both have leaf-list $\{4, 12\}$. Thus, $GC \diamond TA$ is a repeat with 1 don't care. Since it is the longest such repeat, $lr_1(x)$ equals 5. The list of occurrence positions of the longest repeat with one don't care of x is $\{1, 9\}$.

4   Algorithm

The initialization phase of the algorithm consists of two main steps. In the first step, the suffix tree of x is constructed and then traversed in a preorder manner

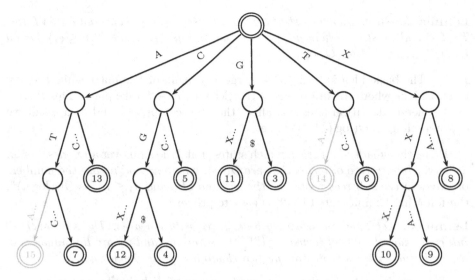

Fig. 2. The suffix tree of `ATACGXXXATCCG`. Each index i is replaced by $16 - i$. The gray nodes may be omitted.

where a number is assigned to each node. For each index i, $no(i)$ is the preorder number assigned to the leaf node v labelled with i in $\mathcal{T}(x)$. This is done during the tree depth-first traversal. For example, if $\mathcal{T}(x)$ is the tree of Fig. 1 then

i	1	2	3	4	5	6	7	8	9	10	11	12	13
$no(i)$	11	8	9	14	4	18	19	20	12	7	3	15	5

In the second step, the suffix tree of $\overleftarrow{x}$ is built. In addition, a list is associated with each leaf node v. For each leaf node v labelled with the i-th suffix of $\overleftarrow{x}$, this list is initialized with the element $no(n + 1 - i + (k - 1))$.

For each internal node v, the list is the sorted union of the disjoint lists of the children of v. The computation of the lists for the internal nodes can be done during a depth-first traversal of the tree. However, in order to guarantee an efficient merge of the lists associated with the children of a node, $\mathcal{T}(\overleftarrow{x})$ is transformed into a binary suffix tree $\mathcal{B}(\overleftarrow{x})$. Furthermore, to maintain these lists efficiently, these lists were implemented using AVL-trees [1]. Although this implementation is similar to the one used in [2] and [6], any other type of balanced search trees may be used. Note that the efficient merging of two AVL trees is essential to our method. The results on the merge operations of two height-balanced trees stated in [3] are summarized in the following lemmas.

Lemma 1. *Two AVL trees of size at most n and m can be merged in time* $O(\log \binom{n+m}{n})$.

Lemma 2. *Given a sorted list of elements $e_1 \leq e_2 \leq \cdots \leq e_n$, and an AVL tree T of size at most m, where $m \geq n$, we can find $q_i = max\{x \in T | x \leq e_i\}$ for all $i = 1, 2, \ldots, n$ in time $O(\log \binom{n+m}{n})$.*

Proof. The basic idea is to use the merge algorithm of Lemma 1 while keeping the positions where the insertions of the elements $e_i \in T$ take place. This change in the merge algorithm does not affect the time complexity and as a result we can find all q_i in $O(\log \binom{n+m}{n})$ time.

Using the *smaller-half trick*, which states that *"the sum over all nodes v of an arbitrary binary tree of terms that are $O(n_1)$, where n_1 and n_2 are the numbers of leaves in the subtrees rooted at the children of v and $n_1 \leq n_2$, is $O(n \log n)$"*, the following lemma stated in [2] is easy to prove:

Lemma 3. *Let T be an arbitrary binary tree with n leaves. The sum over all internal nodes v in T of terms $\log \binom{n_1+n_2}{n_1}$, where n_1 and n_2 are the numbers of leaves in the subtrees rooted at the two children of v, is $O(n \log n)$.*

The algorithm for finding all longest repeats with k don't cares is given in Fig. 3. Recall that at every node v in $\mathcal{B}(\overleftarrow{x})$ we construct a sorted list, stored in an AVL tree $\mathcal{A}$, of all the preorder numbers associated with the elements in $LL(v)$. This list can be considered as a leaf-list sorted according to the preorder numbers associated to the indices in $\mathcal{T}(x)$. If v is a leaf, then $\mathcal{A}$ is constructed directly (Line 5). If v is an internal node, then $\mathcal{A}$ is constructed by merging $\mathcal{A}_1$ and $\mathcal{A}_2$ (Line 27), where $\mathcal{A}_1$ and $\mathcal{A}_2$ are the AVL trees associated with the two children of v and $|\mathcal{A}_1| \leq |\mathcal{A}_2|$. Before constructing $\mathcal{A}$, we use $\mathcal{A}_1$ and $\mathcal{A}_2$ to check for an occurrence of longest repeat with k don't cares. If a number a in $\mathcal{A}_1$ is going to be inserted between b and c in $\mathcal{A}_2$, then b and c are efficiently obtained (Lemma 2). Let max be the length of the current longest repeat with k don't cares. And let u and v be the nodes representing this longest repeat, where $u \in \mathcal{T}(x)$ and $v \in \mathcal{B}(\overleftarrow{x})$. Since we are moving upward in $\mathcal{B}(\overleftarrow{x})$ minimizing the depth of v, the only way to find a longer repeat with k don't cares is by replacing node u in $\mathcal{T}(x)$ with a node that has greater depth. Clearly, this node should be a lowest common ancestor of a pair of nodes that has not been considered so far, *i.e.* a pair consisting of an element in $\mathcal{A}_1$ and an element in $\mathcal{A}_2$. It follows from Lemma 4, that we do not need to consider all the possible new pairs. In other words, only the pairs of the form (a,b) or (a,c) are the ones that need to be considered. For each pair of nodes considered by the algorithm, the algorithm checks whether the sum of the depth of both nodes is greater than or equal to max. If so, the algorithm uses list M to store the pair. Note that the longest repeat with k don't cares may not be unique. So, each pair (x, y) in M represents a longest repeat obtained by a concatenation of $\overleftarrow{\ell_y}$, k don't cares, and ℓ_x. Where $lr_k(x)$ equals $d_x + k + d_y$ for all pairs $(x, y) \in M$.

Lemma 4. *Let i, j and k be the preorder numbers given to three leaves u, v and w during a preorder traversal of a rooted tree T. If $i < j < k$, then the depth of $lca(u, v)$ cannot be less than the depth of $lca(u, w)$, where lca is the lowest common ancestor of two nodes.*

Algorithm *Longest-Repeat-Don't-Cares*(x, k)
Input: A string x of length n
Output: All longest repeats with k contiguous don't cares

1.   Build the suffix tree $\mathcal{T}(x)$ and traverse the tree in preorder manner numbering all the nodes.
2.   **for** each leaf $v \in \mathcal{T}(x)$
3.   **if** v is labelled with i
4.   **then** $no(i) \leftarrow$ the preorder number of v
5.   Build the binary suffix tree $\mathcal{B}(\overleftarrow{x})$ and create at each leaf an AVL tree of size one that stores $no(n + 1 - i + (k + 1))$, where i is the index associated with the leaf.
6. $(max, u, v) \leftarrow (0, root(\mathcal{T}(x)), root(\mathcal{B}(\overleftarrow{x})))$
7. $M \leftarrow \emptyset$
8.   **for** each node $v \in \mathcal{B}(\overleftarrow{x})$ in bottom-up (depth-first) manner
9. $\mathcal{A}_1, \mathcal{A}_2 \leftarrow$ the AVL trees of the two children of v where $|\mathcal{A}_1| \leq |\mathcal{A}_2|$
10.   **for** $a \in \mathcal{A}_1$ in ascending order
11. $b \leftarrow \max\{x \in \mathcal{A}_2 \mid x \leq a\}$
12. $ab \leftarrow lca(no^{-1}(a), no^{-1}(b))$ in $\mathcal{T}(x)$
13.   **if** $d_{ab} + d_v = max$
14.   **then** $(u, v) \leftarrow (ab, v)$
15. $M \leftarrow M \cup (ab, v)$
16.   **else if** $d_{ab} + d_v > max$
17.   **then** $(max, u, v) \leftarrow (d_{ab} + d_v, ab, v)$
18. $M \leftarrow (ab, v)$
19. $c \leftarrow next(T_2, b)$
20. $ac \leftarrow lca(no^{-1}(a), no^{-1}(c))$ in $\mathcal{T}(x)$
21.   **if** $d_{ac} + d_v = max$
22.   **then** $(u, v) \leftarrow (ac, v)$
23. $M \leftarrow M \cup (ac, v)$
24.   **else if** $d_{ac} + d_v > max$
25.   **then** $(max, u, v) \leftarrow (d_{ac} + d_v, ac, v)$
26. $M \leftarrow (ac, v)$
27. $\mathcal{A} \leftarrow merge(\mathcal{A}_1, \mathcal{A}_2)$
28. $lr_k \leftarrow max + k$
29.   **return** (lr_k, M)

Fig. 3. All longest repeats with k don't cares algorithm.

Proof. The proof is by contradiction. Let x and y be $lca(u, v)$ and $lca(u, w)$, respectively. Assume that the depth of x is less than the depth of y. Since i is less than j, k also must be less than j, which contradicts the condition that $i < j < k$.

The depth of a node in the Lemma 4 is the length of the path from the root to this node. It is quite easy to see that the Lemma can be extended to suffix trees where the depth of a node is the length of the substring spelled by the path from the root to this node.

5   Time Complexity

In this section, we analyze the running time of the algorithm. Recall that, for constant size alphabet, a suffix tree can be built in linear time. Thus, Creating $\mathcal{T}(x)$ and performing the preorder traversal at Line 1 requires $O(n)$ time. The loop on Lines 2-4 takes $O(n)$ time. Building $\mathcal{B}(\overleftarrow{x})$ also takes $O(n)$ time. Creating an AVL tree of size one can be done in constant time. Thus, doing so at each of the n leaves of $\mathcal{B}(\overleftarrow{x})$ at Line 5 requires total of $O(n)$ time. Lines 6,7 take $O(1)$ time.

The algorithm then traverses $\mathcal{B}(\overleftarrow{x})$ in depth-first manner (Lines 8-27). At every internal node v, the algorithm runs a search loop on Lines 10-26 and then performs a merge at Line 27. Let $\mathcal{A}_1$ and $\mathcal{A}_2$ be the two AVL trees associated with the two children of v where $|\mathcal{A}_1| \leq |\mathcal{A}_2|$. During the search loop (Lines 10-26), for each $a \in \mathcal{A}_1$, the algorithm searches $\mathcal{A}_2$ to find b and c. According to Lemma 2, the time required to complete the search loop at each node is $O(\log \binom{|\mathcal{A}_1|+|\mathcal{A}_2|}{|\mathcal{A}_1|}))$. Additionally, Lemma 1 states that the merge at Line 27 takes also $O(\log \binom{|\mathcal{A}_1|+|\mathcal{A}_2|}{|\mathcal{A}_1|}))$ time. Summing these terms over all the internal nodes of $\mathcal{B}(\overleftarrow{x})$ gives the total running time of the tree traversal (Lines 8-27), that is $O(n \log n)$ (Lemma 3). Thus, the total running time of the algorithm is $O(n \log n)$ time. The following theorem states the result.

Theorem 1. *Algorithm Longest-Repeats-Don't-Cares extracts all longest repeats with k don't cares from a given string in $O(n \log n)$ time.*

References

1. Adel'son-Vel'skii, G.M., Landis, Y.M.: An Algorithm for the Organisation of Information. Doklady Akademii Nauk SSSR, Vol. 146 (1962) 263–266
2. Brodal, G.S., Lyngsø, R.B., Pedersen, C.N.S., Stoye, J.: Finding Maximal Pairs with Bounded Gaps. Journal of Discrete Algorithms,, Special Issue of Matching Patterns, Vol. 1 1 (2000) 77–104
3. Brown, M.R., Tarjan, R.E.: A Fast Merging Algorithm. Journal of the ACM, Vol. 26 2 (1979) 211–226
4. Crochemore, M., Rytter, W.: Jewels of Stringology. World Scientific (2002)
5. Gusfield, D.: Algorithms on Strings, Trees and Sequences: Computer Science and Computational Biology. Cambridge Univesity Press (1997)
6. Iliopoulos, C.S., Markis, C., Sioutas, S., Tsakalidis, A., Tsichlas, K.: Identififying Ocuurences of Maximal Pairs in Multiple Strings. CPM (2002) 133–143. LNCS Vol. 2373 (2002) 133–143
7. Kolpakov, R., Kucherov, G.: Finding Repeats with Fixed Gap. SPIRE (2002) 162–168
8. Schieber, B., Vishkin, U.: On Finding Lowest Common Ancestors: Simplifications and Parallization. SIAM Journal of Computation, Vol. 17 (1988) 1253–1262
9. Stoye, J.: Affix Trees. Diploma Thesis, Universität Bielefeld, Forschungsbericht der Technischen Fakultät, Abteilung Informationstechnik, Report 2000-04 (2000) (ISSN 0946-7831)

Join Irreducible Pseudovarieties, Group Mapping, and Kovács-Newman Semigroups

John Rhodes[1] and Benjamin Steinberg[2*]

[1] Department of Mathematics
University of California at Berkeley
Berkeley, CA 94720, USA
rhodes@math.berkeley.edu
[2] School of Mathematics and Statistics
Carleton University
Ottawa, ON K1S 5B6, Canada
bsteinbg@math.carleton.ca

Abstract. We call a pseudovariety finite join irreducible if

$$\mathbf{V} \leq \mathbf{V}_1 \vee \mathbf{V}_2 \implies \mathbf{V} \leq \mathbf{V}_1 \text{ or } \mathbf{V} \leq \mathbf{V}_2.$$

We present a large class of group mapping semigroups generating finite join irreducible pseudovarieties. We show that many naturally occurring pseudovarieties are finite join irreducible including: $\mathbf{S}$, $\mathbf{DS}$, $\mathbf{CR}$, $\mathbf{CS}$ and $\overline{\mathbf{H}}$, where $\mathbf{H}$ is a group pseudovariety containing a non-nilpotent group.

1 Introduction

The following results, appearing here for the first time, are part of the authors' forthcoming book, "The q-theory of finite semigroups [11];" further results shall appear therein.

All semigroups in this paper are assumed to be finite. Recall that a pseudovariety of semigroups [4] is a class of semigroups closed under finite direct products, subsemigroups and homomorphic images. They play an important role in Formal Language Theory thanks to Eilenberg's Variety Theorem [4], which establishes an isomorphism between the complete lattices $\mathbf{PV}$ of pseudovarieties of semigroups and $\mathbf{RAT}$ of varieties of rational (=regular) languages. Since the join operation on $\mathbf{RAT}$ corresponds to closing under the Boolean operations, it is quite natural, from the Formal Language Theory point-of-view, to want to study these lattices.

We now recall some basic notions from lattice theory [11] in order to state our results. Fix a *complete lattice* L. We say that $l \in L$ is:

(ji) *join irreducible* if $l \leq \bigvee_{i \in I} l_i \implies l \leq l_i$, some $i \in I$;

* The second author was supported in part by NSERC and by the FCT and POCTI approved project POCTI/32817/MAT/2000 in participation with the European Community Fund FEDER

M. Farach-Colton (Ed.): LATIN 2004, LNCS 2976, pp. 279–291, 2004.

(sji) *strictly join irreducible* if $l = \bigvee_{i \in I} l_i$ implies $l = l_i$, some $i \in I$;

(fji) *finite join irreducible* if $l \leq l_1 \vee l_2 \implies l \leq l_i$, some i;

(sfji) *strictly finite join irreducible* if $l = l_1 \vee l_2 \implies l = l_i$, some i.

The reader is cautioned that sfji is sometimes called join irreducible in the literature, while fji is sometimes called co-prime.

Let S be a semigroup; then $\mathbb{HSP}_f(S)$ denotes the pseudovariety generated by S. If $\mathbf{V} \in \mathbf{PV}$, then $\mathbf{V} = \bigvee_{S \in \mathbf{V}} \mathbb{HSP}_f(S)$ and so any sji (and hence ji) pseudovariety must be generated by a single semigroup. It is straightforward to verify that a one-generated pseudovariety is ji (respectively, sji) if and only if it is fji (respectively, sfji).

One reason that sfji is important is the following: An sfji (and hence fji) pseudovariety that is not one-generated cannot have a maximal proper subpseudovariety. Indeed, if $\mathbf{W}$ is a maximal proper subpseudovariety and $S \in \mathbf{V} \setminus \mathbf{W}$, then $\mathbf{V} = \mathbf{W} \vee \mathbb{HSP}_f(S)$ implies $\mathbf{V} = \mathbb{HSP}_f(S)$. On the other hand, one can show that $\mathbb{HSP}_f(S)$ is sfji if and only if it has a unique maximal proper subpseudovariety [11]. We mention that there are naturally occuring pseudovarieties that are not sfji, such as the pseudovariety of $\mathcal{J}$-trivial semigroups [1,11]. Also, for example, the pseudovariety of nilpotent semigroups is sfji but not fji [11].

Key to this paper is the notion of a Kovács-Newman semigroup. Roughly speaking a Kovács-Newman semigroup is a semigroup S such that every division from S onto a subdirect product factors through a projection. Each such semigroup generates a *distinct* join irreducible pseudovariety. We mention in passing that we do not distinguish between isomorphic semigroups.

Our main results include the following. Let $\mathbf{H}$ be a pseudovariety of groups containing a non-nilpotent group. Then $\overline{\mathbf{H}}$ (subgroups in $\mathbf{H}$) is finite join irreducible as well as its intersections with $\mathbf{DS}$ (regular $\mathcal{J}$-classes are semigroups) and $\mathbf{CR}$ (union of groups). Hence none of these pseudovarieties of semigroups have maximal proper subpseudovarieties. In particular, the pseudovarieties $\mathbf{S}$ of all semigroups, $\mathbf{DS}$ and $\mathbf{CR}$ are finite join irreducible. The case of $\mathbf{S}$ was first obtained by Margolis *et. al* [10] using profinite methods. They, in fact, showed that $\overline{\mathbf{H}}$ is sfji whenever $\mathbf{H}$ is closed under semidirect product. Auinger and the second author [3,2] have shown that a sufficient condition for a pseudovariety of groups $\mathbf{H}$ to be fji is the following: For each $G \in \mathbf{H}$, there exists a group H so that the wreath product $H \wr G$ belongs to $\mathbf{H}$. This is a relatively mild condition and there are uncountably many such pseudovarieties; see [3]. Our methods give an elementary proof for a large subclass of such pseudovarieties including the pseudovariety of groups $\mathbf{G}$.

2   The Group Case

2.1   Preliminaries

Recall that a non-empty partially ordered set D is said to be *directed* if any two elements of D have an upper bound.

Lemma 2.1. *Let L be a complete lattice and suppose that $D \subseteq L$ is a directed subset of fji elements. Then $\bigvee D$ is fji.*

Proof. Let $d = \bigvee D$ and suppose $d \leq l_1 \vee l_2$. Suppose $d \not\leq l_1$; then $l \not\leq l_1$ for some $l \in D$. Since l is fji, $l \leq l_2$. Let $l' \in D$ and let $k \in D$ be an upper bound for l and l'. If $k \leq l_1$, then $l \leq l_1$, a contradiction. Since k is fji, we conclude that $k \leq l_2$. Hence $l' \leq l_2$. Since l' was arbitrary, we conclude $d \leq l_2$, as desired. □

We shall also need a variant of this lemma.

Lemma 2.2. *Let $\mathbf{V}$ be a pseudovariety such that, for each $S \in \mathbf{V}$, there exists $S' \in \mathbf{V}$ such that:*

1. *$S \in \mathbb{HSP}_f(S')$*
2. *$S' \in \mathbf{V}_1 \vee \mathbf{V}_2 \implies S \in \mathbf{V}_1$ or $S \in \mathbf{V}_2$.*

Then $\mathbf{V}$ is fji.

Proof. Let $S_0, S_1, S_2, \ldots$ be an enumeration of the elements of $\mathbf{V}$ with S_0 trivial. Set $T_0 = S_0$ and define, for $i > 0$, $T_i = (T_{i-1} \times S_i)'$. Then, by (1), we have

$$\forall i, \; T_i \in \mathbb{HSP}_f(T_{i+1}) \text{ and } S_i \in \mathbb{HSP}_f(T_i) \tag{2.1}$$

Suppose $\mathbf{V} \leq \mathbf{V}_1 \vee \mathbf{V}_2$ and $\mathbf{V} \not\leq \mathbf{V}_1$. Then $S_i \notin \mathbf{V}_1$ for some i. Let $j \geq i$; then $S_i \in \mathbb{HSP}_f(T_j)$ by (2.1), so $T_j \notin \mathbf{V}_1$. Since $T_j \in \mathbf{V}_1 \vee \mathbf{V}_2$, we deduce from (2) that $T_{j-1} \times S_j \in \mathbf{V}_2$, $j \geq i$. Hence $T_k \in \mathbf{V}_2$ for $k \geq i - 1$. We conclude, using (2.1), that $\mathbf{V} \leq \mathbf{V}_2$, as desired. □

Let $\mathbf{GPV}$ be the lattice of group pseudovarieties. The following is a straightforward consequence of the fact that groups "lift" under surjective homomorphisms [5,4]; the proof is left as an exercise.

Proposition 2.3. *If $\mathcal{P}$ is any of the properties ji, sji, fji or sfji, then $\mathbf{H} \in \mathbf{GPV}$ is $\mathcal{P}$ in $\mathbf{GPV}$ if and only if it is $\mathcal{P}$ in $\mathbf{PV}$.*

2.2 Kovács-Newman Groups

We refine slightly a result of Kovács and Newman [6,7,8,9] showing that certain pseudovarieties of groups are ji. By Proposition 2.3, we may restrict our attention to the group setting.

Recall that a semigroup S *divides* a semigroup T if S is a quotient of a subsemigroup of T [5,4].

A semigroup T is said to be a *subdirect product* of T_1 and T_2, written $T <\!< T_1 \times T_2$, if the projections from T to the T_i are surjective. A semigroup T is called *subdirectly indecomposable* if $T <\!< T_1 \times T_2$ implies that at least one of the projections $\pi_i : T \twoheadrightarrow T_i$ is an isomorphism. It is not hard to see [9] that a group G is subdirectly indecomposable if and only if it has a unique minimal (non-trivial) normal subgroup M, called its *monolith* [9]; sometimes G is called *monolithic* [9]. A pseudovariety is always generated by its subdirectly indecomposable members.

In what follows, we identify the direct factors G_1 and G_2 of $G_1 \times G_2$ with $G_1 \times 1$, $1 \times G_2$, respectively.

We call a non-trivial group G a *Kovács-Newman group* (or KN group for short) if it has the following property: Whenever there is a diagram

$$G \overset{\varphi}{\leftarrow} H << G_1 \times G_2 \qquad (2.2)$$

φ factors through one of the projections. Since it is clear that $G \in \mathbf{H}_1 \vee \mathbf{H}_2$ if and only if there is a diagram as in (2.2) with $G_1 \in \mathbf{H}_1$ and $G_2 \in \mathbf{H}_2$, it follows that if G is KN, then $\mathbb{HSP}_f(G)$ is fji and hence ji. Observe that if G is a KN group and $G \in \mathbb{HSP}_f(H)$, then G divides H. Indeed, G must divide a product of copies of H and so, being a KN group, it divides H. In particular, *two non-isomorphic KN groups cannot generate the same pseudovariety*.

We remark that there are ji pseudovarieties that are not generated by KN groups. In fact, we show that if p is a prime, then $G = \mathbb{Z}_p$ is not a KN group, but $\mathbb{HSP}_f(\mathbb{Z}_p)$ is ji. To see that $\mathbb{HSP}_f(\mathbb{Z}_p)$ is ji, suppose $\mathbb{Z}_p$ divides $G_1 \times G_2$ with $G_i \in \mathbf{H}_i$, $i = 1, 2$. Then p divides $|G_1| \cdot |G_2|$ and hence $|G_i|$ for some i. Thus $\mathbb{Z}_p$ is a subgroup of G_i and so $\mathbb{Z}_p \in \mathbf{H}_i$; we conclude $\mathbb{HSP}_f(\mathbb{Z}_p)$ is ji. The following proposition shows that KN groups are centerless (and hence no element of $\mathbb{HSP}_f(\mathbb{Z}_p)$ can be a KN group).

Proposition 2.4. *Let G be a KN group. Then the center $Z(G)$ is trivial. In particular, no nilpotent group is a KN group.*

Proof. Suppose $Z(G) \neq 1$ and let $H = G \times Z(G)$. Consider the onto homomorphism $\varphi : H \twoheadrightarrow G$ given by $(g, z)\varphi = gz^{-1}$; it's straightforward to verify φ is a homomorphism. However, φ does not factor through either projection; indeed, $\ker \varphi = \{(g, g) \mid g \in Z(G)\}$ is the diagonal embedding of $Z(G)$ and intersects the two factors trivially. The last statement follows since nilpotent groups have non-trivial centers. $\square$

Kovács and Newman essentially showed that there is a large number of KN groups. Clearly any KN group must be monolithic. The following proposition about subdirect products is straightforward [9]; we omit the proof.

Proposition 2.5. *Suppose $H << G_1 \times G_2$. Then a subgroup $N \leq G_i$ is normal in G_i if and only if it is normalized by H. In particular, if $K \lhd H$ and $N \lhd G_i$, then $K \cap N$ is normal in G_i.*

Let us establish some notation. If $\psi : H \to G$ is a homomorphism and $N \lhd G$ is a normal subgroup, then H acts on N by first applying ψ and then acting by conjugation. Let $C_H(N)$ denote the centralizer of N under this action. Notice that $\ker \psi \leq C_H(N)$, in fact, $C_H(N) = C_G(N)\psi^{-1}$. Centralizers of minimal normal subgroups shall play an important role in this paper thanks to Theorem 2.8 below.

Lemma 2.6. *Suppose G is a monolithic group with non-Abelian monolith M. Then $C_G(M)$ is trivial.*

Proof. If $C_G(M)$ is non-trivial, then $M \leq C_G(M)$. But this implies that M is Abelian.   □

Another useful fact about minimal normal subgroups is the following.

Lemma 2.7. *Let G be a group and $M \lhd G$ be a minimal normal subgroup. Then there exists a normal subgroup $N \lhd G$ such that $N \cap M = 1$ and G/N is monolithic with monolith $MN/N \cong M$. Moreover, $C_G(M) = C_G(MN/N)$.*

Proof. Let N be a maximal normal subgroup such that $N \cap M = 1$. Let $N < K \lhd G$; then $K \cap M \neq 1$. Since M is minimal, we conclude $M \leq K$. It follows that MN/N is the unique minimal normal subgroup of G/N. Since $M \cap N = 1$, $M \cong MN/N$.

Clearly $C_G(M) \leq C_G(MN/N)$. For the converse, suppose $m \in M$ and $g \in C_G(MN/N)$. Then $g^{-1}mg = mn$ with $n \in N$. So $n = m^{-1}(g^{-1}mg) \in M \cap N = 1$. We conclude $g \in C_G(M)$.   □

The following technical theorem on lifting minimal normal subgroups and their centralizers is essentially due to Kovács and Newman [6,7,8]; our proof is adapted from [9, Chpt. 5, §3].

Theorem 2.8. *Let G be a monolithic group with monolith M. Suppose that one has a diagram as in (2.2). Let $\tau : G \to G/C_G(M)$ be the natural quotient map. Then $\varphi\tau$ factors through one of the direct product projections.*

Proof. Let $K = \ker\varphi$. To each pair (N_1, N_2) of normal subgroups $N_1 \lhd G_1$, $N_2 \lhd G_2$, we associate the positive number $|G_1/N_1| + |G_2/N_2|$; this number is called the *weight* of the pair. Suppose (N_1, N_2) is of minimal weight with the property that there is a factorization:

$$H \xrightarrow{\psi} H\psi << G_1/N_1 \times G_2/N_2$$
$$\varphi \downarrow \quad \nearrow \Phi$$
$$G$$

$$(2.3)$$

That is choose (N_1, N_2) of minimal weight so that for ψ, as defined in (2.3), $\ker\psi \leq K$. Set $D_i = G_i/N_i$. Notice that if D_1 (respectively, D_2) is trivial, then φ (and hence $\varphi\tau$) factors through the projection to G_2 (respectively, G_1) and we are done. So we assume from now on that $D_1, D_2 \neq 1$. Let us fix the following notation: $H' = H\psi$, $K' = K\psi$.

Fact 2.9. *Every non-trivial normal subgroup of D_i intersects H' non-trivially.*

Proof. Suppose, without loss of generality, $N = N'/N_1$ (with $N' \supseteq N_1$) is a non-trivial normal subgroup of D_1 with $H' \cap N = 1$. Then $H' << D_1/N \times D_2$ under the natural map and so we have a factorization

$$H \longrightarrow H' << G_1/N' \times G_2/N_2$$
$$\varphi \downarrow \quad \nearrow$$
$$G$$

where (N', N_2) has strictly smaller weight than (N_1, N_2) – contradiction.   □

Fact 2.10. *The normal subgroup K' of H' intersects D_1 and D_2 trivially.*

Proof. Suppose, without loss of generality, that $N = K' \cap D_1 \neq 1$. By Proposition 2.5, N is normal in D_1. Thus, recalling $N \leq K'$, we obtain a factorization

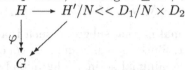

contradicting the minimality of the weight of (N_1, N_2).   □

We need one last fact before completing the proof.

Fact 2.11. *Let $\pi_i : H \twoheadrightarrow D_i$ be the projection and $L_i = \ker \pi_i$. Suppose N is a minimal normal subgroup of D_i. Then $N \leq H'$ and Φ takes N isomorphically onto M. Moreover,*

$$L_i \leq C_H(N) = C_H(M) = C_G(M)\varphi^{-1}. \tag{2.4}$$

Proof. By Fact 2.9, $N \cap H' \neq 1$. By Proposition 2.5, $N \cap H'$ is normal in D_i. Hence, by minimality of N, $N \leq H'$. Also Proposition 2.5 immediately implies that N is a minimal normal subgroup of H'.

Since $N \leq D_i$, Fact 2.10 shows that $K' \cap N = 1$. Thus $\Phi : H' \twoheadrightarrow H'/K' = G$ is injective on N. Hence $N\Phi = N'$ is a minimal normal subgroup of G. Indeed, if N_0 is a normal subgroup of G properly contained in N', then $N_0\Phi^{-1} \cap N$ is a normal subgroup of H' properly contained in N, a contradiction. Since G is monolithic, it follows $N\Phi = M$.

We are now left with proving (2.4). Without loss of generality, let us take $i = 1$. Notice that there is an ambiguity in the notation $C_H(N)$ since we can either view H as acting on N by first doing ψ and then conjugating, or by first doing π_1 and then conjugating. We show that the centralizers under either interpretation are the same. Indeed, let $h \in H$ and $n \in N$. Set $h\psi = (h_1, h_2) = (h\pi_1, h\pi_2)$. Then

$$(h\psi)^{-1}nh\psi = (h_1^{-1}, h_2^{-1})(n, 1)(h_1, h_2) = (h_1^{-1}nh_1, 1),$$

so $h\psi$ centralizes N if and only if $h_1 = h\pi_1$ centralizes N.

It now follows that $L_1 = \ker \pi_1 \leq C_H(N)$. Clearly if $h\psi$ centralizes N, then $h\psi\Phi$ centralizes $N\Phi = M$. Thus $C_H(N) \leq C_H(M)$. Suppose $h \in C_H(M)$. Let $h\psi = (h_1, h_2)$ and let $n \in N$. Since $h\psi\Phi$ centralizes $n\Phi$,

$$(h_1^{-1}nh_1, 1) = (h_1, h_2)^{-1}(n, 1)(h_1, h_2) = (h\psi)^{-1}n(h\psi) = nk$$

some $k \in K'$. Hence $k = n^{-1}h_1^{-1}nh_1 \in K' \cap D_1 = 1$, by Fact 2.10. Thus $h \in C_H(N)$. The equality $C_H(M) = C_G(M)\varphi^{-1}$ is elementary.   □

Fact 2.11 completes the proof of the theorem since if $\rho_i : H \to G_i$ is the projection, then $\ker \rho_i \leq L_i \leq C_H(M) = \ker \varphi\tau$, as desired.   □

Hence we obtain one of our principal results[1].

Theorem 2.12. *Let G be a monolithic group with non-Abelian monolith M. Then G is a Kovács-Newman group.*

Proof. Suppose one has a diagram as per (2.2). Then, by Theorem 2.8, $\varphi\tau$ factors through a direct product projection (retaining the notation of that theorem). But, by Lemma 2.6, τ is the identity map; we deduce that G is a KN group. $\square$

Corollary 2.13 (Kovács-Newman [6,7,8,9]). *Each distinct monolithic group with non-Abelian monolith generates a **distinct** join irreducible pseudovariety.*

2.3   Applications

We use the above results to provide some fji group pseudovarieties, including a weaker version of the result of Auinger and the second author [2,3] mentioned earlier (but with a more elementary proof).

Let A_n be the alternating group on n letters. It is well known that every finite group embeds in A_n for some $n \geq 5$ and that A_n is simple non-Abelian, $n \geq 5$, and hence a KN group by Theorem 2.12. Thus $\mathbf{G} = \bigvee \mathbb{HSP}_f(A_n)$ is fji by Lemma 2.1.

The authors are indebted to John Dixon for pointing out the following example. Set $G_i = \mathbb{PSL}(2, 2^i)$; the G_i are simple non-Abelian groups and $G_i \leq G_{i+1}$, so Lemma 2.1 shows $\mathbf{H} = \bigvee \mathbb{HSP}_f(G_i)$ is fji. Since, for any $q > 2$, the q-subgroups of G_i are Abelian, $\mathbf{H}$ is a proper fji pseudovariety of groups.

Corollary 2.14. *Suppose $\mathbf{H}$ is a pseudovariety of groups such that, for each $G \in \mathbf{H}$, there is a simple non-Abelian group H such that $H \wr G \in \mathbf{H}$. Then $\mathbf{H}$ is finite join irreducible.*

Proof. We use Lemma 2.2. Let $G \in \mathbf{H}$ and H be a simple non-Abelian group such that $W = H \wr G \in \mathbf{H}$. Set $M = H^G \lhd W$; we claim that M is a minimal normal subgroup. Indeed, suppose $1 \neq f \in M$; we show that the normal closure N of f is M. Conjugating by an element of G, we may assume $1f \neq 1$. Since H has trivial center, there exists $h \in H$ such that $h^{-1}(1f)h \neq 1f$. Define $g \in M$ by $1g = h$, $h'g = 1$ for $h' \in H \setminus 1$. Set $k = f(g^{-1}fg)^{-1}$; then $1k \neq 1$, $h'k = 1$ all $h' \in H \setminus 1$ and $k \in N$. Since H is simple, it now follows that $K = \{f \in M \mid hf = 1, \forall h \in H \setminus 1\} \leq N$. But $\langle g^{-1}Kg \mid g \in G \rangle = M$.

By Lemma 2.7, there is a normal subgroup $N \lhd W$ such that: $N \cap M = 1$, $G' = W/N$ is monolithic with monolith $MN/N \cong M$ and $C_W(MN/N) = C_W(M)$. Since G acts faithfully on M, $G \cap C_W(M) = 1$. Thus

$$G \cap N \leq G \cap C_W(MN/N) = G \cap C_W(N) = 1,$$

and so $G \leq G'$. Clearly (1) of Lemma 2.2 is satisfied; since M is non-Abelian, G' is a KN group and so (2) is also satisfied. $\square$

---

[1] After being shown a preprint of this paper, L. G. Kovács (private communication) was able to prove the converse of Theorem 2.12.

3   The Semigroup Case

3.1   Preliminaries

Before defining KN semigroups, we review some preliminary notions concerning simple semigroups and minimal ideals. Each semigroup S has a minimal ideal $K(S)$, sometimes called its *kernel*, which is a (completely) simple semigroup [5]. Moreover, if $\varphi : S \twoheadrightarrow T$ is an onto homomorphism, then $K(S)\varphi = K(T)$ [5].

Notice that S acts on the left and right of $K(S)$. Following [5, Chpt. 8], S is said to be *generalized group mapping* over its kernel if S acts faithfully on both the left and right of $K(S)$. It is shown [5, Chpt. 8] that S is generalized group mapping over $K(S)$ if and only if the following congruence on S is the equality relation:

$$s_1 \equiv s_2 \iff \forall k_1, k_2 \in K(S), \ k_1 s_1 k_2 = k_1 s_2 k_2. \tag{3.1}$$

One says that S is *group mapping* over $K(S)$ if either $S = 1$ or S is generalized group mapping over $K(S)$ and $K(S)$ contains a non-trivial group.

Recall that a simple semigroup S is always isomorphic to a regular Rees matrix semigroup $\mathcal{M}(G, A, B, C)$, where A, B are sets, G is a group, called the maximal subgroup of S, and $C : B \times A \to G$ is a matrix, c.f. [5, Chpt. 7]. It follows easily from [5, Chpt. 8, Fact 2.22] that a simple semigroup S is group mapping if and only if G is non-trivial and no two rows of C are proportional on the left and no two columns of C are proportional on the right. In particular, any group is group mapping.

3.2   Kovács-Newman Semigroups

We now define a KN semigroup: A non-trivial semigroup S is a *Kovács-Newman semigroup* (KN semigroup) if whenever there is a diagram

$$S \overset{\varphi}{\leftarrow} T << T_1 \times T_2 \tag{3.2}$$

φ factors through one of the projections. As in the case of groups, it is clear that if S is a KN semigroup, then $\mathbb{HSP}_f(S)$ is ji; moreover, *non-isomorphic KN semigroups generate distinct pseudovarieties*.

It is not clear *a priori* that a KN group is a KN semigroup. This will be a consequence of our main result stating: if S is group mapping over $K(S)$ and the maximal subgroup of $K(S)$ is a KN group, then S is a KN semigroup. In particular, each KN group is a KN semigroup.

To prove this, we shall need a special case of [5, Chpt. 8, Prop. 3.28], which highlights the importance of group mapping semigroups by saying that homomorphisms to group mapping semigroups "have kernels."

Proposition 3.1. *Suppose that $\varphi : T \twoheadrightarrow S$ is a surjective homomorphism and that S is group mapping over $K(S)$. Let H be a maximal subgroup of $K(T)$ and suppose $\psi : T \twoheadrightarrow T'$ is a surjective homomorphism such that* $\ker \psi|_H \leq \ker \varphi|_H$. *Then φ factors through ψ.*

Proof. Suppose that $t_1, t_2 \in T$ and $t_1\psi = t_2\psi$; we need to show that $t_1\varphi = t_2\varphi$. Since S is group mapping, to do this it suffices, by (3.1), to show that, for all $k_1, k_2 \in K(S)$,

$$k_1(t_1\varphi)k_2 = k_1(t_2\varphi)k_2.$$

Since $K(S) = K(T)\varphi$, this is equivalent to showing, for all $j_1, j_2 \in K(T)$,

$$(j_1t_1j_2)\varphi = (j_1t_2j_2)\varphi.$$

Set $t_i = j_1t_ij_2$, $i = 1, 2$. First observe that $t_1 \mathcal{H} t_2$ (by finiteness [5]) and so they belong to the same maximal subgroup G of $K(T)$. By, say Green's Lemma or Rees's Theorem [5], there exist $x, y, x', y' \in K(T)$ such that $u \mapsto xuy$ is a bijection from G to H with inverse given by $v \mapsto x'vy'$. Let $K = \ker\varphi|_H$ and $N = \ker\psi|_H$: so $N \leq K$ by hypothesis. Since $t_1\psi = t_2\psi$, $(xt_1y)\psi = (xt_2y)\psi$ and so $xt_1yN = xt_2yN$. Since $N \leq K$, we have $xt_1yK = xt_2yK$, that is, $(xt_1y)\varphi = (xt_2y)\varphi$. Thus

$$t_1\varphi = (x'xt_1yy')\varphi = (x'xt_2yy')\varphi = t_2\varphi,$$

completing the proof. □

The following theorem, along with Theorems 2.12 and 3.6, can be viewed as the principal results of this paper.

Theorem 3.2. *Let S be semigroup that is group mapping over $K(S)$ and such that $K(S)$ has a maximal subgroup G that is a Kovács-Newman group. Then S is a Kovács-Newman semigroup. In particular, every KN group is a KN semigroup.*

Proof. Suppose we have a diagram as in (3.2); let $\pi_i : T \to T_i$ be the projection. Since $K(S) = K(T)\varphi$, standard results about simple semigroups (c.f. [5, Chpt. 7]) show that there is a maximal subgroup $H \leq K(T)$ such that $H\varphi = G$. Let $K = \ker\varphi|_H$; then $G = H/K$. Clearly $H \lessdot\lessdot H\pi_1 \times H\pi_2$. Setting $N_i = \ker\pi_i|_H$, we have, since G is a KN group, $N_i \leq K$ for some i. Proposition 3.1 then implies φ factors through π_i. □

Corollary 3.3. *Let S be a semigroup that is group mapping over $K(S)$ and such that the maximal subgroup of $K(S)$ is monolithic with non-Abelian monolith. Then S is a KN semigroup.*

An open question is to describe all KN semigroups[2].

3.3 Applications

Our first application is the following theorem.

Theorem 3.4. *The pseudovariety **CS** of (completely) simple semigroups is finite join irreducible.*

---

[2] Since this paper was submitted, the authors were able to prove (using in part L. G. Kovács solution to the group case) the converse of Corollary 3.3, thus resolving this question.

Proof. It is well known [5, Chpt. 8] that every simple semigroup is a subdirect product of a right zero semigroup, a left zero semigroup and a group mapping simple semigroup. A non-trivial group G can be embedded into the group mapping simple semigroup $S = \mathcal{M}(G, 2, 2, C)$ with the structure matrix

$$C = \begin{pmatrix} 1 & 1 \\ 1 & g \end{pmatrix}$$

where $1 \neq g \in G$. Moveover, the two element left and right zero semigroups divide S, so we may conclude that **CS** is generated by the collection of all group mapping simple semigroups that are not groups.

Let $\mathcal{M}(G, A, B, C)$ be a group mapping simple semigroup, which is not a group, and suppose $G \leq A_n$ with $n \geq 5$. Then

$$\mathcal{M}(G, A, B, C) \leq \mathcal{M}(A_n, A, B, C)$$

(where we now view C as a matrix over A_n) and the latter semigroup is group mapping. It follows that **CS** is generated by group mapping simple semigroups with structure group A_n, $n \geq 5$. Each such semigroup generates a ji pseudovariety by Corollary 3.3.

We now show that given two such semigroups

$$S_1 = \mathcal{M}(A_n, A, B, C), \quad S_2 = \mathcal{M}(A_j, A', B,' C')$$

there is such a semigroup containing them both. First observe, that we may assume $n = j$ by replacing the smaller index by the larger one.

Now construct a matrix P as follows. Without loss of generality, we may assume that C has at least as many rows of C'. Then add to C' as many rows of 1's as needed in order to obtain a matrix C'' with the same number of rows as C. Let $P = (C \; C'')$. No two rows of P are proportional on the left, since no two rows of C are proportional on the left; however P may have some columns proportional on the right. So we identify the proportional columns of P to obtain a new matrix P'. Since multiplying a column on the right by a scalar and changing the order of the columns does not change a Rees matrix semigroup, the resulting Rees matrix semigroup over A_n with structure matrix P' contains a copy of S_1 and S_2.

We may conclude that **CS** is the directed supremum of ji pseudovarieties and so an application of Lemma 2.1 establishes the theorem. □

Recall that a semigroup S is said to be an *ideal extension* of B by T if B is an ideal of S and $S/B = T$. For a semigroup S, we use S^0 to denote S with an adjoined zero and S^I to denote S with an adjoined identity (in both cases, even if S already had one).

Lemma 3.5. *Let S be a semigroup, G be a group and $1 \neq g \in G$. Then there is a semigroup $S(G, g)$ such that:*

1. $S(G, g)$ is an ideal extension of $K(S(G, g))$ by S^0;

2. G is the maximal subgroup of $K(S(G,g))$;

3. $S(G,g)$ is group mapping over $K(S(G,g))$.

Moreover, if $\varphi : G \to H$ is a surjective homomorphism and $g\varphi \neq 1$, then there is a natural surjective morphism $\widetilde{\varphi} : S(G,g) \twoheadrightarrow S(H,g\varphi)$ such that $\widetilde{\varphi}$ is injective on S and $\widetilde{\varphi}|_G = \varphi$.

Proof. We first construct a semigroup $S_0(G,g)$ meeting the requirements 1-2. Let $K(G,g) = \mathcal{M}(G,A,B,P_0)$ be constructed as follows. We set $B = S^I$. Choose $1 \neq g \in G$. Construct the $(n+1) \times (n+1)$ matrix

$$P = \begin{pmatrix} g & 1 & \cdots & 1 \\ 1 & g & 1 & 1 \\ \vdots & 1 & \ddots & \vdots \\ 1 & 1 & \cdots & g \end{pmatrix}$$

The rows of P shall be indexed by S^I and the columns by $\{a_{I,1}, \ldots, a_{I,n+1}\}$. For each $s \in S$, we create $n+1$ new columns as follows: Column $a_{s,j}$ has entry in row $s_i \in S^I$ equal to the entry in row $s_i s$, column $a_{I,j}$. The resulting matrix is denoted P_0.

We now form $S_0(G,g) = S \cup K(G,g)$ where multiplication of elements of $K(G,g)$ by elements of S is defined as follows:

$$(a_{s_1,j}, h, s_2)s = (a_{s_1,j}, h, s_2 s), \text{with } s_1, s_2 \in S^I, \ s \in S, \ h \in H$$
$$s(a_{s_1,j}, h, s_2) = (a_{ss_1,j}, h, s_2), \text{with } s_1, s_2 \in S^I, \ s \in S, \ h \in H$$

It is an exercise in the linked equations [5, Chpt. 7] to verify $S_0(G,g)$ is a semigroup with $K(S_0(G,g)) = K(G,g)$. Define a congruence $\equiv$ on $S_0(G,g)$ by $t_1 \equiv t_2$ if and only if $k_1 t_1 k_2 = k_1 t_2 k_2$ for all $k_1, k_2 \in K(G,g)$. Then $S(G,g) = S_0(G,g)/\equiv$ satisfies 2 and 3; see [5]. We need to show that 1 is satisfied. To prove this, it suffices to show that $\equiv$ does not identify elements of S. Suppose $s, s' \in S$ are distinct. By construction of P_0, there is a column $a_{I,j}$ such that the entry in row s is g and the entry in row s' is 1. Then

$$(a_{I,1}, 1, I)s(a_{I,j}, 1, I) = (a_{I,1}, 1, s)(a_{I,j}, 1, I) = (a_{I,1}, g, I)$$
$$(a_{I,1}, 1, I)s'(a_{I,j}, 1, I) = (a_{I,1}, 1, s')(a_{I,j}, 1, I) = (a_{I,1}, 1, I)$$

and so $s \not\equiv s'$, as desired.

Suppose now that $\varphi : G \to H$ is a surjective homomorphism and $g\varphi \neq 1$. The map $\Phi : S_0(G,g) \twoheadrightarrow S_0(H,g\varphi)$ defined by $s\varphi = s$ for $s \in S$ and $(a, g', b)\Phi = (a, g'\varphi, b)$ for $(a, g', b) \in K(G,g)$ is clearly a surjective homomorphism and $\Phi|_G = \varphi$. Since $K(S_0(G,g))$ is the minimal $\mathcal{J}$-class mapping onto $K(S_0(H,g\varphi))$, the results of [5, Chpt. 8] immediately imply that there is an onto homomorphism $\widetilde{\varphi} : S(G,g) \twoheadrightarrow S(H,g\varphi)$ such that

$$S_0(G,g) \xrightarrow{\Phi} S_0(H, g\varphi)$$

$$\downarrow \qquad \qquad \downarrow \qquad \qquad (3.3)$$

$$S(G,g) \xrightarrow[\widetilde{\varphi}]{} S(H, g\varphi)$$

commutes. From the commutativity of (3.3) and (1) for $S(H, g\varphi)$, it immediately follows that $\widetilde{\varphi}$ is injective on S and $\widetilde{\varphi}|_G = \varphi$.   □

Let $\mathbf{H}$ be a group pseudovariety; denote by $\overline{\mathbf{H}}$ the pseudovariety of semigroups with subgroups in $\mathbf{H}$. If $\mathbf{V}$ is a pseudovariety, set $\mathbf{V}(\mathbf{H}) = \mathbf{V} \cap \overline{\mathbf{H}}$. Let $\mathbf{CR}$ be the pseudovariety of completely regular semigroups and let $\mathbf{DS}$ be the pseudovariety of semigroups whose regular $\mathcal{J}$-classes are subsemigroups. The main application of this paper is the next theorem and its corollaries.

Theorem 3.6. *Suppose that $\mathbf{H}$ is a pseudovariety of groups containing a non-nilpotent group. Let $\mathbf{V}$ be a pseudovariety of semigroups containing a non-trivial semilattice and closed under ideal extensions of elements of $\mathbf{CS}(\mathbf{H})$ by elements of $\mathbf{V}$. Then $\mathbf{V}$ is finite join irreducible.*

Proof. First note that $\mathbf{V}$ is closed under the operation of adjoining a (new) zero since it contains a non-trivial semilattice [4].

We claim that $\mathbf{H}$ contains a monolithic group G with non-central monolith M. Indeed, let $G \in \mathbf{H}$ be a non-nilpotent group of minimal order. Clearly G must be subdirectly indecomposable; let M be its monolith. By choice of G, G/M is nilpotent. If M were central, then G would be a central extension by a nilpotent group and hence nilpotent, contradicting the choice of G.

Let $S \in \mathbf{V}$ and let $g \notin C_G(M)$. Set $S' = S(G, g)$. By hypothesis, $S' \in \mathbf{V}$; clearly (1) of Lemma 2.2 holds; we show that (2) holds.

Suppose $S' \in \mathbf{V}_1 \vee \mathbf{V}_2$. Then there is a diagram $S' \xleftarrow{\varphi} T << T_1 \times T_2$ with $T_i \in \mathbf{V}_i$, $i = 1, 2$. As in the proof of Theorem 3.2, $K(T)\varphi = K(S')$ and there is a maximal subgroup H of $K(T)$ with $H\varphi = G$. Also $H << H\pi_1 \times H\pi_2$, where $\pi_i : T \twoheadrightarrow T_i$ is the projection, $i = 1, 2$. Set $N_i = \ker \pi_i|_H$, $i = 1, 2$. By Theorem 2.8, $N_i \leq C_H(M)$, some i, say $i = 1$.

Let $\rho : G \twoheadrightarrow G/C_G(M)$ be the projection and consider the map $\widetilde{\rho} : S(G, g) \rightarrow S(G/C_G(M), g\rho)$ as per Lemma 3.5; note that $g\rho \neq 1$. Then $N_1 \leq C_H(M) = \ker(\varphi\widetilde{\rho})|_H$, so, by Proposition 3.1, the quotient $\varphi\widetilde{\rho} : T \twoheadrightarrow S(G/C_G(M), g\rho)$ factors through π_1. Since $S \leq S(G/C_G(M), g\rho)$, it follows S divides T_1 and so $S \in \mathbf{V}_1$. This completes the proof that (2) of Lemma 2.2 holds, establishing the theorem.   □

Corollary 3.7. *The pseudovarieties $\mathbf{S}$, $\mathbf{CR}$ and $\mathbf{DS}$ are all finite join irreducible. More generally, suppose $\mathbf{H}$ is a pseudovariety of groups containing a non-nilpotent group. Then $\overline{\mathbf{H}}$, $\mathbf{CR}(\mathbf{H})$ and $\mathbf{DS}(\mathbf{H})$ are all finite join irreducible. Hence none of these pseudovarieties has a maximal proper subpseudovariety.*

Since the only nilpotent pseudovarieties of groups closed under semidirect product are **1** and $\mathbf{G}_p$ (p-groups), we have recovered all of the join results of [10] with the exceptions of $\mathbf{A} = \overline{\mathbf{1}}$ (aperiodic semigroups) and $\overline{\mathbf{G}_p}$. In fact, our results are stronger since we prove fji rather than sfji.

References

1. Almeida, J.: Finite Semigroups and Universal Algebra. World Scientific, Singapore, 1994.
2. Auinger, K., Steinberg, B.: The geometry of profinite graphs with applications to free groups and finite monoids. Trans. Amer. Math. Soc. **356** (2004) 805–851.
3. Auinger, K., Steinberg, B.: On power groups and embedding theorems for relatively free profinite monoids. Proc. Cambridge Philos. Soc., to appear.
4. Eilenberg, S.: Automata, Languages and Machines. Academic Press, New York, Vol B, 1976.
5. Krohn, K., Rhodes, J., Tilson, B.: Lectures on the algebraic theory of finite semigroups and finite-state machines. Chapters 1, 5-9 (Chapter 6 with Arbib, M. A.) of The Algebraic Theory of Machines, Languages, and Semigroups (Arbib, M. A. ed.), Academic Press, New York, 1968.
6. Kovács, L. G., Newman, M. F.: Cross varieties of groups. Proc. Roy. Soc. (London) A **292** (1966) 530–536.
7. Kovács, L. G., Newman, M. F.: Minimal verbal subgroups. Proc. Cambridge Phil. Soc. **62** (1966) 347–350.
8. Kovács, L. G., Newman, M. F.: On critical groups. J. Austral. Math. Soc. **6** (1966) 237–250.
9. Neumann, H.: Varieties of Groups. Springer, Berlin Heidelberg New York, 1967.
10. Margolis, S. W., Sapir, M., Weil P.: Irreducibility of certain pseudovarieties. Comm. Algebra **26** (1998) 779–792.
11. Rhodes, J., Steinberg, B.: The q-theory of finite semigroups. In preparation, http://www.mathstat.carleton.ca/~bsteinbg/qtheor.

Complementation of Rational Sets on Scattered Linear Orderings of Finite Rank

Olivier Carton[1] and Chloé Rispal[2]

[1] LIAFA, Université Paris 7, 2, place Jussieu, F-75251 Paris cedex 05, France,
Olivier.Carton@liafa.jussieu.fr,
[2] IGM, Université de Marne-la-Vallée, 5 boulevard Descartes, F-77454
Marne-la-Vallée Cedex 2, France,
chloe.rispal@univ-mlv.fr

Abstract. In a preceding paper (Bruyère and Carton, automata on linear orderings, MFCS'01), automata have been introduced for words indexed by linear orderings. These automata are a generalization of automata for finite, infinite, bi-infinite and even transfinite words studied by Büchi. Kleene's theorem has been generalized to these words. We show that deterministic automata do not have the same expressive power. Despite this negative result, we prove that rational sets of words of finite ranks are closed under complementation.

1 Introduction

Automata were first introduced by Kleene who showed that they have the same expressive power as rational expressions [13]. Since then, many extensions of this deep result have been proved. Different kinds of structures have been considered like infinite words [7,14], bi-infinite words [11,15] and transfinite words [9,10,22], finite and infinite trees [18], finite and infinite traces, pictures, *etc.*

In [2,3], have been introduced automata that accept linearly-ordered stuctures. These automata are a simple and natural generalization of usual automata with additional limit transitions of the form $P \to q$ and $q \to P$ where P is subset of states. They allow to treat in the same framework finite, infinite words, bi-infinite words and transfinite words. These automata were proved to be equivalent to some rational expressions when the orderings are restricted to scattered orderings. Recall that scattered orderings are those orderings which do not contain a dense sub-orderering like $\mathbb{Q}$. They include the ordinals and their mirrors.

One main property of rational sets is the closure under complementation. It means that for any automaton $\mathcal{A}$, there is another automaton $\mathcal{B}$ accepting exactly the structures that are not accepted by $\mathcal{A}$. This property holds for almost all structures: finite and infinite words, finite and infinite trees and even for transfinite words on ordinals.

This property is important both from the pratical and the theoretical point of view. It means that the class of rational sets forms an effective boolean algebra. It is used whenever some logic is translated into automata. For instance, in both

M. Farach-Colton (Ed.): LATIN 2004, LNCS 2976, pp. 292–301, 2004.
© Springer-Verlag Berlin Heidelberg 2004

proofs of the decidability of the monadic second-order theory of the integers by Büchi [8] and the decidability of the monadic second-order theory of the infinite binary tree by Rabin [18], the closure under complementation of automata is the key property. It is well known that automata have the same expressive power as the monadic second order theory on many structures like finite, infinite and transfinite words and trees. A nice result would be to extend this equivalence to linear orderings. Proving the closure under complementation is one step towards this result.

In [3], the closure under complementation was left open. In this paper, we address this problem and we solve it for a subclass of scattered linear orderings. Namely, we prove that rational sets of words on scattered orderings of finite ranks are closed under complementation. Recall that Hausdorff's result [12] states that scattered orderings can be obtained from the finite orderings by repetitive applications of ω-sums and $-\omega$-sums (see Theorem 1). The rank of a scattered linear ordering is the number of nested ω-sums and $-\omega$-sums needed to obtain it. The ranks of all countable scattered linear orderings range over all countable ordinals. It can be seen as a measure of its complexity. For instance, ω and ζ are scattered orderings of rank 1. Our result generalizes both the complementation of infinite and bi-infinite words. The class of scattered orderings of finite rank includes ordinals smaller than ω^ω. Therefore, our result holds for sets of transfinite words studied by Choueka [10].

The classical method to get an automaton for the complement of a set of finite words accepted by an automaton $\mathcal{A}$ is through determinization [1]. Another method uses algebraic objects like semigroups [17]. The determinization method can still be used for infinite words but it becomes more involved [21,4]. This method has been pushed further by Büchi for countable transfinite words but it is then very complex [9]. The algebraic method can also be extended to ordinals [5,6]. In our case, this method can not be applied since automata can not be made deterministic. In this paper, we give an example of a rational set of words that cannot be accepted by a derterministic automaton. Therefore, we use another method which was introduced by Büchi for infinite words. It is based on an equivalence relation on words whose classes are shown to be rational.

The paper is organized as follows. In Section 2, we introduce words indexed by linear orderings and recall the Hausdorff characterization of countable scattered linear orderings. Then rational sets of words are defined from rational operators and automata in section 3. We finally prove in section 4 that rational sets of words indexed by countable scattered linear orderings of finite ranks are closed under complementation.

2 Words on Linear Orderings

In this section, we recall some definitions and operations on linear orderings but we refer the reader to [20] for a complete introduction to linear orderings. We give the Hausdorff's characaterization of countable scattered linear orderings and introduce words indexed by linear orderings.

Let J be a set equipped with an order $<$. The ordering J is *linear* if for any j and k in J, either $j < k$ or $k < j$. A linear ordering J is *dense* if for any j and k in J such that $j < k$, there exists an element i of J such that $j < i < k$. It is *scattered* if it contains no dense subordering. The ordering ω of natural integers and the ordering ζ of relative integers are scattered. More generally, ordinals are scattered orderings.

Let A be a finite alphabet. A *word* $x = (a_j)_{j \in J}$ indexed by a linear ordering J is a function from J to A. J is called the *length* of x. For instance ω is the length of right-infinite words $a_0 a_1 \ldots$ and ζ is the length of bi-infinite words $\ldots a_{-1} a_0 a_1 \ldots$.

In order to define the rank of scattered linear orderings, we recall operators.

2.1   Operations on Linear Orderings

For any linear ordering J, we denote by $-J$ the backward linear ordering that is the set J equipped with the reverse ordering. For instance, $-\omega$ is the linear ordering of negative integers.

The sum $J + K$ of two linear orderings is the set $J \cup K$ equipped with the ordering $<$ extending the orderings of J and K by setting $j < k$ for any $j \in J$ and $k \in K$. For instance, $\zeta = -\omega + \omega$. Formally, the *sum* $\sum_{j \in J} K_j$ is the set of all pairs (k, j) such that $k \in K_j$ equipped with the ordering defined by $(k_1, j_1) < (k_2, j_2)$ if and only if $j_1 < j_2$ or $(j_1 = j_2$ and $k_1 < k_2$ in $K_{j_1})$.

The sum of linear orderings helps to define the lengths of the products of words. Let J be a linear ordering and let $(x_j)_{j \in J}$ be words of respective length K_j for any $j \in J$. The word $x = \prod_{j \in J} x_j$ obtained by concatenation of the words x_j with respect to the ordering on J is of length $L = \sum_{j \in J} K_j$. We call *J-product* a product indexed by the ordering J. For instance, the ω-product of the word a^ω is the word $(a^\omega)^\omega$ of length $\sum_\omega \omega$. The sequence $(x_j)_{j \in J}$ of words is a *J-factorization* of the word $x = \prod_{j \in J} x_j$.

2.2   Construction of Countable Scattered Linear Orderings

Countable scattered linear orderings are defined through a forbidden pattern, namely that they do not contain a dense subordering. Hausdorff's theorem states that they can be constructed from finite orderings.

We denote by $\mathcal{N}$ the subclass of finite linear orderings, $\mathcal{O}$ the class of countable ordinals and $\mathcal{S}$ the class of countable scattered linear orderings.

Theorem 1. [12] *A countable linear ordering J is scattered if and only if J belongs to* $\bigcup_{\alpha \in \mathcal{O}} V_\alpha$ *where the classes V_α are inductively defined by:*

1. $V_0 = \{\mathbf{0}, \mathbf{1}\}$

2. $V_\alpha = \left\{ \sum_{j \in J} K_j \mid J \in \mathcal{N} \cup \{\omega, -\omega, \zeta\} \text{ and } K_j \in \bigcup_{\beta < \alpha} V_\beta \right\}.$

where $\mathbf{0}$ *and* $\mathbf{1}$ *are respectively the orderings of zero and one element.*

Intuitively, the rank of a linear ordering is the maximum number of nested ω and $-\omega$. It is linked to its Hausdorff's class. For instance the orderings ω of rank 1 and ω^2 of rank 2 belong respectively to V_1 and V_2. Nevertheless, the class V_α is not exactly the set of orderings of rank α. For instance, the ordering $\omega + \omega$ is of rank 1 and belongs to V_2. Therefore, we work on slightly different inductive classes. For any $\alpha \in \mathcal{O}$, we define the class W_α by :

$$W_\alpha = \left\{ \sum_{j \in J} K_j \mid J \in \mathcal{N} \text{ and } K_j \in V_\alpha \right\}.$$

Those classes are strictly intermediate to the Hausdorff's ones: the inclusions $V_\alpha \subset W_\alpha \subset V_{\alpha+1}$ hold for any ordinal α. For instance, the ordering $\omega^\alpha + \omega^\alpha$ belongs to W_α but does not belong to V_α and the ordering $\omega^{\alpha+1}$ belongs to $V_{\alpha+1}$ but does not belong to W_α. Formally, the *rank* of a linear ordering J is the smallest ordinal α such that $J \in W_\alpha$. For instance the orderings of rank 0 are the finite ones. In this paper, we restrict to linear orderings of finite ranks that is the set $\bigcup_{n<\omega} W_n = \bigcup_{n<\omega} V_n$.

By extension, the rank of a word is the rank of its length and the rank of a set of words is the upper bound of the ranks of its elements.

We denote by $A^\diamond$ the set of all words indexed by countable scattered linear orderings and we also denote by A^{W_r} (respectively A^{V_r}) the set of words whose length is an ordering in W_r (respectively V_r) for some integer r. Thus the words of A^{W_r} have a rank lower than or equal to r.

3   Rational Sets of Words on Linear Orderings

Bruyère and Carton [2] have introduced rational expressions and automata for words indexed by countable scattered linear orderings. They have proved that a set of words is rational if and only if it is recognizable extending Kleene's theorem. More precisely, they have defined a whole hierarchy of rational sets [3]. For each subset of rational operations, they consider the class of corresponding rational languages and define transitions of automata capturing the same languages. In the following section, the characterization of rational sets of words of finite rank is notified.

3.1   Rational Expressions

The rational sets of finite rank can be obtained from finite sets of finite words using the union $+$, the concatenation $\cdot$, the star $*$, the omega iteration ω and the backwards omega iteration $-\omega$. Let X and Y be two sets of words, we define:

$$X + Y = \{z|\ z \in X \cup Y\}$$
$$X \cdot Y = \{x \cdot y|\ x \in X, y \in Y\}$$
$$X^* = \{\prod_{j=1}^{n} x_j|\ n \in \mathcal{N}, x_j \in X\}$$
$$X^\omega = \{\prod_{j \in \omega} x_j|\ x_j \in X\}$$
$$X^{-\omega} = \{\prod_{j \in -\omega} x_j|\ x_j \in X\}$$

To define rational sets of words indexed by all linear orderings, three more operations are needed : the ordinal iteration $\#$, the backwards ordinal iteration $-\#$ and the iteration for all linear countable scattered orderings $\diamond$.

$$X^{\#} = \{\prod_{j \in J} x_j|\ J \in \mathcal{O}, x_j \in X\}$$
$$X^{-\#} = \{\prod_{j \in -J} x_j|\ J \in \mathcal{O}, x_j \in X\}$$
$$X \diamond Y = \{\prod_{j \in J \cup \hat{J}^*} z_j|\ J \in \mathcal{S} \setminus \emptyset, z_j \in X \text{ if } j \in J \text{ and } z_j \in Y \text{ if } j \in \hat{J}^*\}$$

In this paper, we are only interested in languages which are defined using $+$, $\cdot$, ω and $-\omega$. We refer the reader to [2] for a precise definition of other rational operations. A set of words on linear orderings is *rational* if it is obtained from finite sets of finite words using the rational operations defined above.

3.2   Automata on Linear Orderings

Let (Q, A, E, I, F) be a classical automaton on finite words with usual notations. As the set E of transitions is a subset of $Q \times A \times Q$, the paths of such an automaton are finite. In Büchi automata, a word is accepted if it is the label of a path going infinitely times through a given set of states. The problem is that this accepting condition does not even allow to recognize the concatenation of infinite words. To cope with this difficulty, a set of limit transitions included in $\mathcal{P}(Q) \times Q$ is introduced. This way, if an infinite path goes infinitely many times through the states of a set P and that the transition (P, q) exists, then the next state of the path may be q.

Example 1. : Let $\mathcal{A} = (Q, A, E, I, F)$ be the automaton of Figure 1 where $Q = \{1, 2, 3\}$, $A = \{a, b\}$, $I = \{1\}$, $F = \{3\}$.

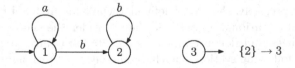

Fig. 1. Automaton recognizing $a^* b^\omega$

A limit transition $\{2\} \to 3$ is added to E. Intuitively, an infinite path going through the state 2 infinitely many times leads to state 3 and a path in $\mathcal{A}$ leading from state 2 to state 3 is labelled b^ω. Finally, this automaton recognizes the language $a^* b^\omega$.

The previous limit transitions called *left limit transitions* allow to recognize sets of words indexed by countable ordinals. In order to get words indexed by linear scattered orderings, we also need *right limit transitions*.

Definition 1. *An automaton $\mathcal{A}$ on linear orderings is defined by $\mathcal{A} = (Q, A, E, I, F)$ where Q is a finite set of states, A is a finite alphabet, $E \subseteq (Q \times A \times Q) \cup (\mathcal{P}(Q) \times Q) \cup (Q \times \mathcal{P}(Q))$ is the set of transitions and $I \subseteq Q$ and $F \subseteq Q$ are respectively the sets of initial and final states.*

Right limit transitions are used symetrically when a path has a limit length on the left. In order to use nested limit transitions, it is needed to define the left (respectively right) limit sets of states in a given point of the path.

Consider a finite path $q_0 \xrightarrow{a_1} q_1 \xrightarrow{a_2} \ldots \xrightarrow{a_n} q_n$ labelled $x = a_1 \ldots a_n$. Note that a state is inserted between any two consecutive letters of x. In other words, to any two-factorization $x = (a_1 \ldots a_k)(a_{k+1} \ldots a_n)$ of x is associated a state q_k. This definition of paths is generalized to automata on linear orderings in the following way: Let x be a word indexed by a linear scattered ordering J. To any two-factorization $x = yz$ of x, one can associate a partition of J into two intervals (K, L) such that $|y| = K$ and $|z| = L$. Then, a path labelled x is a function from the set $\hat{J} = \{(K, L) | K \cup L = J \wedge \forall k \in K, \forall l \in L, k < l\}$ into the set of states. As the set $\hat{J}$ is naturally equipped with the ordering $(K_1, L_1) < (K_2, L_2)$ if and only if $K_1 \subset K_2$, a path labelled by a word of length J is a word over Q of length $\hat{J}$. An element of $\hat{J}$ is called a *cut*.

Let $\gamma = (q_c)_{c \in \hat{J}}$ be a word of length $\hat{J}$ over Q, we are now able to define the limit sets of states of γ in a given cut c of $\hat{J}$:

$$\lim_{c^-} \gamma = \{q \in Q | \ \forall c' < c, \ \exists c' < c'' < c \text{ such that } q = q_{c''}\}$$

$$\lim_{c^+} \gamma = \{q \in Q | \ \forall c' > c, \ \exists c < c'' < c' \text{ such that } q = q_{c''}\}$$

For instance, in example 1, the word $\gamma = (q_c)_{c \in \hat{\omega}}$ defined by $q_{(\emptyset, \omega)} = 1$, $q_{(\{0, 1, \ldots, n\}, \{n+1, \ldots\})} = 2$ for any positive integer n and $q_{(\omega, \emptyset)} = 3$ has the following nonempty limit $\lim_{(\omega, \emptyset)-} \gamma = \{2\}$.

Finally, a path has to be compatible with the automata transitions:

Definition 2. *Let $\mathcal{A} = (Q, A, E, I, F)$ be an automaton on linear orderings and let $x = (a_j)_{j \in J}$ be a word of length J on A.*
A path γ of label x in $\mathcal{A}$ is a word $\gamma = (q_c)_{c \in \hat{J}}$ of length $\hat{J}$ over Q such that for any $(K, L) \in \hat{J}$:

- *If there exists $l \in L$ such that $(K \cup \{l\}, L \setminus \{l\}) \in \hat{J}$*
  then $q_{(K,L)} \xrightarrow{a_l} q_{(K \cup \{l\}, L \setminus \{l\})} \in E$ else $q_{(K,L)} \to \lim_{(K,L)-} \gamma \in E$.

- *If there exists $k \in K$ such that $(K \setminus \{k\}, L \cup \{k\}) \in \hat{J}$*
  then $q_{(K \setminus \{k\}, L \cup \{k\})} \xrightarrow{a_k} q_{(K,L)} \in E$ else $\lim_{(K,L)+} \gamma \to q_{(K,L)} \in E$.

Thus, if a cut has a predecessor or a successor, usual transitions are used, else the path is built on limit transitions. As $\hat{J}$ has the least element $(\emptyset, J)$ and the greatest element $(J, \emptyset)$ for any linear ordering J, a path has always a first and a last state. It is said to be *successful* if it leads from an initial state to a final state. A word is *recognized* by an automata if it is the label of a successful path.

We denote by $p \overset{x}{\Longrightarrow} q$ the existence of a path leading from state p to q of label x. The *content* of a path is the set of states occuring in the path and $p \overset{x}{\underset{P}{\Longrightarrow}} q$ denotes a path leading from p to q of label x and of content P.

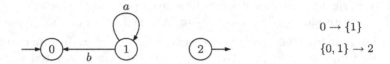

$$0 \to \{1\}$$
$$\{0, 1\} \to 2$$

Fig. 2. Automaton on linear orderings recognizing $(a^{-\omega}b)^{\omega}$

3.3   Generalisations of Kleene's Theorem

Bruyère and Carton have generalized Kleene's theorem on words indexed by countable scattered linear orderings:

Theorem 2. *[2] A set of words indexed by countable scattered linear orderings is rational if and only if it is recognizable.*

Moreover, they have defined a subclass of automata on linear orderings which recognizes rational languages of finite ranks.

Theorem 3. *[3] A set of words of finite rank is rational if and only if it is recognized by an automata on linear orderings where limit transitions $P \to q$ or $q \to P$ verify $q \notin P$.*

4   Complement of a Rational Set of Finite Rank

In the case of finite words, it is known that rational sets are closed under complementation. Given an automaton on finite words recognizing a language L, the construction of an automaton recognizing the complement $A^* \setminus L$ is based on the property that any finite automaton on finite words can be determinized. Büchi has generalized this result for sets of words indexed by countable ordinals of finite ranks [9]. This property does not hold any longer for automata on linear orderings. An automaton on linear orderings $\mathcal{A} = (Q, A, E, I, F)$ is *deterministic* if for any state $q \in Q$ and any word $u \in A^{\diamond}$, there exists at most one path labelled u starting from q.

Proposition 1. *The language $(a^{-\omega})^{-\omega}$ can not be recognized by a derterministic automaton.*

To cope with this difficulty of determinism, we use a different method based on equivalence classes to prove the closure of rational sets under complementation. Up to now, we are only able to prove this result in the case of rational sets of words of finite ranks.

Theorem 4. *Let L be a rational set of words on linear orderings and let r be a finite integer. The complement $A^{W_r} \setminus L$ is rational.*

In the case of finite words, Büchi has given a different proof of the closure under complement of rational sets. It does not need the property of determinizability but it is based on the following equivalence relation defined for any finite automaton $\mathcal{A} = (Q, A, E, I, F)$ on finite words:

$$u \sim v \text{ if and only if } \forall p \in Q, \forall q \in Q, p \overset{u}{\Longrightarrow} q \iff p \overset{v}{\Longrightarrow} q$$

Note that if a word u is the label of a successful path in $\mathcal{A}$, it holds for any equivalent word. So any equivalence class is either contained in the language L recognized by $\mathcal{A}$ or disjoint from L. Moreover, equivalence classes are rational thus the complement of L is rational as a finite union of equivalence classes. We extend this proof to automata on linear orderings of finite ranks. Let $\mathcal{A} = (Q, A, E, I, F)$ be an automaton on linear orderings recognizing L. Recall that a path from p to q with label u and content P is denoted by $p \overset{u}{\underset{P}{\Longrightarrow}} q$. As the contents of paths are needed in limit transitions, we define the equivalence relation $\sim$ by:

$$u \sim v \text{ if and only if } \forall p \in Q, \forall q \in Q, \forall P \subseteq Q, p \overset{u}{\underset{P}{\Longrightarrow}} q \iff p \overset{v}{\underset{P}{\Longrightarrow}} q$$

Note first that the equivalence relation has finitely many classes. Indeed the class of a word u depends on whether there is a path from p to q with content P for each triple (p, q, P). Since there are $n^2 2^n$ such triples, the relation $\sim$ has at most $2^{n^2 2^n}$ equivalence classes. We denote by $\mathcal{C}$ the set of all equivalence classes of $\sim$. For each integer r, we denote by $\mathcal{C}_r = \{C \cap A^{W_r} | C \in \mathcal{C}\}$ the set of equivalence classes of rank r. The cardinality of $\mathcal{C}_r$ is at most the cardinality of $\mathcal{C}$. As in the case of finite words, each class C is either contained in L or disjoint from L. Therefore we have both equalities

$$L = \bigcup_{C \in \mathcal{C}, C \cap L \neq \emptyset} C \text{ and } \bar{L} = A^\diamond \setminus L = \bigcup_{C \in \mathcal{C}, C \cap L = \emptyset} C$$

The same holds for words of rank less than r.

$$L \cap A^{W_r} = \bigcup_{C \in \mathcal{C}_r, C \cap L \neq \emptyset} C \text{ and } A^{W_r} \setminus L = \bigcup_{C \in \mathcal{C}_r, C \cap L = \emptyset} C.$$

For each integer r, the family $\mathcal{C}_r$ contains finitely many classes. To prove that $A^{W_r} \setminus L$ is rational, it suffices to prove that each $C \in \mathcal{C}_r$ is rational. We prove that claim by induction on r. The result holds obviously for $r = 0$ and the induction step is based on the following idea. Suppose that $\mathcal{C}_r$ contains the classes $\{C_1, ..., C_m\}$. We define rational expressions using the C_i as letters. An elementary expression is an expression of the form C_i, C_i^{ω} or $C_i^{-\omega}$ where C_i is a class of $\mathcal{C}_r$. We denote by B the set of elementary expressions. We consider the set B^* of all expressions obtained by concatenation of elementary expressions. Suppose for instance that $\mathcal{C}_r = \{C_1, C_2\}$. The set of elementary expressions is $B = \{C_1, C_1^{\omega}, C_1^{-\omega}, C_2, C_2^{\omega}, C_2^{-\omega}\}$ and a typical example of element of B^* is $C_2^{\omega} C_1 C_2^{-\omega} C_1 C_2^{-\omega}$. We consider each element of B^* as a rational expression over the letters C_i. Each expression of B^* denotes a set of words of rank at most $r+1$. By a slight abuse of language, we say that a word belongs to an expression R in B^* if it actually belongs to the set denoted by R. The two following lemmas are needed in the proof of proposition 2. Their proofs are not detailed in this paper because of the lack of space. In Lemma 1, we first prove that each word of rank at most $r + 1$ belongs to at least one expression in B^*.

Lemma 1. $A^{W_{r+1}} = \bigcup_{R \in B^*} R.$

In Lemma 2, we prove that two words belonging to the same expression are $\sim$-equivalent. This means that each set denoted by an expression of B^* is included in a single $\sim$-class.

Lemma 2. *If two words x, y of rank at most $r + 1$ belong to the same expression R of B^*, then they satisfy $x \sim y$.*

It follows from Lemmas 1 and 2 that each class C in $\mathcal{C}_{r+1}$ satisfies

$$C = \bigcup_{R \in B^*, C \cap R \neq \emptyset} R$$

However, this is not a rational expression since there are infinitely many such expressions R included in C. In the following proposition, we show that the set of rational expressions included in some class C can be described by a rational expression over the elementary expressions.

Proposition 2. *Each equivalence class in $\mathcal{C}_r$ is rational.*

The proof by induction on the rank r is not detailed in this paper. We come back to the proof of Theorem 4.

Proof. Let $\mathcal{A}$ be an automaton on linear orderings recognizing L and let r be a finite rank. Let $\mathcal{C}_r$ be the set of equivalence classes of rank r according to $\mathcal{A}$. From proposition 2, we have that each class of $\mathcal{C}_r$ is rational. Moreover, considering the definition of $\sim$, we note that if a word u is the label of a successful path in $\mathcal{A}$, it holds for any equivalent word. So an equivalence class is either contained in L or disjoint of L. We deduce a rational expression of $A^{W_r} \setminus L$ as a finite union of classes of $\mathcal{C}_r$:

$$A^{W_r} \setminus L = \bigcup_{C \in \mathcal{C}_r, C \cap L = \emptyset} C$$

$\square$

As a conclusion, we mention a question that is left open by this paper. A generalization of our result is that the class of rational sets of countable scattered linear orderings is closed under complementation.

References

1. A. V. Aho, J. E. Hopcroft, and J. D. Ullman. *The Design and Analysis of Computer Algorithms.* Addison-Wesley, London, 1974.
2. V. Bruyère and O. Carton. Automata on linear orderings. In J. Sgall, A. Pultr, and P. Kolman, editors, *MFCS'2001*, volume 2136 of *Lect. Notes in Comput. Sci.*, pages 236–247, 2001. IGM report 2001-12.
3. V. Bruyère and O. Carton. Hierarchy among automata on linear orderings. In *IFIP TCS'2002*, pages 107–118, 2002.
4. N. Bedon. Finite automata and ordinals. *Theoret. Comput. Sci.*, 156:119–144, 1996.
5. N. Bedon. Automata, semigroups and recognizability of words on ordinals. *Int. J. Alg. Comput.*, 8:1–21, 1998.
6. N. Bedon and O. Carton. An Eilenberg theorem for words on countable ordinals. In Cláudio L. Lucchesi and Arnaldo V. Moura, editors, *Latin'98: Theoretical Informatics*, volume 1380 of *Lect. Notes in Comput. Sci.*, pages 53–64. Springer-Verlag, 1998.
7. J. R. Büchi. Weak second-order arithmetic and finite automata. *Z. Math. Logik und grundl. Math.*, 6:66–92, 1960.
8. J. R. Büchi. On a decision method in the restricted second-order arithmetic. In *Proc. Int. Congress Logic, Methodology and Philosophy of science, Berkeley 1960*, pages 1–11. Stanford University Press, 1962.
9. J. R. Büchi. Transfinite automata recursions and weak second order theory of ordinals. In *Proc. Int. Congress Logic, Methodology, and Philosophy of Science, Jerusalem 1964*, pages 2–23. North Holland, 1965.
10. Y. Choueka. Finite automata, definable sets, and regular expressions over ω^n-tapes. *J. Comput. System Sci.*, 17(1):81–97, 1978.
11. D. Girault-Beauquier. Bilimites de langages reconnaissables. *Theoret. Comput. Sci.*, 33(2–3):335–342, 1984.
12. F. Hausdorff. Set theory. In *Chelsea*, New York, 1957.
13. S. C. Kleene. Representation of events in nerve nets and finite automata. In C.E. Shannon, editor, *Automata studics*, pages 3–41. Princeton university Press, Princeton, 1956.
14. D. Muller. Infinite sequences and finite machines. In Proc. of Fourth Annual IEEE Symp., editor, *Switching Theory and Logical Design*, pages 3–16, 1963.
15. M. Nivat and D. Perrin. Ensembles reconnaissables de mots bi-infinis. In *Proceedings of the Fourteenth Annual ACM Symposium on Theory of Computing*, pages 47–59, 1982.
16. D. Perrin and J.E. Pin. Infinite words. In Elsevier, editor, *Academic Press*, 2003.
17. J.-E. Pin. *Handbook of formal languages*, volume 1, chapter Syntactic semigroups, pages 679–746. Springer-Verlag, 1997.
18. M. O. Rabin. Decidability of second-order theories and automata on infinite trees. 141:1–35, 1969.
19. F. D. Ramsey. On a problem of formal logic. *Proc. of the London math. soc.*, 30:338–384, 1929.
20. J. G. Rosenstein. *Linear ordering.* Academic Press, New York, 1982.
21. S. Safra. On the complexity of ω-automata. In *29th Annual Symposium on Foundations of computer sciences*, pages 24–29, 1988.
22. J. Wojciechowski. Finite automata on transfinite sequences and regular expressions. *Fundamenta informaticæ*, 8(3-4):379–396, 1985.

Expected Length of the Longest Common Subsequence for Large Alphabets

Marcos Kiwi[1*], Martin Loebl[2**], and Jiří Matoušek[2***]

[1] University of Chile, Dept. de Ing. Matemática and Ctr. de Modelamiento Matemático, UMR–UChile 2071, Correo 3, Santiago 170–3, Chile, mkiwi@dim.uchile.cl
[2] Charles University, Dept. of Applied Mathematics and Institute of Theoretical Computer Science (ITI), Malostranské nám. 25, 118 00 Praha 1, Czech Republic {loebl,matousek}@kam.mff.cuni.cz

Abstract. We consider the length L of the longest common subsequence of two randomly uniformly and independently chosen n character words over a k-ary alphabet. Subadditivity arguments yield that $\mathbf{E}[L]/n$ converges to a constant γ_k. We prove a conjecture of Sankoff and Mainville from the early 80's claiming that $\gamma_k \sqrt{k} \to 2$ as $k \to \infty$.

1 Introduction

Consider two sequences of length n, with letters from a size k alphabet Σ, say μ and ν. The longest common subsequence (LCS) problem is that of finding the largest value L for which there are $1 \le i_1 < i_2 < \ldots < i_L \le n$ and $1 \le j_1 < j_2 < \ldots < j_L \le n$ such that $\mu_{i_t} = \nu_{j_t}$, for all $t = 1, 2, \ldots, L$.

The LCS problem has emerged more or less independently in several remarkably disparate areas, including the comparison of versions of computer programs, cryptographic snooping, and molecular biology. The biological motivation of the problem is that long molecules such as proteins and nucleic acids like DNA can be schematically represented as sequences from a finite alphabet. Taking an evolutionary point of view, it is natural to compare two DNA sequences by finding their closest common ancestors. If one assumes that these molecules evolve only through the process of inserting new symbols in the representing strings, then ancestors are substrings of the string that represent the molecule. Thus, the length of the longest common subsequence of two strings is a reasonable measure of how close both strings are. In the mid 1970's, Chvátal and Sankoff [6] proved that the expected length of the LCS of two random k-ary sequences of length n when normalized by n converges to a constant. The value of this constant γ_k is unknown although much effort has been spent in finding good upper an lower bounds for it (see, for example, [3] and references therein). The best known upper and lower bounds for γ_k do not have a closed form. There are obtained either as numeric

---

* Gratefully acknowledges the support of ICM P01–05 and Fondecyt 1010689.
** This work was done while visiting the Dept. de Ing. Matemática, University of Chile, supported by ICM-P01-05.
*** This research was done while visiting the Ctr. de Modelamiento Matemático, UMR–UChile 2071, supported by Fondap in Applied Mathematics.

M. Farach-Colton (Ed.): LATIN 2004, LNCS 2976, pp. 302–311, 2004.
© Springer-Verlag Berlin Heidelberg 2004

approximation to the solutions of a nonlinear equation or as a numeric evaluation of some series expansion (see [7] for a survey of such results).

Although the problem of determining γ_k has a simple statement, it has turned out to be a challenging mathematical endeavor. Moreover, its quite naturally motivated. Indeed, a claim that two DNA sequences of length n are somewhat related makes sense provided their LCS differs significantly from $\gamma_4 n$ (since DNA sequences have 4 basis elements). We analyze the behavior of γ_k for k tending to infinity, and more generally, we consider the expected length of the LCS when k is an (arbitrarily slowly growing) function of n and $n \to \infty$. We confirm a conjecture of Sankoff and Mainville from the early 80's [20] stating that

$$\lim_{k \to \infty} \gamma_k \sqrt{k} = 2. \tag{1}$$

(See [19, § 6.8] for a discussion of a stronger version, due to Arratia and Steele, of the above stated conjecture.)

The constant 2 in (1) arises from a connection with the famous longest increasing sequence (LIS) problem. An increasing subsequence of length L of a permutation π of $\{1, \ldots, n\}$ is a sequence $1 \leq i_1 < i_2 < \ldots < i_L \leq n$ such that $\pi(i_1) < \pi(i_2) < \ldots < \pi(i_L)$. A LIS is an increasing subsequence of maximum length. The LIS problem concerns the determination of the asymptotic, on n, behavior of the length of a LIS of a randomly and uniformly chosen permutation π. The LIS problem is also referred to as "Ulam's problem." (e.g., in [14,4,18]). Ulam is often credited for raising it in [23] where he mentions (without reference) a "well–known theorem" asserting that given $n^2 + 1$ integers in any order, it is always possible to find among them a monotone subsequence of $n + 1$. The theorem is due to Erdős and Szekeres [8]. The discussion in [23] solely concerns monotonic subsequences of a randomly and uniformly chosen permutation of $n^2 + 1$ elements. Monte Carlo simulations are reported in [2], where it is observed that over the range $n \leq 100$, the limit of the length of the LIS of $n^2 + 1$ randomly chosen elements, when normalized by n, approaches 2. Hammersley [11] gave a rigorous proof of the existence of the limit and conjectured it was equal to 2. Later, Logan and Shepp [17] based on a result by Schensted [21] proved that $\gamma \geq 2$; finally, Vershik and Kerov [24] showed that $\gamma \leq 2$. In a major recent breakthrough due to Baik, Deift, Johansson [4] the asymptotic distribution of the longest increasing sequence random variable has been determined. For a detailed account of these results, history and related work see the surveys of Aldous and Diaconis [1] and Stanley [22].

2 Statement of Results

Henceforth we denote by A and B two disjoint totally ordered sets. We assume that the elements of A and B are numbered $1, 2, \ldots, |A|$ and $1, 2, \ldots, |B|$ respectively. We denote by r and s the size of $|A|$ and $|B|$, respectively. Typically, we have $r = s = n$.

Throughout the paper we follow standard graph theory notation (for a reference the reader might consult Bollobás [5]). We let G denote a bipartite graph with color classes A and B. Two distinct edges ab and $a'b'$ of G are said to be *noncrossing* if a and a' are in the same order as b and b'; in other words, if $a < a'$ and $b < b'$ or $a' < a$ and $b' < b$. A matching of G is called *planar* if every distinct pair of its edges is noncrossing. We let $L(G)$ denote the number of edges of a maximum size planar matching in G (note that $L(G)$ depends on the graph G *and* on the ordering of its color classes). We want to understand $L(G)$'s behavior for random choices of G. Primarily choices according to the following two models of random graphs:

- The *random words model* $\Sigma(K_{n,n}; k)$: the distribution over the set of subgraphs of $K_{n,n}$ obtained by uniformly and independently assigning each node of $K_{n,n}$ one of k characters and keeping those edges whose end-points are associated to equal characters.
- The *binomial random graph model* $G(K_{n,n}; p)$: the distribution over the set of subgraphs of $K_{n,n}$ where each edge of $K_{n,n}$ is included with probability p, and these events are mutually independent. (This is an obvious modification of the usual $G(n, p)$ model for bipartite graphs with ordered color classes.)

For a bipartite graph G over color classes A and B, let

$$L(G) = \max\{L : \exists a_1 < \ldots < a_L, b_1 < \ldots < b_L, a_i b_i \in E(G), 1 \le i \le L\},$$

Observe that $L(G)$, when G is chosen according to $\Sigma(K_{n,n}; k)$, is precisely the length of the LCS of the two words, one for each of the color classes of G, corresponding to the characters associated to $K_{n,n}$'s nodes. Also note that the latter words are uniformly and independently distributed length n sequences of characters over a k size alphabet. In other words, the study of $L(\Sigma(K_{n,n}; k))$ is just a re-wording of a similar study of the LCS of two randomly chosen n length sequences over a size k alphabet. Nevertheless, it will be more convenient to cast our discussion in the language of graph theory.

We now argue that $L(G)$ is "subadditive" and from it draw an important conclusion about its expected asymptotic behavior. Indeed, consider two bipartite graphs G and G' over disjoint color classes A-B and A'-B', respectively. Denote by $G \oplus G'$ the bipartite graph over color classes $A \cup A'$-$B \cup B'$. In order for $L(G \oplus G')$ to be well defined we adopt the convention that the elements of G''s color classes are strictly larger than those of G's color classes. It follows immediately that $L(\cdot)$ is subadditive, i.e., $L(G \oplus G') \ge L(G) + L(G')$. Thus, for G and G' chosen according to $\Sigma(K_{n,n}; k)$ and $\Sigma(K_{m,m}; k)$ respectively

$$\mathbf{E}\left[L(G \oplus G')\right] \ge \mathbf{E}\left[L(G)\right] + \mathbf{E}\left[L(G')\right].$$

A standard subadditivity argument implies existence of $\lim_{n \to \infty} \mathbf{E}\left[L(\Sigma(K_{n,n}; k))\right]/n]$. The same claim holds for the binomial random graph model.

Our main result essentially says that $L(\Sigma(K_{n,n}; k)) \cdot \sqrt{k}/n$ converges to 2 as $k \to \infty$, provided that n is sufficiently large in terms of k. Specifically,

Theorem 1. *For every $\varepsilon > 0$ there exist k_0 and C such that for all $k > k_0$ and all n with $n/\sqrt{k} > C$ we have*

$$(1 - \varepsilon) \cdot \frac{2n}{\sqrt{k}} \le \mathbf{E}\left[L(\Sigma(K_{n,n}; k))\right] \le (1 + \varepsilon) \cdot \frac{2n}{\sqrt{k}}.$$

Moreover, there is an exponentially small tail bound; namely, for every $\varepsilon > 0$ there exists $c > 0$ such that for k and n as above,

$$\mathbf{P}\left[\left|L(\Sigma(K_{n,n}; k)) - \frac{2n}{\sqrt{k}}\right| \ge \varepsilon \frac{2n}{\sqrt{k}}\right] \le e^{-cn/\sqrt{k}}.$$

Corollary 1. *The limit* $\gamma_k = \lim\limits_{n\to\infty} \mathbf{E}\left[L(\Sigma(K_{n,n}; k))/n\right]$ *exists, and* $\lim\limits_{k\to\infty} \gamma_k \sqrt{k} = 2$.

The focus on the case where $k \to \infty$ is partly inspired by [15]. There, it is shown that $L(G)/\sqrt{dn} \to 2$ in probability as $n \to \infty$ provided $d = o(n^{1/4})$ and G is a uniformly chosen d-regular subgraph of $K_{n,n}$. Under the $d = o(n^{1/4})$ condition, any node of the d-regular bipartite graph can potentially be matched to a $d/n \to 0$ fraction of the other color class nodes. In the case of interest here, that is the LCS problem with $k \to \infty$, it also happens that any sequences' character can be matched to an expected $1/k \to 0$ fraction of the other sequence's characters. Both for this work and in [15], the vanishing fraction of (expected) potential matches is a key issue. Indeed, this is where the connection with the LIS problem arises. To clarify this point, suppose G is chosen according to $\Sigma(K_{n,n}; k)$ and assume $n \ll k$. Then, an easy calculation shows that the expected number of edges of G is n^2/k and that the average degree of a node is $n/k \ll 1$. For $n^2/k \gg 1$, it turns out that disregarding degree 0 nodes, G is essentially a perfect matching on approximately n^2/k nodes. In other words, essentially a permutation π on approximately n^2/k elements! Moreover, a LIS of π is a planar matching in the original graph G. It turns out that the length of a LIS of π is in fact very close to $L(G)$.

Here is an outline of the paper. First, we state in Section 3, the estimate for the length of a LIS of a uniformly chosen permutation on which we rely. Then in Section 4, we formalize the claim of the previous paragraph. In Sections 5 and 6, we handle the case were n is not small in comparison with k. In these sections we re-establish, respectively, the lower and upper bounds derived in Section 4, but lifting the constraint that n be "small". This completes the proof of our main result. The gist of the paper is the material of Section 6 showing how the upper bound for "small" values of n are used to obtain upper bounds without this latter restriction on n. In order to simplify the exposition, we focus mainly on the random words model. Nevertheless, in Section 7 we discuss other models, among them the binomial random graph model. Thus we hope to convey, to some extent, that our proof arguments can be successfully adapted to a wider class of probabilistic distributions over bipartite graphs. Due to space considerations, basically all proofs are omitted from this extended abstract. Full proofs can be found in [16].

3 Tools

The crucial ingredient in our proofs is a sufficiently precise result on the distribution of the length of the longest increasing subsequence in a random permutation. We state a remarkable strong result of Baik, Deift and Johansson [4, eqn. (1.7) and (1.8)] (our formulation slightly weaker than theirs, in order to make the statement simpler). A much weaker tail bound than provided by them would actually suffice for our proof (e.g., Frieze's [9] LIS concentration result).

Theorem 2. *Let* LIS_N *be the length of the longest increasing subsequence of a randomly chosen permutation of* $\{1, \ldots, N\}$. *There are positive constants* B_0, B_1, *and* c *such that for every* λ *with* $B_0/N^{1/3} \leq \lambda \leq \sqrt{N} - 2$,

$$\mathbf{P}\left[\mathrm{LIS}_N \geq 2\sqrt{N} + \lambda\sqrt{N}\right] \leq B_1 \exp\left(-c\lambda^{3/5} N^{1/5}\right),$$

and for every λ *with* $B_0/N^{1/3} \leq \lambda \leq 2$,

$$\mathbf{P}\left[\mathrm{LIS}_N \leq 2\sqrt{N} - \lambda\sqrt{N}\right] \leq B_1 \exp\left(-c\lambda^3 N\right).$$

4   Small Graphs

In this section we derive a result essentially saying that Theorem 1 holds if k is sufficiently large in terms of n. For technical reasons, we also need to consider bipartite graphs with color classes of unequal sizes.

Proposition 1. *For every* $\delta > 0$, *there exists a (large) positive constant* C *such that:*

(i) *If* $rs \geq Ck$ *and* $(r + s)\sqrt{rs} \leq \delta k^{3/2}/6$, *then with* $m_u = m_u(r, s) = 2(1 + \delta)\sqrt{rs/k}$, *for all* $t \geq 0$,

$$\mathbf{P}[L(\Sigma(K_{r,s}; k)) \geq m_u + t] \leq 2e^{-t^2/8(m_u+t)}.$$

(ii) *If* $rs \geq Ck$ *and and* $r + s \leq \delta k/6$, *then with* m_u *as above and* $m_l = m_l(r, s) = 2(1 - \delta)\sqrt{rs/k}$, *for all* $t \geq 0$,

$$\mathbf{P}[L(\Sigma(K_{r,s}; k)) \leq m_l - t] \leq 2e^{-t^2/8m_u}.$$

The idea behind the proof of Proposition 1 is simple: we show that (ignoring degree 0 nodes) for G chosen according to $\Sigma(K_{r,s}; k)$ is "almost" a matching, but the size of the largest planar matching in a random matching corresponds precisely to the length of a LIS in a randomly chosen permutation.

First, we deal with the (usually few) nodes of G of degree larger than one. To this end, we define a graph G' obtained from G by removing all edges incident to nodes of degree at least 2. Throughout, E and E' denote $E(G)$ and $E(G')$, respectively.

Ignoring degree 0 nodes, G' is clearly a matching on its end-points — equivalently its a permutation of $\{1, \ldots, |E'|\}$. Theorem 2 thus gives us an estimation of $L(G')$ in terms of $|E'| = |E| - |E \setminus E'|$. But, $L(G') \leq L(G) \leq L(G') + |E \setminus E'|$. Hence, good estimates on $|E|$ and $|E \setminus E'|$ coupled with the aforementioned estimate of $L(G')$ yields the sought after bounds on $L(G)$.

It follows easily that $\mathbf{E}[|E|] = rs/k$. A simple second-moment argument (Chebyshev's inequality) suffices to obtain an estimate of $|E|$.

Lemma 1. *For every* $\eta > 0$, *we have* $\mathbf{P}\left[\left||E| - \dfrac{rs}{k}\right| \geq \eta \cdot \dfrac{rs}{k}\right] \leq \dfrac{1}{\eta^2(rs/k)}.$

Based on a Markov bound we estimate $|E \setminus E'|$. Thus we require the following

Lemma 2. *We have* $\mathbf{E}[|E \setminus E'|] \leq (r + s)rs/k^2.$

Although the underlying idea of the proof of Proposition 1 should be clear from this section's discussion, there are technicalities involved in it (see [16] for details). One key technical aspect is the use of Talagrand's inequality [12, Theorem 2.29] which provides a concentration for $L(\Sigma(K_{r,s}; k))$ around one of its medians. The remaining part of the argument consists in estimating, based on Theorem 2 and Lemmas 1 and 2, the magnitude of one such median.

5   The Lower Bound in Theorem 1

In this section we establish the lower bound on the expectation of $L(\Sigma(K_{n,n}; k))$ and the lower tail bound for its distribution.

Given ε, let $\delta > 0$ be such that $(1 - 2\delta)^2 = 1 - \varepsilon$, and let $C = C(\delta)$ be as in Proposition 1. Fix $\widetilde{C} \geq \sqrt{C}$ large enough so that

$$\exp\left(-\frac{\delta^2}{4(1+\delta)} \cdot \widetilde{C}\right) \leq \delta.$$

Let $\tilde{n}(k) = \tilde{n} = \lfloor \delta k/12 \rfloor$. Proposition 1 applies for $k \geq k_0$ where k_0 is such that $\tilde{n}(k_0) \geq \widetilde{C}\sqrt{k_0}$. It follows that

$$\mathbf{E}\left[L(\Sigma(K_{\tilde{n},\tilde{n}}; k))\right] \geq (1 - 2\delta) \cdot \frac{2\tilde{n}}{\sqrt{k}} \cdot \mathbf{P}\left[L(G) \geq 2(1 - 2\delta)\frac{\tilde{n}}{\sqrt{k}}\right]$$

$$\geq (1 - 2\delta) \cdot \frac{2\tilde{n}}{\sqrt{k}}\left(1 - 2\exp\left(-\frac{\delta^2}{4(1+\delta)} \cdot \frac{\tilde{n}}{\sqrt{k}}\right)\right) \geq (1 - \varepsilon) \cdot \frac{2\tilde{n}}{\sqrt{k}}.$$

Since, as already mentioned $\mathbf{E}\left[L(\Sigma(K_{n,n}; k)\right]$ is subadditive, $\mathbf{E}\left[L(\Sigma(K_{n,n}; k)/n\right]$ is non-decreasing. The desired lower bound on the expectation follows.

Now we establish the lower tail bound. First, we redefine $\tilde{n} = \lceil C\sqrt{k} \rceil$ and let $q = \lfloor n/\tilde{n} \rfloor$. Moreover, we let G be chosen according to $\Sigma(K_{n,n}; k)$ and let G_i be the subgraph induced in G by the nodes $(i-1) \cdot \tilde{n} + 1, \ldots, i \cdot \tilde{n}$ in each color class, $i = 1, \ldots, q$. We observe that $L(G_1), \ldots, L(G_q)$ are independent identically distributed with distribution $\Sigma(K_{\tilde{n},\tilde{n}}; k)$ and $L(G) \geq L(G_1) + \cdots + L(G_q)$. Let $\mu = \mathbf{E}\left[L(G_i)\right]$ and $t = \varepsilon(2n/\sqrt{k})$. Since $n \leq (q+1)\tilde{n}$, the lower bound on μ proved above yields that

$$\mathbf{P}\left[L(G) \leq (1 - 3\varepsilon) \cdot \frac{2n}{\sqrt{k}}\right] \leq \mathbf{P}\left[\sum_{i=1}^{q} L(G_i) \leq q\mu - t + (\mu - t)\right].$$

An argument similar to the one used above to derive the bound $\mu \geq (1 - \varepsilon)2\tilde{n}/\sqrt{k}$ can be used to obtain $\mu \leq (1 + \varepsilon)2\tilde{n}/\sqrt{k}$ from Proposition 1. Let n be large enough so that $n \geq \tilde{n}(1 + 2\varepsilon)/\varepsilon$. Thus, $q \geq (1 + \varepsilon)/\varepsilon$ and $t \geq \varepsilon q\mu/(1 + \varepsilon) \geq \mu$. Hence, a standard Chernoff bound [12, Theorem 2.1] implies that

$$\mathbf{P}\left[L(G) \leq (1 - 3\varepsilon) \cdot \frac{2n}{\sqrt{k}}\right] \leq \mathbf{P}\left[\sum_{i=1}^{q} L(G_i) \leq q\mu - t\right]$$

$$\leq \exp\left(-\frac{t^2}{2q\mu}\right) \leq \exp\left(-\frac{\varepsilon^2}{2(1+\varepsilon)} \cdot \frac{2n}{\sqrt{k}}\right).$$

This establishes the sought after lower tail bound.

6   The Upper Bound in Theorem 1

We will only discuss the tail bound since $L(\Sigma(K_{n,n}; k)) \leq n$ always, and so the claimed estimate for the expectation follows from the tail bound.

Let $\varepsilon > 0$ be fixed. We choose a sufficiently small $\delta = \delta(\varepsilon) > 0$, much smaller than ε. Requirements on δ will be apparent from the subsequent proof.

Henceforth, we fix constants $1/2 < \alpha < \beta < 3/4$ (any choice of α and β in the specified range would suffice for our purposes). In this section, we will always assume that $k \geq k_0$ for a sufficiently large integer $k_0 = k_0(\varepsilon)$, and that n is sufficiently large compared to k: $n \geq k^\beta$, say. Note that for $n \leq k^\beta$ (and k sufficiently large), the tail bound of Theorem 1 follows from Proposition 1.

Below, we first introduce the notion of a block partition associated to a "large" planar matching. We then classify block partitions into different types. Finally, we show that there are not too many different types, and that there is a very small probability that a graph chosen according to $L(\Sigma(K_{n,n};k))$ is of a given fixed type. A bound on the probability of a "large" planar matching occurring immediately follows. This provides us with the sought after upper tail bound.

Block partitions. Let us write $m_{\max} = (1+\varepsilon) \cdot (2n/\sqrt{k})$ for the upper bound on the expected size of a planar matching as in Theorem 1. We also define an auxiliary parameter $\ell = k^\alpha$. This is a somewhat arbitrary choice (but given by a simple formula). The essential requirements on ℓ are that ℓ be much larger than $\sqrt{k}$ and much smaller than $k^{3/4}$. We note that n/ℓ is large by our assumption $n \geq k^\beta$.

Let M be a planar matching with $m_{\max}$ edges on the sets A and B, $|A| = |B| = n$. We define a partition of M into blocks of consecutive edges. There will be roughly n/ℓ blocks, each of them containing at most

$$e_{\max} = \left\lfloor \frac{1}{\delta} \cdot \frac{\ell}{n} \cdot m_{\max} \right\rfloor$$

edges of M. So $e_{\max}$ is of order $\ell/\sqrt{k}$, which by our assumptions can be assumed to be larger than any prescribed constant. Moreover, we require that no block is "spread" over more than ℓ consecutive nodes in A or in B.

Formally, the ith block of the partition will be specified by nodes $a_i, a_i' \in A$ and $b_i, b_i' \in B$; $a_i b_i \in M$ is the first edge in the block and $a_i' b_i' \in M$ is the last edge (the block may contain only one edge, and so $a_i b_i = a_i' b_i'$ is possible). The edge $a_1 b_1$ is the first edge of M, and $a_{i+1} b_{i+1}$ is the edge of M immediately following $a_i' b_i'$. Finally, given $a_i b_i$, the edge a_i', b_i' is taken as the rightmost edge of M such that

- the ith block has at most $e_{\max}$ edges of M, and
- $a_i' - a_i \leq \ell$ and $b_i' - b_i \leq \ell$ (here and in the sequel, with a little abuse of notation, we regard the nodes in A and those in B as natural numbers $1, 2, \ldots, n$, although of course, the nodes in A are distinct from those of B).

Let q denote the number of blocks obtained in this way. It is easily seen that $q = O(n/\ell)$. A block partition is schematically illustrated in Fig. 1.

Counting the types. Let e_i be the number of edges of M in the ith block. Let us call the $5q$-tuple $T = (a_1, a_1', b_1, b_1', e_1, \ldots, a_q, a_q', b_q, b_q', e_q)$ the *type* of the block partition of M, and let us write $T = T(M)$. Let $\mathcal{T}$ denote the set of all possible types of block partitions of planar matchings as above.

Lemma 3. *For a suitable absolute constant C_1, we have $|\mathcal{T}| \leq \exp\left(C_1 \frac{n}{\ell} \log \ell\right)$.*

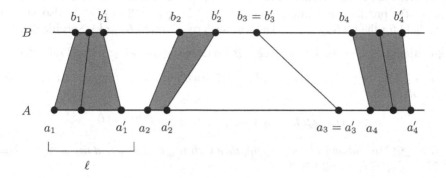

Fig. 1. A block partition.

The probability of a matching with a given type of block partition. Relying on the fact that blocks are small graphs similar to those dealt with in Section 4, we show that for every fixed type T, the probability that our random choice contains a planar matching of size $m_{\max}$ with that type of block partition is very small.

Lemma 4. *Let n and k be as above. For any given type $T \in \mathcal{T}$, the probability p_T that the random graph $\Sigma(K_{n,n}; k)$ contains a planar matching M with $m_{\max}$ edges and with $T(M) = T$ satisfies*

$$p_T \leq \exp\left(-c\varepsilon^2\delta \cdot \frac{n}{\sqrt{k}}\right)$$

with a suitable absolute constant $c > 0$.

Proof of Theorem 1. We have

$$\mathbf{P}[L(\Sigma(K_{n,n}; k)) \geq m_{\max}] \leq \sum_{T \in \mathcal{T}} p_T \leq |\mathcal{T}| \cdot \max_T p_T.$$

From Lemmas 3 and 4 follows the sought after estimate.   □

7   Extensions

One can prove results for the Erdős model analogous to those obtained in previous sections (essentially, k is now replaced by $1/p$):

Theorem 3. *For every $\varepsilon > 0$ there exist constants $p_0 \in (0, 1)$ and C such that for all $p < p_0$ and all n with $n\sqrt{p} > C$ we have*

$$(1 - \varepsilon) \cdot 2n \cdot \sqrt{p} \leq \mathbf{E}\left[L(G(K_{n,n}; p))\right] \leq (1 + \varepsilon) \cdot 2n \cdot \sqrt{p}.$$

Moreover, there is an exponentially small tail bound; namely, for every $\varepsilon > 0$ there exists $c > 0$ such that for p and n as above,

$$\mathbf{P}[|L(G(K_{n,n}; p)) - 2n\sqrt{p}| \geq \varepsilon 2n\sqrt{p}] \leq e^{-cn\sqrt{p}}.$$

Subadditivity arguments yield that $\mathbf{E}\left[L(G(K_{n,n};p))/n\right]$ converges to a constant Δ_p as $n \to \infty$. The previous theorem thus implies that $\Delta_p/\sqrt{p} \to 2$ as $p \to 0$. Also, similar results hold for the $G(K_{r,s};p)$ model as those derived for $\Sigma(K_{r,s};k)$. Specifically,

Proposition 2. *For every $\delta > 0$, there exists a (large) positive constant C such that:*

(i) *If $rs \geq C/p$ and $(r+s)\sqrt{rs} \leq \delta/6p^{3/2}$, then with $m_u = m_u(r,s) = 2(1+\delta)\sqrt{rsp}$, for all $t \geq 0$,*

$$\mathbf{P}[L(G(K_{r,s};p)) \geq m_u + t] \leq 2e^{-t^2/8(m_u+t)}.$$

(ii) *If $rs \geq C/p$ and and $r+s \leq \delta/6p$, then with m_u as above and $m_l = m_l(r,s) = 2(1-\delta)\sqrt{rsp}$, for all $t \geq 0$,*

$$\mathbf{P}[L(G(K_{r,s};p)) \leq m_l - t] \leq 2e^{-t^2/8m_u}.$$

In [13], Johansson implicitly considers a model somewhat related to the $G(K_{n,n};p)$ model. Specifically, a distribution $G^*(K_{n,n};p)$ over weighted instances of $K_{n,n}$. The weight of each edge is a geometrically distributed random variable taking the value $k \in \mathbf{N}$ with probability $(1-p)^k p$, and the edge weights are mutually independent. Denoting the maximum weight planar matching of an instance drawn according to $G^*(K_{n,n};p)$ by $L(G^*(K_{n,n};p))$, Johansson's result [13, Theorem 1.1] says that for all $p \in (0,1)$,

$$\lim_{n \to \infty} \frac{1}{n} \cdot \mathbf{E}\left[L(G^*(K_{n,n};p))\right] = \frac{1}{p} \cdot (1 + \sqrt{1-p})^2.$$

Note that an instance G of $G(K_{n,n};p)$ can be obtained from one drawn according to $G^*(K_{n,n};p)$ by including in G only those edges of $K_{n,n}$ with nonzero weight. Hence,

$$\mathbf{E}\left[L(G(K_{n,n};p))\right] \leq \mathbf{E}\left[L(G^*(K_{n,n};p))\right].$$

It follows that $\Delta_p \leq (1 + \sqrt{1-p})^2/p$ for all $p \in (0,1)$. We shall see below that known results imply a much stronger bound on Δ_p for not too large values of p.

Gravner, Tracy and Widom [10] consider processes associated to random $(0,1)$–matrices where each entry takes the value 1 with probability p, independent of the values of other matrix entries. In particular they study a process called *oriented digital boiling* (ODB) and analyze the behavior of a so called *height function* which equals, in distribution, the longest sequence (i_l, j_l) of positions in a random $(0,1)$–matrix of size $n \times n$ which have entry 1 such that the i_l's are increasing and the j_l's are non-decreasing. In contrast, $L(G(K_{n,n};p))$ equals in distribution the longest such sequence with both i_l's and j_l's increasing. This latter model is referred to as *strict oriented digital boiling* in [10], but no results are claimed for it. Clearly, an ODB process dominates that of a strict ODB process. Hence, [10, §3, (1)] implies that for any $p < 1/2$,

$$\Delta_p \leq \kappa_p := \lim_{n \to \infty} \frac{1}{n} \cdot \mathbf{E}\left[L(G(K_{n,n};p))\right] = 2\sqrt{p(1-p)},$$

which in turn implies that $\limsup_{p \to 0} \Delta_p/\sqrt{p} \leq 2$. Nevertheless, our derivation of this latter limit value is elementary in comparison with the highly technical nature of [10].

Acknowledgments. We thank Ricardo Baeza for calling to our attention reference [20].

References

1. D. Aldous and P. Diaconis. Longest increasing subsequences: From patience sorting to the Baik–Deift–Johansson theorem. *Bull. of the AMS*, 36(4):413–432, 1999.
2. R. Baer and P. Brock. Natural soritng over permutation spaces. *Mathematics of Computation*, pages 385–410, 1967.
3. R. Baeza-Yates, G. Navarro, R. Gavaldá, and R. Schehing. Bounding the expected length of the longest common subsequences and forests. *Theory of Computing Systems*, 32(4):435–452, 1999.
4. J. Baik, P. Deift, and K. Johansson. On the distribution of the length of the longest increasing subsequence of random permutaions. *J. Amer. Math. Soc.*, 12:1119–1178, 1999.
5. B. Bollobás. *Modern Graph Theory*, volume 184 of *Graduate Text in Mathematics*. Springer, 1998.
6. V. Chvátal and D. Sankoff. Longest common subsequences of two random sequences. *J. Appl. Prob*, 12:306–315, 1975.
7. V. Dančík. *Expected Length of Longest Common Subsequences*. PhD thesis, Department of Computer Science, University of Warwick, September 1994.
8. P. Erdös and G. Szekeres. A combinatorial problem in geometry. *Compositio Math.*, 2:463–470, 1935.
9. A. Frieze. On the length of the longest monotone subsequence in a random permutation. *The Annals of Applied Prob.*, 1(2):301–305, 1991.
10. J. Gravner, C. Tracy, and H. Widom. Limit theorems for height fluctuations in a class of discrete space time growth models. *J. Stat. Phys.*, 102:1085–1132, 2001.
11. J. M. Hammersley. A few seedlings of research. In *Proc. Sixth Berkeley Sympos. Math. Stat. Prob.*, pages 345–394, Berkeley, Calif., 1972. Univ. of California Press.
12. S. Janson, T. Luczak, and A. Rucinski. *Random Graphs*. Wiley, 2000.
13. K. Johannson. Shape fluctuations and random matrices. *Commun. Math. Phys.*, 209:437–476, 2000.
14. J. F. C. Kingman. Subadditive ergodic theory. *The Annals of Prob.*, 1(6):883–909, 1973.
15. M. Kiwi and M. Loebl. Largest planar matching in random bipartite graphs. *Random Structures and Algorithms*, 21(2):162–181, 2002.
16. M. Kiwi, M. Loebl, and J. Matoušek. Expected length of the longest common subsequence for large alphabets. Technical Report math.CO/0308234, ArXiv.org, August 2003.
17. B. Logan and L. Shepp. A variational problem or random Young tableaux. *Adv. in Math.*, 26:206–222, 1977.
18. A. Okounkov. Random matrices and ramdom permutations. *International Mathematics Research Notices*, pages 1043–1095, 2000.
19. P. Pevzner. *Computational Molecular Biology: An Algorithmic Approach*. MIT Press, 2000.
20. D. Sankoff and J. Kruskal, editors. *Common subsequences and monotone subsequences*, chapter 17, pages 363–365. Addison–Wesley, Reading, Mass., 1983.
21. C. Schensted. Longest increasing and decreasing subsequences. *Canad. J. Math.*, 13:179–191, 1961.
22. R. Stanley. Recent progress in algebraic combinatorics. *Bull. of the AMS*, 40(1):55–68, 2002.
23. S. Ulam. Monte Carlo calculations in problems of mathematical physics. In *Modern Mathematics for the Engineers*, pages 261–281. McGraw-Hill, 1961.
24. A. Vershik and S. Kerov. Asymptotics of the Plancherel measure of the symmetric group and the limiting form of Young tableaux. *Dokl. Akad. Nauk SSSR*, 233:1024–1028, 1977.

Universal Types and Simulation of Individual Sequences

Gadiel Seroussi

Hewlett-Packard Laboratories
1501 Page Mill Road
Palo Alto, CA 94304, USA
gadiel.seroussi@hp.com

Abstract. We define the universal type class of an individual sequence x_1^n, in analogy to the classical notion used in the method of types of information theory. Two sequences of the same length are said to be of the same universal (LZ) type if and only if they yield the same set of phrases in the incremental parsing of Ziv and Lempel (1978). We show that the empirical probability distributions of any finite order k of two sequences of the same universal type converge, in the variational sense, as the sequence length increases. Consequently, the logarithms of the probabilities assigned by any k-th order probability assignment to two sequences of the same universal type converge, for any k. We estimate the size of a universal type class, and show that its behavior parallels that of the conventional counterpart, with the LZ78 code length playing the role of the empirical entropy. We present efficient procedures for enumerating the sequences in a universal type class, and for drawing a sequence from the class with uniform probability. As an application, we consider the problem of universal simulation of individual sequences. A sequence drawn with uniform probability from the universal type class of x_1^n is a good simulation of x_1^n in a well defined mathematical sense.

1 Introduction

Let A be a finite alphabet of cardinality $|A| \geq 2$. We denote by x_j^k the sequence $x_j x_{j+1} \ldots x_k$, $x_i \in A$, $j \leq i \leq k$, with the subscript j sometimes omitted from x_j^k when $j = 1$. If $j > k$, $x_j^k = \lambda$, the null string. The terms "string" and "sequence" are used interchangeably; we denote by $|w|$ the length of a string $w \in A^*$, and by vw the concatenation of $v, w \in A^*$.

The *method of types* [1,2,3] has proven very useful in deriving results in source and channel coding. Although often discussed in the memoryless setting, the method generalizes readily to wider classes of parametric probability distributions on sequences over discrete alphabets. Specifically, consider a class $\mathbb{P}$ of probability distributions P_Θ on A^n, $n \geq 1$, parametrized by a finite-dimensional vector Θ of real-valued parameters. The *type class* of x^n with respect to $\mathbb{P}$ is the set of all sequences y^n such that $P_\Theta(x^n) = P_\Theta(y^n)$ for all admissible values of the parameter vector Θ. Generally, type classes are characterized by a set of

M. Farach-Colton (Ed.): LATIN 2004, LNCS 2976, pp. 312–321, 2004.

empirical statistics, whose structure is determined by the class $\mathbb{P}$. For example, in the case where the components of x^n are independent and identically distributed (i.i.d.), and the class $\mathbb{P}$ is parametrized by the $|A| - 1$ free parameters corresponding to the probabilities of individual symbols from A, the type class of x^n consists of all sequences that have the same single-symbol empirical distribution as x^n [1]. Type classes for families of memoryless distributions with more elaborate parametrizations are discussed in [4]. In the case of finite memory (Markov) distributions of a given order k, empirical joint distributions of order $k + 1$ determine the type classes [3].

In all the cases mentioned, to define the type classes, one needs knowledge on the structure (e.g., number of parameters) of $\mathbb{P}$. In this paper, we define a notion of *universal type* that does not require such knowledge. The universal type class of x^n will be characterized, as in the conventional case, by the combinatorial structure of x^n. Rather than explicit symbol counts, however, we will base the characterization on the data structure built by a universal data compression scheme, namely, the variant of Lempel-Ziv compression described in [5], often referred to as LZ78.[1]

The *incremental parsing rule* [5] parses the string x^n as $x^n = \mathbf{p}_0\mathbf{p}_1\mathbf{p}_2\cdots\mathbf{p}_c\mathbf{t}_x$, where $\mathbf{p}_0 = \lambda$, and the *phrase* $\mathbf{p}_i$, $1 \leq i \leq c$, is the shortest substring of x^n starting at the point following $\mathbf{p}_{i-1}$ such that $\mathbf{p}_i \neq \mathbf{p}_j$ for all $j < i$ (x_1 is assumed to follow $\mathbf{p}_0$). The substring $\mathbf{t}_x$, referred to as the *tail* of x^n, is a (possibly empty) suffix for which the parsing rule was truncated due to the end of the string x^n. Conversely, we refer to the prefix $\mathbf{p}_1\mathbf{p}_2\cdots\mathbf{p}_c$ as the *head* of x^n. Notice that all the phrases are distinct, and $\mathbf{t}_x$ must be equal to one of the phrases, for otherwise an additional phrase could have been parsed. Clearly, the number of phrases is a function $c(x^n)$ of the input sequence, but we shall omit its argument when clear from the context.

Let $\Phi_{x^n} = \{\mathbf{p}_1, \mathbf{p}_2, \ldots, \mathbf{p}_c\}$ denote the set of phrases in the parsing of x^n. We define the *universal (LZ) type class* (in short, UTC) of x^n, denoted $\mathcal{T}_{x^n}$, as the set

$$\mathcal{T}_{x^n} = \{ y^n \in A^n : \Phi_{y^n} = \Phi_{x^n} \}.$$

For arbitrary strings u^k, v^m, $m \geq k \geq 1$, let

$$N(u^k, v^m) = |\{i : v_i^{i+k-1} = u^k, \ 1 \leq i \leq m - k + 1\}|$$

denote the number of (possibly overlapping) occurrences of u^k in v^m. Denote the empirical (joint) distribution of order k, $1 \leq k \leq n$, of x^n by $\hat{P}_{x^n}^{(k)}$, with $\hat{P}_{x^n}^{(k)}(u^k) = N(u^k, x^n)/(n-k+1)$, $u^k \in A^k$. A fundamental property of the UTC is given in the following theorem, proved in Section 2.

---

[1] Similar notions of universal type can be defined also for other universal compression schemes, e.g. Context [6]. Presently, however, the LZ78 scheme appears more amenable to a combinatorial characterization of its type classes.

Theorem 1. *Let x^n be an arbitrary sequence of length n, and k a fixed positive integer. If $y^n \in \mathcal{T}_{x^n}$, then, for all $u^k \in A^k$, we have*[2]

$$\hat{P}_{x^n}^{(k)}(u^k) - \hat{P}_{y^n}^{(k)}(u^k) = o(1) \quad as \ n \to \infty. \tag{1}$$

A k-th order (finite-memory) probability assignment Q_k is defined by a set of conditional probability distributions $Q_k(u^{k+1}|u_1^k)$, $u^{k+1} \in A^{k+1}$, and a distribution $Q_k(x_1^k)$ on the initial state, so that $Q_k(x_1^n) = Q_k(x_1^k) \prod_{i=k+1}^{n} Q_k(x_i|x_{i-k}^{i-1})$. In particular, Q_k could be defined by the k-th order approximation of an ergodic measure [2]. The following is an immediate consequence of Theorem 1.

Corollary 1. *Let x^n and y^n be sequences such that $y^n \in \mathcal{T}_{x^n}$. Then, for any nonnegative integer k, and any k-th order probability assignment Q_k such that $Q_k(x^n) \neq 0$ and $Q_k(y^n) \neq 0$, we have $\frac{1}{n} \log \frac{Q_k(x^n)}{Q_k(y^n)} = o(1)$ as $n \to \infty$.*

Theorem 1 and Corollary 1 are universal analogues of well known properties of classical types. In a classical type class, all the sequences in the class have the same empirical distribution (relative to the model class defining the type, e.g., k-th order joint empirical distributions for $k-1$st order finite-memory), and they are assigned identical probabilities by any distribution from the model class. In a sense, both properties mean that sequences from the same type class are statistically "indistinguishable" by distributions in the model class. In the universal type case, "same empirical distribution" and "identical probabilities" are weakened to asymptotic notions, i.e. "equal in the limit," but they hold for any model order. The weakened "indistinguishability" is the price paid for universality.

For simplicity, we will focus on the case of binary sequences, i.e., $A = \{0, 1\}$. All the principles and main results presented carry without difficulty (albeit with increased notational complexity) to other finite alphabets. The rest of this extended summary is organized as follows. In Section 2 we prove Theorem 1, and we analyze the structure and size of UTCs. In Section 3, we present a procedure for drawing a random element with uniform probability from a UTC, and describe an application of UTCs to the universal simulation of individual sequences. Proofs for all the results in this extended summary are presented in [7]. The full version also discusses additional properties of universal types, such as the number of UTCs for a given sequence length n, which, contrary to the conventional finite-parametric case, is not polynomial in n. These discussions are omitted here due to length constraints.

2   The Universal Type Class of x^n

If $y^n \in \mathcal{T}_{x^n}$, then y^n parses into the same set of phrases as x^n. However, the phrases need not (and, except when $y^n = x^n$, *will not*) occur in the same order

---

[2] The asymptotic language in this statement, and others in the sequel, should be interpreted as relating to any infinite sequence of sequences x^n, one for each length n, and not necessarily related by prefix relations.

in y^n as they do in x^n. Also, the tail of y^n may be any phrase $\mathbf{t}_y \in \Phi_{x^n}$, of length $|\mathbf{t}_y| = |\mathbf{t}_x| = n - n_T(x^n)$, where

$$n_T(x^n) = |\mathbf{p}_1| + |\mathbf{p}_2| + \cdots + |\mathbf{p}_c|. \tag{2}$$

Example 1. Consider the string $x^8 = 10101100$, with $n = 8$. The incremental parsing for x^8 is $1, 0, 10, 11, 00$, with $c = 5$, $n_T(x^8) = 8$, and a null tail. The sequence $y^8 = 01001011$ is parsed into $0, 1, 00, 10, 11$, defining the same set of phrases as x^8. Thus, x^8 and y^8 are in the same UTC.

2.1   Proof of Theorem 1

Proof. We claim that the following inequalities hold for x^n:

$$\sum_{j=1}^{c} N(u^k, \mathbf{p}_j) \le N(u^k, x^n) \le \sum_{j=1}^{c} N(u^k, \mathbf{p}_j) + (k-1)c + |\mathbf{t}_x|. \tag{3}$$

The first inequality follows from the fact that the phrases are distinct, and they parse the head of x^n. The second inequality follows from the fact that an occurrence of u^k is either completely contained in a phrase of the parsing, or it spans a phrase boundary, or it is contained in the tail of x^n. To span a boundary, an occurrence of u^k must start at most $k-1$ locations before the end of a phrase. Clearly, the inequalities in (3) hold also for any sequence $y^n \in \mathcal{T}_{x^n}$, since they all define the same set of phrases as x^n, and have tails of the same length. Thus, it follows from (3) that

$$|N(u^k, x^n) - N(u^k, y^n)| \le (k-1)c + |\mathbf{t}_x|, \quad \forall y^n \in \mathcal{T}_{x^n}. \tag{4}$$

It is well known (cf. [5,2]) that for the LZ78 incremental parsing, the number of phrases satisfies $c \le n/(\log n - o(\log n))$, and the length ℓ of any phrase (and, thus, also of $|\mathbf{t}_x|$) satisfies $\ell \le \sqrt{2n} - o(\sqrt{n})$. Hence, we have $(k-1)c + |\mathbf{t}_x| = o(n)$ for fixed k, and the claim of the theorem follows from (4), after normalization by $n - k + 1$.   □

2.2   The Size of the Universal Type Class

The set of phrases in the incremental parsing of x^n is best represented by means of a rooted, ordered binary *parsing tree* T_{x^n}, where each node represents a phrase, and each branch is labeled with a binary symbol. The phrase associated with a node is the concatenation of the edge labels on the path from the root (associated with λ) to the node. The number of nodes in the tree is $c(x^n) + 1$, and its *path length* [8] is $n_T(x^n)$ as defined in (2), which depends only on the tree, a fact we will emphasize by omitting the argument x^n. All the sequences in a UTC share the same parsing tree T, which can serve as a canonical representation of the type class. In general, a complete specification of the UTC requires also the sequence length n, since the same parsing tree T might result from parsing sequences of different lengths, due to possibly different tail lengths. For a given tree T, n can vary from n_T to $n_T + \ell_{\max}$, where $\ell_{\max}$ is the maximal depth of a leaf of T. When $n_T = n$ (i.e., x^n has a null tail), we say that $\mathcal{T}_{x^n}$ is the *natural* UTC associated with T_{x^n}. If T is an arbitrary parsing tree, we denote the associated natural UTC by $\mathcal{T}(T)$, without reference to a specific string in the type class. We call a valid pair (T, n), the *universal type* (UT) of the sequences in the corresponding UTC. When n is not specified, the natural UT is assumed. An example of a parsing tree, corresponding to the string x^8 of Example 1, is shown in Figure 1. In the example, $n = n_T$, and, thus, $\mathcal{T}(T) = \mathcal{T}_{x^8}$.

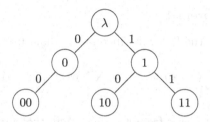

Fig. 1. Parsing tree for $x^8 = 10101100$

Each sequence in the UTC of x^n is determined by some permutation of the order of the phrases in Φ_{x^n}. Not all the possible permutations are allowed, though, since a phrase $\mathbf{p}_i$ that is a prefix of another phrase $\mathbf{p}_j$ must always precede $\mathbf{p}_j$. Thus, only permutations that respect the prefix partial order are valid. Notice that, in general, the parsing tree is not necessarily complete.[3] We call a node (or the corresponding phrase) a *bridge* if it has exactly one child. For example, the phrase '0' in the example of Figure 1 is a bridge.

Let T^0 and T^1 denote the subtrees of T_{x^n} rooted at the respective children of the root of T_{x^n}. Each subtree, in turn, defines the UT of a subsequence of x^n; namely, T^a defines the UTC $\mathcal{T}(T^a)$ of the subsequence resulting from the concatenation of all phrases in Φ_{x^n} starting with the symbol $a \in \{0, 1\}$. Notice that either subtree might be missing if the root of T_{x^n} is a bridge or a leaf. When both children are missing, we have the trivial case where $x^n = \lambda$, and $|\mathcal{T}_{x^n}| = 1$. We denote the number of nodes in T^a by $c_a + 1$, $a = 0, 1$. In analogy with the notation for T_{x^n}, c_a denotes the number of non-root nodes in the respective subtree. When T^a is missing, we set $c_a = -1$, and $|\mathcal{T}_a| = 1$. We have $c(x^n) = c_0 + c_1 + 2$. A valid permutation defining a sequence $y^n \in \mathcal{T}_{x^n}$ must result from a valid permutation of the phrases in T^0 and a valid permutation of the phrases in T^1. The resulting ordered sub-lists of phrases can be freely interleaved to form a valid ordered list of phrases for y^n, since there is no order constraint

---

[3] We call a binary tree *complete* if every node has either two children or none. There is a remarkable lack of terminology consensus for this notion in the literature; see a footnote in [9] for a sample of different terms authors have used.

between phrases in different subtrees. The number of possible interleavings is, therefore,

$$M(c_0, c_1) = \begin{pmatrix} c_0 + c_1 + 2 \\ c_0 + 1 \end{pmatrix},$$

which is the well known formula for the number of ways to merge an ordered list of size $c_0 + 1$ with one of size $c_1 + 1$, while preserving the respective orders.

Given the sub-lists and their interleaving, to completely specify y^n, we must also define its tail, which can be any phrase of length $|\mathbf{t}_x|$ (at least one such phrase exists, namely, $\mathbf{t}_x$ itself). Therefore, we can write the following recursion for the size of $\mathcal{T}_{x^n}$:

$$|\mathcal{T}_{x^n}| = |\mathcal{T}(T^0)| \cdot |\mathcal{T}(T^1)| \cdot M(c_0, c_1) \cdot m(|\mathbf{t}_x|), \tag{5}$$

where $m(|\mathbf{t}_x|)$ denotes the number of nodes at level $|\mathbf{t}_x|$ in $\mathcal{T}_{x^n}$. Notice that when (5) is used recursively, all recursion levels except the outermost deal with natural UTCs. Therefore, a nontrivial term $m(|\mathbf{t}_x|)$ occurs only at the outermost application of (5).

Example 2. For the tree T in Figure 1, we have $c_0=1$, $c_1=2$, $|\mathbf{t}_x|=0$. Therefore,

$$|\mathcal{T}(T)| = |\mathcal{T}(T^0)||\mathcal{T}(T^1)|\begin{pmatrix} 5 \\ 2 \end{pmatrix} = (1) \cdot (1 \cdot 1 \cdot \begin{pmatrix} 2 \\ 1 \end{pmatrix}) \cdot \begin{pmatrix} 5 \\ 2 \end{pmatrix} = 20,$$

which is the number of ways to merge the list $[0, 00]$ with either the list $[1, 10, 11]$ or the list $[1, 11, 10]$ while preserving the order of each list.

Let $\hat{T}$ denote the tree obtained from a parsing tree T by collapsing paths of the form $v_1 \xrightarrow{a_1} v_2 \xrightarrow{a_2} v_3$, where v_2 is a bridge node, to single edges $v_1 \xrightarrow{a_1} v_3$, or, if the root is a bridge node, eliminating the root and its outgoing edge (with the node at the end of this edge becoming the new root). Bridge nodes are eliminated sequentially, one bridge at a time, until none are left. By construction, $\hat{T}$ is a complete tree.

Lemma 1. *We have* $|\mathcal{T}(T)| = |\mathcal{T}(\hat{T})|$.

We denote by $c_B(x^n)$ the number of bridges in the parsing tree of x^n.

Theorem 2. *Let* x^n *be an arbitrary sequence of length* n, *and* $\hat{c} = c(x^n) - c_B(x^n)$. *Then,*

$$(1 - \beta)\hat{c}\log\hat{c} \leq \log|\mathcal{T}_{x^n}| \leq \hat{c}\log\hat{c}, \tag{6}$$

where, as $n \to \infty$, β *can be bounded away from zero only if* $\frac{\log\hat{c}}{\log n}$ *is bounded away from one.*

The bounds are expressed in terms of $\hat{c}$ rather than c, due to Lemma 1. The upper bound is a straightforward consequence of the fact that the UTC size for a sequence with $\hat{c}$ phrases is upper-bounded by $\hat{c}!$. A direct combinatorial proof of the lower bound is presented in [7]. However, the necessity of a lower bound of this kind follows from the optimality of the LZ78 algorithm, and the fact that UTs can be used to define an asymptotically optimal *enumerative coding* procedure [10] for sequences of length n, in which the code length assigned to x^n is $\log|\mathcal{T}_{x^n}| + o(n)$. Theorem 2 shows another fundamental parallel between universal types and conventional types: the size of a conventional type is known to be $2^{n\hat{H}(x^n)+o(n)}$, where $\hat{H}(x^n)$ denotes the empirical entropy rate of x^n with respect to the model class defining the types (cf. [2,3]). Theorem 2 states that a similar statement is true for UTs, with the normalized LZ78 code length $(c\log c)/n$ playing the role of the empirical entropy rate. Notice that sequences for which $\log c/\log n$ is bounded away from one have vanishing LZ78 compressibility rate, so the analogy does not break when (6) is not tight.

3 Random Sequences from Universal Types

The recursion (5) is helpful for deriving efficient procedures for enumerating $\mathcal{T}_{x^n}$, and for drawing a random sequence from it with uniform probability. We present the random selection procedure here. Enumeration procedures are presented in [7], and will provide an alternative way of drawing a sequence from the UTC with uniform probability.

3.1 Random Selection Algorithm

The algorithm in Figure 2 draws a random sequence from $\mathcal{T}_{x^n}$. In the algorithm, we label nodes of T with their associated phrases, we mark nodes as *used* or *unused*, and we denote by $U(v)$ the number of *unused* nodes in the subtree rooted at v. For a node v, and $b \in \{0,1\}$, we say that the path from v to vb is *blocked* if either there is no node labelled vb in the tree, or $U(vb) = 0$.

Theorem 3. *The algorithm in Figure 2 outputs a sequence drawn with uniform probability from $\mathcal{T}_{x^n}$. It requires a total of $\log|\mathcal{T}_{x^n}| + o(n)$ random bits to do so.*

Table 1 shows the steps taken in drawing a random sequence from the UTC associated with the sequence x^8 and the parsing tree of Figure 1. The output of the run is the string $y^8 = 0\,00\,1\,10\,11$, which is in $\mathcal{T}_{x^8}$. Checking the probabilities of random choices made in Step 5, we observe that the execution path taken had overall probability $\frac{2}{5} \cdot \frac{1}{4} \cdot \frac{1}{2} = \frac{1}{20}$, consistent with our previous determination of $|\mathcal{T}_{x^8}| = 20$ and a uniform distribution.

3.2 Application to Universal Simulation of Individual Sequences

Informally, given an individual sequence x^n, we call a random sequence y^n a "good simulation" of x^n if the following conditions hold:

1. y^n is "statistically similar" to x^n;
2. given that y^n satisfies Condition 1, it has the maximum possible entropy.

Input: x^n, parsing tree $T = T_{x^n}$.
Output: Sequence y^n, drawn with uniform probability from $\mathcal{T}_{x^n}$.

---

1. Mark the root of T as *used*, and all other nodes as *unused*.
2. Set $v \leftarrow \lambda$ (the root). If $U(v) = 0$, go to Step 6.
   Otherwise, proceed to Step 3.
3. If v is *unused*, output v as the next phrase of y^n, mark v as *used*,
   and go to Step 2. Otherwise, proceed to Step 4.
4. If the path from v to $v0$ is blocked, set $v \leftarrow v1$ and go to Step 3.
   Else, if the path from v to $v1$ is blocked, set $v \leftarrow v0$ and go to Step 3.
   Otherwise, proceed to Step 5.
5. Draw a random bit b with $\text{Prob}(b = 1) = \frac{U(v\,1)}{U(v\,0)+U(v\,1)}$.
   Set $v \leftarrow vb$, and go to Step 3.
6. Output a random phrase of length $|t_x|$ as the tail of y^n. **Stop.**

Fig. 2. Algorithm for drawing a random sequence from $\mathcal{T}_{x^n}$

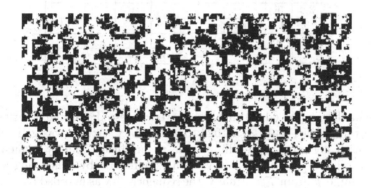

Fig. 3. Texture simulation

Condition 1 is stated in a purposely vague fashion, as the desired similarity criterion may vary from setting to setting. In [4], for example, x^n is assumed to have been emitted by a source from a certain parametric class, and a strict criterion is used, where y^n must be assigned exactly the same probability as x^n by all sources in the class. We will be satisfied with a less stringent requirement (since, among other things, we do not assume x^n was generated by a probabilistic source), as given by the property stated in Theorem 1. Condition 2, on the other hand, is necessary to avoid, in the extreme case, a situation where x^n (which certainly satisfies Condition 1) is returned as its own "simulation." We wish to have as much variety as possible in the space of simulations of x^n. It is proved in [7] that a simulation procedure based on drawing y^n uniformly at random from $\mathcal{T}_{x^n}$ satisfies both conditions in a well defined mathematical sense.

Table 1. Execution of the random selection algorithm

Step	v	status	$U(v0), U(v1)$	choice	Prob	output
2	root					
3	root	used				
4	root		2,3			
5	root		2,3	0	2/5	
3	0	unused				0
2,3,4	root		1,3			
5	root		1,3	0	1/4	
3	0	used	1,0			
4	0		1,0	0		
3	00	unused				00
2,3,4	root		0,3	1		
3	1	unused				1
2,3,4	root		0,2	1		
3,4	1	used	1,1			
5	1		1,1	0	1/2	
3	10	unused				10
2,3,4	root			1		
3,4	1			1		
3	11	unused				11
2	root		0,0			
6			**stop**			

The simulation procedure outlined above was tested on some binary textures. For the example in Figure 3, a 1024×1024 binary texture was generated, and scanned with a Peano plane-filling scan, to produce a binary sequence x^n of $n = 2^{20}$ samples. The sequence was then "simulated" by generating a uniform random sample y^n from $\mathcal{T}_{x^n}$. Finally, the sequence y^n was mapped back, reversing the same Peano scan order, to a 1024×1024 image. The left half of Figure 3 shows a 512×512 patch of the original texture, while the right half shows a 512×512 patch of the simulated one (the smaller patches were used to comply with page size limitations without sub-sampling or altering the visual quality of the images). It is evident from the figure that the right half indeed "looks like" the left half, and the seam between the images appears unnoticeable. Yet, the right half is completely different from the left half, and was selected from a very large class of possible simulation images. In fact, the size of $\mathcal{T}_{x^n}$ in this example was estimated using the recursion (5), resulting in $\log |\mathcal{T}_{x^n}| \approx 109,700$.

Acknowledgments. Thanks to Erik Ordentlich, Wojciech Szpankowski, Alfredo Viola, Marcelo Weinberger, and Tsachy Weissman for very useful discussions.

References

1. Csiszár, I., Körner, J.: Information Theory: Coding Theorems for Discrete Memoryless Systems. Academic, New York (1981)
2. Cover, T.M., Thomas, J.A.: Elements of Information Theory. John Wiley & Sons, Inc., New York (1991)
3. Csiszár, I.: The method of types. IEEE Trans. Inform. Theory **IT-44** (1998) 2505–2523
4. Merhav, N., Weinberger, M.J.: On universal simulation of information sources using training data. Technical Report HPL-2002-263, Hewlett-Packard Laboratories (2002) To appear in *IEEE Transactions on Information Theory*.
5. Ziv, J., Lempel, A.: Compression of individual sequences via variable-rate coding. IEEE Trans. Inform. Theory **IT-24** (1978) 530–536
6. Rissanen, J.: A universal data compression system. IEEE Trans. Inform. Theory **IT-29** (1983) 656–664
7. Seroussi, G.: Universal types and simulation of individual sequences. in preparation (2003)
8. Knuth, D.E.: The Art of Computer Programming. Seminumerical Algorithms. third edn. Volume 2. Addison-Wesley, Reading, MA (1998)
9. Freiling, C., Jungreis, D., Theberge, F., Zeger, K.: Almost all complete binary prefix codes have a self-synchronizing string. IEEE Trans. Inform. Theory **IT-49** (2003) 2219–2225
10. Cover, T.M.: Enumerative source encoding. IEEE Transactions on Information Theory **IT-19** (1973) 73–77

Separating Codes: Constructions and Bounds

Gérard Cohen[1] and Hans Georg Schaathun[2]

[1] Ecole Nationale Supérieure des Télécommunications
46 rue Barrault
F-75634 Paris Cedex, France
cohen@enst.fr
[2] Department of Informatics
University of Bergen
Høyteknologisenteret
N-5020 Bergen, Norway
georg@ii.uib.no

Abstract. Separating codes, initially introduced to test automaton, have revived lately in the study of fingerprinting codes, which are used for copyright protection. Separating codes play their role in making the fingerprinting scheme secure against coalitions of pirates. We provide here better bounds, constructions and generalizations for these codes.

1 Introduction

Separating codes were introduced in 1969 and have been the topic of several papers with various motivations. Many initial results are due to Sagalovich; see [3] for a survey, and also [2,5]. New applications of separating codes have appeared during the last decade, namely *traitor tracing* and *fingerprinting*.

Fingerprinting is a proposed technique for copyright protection. The vendor has some copyrighted work of which he wants to sell copies to customers. If he is not able to prevent the customer from duplicating his copy, he may individually mark every copy sold with a unique fingerprint. If an illegal copy (for which the vendor has not been paid) subsequently appears, it may be traced back to one legal copy and one pirate via the fingerprint. A pirate is here any customer guilty of illegal copying of the copyrighted work.

Traitor tracing is the same idea applied to broadcast encryption keys. E.g. the vendor broadcasts encrypted pay-TV, and each customer buys or leases a decoder box to be able to decrypt the programmes. If the vendor is not able to make the decoder completely tamperproof, he may fingerprint the decryption keys which are stored in the box.

The set of fingerprints in use, is called the fingerprinting code. Separating codes are used in the study of *collusion secure* fingerprinting codes. If several pirates collude, they posess several copies with different fingerprints. By comparing their copies, they will find differences which must be part of the fingerprint. These identified "marks" may be changed to produce a false fingerprint. A collusion secure code should aim to identify at least one of the pirates from this false fingerprint.

M. Farach-Colton (Ed.): LATIN 2004, LNCS 2976, pp. 322–328, 2004.

We shall introduce two useful concepts regarding collusion secure codes. If the code is t-frameproof, it is impossible for any collusion of at most t pirates to produce a false fingerprint which is also a valid fingerprint of an innocent user. In other words, no user may be framed by a coalition of t pirates or less. A t-frameproof code is the same as a $(t, 1)$-separating code, which will be defined formally in the next section.

If the code is t-identifying, the vendor is always able to identify at least one pirate from any coalition of size at most t, given a false fingerprint created by the coalition. A first step towards identification is (t, t)-separation (see, e.g. [4]), which we study and generalize here.

2   Definitions

For any positive real number x we denote by $\lceil x \rceil$ the smallest integer at least equal to x. Let A be an alphabet of q elements, and A^n the set of sequences of length n over it. A subset $C \subseteq A^n$ is called an $(n, M)_q$ or (n, M)-*code* if $|C| = M$. Its *rate* is defined by $R = (\log_q M)/n$. For any $\mathbf{x} \in A^n$, we write x_i for the i-th component, so that $\mathbf{x} = (x_1, x_2, \dots, x_n)$. The minimum Hamming distance between two elements (codewords) of C is denoted by $d(C)$ or d, and the normalised quantity d/n by δ.

Consider a subset $\mathcal{C} \subseteq C$. For any position i, we define the *projection* $P_i(\mathcal{C}) = \bigcup_{\mathbf{a} \in \mathcal{C}}\{a_i\}$. The *feasible set* of $\mathcal{C}$ is

$$F(\mathcal{C}) = \{\mathbf{x} \in A^n : \forall i, x_i \in P_i(\mathcal{C})\}.$$

If $\mathcal{C}$ is the fingerprints held by some pirate coalition, then $F(\mathcal{C})$ is the set of fingerprints they may produce. If two non-intersecting coalitions can produce the same descendant, i.e., if their feasible sets intersect, it will be impossible to trace with certainty even one pirate. This motivates the following definition.

Definition 1. *A code C is (t, t')-separating if, for any pair (T, T') of disjoint subsets of C where $|T| = t$ and $|T'| = t'$, the feasible sets are disjoint, i.e. $F(T) \cap F(T') = \emptyset$.*

Such codes are also called *separating systems*, abbreviated by SS.

Since the separation property is preserved by translation, we shall always assume that $\mathbf{0} \in C$. The separation property can be rephrased as follows when $q = 2$: For any ordered $t + t'$-tuple of codewords, there is a coordinate where the $t + t'$-tuple $(1..10..0)$ of weight t or its complement occurs.

Given a (t, t')-configuration (T, T') we define the separating set $\Theta(T, T')$ to be the set of coordinate positions where (T, T') is separated. Let $\theta(T, T') := \#\Theta(T, T')$ be the separating weight. Clearly $\theta(T, T') \geq 1$ is equivalent with (T, T') being separated. The minimum (t, t')-separating weight $\theta_{t,t'}(C)$ is the least separating weight of any (t, t')-configuration of C. We abbreviate $\theta_{i,i}(C)$ to $\theta_i(C)$ or θ_i. Clearly $\theta_1(C) = d(C)$. The minimum separating weights have previously been studied by Sagalovich [3].

3 Bounds on $(t, 1)$ Separating Codes

The case $t' = 1$ corresponds to "frameproof" codes introduced in [1]. Körner (personal communication) has a simplified proof of $R \le 1/2$ for (1,2)-separation in the binary case. We generalize it to any t and q, and for bounded separating weight $n\tau$.

A (t, τ)-*coverfree* code is a code with $(t, 1)$-separating weight at least equal to τn. Their study in [11] and [9] is motivated by broadcast encryption.

Partition $\{1, 2, ..n\}$ into t almost equal parts $P_1, \ldots, P_t$ of size approximately n/t. Say a codeword c is *isolated* on P_i if no other codeword projects on P_i on a vector located at distance less than $(n/t)\tau$ from c. Denote by U_i the subset of codewords isolated on P_i.

Lemma 1. *If C is (t, τ)-coverfree, then every codeword c of C is isolated on at least one P_i.*

Proof: Suppose for a contradiction that there is a codeword $\mathbf{c}_0$ which is not isolated. Let $\mathbf{c}_i$ be a codeword which is at distance less than $(n/t)\tau$ when projected onto P_i, for $i = 1, \ldots, t$. Now $\mathbf{c}_0$ is separated from $\{\mathbf{c}_1, \ldots, \mathbf{c}_t\}$ on less than $(n/t)\tau$ coordinates per block, or at most $n\tau - t$ coordinate positions total. This contradicts the assumption on the separating weight τ.

If we let τ tend to zero, we get an upper bound on the size of $(t, 1)$-separating codes, which was found independently in [13] and [12]. The proofs are essentially the same as the one presented here.

Theorem 1. *If C is (t, τ)-coverfree, then $|C| \le t q^{\lceil (1-\tau)n/t \rceil}$.*

For constant t, this asymptotically gives a rate $R \le (1-\tau)/t$ when n increases. A lower bound on the rate can now be obtained by invoking a sufficient condition for C to be (t, τ)-coverfree, based on its minimum distance d: $td \ge (t - 1 + \tau)n$. This is proved in a more general form in Proposition 1. Using algebraic-geometric (AG) codes [7] with $\delta > t^{-1}(1 - \tau)$ and $R \approx 1 - \delta - 1/(q^{1/2} - 1)$ gives the following asymptotically tight (in q):

Theorem 2. *For fixed t and large enough q, the largest possible rate of a q-ary family of (t, τ)-coverfree codes satisfies $R = t^{-1}(1 - \tau)(1 + o(1))$.*

4 Large Separation

Definition 2. *A code C of length n is (t, t', τ)-separating if, for any pair (T, T') of disjoint subsets of C where $|T| = t$ and $|T'| = t'$, $\theta(T, T') \ge \tau n$.*

Proposition 1. *A code with minimum distance d is (t, t', τ)-separating if*

$$tt'd \ge (tt' - 1 + \tau)n.$$

Proof: Consider two disjoints sets T and T' of sizes t and t' respectively and count the sum Σ of pairwise distances between them: on one hand, $\Sigma \geq tt'd \geq (tt' - 1 + \tau)n$. Computing Σ coordinatewise now, we get that the contribution to Σ of at least τn coordinates must be greater than $tt' - 1$, i.e. tt'. Thus, these coordinates separate T and T'.

To construct infinite families of separating codes over small alphabets, we can resort to the classical notion of *concatenation*.

Definition 3 (Concatenation). *Let C_1 be a $(n_1, Q)_q$ and let C_2 be an $(n_2, M)_Q$ code. Then the concatenated code $C_1 \circ C_2$ is the $(n_1 n_2, M)_q$ code obtained by taking the words of C_2 and mapping every symbol on a word from C_1.*

The following result is an easy consequence of the definition.

Proposition 2. *Let Γ_1 be a $(n_1, M)_{M'}$ code with minimum separating weight $\theta_{t,t'}^{(1)}$, and let Γ_2 be a $(n_2, M')_q$ code with minimum separating weight $\theta_{t,t'}^{(2)}$. Then the concatenated code $\Gamma := \Gamma_2 \circ \Gamma_1$ has minimum separating weight $\theta_{t,t'} = \theta_{t,t'}^{(1)} \cdot \theta_{t,t'}^{(2)}$.*

We shall illustrate the concatenation method with $q = 2, t = 2, t' = 1$ in the next section.

5 The Binary Case

5.1 (2, 1)-Separation

In [8], it was pointed out that shortened Kerdock codes $K'(m)$ for $m > 4$ are $(2, 1)$-separating. Take an arbitary subcode of size 11^2 in $K'(4)$ which is a $(15, 2^7)$ $(2, 1)$-SS. Concatenate it with an infinite family of algebraic-geometry codes over $\mathrm{GF}(11^2)$ (the finite field with 11^2 elements) with $\delta > 1/2$ (hence $(2, 1)$-separating by Proposition 1) and $R \approx 1/2 - 1/11$ [7]. After some easy computations, this gives:

Theorem 3. *There is a constructive asymptotic family of binary $(2, 1)$-separating codes with rate $R = 0.1845$.*

This can even be refined if we concatenate with the codes contained in the following proposition from [10].

Proposition 3. *Suppose that $q = p^{2r}$ with p prime, and that t is an integer such that $2 \leq t \leq \sqrt{q} - 1$. Then there is an asymptotic family of $(t, 1)$-separating codes with rate*

$$R = \frac{1}{t} - \frac{1}{\sqrt{q} - 1} + \frac{1 - 2\log_q t}{t(\sqrt{q} - 1)}.$$

Remark 1. If we use the Xing's codes ([10]), we get an improved rate of $R \approx 0.2033$, but at the expense of constructivity.

5.2   A Stronger Property

Definition 4 (Completely Separating Code). *A binary code is said to be* (t, t')-*completely separating* $((t, t')$-*CSS) if for any set ordered set of* $t + t'$ *code-words, there is at least one column with 1 in the* t *upper positions, and 0 elsewhere, and one column with 0 in the* t *upper positions and 1 in the* t' *lower ones.*

We define $R_{SS}(t, t')$ as the largest possible asymptotical rate of a family of (t, t')-SS, and similarly $R_{CSS}(t, t')$ for (t, t')-CSS. We clearly have

$$R_{SS}(t, t') \geq R_{CSS}(t, t') \geq \frac{1}{2} R_{SS}(t, t'). \tag{1}$$

5.3   Improved Upper Bounds on (t, t)-Separating Codes

Theorem 4. *A* (t, t)-*separating* (θ_0, M, θ_1) *code with separating weights* $(\theta_1, \ldots, \theta_t)$ *gives rise to a* (i, i)-*CSS* $(\theta_{t-i}, M - 2t + 2i, 2\theta_{t+1-i})$ *with complete-separating weight* θ_i, *for any* $i < t$.

Proof: Consider a pair of $(t - i)$-tuples of vectors which are separated on θ_{t-i} positions. Pick any vector **c** from the first $(t - i)$-tuple and replace the code C by its translation $C - \mathbf{c}$. Thus all the columns which separates the two tuples have the form $(0 \ldots 01 \ldots 1)$.

Now consider any two i-tuples of vectors. Coupling each i-tuple with a $(t - i)$-tuple, we get two t-tuples which must be separated on θ_t positions, i.e. the two i-tuples must have at least θ_t columns of the form $(0 \ldots 01 \ldots 1)$. Now, observe that we can swap the two $(t - i)$-tuples, and the two resulting t-tuples are still separated. This guarantees at least θ_t columns of the form $(1 \ldots 10 \ldots 0)$.

Deleting all the columns where the two $(t - i)$-tuples are not separated, and the words of these two tuples must this leave us with an (i, i)-CSS with complete-separating weight θ_i and parameters $(\theta_{t-i}, M - 2t + 2i, 2\theta_{t+1-i})$, as required.

Theorem 5. *Any completely* (t, t)-*separating* $(\theta_0, M, 2\theta_1)$ *code with complete-separating weights* $(\theta_1, \ldots, \theta_t)$ *gives rise to a completely* (i, i)-*separating* $(\theta_{t-i}, M - 2t + 2i, 2\theta_{t+1-i})$ *code with complete-separating weight* θ_i, *for any* $i < t$.

This is proved in the same way as the previous theorem.

Theorem 6. *For any* (t, t)-*CSS, the rate* R_t *satisfies*

$$R_t \leq \bar{R}(2R_t/\bar{R}_{t-1}),$$

where $\bar{R}(\delta)$ *is any upper bound on the rate of error-correcting codes in terms of the normalised minimum distance, and* $\bar{R}_{t-1}$ *is the upper bound on the rate of any* $(t - 1, t - 1)$-*CSS.*

Proof: Let C_{t-1} be the $(t-1, t-1)$-CSS which exists by Theorem 5, and let R_{t-1} be its rate. We have that

$$\delta_t = 2\frac{\theta_1}{\theta_0} = 2\frac{\log M}{\theta_0}\frac{\theta_1}{\log M} = 2R_t/R_{t-1}.$$

Now, obviously $R_t \leq \bar{R}(\delta_t)$, which is decreasing in δ_t, and this gives the result.
With a completely analogous proof, we also get the following.

Theorem 7. *For any (t,t)-SS, the rate R satisfies*

$$R \leq \bar{R}(R/\bar{R}_{t-1}),$$

where $\bar{R}(\delta)$ is any upper bound on the rate of error-correcting codes in terms of the normalised minimum distance, and $\bar{R}_{t-1}$ is the upper bound on the rate of any $(t-1, t-1)$-CSS.

Table 1. Rate bounds on CSS and SS.

(t,t)	Bound 1 CSS rate	Bound 1 SS rate	D'yachkov et al. CSS rate	Bound 2 CSS rate	Bound 2 SS rate
(1,1)	1	1	1	1	1
(2,2)	0.1712	0.2835	0.161	–	–
(3,3)	0.03742	0.06998	0.0445	0.0354	0.0663
(4,4)	0.008843	0.01721	0.0123	0.00837	0.0163
(5,5)	0.002156	0.004261	0.00333	0.00204	0.00404

Setting equality in the bounds and solving, we get the upper bounds given as 'Bound 1' in Table 1. Comparing with the CSS bounds of [6] shows an improvement from $(3,3)$-CSS onwards. However, [6] has a good bound on $(2,2)$-CSS, used as a seed for the recursive bounds of our theorems to obtain 'Bound 2' in the table.

Example 1. Let C_1 be an asymptotic class of $(\theta_0, 2^k, \theta_1)$ $(3,3)$-SS. Then there is an asymptotic class C_2 of $(\theta_1, 2^k, \theta_2)$ $(2,2)$-CSS. We have that $R_2 = k/\theta_1 \leq 0.161$, and

$$R_1 = k/\theta_0 = R_2\delta_1 \leq 0.161\delta_1,$$

which is equivalent to $\delta_1 \geq R_1/0.161$. We can use any upper bound $\bar{R}(\delta)$ on R_1, and get

$$R_1 \leq \bar{R}(\delta_1) \leq \bar{R}(R_1/0.161),$$

and $R_1 \leq 0.0663$ by the linear programming bound.

References

1. D. Boneh and J. Shaw, "Collusion-secure fingerprinting for digital data", *IEEE Trans. on Inf. Theory*, **44** (1998), pp. 480–491.
2. J. Körner and G. Simonyi, "Separating partition systems and locally different sequences," *SIAM J. Discrete Math.*, **1** No 3 (1988) pp. 355–359.
3. Yu. L. Sagalovich, "Separating systems", *Probl. Inform. Trans.* **30** No 2 (1994) pp. 105–123.
4. A. Barg, G. Cohen, S. Encheva, G. Kabatiansky, and G. Zémor, "A hypergraph approach to the identifying parent property", *SIAM J. Disc. Math.*, vol. 14, 3 (2001) pp. 423-431.
5. G. Cohen, S. Encheva, and H.G. Schaathun, "More on $(2, 2)$-separating systems", *IEEE Trans. Inform. Theory*, vol. 48, 9 (2002) pp. 2606-2609.
6. A. D'yachkov, P. Vilenkin, A. Macula, and D. Torney, "Families of finite sets in which no intersection of ℓ sets is covered by the union of s others", *J. Combinatorial Theory*, vol. 99, 195-208 (2002).
7. M.A. Tsfasmann, "Algebraic-geometric codes and asymptotic problems", *Discrete Applied Math.*, vol. 33 (1991) pp. 241-256.
8. A. Krasnopeev and Yu. Sagalovitch, "The Kerdock codes and separating systems", *Eighth International Workshop on Algebraic and Combinatorial Theory*, 8-14 Sept. 2002, pp. 165-167.
9. R. Kumar, S. Rajagopalan and A. Sahai, "Coding constructions for blacklisting problems without computational assumptions", Crypto'99 *Springer LNCS* 1666 (1999) 609-623.
10. Chaoping Xing, "Asymptotic bounds on frameproof codes", *IEEE Trans. Inform. Th.* 40 (2002) 2991-2995.
11. J. Garay, J. Staddon and A. Wool, "Long-lived broadcast encryption", Crypto 2000 *Springer LNCS* 1880 (2000) pp. 333–352.
12. S. R. Blackburn, "Frameproof codes", *SIAM J. Discrete Math.*, vol. 16 (2003) 499-510.
13. G. Cohen and H.- G. Schaathun, "New upper bounds on separating codes", *2003 International Conference on Telecommunications*, February 2003.

Encoding Homotopy of Paths in the Plane

Sergei Bespamyatnikh

Department of Computer Science,
University of Texas at Dallas, Box 830688,
Richardson, TX 75083,USA. besp@utdallas.edu

Abstract. We study the problem of encoding homotopy of simple paths in the plane. We show that the homotopy of a simple path with k edges in the presence of n obstacles can be encoded using $O(n \log(n + k))$ bits. The bound is tight if $k = \Omega(n^{1+\varepsilon})$. We present an efficient algorithm for encoding the homotopy of a path. The algorithm can be applied to find homotopic paths among a set of simple paths. We show that the homotopy of a general (not necessary simple) path can be encoded using $O(k \log n)$ bits. The bound is tight. The code is based on a homotopic minimum-link path and we present output-sensitive algorithms for computing a path and the code.

1 Introduction

A fundamental problem is to find shortest paths in a geometric domain [12]. Chazelle [4] and Lee and Preparata [10] gave a funnel algorithm that computes the shortest path between two points in a simple polygon. Hershberger and Snoeyink [8] simplified the funnel algorithm and studied various optimizations of a given path among obstacles under the Euclidean and link metrics and under polygonal convex distance functions.

The topological concept of homotopy captures the notion of deforming paths. A *path* is a continuous map $\pi : [0, 1] \to \mathbb{R}^2$. Let $\alpha, \beta : [0, 1] \to \mathbb{R}^2$ be two paths that share starting and ending endpoints, $\alpha(0) = \beta(0)$ and $\alpha(1) = \beta(1)$. Let $B \subset \mathbb{R}^2$ be a set of *barriers* that includes the endpoints of α and β. We assume that the interiors of the paths α and β avoid B, i.e. $\{t \mid \alpha(t) \in B\} = \{t \mid \beta(t) \in B\} = \{0, 1\}$. The paths α and β are *homotopic* with respect to the barrier set B if α can be continuously transformed into β avoiding B.

Problems related to the homotopy of paths in the plane received attention very recently [3,6,2]. In this paper we consider the following questions. How the homotopy of a path can be represented in a computer? What is the minimum number of bits needed to encode the homotopy of a path with k edges in the presence of n obstacles?

Homotopy Encoding. Let B be a set of n barrier points in the plane and s and t be two barrier points. Let Π be a class of st-paths with at most k edges such that each path intersects B by the endpoints only. Find an integer number N and a map $\tau : \Pi \to [0..N]$ such that two

M. Farach-Colton (Ed.): LATIN 2004, LNCS 2976, pp. 329–338, 2004.

paths $\pi_1, \pi_2 \in \Pi$ are homotopic iff their codes are equal, $\tau(\pi_1) = \tau(\pi_2)$. Minimize the number of bits $\log N$ as a function of n and k.

To the best of our knowledge this paper is the first to study the problem of homotopy encoding *HEP*. Homotopy encoding can be used to solve a problem of testing homotopy of two paths [3]. One can test homotopy of two paths by, first, encoding their homotopies and then comparing the codes. Cabello *et al.* [3] established a criteria for two simple paths to be homotopic. They introduced a *canonical sequence* for a path and proved that two simple paths are homotopic if their canonical sequences are equal. Unfortunately, the canonical sequences can have $\Omega(nk)$ length which makes them ineffective for testing homotopy. Cabello *et al.* [3] found a way avoiding the computation of canonical sequences to test homotopy in $O(m \log m)$ time where $m = k + n$.

We focus on two classes of paths: simple paths and general (not necessarily simple) paths. For the simple paths we show a lower bound of $\Omega(n \log k)$ for the number of bits in a homotopy code if $k = \Omega(n^{1+\varepsilon})$. We introduce a *spanning homotopic graph* and show that it can be used to recognize the path homotopy. It can be used to encode the homotopy of a simple path using $n(\log n + 3 \log k + 3 \log 12) + o(n)$ bits. The bound is tight if $k = \Omega(n^{1+\varepsilon})$.

The main difficulty in homotopy encoding lies in computation of the spanning homotopic graph. Our approach is based on recently developed techniques for computing shortest homotopic paths [3,6,2]. We are interested in a special case of the problem (single path): given a path with k edges and a set of n barriers, find the shortest homotopic path. Very recently Efrat *et al.* [6] presented output sensitive algorithm for computing the shortest homotopic paths. The algorithm runs in $O(n^{3/2} + k \log n + K)$ time where K is the size of the output path. Note that K can be $\Omega(kn)$ in the worst case. They also gave a randomized algorithm with $O(n \log^{1+\varepsilon} n + k \log n + K)$ running time. In [2] the deterministic algorithm was improved and running time $O(n \log^{1+\varepsilon} n + k \log n + K)$ is achieved. We show that the path homotopy can be encoded in $O(m \log n)$ time where $m = \max(n, k)$. We show that the shortest homotopic path can retrieved (or decoded) from the homotopy code in $O(n + K)$ time.

For non-simple paths, we show a lower bound of $O(k \log n)$ for any homotopy encoding. We provide a homotopy code that achieves this bound. We introduce *canonical minimum-link path* homotopic to a path and show that it can be computed in $O(n^{2+\varepsilon} + k \log^2 n)$ time. The path can be used for the homotopy code. We also show that using space-time tradeoff the running time can be improved in the case $k < n^{2+\varepsilon}$ to $O((n + k + (nk)^{2/3}) \text{polylog}(n))$. This improves an algorithm by Hershberger and Snoeyink [8] that computes a minimum-link path homotopic to a path in $O(nk)$ time.

Our algorithms for homotopy encoding can be used for testing homotopy among multiple paths. This can be applied both for simple paths and general paths. To the best of our knowledge the problem with multiple paths was not considered. It can be stated as *homotopy classification*: Given simple/non-simple paths $\Pi = \{\pi_1, \ldots, \pi_l\}$ in the plane avoiding a set of n barriers, partition Π into classes of homotopy equivalent paths. For simple paths, the problem can be solved in $O(M \log n)$ where $M = \max(n, K)$ and K is the total complexity of the paths in Π. For non-simple paths the problem can be solved in $O((n + K + (nK)^{2/3})\text{polylog}(n))$ time. The solution is based on computing homotopy code for each path and sorting the codes.

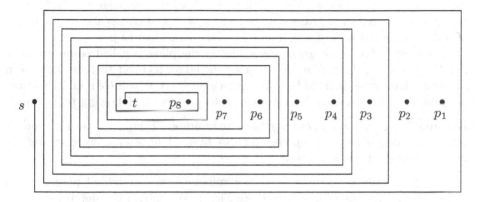

Fig. 1. Lower bound. $\alpha = (1, 0, 1, 2, 0, 3, 1, 2)$. The path winds around a point p_i $\alpha_1 + \ldots + \alpha_i$ times.

2 Lower Bound for Simple Paths

In this Section we show a lower bound for a homotopy code of simple paths.

Theorem 1. *Suppose that $k > n^{1+\varepsilon}$. Then any homotopy code for simple paths has at least $\Omega(n \log k)$ bits.*

Proof. Let $p_1, p_2, \ldots, p_n$ be barrier points located on a horizontal line in decreasing order of x-coordinates, see Fig. 1. We put the points s and t on the same line so that all the barrier points are on the right side. Let $m = \lfloor k/4 \rfloor$ and let α be a vector from $\mathbb{Z}^n$ such that all $\alpha_i \geq 0$ and $\sum_{1 \leq i \leq n} \alpha_i = m$. We construct a simple path such that, for each i, the path winds around the point p_i $\sum_{1 \leq j \leq i} \alpha_j$ times, see Fig. 1. This property holds for the shortest homotopic path as well. Therefore distinct vectors α and α' correspond to non-homotopic paths. Thus the total number of pairwise non-homotopic paths of length k is at least the number $t(n, k)$ of distinct vectors α.

Suppose that $k => n^{1+\varepsilon}$. A simple combinatorial observation is that $t(n, k) = \binom{m+n}{n} = (m + n)!/(n!m!)$. Using Stirling's formula one can obtain $\log t(n, k) = \Omega(n \log k)$.

3 Homotopy Code for Simple Paths

In this Section we introduce a fairly simple schema to encode the homotopy of simple paths that achieves optimal size if k is larger than $n^{1+\varepsilon}$. The idea is based on the shortest homotopic path. We define a *spanning homotopic graph* G as follows. Let π be a simple path in the presence of obstacles B. Let π' be its shortest homotopic path. The graph G has $B \cup \{s, t\}$ as the set of vertices and two vertices a and b are adjacent if the segment ab is in π'. Abusing notation we treat the vertices of G as points in $\mathbb{R}^2$ as well. An edge can be traversed many times if one walks along π'. We assign a weight $w(e)$ to an edge $e = (a, b)$ to be the number of times e is traversed. Let p be a vertex of G. Let $w(p)$ be a *weight* of p defined as the sum of weights of edges in G incident to p.

A vertex v of a planar graph embedded in the plane is called *pointed* if there is a line l passing through v such that all the edges incident to v are located in one of the halfplanes defined by l. We call a planar graph *pointed* if all its vertices are pointed. The spanning homotopic graphs possess the following properties.

Lemma 1. *Let G be a spanning homotopic graph. Then G is pointed and its vertices satisfy a* parity *property: vertices from B have even weights and the vertices s and t have odd weights.*

The properties in Lemma 1 are not sufficient for a weighted planar graph to be a spanning homotopic graph, see Fig. 2. We extend the definition of the spanning homotopic graph to a set of disjoint and simple paths, open or closed (note that, unlike open paths, a closed path does not contribute to the set of graph vertices). Let Π be a collection of disjoint simple paths $\pi_1, \pi_2, \ldots, \pi_m$, each path is either open or closed. We assume that the endpoints of an open path π_i are barrier points (note that any two open paths have disjoint enpoints). For each path $\pi_i \in \Pi$ we find its shortest homotopic path in the presence of obstacles B. A vertex of the spanning homotopic graph G of Π is either a barrier point or an endpoint of an open path π_i. A pair (p, q) is an edge of G if the shortest path of a path π_i has the line segment pq. The weight of an edge (p, q) is defined as the number of times the segment pq is traversed by the sortest paths.

(a)   (b)

Fig. 2. (a) Graph G is pointed and satisfies the parity property but (b) has no underlying simple path.

Theorem 2. *Let Π be a set of disjoint simple paths $\pi_1, \pi_2, \ldots, \pi_m$. Let Π' be a set of disjoint simple paths $\pi'_1, \pi'_2, \ldots, \pi'_{m'}$. If the spanning homotopic graphs of Π and Π' are equal then $m' = m$ and there is a permutation $\sigma_1, \sigma_2, \ldots, a_m$ such that, for every i, the paths π_i and π'_{σ_i} are either*

(i) *both closed paths that are homotopic in presence of obstacles B, or*

(ii) *both open paths with the same endpoints from B and are homotopic in presence of obstacles B.*

We apply Theorem 2 to a single path.

Corollary 1. *Two st-paths induce the same spanning homotopic graph iff they are homotopic.*

Let $v_1, v_2, \ldots$ be the list of vertices from V. We define *homotopy code* $\chi(\pi, B)$ of a simple path $\pi \in \Pi$ as the sequence of triples $(v_i, v_j, w(v_i, v_j))$ in the lexicographical order. By Corollary 1 two paths are homotopic iff their homotopy codes χ are equal.

We mention for completeness that, for every pointed weighted graph G, there is a set of paths whose spanning homotopic graph is G.

3.1 Succinct Homotopy Code

The number of triples $(v_i, v_j, w(v_i, v_j))$ in a homotopy code is $O(n)$. The explicit representation of a homotopy code requires $O(n \log(n+k))$ bits since the indices i and j need $O(\log n)$ bits (this is the widely used adjacency-list encoding) and the weight $w(v_i, v_j)$ need $O(\log k)$ bits. For small values of k this representation can exceed $O(n \log k)$ bound. We apply succinct encoding of labeled planar graphs [9,13] where the vertices of a planar graph are embedded into the plane and labeled. Keeler and Westbrook [9] proved that a labeled planar graph with n vertices and m edges can be encoded using $n \log n + 3n \log 12 + o(n)$ bits.

We encode the spanning homotopic graph without weights and the weights separately. We label the spanning homotopic graph according to the lexicographical order of the vertex coordinates. The weight components of χ can be encoded in $m \log k \leq 3n \log k$ bits. This imples the following theorem.

Theorem 3. *The homotopy of a simple path can be encoded using $n(\log n + 3 \log k + 3 \log 12) + o(n)$ bits.*

4   Shortest Homotopic Paths

In this Section we briefly describe the construction of canonical paths [3], the bundling [6] and an algorithm [2,3,6]. As in [6] we use the canonical paths to shortcut the given path and divide it into *x-monotone* paths. We can treat the monotone paths as horizontal segments and obtain rectified paths [3]. To rectify the paths one needs "aboveness" relation between the monotone paths and the barriers. This can be computed using a triangulation or trapezoidization [6] by an algorithm of Bar-Yehuda and Chazelle [1]. The running time is $O(n \log^{1+\varepsilon} n + k_{in})$ for any fixed $\varepsilon > 0$. The rectified paths can be shortcut by vertical segments. Applying the segment dragging queries by Chazelle [5] shortcuts can be done in $O(k_{in} \log n)$ time using $O(n \log n)$ preprocessing and $O(n)$ space.

The number homotopically different paths produced by shortcutting is at most $2n$ [6]. The homotopic paths can be bundled reducing the problem to the case $k_{in} \leq 2n$.

The problem is further reduced to finding a shortest monotone path in a simple polygon with barriers colored in two colors [2]. A hierarchical data structure is constructed to compute the paths efficiently. It stores the paths in a compact way by partitioning them into so called λ-paths. We modify the data structure to store *weighted λ-paths* where the weight is a number of times the λ-path is traversed. The weights of the paths can be used to compute the weights of edges of the spanning homotopic graph. We leave details for final version.

Theorem 4. *The homotopy code $\chi(\pi, B)$ of a simple path π can be computed in $O((n + k) \log n)$ time using $O(n + k)$ space.*

We show how the shortest homotopic path can be extracted from the homotopy code χ. We assume that the weights of edges take $O(1)$ space per weight. This is reasonable assumption since weights are bounded by k. The running time is $O(n + K)$.

5   Non-simple Paths

5.1   Lower Bound

We show a lower bound for general paths and design a homotopy code achieving this bound.

Lemma 2. *Any homotopy code for general paths has at least $\Omega(k \log n)$ bits.*

Proof. To show the lower bound we provide different homotopy paths from s to t. Let $\alpha = (\alpha_1, \alpha_2, \ldots, \alpha_m), m = \lceil k/3 \rceil$ be a sequence of integers from $1 \leq \alpha_i \leq n$ such that there are no repetitions $\alpha_i \neq \alpha_{i+1}$ and $p_{\alpha_1} \neq s, p_{\alpha_m} \neq t$. We generate a path for each sequence α. The idea is that, for three consecutive numbers $\alpha_{i-1}\alpha_i\alpha_{i+1}$, we can make a path $p_{\alpha_{i-1}}p_{\alpha_{i+1}}$ using two or three links such that p_{α_i} supports the path, see Fig. 3. The path has at most $3m \leq k$ edges (each point p_{α_i} contributes at most three edges to the path).

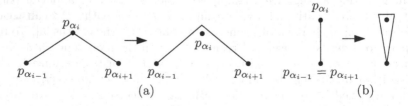

Fig. 3. Generating a path. (a) Distinct $p_{\alpha_{i-1}}$ and $p_{\alpha_{i+1}}$. (b) $p_{\alpha_{i-1}} = p_{\alpha_{i+1}}$.

All the generated paths have different homotopy since the shortest homotopic path for α is $sp_{\alpha_1} \ldots p_{\alpha_m} t$. The number of paths is $s(n, m) = n(n-1)^{m-2}$. The lower bound follows since $\log s(n, m) = \Omega(m \log n) = \Omega(k \log n)$.

5.2   Supported Paths

To show the upper bound we need some notations and properties and then introduce a *canonical homotopic path* for a general path π. The key idea is based on a *minimum-link path* that is defined as a homotopic path with minimum number of edges. A minimum-link path is not unique but it preserves the path complexity. Hershberger and Snoeyink [8] designed an algorithm for finding minimum-link path in a simple polygon with holes. The polygon is triangulated. They proved that the minimum-link path, π', homotopic to a given path π can be found in time $O(C_\pi + \Delta_\pi + \Delta_{\pi'})$ where C_π is the path complexity, Δ_π is the number of times that α crosses a triangulation edge. In terms of n and k, this bound is $O(nk)$. The algorithm exploits useful properties of minimum-link paths.

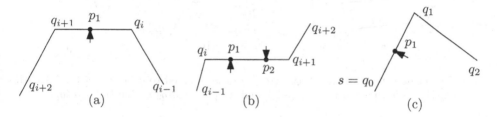

Fig. 4. Support points. Arrows show the sides of support. (a) Point p_1 supports the edge $q_i q_{i+1}$ from below, (b) points p_1 and p_2 support the inflection edge $q_i q_{i+1}$ from below and above, and (c) point p_1 supports the start edge $q_0 q_1$.

The approach by Ghosh [7] and Hershberger and Snoeyink [8] is based on computing the shortest homotopic path. This approach in our setting leads to $\Omega(nk)$ algorithm since the complexity of the shortest homotopic path can be $\Omega(nk)$. In order to make a faster algorithm we apply another approach using *supported paths*. Let $\pi = q_0 = s, q_1, \ldots, q_k = t$ be a path from s to t. Traversing π from s to t, we can label each vertex as a *left* or *right* turn. An edge $q_i q_{i+1}, 1 \le i < k$ whose endpoints make left turns is *supported* if it touches a barrier point on its left side, see Fig. 4 (a). The barrier point is called *support point* of the edge. The support point can be one of the endpoints of the edge. Similarly we define supported edge whose endpoints make right turns. As in [8] we call such an edge $q_i q_{i+1}, 1 \le i < k$ *inflection edge* if its endpoints make different turns. We require two support points for the inflection edge $q_i q_{i+1}$ to be supported and the barrier points should be on different sides of the edge and their order should correspond the turns at q_i and q_{i+1}, see Fig 4 (b). We call the edge incident to the support points of an inflection edge *support edge*. We also define conditions for the first and last edge to be supported. The edge $q_0 q_1$ is supported if q_1 makes left/right turn and $q_0 q_1$ has a barrier on its left/right side, see Fig. 4 (c). Similarly the support of the last edge is defined. A path is *supported* if all its edges are supported. An *extension* of a line segment ab is any line segment containing ab. Note that a supported inflection edge is an extension of its support edge.

5.3   Normalization

We introduce normalization tools to update a path. We show that a path π that is not supported can be normalized. If an edge $q_i q_{i+1}$ whose endpoints make left turns is not supported, then we slide $q_i q_{i+1}$ until either it hits a barrier point or it reaches one of the endpoints q_{i-1} or q_{i+2}, see Fig. 5 (a) and (b). The start edge can be normalized by sliding q_1 toward q_2, see Fig. 5 (c).   Normalization

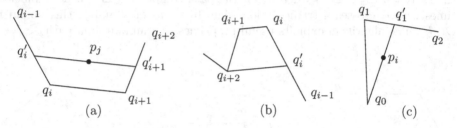

Fig. 5. Normalization. (a) Path $q_{i-1} q_i q_{i+1} q_{i+2}$ is changed to $q_{i-1} q_i' q_{i+1}' q_{i+2}$, (b) path $q_{i-1} q_i q_{i+1} q_{i+2}$ is changed to $q_{i-1} q_i' q_{i+2}$, and (c) path $q_0 q_1 q_2$ is changed to $q_0 q_1' q_2$.

of an edge with different turns at endpoints is more complicated. Let $q_i q_{i+1}$ be such an edge. We can assume that both the triangle $q_{i-1} q_i q_{i+1}$ and the triangle $q_i q_{i+1} q_{i+2}$ contain barrier points, otherwise we shortcut the path by trading two edges for one. If the union of two triangles is a convex quadrangle, then we change the edge $q_i q_{i+1}$ by common tangent of two barrier sets in the triangles $q_{i-1} q_i q_{i+1}$ and $q_i q_{i+1} q_{i+2}$, see Fig. 6 (a) (note that if one or both barrier sets are missing we can reduce the number of edges by one or two). If the quadrangle $q_{i-1} q_i q_{i+1} q_{i+2}$ is not convex then we truncate one of the triangles and use the above argument, for example, we consider the triangles $q_{i-1} q_i q_{i+1}$ and $q_i q_{i+1} q_{i+1}'$ in the case depicted in Fig. 6 (b).

A path is *normalized* if none of normalization tools can be applied to it.

Lemma 3. *Let π' be a normalized path obtained from a path π. Then π' is supported and has no more edges than π.*

5.4   Inflection Edges

Let $\mathcal{S}(\alpha)$ denote the set of support edges of inflection edges of a path α.

Theorem 5. *Let π' and π'' be two supported paths homotopic to π. They have the same set of support edges of inflection edges $\mathcal{S}(\pi') = \mathcal{S}(\pi'')$ and the edges of $\mathcal{S}(\pi')$ occur in the same order in the paths π' and π''.*

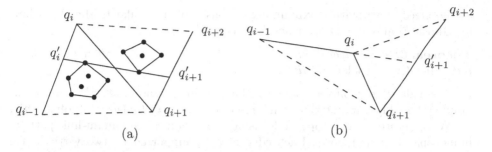

Fig. 6. Normalization. (a) Common tangent and the path $q_{i-1}q_i'q_{i+1}'q_{i+2}$, (b) changing triangle $q_iq_{i+1}q_{i+2}$ by $q_iq_{i+1}q_{i+1}'$.

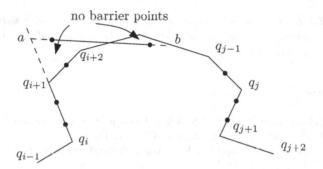

Fig. 7. Canonical path.

We show that the paths between inflection edges of π' can be further normalized producing a unique path.

5.5 Canonical Minimum-Link Path and Homotopy Code

Let q_iq_{i+1} and $q_jq_{j+1}, i < j$ be two inflection edges of a supported path. Let $\alpha = q_{i+1}q_{i+2}\ldots q_{j-1}q_j$ be the path between two edges. We also assume that there are no inflection edges in α. The vertices $q_{i+1},\ldots,q_j$ make the same turn and we assume that it is right turn, see Fig. 7. We find a point a on the ray q_iq_{i+1} and a point b on α such that the paths $q_{i+1}q_{i+2}\ldots b$ and $q_{i+1}ab$ are homotopic. In other words there are no barrier points between these paths, see Fig. 7. We optimize ab so that the path $q_{i+1}b$ has maximum length. We can view this as a travel of b along α starting from q_{i+2}. The travel of the point b is can be stopped if two barrier points on ab are found as in Fig. 7. The travel of b is restricted by a point on α such that a line parallel to q_iq_{i+1} is tangent to b (this may happen if α winds several times). We substitute the path $q_{i+1}q_{i+2}\ldots b$ of α by $q_{i+1}ab$ without increasing the number of edges. If the path ab can extended to reach the ray $q_{j+1}q_j$ preserving homotopy then we use the extension. Otherwise we continue this process on the path bq_j. This gives *canonical minimum-link path* homotopic to π. The construction can be viewed as the greedy path with maximum extensions. This is similar to the algorithms by Ghosh [7] and Hershberger and Snoeyink [8].

Theorem 6. *Canonical minimum-link path homotopic to a path π can be computed in $O(n^{2+\varepsilon} + k \log^2 n)$ time.*

A standard tradeoff technique can be used in the case $k < n^{2+\varepsilon}$, see for example [11,2]. The running time can be reduced to $O((n+k+(nk)^{2/3})\text{polylog}(n))$.

We generate a homotopy code using the canonical minimum-link path π' homotopic to π as follows. Each edge of π' is supported by two vertices. We store the vertices and the corresponding sides of the edge in a list. Each item takes $O(\log n)$ bits using indices of barriers (the sides takes two bits). The total size is $O(k \log n)$. We conclude the following theorem.

Theorem 7. *There is a homotopy code for paths in the plane using $O(k \log n)$ bits. The homotopy code of a path can be computed in $O((n + k + (nk)^{2/3})\text{polylog}(n))$ time.*

References

1. R. Bar-Yehuda and Bernard Chazelle. Triangulating disjoint Jordan chains. *Internat. J. Comput. Geom. Appl.*, 4(4):475–481, 1994.
2. S. Bespamyatnikh. Computing homotopic shortest paths in the plane. *J. Algorithms*, 49(2):284–303, 2003.
3. S. Cabello, Y. Liu, A. Mantler, and J. Snoeyink. Testing homotopy for paths in the plane. In *Proc. 18th Annu. ACM Sympos. Comput. Geom.*, pages 160–169, 2002.
4. Bernard Chazelle. A theorem on polygon cutting with applications. In *Proc. 23th Annu. IEEE Sympos. Found. Comput. Sci.*, pages 339–349, 1982.
5. Bernard Chazelle. An algorithm for segment-dragging and its implementation. *Algorithmica*, 3:205–221, 1988.
6. A. Efrat, S.G. Kobourov, and A. Lubiw. Computing homotopic shortest paths efficiently. In *Proc. 10th Annu. European Sympos. Algorithms*, pages 411–423, 2002.
7. S. K. Ghosh. Computing visibility polygon from a convex set and related problems. *J. Algorithms*, 12:75–95, 1991.
8. J. Hershberger and J. Snoeyink. Computing minimum length paths of a given homotopy class. *Comput. Geom. Theory Appl.*, 4:63–98, 1994.
9. K. Keeler and J. Westbrook. Short encodings of planar graphs and maps. *Discrete Applied Mathematics*, 58(3):239–252, 1995.
10. D. T. Lee and F. P. Preparata. Euclidean shortest paths in the presence of rectilinear barriers. *Networks*, 14:393–410, 1984.
11. J. Matoušek. Range searching with efficient hierarchical cuttings. *Discrete Comput. Geom.*, 10(2):157–182, 1993.
12. Joseph S. B. Mitchell. Geometric shortest paths and network optimization. In Jörg-Rüdiger Sack and Jorge Urrutia, editors, *Handbook of Computational Geometry*, pages 633–701. Elsevier Science Publishers B.V. North-Holland, Amsterdam, 2000.
13. J. Ian Munro and V. Raman. Succinct representation of balanced parentheses and static trees. *SIAM J. Comput.*, 31(3):762–776, 2001.

A Unified Approach to Coding Labeled Trees

Saverio Caminiti[1], Irene Finocchi[2], and Rossella Petreschi[1]

[1] DSI, Università degli Studi di Roma "La Sapienza"
Via Salaria 113, 00198 Roma, Italy
{caminiti, petreschi}@dsi.uniroma1.it
[2] DISP, Università degli Studi di Roma "Tor Vergata"
Via del Politecnico 1, 00133 Roma, Italy
finocchi@disp.uniroma2.it

Abstract. We consider the problem of coding labeled trees by means
of strings of node labels and we present a unified approach based on a
reduction of both coding and decoding to integer (radix) sorting. Applying this approach to four well-known codes introduced by Prüfer [18],
Neville [17], and Deo and Micikevicius [5], we close some open problems.
With respect to coding, our general sequential algorithm requires optimal linear time, thus solving the problem of optimally computing the
second code presented by Neville. The algorithm can be parallelized on
the EREW PRAM model, so as to work in $O(\log n)$ time using $O(n)$ or
$O(n\sqrt{\log n})$ operations, depending on the code.
With respect to decoding, the problem of finding an optimal sequential
algorithm for the second Neville code was also open, and our general
scheme solves it. Furthermore, in a parallel setting our scheme yields the
first efficient decoding algorithms for the codes in [5] and [17].

1 Introduction

Labeled trees are of interest in practical and theoretical areas of computer science. For example, Ethernet has a unique path between terminal devices, thus
being a tree: labeling the tree nodes is necessary to uniquely identify each device
in the network. An interesting alternative to the usual representations of tree
data structures in computer memories is based on coding labeled trees by means
of strings of node labels. This representation was first used in the proof of Cayley's theorem [2,18] to show a one-to-one correspondence between free labeled
trees on n nodes and strings of length $n - 2$. In addition to this purely mathematical use, string-based codings of trees have many practical applications. For
instance, they make it possible to generate random uniformly distributed trees
and random connected graphs [14]: the generation of a random string followed
by the use of a fast decoding algorithm is typically more efficient than random
tree generation by the addition of edges, since in the latter case one must pay
attention not to introduce cycles. In addition, tree codes are employed in genetic
algorithms, where chromosomes in the population are represented as strings of
integers, and in heuristics for computing minimum spanning trees with additional constraints, e.g., on the number of leaves or on the diameter of the tree

M. Farach-Colton (Ed.): LATIN 2004, LNCS 2976, pp. 339–348, 2004.
© Springer-Verlag Berlin Heidelberg 2004

itself [7,8,20]. Not last, tree codes are used for data compression [19] and for computing the tree and forest volumes of graphs [13].

Tree codes. We now survey the main tree codes known in the literature, referring the interested reader to the taxonomy in [6] for further details. We assume to deal with a labeled n-node rooted tree T whose nodes have distinct labels from $[1, n]$. All the codes that we discuss are obtained by progressively *updating* the tree through the *deletion of leaves*: when a leaf is eliminated, the label of its parent is added to the code.

The oldest and most famous code is due to Prüfer [18] and always deletes the leaf with smallest label. In 1953, Neville [17] presented three different codes, the first of which coincides with Prüfer's one. The second Neville's code, before updating T, eliminates *all* the leaves ordered by increasing labels. The third Neville's code works by deleting chains. We call *pending chain* a path $u_1, \ldots, u_k$ of the tree such that the starting point u_1 is a leaf, and, for each $i \in [1, k-1]$, the elimination of u_i makes u_{i+1} a leaf: the code works by iteratively eliminating the pending chain with the smallest starting point. Quite recently, Deo and Micikevicius [5] suggested the following coding approach: at the first iteration, eliminate all tree leaves as in the second Neville's code, then delete the remaining nodes in the order in which they assume degree 1. For brevity, we will denote the codes introduced above with PR, N2, N3, and DM, respectively.

All these codes have length $n - 1$ and the last element is the root of the tree. If the tree is unrooted, the elimination scheme implicitly defines a root for it. Actually, it is easy to see that the last element in the code is: a) the maximum node label (i.e., n) for Prüfer's code; b) the maximum label of a center of the tree for the second Neville's code; c) the label of the maximum leaf for the third Neville's code; d) the label of any tree center for Deo and Micikevicius's code. In cases a), c), and d), the value of the last element can be univocally determined from the code and thus the code length can be reduced to $n - 2$. We remark that codes PR and DM have been originally presented for free trees, while Neville's codes have been generalized for free trees by Moon [16].

Related work. A linear time algorithm for computing Prüfer codes is presented in [3]. The algorithm can be easily adapted to the third Neville's code. Deo and Micikevicius give a linear time algorithm for code DM based on a quite different approach. As stated in [6], no $O(n)$ time algorithm for the second Neville's code was known so far: sorting the leaves before each tree update yields indeed an $O(n \log n)$ bound. An optimal parallel algorithm for computing Prüfer codes, which improves over a previous result due to Greenlaw and Petreschi [10], is given in [9]. A few simple changes make the algorithm work also for code N3. Efficient, but not optimal, parallel algorithms for codes N2 and DM are presented in [7]. A simple – non optimal – scheme for constructing a tree T from a Prüfer code is presented in [6]. The scheme can be promptly generalized for building T starting from any of the other codes: this implies decoding in linear time codes N3 and DM, and in $O(n \log n)$ time codes PR and N2. An $O(n)$ time decoding algorithm for PR is described in [9]. In a parallel setting, Wang, Chen, and Liu [19] propose an $O(\log n)$ time decoding algorithm for Prüfer codes using $O(n)$ processors on the EREW PRAM computational model. At the best of our knowledge, parallel decoding algorithms for the other codes were not known in the literature until this paper.

Our contribution. We show that both coding and decoding can be reduced to integer (radix) sorting. Based on this reduction, we present a unified approach that works for all the codes introduced so far and can be applied both in a sequential and in a parallel setting. The coding scheme is based on the definition of pairs associated to the nodes of T according to criteria dependent on the specific code: the coding problem is then reduced to the problem of sorting these pairs in lexicographic order. The decoding scheme is based on the computation of the rightmost occurrence of each label in the code: this is also reduced to integer radix sorting.

Concerning coding, our general sequential algorithm requires optimal linear time for all the presented codes; in particular, it solves the problem of computing the second Neville code in time $O(n)$, which was still open [6]. The algorithm can be parallelized, and its parallel version either matches or improves by a factor $O(\sqrt{\log n})$ the performances of the best ad-hoc approaches known so far. Concerning decoding, we design the first parallel algorithm for codes N2, N3, and DM, working on the EREW PRAM model in $O(\log n)$ time and $O(n\sqrt{\log n})$ operations (with respect to PR, our algorithm matches the performances of the best previous result). Our parallel results both for coding and for decoding are summarized in the following table:

	Coding		Decoding	
	before	this paper	before	this paper
PR	$O(n)$ [9]	$O(n)$	$O(n \log n)$ [19]	$O(n \log n)$
N2	$O(n \log n)$ [7]	$O(n\sqrt{\log n})$	open	$O(n\sqrt{\log n})$
N3	$O(n)$ [7,9]	$O(n)$	open	$O(n\sqrt{\log n})$
DM	$O(n \log n)$ [7]	$O(n\sqrt{\log n})$	open	$O(n\sqrt{\log n})$

where costs are expressed in terms of number of operations. We remark that the problem of finding an optimal sequential decoding algorithm for code N2 was also open, and our general scheme solves it optimally. Hence, we show that labeled trees can be coded and decoded in linear sequential time independently of the specific code. Due to lack of space, we omit many details in this extended abstract.

2 A Unified Coding Algorithm

Many sequential and parallel coding algorithms have been presented in the literature [3,5,6,9,10,19], but all of them strongly depend on the properties of the code which has to be computed and thus are very different from each other. In this section we show a unified approach that works for all the codes introduced in Section 1 and can be used both in a sequential and in a parallel setting.

Table 1. Pair (x_v, y_v) associated to node v for different codes.

	PR	N2	N3	DM
x_v	$\mu(v)$	$l(v)$	$\lambda(v)$	$l(v)$
y_v	$d(\mu(v),v)$	v	$d(\lambda(v),v)$	$\gamma(v)$

Namely, we associate each tree node with a pair of integer numbers and we sort nodes using such pairs as keys. The obtained ordering corresponds to the order in which nodes are removed from the tree and can be thus used to compute the code. In the rest of this section we show how different pair choices yield Prüfer, Neville, and Deo and Micikevicius codes, respectively. We then present a linear time sequential coding algorithm and its parallelization on the EREW PRAM model. The parallel algorithm works in $O(\log n)$ time and requires either $O(n)$ or $O(n\sqrt{\log n})$ operations, depending on the code.

Coding by sorting pairs. Let T be a rooted labeled n-node tree. If T is not rooted, we choose a root r as in points a) – d) in Section 1. Let u, v be any two nodes of tree T. Let us call: T_v, the subtree of T rooted at v; $d(u, v)$, the distance between any two nodes u and v ($d(v, v) = 0$); $l(v)$, the level of a node v, i.e., the maximum distance of v from a leaf in T_v; $\mu(v)$, the maximum label among nodes in T_v; $\lambda(v)$, the maximum label among leaves in T_v; $\gamma(v)$, the maximum label among the leaves in T_v at maximum distance from v; (x_v, y_v), a pair associated to node v according to the specific code as shown in Table 1; P, the set of pairs (x_v, y_v). The following lemma establishes a correspondence between the set P of pairs and the order in which nodes are removed from the tree. Due to lack of space, we defer the proof to the extended version of this paper.

Lemma 1. *For each code, the lexicographic ordering of the pairs (x_v, y_v) in set P corresponds to the order in which nodes are removed from tree T according to the code definition.*

Before describing the sequential and parallel algorithms, note that it is easy to sort the pairs (x_v, y_v) used in the coding scheme. Indeed, independently of the code, each element in such pairs is in the range $[1, n]$. A radix-sort like approach [4] is thus sufficient to sort them according to y_v, first, and x_v, later. In Figure 1 the pairs relative to the four codes are presented. The tree used in the example is the same in the four cases and is rooted according to points a) – d) in Section 1. Bold arcs in the trees related to codes PR and N3 indicate chains [1] and pending chains, respectively; dashed lines in the trees related to codes N2 and DM separate nodes at different levels. In each figure the string representing the generated code is also shown.

Sequential algorithm. Using the pairs defined in Table 1, an optimal sequential coding algorithm is now straightforward:

---

[1] According to the definition of Prüfer's code, when the node $\mu(v)$ is chosen for removal, the only remaining subtree of T_v consists of a *chain* from $\mu(v)$ to v.

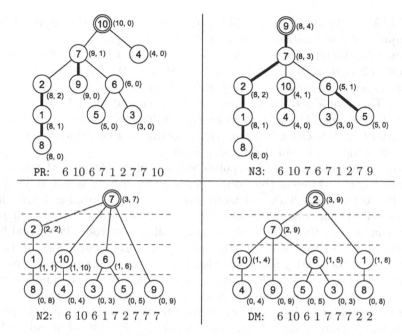

Fig. 1. Pair associated to each tree node as specified in Table 1.

UNIFIED CODING ALGORITHM:
1. **for each** node v, **compute** the pair (x_v, y_v)
2. **sort** the tree nodes according to pairs (x_v, y_v)
3. **for** $i = 1$ **to** $n - 1$ **do**
4.   **let** v be the i-th node in the ordering
5.   **append** $parent(v)$ to the code

The UNIFIED CODING ALGORITHM clearly requires linear time: the set of pairs can be easily computed in $O(n)$ time using a post-order visit of the tree, and two bucket-sorts can be used to implement step 2. Hence we have the following theorem:

Theorem 1. *Let T be a n-node tree and let the pair (x_v, y_v) associated to each node v of T be defined as in Table 1. The* UNIFIED CODING ALGORITHM *computes codes* PR, N2, N3, *and* DM *in* $O(n)$ *worst-case running time.*

Parallel algorithm. We now show how to parallelize each step of the sequential algorithm presented above. We work in the simplest PRAM model with exclusive read and write operations (EREW [12]). The Euler tour technique makes it possible to root the tree at the node r specified in Section 1 in $O(\log n)$ time with cost $O(n)$ [12]. The node r can be easily identified; in particular, when r is the center of the tree, we refer to the approach described in [15]. The pairs given in Table 1 can be computed in $O(\log n)$ time with cost $O(n)$ using standard techniques, such as Euler tour, rake, and list ranking [12].

Step 3 can be trivially implemented in $O(1)$ time with cost $O(n)$. The sorting in step 2 is thus the most expensive operation. We follow a radix-sort like approach and use the stable integer-sorting algorithm presented in [11] as a subroutine. This requires $O(\log n)$ time, linear space and $O(n\sqrt{\log n})$ cost on an EREW PRAM with $O(\log n)$ word length. Under the hypothesis that the machine word length is $O(\log^2 n)$, the cost of sorting can be reduced to $O(n)$ [11], and so does the cost of our coding algorithm.

We remark that our algorithm solves within a unified framework the problem of computing four different tree codes. In addition, with respect to codes N2 and DM, it improves of an $O(\sqrt{\log n})$ factor over the best approaches known in the literature [7]. Unfortunately, as far as we have described it, it does not match the perfomances of the optimal algorithms available for codes PR [9] and N3 [7]. However, in these cases we can further reduce the cost of our algorithm to $O(n)$ by using an ad-hoc sorting procedure that benefits from the partition into chains.

Let us consider Prüfer codes first. As observed in [10], the final node ordering can be obtained by sorting chains among each other and nodes within each chain. In our framework, the chain ordering is given by the value $\mu(v)$, and the position of each node within its chain by the distance $d(\mu(v), v)$. Instead of using a black-box integer sorting procedure, we exploit the fact that we can compute optimally the size of each chain, i.e., the number of nodes with the same $\mu(v)$. A prefix sum computation gives, for each chain head, the number of nodes in the preceding chains, i.e., its final position. At last, the position of the remaining nodes is univocally determined summing up the position of the chain head $\mu(v)$ with the value $d(\mu(v), v)$. Similar considerations can be applied to the third Neville's code. The following theorem summarizes our results on parallel coding:

Theorem 2. *Let T be a n-node tree and let the pair (x_v, y_v) associated to each node v of T be defined as in Table 1. On the EREW PRAM model, the* UNIFIED CODING ALGORITHM *computes codes* PR *and* N3 *optimally, i.e., in $O(\log n)$ time with cost $O(n)$, and codes* N2 *and* DM *in $O(\log n)$ time with cost $O(n\sqrt{\log n})$.*

3   Decoding Algorithms

In this section we present sequential and parallel algorithms for decoding, i.e., for building the tree T corresponding to a given code C. As far as C is computed, each node label in it represents the parent of a leaf eliminated from T. Hence, in order to reconstruct T, it is sufficient to compute the ordered sequence of labels of the eliminated leaves, say S: for each $i \in [1, n-1]$, the pair (C_i, S_i) will thus be an arc in the tree. Before describing the algorithms, we argue that computing the rightmost occurrence of a node in the code is very useful for decoding, and we show how to obtain such an information both in a sequential and in a parallel setting.

Decoding by rightmost occurrence computation. We first observe that the leaves of T are exactly those nodes that do not appear in the code, as they are not parents of any node. Each internal node, say v, in general may appear in C more than once; each appearance corresponds to the elimination of one of its children, and therefore to decreasing the degree of v by 1. After the rightmost occurrence in the code, v is clearly a leaf and thus becomes a candidate for being eliminated. More formally:

$$\forall v \neq r, \quad \exists \; unique \; j > rightmost(v, C) \; such \; that \; S_j = v$$

where r is the tree root (i.e., the last element in C) and $rightmost(v, C)$ denotes the index of the rightmost occurrence of node v in C. We assume that $rightmost(v, C) = 0$ if v is a leaf of T.

It is easy to compute the rightmost occurrence of each node sequentially by simply scanning code C. In parallel, we can reduce the rightmost occurrence computation problem to a pair sorting problem. Namely, we sort in increasing order the pairs (C_i, i), for $i \in [1, n-1]$. Let us now consider the sub-sequences of pairs with the same first element C_i: the second element of the last pair in each sub-sequence is the index of the rightmost occurrence of node C_i in the code. Since each pair value is an integer in $[1, n]$, we can use twice the stable integer-sorting algorithm of [11]: this requires $O(\log n)$ time and $O(n\sqrt{\log n})$ cost in the EREW PRAM model. Then, each processor p_i in parallel compares the first element of the i-th pair in the sorted sequence to the first element of the $(i+1)$-th pair, deciding if this is the end of a sub-sequence or not. This requires additional $O(1)$ time and linear cost with exclusive read and write operations.

A unified decoding algorithm. We now describe a decoding algorithm for codes N3, PR, and DM that works on the rightmost occurrences and can be used both in a sequential and in a parallel setting. First, for each code, we show how the position of a node in the sequence S that we want to construct can be expressed as a function of *rightmost*.

Third Neville code. By definition of code N3, each internal node v is eliminated as soon as it becomes a leaf. Thus, the position of v in sequence S is exactly $rightmost(v, C) + 1$. The entries of S which are still free after positioning all the internal nodes are occupied by the leaves of T in increasing order.

Prüfer code. Differently from the third Neville's code, in code PR an internal node v is eliminated as soon as it becomes a leaf if and only if there is no leaf with label smaller than v. In order to test this condition, following [19], we introduce the number of nodes with label smaller than v that become leaves before v: $prev(v, C) = |\{u : u < v \text{ and } rightmost(u, C) < rightmost(v, C)\}|$. Thus, the position of v in sequence S is $rightmost(v, C) + 1$ if and only if $rightmost(v, C) \geq prev(v, C)$. All the other nodes are assigned to the remaining entries of S by increasing label order.

Deo and Micikevicius code. All the leaves of T, sorted by increasing labels, are at the beginning of sequence S. Then, all the internal nodes appear in the order in which they become leaves, i.e., sorted by increasing *rightmost*. It is possible to get a closed formula giving the position of each node. For each $i \in [1, n - 1]$, let $\rho(i)$ be 1 if i is the rightmost occurrence of node C_i, and 0 otherwise. Let $\sigma(i) = \sum_{j \leq i} \rho(j)$. The position of an internal node v is exactly $|leaves(T)| + \sigma(rightmost(v, C))$.

Our unified decoding algorithm is as follows:

DECODING ALGORITHM:
1.   **for each node** v compute $rightmost(v, C)$
2.   **for each node** v **except for the root do**
3.   **if** $(test(v) = $ **true) then** $S[position(v)] \leftarrow v$
4.   **let** L **be the list of nodes not yet assigned in increasing order**
5.   **let** P **be the set of positions of** S **which are still empty**
6.   **for each** $i = 1$ **to** $|P|$ **do**
7. $S[P[i]] \leftarrow L[i]$

where $test(v)$ and $position(v)$ are specified in Table 2. With respect to Prüfer's code, the algorithm is essentially the same as the one described in [19], and we refer to [19,9] for a detailed parallel analysis. As observed in Section 1, a linear sequential decoding algorithm for Prüfer codes is presented in [9], while the straightforward sequential implementation of our algorithm would require $O(n \log n)$ time. This can be easily reduced to $O(n)$ time by adapting the DE-CODING ALGORITHM in such a way that the $prev$ computation can be avoided. With respect to codes N3 and DM, the DECODING ALGORITHM runs in linear time.

In parallel, $\sigma(i)$ can be computed for each i using a prefix sum operation [12]. In order to get set L in step 4, we can mark each node not yet assigned to S and obtain its rank in L by computing prefix sums. Similarly for set P. Hence, the most expensive step is the rightmost computation, which requires integer sorting. This implies the following result:

Theorem 3. *Let C be a string of $n-1$ integers in $[1, n]$. Let $\mathcal{C}$ be the set of codes* PR, N3, *and* DM. *For each $i \in [1, n - 1]$, let* test(C[i]) *and* position(C[i]) *be defined as in Table 2. For each code in $\mathcal{C}$, the* DECODING ALGORITHM *computes the tree corresponding to string C in $O(n)$ sequential time. Decoding on the EREW PRAM model requires $O(\log n)$ time with cost $O(n \log n)$ for code* PR *and $O(\log n)$ time with cost $O(n\sqrt{\log n})$ for codes* N3 *and* DM.

Second Neville code. Differently from the other codes, in code N2 the right-most occurrence of each node in C gives only partial information about sequence S. Thus, we treat N2 separately in this section. We first observe that if all nodes were assigned with a level, an ordering with respect to pairs $(l(v), v)$ would give sequence S, and thus the tree. We refer to Section 2 for details on the correctness of this approach. We now show how to compute $l(v)$.

Let x be the number of leaves of T, which have level 1 and rightmost occurrence 0. Consider the first x elements of code C, say $C[1], \ldots, C[x]$. For each i, $1 \leq i \leq x$, such that i is the rightmost occurrence of $C[i]$, we know that node $C[i]$ has level 2. The same reasoning can be applied to get level-3 nodes from level-2 nodes, and so on. With respect to the running time, a sequential scan

Table 2. Condition on node v that is checked in the DECODING ALGORITHM and position of v as a function of $rightmost(v, C)$.

	$test(v)$	$position(v)$
N3	true	$rightmost(v, C) + 1$
PR	$rightmost(v, C) \geq prev(v, C)$	$rightmost(v, C) + 1$
DM	true	$|leaves(T)| + \sigma(rightmost(v, C))|$

of code C is sufficient to compute the level of each node in linear time. Integer sorting does the rest. Unfortunately, this approach is inherently sequential and thus inefficient in parallel.

Before describing our parallel approach, we note that the procedure for level computation described above can be applied also for code DM. Indeed, let T' be the tree obtained interpreting C as the code by Deo and Micikevicius and let S' be the corresponding sequence: although T and T' are different, they have the same nodes at the same levels and, both in S and S', nodes at level $i + 1$ appear after nodes at level i, but are differently permuted within the level. In view of these considerations, we are able to solve our problem in parallel, using T' to get our missing level information. Namely, first we build tree T' using the DECODING ALGORITHM, then we compute node levels applying the Euler tour technique, and finally we obtain sequence S (corresponding to tree T) by sorting the pairs $(l(v), v)$. It is to remark that the Euler tour technique requires a particular data structure [12] that can be built as described in [10]. The bottleneck of this procedure is sorting of pairs of integers in $[1, n]$, and thus we can use the parallel integer sorting presented in [11]. We can summarize the results concerning code N2 as follows.

Theorem 4. *Let C be a string of $n - 1$ integers in $[1, n]$. The tree corresponding to C according to code N2 can be computed in $O(n)$ sequential time and in $O(\log n)$ time with cost $O(n\sqrt{\log n})$ on the EREW PRAM model.*

4   Conclusions and Open Problems

We have presented a unified approach for coding labeled trees by means of strings of node labels and have applied it to four well-known codes: PR [18], N2 [17], N3 [17], and DM [5]. The coding scheme is based on the definition of pairs associated to the nodes of the tree according to some criteria dependent on the specific code. The coding problem is reduced to the problem of sorting these pairs in lexicographic order. The decoding scheme is based on the computation of the rightmost occurrence of each label in the code: this is also reduced to radix sorting. We have applied these approaches both in a sequential and in a parallel setting. We have completely closed the sequential coding and decoding problem, showing that both operations in all the four codes can be done in linear time. In the parallel setting, further work is still needed in order to improve all the non optimal coding and decoding algorithms. We remark that any improvement on the computation of integer sorting would yield better results for our parallel algorithms.

References

1. ATALLAH, M.J., COLE, R., AND GOODRICH, M.T.: Cascading divide-and-conquer: a technique for designing parallel algorithms. *SIAM Journal of Computing*, 18(3), pp. 499–532, 1989.
2. CAYLEY, A.: A theorem on trees. *Quarterly Journal of Mathematics*, 23, pp. 376–378, 1889.
3. CHEN, H.C. AND WANG, Y.L.: An efficient algorithm for generating Prüfer codes from labelled trees. *Theory of Computing Systems*, 33, pp. 97–105, 2000.
4. T.H. CORMEN, C.E. LEISERSON, R.L. RIVEST, AND C. STEIN: *Introduction to algorithms*. McGraw-Hill, 2001.
5. DEO, N. AND MICIKEVICIUS, P.: A new encoding for labeled trees employing a stack and a queue. *Bulletin of the Institute of Combinatorics and its Applications*, 34, pp. 77–85, 2002.
6. DEO, N. AND MICIKEVICIUS, P.: Prüfer-like codes for labeled trees. *Congressus Numerantium*, 151, pp. 65–73, 2001.
7. DEO, N. AND MICIKEVICIUS, P.:æ Parallel algorithms for computing Prüfer-like codes of labeled trees.æ *Computer Science Technical Report* CS-TR-01-06, 2001.
8. EDELSON, W. AND GARGANO, M.L.: Feasible encodings for GA solutions of constrained minimal spanning tree problems. *Proceedings of the Genetic and Evolutionary Computation Conference*, Morgan Kaufmann Publishers, page 754, 2000.
9. GREENLAW, R., HALLDORSSON, M.M. AND PETRESCHI, R.: On computing Prüfer codes and their corresponding trees optimally. *Proceedings Journees de l'Informatique Messine, Graph algorithms*, 2000.
10. GREENLAW, R. AND PETRESCHI, R.: Computing Prüfer codes efficiently in parallel. *Discrete Applied Mathematics*, 102, pp. 205-222, 2000.
11. HAN, Y. AND SHEN, X.: Parallel integer sorting is more efficient than parallel comparison sorting on exclusive write PRAMS. *SIAM Journal of Computing*, 31(6), pp. 1852–1878, 2002.
12. JÁJÁ, J.: *An Introduction to parallel algorithms*. Addison Wesley, 1992.
13. KELMANS, A., PAK, I., AND POSTNIKOV, A.: Tree and forest volumes of graphs. *DIMACS Technical Report* 2000-03, 2000.
14. KUMAR, V., DEO, N., AND KUMAR, N.: Parallel generation of random trees and connected graphs. *Congressus Numerantium*, 130, pp. 7–18, 1998.
15. LO, W.T., AND PENG, S.: The optimal location of a structured facility in a tree network. *Journal of Parallel Algorithms and Applications*, 2, pp. 43–60, 1994.
16. MOON, J.W.: *Counting labeled trees*. William Clowes and Sons, London, 1970.
17. NEVILLE, E.H.: The codifying of tree structures. *Proceedings of Cambridge Philosophical Society*, 49, pp. 381–385, 1953.
18. PRÜFER, H.: Neuer Beweis eines Satzes über Permutationen. *Archiv für Mathematik und Physik*, 27, pp. 142–144, 1918.
19. WANG, Y.L., CHEN, H.C. AND LIU, W.K.: A parallel algorithm for constructing a labeled tree. *IEEE Transactions on Parallel and Distributed Systems*, 8(12), pp. 1236–1240, 1997.
20. ZHOU, G. AND GEN, M.: A note on genetic algorithms for degree-constrained spanning tree problems. *Networks*, 30, pp. 91–95, 1997.

Cost-Optimal Trees for Ray Shooting[*]

Hervé Brönnimann[1][**] and Marc Glisse[2][***]

[1] Computer and Information Science, Polytechnic University, Six Metrotech Center, Brooklyn, NY 11201, USA.
[2] Ecole Normale Supérieure Paris, 45 Rue d'Ulm, 75230 Paris, Cedex 5, France.

Abstract. Predicting and optimizing the performance of ray shooting is a very important problem in computer graphics due to the severe computational demands of ray tracing and other applications, e.g., radio propagation simulation. Aronov and Fortune were the first to guarantee an overall performance within a constant factor of optimal in the following model of computation: build a triangulation compatible with the scene, and shoot rays by locating origin and traversing until hit is found. Triangulations are not a very popular model in computer graphics, but space decompositions like kd-trees and octrees are used routinely. Aronov et al. [1] developed a cost measure for such decompositions, and proved it to reliably predict the average cost of ray shooting.

In this paper, we address the corresponding optimization problem, and more generally d-dimensional trees with the cost measure of [1] as the optimizing criterion. We give a construction of quadtrees and octrees which yields cost $O(M)$, where M is the infimum of the cost measure on all trees, for points or for $(d-1)$-simplices. Sometimes, a balance condition is important. (Informally, balanced trees ensures that adjacent leaves have similar size.) We also show that rebalancing does not affect the cost by more than a constant multiplicative factor, for both points and $(d-1)$-simplices. To our knowledge, these are the only results that provide performance guarantees within approximation factor of optimality for 3-dimensional ray shooting with the octree model of computation.

1 Introduction

Given a set S of objects, called a *scene*, the ray-shooting problem asks, given a ray, what is the first object in S intersected by this ray. Solving this problem is essential in answering visibility queries. Such queries are used in computer

---

[*] This research was initiated at the McGill-INRIA Workshop on Computational Geometry in Computer Graphics, February 9-15, 2002, co-organized by H. Everett, S. Lazard, and S. Whitesides, and held at the Bellairs Research Institute of McGill University.

[**] Work on this paper has been supported by NSF ITR Grant CCR-0081964 and NSF CAREER Grant CCR-0133599.

[***] Work by the second author was performed while spending a semester at Polytechnic University, on leave from École Normale Supérieure, Paris, France.

M. Farach-Colton (Ed.): LATIN 2004, LNCS 2976, pp. 349–358, 2004.
© Springer-Verlag Berlin Heidelberg 2004

graphics (e.g., ray tracing and radiosity techniques for photo-realistic 3D rendering), radio- and wave-propagation simulation, and a host of other practical problems.

A popular approach to speed up ray-shooting queries is to construct a space decomposition such as a quadtree in 2D and an octree in 3D. The query is then answered by traversing the leaves of the tree as they are intersected by the ray, and for each cell in turn, testing for an intersection between the ray and the subset of objects intersecting that cell. The performance of such an approach greatly depends on the quality of that space decomposition.

Unfortunately, not much is understood about how to measure this quality. Practioners use a host of heuristics and parameters of the scene, of which the object count is less important than, e.g., the size of the objects in the scene, and other properties of the object distribution (density, depth complexity, surface area of the subdivision). Those parameters are used to develop automatic termination criteria for recursively constructing the decompositions.While they perform acceptably well most of the time, none of these heuristics performs better than the brute-force method in the worst case. More importantly, occasionally the termination criteria will produce a bad decomposition, and in any case there is no way to know the quality of the decomposition because lower bounds are hard to come by.

Our results. In [1], we proposed a measure for bounded-degree space decompositions, based on the surface area heuristic, which is a simplification (for practicality) of a more complicated but theoretically sound cost measure: under certain assumptions on the ray distribution, the cost measure provably reflects the cost of shooting an average ray using the space decomposition. This has been experimentally verified [1, 2]

In [6] and in this paper, we are interested in constructing trees with cost as low as possible, with a guaranteed approximation ratio. The only objects we consider are simplices (points and segments inside the unit square $[0, 1]^2$ in $\mathbb{R}^2$, or points, segments and triangles inside the unit cube $[0, 1]^3$ in $\mathbb{R}^3$). We however assume the Real-RAM model so as to avoid a discussion on the bit-length of the coordinates. We give and analyze algorithms that produce trees with cost $O(M)$, where M is a lower bound on the cost of any tree. The novelty from [6] is the extension to d=3 and higher of the results. we also examine the effect of rebalancing the tree on the cost measure, and prove that rebalancing only increases the cost by a constant multiplicative factor.

Related work. The work on quadtrees and octrees in the mesh generation and graphics community (see the book by Samet [10], the thesis of Moore [8], or the survey by Bern and Eppstein [5] for references) is usually concerned with a tradeoff between the size of the tree and their accuracy with respect to a certain measure (that usually evaluates a maximum interpolation error). It is not relevant here.

There is, however, a rich history of data-structure optimization for ray shooting in computer graphics. Cost measures have been proposed for ray shooting in octrees by McDonald and Booth [7], Reinhard and coll. [9], Whang and coll. [12], and for other structures, such as bounding volume hierarchies BSP-trees uniform grids and hierarchical uniform grids (see [1] and refs. therein). All of these approaches use heuristic criterion (sometimes very effectively) but none offer theoretical guarantees.

2   General Cost Measure Results

The following cost measure was introduced by Aronov *et al.* [1] for the purpose of predicting the traversal cost of shooting a ray in $\mathbb{R}^d$, while using a quadtree (for $d = 2$) or an octree (for $d = 3$) to store S:

$$c_S(\mathcal{T}) = \sum_{\sigma \in \mathcal{L}(\mathcal{T})} (\gamma + |S \cap \sigma|) \times \lambda_{d-1}(\sigma), \tag{1}$$

where $\mathcal{L}(\mathcal{T})$ is the set of leaves of the quadtree, $S \cap \sigma$ is the set of scene objects meeting a leaf σ, and $\lambda_{d-1}(\sigma)$ is the perimeter length (if $d = 2$) or surface area (if $d = 3$) of σ.

This cost function provably models the cost of finding all the objects intersected by a random line, with respect to the rigid-motion invariant distribution of lines [1]. Here's an overly simplified explanation why: when shooting a ray, the octree is traversed and all the objects in a traversed leaf are tested against the ray to find the first hit. The cost in a leaf σ is thus $O(\gamma + |S \cap \sigma|)$. The coefficient γ depends on the implementation, and models the ratio of the cost of the tree traversal (per cell) to that of a ray-object intersection test (per test).[1] Integral geometry tells us that a random ray will intersect σ with probability $\lambda_{d-1}(\sigma)$ (this is not quite true; read [1] for the niceties). Hence the average cost of ray shooting is given by (1) as claimed.

Tree and object costs. Observe that the cost measure can be split into two terms: $c_t(\mathcal{T}) = \gamma\lambda_{d-1}(\mathcal{T}) = \gamma\sum_{\sigma \in \mathcal{L}(\mathcal{T})} \lambda_{d-1}(\sigma)$ (the *tree cost*), and $c_o(\mathcal{T}) = \sum_{s \in S} \lambda_{d-1}(s \cap \mathcal{T})$ (the *object cost*), where $s \cap \mathcal{T}$ denote the set of leaves of $\mathcal{T}$ crossed by s and λ_{d-1} is extended to sets of leaves by summation. It is useful to keep in mind the following simple observations: when subdividing a cell σ, the total tree cost of its children is twice the tree cost of σ, and the object cost of an object is multiplied by $m/2^{d-1}$ where $m \in [1 \ldots 2^d]$ is the number of children intersected by the object. Note that $m \leq 3$ for a segment in 2D and $m \leq 7$ for a triangle in 3D (unless they pass through the center of the cell). As the tree grows finer, the tree cost increases while the object cost presumably decreases.

Lemma 1. *For any set S of simplices in $[0,1]^d$, $c(\mathcal{T}) \geq 2d\gamma + d\sqrt{2}\sum_{s \in S} \lambda_{d-1}(s)$, for any $d \geq 2$.*

---

[1] In [2], we show how to choose γ to reliably get the best cost possible.

Proof. The tree cost cannot be less than $\lambda_{d-1}([0,1]^d)\gamma = 2d\gamma$, and the object cost cannot be less than $\sum_{s \in S} \lambda_{d-1}(s)S$. We can improve this lower bound further by noting that any leaf σ that is intersected by an object s has area at least $d\sqrt{2}$ times $\lambda_{d-1}(s \cap \sigma)$. Indeed, the smallest ratio $\lambda_{d-1}(\sigma)/\lambda_{d-1}(s \cap \sigma)$ happens when s maximizes $\lambda_{d-1}(s \cap \sigma)$; this happens for a diagonal segment of length $\sqrt{2}$ for the unit square (of perimeter 4), and for a maximal rectangular section of area $\sqrt{2}$ for the unit cube (of area 6). In fact, the maximal section of the unit d-cube is $\sqrt{2}$ [4], hence the ratio is at least $2d/\sqrt{2} = d\sqrt{2}$ in any dimension. ∎

3 Tree Construction Schemes

All we said so far was independent of the particular algorithm used to construct the tree. In this section, we introduce several construction schemes and explore their basic properties.

Terminology and notation. We follow the same terminology as [6], and generalize it to encompass any dimension. For the d-cube $[0,1]^d$ and the cells of the decomposition, we borrow the usual terminology of polytopes (vertex, facet, h-face, etc.). The *square* is a quadtree that has a single leaf (no subdivision), the *cube* is an octree with a single leaf, and the d-*cube* is a single-leaf tree (for any d). We call this tree the *unit cell* and denote it by $\mathcal{T}^{(\text{empty})}$. If we subdivide this leaf recursively until depth k, we get a *complete tree (of depth k)*, denoted by $\mathcal{T}_k^{(\text{complete})}$, and its leaves form a regular d-dimensional grid with 2^k cells on each side. In a quadtree, if only the cells incident to one facet (resp. d facets sharing a vertex, or touching any of the $2d$ facets) of a cell are subdivided, and this recursively until depth k, the subtree rooted at that cell is called a k-*side* (resp. k-*corner* and k-*border*) tree, and denoted by $\mathcal{T}_k^{(\text{side})}$ (resp. $\mathcal{T}_k^{(\text{corner})}$ and $\mathcal{T}_k^{(\text{border})}$); see Figure 1 for an illustration of the 2D case. In higher dimensions, there are other cases (one for each dimension between 1 and $d-2$). All this notation is extended to starting from a cell σ instead of a unit cell, by substituting σ for $\mathcal{T}$: for instance, the complete subtree of depth k subdividing σ is denoted by $\sigma_k^{(\text{complete})}$.

The subdivision operation induces a partial ordering $\prec$ on trees, whose minimum is the unit cell. Again, this partial ordering is extended to subtrees of a fixed cell σ.

We consider recursive algorithms for computing a tree of a given set S of objects, which subdivide each cell until some given termination criterion is satisfied. In particular, we may recursively subdivide the unit cell until each leaf meets at most one object. We call this the *separation criterion*, and the resulting tree the *minimum separating tree*, denoted $\mathcal{T}^{(\text{sep})}(S)$, with variants where the recursion stops at depth k, denoted $\mathcal{T}_k^{(\text{sep})}(S)$. (Note that the depth of $\mathcal{T}^{(\text{sep})}$ is always infinite if any two simplices intersect.) In 3D, for non-intersecting triangles, a variant of [3] stops the recursive subdivision also when no triangle edge

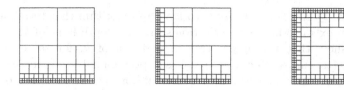

Fig. 1. The k-side quadtree $\mathcal{Q}_k^{(\text{side})}$ (left), a corner $\mathcal{Q}_k^{(\text{corner})}$ (center), and a border $\mathcal{Q}_k^{(\text{border})}$ (right).

intersects the leaf (but any number of non-intersecting triangle interiors may slice the leaf). We will not analyze this variant in this paper.

Dynamic programming and greedy strategies. As introduced in [6], the *dynamic programming algorithm* finds the tree that minimizes the cost over all trees with depth at most k, which we denote by $\mathcal{T}_k^{(\text{opt})}(S)$ (or $\sigma_k^{(\text{opt})}(S)$ if we start from a cell σ instead of the unit cell): the algorithm starts with the complete tree $\mathcal{T}_k^{(\text{complete})}$, and simply performs a bottom-up traversal of all the nodes, while maintaining the optimum cost of a tree rooted at that node. The decision whether to keep the subtree of a cell or prune it is based on the cost of the cell vs. the sum of the optimum costs of the subtrees rooted at its 2^d children.

Unfortunately, the memory requirements of this algorithm are huge for large values of k (although they remain polynomial if $k = \Theta(\log n)$; see next section). Therefore we also propose a greedy strategy with bounded lookahead: the algorithm proceeds by recursively subdiving the nodes with a greedy termination criterion: when examining a cell σ, we run the dynamic programming within σ with depth k (k is a parameter called *lookahead*). If the best subtree $\sigma_k^{(\text{opt})}(S)$ does not improve the cost of the unsubdivided node σ, then the recursion terminates. Otherwise, we replace σ by the subtree $\sigma_k^{(\text{opt})}(S)$ and recursively evaluate the criterion for the leaves of $\sigma_k^{(\text{opt})}(S)$. We call this the k-greedy strategy, and denote the resulting tree by $\mathcal{T}^{(k\text{-greedy})}(S)$ (or $\sigma^{(k\text{-greedy})}(S)$ if we start from a cell σ instead of the unit cell). Note that unlike all the other quadtrees constructed up to now, that tree could be infinite. We use the notation $\mathcal{T}_\ell^{(k\text{-greedy})}$ to denote the tree constructed with the k-greedy lookahead criterion combined with a maximum depth of ℓ.

With no lookahead ($k = 1$), the *greedy strategy* simply examines whether one subdivision decreases the cost measure. Below, we show that this does not yield good trees in general. We will analyze the greedy strategies without and with lookahead, first for points, then for simplices. But first, we must grapple with the issue of infinite depth.

Pruning beyond a given depth. The "optimal" tree may not have finite depth (it is conceivably possible to decrease the cost by subdividing ad infinitum), so we let M denote the infimum of $c(\mathcal{T})$ over all trees $\mathcal{T}$ for S. (As a consequence of Lemma 1, $M \geq 2d\gamma > 0$.) In order to have an algorithm that terminates, we usually add an extra termination criterion such as a depth limit of k.

We now show that pruning a tree at a depth of k, for some choice of $k = \Theta(\log n)$ (to ensure that the tree has a polynomial size), increases cost at most by a constant factor. We first show it for arbitrary convex obstacles (simplices in particular). Then we improve on the result for the case of points. The proofs are no difficult and omitted for space consideration. Nevertheless, these considerations of depths are necessary to ensure that the computation is meaningful.

Lemma 2. *Let $\mathcal{T}$ be a d-dimensional tree which stores a set S of n convex objects of dimensions not more than $d - 1$. For $k = \log_2 n + C$, let $\mathcal{T}_k$ be the tree obtained from $\mathcal{T}$ by removing every cell of depth greater than k. Then $c(\mathcal{T}_k) = O(c(\mathcal{T}))$ and the constant does not depend on n nor S.*

Remark. A choice of $k = \log_2 n + C$ ensures that $\mathcal{T}_k$ has at most $(2^k)^d = O(n^d)$ leaves, for any fixed d. Hence the algorithm which computes the full subdivision at depth k and then applies the dynamic programming heuristic provably computes a tree whose cost is $O(M)$ in polynomial time, as a consequence of Lemma 2.

As a side note, with slightly more restrictive hypotheses on $\mathcal{T}$, the depth k can be reduced for points so that $\mathcal{T}_k$ has size at most $O(n^{1+\frac{1}{d-1}}) = O(n^2)$ (for any $d \geq 2$) and cost as close as desired to that of $\mathcal{T}$.

Lemma 3. *Let $\mathcal{T}$ be a d-dimensional tree, which stores a set S of n simplices. Assume that $\mathcal{T}$ does not contain empty internal nodes (i.e. that are subdivided but do not contain any object). Let $\mathcal{T}_k$ be the tree obtained from $\mathcal{T}$ by removing every cell of depth greater than k. Then, for every $\varepsilon > 0$ there exists a C (that depends only on ε and γ but not on n) such that, for $k = \frac{1}{d-1}\log_2 n + C$, we have*

$$c(\mathcal{T}_k) \leq (1 + \varepsilon)c(\mathcal{T}).$$

4   Cost-Optimal Trees

The following lemma was proven in [6] for the case $d = 2$. Its statement and proof extend straightforwardly to higher dimensions.

Lemma 4. *The lookahead greedy strategy does not always give a cost-optimal tree. Specifically, for any k, there is a set S of n objects such that no tree of depth at most k has cost less than $2d(\gamma + n)$, but some quadtree of depth at least $k + 1$ has cost less than $2d(\gamma + n)$.*

Although the lookahead greedy strategy does not produce the optimal solution, in the counter-example of the lemma (given for $d = 2$ in [6]) it does give a good approximation. In fact, this can be proven for all scenes.

Theorem 1. *Given a set S of flat objects in the unit cube, and let M be the infimum of $c(\mathcal{T})$ over all trees $\mathcal{T}$. There is an integer p (for $d = 2$ or $d = 3$,*

$p = 3$) such that the tree $\mathcal{T}^{(p\text{-greedy})}$ constructed by the p-greedy strategy has cost $c(\mathcal{T}^{(p\text{-greedy})}) = O(M)$.

Proof. The intuition is that small objects behave well, and the cost of a big object is bounded below by a constant times its size so it cannot be reduced by very much. Let us look at a cell σ of the tree $\mathcal{T}^{(p\text{-greedy})}$: we are going to show that, when the optimal decomposition of depth at most p of a cell σ does not improve on the cost of σ, then the cost of σ is $O(M_\sigma)$ where M_σ is the infimum cost of all the possible tree subdivisions of σ. If this holds true for every leaf σ of the p-greedy strategy, then $c(\mathcal{Q}^{(p\text{-greedy})}) = O(M)$ as well.

Assume there are a objects meeting at most $C_d(2^p)$ cells, and b other objects. The cost of σ is $(\gamma + a + b)\lambda_1(\sigma)$. Since $\sigma_p^{(\text{complete})}$ has cost at most $\left(2^p\gamma + a\frac{C_d(2^p)}{2^{p(d-1)}} + 2^p b\right)\lambda_1(\sigma)$, which we assumed to be at least $c(\sigma)$, we have $c(\sigma) = (\gamma + a + b)\lambda_1(\sigma) \leq \left(2^p\gamma + a\frac{C_d(2^p)}{2^{p(d-1)}} + 2^p b\right)\lambda_1(\sigma)$, which implies that $a \leq (\gamma + b)(2^p - 1)\left(1 - \frac{C_d(2^p)}{2^{p(d-1)}}\right)^{-1}$. We will need a technical lemma:

Lemma 5. *For every d and k, there exist constants $C_d(k)$ and $S_d(k) > 0$ such that for any convex object s of dimension at most $d - 1$, either s intersects at most $C_d(k)$ cells of the regular grid of side k, or else $\lambda_{d-1}(s) \geq S_d(k)$. We may take $C_d(k) \leq d^2 k^{d-2}$ and $C_3(k) = 7k - 6$.*

Let p be the smallest integer such that $C_d(2^p) < (2^p)^{d-1}$. By lemma 5, an object which belongs to more than $C_d(2^p)$ cells has measure at least $S_d(2^p)$ so its contribution to the cost is at least $(d\sqrt{2}S_d(2^p))\lambda_{d-1}(\sigma)$. The optimal cost M_σ is then greater than $(\gamma + bd\sqrt{2}S_d(2^p))\lambda_{d-1}(\sigma)$. We have then proved that M_σ is at least a fixed fraction of the cost of σ, and the lemma follows. ∎

Already in 2D, the separating quadtree strategy does not work as well for segments as for points, especially since it is not able to distinguish between a segment that barely intersects the corner of the square and the diagonal (in the first case it is usually good to subdivide, and in the second case it is not). The lookahead strategy is then a true improvement.

The case of points. Arguably, the case of points is of theoretical interest only, but has relevance since simplices are usually very small in a graphics scene (when they come from a subdivision surface), and can be thought of as points. This is lent credence by a recent trend: point cloud data (PCD) is becoming an important primitive in computer graphics, and several algorithms for rendering them have been given of late, which are amenable our cost measure.

In the plane, the 1- and 2-greedy strategy may produce a quadtree of cost $\Theta(n)$ times the optimal cost, and so does 1-greedy in higher dimensions. Nevertheless, with one more level of lookahead, everything works near-optimally. We simply state the lemma and omit the proof, similar the the one given above for simplices.

Lemma 6. *The 3-greedy strategy in the plane, and the 2-greedy strategy in d dimensions (d ≥ 3) produce near-optimal trees for points. Namely, if S be a set*

*of n points in the unit d-cube, and M is the infimum of $c(\mathcal{T})$ over all trees $\mathcal{T}$,
then $c(\mathcal{T}^{(3\text{-greedy})}) = O(M)$ for all $d \geq 2$, and $c(\mathcal{T}^{(2\text{-greedy})}) = O(M)$ for all
$d \geq 3$.*

As for finding the optimal quadtree, the question is still open whether for
given values of γ (and maybe of n) there exists a k such that the lookahead
strategy yields the optimal result. All we know is that if $n = 5$ and $\gamma < 1$ tends
to 1, the required k tends to infinity. We can also mention that if every point
belongs to at most one cell, then $k = 1$ leads to the optimal tree.

5   Rebalancing Quadtrees and Octrees

Quadtrees are used in meshing for computer graphics, and octrees are used as
a space subdivision method for ray casting and radiosity methods, for instance.
Both quadtrees and octrees are used in scientific computing numerical simulations (e.g., finite element methods). These are but a few applications where
quadtrees and octrees have appeared. In all these applications, it can be important to maintain aspect ratio (hence starting with a unit cell) and to ensure
that two neighboring cells don't have wildly differing sizes. This has led several
authors to propose balancing for trees. Also from our perspective, since the cost
measure of [1] provably relates to the cost of traversal only for balanced trees,
we are interested in balancing trees as well. In this section, we prove that rebalancing does not affect the cost by more than a multiplicative constant factor.

Definitions. Two leaves are k-*adjacent* if they intersect in a convex portion of
dimension k. A tree is called k-*balanced* if the depths of any two k-adjacent leaves
differ by at most one. Notice that when considering two k-balanced trees, their
intersection, constructed from the unit cell by subdividing all and only cells that
are subdivided in both trees, is k-balanced. Thus for a tree $\mathcal{T}$, there is a unique
balanced tree $\mathrm{bal}_k(\mathcal{T}) = \min\{\mathcal{T}' \succ \mathcal{T} : \mathcal{T}' \text{ is } k\text{-balanced}\}$, which is called the
k-rebalancing of $\mathcal{T}$.

For instance, 0-balanced quadtrees are what Moore called *smooth*
quadtrees [8], and 1-balanced what he called *balanced* and others called 1-
irregular or *restricted*.

Cost analysis. While the size of $\mathrm{bal}_k(\mathcal{T})$ is known to increase by at most a
constant factor from the size of $\mathcal{T}$, the final cost $c(\mathrm{bal}_k(\mathcal{T}))$ is unknown, however.
Our main result concerning cost analysis is the following, when objects in $\mathcal{S}$ are
either points or segments.

Theorem 2. *Let $\mathcal{T}$ be a tree storing both points and/or simplices in the unit
cube. Then for any k, $0 \leq k < d$, and $d \leq 3$,*

$$c(\mathrm{bal}_k(\mathcal{T})) \leq 3^d c(\mathcal{T}).$$

The result becomes $c(\mathrm{bal}_k(\mathcal{T})) = O(d^2 4^d)c(\mathcal{T})$ in higher dimensions.

The following lemma was first proven by Weiser in 2D, and by Moore for any dimension $d \geq 2$.

Lemma 7 ([8]). *Let $\mathcal{T}$ be a tree. There is a 0-balanced tree $\mathcal{T}' \succ \mathcal{T}$ such that $c_t(bal_0(\mathcal{T})) \leq 3^d c_t(\mathcal{T})$.*

Since $\mathcal{T}'$ must be a refinement of $bal_0(\mathcal{T})$, which is itself a refinement of $bal_k(\mathcal{T})$, for any $k > 0$, this implies that $c_t(bal_k(\mathcal{T})) \leq c_t(bal_0(\mathcal{T})) \leq 3^d c_t(\mathcal{T})$. Note that the same construction also implies the factor 3^d on the number of leaves.

Next we prove in Lemma 8 that the object cost of $bal_k(\mathcal{T})$ is at most twice (for points) and some constant $B_d < S^d$ (for simplices) times that of $\mathcal{T}$. The proof is omitted for lack of space.

Lemma 8. *Let $\mathcal{T}$ be a tree, and consider the object cost of a single object $s \in S$ both in $\mathcal{T}$ and in $bal_k(\mathcal{T})$. If s is a point, then $\lambda_{d-1}(s \cap bal_k(\mathcal{T})) \leq 2\lambda_{d-1}(s \cap \mathcal{T})$. If s is a convex object of dimension at most $d-1$ (e.g. a $(d-1)$-simplex), then $\lambda_{d-1}(s \cap bal_k(\mathcal{T})) \leq B_d \lambda_{d-1}(s \cap \mathcal{T})$.*

Compounding all these costs together, we have $c_o(bal_k(\mathcal{T})) \leq 3^d c_o(\mathcal{T})$ for any $0 \leq k < d$, which ends the proof of the theorem.

Remark. It could also very well be that rebalancing actually decreases the cost. We don't know that, and we don't need it since we are mostly interested in trees for which $c(\mathcal{T}) = O(M)$. In any case, we can ask if there is a reverse theorem (lower bound on $c(bal_k(\mathcal{T}))$ in terms of $c(\mathcal{T})$).

6   Conclusion

In this paper we have proved that instead of considering the optimal octree, and without increasing the cost too much, we may consider the octree given by the lookahead strategy. Still, this may yield an infinite subdivision. In order to have an effective algorithm, we need to add a termination criteria such as a depth limit of $\log_2 n$. As we have also proven, this increases again the cost at most by a constant factor. In practice, we find that greedy with or without lookahead yield near-optimal octrees, hence the approximation ratio seems close to one.[2]

All the results stated in this paper should extend easily to recursive grids and simplicial trees as well, in two and higher dimensions, with only small differences. However, the constants involved in the analysis would be even higher than they are here.

We conclude with a few open problems: first, is it true that by pruning at depth $k = \Theta(\log n)$, we can approach the cost to within $1 + \varepsilon$? Since the optimal tree might be infinite, there is little sense in asking for an algorithm

---

[2] Actually, they both yield octrees of same cost which are the lowest cost we observe with other heuristics; we find it hard to believe that they would all be c times optimal, for some constant $c > 1$.

that constructs the optimal tree. But if the answer to the first question were true, it would be nice to have a PTAS with respect to the cost measure. We don't know if the greedy strategy for high enough lookahead would fit the bill.

Lastly, the cost measure considered here is simple but does not model the average traversal cost during ray shooting. For this, the following cost measure should be considered [1]:

$$c^*(\mathcal{T}) = \sum_\sigma (\gamma + |S \cap \sigma|) \times (\lambda_{d-1}(\sigma) + \lambda_{d-1}(S \cap \sigma)), \qquad (2)$$

x where $\lambda_{d-1}(S \cap \sigma)$ measures the portion of the objects within σ. Our only result here is that the greedy strategy does not work.

References

1. B. Aronov, H. Brönnimann, A.Y. Chang, and Y.-J. Chiang. Cost prediction for ray shooting. In *Proc. of Eighteenth ACM Symp. on Geom. Comput.*, pages 293–302, 2002, Barcelona, Spain.
2. B. Aronov, H. Brönnimann, A.Y. Chang, and Y.-J. Chiang. Cost-driven octree construction schemes: an experimental study. In *Proc. of Nineteenth ACM Symp. on Geom. Comput.*, 2003, San Diego, CA.
3. B. Aronov and S. Fortune. Approximating minimum weight triangulations in three dimensions. *Discrete Comput. Geom.*, 21(4):527–549, 1999.
4. K. Ball. *Cube slicing in* $\mathbb{R}^n$. Proc. Amer. Math. Soc. 97 (1986), no. 3, 465–473.
5. M. Bern and D. Eppstein. Mesh generation and optimal triangulation. In *Computing in Euclidean Geometry*, Lecture Notes Series on Computing, Volume 4, pages 47–123, World Scientific, Singapore, 1992.
6. H. Brönnimann, M. Glisse, and D. Wood. Cost-optimal quadtrees for ray shooting. In *Proc. Canad. Conf. on Comput. Geom.* (CCCG'02), Lethbridge, Alberta.
7. J.D. MacDonald and K.S. Booth. Heuristics for ray tracing using space subdivision. *The Visual Computer*, 6:153–166, 1990.
8. D.W. Moore. *Simplicial Mesh Generation with Applications*. Ph.D dissertation, Cornell University, 1992.
9. E. Reinhard, A.J.F. Kok, and F.W. Jansen. Cost prediction in ray tracing. In P. Hanrahan and W. Purgathofer et al., editors, *Rendering Techniques '97*, pages 42–51. Porto, Portugal, 1996.
10. H. Samet. *Design and Analysis of Spatial Data Structures*. Addison-Wesley, 1990.
11. K.R. Subramanian and D.S. Fussell. Automatic termination criteria for ray tracing hierarchies. In *Proc. of Graphics Interface '91*, June 3-7 1991.
12. K. Y. Whang, J. W. Song, J. W. Chang, J. Y. Kim, W. S. Choand, C. M. Park, and I. Y. Song. Octree-R: An adaptive octree for efficient ray tracing. *IEEE Trans. Visual and Comp. Graphics*, 1:343–349, 1995.

Packing Problems with Orthogonal Rotations*

Flavio Keidi Miyazawa[1]** and Yoshiko Wakabayashi[2]

[1] Instituto de Computação, Universidade Estadual de Campinas
Caixa Postal 6176, 13084-971, Campinas-SP, Brazil
`fkm@ic.unicamp.br`
[2] Instituto de Matemática e Estatística, Universidade de São Paulo
Rua do Matão, 1010-05508-090, São Paulo-SP, Brazil
`yw@ime.usp.br`

Abstract. In this extended abstract, we present approximation algorithms for the following packing problems: the *strip packing problem*, the *two-dimensional bin packing problem*, the *three-dimensional strip packing problem*, and the *three-dimensional bin packing problem*. For all these problems, we consider orthogonal packings where 90° rotations are allowed. The algorithms we show for these problems have asymptotic performance bounds 1.613, 2.64, 2.76 and 4.89, respectively. We also present an algorithm for the *z-oriented three-dimensional packing problem* with asymptotic performance bound 2.64. To our knowledge the bounds presented here are the best known for each problem.

1 Introduction

We present approximation algorithms for packing problems allowing orthogonal rotations. We consider the following problems: the *strip packing problem*, the *two-dimensional bin packing problem*, the *three-dimensional strip packing problem*, and the *three-dimensional bin packing problem*. The packings must be orthogonal and 90° rotations are allowed. The algorithms we show for these problems have asymptotic performance bounds 1.613, 2.64, 2.76 and 4.89, respectively. We also present an algorithm for the *z-oriented three-dimensional packing problem* with asymptotic performance bound 2.64. This result improves our previous results in [10,11].

Approximation algorithms for the oriented version of these packing problems have been extensively considered, but there are very few results when rotations are considered. For a survey on approximation algorithms for packing problems, see [3,4].

In all problems considered in this paper, all items must be packed into recipients and they may not overlap. We consider orthogonal packings where orthogonal rotations are allowed, i.e. 90° rotations, around any axis.

There are many applications in which orthogonal rotations are allowed: cutting of hardboard, glass, cloth (when there is no oriented pattern), foam, etc. [3]. Interesting applications also occur in job scheduling problems [9,11].

---

* This work has been partially supported by MCT/CNPq – Project ProNEx (Proc. 664107/97-4), and CNPq (Proc. 300301/98-7, 304527/89-0, 464114/00-4, 470608/01-3, 478818/03-3).
** Corresponding author.

M. Farach-Colton (Ed.): LATIN 2004, LNCS 2976, pp. 359–368, 2004.

The *strip packing problem*, SPP, is the following: given a list of rectangles $L = (r_1, \ldots, r_n)$, where $r_i = (x_i, y_i)$, and a rectangle $R = (a, \infty)$, find a packing of the rectangles of L into R minimizing the size of the packing in the unlimited direction of R.

In the *two-dimensional bin packing problem*, 2BP, we are given a list of rectangles $L = (r_1, \ldots, r_n)$, where $r_i = (x_i, y_i)$, and two-dimensional bins $R = (a, b)$, and we wish to pack the rectangles of L into bins R using the smallest number of bins.

In the *three-dimensional strip packing problem*, TPP, we are given a list of boxes $L = (b_1, \ldots, b_n)$, where $b_i = (x_i, y_i, z_i)$, and a box $B = (a, b, \infty)$, pack the boxes of L into B, minimizing the size of the packing in the unlimited direction of the recipient.

The *three-dimensional bin packing problem*, 3BP, is defined as follows: given a list of boxes $L = (b_1, \ldots, b_n)$, where $b_i = (x_i, y_i, z_i)$, and three-dimensional bins $B = (a, b, c)$, find a packing of the boxes of L into the smallest number of bins B.

We also consider a special version of the three-dimensional strip packing problem, called *z-oriented three-dimensional strip packing problem*, TPP^z, which is the same as in the definition of TPP, except that boxes are oriented in the z-axis. That is, a box can be rotated around the z-axis (height direction), but cannot be laid down.

In section 2, we present some notation and discuss some results when orthogonal rotations are allowed. In section 3, we present the approximation algorithm for the strip packing problem and in the section 4, the approximation algorithm for the two-dimensional bin packing problem. In section 5, we present the results we have obtained for the other packing problems. In section 6, we present some concluding remarks.

2 Preliminaries

To define the packings, we consider the Euclidean space $\mathbb{R}^3$, with the xyz coordinate system. An item e to be packed has its dimensions defined as $x(e)$, $y(e)$ and $z(e)$, also denoted as the length, width and height of item e, where each of these dimensions is the measure in the corresponding axis of the xyz system. For the one- and the two-dimensional cases, some of these values are not defined.

We denote by $\mathrm{SPP}(a)$, $2\mathrm{BP}(a, b)$, $\mathrm{TPP}(a, b)$, $\mathrm{TPP}^z(a, b)$ and $3\mathrm{BP}(a, b, c)$ the corresponding problems versions with the recipient sizes defined by values a, b and c. We denote by $S(r_i)$ the area of the rectangle $r_i = (x_i, y_i)$ and $V(b_i)$ the volume of the box $b_i = (x_i, y_i, z_i)$. Given a function $f : C \to \mathbb{R}$ and a subset $C' \subseteq C$, we denote by $f(C')$ the sum $\sum_{e \in C'} f(e)$. For all algorithms, we consider that the items are given in an initial configuration that can be packed. That is, given a box e we have $x(e) \leq a$, $y(e) \leq b$ and $z(e) \leq c$. Given an item $e = (a, b)$, we denote by $\rho(e) := (b, a)$. We also consider that each item dimension is not greater than a constant Z.

The following is a convenient notation to define and restrict the input list of items:
$$\mathcal{X}[p, q] := \{e : \ p \cdot a < x(e) \leq q \cdot a\}, \quad \mathcal{Y}[p, q] := \{e : \ p \cdot b < y(e) \leq q \cdot b\},$$
$$\mathcal{C}[p_1, q_1 ; p_2, q_2] := \mathcal{X}[p_1, q_1] \cap \mathcal{Y}[p_2, q_2], \quad \mathcal{C}_m := \mathcal{C}\left[0, \tfrac{1}{m} ; 0, \tfrac{1}{m}\right],$$
$$\wp_1 := \mathcal{C}\left[0, \tfrac{1}{2} ; 0, \tfrac{1}{2}\right], \wp_2 := \mathcal{C}\left[0, \tfrac{1}{2} ; \tfrac{1}{2}, 1\right], \wp_3 := \mathcal{C}\left[\tfrac{1}{2}, 1 ; 0, \tfrac{1}{2}\right], \wp_4 := \mathcal{C}\left[\tfrac{1}{2}, 1 ; \tfrac{1}{2}, 1\right].$$

Given a list L of items to be packed, and an algorithm $\mathcal{A}$, we denote by $\mathcal{A}(L)$ the size of the packing generated by algorithm $\mathcal{A}$ when applied to list L. Such size can be the height of the packing or the number of bins used in the packing, depending on which version of packing problem we are considering. For a packing $\mathcal{P}$, we denote by $H(\mathcal{P})$ its height, and by $\#(\mathcal{P})$ the number of bins that is used by $\mathcal{P}$. We denote by $\mathrm{OPT}(L)$ the size of an optimum packing of L. We say that an algorithm $\mathcal{A}$ has *asymptotic performance bound* α if there exists a constant β such that $\mathcal{A}(L) \leq \alpha \cdot \mathrm{OPT}(L) + \beta$, for all input list L.

Since the one-dimensional bin-packing problem is a particular case of all problems considered in this paper, it follows that each problem considered here is $\mathcal{NP}$-hard.

One idea to solve problems allowing orthogonal rotations is to simply apply the algorithms developed for the oriented case, ignoring any possible rotation. It can be shown that there is no algorithm, developed in the way we described above, for the strip packing problem with asymptotic performance bound less than 2, and for the two-dimensional bin packing problem and the three-dimensional bin packing problem there is no known algorithm with asymptotic performance bound less than 3 (see [11]).

Most of the results concerning approximation results do not consider rotations and some questions where posed in the early 80's. The following quote where extracted from a paper of F.R. Chung, M.R. Garey and D.S. Johnson [2] on the two-dimensional bin packing problem:

> "A second line of attack would be to design and analyze algorithms which could make use of the fact that, in some applications, 90° rotations of rectangles might be allowable.
> Algorithms which consider the possibility of rotations might well yield improvements. Can one prove worst case bounds that reflect these improvements?"

There are other papers in the literature that raise questions about orthogonal rotations, as for example, [3,5]. Although these papers are from the early 80's, very few has been done about orthogonal rotations. In fact, when the scale does not affect the problem, we can show that for any of the general packing problems considered, the version allowing orthogonal rotations is as hard to approximate as the oriented version.

Theorem 1. *Let* PROB^r *be one of the problems defined previously, considering orthogonal rotations around some of the axes* x *or* y *or* z *(maybe in several axes),* α *and* β *constants and* A^r *an algorithm such that,* $A^r(L) \leq \alpha \cdot \mathrm{OPT}(L) + \beta$ *for any instance* L *of* PROB^r. *Then, we can adapt this algorithm to another algorithm* $\mathcal{A}$ *for a variant of* PROB^r, *called* PROB, *where we fix the orientation of the items (with respect to some axis), in such a way that the following relation holds:* $\mathcal{A}(L) \leq \alpha \cdot \mathrm{OPT}(L) + \beta$ *for any instance* L *of* PROB.

3   Strip Packing Problem

In this section we consider the strip packing problem. In [5], Coffman, Garey, Johnson and Tarjan present the algorithms NFDH (Next Fit Decreasing Height) and FFDH (First Fit Decreasing Height) for the oriented case and prove that their asymptotic performance bounds are 2 and 1.7, respectively. Another algorithm with asymptotic performance bound 2 is the BLDW, Bottom Leftmost Decreasing Width, algorithm presented by Baker, Coffman and Rivest [1]. Kenyon and Remila [8], presented an asymptotic approximation scheme for the oriented strip packing problem.

When orthogonal rotations are allowed, the bound 2 of the algorithms BLDW and NFDH are also valid, since the proofs of the bounds are based only in area arguments.

The algorithm we present in this section is called ASP. It uses the critical set combination strategy presented in [10,11]. The idea is to combine items that do not lead to packings with good space filling, if considered independently. We call these items *critical items*. We take two sets with critical items, called here *critical sets*, and generate a combined packing of the items in these two sets. Each algorithm that combines items of two critical sets has the property that at the end of this combination, one of these sets is totally packed in the combined packing. Moreover, the combined packing has a better configuration than the one we can obtain for each critical set.

Before presenting the algorithm, we describe an algorithm used as subroutine called TC (Two Column) which builds a packing with two columns, each one is a stack of rectangles packed one on top of the other. Each column is associated with only one critical set. The algorithm TC is called with parameters (L_1, L_2, x_1, x_2), where L_1 and L_2 are two critical sublists and x_1 and x_2 are positions where the columns are built aligned to the left, from the bottom of the strip. We call *height of the column* the sum of the rectangles heights in the corresponding column. To pack a rectangle, the algorithm chooses the first column with smallest height. Let h be the height of this column and L_i the list associated with this column. If not all rectangles of L_i have been packed, then the next rectangle of L_i, say r, is packed in the position (x_i, h). Then, the list L_i is updated (by removing the rectangle r). This process is repeated until one of the lists, L_1 or L_2 is totally packed. We assume the positions x_1 and x_2 and the lists L_1 and L_2 are such that they do not generate infeasible packings. Any algorithm that combines critical sets returns a pair $(\mathcal{P}', L')$, where $\mathcal{P}'$ is the packing generated and L' is the set of rectangles packed in $\mathcal{P}'$. The following lemma can be proved for the algorithm TC.

Lemma 1. *Let $\mathcal{P}$ be a packing of $L' \subseteq L_1 \cup L_2$ generated by the algorithm TC when applied to lists L_1 and L_2 for $\mathrm{SPP}(a)$. If $x(r) > l_i \cdot a$ for all $r \in L_i$, $i = 1, 2$, then $H(\mathcal{P}) \leq \frac{1}{l_1+l_2} \frac{S(L')}{a} + Z$.*

We denote the value s in inequalities of the form $H(\mathcal{P}) \leq \frac{1}{s} \frac{S(L)}{a} + Z$ as an *area guarantee* of the packing $\mathcal{P}$. The idea used in the algorithm ASP is to generate the final packing consisting of two parts: one associated with a partial optimum packing generated with rectangles with width greater than $\frac{a}{2}$, and the other associated with a packing with better area guarantee.

Note that in an oriented packing, if L_1 has only rectangles with width greater than $\frac{a}{2}$ then we can obtain an optimum packing only by placing one rectangle on top of the other. When rotations are allowed, we can also have a similar result. In this case, we first reorient and remove all rectangles of L_1 with width greater than $\frac{a}{2}$ but height at most $\frac{a}{2}$. In this way, the remaining rectangles are the ones that cannot be reoriented, or if they can, they continue with width greater than $\frac{a}{2}$. Clearly, the only way to pack these remaining rectangles is to pack one on top of the other. After this rotation step, we introduce another reorientation step for the new rectangles in L_1 in such a way that each rectangle stays with the lowest possible height, maintaining a feasible orientation. Thus, we can continue having a partial optimum packing only by packing one rectangle on top of the other.

The algorithm TC is used to pack the critical rectangles of L_1 and critical rectangles of the remaining part. The packing of the remaining non-packed rectangles is done using NFDH strategy, which consists in first sorting the rectangles in non-decreasing order of their height and then packing the rectangles in this order, placing rectangles side by side generating levels. When a rectangle cannot be packed in a level it is packed in a new consecutive parallel level (for more details of the algorithm NFDH, see [5]). The following result is valid for NFDH.

Lemma 2. *Let $N_1, \ldots, N_v$ be the levels generated by NFDH for a list L, in the order they are generated. If $w(N_i)$ is the total sum of the width of the rectangles in N_i, and there exists a constant s such that $w(N_i) \geq s \cdot a$, for $1 \leq i \leq v - 1$, then we have $\mathrm{NFDH}(L) \leq \frac{1}{s} \frac{S(L)}{a} + Z$.*

Algorithm ASP(L)
 Input: List of rectangles L for SPP(a)
1 Rotate all rectangles $r \in L$ with $x(r) > \frac{a}{2}$ and $y(r) \leq \frac{a}{2}$.
2 Rotate all rectangles $r \in L$ with $x(r) > \frac{a}{2}$ and $x(r) < y(r) \leq a$.
3 Let $p \leftarrow 1/\sqrt{6}$ and $L_i' \leftarrow \{s \in L : \frac{a}{i+1} < x(s) \leq \frac{a}{i}\}$, $i = 1, \ldots, 4$,
      $L_5' \leftarrow \{s \in L : (1-2p) \cdot a < x(s) \leq \frac{a}{5}\}$, $L_6' \leftarrow \{s \in L : x(s) \leq (1-2p) \cdot a\}$,
      $L_A \leftarrow \{r \in L_1' : x(r) \leq (1-p) \cdot a\}$, $L_B \leftarrow \{r \in L_2' \cup \ldots \cup L_5' : x(r) \leq p \cdot a\}$.
4 $(\mathcal{P}_{AB}, L_{AB}) \leftarrow \mathrm{TC}(L_A, L_B, 0, 1-p)$; $L_i \leftarrow L_i' \setminus L_{AB}$,   for $i = 1, \ldots, 6$;
5 $\mathcal{P}_i \leftarrow \mathrm{NFDH}(L_i)$, $i = 1, \ldots, 6$;
6 $\mathcal{P}_{opt} \leftarrow \mathcal{P}_1 \| \mathcal{P}_{AB}$; $\mathcal{P}_{aux} \leftarrow \mathcal{P}_2 \| \ldots \| \mathcal{P}_6$.
7 Return $\mathcal{P}_{opt} \| \mathcal{P}_{aux}$.

Theorem 2. *For any input list L for the SPP(a), where the rectangles of L have dimensions at most Z, we have $\mathrm{ASP}(L) \leq 1.613 \cdot \mathrm{OPT}(L) + 6Z$.*

Proof. Since each rectangle of L_A has width at least $\frac{a}{2}$ and each rectangle of L_B has width at least $(1-2p)a$, from Lemma 1 we conclude that the following inequality holds.

$$H(\mathcal{P}_{AB}) \leq \frac{1}{1/2 + (1-2p)} \frac{S(L_{AB})}{a} + Z \leq \frac{1}{1-p} \frac{S(L_{AB})}{a} + Z. \tag{1}$$

Denote by L_{opt} the set of rectangles packed in $\mathcal{P}_{opt}$. It is easy to see that $\mathcal{P}_{opt}$ is an asymptotic optimum packing of L_{opt} since the *large* rectangles (with $x(r) > \frac{a}{2}$) of $L_1 \cup \ldots \cup L_6 \cup L_{AB}$ either cannot be rotated, or if they can, they remain in the set defined for L_1. Moreover, the large rectangles of L_{opt} are packed with the smallest possible height. Therefore, we have

$$H(\mathcal{P}_{opt}) \leq \mathrm{OPT}(L) + Z. \tag{2}$$

Now, we analyze two cases. In the first (second), we consider that all rectangles of L_A (all rectangles of L_B) are totally packed in $\mathcal{P}_{AB}$.
Case 1. $L_A \subseteq L_{AB}$. In this case, we have all rectangles of $L_1 := L_1' \setminus L_{AB}$ with width greater than $(1-p) \cdot a$. Using (1), we obtain $S(L_{opt}) \geq (H(\mathcal{P}_{opt}) - Z) \cdot (1-p) \cdot a$. That is,

$$H(\mathcal{P}_{opt}) \leq \frac{1}{1-p} \frac{S(L_{opt})}{a} + Z. \tag{3}$$

For each list L_i $(i = 2, \ldots, 6)$ the area guarantee in each level of $\mathcal{P}_i$, except perhaps in the last, is at least $\frac{2}{3}$. This area guarantee is obtained by considering the width occupation in each level of the packing generated by the algorithm NFDH. Since the packing $\mathcal{P}_{aux}$ is the concatenation of packings $\mathcal{P}_2, \ldots, \mathcal{P}_6$, from Lemma 2 we have

$$H(\mathcal{P}_{aux}) \leq \frac{1}{2/3} \frac{S(L_{aux})}{a} + 5Z. \tag{4}$$

Defining $h_1 := H(\mathcal{P}_{opt}) - Z$ and $h_2 := H(\mathcal{P}_{aux}) - 5Z$, we have

$$\mathrm{OPT}(L) \geq \frac{S(L)}{a} = \frac{S(L_{opt})}{a} + \frac{S(L_{aux})}{a} \geq (1 - p) \cdot h_1 + \frac{2}{3} \cdot h_2 . \tag{5}$$

From (2) and (5) we have $\mathrm{OPT}(L) \geq \max\{h_1, (1 - p) \cdot h_1 + \frac{2}{3} \cdot h_2\}$, and therefore, we obtain $H(\mathcal{P}) \leq \alpha_1 \cdot \mathrm{OPT}(L) + 6Z$, where $\alpha_1 = \frac{h_1 + h_2}{\max\{h_1, (1-p) \cdot h_1 + \frac{2}{3} \cdot h_2\}} \leq \frac{2 + 3p}{2}$. This last inequality can be proved by analyzing the two possible values attained by the maximum.

Case 2. $L_B \subseteq L_{AB}$. The analysis of this case uses the same arguments used in case 1. Therefore, we present only the inequality one could obtain

$$H(\mathcal{P}) \leq \alpha_2 \cdot \mathrm{OPT}(L) + 6Z,$$

where $\alpha_2 = \frac{h_1 + h_2}{\max\{h_1, \frac{1}{2}h_1 + 2ph_2\}} \leq \frac{4p+1}{4p}$.

Substituting the value of p in the bounds obtained in the two cases above, we obtain $H(\mathcal{P}) \leq 1.613 \cdot \mathrm{OPT}(L) + 6Z$. □

4   Two-Dimensional Bin Packing Problem

In 1982, Chung et al. [2] presented an algorithm with asymptotic performance bound 2.125, which is the best bound known for the oriented two-dimensional bin packing problem.

In this section, we present an algorithm using orthogonal rotations, called $\mathrm{BI}_{k,\epsilon}$, with asymptotic performance bound that can be made not larger than 2.64. This algorithm follows the same technique used in the algorithm ASP. It also defines critical sets and combination of them, although in more steps and phases.

Before presenting the algorithm, we describe some algorithms used as subroutines by the main algorithm: the first algorithm is the algorithm NFDH^x, used for the packing of simple sublists. The algorithm uses two subroutines to combine critical sets: the algorithms COMBINE-AB$_k$ and COMBINE-CD. The algorithms FL_ϵ and the algorithm FFD, both for one-dimensional bin packing problem are used to generate packings of special sets that can be interpreted as one-dimensional packing problems.

The algorithm FL_ϵ was developed by Fernandez de la Vega and Lucker [6]. The algorithm presented by these authors is a linear time algorithm with additive constant $O(1/\epsilon)$. For our purposes, we consider a polynomial time version of it with additive constant 1, which can be found in [12]:

Theorem 3. *[6,12] For any $\epsilon > 0$, there exists a polynomial time algorithm FL_ϵ for the bin-packing problem such that $\mathrm{FL}_\epsilon(L) \leq (1 + \epsilon) \cdot \mathrm{OPT}(L) + 1$.*

The algorithm FFD first sorts the items of L in non-increasing order of their lengths, and then packs items in the order given by L. To pack an item, the algorithm FFD tries to pack the new item into one of the previous bins, considering the order they were generated. If it is not possible to pack in the previous bins, it packs in a new bin. Johnson [7] proved that the asymptotic performance bound of the algorithm FFD is $11/9$.

The algorithm $NFDH^x$ *(Next Fit Decreasing Height)* first sorts the input list L in non-decreasing order of height, then packs the rectangles side by side (first horizontally and then vertically) generating levels. When a rectangle cannot be packed in the current level, a new level is generated parallel to the last one, which becomes the current level. When a level cannot be generated in the current bin, it is generated at the bottom of a new bin, which becomes the current bin.

The variant of this algorithm that generates levels in the y-direction is denoted by $NFDH^y$. Subdividing a list $L \subseteq C_m$ into more sublists, and applying an appropriate variant of the algorithm $NFDH^x$, we can obtain an algorithm, denoted by BI_m, for which the following lemma holds.

Lemma 3. *For any list of rectangles $L = (r_1, \ldots, r_n)$, where $x(r_i) \leq \frac{1}{m}$ and $y(r_i) \leq \frac{1}{m}$, we have $BI_m(L) \leq \left(\frac{m+1}{m}\right)^2 S(L) + 6$.*

The following lemma gives an upper bound for the number of bins used in a packing that has a minimum area occupation in each bin.

Lemma 4. *If $\mathcal{P}$ is a packing of a list of rectangles L such that all bins (a, b), except perhaps k of them, have an area occupation of at least f, then $\#(\mathcal{P}) \leq \frac{1}{f} \frac{S(L)}{ab} + k$.*

The critical sets used by the algorithm COMBINE-AB$_k$ are defined by the numbers $r_i^{(k)}$ and $s_i^{(k)}$, presented in [10,11], defined as follows.

Definition 1. *Let $r_1^{(k)}, r_2^{(k)}, \ldots, r_{k+15}^{(k)}$ and $s_1^{(k)}, s_2^{(k)}, \ldots, s_{k+14}^{(k)}$ be real numbers such that $r_1^{(k)} < \frac{4}{9}$; $r_1^{(k)} \frac{1}{2} = r_2^{(k)}(1 - r_1^{(k)}) = r_3^{(k)}(1 - r_2^{(k)}) = \ldots = r_k^{(k)}(1 - r_{k-1}^{(k)}) = \frac{1}{3}(1 - r_k^{(k)})$; $r_{k+1}^{(k)} = \frac{1}{3}$, $r_{k+2}^{(k)} = \frac{1}{4}, \ldots, r_{k+15}^{(k)} = \frac{1}{17}$; $s_i^{(k)} = 1 - r_i^{(k)}$ for $i = 1, \ldots, k$; and $s_{k+i}^{(k)} = 1 - \left(\frac{2i+4-\lfloor \frac{i+2}{3} \rfloor}{4i+10}\right)$ for $i = 1, \ldots, 14$.*

For simplicity we omit the superscripts $^{(k)}$ of the notation $r_i^{(k)}$, $s_i^{(k)}$ when k is clear from the context. Using a continuity argument, we can prove that the numbers $r_1, r_2, \ldots, r_k$ are such that $r_1 > r_2 > \cdots > r_k > \frac{1}{3}$ and $r_1 \to \frac{4}{9}$ as $k \to \infty$. Now, we can define the following critical sets.

$$\mathcal{A}_i = \mathcal{C}\left[r_{i+1}, r_i; \tfrac{1}{2}, s_i\right], \quad \mathcal{B}_i = \mathcal{C}\left[\tfrac{1}{2}, s_i; r_{i+1}, r_i\right], \quad \mathcal{A} = \bigcup_{i=1}^{k+14} \mathcal{A}_i, \quad \mathcal{B} = \bigcup_{i=1}^{k+14} \mathcal{B}_i.$$

Lemma 5. *Given a list L of rectangles for the $2BP(a, b)$, the algorithm COMBINE-AB packs all rectangles of type $\mathcal{A}$ or all rectangles of type $\mathcal{B}$. Moreover, if $\mathcal{P}_{AB}$ is the packing generated and L_{AB} the rectangles in $\mathcal{P}_{AB}$, then $\#(\mathcal{P}_{AB}) \leq \frac{36}{17} \frac{S(L_{AB})}{ab} + 2k + 41$.*

After applying the algorithm COMBINE-AB, suppose that all rectangles of type $\mathcal{B}$ have been packed in $\mathcal{P}_{AB}$. Consider the lists $L_1, \ldots, L_{23}$, defined in step 4.3 of the algorithm $BI_{k,\epsilon}$. The packing of lists L_1 and L_{18}, generated by the algorithm NFDH, has an area guarantee close to $\frac{4}{9}$, but for sublists $L_2, \ldots, L_{17}, L_{19}, \ldots, L_{23}$ the NFDH strategy generates packings with volume guarantee better than $\frac{4}{9}$ (namely at least $17/36$). Therefore, we define the critical set $L_D := L'_D \cup L''_D$ (see step 4.3) as the set of rectangles of $L_1 \cup L_{18}$ which leads to packings with area guarantee close to $4/9$. The area guarantee we can obtain for rectangles in $L \cap (\wp_2 \cup \wp_4) \setminus L_{AB}$ is $\frac{1}{4}$. Therefore, we define the critical set L_C as the critical rectangles in $L \cap \wp_4$ (see step 4.3). Let us denote the algorithm that combines the sets L_C and L_D by COMBINE-CD. The packing that is generated has bins with one rectangle of L_C and one rectangle of L'_D, or one rectangle of L_C and two rectangles of L''_D. If $\mathcal{P}_{CD}$ is a packing generated by COMBINE-CD, and L_{CD} is the set of rectangles packed in $\mathcal{P}_{CD}$, then or $L_C \subseteq L_{CD}$ or $L_D \subseteq L_{CD}$. Moreover, the packing $\mathcal{P}_{CD}$ has an area guarantee close to $17/36$. More precisely,

Lemma 6. *If $\mathcal{P}_{CD}$ is a packing generated by the algorithm COMBINE-CD and L_{CD} is the set of rectangles in $\mathcal{P}_{CD}$, then $\#(\mathcal{P}_{CD}) \leq \frac{1}{1/4 + r_1/2} \frac{S(L_{CD})}{ab} + 2$.*

The packings $\mathcal{P}_{AB}$ and $\mathcal{P}_{CD}$ have area guarantee close to $\frac{17}{36}$. Depending on which set is totally packed in $\mathcal{P}_{CD}$, we can improve either the area guarantee of $\frac{1}{4}$, of $L \cap (\wp_2 \cup \wp_4) \setminus (L_{AB} \cup L_{CD})$ packing, or the area guarantee of $\frac{4}{9}$, of the $L \cap (\wp_1 \cup \wp_3) \setminus (L_{AB} \cup L_{CD})$ packing. Now, it is possible to obtain an algorithm for 2BP with asymptotic performance bound close to 2.64. For simplicity, we denote by Update(L) the procedure that removes from L all rectangles previously packed.

Algorithm $BI_{k,\epsilon}(L)$

Input: List of rectangles L for the 2BP(a, b)

1 Rotate all rectangles $r \in L \cap \wp_4$, where $\rho(r) \in \wp_1 \cup \wp_2 \cup \wp_3$.

2 $t \leftarrow 0.4574271$.

3 $(\mathcal{P}_{AB}, L_{AB}) \leftarrow$ COMBINE-AB$_k(L)$. Update(L).

4 If all rectangles of type $\mathcal{B}$ were packed then

    4.1 Rotate each rectangle of $L \cap \wp_2$ that fits in $\wp_1 \cup \wp_3$.

    4.2 Rotate each rectangle of $L \cap (\wp_2 \cup \wp_4)$, so that if $b \in L \cap (\wp_2 \cup \wp_4)$ then $x(b) \leq y(b)$ or $\rho(b) \notin \mathcal{C}_1$.

    4.3 Subdivide the list L into sublists $L_1, \ldots, L_{23}$ as follows.

$L_i \leftarrow L \cap \mathcal{C}\left[\frac{1}{2}, 1; \frac{1}{i+2}, \frac{1}{i+1}\right]$, for $i = 1, \ldots, 16$, $L_{17} \leftarrow L \cap \mathcal{C}\left[\frac{1}{2}, 1; 0, \frac{1}{18}\right]$,

$L_{18} \leftarrow L \cap \mathcal{C}\left[\frac{1}{3}, \frac{1}{2}; \frac{1}{3}, \frac{1}{2}\right]$, $L_{19} \leftarrow L \cap \mathcal{C}\left[\frac{1}{3}, \frac{1}{2}; \frac{1}{4}, \frac{1}{3}\right]$, $L_{20} \leftarrow L \cap \mathcal{C}\left[\frac{1}{3}, \frac{1}{2}; 0, \frac{1}{4}\right]$,

$L_{21} \leftarrow L \cap \mathcal{C}\left[\frac{1}{4}, \frac{1}{3}; \frac{1}{3}, \frac{1}{2}\right]$, $L_{22} \leftarrow L \cap \mathcal{C}\left[0, \frac{1}{4}; \frac{1}{3}, \frac{1}{2}\right]$, $L_{23} \leftarrow L \cap \mathcal{C}\left[0, \frac{1}{3}; 0, \frac{1}{3}\right]$,

$L_C \leftarrow L \cap \mathcal{C}\left[\frac{1}{2}, 1; \frac{1}{2}, 1-t\right]$, $L'_D \leftarrow L_1 \cap \mathcal{C}[0, t; 0, 1]$, $L''_D \leftarrow L_{18} \cap \mathcal{C}[0, t; 0, 1]$,

$L_D \leftarrow L'_D \cup L''_D$.

    4.4 Generate packing $\mathcal{P}_{CD}$ as follows.

       $(\mathcal{P}_{CD}, L_{CD}) \leftarrow$ COMBINE-CD(L); $L_1 \leftarrow L_1 \setminus L_{CD}$; $L_{18} \leftarrow L_{18} \setminus L_{CD}$.

    4.5 Generate packings $\mathcal{P}_1, \ldots, \mathcal{P}_{23}$ as follows.

       $\mathcal{P}_i \leftarrow$ NFDH$^y(L_i)$ for $i = 1, \ldots, 21$; $\mathcal{P}_i \leftarrow$ NFDH$^x(L_i)$ for $i = 22$;

       $\mathcal{P}_{23} \leftarrow$ BI$_3(L_{23})$.   Update(L). Now, we have $L \subseteq \wp_2 \cup \wp_4$.

    4.6 Consider each rectangle of L as a one-dimensional item of length $x(r)$ and each two-dimensional bin as a one-dimensional bin of length a.

       Let $\mathcal{P}_{FL_\epsilon}$ be the packing obtained by $FL_\epsilon(L)$ and let $\mathcal{P}_{FFD}$ be the packing

$\text{FFD}(L \cap \mathcal{X}[0, \frac{1}{3}]) \| \text{FFD}(L \cap \mathcal{X}[\frac{1}{3}, \frac{1}{2}]) \| \text{FFD}(L \cap \mathcal{X}[\frac{1}{2}, 1]).$

Let $\mathcal{P}_{\text{UNI}}$ be the smallest packing in $\{\mathcal{P}_{\text{FL}_\epsilon}, \mathcal{P}_{\text{FFD}}\}$.

4.7 $\mathcal{P}_{aux} \leftarrow \mathcal{P}_{AB} \| \mathcal{P}_{CD} \| \mathcal{P}_1 \| \ldots \| \mathcal{P}_{23};$

4.8 $\mathcal{P} \leftarrow \mathcal{P}_{\text{UNI}} \| \mathcal{P}_{aux}.$

5 If all rectangles of type $\mathcal{A}$ were packed, then generate a packing $\mathcal{P}$ of L as in step 4 (in a symmetric way).

6 Return $\mathcal{P}$.

Theorem 4. *For any instance L of* 2BP, *we have* $\text{BI}_{k,\epsilon}(L) \leq \alpha_{k,\epsilon} \cdot \text{OPT}(L) + \mathcal{O}(k)$, *where* $\alpha_{k,\epsilon} \to 2.63 \ldots$ *as* $k \to \infty$ *and* $\epsilon \to 0$.

Proof. We present the proof for the case in which all rectangles of type $\mathcal{B}$ are packed in step 3. The proof for the other case (all rectangles of type $\mathcal{A}$ are packed) is analogous. This proof is divided in two cases, according to step 4.4 ($L_C \subset L_{CD}$ or $L_D \subseteq L_{CD}$).

Analyzing the configuration of each bin in the packing $\mathcal{P}_i$, $i \in \{1, \ldots, 23\} \setminus \{1, 18\}$, we can show that $\mathcal{P}_i$ have an area occupation of at least $\frac{17}{36} \cdot a \cdot b$ in each bin, except perhaps in a constant number of them. Therefore, applying lemmas 4, 3, 5 and 6 we have

$$\#(\mathcal{P}_i) \leq \frac{36}{17} \frac{S(L_i)}{ab} + 1, \quad \text{for } i \in \{1, \ldots, 23\} \setminus \{1, 18\}. \tag{6}$$

$$\#(\mathcal{P}_{AB}) \leq \frac{36}{17} \frac{S(L_{AB})}{ab} + (2k + 41), \quad \#(\mathcal{P}_{CD}) \leq \frac{1}{(\frac{1}{4} + \frac{r_1}{2})} \frac{S(L_{CD})}{ab} + 2. \tag{7}$$

Case 1. $L_C \subseteq L_{CD}$. For packings $\mathcal{P}_1$ and $\mathcal{P}_{18}$, we have

$$\#(\mathcal{P}_1) \leq \frac{1}{r_1} \frac{S(L_1)}{ab} + 1, \quad \#(\mathcal{P}_{18}) \leq \frac{1}{\frac{4}{9}} \frac{S(L_{18})}{ab} + 1. \tag{8}$$

By Theorem 3, and analyzing the area guarantee of $\mathcal{P}_{\text{FFD}}$, we have

$$\#(\mathcal{P}_{\text{UNI}}) \leq \#(\mathcal{P}_{\text{FL}_\epsilon}) \leq (1 + \epsilon) \cdot \text{OPT}(L_{\text{UNI}}) + 1. \tag{9}$$

$$\#(\mathcal{P}_{\text{UNI}}) \leq \#(\mathcal{P}_{\text{FFD}}) \leq \frac{1}{(1-t) \cdot \frac{1}{2}} \cdot \frac{S(L_{\text{UNI}})}{ab} + 3. \tag{10}$$

Now, for the packing $\mathcal{P}_{aux} = \mathcal{P}_{AB} \| \mathcal{P}_1 \| \ldots \| \mathcal{P}_{23}$, using the inequalities (6)–(8) and the fact that $r_1 = \min \{\frac{17}{36}, r_1, \frac{1}{4} + \frac{r_1}{2}, \frac{4}{9}\}$, we obtain

$$\#(\mathcal{P}_{aux}) \leq \frac{1}{r_1} \frac{S(L_{aux})}{ab} + (2k + 68),$$

where L_{aux} denotes the set of rectangles in $\mathcal{P}_{aux}$. Let $n_1 := \#(\mathcal{P}_{\text{UNI}}) - 3$ and $n_2 := \#(\mathcal{P}_{aux}) - (2k + 68)$. From inequality (9) and the fact that $\text{OPT} \geq \frac{S(L)}{ab}$, we have $\text{OPT}(L) \geq \max \{\frac{1}{1+\epsilon} n_1, \frac{1-t}{2} n_1 + r_1 \cdot n_2\}$. Since $\#(\mathcal{P}) = \#(\mathcal{P}_{aux}) + \#(\mathcal{P}_{\text{UNI}})$, now we have $\#(\mathcal{P}) = (n_2 + (2k + 68) + n_1 + 3) = n_1 + n_2 + (2k + 71)$. Therefore,

$$\text{BI}_{k,\epsilon}(L) \leq \alpha'_{k,\epsilon} \cdot \text{OPT}(L) + (2k + 71),$$

where $\alpha'_{k,\epsilon} = (n_1 + n_2)/\max \{\frac{1}{1+\epsilon} n_1, \frac{(1-t)}{2} n_1 + r_1 \cdot n_2\} \leq \frac{1}{r_1} - \frac{(1-t)}{2r_1} + (1 + \epsilon)$.

Case 2. $L_D \subseteq L_{CD}$. In this case, the proof is analogous and we can obtain that $\text{BI}_{k,\epsilon}(L) \leq \alpha''_{k,\epsilon} \cdot \text{OPT}(L) + (2k + 71)$, where $\alpha''_{k,\epsilon} = \frac{(n_1 + n_2)}{\max\{\frac{1}{1+\epsilon} n_1, \frac{1}{4} n_1 + t \cdot n_2\}} \leq \frac{1}{t} - \frac{(1+\epsilon)}{4t} + (1 + \epsilon)$.

From cases 1 and 2, we conclude that for $k \to \infty$ and $\epsilon \to 0$ the theorem follows. $\square$

5   Three-Dimensional Packing Problems

Due to space limitations, we only present the results for the algorithms ATP, ATP^z and A3D we obtained by applying the same techniques to the problems TPP, TPP^z and 3BP, respectively.

Theorem 5. *For any instance L of* TPP, *we have* $\text{ATP}_{k,\epsilon}(L) \leq \alpha_{k,\epsilon} \cdot \text{OPT}(L) + \mathcal{O}\left(k + \frac{1}{\epsilon}\right) \cdot Z$, *where* $\alpha_{k,\epsilon} \rightarrow 2.76\ldots$ *as* $k \rightarrow \infty$ *and* $\epsilon \rightarrow 0$.

Theorem 6. *For any instance L of* TPP^z, *we have* $\text{ATP}^z_{k,\epsilon}(L) \leq \alpha_{k,\epsilon} \cdot \text{OPT}(L) + \mathcal{O}\left(k + \frac{1}{\epsilon}\right) \cdot Z$, *where* $\alpha_{k,\epsilon} \rightarrow 2.63\ldots$ *as* $k \rightarrow \infty$ *and* $\epsilon \rightarrow 0$.

This last result improves our previous result in [10,11].

Theorem 7. *For any list of boxes L for* 3BP, *we have* $\text{A3D}(L) \leq \alpha_{k,\epsilon} \cdot \text{OPT}(L) + \beta_{k,\epsilon}$, *where* $\lim_{k \rightarrow \infty, \epsilon \rightarrow 0} \alpha_{k,\epsilon} \leq 4.88\ldots$, *and* $\beta_{k,\epsilon}$ *is constant for constant values of k and ϵ.*

6   Concluding Remarks

We presented several approximation algorithms for packing problems where orthogonal rotations are allowed. These problems have been less investigated in the literature. To our knowledge, the bounds presented are the best ones known for each problem. We would like to thank David S. Johnson, for his comments about the status of these problems.

References

1. B.S. Baker, E.G. Coffman Jr., and R.L. Rivest. Orthogonal packings in two-dimensions. *SIAM Journal on Computing*, 9:846–855, 1980.
2. F.R.K. Chung, M.R. Garey, and D.S. Johnson. On packing two-dimensional bins. *SIAM Journal on Algebraic and Discrete Methods*, 3:66–76, 1982.
3. E.G. Coffman, Jr., M.R. Garey, and D.S. Johnson. Approximation algorithms for bin packing - an updated survey. In G. Ausiello, M. Lucertini, and P. Serafini, editors, *Algorithms design for computer system design*, pages 49–106. Spring-Verlag, New York, 1984.
4. E.G. Coffman, Jr., M.R. Garey, and D.S. Johnson. *Approximation algorithms (ed. D. Hochbaum)*, chapter Approximation algorithms for bin packing - a survey. PWS, 1997.
5. E.G. Coffman, Jr., M.R. Garey, D.S. Johnson, and R.E. Tarjan. Performance bounds for level oriented two-dimensional packing algorithms. *SIAM J. on Computing*, 9:808–826, 1980.
6. W. Fernandez de la Vega and G.S. Lueker. Bin packing can be solved within $1 + \epsilon$ in linear time. *Combinatorica*, 1(4):349–355, 1981.
7. D.S. Johnson. *Near-optimal bin packing algorithms*. PhD thesis, Massachusetts Institute of Technology, Cambridge, Mass., 1973.
8. C. Kenyon and E. Remila. Approximate strip packing. In *37th Annual Symposium on Foundations of Computer Science*, pages 31–36, 1996.
9. K. Li and K-H. Cheng. Static job scheduling in partitionable mesh connected systems. *Journal of Parallel and Distributed Computing*, 10:152–159, 1990.
10. F.K. Miyazawa and Y. Wakabayashi. An algorithm for the three-dimensional packing problem with asymptotic performance analysis. *Algorithmica*, 18(1):122–144, 1997.
11. F.K. Miyazawa and Y. Wakabayashi. Approximation algorithms for the orthogonal z-oriented 3-D packing problem. *SIAM Journal on Computing*, 29(3):1008–1029, 2000.
12. V.V. Vazirani. *Approximation Algorithms*. Springer-Verlag, 2001.

Combinatorial Problems on Strings with Applications to Protein Folding

Alantha Newman[1] and Matthias Ruhl[2]

[1] MIT Laboratory for Computer Science
Cambridge, MA 02139
alantha@theory.lcs.mit.edu
[2] IBM Almaden Research Center
San Jose, CA 95120
ruhl@almaden.ibm.com

Abstract. We consider the problem of protein folding in the HP model on the 3D square lattice. This problem is combinatorially equivalent to folding a string of 0's and 1's so that the string forms a self-avoiding walk on the lattice and the number of adjacent pairs of 1's is maximized. The previously best-known approximation algorithm for this problem has a guarantee of $\frac{3}{8} = .375$ [HI95]. In this paper, we first present a new $\frac{3}{8}$-approximation algorithm for the 3D folding problem that improves on the absolute approximation guarantee of the previous algorithm. We then show a connection between the 3D folding problem and a basic combinatorial problem on binary strings, which may be of independent interest. Given a binary string in $\{a,b\}^*$, we want to find a long subsequence of the string in which every sequence of consecutive a's is followed by at least as many consecutive b's. We show a non-trivial lower-bound on the existence of such subsequences. Using this result, we obtain an algorithm with a slightly improved approximation ratio of at least .37501 for the 3D folding problem. All of our algorithms run in linear time.

1 Introduction

We consider the problem of protein folding in the HP model on the three-dimensional (3D) square lattice. This optimization problem is combinatorially equivalent to folding a string of 0's and 1's, i.e. placing adjacent elements of the string on adjacent lattice points, so that the string forms a self-avoiding walk on the lattice and the number of adjacent pairs of 1's is maximized. Figure 1 shows an example of a 3D folding of a binary string.

Background. The widely-studied HP model was introduced by Dill [Dil85, Dil90]. A protein is a chain of amino acid residues. In the HP model, each amino acid residue is classified as an H (hydrophobic or non-polar) or a P (hydrophilic or polar). An optimal configuration for a string of amino acids in this model is one that has the lowest energy, which is achieved when the number of H-H contacts (i.e. pairs of H's that are adjacent in the folding but not in the string) is maximized. The *protein folding* problem in the hydrophobic-hydrophilic (HP)

M. Farach-Colton (Ed.): LATIN 2004, LNCS 2976, pp. 369–378, 2004.
© Springer-Verlag Berlin Heidelberg 2004

model on the 3D square lattice is combinatorially equivalent to the problem we just described: we are given a string of P's and H's (instead of 0's and 1's) and we wish to maximize the number of adjacent pairs of H's (instead of 1's). An informative discussion on the HP model and its applicability to protein folding is given by Hart and Istrail [HI95].

Related Work. Berger and Leighton proved that this problem is NP-hard [BL98]. On the positive side, Hart and Istrail gave a simple algorithm with an approximation guarantee of $\frac{3}{8}OPT - \Theta(\sqrt{OPT})$ [HI95]. Folding in the HP model has also been studied for the 2D square lattice. This variant is also NP-hard [CGP+98]. Hart and Istrail gave a $\frac{1}{4}$-approximation algorithm for this problem [HI95], which was recently improved to a $\frac{1}{3}$-approximation algorithm [New02].

Our Contribution. Improving on the approximation guarantee of $\frac{3}{8}$ for the 3D folding problem has been an open problem for almost a decade. In this paper, we first present a new 3D folding algorithm (Section 2.1). Our algorithm produces a folding with $\frac{3}{8}OPT - \Theta(1)$ contacts, improving the absolute approximation guarantee. We then show that if the input string is of a certain special form, we can modify our algorithm to yield $\frac{3}{4}OPT - O(\delta(S))$ contacts, where $\delta(S)$ is the number of transitions in the input string S from sequences of 1's in odd positions in the string to sequences of 1's in even positions. This is described in Section 2.2.

In Section 3, we reduce the general 3D folding problem to the special case above, yielding a folding algorithm producing $.439 \cdot OPT - O(\delta(S))$ contacts. This reduction is based on a simple combinatorial problem for strings, which may be of independent interest.

We call a binary string from $\{a, b\}^*$ *block-monotone* if every maximal sequence of consecutive a's is immediately followed by a block of at least as many b's. Suppose we are given a binary string with the following property: every suffix of the string (i.e. every sequence of consecutive elements that ends with the last element of the string) contains at least as many b's as a's. What is the longest block-monotone subsequence of the string? It is easy to see that we can find a block-monotone subsequence with length at least half the length of the string by removing all the a's. In Section 3.1, we show that there always is a block-monotone subsequence containing at least a $(2 - \sqrt{2}) \approx .5857$ fraction of the string's elements.

Finally, we combine our folding algorithm with a simple, case-based algorithm that achieves $.375 \cdot OPT + \Omega(\delta(S))$ contacts, which is described in the full version of this paper. We thereby remove the dependence on $\delta(S)$ in the approximation guarantee and obtain an algorithm with a slightly improved approximation guarantee of $.37501$ for the 3D folding problem. Due to space restrictions, all proofs are omitted and can be found in the full version of this paper.

2   A New 3D Folding Algorithm

Let $S \in \{0,1\}^n$ represent the string we want to fold. We refer to each 0 or 1 as an *element*. We let s_i represent the ith element of S, i.e. $S = s_1 s_2 \ldots s_n$. We refer to a 1 in an odd position (i.e. $s_i = 1$ with odd index i) as an *odd-1* and a 1 in an even position (i.e. $s_i = 1$ with even index i) as an *even-1*. An *odd* or *even* label is determined by an element's position in the input string and does not change at any stage of the algorithm. We will use $\mathcal{O}[S]$ and $\mathcal{E}[S]$ to denote the number of odd-1's and even-1's, respectively, in a string S. For example, for $S = 10111100101101$, we have $\mathcal{O}[S] = 5$ and $\mathcal{E}[S] = 4$.

Note that because the square lattice is bipartite, the odd/even label determines the set of lattice points on which an element can be placed. For example, suppose we divide the lattice points into two bipartite sets, one red and one blue. If the first element of the string is placed on a red lattice point, then all the elements in odd positions in the string will be placed on red lattice points and all the elements in even positions in the string will be placed on blue lattice points.

A contact between two elements placed on the square lattice can therefore only occur between an odd-1 and an even-1. Each lattice point is adjacent to six neighboring lattice points. In any folding, if an odd-1 is placed on a particular lattice point, two neighboring lattice points will be occupied by preceding and succeeding (even) elements of the string unless the element is one of the two endpoints of the string. Therefore, there are four remaining adjacent lattice points with which contacts can be formed. Thus, an upper bound on the size of an optimal solution is $OPT \leq 4 \min\{\mathcal{O}[S], \mathcal{E}[S]\} + 2$.

2.1   The Diagonal Folding Algorithm

We now present an algorithm that produces a folding with at least $\frac{3}{8}OPT - \Theta(1)$ contacts in the worst case, thereby improving the *absolute* approximation guarantee of the algorithm of Hart and Istrail [HI95]. Our algorithm is based on *diagonal folds*. The algorithm guarantees that contacts form on and between two adjacent 2D planes. Each point in the 3D lattice has an (x, y, z)-coordinate, where x, y, and z are integers. We will fold the string so that all contacts occur on or between the planes $z = 0$ and $z = 1$. The DIAGONAL FOLDING ALGORITHM is described on the next page and illustrated in Figure 1.

Lemma 1. *The* DIAGONAL FOLDING ALGORITHM *produces a folding with at least* $\frac{3}{8}OPT - O(1)$ *contacts.*

2.2   Relating Folding to String Properties

As the number of 1's placed on the diagonal in the DIAGONAL FOLDING ALGORITHM increases, the length (i.e. $\frac{1}{2}\min\{\mathcal{O}[S], \mathcal{E}[S]\}$) of the resulting folding increases in a direction parallel to the line $x = y$. The height of the folding may also increase depending on the maximum distance between consecutive odd-1's

---

DIAGONAL FOLDING ALGORITHM

Input: a binary string S.
Output: a folding of the string S.
1. Let $k = \min\{\mathcal{O}[S], \mathcal{E}[S]\}$.
2. Divide S into two strings such that $S_\mathcal{O}$ contains at least half the odd-1's and $S_\mathcal{E}$ contains at least half the even-1's. We can do this by finding a point on the string such that half of the odd-1's are on one side of this point and half the odd-1's are on the other side. One of these sides contains at least half of the even-1's. We call this side $S_\mathcal{E}$ and the remaining side $S_\mathcal{O}$. Then we replace all the even-1's in $S_\mathcal{O}$ with 0's and replace all the odd-1's in $S_\mathcal{E}$ with 0's.
3. Place the first odd-1 in $S_\mathcal{O}$ on lattice point $(1,1,1)$ and the next odd-1 in $S_\mathcal{O}$ on lattice point $(2,2,1)$ and so on. For the first $\frac{k}{4}$ of the odd-1's in $S_\mathcal{O}$, place the ith odd-1 on lattice point $(i,i,1)$. Then place the $(k/4+1)$ odd-1 on lattice point $(k/4-1, k/4+1, 1)$. For the first $\frac{k}{4} - 1$ of the even-1's in $S_\mathcal{E}$, place the ith even-1 on lattice point $(i, i+1, 1)$. Use the dimensions $z \geq 1$ to place the strings of 0's between consecutive odd-1's in $S_\mathcal{O}$ and the strings of 0's between consecutive even-1's in $S_\mathcal{E}$.
4. Place the $(k/4+2)$ odd-1 in $S_\mathcal{O}$ on lattice point $(k/4-2, k/4+1, 0)$. Then place the $(k/4+i)$ odd-1 in $S_\mathcal{O}$ on lattice point $(k/4 - i + 1, k/4 - i + 2, 0)$. Place the $(k/4)$ even-1 in $S_\mathcal{E}$ on lattice point $(k/4-1, k/4-1, 0)$. Place the $(k/4+i)$ even-1 in $S_\mathcal{E}$ on lattice point $(k/4 - i - 1, k/4 - i - 1, 0)$. Use the dimensions $z \leq 0$ to place the strings of 0's between consecutive 1's in $S_\mathcal{O}$ or $S_\mathcal{E}$.

---

in $S_\mathcal{O}$ or consecutive even-1's in $S_\mathcal{E}$. However, regardless of the input string, the resulting folding has the same constant width in the direction parallel to the line $x = -y$. In other words, although the algorithm produces a three-dimensional folding, with increasing k and n, the folding may increase in length and height but not in width. We will explain how we can use this unused space to improve the algorithm for a special class of strings.

By *consecutive odd-1's* we mean odd-1's that are not separated by even-1's and similarly for consecutive even-1's. For example, in the string 1010001100011, there is a sequence of 3 consecutive odd-1's followed by two consecutive even-1's followed by an odd-1.

Definition 2. *A string $S_\mathcal{O}$ is called* odd-monotone *if every maximal sequence of consecutive even-1's is immediately preceded by at least as many consecutive odd-1's.*

An *even-monotone* string is defined analogously. For example, the string 10101100011 is odd-monotone and the string 0100010101101101011 is even-monotone. We define a *switch* as follows:

Definition 3. *A* switch *is an odd-1 followed by an even-1 (separated only by 0's). We denote the number of switches in S by $\delta(S)$.*

For example, for the string $S = 1001000101011011\underline{1}01011$, $\delta(S) = 2$ since there are two transitions (underlined) from a maximal sequence of consecutive odd-1's to a sequence of even-1's. We use these definitions in the following theorem.

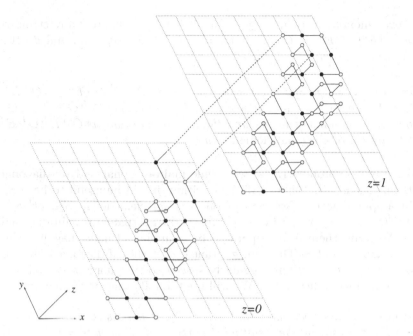

Fig. 1. This figure illustrates Steps 3 and 4 of the DIAGONAL FOLDING ALGORITHM. In the folding resulting from this algorithm, all contacts are formed on or between the 2D planes $z = 0$ (lower) and $z = 1$ (upper). Black dots represent 1's and white dots represent 0's.

Theorem 4. *Let $S = S_O S_\mathcal{E}$ and let S_O be an odd-monotone string and $S_\mathcal{E}$ be an even-monotone string such that $\mathcal{O}[S_O] = \mathcal{E}[S_\mathcal{E}]$ and $\mathcal{E}[S_O] = \mathcal{O}[S_\mathcal{E}]$. Then there is a linear time algorithm that folds these two strings achieving $\frac{3}{4}OPT - 16\delta(S) - O(1)$ contacts.*

The main idea behind the proof of Theorem 4 is that we partition the elements in S_O and $S_\mathcal{E}$ into *main-diagonal elements* and *off-diagonal elements*. We then use the DIAGONAL FOLDING ALGORITHM to fold the main-diagonal elements along the direction $x = y$ and the off-diagonal elements into branches along the direction $x = -y$ (see Figure 2). All 1's will receive 3 contacts except for a constant number of 1's for each off-diagonal branch, which correspond to switches in the strings S_O and $S_\mathcal{E}$, and a constant number at the ends of the main diagonal. This yields the claimed number of $\frac{3}{4}OPT - O(\delta(S)) - O(1)$ contacts.

To precisely define *main-diagonal* and *off-diagonal* elements, we use additional notation. We use 0^k and 1^k (for some integer $k \geq 0$) to refer to the strings consisting of k 0's and k 1's, respectively. By writing $S = E^k$ for some integer k, we mean that S is of the form $S = 0^{2i_0+1}10^{2i_1+1}10^{2i_2+1}10^{2i_3+1} \ldots 0^{2i_{k-1}+1}10^{i_k}$ for integers $i_j \geq 0$, and all the 1's in S are even-1's. Likewise, we write $S = O^k$ to refer to a string of the same form where all 1's are odd-1's, i.e. $S = 10^{2i_1+1}10^{2i_2+1}\,10^{2i_3+1} \ldots 0^{2i_{k-1}+1}10^{i_k}$. So we can express any string $S_\mathcal{E}$ as $S_\mathcal{E} = E^{a_1}O^{b_1}E^{a_2}O^{b_2} \ldots E^{a_k}O^{b_k}$ for $k = \delta(S_\mathcal{E})$ and integers a_i and b_i. If

$S_{\mathcal{E}}$ is even-monotone, then $a_i \geq b_i$ for all i. We can express any string $S_{\mathcal{O}}$ as $S_{\mathcal{O}} = O^{c_1} E^{d_1} O^{c_2} E^{d_2} \ldots O^{c_\ell} E^{d_\ell}$ for $\ell = \delta(S_{\mathcal{O}})$ and integers c_i and d_i. If $S_{\mathcal{O}}$ is even-monotone, then $c_i \geq d_i$ for all i.

Definition 5. *For an odd-monotone string $S_{\mathcal{O}} = O^{c_1} E^{d_1} O^{c_2} E^{d_2} \ldots O^{c_\ell} E^{d_\ell}$, the first set of $c_i - d_i$ odd-1's in each block, i.e. the elements $O^{c_1 - d_1} O^{c_2 - d_2} \ldots O^{c_\ell - d_\ell}$, are the* main-diagonal elements *and the remaining elements $O^{d_1} E^{d_1} O^{d_2} E^{d_2} \ldots O^{d_\ell} E^{d_\ell}$ are the* off-diagonal elements *in $S_{\mathcal{O}}$.*

For even-monotone strings, we define main-diagonal and off-diagonal elements analogously. In our modified algorithm, it will be useful to have $S_{\mathcal{E}}$ and $S_{\mathcal{O}}$ in a special form. Two sets of off-diagonal elements in $S_{\mathcal{O}}$, $O^{d_i} E^{d_i}$ and $O^{d_{i+1}} E^{d_{i+1}}$, are separated by $c_{i+1} - d_{i+1}$ odd-1's that are main-diagonal elements. We want them to be separated by a number of main-diagonal elements that is a multiple of 8. This will guarantee that the off-diagonals used to fold the off-diagonal elements are regularly spaced so that none of the off-diagonal folds interfere with each other. We will use the following simple lemma.

Lemma 6. *For any odd-monotone string $S_{\mathcal{O}}$ it is possible to change at most $8\delta(S_{\mathcal{O}})$ 1's to 0's so that the resulting string S' is of the form $S' = O^{a_1} E^{b_1} O^{a_2} E^{b_2} \ldots O^{a_k}$, where $a_i - b_i$ is a positive multiple of 8 for $1 \leq i < k$.*

We note that there is an analogous version of Lemma 6 for even-monotone strings. With this preparation, we can now state our folding algorithm.

OFF-DIAGONAL FOLDING ALGORITHM

Input: A binary string $S = S_{\mathcal{O}} S_{\mathcal{E}}$, such that $S_{\mathcal{O}}$ is odd-monotone, $S_{\mathcal{E}}$ is even-monotone, $\mathcal{O}[S_{\mathcal{O}}] = \mathcal{E}[S_{\mathcal{E}}]$ and $\mathcal{E}[S_{\mathcal{O}}] = \mathcal{O}[S_{\mathcal{E}}]$.
Output: A folding of the string S.

1. Change at most $8\delta(S)$ 1's to 0's in $S_{\mathcal{O}}$ and $S_{\mathcal{E}}$ to yield the form specified in Lemma 6.
2. Run DIAGONAL FOLDING ALGORITHM on *main-diagonal* elements along the direction $x = y$ and change from plane $z = 0$ to $z = 1$ when the length of the main diagonal equals $4 \cdot \lfloor \mathcal{O}[S_{\mathcal{O}}]/8 \rfloor + 2$.
3. Run DIAGONAL FOLDING ALGORITHM on the *off-diagonal* elements along the direction $x = -y$. The *off-diagonal* elements attached to the *main-diagonal* elements on the plane $z = 1$ are folded along the diagonals $x = -y + 8k$. The *off-diagonal* elements attached to the *main-diagonal* elements on the plane $z = 0$ are folded along the diagonals $x = -y + 8k + 4$. (See Figure 2.)

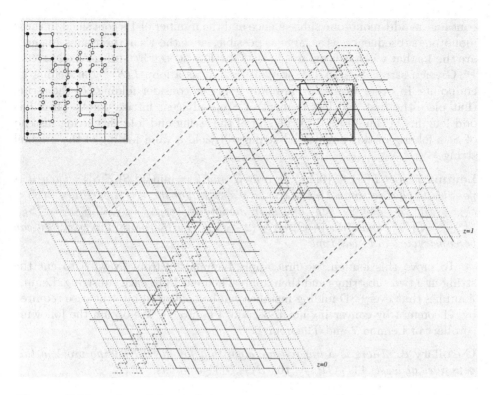

Fig. 2. Folding the *off-diagonal* elements in Step 3 of the OFF-DIAGONAL FOLD-ING ALGORITHM. The *main-diagonal* elements are represented by the dashed lines on the main diagonal. The *off-diagonal* elements are represents by the solid lines on the off-diagonals. This figure shows how the repetitions of the DIAGONAL FOLDING AL-GORITHM on the off-diagonals interleave and thus so not interfere with each other. The closeup gives an example of how the off-diagonal folds are connected to the main diagonal.

3 Combinatorial Problems on Strings

In this section, we present a combinatorial theorem about binary strings that allows us to use the algorithm from Section 2.2 for the general 3D folding problem. The binary strings that we consider in this section are from the set $\{a, b\}^*$. Given a string to fold in $\{0, 1\}^*$, we map it to a corresponding string in $\{a, b\}^*$ by representing each odd-1 by an a and each even-1 by a b. For example, the string 10100101 would be mapped to the string $aabb$. We will use theorems about the strings in $\{a, b\}^*$ to prove theorems about *subsequences* of the strings in $\{0, 1\}^*$ that we want to fold.

The combinatorial problem that we want to solve is the following: given a string $S \in \{0, 1\}^*$ such that $\mathcal{E}[S] = \mathcal{O}[S]$, we want to divide the string into two substrings such that one contains an even-monotone subsequence and the other contains an odd-monotone subsequence and the number of 1's contained in these monotone subsequences is as large as possible, since the 1's in these subsequences are the 1's that will have contacts in the OFF-DIAGONAL FOLDING ALGORITHM.

Given a string $S \in \{0,1\}^*$, we will treat it as a loop $L(S)$ by attaching its endpoints. In other words, we are only going to consider foldings of the string that place the first and last element of S on adjacent lattice points. (If S has odd length, we can add a 0 to the end of the string and fold this string instead of S; a folding of this augmented string will yield a valid folding of the original string.)

Lemma 7. *Let $L(S) \in \{0,1\}^*$ be a loop, and $k = \min\{\mathcal{O}[S], \mathcal{E}[S]\}$. Then it is possible to change some 1's of $L(S)$ to 0's such that there is a partition $L(S) = S_{\mathcal{O}} S_{\mathcal{E}}$ with $S_{\mathcal{O}}$ and $S_{\mathcal{E}}$ odd- and even-monotone, respectively, $\mathcal{O}[S_{\mathcal{O}}] = \mathcal{E}[S_{\mathcal{E}}]$, $\mathcal{E}[S_{\mathcal{O}}] = \mathcal{O}[S_{\mathcal{E}}]$, and $\mathcal{O}[S_{\mathcal{O}}] + \mathcal{O}[S_{\mathcal{E}}] \geq (2 - \sqrt{2})k$. Furthermore, this partition can be constructed in linear time.*

To prove this lemma, we first apply Lemma 2.2 from [New02] to cut the string into two substrings and then apply Theorem 13 to each substring. Lemma 7 implies that every 3D folding instance can be converted into the case required by Theorem 4 by converting not too many 1's into 0's. We obtain the following corollary of Lemma 7 and Theorem 4.

Corollary 8. *There is a linear time algorithm for the 3D folding problem that generates at least $.439 \cdot OPT - 16\delta(S) - O(1)$ contacts.*

3.1   Block-Monotone Subsequences

Let S be a binary string, $S \in \{a, b\}^n$. We will use the following definitions.

Definition 9. *Let $n_a(S)$ and $n_b(S)$ denote the number of a's and b's, respectively, in a string S.*

Definition 10. *A block is a maximal substring of consecutive a's or b's in a binary string.*

Definition 11. *A binary string is block-monotone if every block of a's is immediately followed by a block of at least as many b's.*

For example, the string $bbbbaaabb$ has two blocks of b's (of length four and two) and one block of a's (of length three). An example of a block-monotone string is $baaabbbaaabbbb$. The string $aabbaaabb$ is not block-monotone.

Given a binary string S, our goal is to find a long block-monotone subsequence. It is easy to see that S contains a block-monotone subsequence of length at least $n_b(S)$ since the subsequence of b's is trivially block-monotone. It is also easy to see that there are strings for which we cannot do better than this. For example, consider the string $b^i a^i$. In this string, there is no block monotone subsequence that contains any of the a's. Thus, we will put a stronger condition on the binary strings in which we want to find long block-monotone subsequences.

Notation. $\alpha := 1 - \frac{1}{\sqrt{2}} \approx 0.2929$

Definition 12. *A binary string* $S = s_1 \ldots s_n$ *is* suffix-monotone *if for every suffix* $\overline{S}_k = s_{k+1} \ldots s_n$, $0 \leq k < n$, *we have* $n_b(\overline{S}_k) \geq \alpha \cdot (n - k)$.

For example if every suffix of S has at least as many b's as a's, the string is suffix-monotone. We will give an algorithm to prove the following theorem.

Theorem 13. *Suppose* S *is a suffix-monotone string of length* n. *Then there is a block-monotone subsequence of* S *with length at least* $n - n_a(S)(2\sqrt{2} - 2)$. *Furthermore, such a subsequence can be found in linear time.*

If $n_a(S) \leq \frac{1}{2}n$ and S is suffix-monotone, then Theorem 13 states that we can find a block-monotone subsequence of length at least $(2 - \sqrt{2}) > .5857$ the length of S. This is accomplished by the following algorithm.

BLOCK-MONOTONE ALGORITHM

> *Input*: a suffix-monotone string $S = s_1 \ldots s_n$
> *Output*: a block-monotone subsequence of S
> Let $S_i = s_1 \ldots s_i$, $\overline{S}_i = s_{i+1} \ldots s_n$ for $i: 1 < i \leq n$
> 1. If $s_1 = b$:
>> (i) Find the largest index k such that S_k is a block of b's and output S_k
> 2. If $s_1 = a$:
>> (i) Find the smallest index k such that:
>> $\quad n_b(S_k) \geq \alpha k$
>> (ii) Let $S'_\ell = s_{\ell+1} \ldots s_k$ for $\ell: 1 \leq \ell < k$
>> (iii) Find ℓ such that:
>> $\quad n_a(S_\ell) \leq n_b(S'_\ell)$
>> $\quad n_a(S_\ell) + n_b(S'_\ell)$ is maximized
>> (iv) Remove all the b's from S_ℓ and output S_ℓ
>> (v) Remove all the a's from S'_ℓ and output S'_ℓ
> 3. Repeat algorithm on string $\overline{S}_k$

---

4   Conclusion

We conclude by stating an approximation guarantee independent of $\delta(S)$. In the full version of this paper, we give a case-based algorithm whose approximation guarantee is $\frac{3}{8}OPT + O(\delta(S))$. This algorithm is based on the following idea: Suppose $S_\mathcal{O}$ and $S_\mathcal{E}$ contain half the odd-1's and half the even-1's, respectively. We use the DIAGONAL FOLDING ALGORITHM, but for each switch in $S_\mathcal{O}$, we use different local foldings to obtain an additional (constant) number of contacts, e.g. we use an even-1 in the switch to obtain another contact with an odd-1 placed on the main diagonal. The performance of this algorithm is summarized in Lemma 14, which in combination with Corollary 8 yields Lemma 15.

Lemma 14. *We can modify the* DIAGONAL FOLDING ALGORITHM *to create a folding with* $\frac{3}{8}OPT + \frac{\delta(S)}{256} - O(1)$ *contacts for any binary string* S.

Lemma 15. *There is a linear time algorithm for the 3D folding problem that creates a folding with* $.37501 \cdot OPT - O(1)$ *contacts for any binary string* S.

We have described an algorithm for protein folding in the HP model on the 3D square lattice that slightly improves on the previously best-known algorithm to yield an approximation guarantee of .37501. The contribution of this paper is not so much the actual gain in the approximation ratio, but the demonstration that the previously best-known algorithm is not optimal, even though there have been no improvements for almost a decade.

In closing, we discuss the problem of finding block-monotone subsequences of binary strings. One way to improve the approximation ratio of our algorithm is to improve the guarantee given by Theorem 13. We note that we only apply Theorem 13 to binary strings in which every suffix contains at least as many b's as a's—a stronger condition than our definition of block-monotone. Theorem 13 implies that such strings contain block-monotone subsequences of at least .5857 their length. We conjecture that the real lower bound is actually $\frac{2}{3}$ their length. Currently, the best upper bound we are aware of is the string:

$$aaaaabaaaabaaabaababbbaaabaaababababaababbbbbbbbbbbbbb$$

whose longest block-monotone subsequence is $a^{18}b^{19}$, which is $\frac{37}{52} \approx 71.15\%$ of the length of the original string.

Acknowledgments. We thank Santosh Vempala for many helpful discussions and suggestions and comments on the presentation. We thank Edith Newman for drawing Figures 1 and 2.

References

[BL98]   Bonnie Berger and Tom Leighton. Protein Folding in the Hydrophobic-Hydrophilic (HP) Model is NP-Complete. In *Proceedings of the 2nd Conference on Computational Molecular Biology (RECOMB)*, 1998.

[CGP+98] P. Crescenzi, D. Goldman, C. Papadimitriou, A. Piccolboni, and M. Yannakakis. On the Complexity of Protein Folding. In *Proceedings of the 2nd Conference on Computational Molecular Biology (RECOMB)*, 1998.

[Dil85]   K. A. Dill. Theory for the Folding and Stability of Globular Proteins. *Biochemistry*, 24:1501, 1985.

[Dil90]   K. A. Dill. Dominant Forces in Protein Folding. *Biochemistry*, 29:7133–7155, 1990.

[HI95]   William E. Hart and Sorin Istrail. Fast Protein Folding in the Hydrophobic-hydrophilic Model within Three-eighths of Optimal. In *Proceedings of the 27th ACM Symposium on the Theory of Computing (STOC)*, 1995.

[New02]   Alantha Newman. A New Algorithm for Protein Folding in the HP Model. In *Proceedings of the 13th ACM-SIAM Symposium on Discrete Algorithms (SODA)*, 2002.

Measurement Errors Make the Partial Digest Problem NP-Hard

Mark Cieliebak[1] and Stephan Eidenbenz[2]

[1] Institute of Theoretical Computer Science, ETH Zurich, cieliebak@inf.ethz.ch
[2] Los Alamos National Laboratory[***], eidenben@lanl.gov

Abstract. The PARTIAL DIGEST problem asks for the coordinates of m points on a line such that the pairwise distances of the points form a given multiset of $\binom{m}{2}$ distances. PARTIAL DIGEST is a well-studied problem with important applications in physical mapping of DNA molecules. Its computational complexity status is open. Input data for PARTIAL DIGEST from real-life experiments are always prone to error, which suggests to study variations of PARTIAL DIGEST that take this fact into account. In this paper, we study the computational complexity of the variation of PARTIAL DIGEST in which each distance is known only up to some error, due to experimental inaccuracies. The error can be specified either by some additive offset or by a multiplicative factor. We show that both types of error make the PARTIAL DIGEST problem strongly NP-complete, by giving reductions from 3-PARTITION. In the case of relative errors, we show that the problem is hard to solve even for constant relative error.

1 Introduction

The PARTIAL DIGEST problem is perhaps *the* classic combinatorial problem from computational biology with applications in DNA sequencing. Despite considerable research efforts in the past twenty years, its computational complexity is still an open problem. In the PARTIAL DIGEST problem we are given a multiset D of distances and are asked to find coordinates of points on a line, i.e., a point set P, such that D is exactly the multiset[1] of all pairwise distances of these points. In this case, we say that D is the distance multiset of point set P. A formal definition of the problem is as follows.

Definition 1 (PARTIAL DIGEST). *Given an integer m and a multiset of $k = \binom{m}{2}$ positive integers $D = \{d_1, \ldots, d_k\}$, is there a set of m integers $P = \{p_1, \ldots, p_m\}$ such that $\{|p_i - p_j| \mid 1 \leq i < j \leq m\} = D$?*

For example, if $D = \{2, 5, 7, 7, 9, 9, 14, 14, 16, 23\}$, then $P = \{0, 7, 9, 14, 23\}$ is one feasible solution (cf. Figure 1).

[***] Work partially done while M. Cieliebak was visiting LANL. LA-UR-03:6621.
[1] We will denote multisets like sets, since the fact of being a multiset is not crucial for our purposes.

M. Farach-Colton (Ed.): LATIN 2004, LNCS 2976, pp. 379–390, 2004.

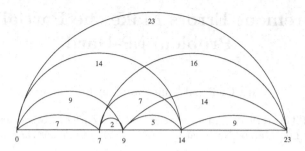

Fig. 1. Example for PARTIAL DIGEST

Previous Work

Intriguingly, the computational complexity of this seemingly straight-forward combinatorial puzzle is a long–standing open problem, and it appears in its pure combinatorial formulation already in the 1930's in the area of X–ray crystallography (acc. to [16]). The problem is also known as "turnpike problem", where we are given the pairwise distances of cities along a highway, and we want to find their ordering along the road [4]. The PARTIAL DIGEST problem can be solved in pseudo–polynomial time [10,13], and there exists a backtracking algorithm (for exact or erroneous data) that has expected running time polynomial in the number of distances [16,17], but exponential worst case running time [20]. The PARTIAL DIGEST problem can be formalized by cut grammars, which have one additional symbol δ, the *cut*, that is neither a non–terminal nor a terminal symbol [14], and the problem is closely related to the theory of homometric sets[2] [16]. Finally, if the points in a solution do not have to be on a line, but only in d–dimensional space, then the problem is NP-hard for some $d \geq 2$ [16]. However, for the original PARTIAL DIGEST problem, neither a polynomial–time algorithm nor a proof of NP-hardness is known [2,4,11,12,15].

Biological Background

PARTIAL DIGEST has several applications; the classical and most prominent is in the study of the structure of DNA molecules. More precisely, given a large DNA molecule (sequence of nucleotides A, C, G, and T), restriction enzymes can be used to generate a physical map of the molecule. A restriction enzyme cuts a DNA molecule at specific patterns, the restriction sites. For instance, the enzyme Eco RI cuts occurrences of the pattern GAATTC into G and AATTC. Under appropriate experimental conditions, *all* fragments between each two restriction sites are created. This process is called *partial digestion*. The lengths of the fragments (i.e., their number of nucleotides) are then measured by using gel electrophoresis, a standard technique in molecular biology. This leaves us with the multiset of distances between all restriction sites, and the objective is to

---

[2] Two (non–congruent) sets of points are homometric if they generate the same multiset of pairwise distances.

reconstruct the original ordering of the fragments in the DNA molecule, which is the PARTIAL DIGEST problem.

Erroneous Data

In real-life, partial digestion experiments cannot be conducted under ideal conditions as outlined above, and thus errors occur in the data. In fact, there is no such thing as error–free data, and typically four types of errors occur [5,6,8, 17]: *additional fragments*, for instance through contamination of the probe with unrelated biological material; *missing fragments*, due to partial cleavage errors, or because of small fragments that remain undetected by gel electrophoresis; *incorrect fragment lengths*, due to the fact that fragment lengths cannot be determined exactly using gel electrophoresis; and, finally, *wrong multiplicities*, due to the intrinsic difficulty to determine the proper multiplicity of a distance by gel electrophoresis[3].

Algorithms for PARTIAL DIGEST with inaccurate data have been studied intensively in the literature [5,8,17], and different error models have been designed, e.g. for measurement errors that are logarithmic in the size of the fragment length [18,19] or for intervals of absolute errors [1,17]. Optimization variations of PARTIAL DIGEST where fragments are either omitted or added in the data, and the number of errors has to be minimized, are known to be NP-hard or hard to approximate, respectively [3].

In this work we will focus on the third type of error, where the lengths of fragments can be erroneous (*measurement errors*). In partial digestion experiments all measurements of fragment lengths are prone to inaccuracies: Using gel electrophoresis, measurement errors within a range of up to 5 percent of the fragment length can occur [5,6,17].

Many experimental variations of partial digest experiments have been studied, see [9] for a survey; and for more detailed discussions on the problem, see [12] and [15].

Definitions and Results

In this paper, we study the computational complexity of PARTIAL DIGEST in the presence of measurement errors, where we allow both additive or multiplicative errors.

We start with additive errors. The PARTIAL DIGEST problem is known to be strongly NP-hard if additive error bounds that can be even zero can be assigned to each distance *individually* [9,16]. However, this does not model reality appropriately, since in real–life data we cannot assume that even one single fragment length can be measured exactly. Therefore, we study the computational complexity of the variation of PARTIAL DIGEST where *all* measurements are prone to some non–zero error. Moreover, we refrain from individual error bounds, and study the variation where all measurements are prone to *the same additive non–zero* error δ. More precisely, we say that value v matches a distance d up to additive error δ if $|v - d| \leq \delta$; moreover, a multiset D is a distance multiset of a point set P up to additive error δ, if there is a bijective function

---

[3] The multiplicity of a fragment is determined from the intensity of the corresponding band in the gel.

$f : D \to \Delta(P)$ such that each distance $d \in D$ matches value $f(d)$ up to error δ; here, $\Delta(P) = \{|p_j - p_i| \mid 1 \leq i < j \leq n\}$ denotes the multiset of pairwise distances in P. The PD–ABSERROR problem is defined as follows.

Definition 2 (PD–ABSERROR). *Given an integer m, a multiset D of $k = \binom{m}{2}$ positive integers, and an integer error bound $\delta > 0$, is there a set P of m points on a line such that D is the distance multiset of P up to additive error δ?*

We show in Section 2 that PD–ABSERROR is strongly NP-complete, by giving a reduction from 3-PARTITION.

We then turn to the case of multiplicative errors. We say that distance d matches a value x up to *multiplicative error $\varepsilon > 0$* if $d(1 - \varepsilon) \leq x \leq d(1 + \varepsilon)$. Observe that this definition is not symmetric, i.e., if d matches x up to error ε, then this does *not* in general imply that x matches d (in contrast to the definition of additive errors, which is symmetric). A multiset D is a distance multiset of point set P up to multiplicative error ε if there is a bijective function $f : D \to \Delta(P)$ such that each distance $d \in D$ matches value $f(d)$ up to multiplicative error ε. The PD–RELERROR problem is defined as follows.

Definition 3 (PD–RELERROR). *Given an integer m, a multiset D of $k = \binom{m}{2}$ positive integers, and a rational error $\varepsilon > 0$, is there a set P of m points on a line such that D is the distance multiset of P up to multiplicative error ε?*

We show in Section 3 that PD–RELERROR is strongly NP-complete, even for constant error, by using a similar reduction as for PD–ABSERROR.

2 Strong NP-Completeness of PD–ABSERROR

In this section, we show that PD–ABSERROR is strongly NP-complete, by giving a reduction from 3-PARTITION, which is the following problem: Given $3n$ positive integers $q_1, \dots, q_{3n}$ and an integer h such that $\sum_{i=1}^{3n} q_i = nh$ and $\frac{h}{4} < q_i < \frac{h}{2}$ for $i \in \{1, \dots, 3n\}$, are there n disjoint triples of q_i's such that each triple adds up to h? The 3-PARTITION problem is NP-complete in the strong sense [7]. Observe that $\frac{h}{4} < q_i < \frac{h}{2}$ already implies that each subset of the q_i's that adds up to h must have exactly three elements.

The idea of the reduction is as follows. Given an instance $q_1, \dots, q_{3n}$ and h of 3-PARTITION, we define a multiset of distances D and an additive error $\delta = \frac{h}{4}$ that form an instance of PD–ABSERROR. Our construction is based on the following observation: If there is a solution for the 3-PARTITION instance, then we can arrange the q_i's such that triples of adjacent q_i's sum up to h. If we sum up, say, 25 adjacent q_i, then we sum over at least 7 complete triples that each have sum h, plus some few (up to four) additional q_i's at the beginning and the end. In the special and trivial case that all q_i's have exactly value $\frac{h}{3}$, we can easily determine the exact sum of the 25 values. However, in a given instance of 3-PARTITION typically not all q_i's will have value $\frac{h}{3}$. However, they have "approximately" value $\frac{h}{3}$, since they satisfy $\frac{h}{4} < q_i < \frac{h}{2}$ by definition. In the proof of the following theorem, we will use additive error δ to "close the gap" between $\frac{h}{3}$ and the true values of the q_i's.

Theorem 4. PD–ABSERROR *is strongly* NP-*complete.*

Proof. The problem PD–ABSERROR is in NP: Given a candidate point set P, we sort all distances between any two points in P, and all distances in D; then P is a solution if error δ is sufficient to match the i-th distance from P to the i-th distance from D.

To prove strong NP-hardness, we give a reduction from 3-PARTITION. Given an instance of 3-PARTITION, i.e., integers $q_1, \ldots, q_{3n}$ and integer h, we define a distance multiset D and an additive error δ that are an instance of PD–ABSERROR. There will be a solution for this instance if and only if there is a solution for the 3-PARTITION instance. Parallel to the definition of D, we show already the "if" direction of the previous statement: To this end, we assume that the 3-PARTITION can be solved, i.e., there are n triples $T_1, \ldots, T_n$ of q_i's that each sum up to h. We show how to construct a point set P that is a solution for the PD–ABSERROR instance, i.e., P matches D up to additive error δ. The opposite direction ("only if") is shown in a second step. We want to stress at this point that although the definition of D and the construction of P are presented simultaneously, the definition of D itself does *not* rely on the fact that there exists a solution for the 3-PARTITION instance.

We assume that $\frac{h}{12}$ is integer. Otherwise, we can achieve this by simply multiplying all values q_i and h by 12. Moreover, we assume w.l.o.g. that the values $q_1, \ldots, q_{3n}$ are ordered such that the three q_i's that belong to the same triple T_j in a solution are adjacent, i.e., $T_1 = (q_1, q_2, q_3), T_2 = (q_4, q_5, q_6)$, and so on. Finally, we assume that the elements in each T_i are sorted in ascending order, i.e., $q_1 \leq q_2 \leq q_3, q_4 \leq q_5 \leq q_6$, and so on. This ordering allows us to derive a set of inequalities for the $q_i's$. Let $(q_{3k+1}, q_{3k+2}, q_{3k+3})$ be a triple that sums up to h, for $0 \leq k \leq n-1$. Then $q_{3k+1} \leq \frac{h}{3}$, since q_{3k+1} is the smallest of the three elements in the triple, and not all of them can be greater than $\frac{h}{3}$. Similarly, $\frac{h}{3} \leq q_{3k+3}$. With $q_{3k+1} + q_{3k+2} = h - q_{3k+3}$, we have $q_{3k+1} + q_{3k+2} \leq h - \frac{h}{3} = \frac{2h}{3}$. In combination with the restriction $\frac{h}{4} < q_i < \frac{h}{2}$ (from the definition of 3-PARTITION) and $H := \frac{h}{12}$, this yields the following inequalities:

$$
\begin{aligned}
3H &< q_{3k+1} & &\leq 4H \\
3H &< q_{3k+2} & &< 6H \\
4H &\leq q_{3k+3} & &< 6H \\
6H &< q_{3k+1} + q_{3k+2} & &\leq 8H \\
8H &\leq q_{3k+2} + q_{3k+3} & &< 12H \\
12H &= q_{3k+1} + q_{3k+2} + q_{3k+3}
\end{aligned}
\tag{1}
$$

We will use these inequalities later to derive upper and lower bounds for the additive error that we need to apply to our distances in order to guarantee the existence of a solution for the PD–ABSERROR instance.

Before we define our distances, we need to introduce the *level* of a distance: For a point set P, we say that a distance d between two points has *level* ℓ if it spans $\ell - 1$ further points, and we say that distance d is an *atom* if it has level 1. E.g. in Figure 1, distance 5 is an atom, and distance 16 has level 3.

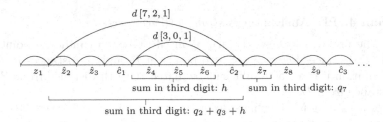

Fig. 2. Atoms and distances in multiset D.

In the following, we will use a vector representation for large numbers that will allow to add up the numbers digit by digit. The numbers are expressed in the number system of some base Z. We denote by $\langle a_1, \ldots, a_n \rangle$ the number $\sum_{i=1}^{n} a_i Z^{n-i}$; we say that a_i is the i-th digit of this number. In our proofs, we will choose base Z large enough such that the additions that we will perform do not lead to carry–overs from one digit to the next. Hence, we can add numbers digit by digit. The same holds for scalar multiplications. For example, having base $Z = 29$ and numbers $\alpha = \langle 3, 5, 1 \rangle$ and $\beta = \langle 2, 1, 0 \rangle$, then $\alpha + \beta = \langle 5, 6, 1 \rangle$ and $3 \cdot \alpha = \langle 9, 15, 3 \rangle$.

We now define our instance of PD–ABSERROR and show at the same time how to construct a solution for this instance. Let $c = n^2 \cdot h^2$. Moreover, define error $\delta := 3H$. The distances are expressed as numbers with base $Z = 10nc$, and each distance consists of three digits. The first digit will denote the *level* of a distance (the meaning of the other two digits will become clear soon).

First we define $4n-1$ distances that will turn out to be atoms in our solution: $z_i = \langle 1, 0, q_i \rangle - \delta$, for $1 \leq i \leq 3n$, and $c_i = \langle 1, c, 0 \rangle - \delta$, for $1 \leq i \leq n-1$. Observe that operation "$-\delta$" only affects the last digit (and in fact, we could have defined z_i by $\langle 1, 0, q_i - \delta \rangle$ instead), since we choose base Z sufficiently large.

Using these distances, we can already define a "solution" P for distance multiset D (although we are not yet finished defining D; in fact, we will construct D in the following such that it matches point set P up to additive error δ): Let $\hat{z}_i = z_i + \delta$ for $1 \leq i \leq 3n$, and $\hat{c}_i = c_i + \delta$ for $1 \leq i \leq n-1$. Observe that each $\hat{z}_i$ has exactly value q_i in its third digit. We call these values z–*pseudoatoms* or c–*pseudoatoms*, respectively, and use them to define a point set $P = \{p_1, \ldots, p_{4n}\}$ by specifying the pairwise distances between the points: Starting in 0, the points have distances $\hat{z}_1, \hat{z}_2, \hat{z}_3, \hat{c}_1, \hat{z}_4, \hat{z}_5, \hat{z}_6, \hat{c}_2, \ldots, \hat{c}_{n-1}, \hat{z}_{3n-2}, \hat{z}_{3n-1}, \hat{z}_{3n}$, i.e., we alternate blocks of three z–pseudoatoms and one c–pseudoatom, starting and ending with a block of three z–pseudoatoms (see Figure 2).

We now show level by level how the distances in D are defined, and that additive error δ (which is $3H$) is sufficient to make all distances from D match some distance between points in P.

By construction of P, the distances of level 1 are the pseudoatoms, and they match the corresponding z_i's and c_i's up to additive error δ.

To denote the distances of higher levels we use notation $d\,[\ell, j, k]$, for appropriate parameters ℓ, j and k. These names already indicate the values of the three digits of a distance: Distance $d[\ell, j, k]$ will have value ℓ in the first digit, which will be the level of the distance in our point set P. The second digit of the distance has value $j \cdot c$, which denotes that this distance will be used to span j c–pseudoatoms (and $\ell - j$ z–pseudoatoms) in our point set P. For instance, in Figure 2 distance $d[7, 2, 1]$ spans the two pseudoatoms $\hat{c}_1$ and $\hat{c}_2$ (and five $\hat{z}_i$'s). Finally, the third digit of distance $d[\ell, j, k]$ has value $k \cdot h$ plus some "small off-set", which will be a multiple of H. Here, k specifies how many *complete* blocks of three adjacent z–pseudoatoms the distance spans in P (recall that such a block corresponds to three q_i's that sum up to exactly h). In the following, we show how to choose these offsets in the third digit such that our point set P matches distance multiset D up to additive error δ.

First consider distances of level 2 in P, i.e., two points $p_i, p_{i+2} \in P$ with one point p_{i+1} in between. There are four possibilities for the two pseudoatoms between these two points, for some $0 \leq k \leq n - 1$: CASE 1: $\hat{z}_{3k+1}$ and $\hat{z}_{3k+2}$; CASE 2: $\hat{z}_{3k+2}$ and $\hat{z}_{3k+3}$; CASE 3: $\hat{z}_{3k+3}$ and $\hat{c}_k$; and CASE 4: $\hat{c}_k$ and $\hat{z}_{3k+1}$.

For the first case, the two pseudoatoms sum up to 2 in the first and to 0 in the second digit. For the third digit of the sum, recall that $\hat{z}_{3k+1}$ has value q_{3k+1} in its third digit, and $\hat{z}_{3k+2}$ has value q_{3k+2} in its third digit. Hence, inequalities (1) yield that the third digit of $\hat{z}_{3k+1} + \hat{z}_{3k+2}$ is bounded below by $6H$ and bounded above by $8H$. We define a distance $d[2, 0, 0] := \langle 2, 0, 9H \rangle$. Obviously, we can span the two pseudoatoms by this distance if we apply at most error δ (recall that $\delta = 3H$). Observe that we could have chosen other values for the third digit of $d[2, 0, 0]$, namely any value between $5H$ and $9H$ (which still allows to match the bounds using additive error δ). Here, we chose value $9H$, since we will use that same distance to cover the two pseudoatoms in Case 2 as well (see below).

Case 1 occurs exactly n times in our point set P, once for each block of three z–pseudoatoms. Hence, we let distance $d[2, 0, 0]$ be n times in our distance multiset D.

Case 2 is similar to Case 1: The third digit of $\hat{z}_{3k+2} + \hat{z}_{3k+3}$ is bounded below by $8H$ and bounded above by $12H$, using again inequalities (1). Like before, this case occurs n times, and we can use n *additional* distances $d[2, 0, 0]$ in D to span such two pseudoatoms up to error δ. Thus, in total we have $2n$ distances $d[2, 0, 0]$ in D that arise from the first two cases.

For the remaining two cases of two pseudoatoms, the last digit of the two pseudoatoms is at least $4H$ and at most $6H$ in Case 3, and at least $3H$ and at most $4H$ in Case 4. Moreover, in both cases the first digit of the sum is 2 and the second digit is c, and both cases occur exactly $n - 1$ times. Hence, we can define distance $d[2, 1, 0] := \langle 2, c, 4H \rangle$ and enclose it $2(n-1)$ times in D, in order to cover these pairs of pseudoatoms, again up to additive error δ.

Before we specify the distances of higher level, we introduce a graphical representation of pseudoatoms: Each z–pseudoatom is represented by a $\bullet$, and each c–pseudoatom by a $|$. This allows us to depict sequences of pseudoatoms without referring to their exact names. E.g. pseudoatoms $\hat{z}_3 \hat{c}_1 \hat{z}_4 \hat{z}_5 \hat{z}_6 \hat{c}_2$ yield $\bullet | \bullet \bullet \bullet |$, and the four cases of two adjacent pseudoatoms above can be represented

level ℓ	pseudoatoms	multiplicity	lower bound	upper bound	distance name	distance value
2	••	n	$6H$	$8H$	$d\,[2,0,0]$	$\langle 2,0,9H \rangle$
	••	n	$8H$	$12H$	$d\,[2,0,0]$	
	•|	$n-1$	$4H$	$6H$	$d\,[2,1,0]$	$\langle 2,c,4H \rangle$
	|•	$n-1$	$3H$	$4H$	$d\,[2,1,0]$	
3	•••	n	$12H$	$12H$	$d\,[3,0,1]$	$\langle 3,0,12H \rangle + \delta$
	|••	$n-1$	$6H$	$8H$	$d\,[3,1,0]$	$\langle 3,c,9H \rangle$
	•|•	$n-1$	$7H$	$10H$	$d\,[3,1,0]$	
	••|	$n-1$	$8H$	$12H$	$d\,[3,1,0]$	
4	••|•	$n-1$	$11H$	$16H$	$d\,[4,1,0]$	$\langle 4,c,13H \rangle$
	•|••	$n-1$	$10H$	$14H$	$d\,[4,1,0]$	
	•••|	$n-1$	$12H$	$12H$	$d\,[4,1,1]$	$\langle 4,c,12H \rangle$
	|•••	$n-1$	$12H$	$12H$	$d\,[4,1,1]$	
5	••|••	$n-1$	$14H$	$20H$	$d\,[5,1,0]$	$\langle 5,c,17H \rangle$
	•••|•	$n-1$	$15H$	$16H$	$d\,[5,1,1]$	$\langle 5,c,16H \rangle$
	•|•••	$n-1$	$16H$	$18H$	$d\,[5,1,1]$	
	|•••|	$n-2$	$12H$	$12H$	$d\,[5,2,1]$	$\langle 5,2c,12H \rangle$
6	•••|••	$n-1$	$18H$	$20H$	$d\,[6,1,1]$	$\langle 6,c,21H \rangle$
	••|•••	$n-1$	$20H$	$24H$	$d\,[6,1,1]$	
	•|•••|	$n-2$	$16H$	$18H$	$d\,[6,2,1]$	$\langle 6,2c,16H \rangle$
	|•••|•	$n-2$	$15H$	$16H$	$d\,[6,2,1]$	
7	•••|•••	$n-1$	$24H$	$24H$	$d\,[7,1,2]$	$\langle 7,c,24H \rangle$
	••|•••|	$n-2$	$20H$	$24H$	$d\,[7,2,1]$	$\langle 7,2c,21H \rangle$
	•|•••|•	$n-2$	$19H$	$22H$	$d\,[7,2,1]$	
	|•••|••	$n-2$	$18H$	$20H$	$d\,[7,2,1]$	

Fig. 3. Distances up to level 7.

by ••, ••, •| and |•. Figure 3 shows the distances, bounds, and multiplicities for level 2 to 7.

Observe that $d\,[2,0,0]$ and $d\,[6,1,1]$ are in a sense "equivalent", since they are used for cases that differ only in one complete block of three z–pseudoatoms and one c–pseudoatom. Hence, we could have written $d\,[6,1,1] = d\,[2,0,0] + \langle 4,c,h \rangle$ instead. Moreover, $d\,[6,2,1] = d\,[2,1,0] + \langle 4,c,h \rangle$ and $d\,[7,2,1] = d\,[3,1,0] + \langle 4,c,h \rangle$. Similarly, distances of level greater than 7 can be decomposed into a distance of low level (4 to 7) and an appropriate number of blocks of three z–pseudoatoms and one c–pseudoatom. We set $\beta := \langle 4,c,h \rangle$ and define in Figure 4 the distances of level 8 to $4n - 5$. In the table, the number of blocks k varies from 1 to $n-3$. Finally, in Figure 5 the distances that have level $4n-4$ to $4n-1$ are shown. Observe that as before they are derived from distances of level 4 to 7, for $k = n-2$. However, not all combinations are necessary for these distances.

Our distance multiset D consists of all atoms z_i and c_i, and all distances specified in Figures 3, 4 and 5, with the corresponding multiplicities. There are $4n - 1$ levels, and for each level ℓ there are $4n - \ell$ distances in D. In total, this yields $\sum_{\ell=1}^{4n-1}(4n - \ell) = \binom{4n}{2}$ distances. The cardinality of D is polynomially bounded in n, and each distance in D is polynomial in h. Hence, multiset D can be constructed in polynomial time from a given instance of 3-PARTITION.

In parallel to the definition of D, we have shown already that a solution for the 3-PARTITION instance yields a solution for the PD–ABSERROR instance. In the following, we show the opposite direction, i.e., we show that a solution for the PD–ABSERROR instance yields a solution for the 3-PARTITION instance. Let $R = \{r_1, \ldots, r_{4n}\}$ be *any* set of $4n$ points on a line that is a solution for the PD–ABSERROR instance, i.e., multiset D is the multiset of pairwise distances

level ℓ	pseudoatoms	multiplicity	distance name	distance value
$4k+4$	●●|...|●	$n-k-1$	$d\,[4+4k,1+k,0+k]$	$d\,[4,1,0]+k\cdot\beta$
	●|...|●●	$n-k-1$	$d\,[4+4k,1+k,0+k]$	
	●●●|...|	$n-k-1$	$d\,[4+4k,1+k,1+k]$	$d\,[4,1,1]+k\cdot\beta$
	|...|●●●	$n-k-1$	$d\,[4+4k,1+k,1+k]$	
$5+4k$	●●|...|●●	$n-k-1$	$d\,[5+4k,1+k,0+k]$	$d\,[5,1,0]+k\cdot\beta$
	●●●|...|●	$n-k-1$	$d\,[5+4k,1+k,1+k]$	$d\,[5,1,1]+k\cdot\beta$
	●|...|●●●	$n-k-1$	$d\,[5+4k,1+k,1+k]$	
	|...|●●●|	$n-k-2$	$d\,[5+4k,2+k,1+k]$	$d\,[5,2,1]+k\cdot\beta$
$6+4k$	●●●|...|●●	$n-k-1$	$d\,[6+4k,1+k,1+k]$	$d\,[6,1,1]+k\cdot\beta$
	●●|...|●●●	$n-k-1$	$d\,[6+4k,1+k,1+k]$	
	●|...|●●●|	$n-k-2$	$d\,[6+4k,2+k,1+k]$	$d\,[6,2,1]+k\cdot\beta$
	|...|●●●|●	$n-k-2$	$d\,[6+4k,2+k,1+k]$	
$7+4k$	●●●|...|●●●	$n-k-1$	$d\,[7+4k,1+k,2+k]$	$d\,[7,1,2]+k\cdot\beta$
	●●|...|●●●|	$n-k-2$	$d\,[7+4k,2+k,1+k]$	$d\,[7,2,1]+k\cdot\beta$
	●|...|●●●|●	$n-k-2$	$d\,[7+4k,2+k,1+k]$	
	|...|●●●|●●	$n-k-2$	$d\,[7+4k,2+k,1+k]$	

Fig. 4. Distances with level 8 to $4n-5$ (with $\beta=\langle 4,c,h\rangle$). Value k varies between 1 and $n-3$.

of R, up to additive error δ for each distance. We assume w.l.o.g. that the points are ordered from left to right, i.e., $r_1 < r_2 < \ldots < r_{4n}$. We will show that R is basically identical to P, the point set that we constructed above.

Obviously, additive error δ can affect only the last digit of each distance, since base Z is sufficiently large. Thus, exactly those distances with value 1 in the first digit are atoms, since all other distances have value greater than 1 in the first digit, and since there must be exactly $4n-1$ atoms. This implies immediately that the first digit of each distance denotes the level of the distance in any solution.

We now show that error $+\delta$ has to be applied to each single atom to make it fit to the distances between adjacent points in R. To see this, first observe that the atoms sum up to $\sum_{i=1}^{3n} z_i + \sum_{i=1}^{n-1} c_i = \langle 4n-1, (n-1)c, nh \rangle - (4n-1)\delta$. On the other hand, $d\,[4n-1,n-1,n] = \langle 4n-1, (n-1)c, nh \rangle + \delta$ is the largest distance in D. Each atom is the distance between two adjacent points in R, up to additive error δ, while $d\,[4n-1,n-1,n]$ is the distance between the first and the last point in R, again up to additive error δ. Hence, the atoms must sum up to the length of the largest distance. This is only possible if we apply error $+\delta$ to each atom, yielding sum $\langle 4n-1, (n-1)c, nh \rangle$, and if we apply error $-\delta$ to the largest distance, yielding $\langle 4n-1, (n-1)c, nh \rangle$ as well. Knowing this, we can again define *pseudoatoms* $\hat{z}_i = z_i + \delta$ and $\hat{c}_i = c_i + \delta$, which represent exactly the distances of adjacent points in R (without error). Observe that if we represented the distances between adjacent points in R in our number representation, then pseudoatom $\hat{z}_i$ would have exactly value q_i in its last digit, for all $1 \le i \le 3n$.

We now show that the ordering of the pseudoatoms arising from R is such that there are n blocks of three pseudoatoms $\hat{z}_i$, and each two blocks are separated by one pseudoatom $\hat{c}_i$. Between any two adjacent c–pseudoatoms there must be exactly three z–pseudoatoms: Since there are no distances of level 4 with value $2c$ in the second digit, no combination $||$ or $|\bullet|$ or $|\bullet\bullet|$ is possible, and there are at least three z–pseudoatoms in between two c–pseudoatoms; moreover, since there

level ℓ	lower bound	upper bound	distance name	distance value
$4n-4$	$(n-2)h+11H$	$(n-2)h+16H$	$d[4n-4,n-1,n-2]$	$d[4,1,0]+(n-2)\cdot\beta$
	$(n-2)h+10H$	$(n-2)h+14H$	$d[4n-4,n-1,n-2]$	
	$(n-1)h$	$(n-1)h$	$d[4n-4,n-1,n-1]$	$d[4,1,1]+(n-2)\cdot\beta$
	$(n-1)h$	$(n-1)h$	$d[4n-4,n-1,n-1]$	
$4n-3$	$(n-1)h+3H$	$(n-1)h+4H$	$d[4n-3,n-1,n-1]$	$d[5,1,1]+(n-2)\cdot\beta$
	$(n-1)h+4H$	$(n-1)h+6H$	$d[4n-3,n-1,n-1]$	
	$(n-2)h+14H$	$(n-2)h+20H$	$d[4n-3,n-1,n-2]$	$d[5,1,0]+(n-2)\cdot\beta$
$4n-2$	$(n-1)h+6H$	$(n-1)h+8H$	$d[4n-2,n-1,n-1]$	$d[6,1,1]+(n-2)\cdot\beta$
	$(n-1)h+8H$	$(n-1)h+12H$	$d[4n-2,n-1,n-1]$	
$4n-1$	nh	nh	$d[4n-1,n-1,n]$	$\langle 4n-1,(n-1)c,nh\rangle+\delta$

Fig. 5. Distances with level $4n-4$ to $4n-1$. Each case occurs once.

are $n-2$ distances of level 5 with value $2c$ in the second digit, there must be at least $n-1$ c–pseudoatoms such that there are always at most 3 z–pseudoatoms in between. Hence, the points in R are such that blocks of three z–pseudoatoms alternate with one c–pseudoatom, starting and ending with a block of three z–pseudoatoms.

Finally, we show that the third digits of each three adjacent z–pseudoatoms sum up to h: Consider those distances of level 3 that have a zero in the second digit. There are n such distances, and their third digits sum up to $nh+n\delta$. Each of these distances must span exactly one of the n blocks of three z–pseudoatoms. The total sum of the last digit of all z–pseudoatoms is exactly $\sum_{i=1}^{3n} q_i = nh$. Since the distances of level 3 that span these blocks do not overlap, they have to sum up to the same total. Hence, the error for each such distance of level 3 must be $-\delta$. This implies that each three q_i's that correspond to one block sum up to exactly h (since we have applied error $+\delta$ to each atom to define the z–pseudoatoms). Thus, these triples yield a solution for the 3-PARTITION instance. □

3 Strong NP-Completeness of PD–RELERROR

In this section, we show that PD–RELERROR is strongly NP-complete by using a reduction from 3-PARTITION similar to the one used to prove strong NP-completeness of PD–ABSERROR (see Theorem 4).

Theorem 5. PD–RELERROR *is strongly* NP-*complete, even if the error is a constant.*

Proof (sketch). The problem is in NP analogously to the proof of Theorem 4. The proof of NP-hardness is also along the lines of the proof of Theorem 4. In fact, the proof has a similar structure overall, but the details are quite different. Given an instance of 3-PARTITION, we define a multiset E of distances which are expressed as numbers with a base Z, with $Z = 10hnc$ and $c = n^2h^2$.

We replace the definition of the atoms as follows: $z_i = \langle 1,0,q_i\rangle \cdot \frac{1}{1+\varepsilon}$, for $1 \le i \le 3n$, and $c_i = \langle 1,c,0\rangle \cdot \frac{1}{1+\varepsilon}$, for $1 \le i \le n-1$. All z_i's and c_i's are part of the distance set E. Note that for a fixed level ℓ, the corresponding distances

$d\,[\ell,\cdot,\cdot]$ from the proof of Theorem 4 are defined for at most two consecutive values of the second digit, say j and $j+1$. Here, we define distances $e\,[\ell,j]$ and $e\,[\ell,j+1]$ for all levels $2 \le \ell \le 4n-1$ and corresponding j or $j+1$, respectively, as follows: $e\,[\ell,j] = \langle \ell, j, B_u(\ell,j)\rangle \cdot \frac{1}{1+\varepsilon}$, and $e\,[\ell,j+1] = \langle \ell, j+1, B_l(\ell,j+1)\rangle \cdot \frac{1}{1-\varepsilon}$, using values $B_u()$ and $B_l()$ as specified below.

The first digit ℓ still indicates the level of the distance (i.e., how many atoms it will span in a solution) and the second digit j or $j+1$ indicates the number of c-atoms it will span. Value $B_u(\ell,j)$ is the maximum upper bound from the corresponding column in Figure 3, Figure 4, or Figure 5, taken over all distances $d\,[\ell,j,\cdot]$ (for Figure 4, these bounds result from Figure 3 by adding appropriate multiples of h); similarly, value $B_l(\ell,j+1)$ is the minimum lower bound from the corresponding column in the figures, taken over all distances $d\,[\ell,j+1,\cdot]$. The multiplicity of distance $e\,[\ell,j]$ is the sum of the multiplicities for all distance values $d\,[\ell,j,\cdot]$ taken from the same figures, likewise for distance $e\,[\ell,j+1]$. Thus, for example $e\,[5,1] = \langle 5,1,20H\rangle \cdot \frac{1}{1+\varepsilon}$ with multiplicity $3(n-1)$, while $e\,[6,2] = \langle 6,2,15H\rangle \cdot \frac{1}{1-\varepsilon}$ with multiplicity $2(n-2)$.

For $d\,[\cdot]$-distances with levels divisible by four (i.e., distances $d\,[4\ell',j,\cdot]$ with integer $\ell' < n$), we only have one possible value j for the second digit. Thus, we define the corresponding $e\,[\cdot]$-distances by $e\,[4\ell',j] = \langle 4\ell',j,B_u(4\ell',j)\rangle \cdot \frac{1}{1-\varepsilon}$. Finally, we define two special distances: $e\,[3,0] = \langle 3,0,h\rangle \cdot \frac{1}{1+\varepsilon}$, with multiplicity n, and $e\,[4n-1,n-1] = \langle 4n-1,(n-1)c,nh\rangle \cdot \frac{1}{1-\varepsilon}$ with multiplicity 1.

All the distances, including the atoms, are put into distance multiset E. We set error $\varepsilon = \frac{1}{100}$. This completes our description of how to construct a PD–RELERROR instance from a given 3-PARTITION instance. The proof that a solution for the 3-PARTITION instance yields a solution for the PD–RELERROR instance, and vice versa, as well as the strategy to transform these distances into integer distances, can be found in the full version of this paper.   □

4   Conclusion

We have shown that PARTIAL DIGEST is NP-complete if all measurements are prone to the same additive or multiplicative error. This answers the question whether PARTIAL DIGEST on real-life data can be solved in polynomial time. However, it also gives rise to new questions: While we have shown NP-hardness for even constant relative error, our proof for absolute error uses error $\frac{h}{4}$, which is not constant. Is PARTIAL DIGEST still NP-complete if we restrict the additive error to some (small) constant? What if we allow only one-sided errors (i.e., if the lengths of the distances are always underestimated)? Moreover, the main open problem is still the computational complexity of PARTIAL DIGEST itself.

Acknowledgments. We would like to thank Claudio Gutiérrez, Fabian Hennecke, Roland Ulber, Birgitta Weber, and Peter Widmayer for helpful discussions, and Riko Jacob, who suggested the graphical presentation used in Section 2.

References

1. L. Allison and C. N. Yee. Restriction site mapping is in separation theory. *Computer Applications in the Biosciences*, 4(1):97–101, 1988.
2. J. Błażewicz, P. Formanowicz, M. Kasprzak, M. Jaroszewski, and W. T. Markiewicz. Construction of DNA restriction maps based on a simplified experiment. *Bioinformatics*, 17(5):398–404, 2001.
3. M. Cieliebak, S. Eidenbenz, and P. Penna. Noisy data make the partial digest problem NP-hard. In *Proc. of the 3rd Workshop on Algorithms in Bioinformatics (WABI 2003)*, pages 111–123, 2003.
4. T Dakić. *On the Turnpike Problem*. PhD thesis, Simon Fraser University, 2000.
5. T. I. Dix and D. H. Kieronska. Errors between sites in restriction site mapping. *Computer Applications in the Biosciences*, 4(1):117–123, 1988.
6. J. Fütterer. Personal communication, 2002. ETH Zurich, Institute of Plant Sciences.
7. M. R. Garey and D. S. Johnson. *Computers and Intractability: A Guide to the Theory of NP-Completeness*. Freeman, 1979.
8. J. Inglehart and P. C. Nelson. On the limitations of automated restriction mapping. *Computer Applications in the Biosciences*, 10(3):249–261, 1994.
9. P. Lemke, S. S. Skiena, and W. Smith. Reconstructing sets from interpoint distances. Technical Report TR2002-37, DIMACS, 2002.
10. P. Lemke and M. Werman. On the complexity of inverting the autocorrelation function of a finite integer sequence, and the problem of locating n points on a line, given the $\binom{n}{2}$ unlabelled distances between them. Preprint 453, Institute for Mathematics and its Application IMA, 1988.
11. G. Pandurangan and H. Ramesh. The restriction mapping problem revisited. *Journal of Computer and System Sciences*, 65(3):526–544, 2002. Special issue on Computational Biology.
12. P. A. Pevzner. *Computational Molecular Biology: An Algorithmic Approach*. MIT Press, 2000.
13. J. Rosenblatt and P. Seymour. The structure of homometric sets. *SIAM Journal of Algorithms and Discrete Mathematics*, 3(3):343–350, 1982.
14. D. B. Searls. Formal grammars for intermolecular structure. In *Proc. of the 1st International Symposium on Intelligence in Neural and Biological Systems (INBS'95)*, pages 30–37, 1995.
15. J. Setubal and J. Meidanis. *Introduction to Computational Molecular Biology*. PWS Boston, 1997.
16. S. S. Skiena, W. Smith, and P. Lemke. Reconstructing sets from interpoint distances. In *Proc. of the 6th ACM Symposium on Computational Geometry (SoCG 1990)*, pages 332–339, 1990.
17. S. S. Skiena and G. Sundaram. A partial digest approach to restriction site mapping. *Bulletin of Mathematical Biology*, 56:275–294, 1994.
18. P. Tuffery, P. Dessen, C. Mugnier, and S. Hazout. Restriction map construction using a 'complete sentence compatibility' algorithm. *Computer Applications in the Biosciences*, 4(1):103–110, 1988.
19. M. S. Waterman. *Introduction to Computational Biology*. Chapman & Hall, 1995.
20. Z. Zhang. An exponential example for a partial digest mapping algorithm. *Journal of Computational Biology*, 1(3):235–239, 1994.

Designing Small Keyboards Is Hard

Jean Cardinal and Stefan Langerman*

Université Libre de Bruxelles
Computer Science Department CP212
B-1050 Brussels Belgium
{jcardin,Stefan.Langerman}@ulb.ac.be

Abstract. We study the problem of placing symbols of an alphabet onto the minimum number of keys on a small keyboard so that any word of a given dictionary can be recognized univoquely only by looking at the corresponding sequence of pressed keys. This problem is motivated by the design of small keyboards for mobile devices. We show that the problem is hard in general, and NP-complete even if we only wish to decide whether two keys are sufficient. We also consider two variants of the problem. In the first one, symbols on a same key must be contiguous in an ordered alphabet. The second variant is a fixed-parameter version of the previous one that minimizes a well-chosen measure of ambiguity in the recognition of the words for a given number of keys. Hardness and approximability results are given.

1 Introduction

Keyboards are by far the most commonly used interfaces for entering textual or numerical data on many communication devices. When this device is small, a complete keyboard is not always available: the situation typically occurs for mobile phones. The solution used in that case is the overloading of keys: each key is associated to more than one symbol of the alphabet. The current standard layout for mobile phone is defined by a 1994 ISO specification (cf. Fig. 1 and [1]). Numerous methods allow the user to specify which symbol is needed among the one corresponding to the pressed key. The multi-tap method is a widely proposed one: the desired symbol is selected by pressing more than once the same key. Other methods use an algorithm that tries to predict the input at a first-order level according to the sequence of pressed keys and using a dictionary of words. A common implementation of such an algorithm that uses maximum probability estimation is the T9 algorithm [4]. A survey of text entry and disambiguation procedures for mobile phones can be found in a recent paper from MacKenzie and Soukoreff [9]. Recently, many authors considered the problem of estimating the achievable word rate using various methods (see e.g. [3]). While many researches related to text entry on mobile devices are conducted in the computer-human interface community, it seems that not many of them treat the problem of redefining the actual keyboard layout used. In this paper we consider

---

* Chargé de Recherches FNRS

M. Farach-Colton (Ed.): LATIN 2004, LNCS 2976, pp. 391–400, 2004.
© Springer-Verlag Berlin Heidelberg 2004

Fig. 1. Usual mobile keypad as recommended by the ISO standard [1]

the problem of defining keyboard layouts with key overloading using an optimal partition of an alphabet Σ, in the sense that the user can type any word of a dictionary D, and that word is always recognized without ambiguity, or a certain measure of ambiguity is minimized. This is, to our knowledge, the first theoretical analysis of this problem.

A similar issue has nevertheless been investigated by Lesher et al. in [8]. They study the problem of arranging characters on a small keyboard with key overloading so that the keystroke efficiency is maximized. A heuristic local optimization algorithm is proposed, based on iterative permutation of a fixed number of characters. The objective function, however, is computed based on the assumption of a character-level disambiguation procedure, and without any reference to a dictionary. Only very superficial considerations on the complexity and approximability of the problem are given.

In section 2 we give a formal definition of the problem and prove that it is NP-hard in general, not approximable within $|\Sigma|^{1/5-\epsilon}$ (unless NP=coRP), and remains complex even if we restrict it to two keys. In section 3 we consider a variant in which letters on the same key must be contiguous in an ordered alphabet. We prove that this variant is NP-hard as well, but admits a $(1+2\ln|D|)$ factor approximation algorithm, which is the best possible within a constant factor. Ambiguous keyboards, in which a well-chosen measure of ambiguity is minimized, are considered in section 4. The ambiguity measure is related to the average number of keys that have to be pressed to resolve an ambiguity. It is useful in practice and can allow for a nonuniform probability distribution over D. We show a constant factor approximation for this version of the problem. Finally, in section 5, we describe a linear-time algorithm for measuring the ambiguity and exhibit optimal ambiguous keyboards for an english dictionary. The optimal layout we found for eight keys is interestingly quite different from the standard one and requires on average less than half the number of keystrokes to resolve an ambiguity.

2   General Formulation

We first formalize the problem of designing a keyboard with key overloading that allows unambiguous recognition of any word in a given dictionary.

Definition 1 (KEYBOARD) *An instance of* KEYBOARD *is composed of an alphabet Σ and a dictionary $D \subset \Sigma^*$. A solution of this instance is a partition of Σ such that for any pair $x, y \in D$ with $x = (x_1, x_2, \ldots, x_{|x|})$ and $y = (y_1, y_2, \ldots, y_{|y|})$ either $|x| \neq |y|$ or there exists an index i such that x_i and y_i are in different subsets of the partition. The objective function to minimize is the size of the partition.*

Using a coloring terminology, this problem can be seen as a minimal coloring of the symbols of an alphabet such that any word of a given dictionary can be recognized univoquely only by looking at the corresponding sequence of colors.

Example 1 *Let $\Sigma = \{a, b, c, d\}$ and $D = \{abcd, dabb, bbcc, addb\}$. The partition of Σ in the two subsets $\{a, b, c\}$ and $\{d\}$ is an optimal solution of this instance of* KEYBOARD. *If we replace each occurence of a symbol in Σ by '1' if it belongs to the first subset, and by '2' if it belongs to the second, we obtain the following set: $\{1112, 2111, 1111, 1221\}$, with four distinct words.*

The following definition is useful in the NP-hardness proof for KEYBOARD.

Definition 2 (GRAPH-COLORING) *An instance of* GRAPH-COLORING *is composed of a graph (V, E). A solution is a partition of the set V of vertices such that any two adjacent vertices are in different subsets. The objective function to minimize is the size of the partition.*

Theorem 1 KEYBOARD *is NP-hard.*

Proof. By reduction of GRAPH-COLORING, as follows. Let Σ be defined as V. Select an edge pq, and to each edge of the graph associate a unique word made of the two symbols p and q, of size $l = \lceil \log_2 |E| \rceil$. For each edge $ab \in E$, let w_{ab} be this word. D is composed of words of equal lengths $l + 1$ of the form $w_{ab}a, w_{ab}b$ for each edge ab. The word pair corresponding to the edge pq is $\{p^{l+1}, p^l q\}$, hence $w_{pq} = p^l$. From this, p and q must be in different subsets, hence the words w_e and $w_{e'}$ for distinct edges e and e' are always distinguishable. On the other hand when a and b are adjacent, $w_{ab}a$ and $w_{ab}b$ belong to D and therefore a and b must be in different subsets. In this reduction, $\Sigma = V$, $|D| = 2|E|$, and D is composed of $2|E|$ words of size $l + 1$.   □

Example 2 *Suppose we want to color the graph on Fig. 2. We encode this instance by setting: $\Sigma = \{p, q, r, s, t, u\}$ and*

$$D = \{pppp, pppq, ppqr, ppqs, pqps, pqpq, pqqs, pqqt,$$
$$qpps, qppu, qpqr, qpqt, qqpr, qqpu, qqqt, qqqu\}.$$

Each edge ab on the graph of Fig. 2(a) is labeled by the word w_{ab}.

Although many other reductions are possible, we believe this one is interesting because it combines two useful properties. First, the size of the alphabet is equal to $|V|$. This means that nonapproximability results for GRAPH-COLORING can be transposed to KEYBOARD. In particular, a recent contribution from Bellare et al. [2] implies the following (assuming NP$\neq$coRP).

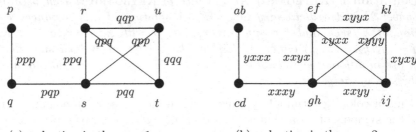

(a) reduction in theorem 1   (b) reduction in theorem 2

Fig. 2. Example graph

Corollary 1 KEYBOARD *is not approximable within* $|\Sigma|^{\frac{1}{5}-\epsilon}$, $\forall \epsilon > 0$.

As a second property, the result also holds in the case where words in D are constrained to have the same size. In general, results presented in this paper are also valid for the case where the words are constrained to have the same size.

This NP-hardness result does not tell us whether testing the existence of a partition of size two is NP-complete, since testing the two-colorability of a graph is a polynomial problem. We provide another reduction using the decision version of GRAPH-COLORING.

Theorem 2 *Asking for the existence of a feasible solution of a given size K in an instance of* KEYBOARD *is NP-complete for any $K \geq 2$.*

Proof. Let us prove this for $K = 2$. We use a reduction of the problem of testing the existence of a 2^M-coloring of a graph, for any $M > 1$. This reduction has the same flavor as the previous one. We use two symbols x and y to define the prefixes of size $l = \lceil \log_2(|E|+1) \rceil$ identifying edges of the graph. The two words x^{l+M} and $x^l y^M$ of size $l + M$ are first included in D. Hence the first prefix x^l is only used to make x and y distinguishable. Then we associate to each vertex $a \in V$ a word v_a of size M made of previously unused symbols. For each edge ab, we include the two words $w_{ab}v_a$ and $w_{ab}v_b$ in D, where w_{ab} is a prefix identifying edge ab, distinct from x^l. In this way, two-coloring symbols of a word v_a corresponds to assigning to vertex a a color in the range $\{0, 1, \ldots, 2^M - 1\}$. In this reduction, $|\Sigma| = 2 + M|V|$, $|D| = 2 + 2|E|$, and D is made of words of equal sizes $l + M$.   □

Example 3 *We consider the graph on Fig. 2 and encode the problem of testing whether this graph has a coloring of size 4 ($M = 2$). We define $\Sigma = \{x, y, a, b, c, d, e, f, g, h, i, j\}$ and*

$$D = \{xxxxxx, xxxxyy, xxxycd, xxxygh,$$

$$xxyxef, xxyxgh, xxyyij, xxyygh, xyxxij, xyxxef,$$

$$xyxyij, xyxykl, xyyxef, xyyxkl, xyyykl, xyyygh, yxxxab, yxxxcd\}$$

The reduction is illustrated on Fig. 2(b).

Again, we point out that the result holds even in the particular case when words have equal lengths.

3   Keyboards with Contiguous Symbols on Each Key

In the previous section, we assumed that symbols of the alphabet could be put anywhere on the keyboard. In other words, the partition of Σ is chosen among all possible partitions. We now consider a more realistic problem in which the alphabet is ordered, and keys of the keyboard are constrained to represent only contiguous alphabet symbols. We show that this constrained variant has very strong connections with the set cover problem.

Definition 3 (CONTIGUOUS-KEYBOARD) *An instance of* CONTIGUOUS-KEYBOARD *is composed of an ordered alphabet Σ and a dictionary $D \subset \Sigma^*$. A solution of this instance is a partition of Σ such that*

1. *each subset of the partition is composed of consecutive symbols of Σ,*
2. *for any pair $x, y \in D$ with $x = (x_1, x_2, \ldots, x_{|x|})$ and $y = (y_1, y_2, \ldots, y_{|y|})$ either $|x| \neq |y|$ or there exists an index i such that x_i and y_i are in different subsets of the partition.*

The objective function to minimize is the size of the partition.

We briefly recall the definition of the set cover problem.

Definition 4 (SET-COVER) *An instance of* SET-COVER *is composed of a ground set S and a set E of subsets of S. A solution is a subset of E that covers each element of S. The objective function to minimize is the size of this subset.*

Theorem 3 *Any instance of* CONTIGUOUS-KEYBOARD *can be encoded as an instance of* SET-COVER.

Proof. Let us first remark that finding a partition of Σ whose subsets are composed of contiguous elements amounts to selecting separators in $\{1, 2, \ldots, |\Sigma| - 1\}$. The partition is then defined as follows: for each selected separator i, all symbols of rank less or equal to i in Σ are in a different subset than those with rank higher than i.

To each separator i in $\{1, 2, \ldots, |\Sigma| - 1\}$, we associate the set

$$C_i = \{\{v, w\} \mid v, w \in D \wedge |v| = |w| \wedge \exists j \, (\mathrm{rank}(v_j) \leq i \wedge \mathrm{rank}(w_j) > i)\},$$

that is, the set of unordered word pairs of equal lengths that are made distinguishable by selecting the separator i. The optimization now consists in finding the minimal set of subsets in $E = \{C_1, C_2, \ldots, C_{|\Sigma|-1}\}$ such that all the unordered word pairs in $S = \{\{v, w\} \mid v, w \in D \wedge |v| = |w|\}$ are covered.   □

Corollary 2 CONTIGUOUS-KEYBOARD *is approximable within $1 + \ln |S| \leq 1 + 2\ln |D|$, where $S = \{\{v, w\} \mid v, w \in D \wedge |v| = |w|\}$.*

Proof. It is well known that SET-COVER is approximable within $1 + \ln |S|$ using the greedy covering algorithm [6,11]. The size of the partition in CONTIGUOUS-KEYBOARD is one more than the number of separators selected in the covering. If we denote by CK (resp. $\mathrm{CK_{OPT}}$) the approximate (resp. optimal) solution of CONTIGUOUS-KEYBOARD and by SC (resp. $\mathrm{SC_{OPT}}$) the approximate (resp. optimal) solution of SET-COVER, we have $\mathrm{SC} \leq (1 + \ln |S|)\mathrm{SC_{OPT}}$, hence

$$\mathrm{CK} - 1 \leq (1 + \ln |S|)(\mathrm{CK_{OPT}} - 1)$$
$$\mathrm{CK} \leq (1 + \ln |S|)\mathrm{CK_{OPT}} - 1 - \ln |S| + 1$$
$$\mathrm{CK} \leq (1 + \ln |S|)\mathrm{CK_{OPT}}.$$

$\square$

So far, it is still not clear whether CONTIGUOUS-KEYBOARD is NP-hard or not. We could imagine that some structure available in CONTIGUOUS-KEYBOARD could be used by a polynomial algorithm to solve it to optimality. The next theorem shows that this is not the case.

Theorem 4 *Any instance of* SET-COVER *can be encoded as an instance of* CONTIGUOUS-KEYBOARD.

Proof. We first remark that the only way to distinguish two consecutive symbols of ranks i and $i + 1$ is to select separator i. It is then possible to encode a SET-COVER problem in a CONTIGUOUS-KEYBOARD problem, by associating a pair of words in D to each element of S, and craft them carefully so that they are contained in only a certain number of subsets in $E = \{C_1, C_2, \ldots, C_{|\Sigma|-1}\}$. First, let $|\Sigma| = |E| + 1$. Let us consider an element x of S and construct a corresponding pair of words $\{v, w\}$ in D. For each i such that x is contained in C_i, we simply append the symbol of rank i to v and the symbol of rank $i + 1$ to w. We also need to always distinguish words of different pairs. To achieve this, we can make the words of different pairs having different lengths by concatenating them with different numbers of copies of themselves. We have $|D| = 2|S|$ and a polynomial reduction. $\square$

Corollary 3 CONTIGUOUS-KEYBOARD *is NP-hard and not approximable within* $c \log |D|$, *for some constant* $c > 0$.

The inapproximability result comes from [10].

Example 4 *Let* $S = \{1, 2, 3, 4\}$ *and* $E = \{\{1, 2\}, \{2, 3\}, \{1, 3, 4\}\}$. *We translate this* SET-COVER *problem into a* CONTIGUOUS-KEYBOARD *problem by letting* $\Sigma = (a, b, c, d)$ *and*

$$D = \{ac, bd, abab, bcbc, bdbdbd, ccccc, c, d\}.$$

In this example, the pair $\{ac, bd\}$ *represents element* $1 \in S$, *found in the first and third subsets. The word pair is therefore separated by the separator 1 between a and b and by the separator 3 between c and d. The distinction between words corresponding to different elements of S is ensured by the variation in length.*

A variant of this reduction in which the words of D are constrained to have the same length could use a system of prefixes, as in the two previous proofs.

4   Ambiguous Keyboards

When dealing with large dictionaries, it is likely that an optimal partition in both of the preceding problems would be quite large, and maybe even of the size of the alphabet itself. It is therefore interesting to consider the problem of an ambiguous keyboard, in which the number of keys is constrained to be at most K, and some well-defined measure of ambiguity between words is minimized.

Definition 5 (Ambiguity) *A partition of Σ defines a confusability relation between words in D:*

$$R = \{\{v, w\} \mid v, w \in D \wedge |v| = |w| \wedge \forall i : v_i \text{ and } w_i \text{ are in the same subsets}\}$$

R is an equivalence relation, hence it partitions D in a set C of equivalence classes. From this observation, we define

- *the number of ambiguous pairs $P = \sum_{c \in C} \binom{|c|}{2}$,*
- *the number of nonambiguous pairs $\bar{P} = |S| - P$.*
- *the ambiguity $A = P/|D|$.*
- *the nonambiguity $\bar{A} = \bar{P}/|D|$.*

The motivation for using this ambiguity measure is the use of a *selection system*. When a user types an ambiguous word, the selection system allows him to select the word he actually wishes to enter among the list of words in the same equivalence class. If the first word in the list is the correct one, no further key needs to be pressed. One click on the "scroll down" key allows him to select the second word. In general, $i - 1$ clicks are necessary for selecting the ith word in the list. Hence the average number of clicks for the selection of a word in an equivalence class c is $\sum_{i=1}^{|c|}(i - 1)/|c| = \binom{c}{2}/|c|$, and the overall average number of clicks needed per word is $A = P/|D|$. This naturally holds only under the assumption of uniform probability distribution of words in D.

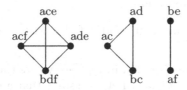

Fig. 3. Graph of a confusability relation between words

Example 5 *Fig. 3 shows the graph of a confusability relation between words obtained when partitioning the alphabet $\Sigma = \{a, b, c, d, e, f\}$ in subsets $\{a, b\}$, $\{c, d\}$ and $\{e, f\}$. We have:*

$$D = \{ace, acf, ade, bdf, ad, ac, bc, be, af\}$$
$$C = \{\{ace, acf, ade, bdf\}, \{ad, ac, bc\}, \{be, af\}\}$$
$$P = \binom{4}{2} + \binom{3}{2} + \binom{2}{2} = 10$$

It is easy to check that $A = P/|D| = 10/9$ is also the average number of clicks per word in the selection system.

For simplicity, we will concentrate on the fixed-parameter version of CONTIGUOUS-KEYBOARD only.

Definition 6 (K-CONTIGUOUS-KEYBOARD) *An instance of K-CONTIGUOUS-KEYBOARD is an instance of CONTIGUOUS-KEYBOARD enriched with an integer K. A solution of this instance is a partition of Σ of size K satisfying the constraints in CONTIGUOUS-KEYBOARD. The problem is parameterized by the (non)ambiguity measure that is to be minimized (maximized).*

To indicate which ambiguity measure is used, we append one of the symbol A or $\bar{A}$ in parentheses. Although the two problems have the same optimal solutions, an approximation algorithm for one problem is not necessarily an approximation algorithm for the other, which is why we distinguish the two.

We now show that this fixed-parameter version of CONTIGUOUS-KEYBOARD corresponds to the fixed-parameter version of SET-COVER.

Definition 7 (MAX-COVERAGE) *An instance of the MAX-COVERAGE problem is an instance of SET-COVER enriched with an integer K. A solution is a subset of E of size K that covers some element of S. The objective function to maximize is the number of elements covered.*

Theorem 5 *Any instance of K-CONTIGUOUS-KEYBOARD($\bar{A}$) can be encoded as an instance of MAX-COVERAGE and any instance of MAX-COVERAGE can be encoded as an instance of K-CONTIGUOUS-KEYBOARD($\bar{A}$).*

Proof. The proofs are the same as those of theorems 3 and 4. We just have to remark that the parameter is not the same: K-CONTIGUOUS-KEYBOARD($\bar{A}$) for a certain K reduces to a MAX-COVERAGE problem with parameter $K - 1$.   □

Corollary 4 (Approximation) K-CONTIGUOUS-KEYBOARD($\bar{A}$) *is approximable within a factor $1 - 1/e$.*

Proof. From the approximation yielded for MAX-COVERAGE by the greedy algorithm, proved in [5,7].   □

The developments above also hold in the case where the probability distribution of the words in D is not uniform. Let us assume that a probability p_v is assigned to each word v in D, with $\sum_{v \in D} p_v = 1$. The average number of clicks per word can be computed easily if we assume that the selection system presents the words in decreasing probability order in each equivalence class of C. We obtain the following generalized objective functions.

Definition 8 (Weighted ambiguity)

$$A = \sum_{c \in C} \sum_{v \in c} p_v \cdot rank_c(v) = \sum_{c \in C} \sum_{\{v,w\} \subseteq c} \min(p_v, p_w)$$

$$\bar{A} = \left(\sum_{v \in S} p_v \cdot rank_D(v) \right) - A$$

The function rank_c sorts the words in a set c: the most probable word has rank 0, the second most probable has rank 1, and so on. There is no need to normalize here, and A is the average number of clicks per word.

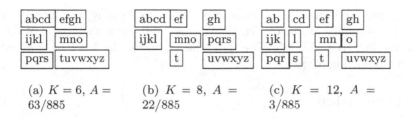

(a) $K = 6$, $A =$ 63/885

(b) $K = 8$, $A =$ 22/885

(c) $K = 12$, $A =$ 3/885

Fig. 4. Some optimal solutions of K-CONTIGUOUS-KEYBOARD with $\Sigma = \{a, b, \dots, z\}$ and a dictionary of 885 frequent words in english

Example 6 *Let us assume that the words ace, acf, ade and bdf are in the same equivalence class c of C, and that $p_{ace} > p_{acf} > p_{ade} > p_{bdf}$. The average number of clicks to select one of the words is $(p_{ace}\cdot 0 + p_{acf}\cdot 1 + p_{ade}\cdot 2 + p_{bdf}\cdot 3)/(\sum_{v\in c} p_v) = (\sum_{\{v,w\}\subseteq c} \min(p_v, p_w))/(\sum_{v\in c} p_v)$.*

By assigning the weight $\min(p_v, p_w)$ to each edge $\{v, w\} \in S$, we can see that the weighted version of K-CONTIGUOUS-KEYBOARD reduces to a weighted maximum coverage problem, hence the corresponding variant K-CONTIGUOUS-KEYBOARD($\bar{A}$) remains approximable within $1 - 1/e$, as in the unweighted case [5,7].

5 Examples of Optimized Keyboards

We now present some examples of optimal keyboards for the latin alphabet and a dictionary of 885 frequent english words. This file was obtained from the Letter-by-Letter Word Games FAQ website. It has been filtered by elimination of uppercase letters.

We concentrate on keyboards with contiguous symbols on each key, more precisely on optimal solutions of K-CONTIGUOUS-KEYBOARD(A), i.e. keyboards that minimize the number of ambiguous word pairs. Exhaustive searching is affordable here: we have at most $\binom{|D|-1}{K-1} = \binom{25}{K-1} \leq \binom{25}{12} \leq 5,200,300$ different partitions. For each possible partition of size K an algorithm for computing the ambiguity measure A is run. We show that this can be done in time linear in $|D|$.

Theorem 6 *Given a dictionary D of words made of symbols in Σ and a partition of Σ, it is possible to check the feasibility condition in KEYBOARD or the ambiguity of the partition in time $O(|D|)$, provided that the maximum length of a word in D is constant.*

Proof. To achieve this complexity in the worst case, we can store the dictionary D in a decision tree and merge the symbols in breadth-first order. In practice, the algorithm can advantageously be implemented using a hash table: for each word of D, the existence of a previously seen word with the same subset sequence can be checked in constant average time.   □

Optimal solutions are shown on Fig. 4. It is interesting to compare Fig. 4(b) with the standard layout of Fig. 1. We computed the ambiguity A of the latter and obtained $A = 57/885$. The individual keys for the letters l, o, s and t are noticeable on Fig 4(c).

6   Conclusion

We proposed an analysis of an original keyboard design problem, formulated as a combinatorial optimization. This is the first theoretical approach of such a problem, and realistic assumptions were made that certainly make this approach directly useful in practice. As a future research, it would be interesting to give other approximability or nonapproximability results for ambiguous keyboards with alternative ambiguity measures or selection systems. It is also likely that this problem appears in other contexts, such as sequence analysis.

References

1. ISO/IEC 9995-8. Information systems – keyboard layouts for text and office systems – part 8: Allocation of letters to keys of a numeric keypad, 1994. International Organisation for standardisation.
2. M. Bellare, M. Goldreich, and M. Sudan. Free bits, PCPs and non-approximability – towards tight results. *SIAM J. Comp.*, 27:804–915, 1998.
3. L. Butts and A. Cockburn. An evaluation of mobile phone text input methods. In *Proc. 3rd Australasian User Interfaces Conference*, 2001.
4. Tegic Communications. T9 text entry. http://www.t9.com.
5. M. Conforti and G. Cornuejols. Submodular functions, matroids and the greedy algorithm: tight worst-case bounds and some generalizations of the Rado-Edmonds theorem. *Discrete Applied Mathematics*, 7:257–275, 1984.
6. M. Garey and D. S. Johnson. *Computers and Intractability : A Guide to the Theory of NP-Completeness*. W. H. Freeman & Co, 1979.
7. D. S. Hochbaum, editor. *Approximation algorithms for NP-hard problems*. PWS Publishing Company, 1997.
8. G. Lesher, B. Moulton, and D. Jeffery Higginbotham. Optimal character arrangements for ambiguous keyboards. *IEEE Trans. on Rehabilitation Engineering*, 6(4), 1998.
9. I. MacKenzie and R. Soukoreff. Text entry for mobile computing: Models and methods, theory and practice. *Human-Computer Interaction*, 17:147–198, 2002.
10. R. Raz and S. Safra. A sub-constant error-probability low-degree test, and a sub-constant error-probability PCP characterization of NP. In *Proceedings of the 29th ACM Symposium on Theory of Computing*, pages 475–484, 1997.
11. V. Vazirani. *Approximation Algorithms, Springer-Verlag, Berlin, 2001.* Springer-Verlag, Berlin, 2001.

Metric Structures in L_1:
Dimension, Snowflakes, and Average Distortion

James R. Lee[1]*, Manor Mendel[3]**, and Assaf Naor[2]

[1] U.C. Berkeley (jrl@cs.berkeley.edu)
[2] Microsoft Research (anaor@microsoft.com)
[3] University of Illinois (mendelma@uiuc.edu)

Abstract. We study the metric properties of finite subsets of L_1. The analysis of such metrics is central to a number of important algorithmic problems involving the cut structure of weighted graphs, including the Sparsest Cut Problem, one of the most compelling open problems in the field of approximation. Additionally, many open questions in geometric non-linear functional analysis involve the properties of finite subsets of L_1.

We present some new observations concerning the relation of L_1 to dimension, topology, and Euclidean distortion. We show that every n-point subset of L_1 embeds into L_2 with average distortion $O(\sqrt{\log n})$, yielding the first evidence that the conjectured worst-case bound of $O(\sqrt{\log n})$ is valid. We also address the issue of dimension reduction in L_p for $p \in (1, 2)$. We resolve a question left open in [1] about the impossibility of *linear* dimension reduction in the above cases, and we show that the example of [2,3] cannot be used to prove a lower bound for the non-linear case. This is accomplished by exhibiting constant-distortion embeddings of snowflaked planar metrics into Euclidean space.

1 Introduction

This paper is devoted to the analysis of metric properties of finite subsets of L_1. Such metrics occur in many important algorithmic contexts, and their analysis is key to progress on some fundamental problems. For instance, an $O(\log n)$-approximate max-flow/min-cut theorem proved elusive for many years until, in [4,5], it was shown to follow from a theorem of Bourgain stating that every metric on n points embeds into L_1 with distortion $O(\log n)$.

The importance of L_1 metrics has given rise to many problems and conjectures that have attracted a lot of attention in recent years. Four basic problems of this type are as follows.

---

* Work partially supported by NSF grant CCR-0121555 and an NSF Graduate Research Fellowship. Part of this work was done while the author was an intern at Microsoft Research.

** Work done while the author was a post-doc fellow at The Hebrew University, and supported in part by the Landau Center and by a grant from the Israeli Science Foundation (195/02).

M. Farach-Colton (Ed.): LATIN 2004, LNCS 2976, pp. 401–412, 2004.

I. Is there an L_1 analog of the Johnson-Lindenstrauss dimension reduction lemma [6]?

II. Are all n-point subsets of L_1 $O\left(\sqrt{\log n}\right)$-embeddable into Hilbert space?

III. Are all squared-ℓ_2 metrics $O(1)$-embeddable into L_1?

IV. Are all planar graphs $O(1)$-embeddable into L_1?

(We recall that a squared-ℓ_2 metric is a space (X,d) for which $(X,d^{1/2})$ embeds isometrically in a Hilbert space.)

Each of these questions has been asked many times before; we refer to [7,8,9, 10], in particular. Despite an immense amount of interest and effort, the metric properties of L_1 have proved quite elusive; hence the name "The mysterious L_1" appearing in a survey of Linial at the ICM in 2002 [9]. In this paper, we attempt to offer new insights into the above problems and touch on some relationships between them.

1.1   Results and Techniques

Euclidean distortion. Our first result addresses problem (II) stated above. We show that the answer to this question is positive on average, in the following sense.

Theorem 1. *For every $f_1,\dots,f_n \in L_1$ there is a linear operator $T : L_1 \to L_2$ such that*

$$\frac{\|T(f_i) - T(f_j)\|_2}{\|f_i - f_j\|_1} \geq \frac{1}{\sqrt{8\log n}}, \qquad 1 \leq i < j \leq n, \ and$$

$$\frac{1}{\binom{n}{2}} \sum_{1 \leq i < j \leq n} \left(\frac{\|T(f_i) - T(f_j)\|_2}{\|f_i - f_j\|_1}\right)^{1/2} \leq 10.$$

In other words, for any n-point subset in L_1, there exists a map into L_2 such that distances are contracted by at most $O(\sqrt{\log n})$ and the average expansion is $O(1)$. This yields the first positive evidence that the conjectured worst-case bound of $O(\sqrt{\log n})$ holds. We remark that a different notion of average embedding was recently studied by Rabinovich [11]; there, one tries to embed (planar) metrics into the line such that the *average distance* does not change too much.

The exponent $1/2$ above has no significance, and we can actually obtain the same result for any power $1-\varepsilon$, $\varepsilon > 0$ (we refer to Section 2 for details). The proof of Theorem 1 follows from the following probabilistic lemma, which is implicit in [12]. We believe that this result is of independent interest.

Lemma 1. *There exists a distribution over linear mappings $T : L_1 \to L_2$ such that for every $x \in L_1 \setminus \{0\}$ the random variable $\frac{\|T(x)\|_2}{\|x\|_1}$ has density $\frac{e^{-1/(4x^2)}}{x^2\sqrt{\pi}}$.*

In contrast to Theorem 1, we show that problem (II) cannot be resolved positively using linear mappings. Specifically, we show that there are arbitrarily large

n-point subsets of L_1 such that any linear embedding of them into L_2 incurs distortion $\Omega(\sqrt{n})$. As a corollary we settle the problem left open by Charikar and Sahai in [1], whether *linear* dimension reduction is possible in L_p, $p \notin \{1, 2\}$. The case $p = 1$ was proved in [1] via linear programming techniques, and it seems impossible to generalize their lower bound to arbitrary L_p. We show that there are arbitrarily large n-point subsets $X \subseteq L_p$ (namely, the same point set used in [1] to handle the case $p = 1$), such that any linear embedding of X into ℓ_p^d incurs distortion $\Omega\left[(n/d)^{|1/p-1/2|}\right]$, thus linear dimension reduction is impossible in any L_p, $p \neq 2$. Additionally, we show that there are arbitrarily large n-point subsets $X \subseteq L_1$ such any linear embedding of X into *any* d-dimensional normed space incurs distortion $\Omega\left(\sqrt{n/d}\right)$. This generalizes the Charikar-Sahai result to arbitrary low dimensional norms.

Dimension reduction. In [2], and soon after in [3], it was shown that if the Newman-Rabinovich diamond graph on n vertices α-embeds into ℓ_1^d then $d \geq n^{\Omega(1/\alpha^2)}$. The proof in [2] is based on a linear programming argument, while the proof in [3] uses a geometric argument which reduces the problem to bounding from below the distortion required to embed the diamond graph in ℓ_p, $1 < p < 2$. These results settle the long standing open problem of whether there is an L_1 analog of the Johnson-Lindenstrauss dimension reduction lemma [6]. (In other words, they show that the answer to problem (I) above is *No.*). In Section 4, we show that the method of proof in [3] can be used to provide an even more striking counter example to this problem.

A metric space X is called *doubling* with constant C if every ball in X can be covered by C balls of half the radius. Doubling metrics with bounded doubling constants are widely viewed as low dimensional (see [13,14] for some practical and theoretical applications of this viewpoint). On the other hand, the doubling constant of the diamond graphs is $\Omega(\sqrt{n})$ (where n is the number of points). Based on a fractal construction due to Laakso [15] and the method developed in [3], we prove the following theorem, which shows a strong lower bound on the dimension required to represent uniformly doubling subsets of L_1.

Theorem 2. *There are arbitrarily large n-point subsets $X \subseteq L_1$ which are doubling with constant 6 but such that every α-embedding of X into ℓ_1^d requires $d \geq n^{\Omega(1/\alpha^2)}$.*

In [16,13] it was asked whether any subset of ℓ_2 which is doubling well-embeds into ℓ_2^d (with bounds on the distortion and the dimension that depend only on the doubling constant). In [13], it was shown that a similar property cannot hold for ℓ_1. Our lower bound exponentially strengthens that result.

Planar metrics. Our final result addresses problems (III) and (IV). Our motivation was an attempt to generalize the argument in [3] to prove that dimension reduction is impossible in L_p for any $1 < p < 2$. A natural approach to this problem is to consider the point set used in [2,3] (namely, a natural realization of the diamond graph, G, in L_1) with the metric induced by the L_p norm instead of the L_1 norm. This is easily seen to amount to proving lower bounds on the distortion required to embed the metric space $(G, d_G^{1/p})$ in ℓ_p^h. Unfortunately, this approach cannot work since we show that, for any planar metric (X, d) and any $0 < \varepsilon < 1$, the metric space $(X, d^{1-\varepsilon})$ embeds in Hilbert space with distortion $O(1/\sqrt{\varepsilon})$, and then using results of Johnson and Lindenstrauss [6], and Figiel, Lindenstrauss and Milman [17], we conclude that this metric can be $O(1/\sqrt{\varepsilon})$ embedded in ℓ_p^h, where $h = O(\log n)$. The proof of this interesting fact is a straightforward application of Assouad's classical embedding theorem [18] and Rao's embedding method [19]. The $O(1/\sqrt{\varepsilon})$ upper bound is shown to be tight for every value $0 < \varepsilon < 1$. We note that the case $\varepsilon = 1/2$ has been previously observed by A. Gupta in his (unpublished) thesis.

2   Average Distortion Euclidean Embedding of Subsets of L_1

The heart of our argument is the following lemma which is implicit in [12], and which seems to be of independent interest.

Lemma 2. *For every $0 < p \le 2$ there is a probability space (Ω, P) such that for every $\omega \in \Omega$ there is a linear operator $T_\omega : L_p \to L_2$ such that for every $x \in L_p \setminus \{0\}$ the random variable $X = \frac{\|T_\omega(x)\|_2}{\|x\|_p}$ satisfies for every $a \in \mathbb{R}$,*
$$\mathbb{E}e^{-aX^2} = e^{-a^{p/2}}. \text{ In particular, for } p = 1 \text{ the density of } X \text{ is } \frac{e^{-1/(4x^2)}}{x^2\sqrt{\pi}}.$$

Proof. Consider the following three sequences of random variables, $\{Y_j\}_{j\ge 1}$, $\{\theta_j\}_{j\ge 1}$, $\{g_j\}_{j\ge 1}$, such that each variable is independent of the others. For each $j \ge 1$, Y_j is uniformly distributed on $[0, 1]$, g_j is a standard Gaussian and θ_j is an exponential random variable, i.e. for $\lambda \ge 0$, $P(\theta_j > \lambda) = e^{-\lambda}$. Set $\Gamma_j = \theta_1 + \cdots + \theta_j$. By Proposition 1.5. in [12], there is a constant $C = C(p)$ such that if we define for $f \in L_p$

$$V(f) = C \sum_{j\ge 1} \frac{g_j}{\Gamma_j^{1/p}} f(Y_j),$$

then $\mathbb{E}e^{iV(f)} = e^{-\|f\|_p^p}$.

Assume that the random variables $\{Y_j\}_{j\ge 1}$ and $\{\Gamma_j\}_{j\ge 1}$ are defined on a probability space (Ω, P) and that $\{g_j\}_{j\ge 1}$ are defined on a probability space (Ω', P'), in which case we use the notation $V(f) = V(f; \omega, \omega')$. Define for $\omega \in \Omega$ a linear operator $T_\omega : L_p \to L_2(\Omega', P')$ by $T_\omega(f) = V(f; \omega, \cdot)$. Since for every fixed $\omega \in \Omega$ the random variable $V(f; \omega, \cdot)$ is Gaussian with variance $\|T_\omega(f)\|_2^2$, for every $a \in \mathbb{R}$, $\mathbb{E}_{P'}e^{iaV(s;\omega,\cdot)} = e^{-a^2\|T_\omega(f)\|_2^2}$. Taking expectation with respect to P we find that, $\mathbb{E}_P e^{-a^2\|T_\omega(f)\|_2^2} = e^{-a^p\|f\|_p^p}$. This implies the required identity. The explicit distribution in the case $p = 1$ follows from the fact that the inverse Laplace transform of $x \mapsto e^{-\sqrt{x}}$ is $y \mapsto \frac{e^{-1/(4y)}}{2\sqrt{\pi y^3}}$ (see for example [20]). $\square$

Theorem 3. *For every $f_1, \ldots, f_n \in L_1$ there is a linear operator $T : L_1 \to L_2$ such that:*

$$\frac{\|T(f_i) - T(f_j)\|_2}{\|f_i - f_j\|_1} \geq \frac{1}{\sqrt{8 \log n}}, \qquad 1 \leq i < j \leq n, \; and$$

$$\frac{1}{\binom{n}{2}} \sum_{1 \leq i < j \leq n} \left(\frac{\|T(f_i) - T(f_j)\|_2}{\|f_i - f_j\|_1} \right)^{1/2} \leq 10.$$

Proof. Using the notation of lemma 2 (in the case $p = 1$) we find that for every $a > 0$, $\mathbb{E} e^{-aX^2} = e^{-\sqrt{a}}$. Hence, for every $a, \varepsilon > 0$ and every $1 < i < j \leq n$,

$$P\left(\frac{\|T_\omega(f_i) - T_\omega(f_j)\|_2}{\|f_i - f_j\|_1} \leq \varepsilon \right) = P\left(e^{-aX^2} \geq e^{-a\varepsilon^2} \right) \leq e^{a\varepsilon^2 - \sqrt{a}}.$$

Choosing $a = \frac{1}{4\varepsilon^4}$ the above upper bound becomes $e^{-1/(4\varepsilon^2)}$. Consider the set

$$A = \bigcap_{1 \leq i < j \leq n} \left\{ \frac{\|T_\omega(f_i) - T_\omega(f_j)\|_2}{\|f_i - f_j\|_1} \geq \frac{1}{\sqrt{8 \log n}} \right\} \subseteq \Omega.$$

By the union bound, $P(A) > \frac{1}{2}$, so that

$$\frac{1}{P(A)} \mathbb{E}\left[\frac{1}{\binom{n}{2}} \sum_{1 \leq i < j \leq n} \left(\frac{\|T_\omega(f_i) - T_\omega(f_j)\|_2}{\|f_i - f_j\|_1} \right)^{1/2} \right]$$

$$\leq 2\mathbb{E} X^{1/2} = \frac{2}{\sqrt{\pi}} \int_0^\infty x^{1/2} \cdot \frac{e^{-1/(4x^2)}}{x^2} dx < 10.$$

It follows that there exists $\omega \subset A$ for which the operator $T = T_\omega$ has the desired properties. $\qquad\square$

Remark 1. There is nothing special about the choice of the the power $1/2$ in Corollary 3. When $p = 1$, $\mathbb{E} X = \infty$ but $\mathbb{E} X^{1-\varepsilon} < \infty$ for every $0 < \varepsilon < 1$, so we may write the above average with the power $1 - \varepsilon$ replacing the exponent $1/2$. Obvious generalizations of Corollary 3 hold true for every $1 < p < 2$, in which case the average distortion is of order $C(p)(\log n)^{1/p - 1/2}$ (and the power can be taken to be 1).

3   The Impossibility of *Linear* Dimension Reduction in L_p, $p \neq 2$

The above method cannot yield a $O\left(\sqrt{\log n}\right)$ bound on the Euclidean distortion of n-point subsets of L_1. In fact, there are arbitrarily large n-point subsets of L_1 on which any *linear* embedding into L_2 incurs distortion at least $\sqrt{\frac{n-1}{2}}$. This follows from the following simple lemma:

Lemma 3. *For every* $1 \leq p \leq \infty$ *there are arbitrarily large n-point subsets of* L_p *on which any linear embedding into* L_2 *incurs distortion at least* $\left(\frac{n-1}{2}\right)^{|1/p-1/2|}$.

Proof. Let $w_1, \ldots, w_{2^k}$ be the rows of the $2^k \times 2^k$ Walsh matrix. Write $w_i = \sum_{j=1}^{2^k} w_{ij} e_j$ where $e_1, \ldots, e_{2^k}$ are the standard unit vectors in $\mathbb{R}^{2^k}$. Consider the set $A = \{0\} \cup \{w_i\}_{i=1}^{2^k} \cup \{e_i\}_{i=1}^{2^k} \subset \ell_p$. Let $T : \ell_p \to L_2$ be any linear operator which is non contracting and L-Lipschitz on A. Assume first of all that $1 \leq p < 2$. Then:

$$2^{k(1+2/p)} = \sum_{i=1}^{2^k} \|w_i\|_p^2 \leq \sum_{i=1}^{2^k} \|Tw_i\|_2^2 = \sum_{i=1}^{2^k} \left\| \sum_{j=1}^{2^k} w_{ij} T(e_j) \right\|_2^2$$

$$= \sum_{i=1}^{2^k} \sum_{j=1}^{2^k} \langle w_i, w_j \rangle \langle T(e_i), T(e_j) \rangle = 2^k \sum_{j=1}^{2^k} \|T(e_j)\|_2^2 \leq 4^k \cdot L^2,$$

which implies that $L \geq 2^{k(1/p-1/2)} = \left(\frac{|A|-1}{2}\right)^{1/p-1/2}$. When $p > 2$ apply the same reasoning, with the inequalities reversed. $\qquad \square$

We remark that the above point set was also used by Charikar and Sahai [1] to give a lower bound on *linear* dimension reduction in L_1. Their proof used a linear programming argument, which doesn't seem to be generalizable to the the case of L_p, $p > 1$. Lemma 3 formally implies their result (with a significantly simpler proof), and in fact proves the impossibility of linear dimension reduction in any L_p, $p \neq 2$. Indeed, if there were a linear operator which embeds A into ℓ_p^d with distortion D then it would also be a $D \cdot d^{|1/p-1/2|}$ embedding into ℓ_2^d. It follows that $D \geq \left(\frac{|A|-1}{2d}\right)^{|1/p-1/2|}$. Similarly, since by John's theorem (see e.g. [21]) any d-dimensional normed space is $\sqrt{d}$ equivalent to Hilbert space, we deduce that there are arbitrarily large n-point subsets of L_1, any linear embedding of which into any d-dimensional normed space incurs distortion at least $\sqrt{\frac{n-1}{2d}}$.

4 An Inherently High-Dimensional Doubling Metric in L_1

This section is devoted to the proof of Theorem 2.

Proof (of Theorem 2). Consider the Laakso graphs, $\{G_i\}_{i=0}^{\infty}$, which are defined as follows. G_0 is the graph on two vertices with one edge. To construct G_i, take six copies of G_{i-1} and scale their metric by a factor of $\frac{1}{4}$. We glue four of them cyclicly by identifying pairs of endpoints, and attach at two opposite gluing points the remaining two copies. See Figure 1 below.

As shown in [15], the graphs $\{G_i\}_{i=0}^{\infty}$ are uniformly doubling (see also [16], for a simple argument showing they are doubling with constant 6). Moreover, since the G_i's are series parallel graphs, they embed uniformly in L_1 (see [22]).

We will show below that any embedding of G_i in L_p, $1 < p \le 2$ incurs distortion at least $\sqrt{1 + \frac{p-1}{4}i}$. We then conclude as in [3] by observing that ℓ_1^d is 3-isomorphic to ℓ_p^d when $p = 1 + \frac{1}{\log d}$, so that if G_i embeds with distortion α in ℓ_1^d then $\alpha \ge \sqrt{\frac{i}{40 \log d}}$. This implies the required result since $i \approx \log |G_i|$.

The proof of the lower bound for the distortion required to embed G_i into L_p is by induction on i. We shall prove by induction that whenever $f : G_i \to L_p$ is non-contracting then there exist two adjacent vertices $u, v \in G_i$ such that $\|f(u) - f(v)\|_p \ge d_{G_i}(u, v)\sqrt{1 + \frac{p-1}{4}i}$ (observe that for $u, v \in G_{i-1}$, $d_{G_{i-1}}(u, v) = d_{G_i}(u, v)$). For $i = 0$ there is nothing to prove. For $i \ge 1$, since G_i contains an isometric copy of G_{i-1}, there are $u, v \in G_i$ corresponding to two adjacent vertices in G_{i-1} such that $\|f(u) - f(v)\|_p \ge d_{G_i}(u, v)\sqrt{1 + \frac{p-1}{4}(i-1)}$. Let a, b be the two midpoints between u and v in G_i. By Lemma 2.1 in [3],

$$\|f(u) - f(v)\|_p^2 + (p-1)\|f(a) - f(b)\|_p^2$$
$$\le \|f(u) - f(a)\|_p^2 + \|f(a) - f(v)\|_p^2 + \|f(v) - f(b)\|_p^2 + \|f(b) - f(u)\|_p^2.$$

Hence:

$$\max\{\|f(u) - f(a)\|_p^2, \|f(a) - f(v)\|_p^2, \|f(v) - f(b)\|_p^2, \|f(b) - f(u)\|_p^2\}$$
$$\ge \frac{1}{4}\|f(u) - f(v)\|_p^2 + \frac{1}{4}(p-1)\|f(a) - f(b)\|_p^2$$
$$\ge \frac{1}{4}\left(1 + \frac{p-1}{4}(i-1)\right) d_{G_i}(u, v)^2 + \frac{p-1}{4}d_{G_i}(a, b)^2$$
$$= \frac{1}{4}\left(1 + \frac{p-1}{4}i\right) d_{G_i}(u, v)^2$$
$$= \left(1 + \frac{p-1}{4}i\right) \max\{d_{G_i}(u, a)^2, d_{G_i}(a, v)^2, d_{G_i}(v, b)^2, d_{G_i}(b, u)^2\}.$$

$\square$

We end this section by observing that the above approach also gives a lower bound on the dimension required to embed expanders in ℓ_∞.

Proposition 1. *Let G be an n-point constant degree expander which embeds in ℓ_∞^d with distortion at most α. Then $d \ge n^{\Omega(1/\alpha)}$.*

Proof. By Matoušek's lower bound for the distortion required to embed expanders in ℓ_p [23], any embedding of G into ℓ_p incurs distortion $\Omega\left(\frac{\log n}{p}\right)$. Since ℓ_∞^d is $O(1)$-equivalent to $\ell_{\log d}^d$, we deduce that $\alpha \ge \Omega\left(\frac{\log n}{\log d}\right)$. $\square$

We can also obtain a lower bound on the dimension required to embed the Hamming cube $\{0, 1\}^k$ into ℓ_∞. Our proof uses a simple concentration argument. An analogous concentration argument yields an alternative proof of Proposition 1.

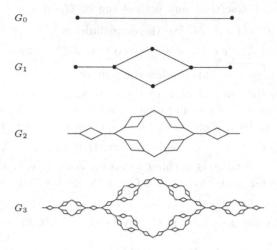

Fig. 1. The Laakso graphs.

Proposition 2. *Assume that $\{0,1\}^k$ embeds into ℓ_∞^d with distortion α. Then $d \geq 2^{k\Omega(1/\alpha^2)}$.*

Proof. Let $f = (f_1, \ldots, f_d) : \{0,1\}^k \to \ell_\infty^d$ be a contraction such that for every $u, v \in \{0,1\}^d$, $\|f(u) - f(v)\|_\infty \geq \frac{1}{\alpha}d(u,v)$ (where $d(\cdot,\cdot)$ denotes the Hamming metric). Denote by P the uniform probability measure on $\{0,1\}^k$. Since for every $1 \leq i \leq k$, f_i is 1-Lipschitz, the standard isoperimetric inequality on the hypercube implies that $P(|f_i(u) - \mathbb{E}f_i| \geq k/(4\alpha)) \leq e^{-\Omega(k/\alpha^2)}$. On the other hand, if $u, v \in \{0,1\}^k$ are such that $d(u,v) = k$ then there exist $1 \leq i \leq d$ for which $|f_i(u) - f_i(v)| \geq k/\alpha$, implying that $\max\{|f_i(u) - \mathbb{E}f_i|, |f_i(v) - \mathbb{E}f_i|\} > k/(4\alpha)$. By the union bound it follows that $de^{-\Omega(k/\alpha^2)} \geq 1$, as required. □

5   Snowflake Versions of Planar Metrics

The problem of whether there is an analog of the Johnson-Lindenstrauss dimension reduction lemma in L_p, $1 < p < 2$, is an interesting one which remains open. In view of the above proof and the proof in [3], a natural point set which is a candidate to demonstrate the impossibility of dimension reduction in L_p is the realization of the diamond graph in ℓ_1 which appears in [2], equipped with the ℓ_p metric. Since this point set consists of vectors whose coordinates are either 0 or 1 (i.e. subsets of the cube), this amounts to considering the diamond graph with its metric raised to the power $\frac{1}{p}$. Unfortunately, this approach cannot work; we show below that any planar graph whose metric is raised to the power $1 - \varepsilon$ has Euclidean distortion $O(1/\sqrt{\varepsilon})$.

Given a metric space (X,d) and $\varepsilon > 0$, the metric space $(X, d^{1-\varepsilon})$ is known in geometric analysis (see e.g. [24]) as the $1-\varepsilon$ snowflake version of (X,d). Assouad's classical theorem [18] states that any snowflake version of a doubling metric space is bi-Lipschitz equivalent to a subset of some finite dimensional Euclidean space. A quantitative version of this result (with bounds on the distortion and the dimension) was obtained in [13]. The following theorem is proved by combining embedding techniques of Rao [19] and Assouad [18]. A similar analysis is also used in [13]. In what follows we call a metric K_r-excluded if it is the metric on a subset of a weighted graph which does not admit a K_r minor. In particular, planar metrics are all K_5-excluded.

Theorem 4. *For any $r \in \mathbb{N}$ there exists a constant $C(r)$ such that for every $0 < \epsilon < 1$, a $1 - \varepsilon$ snowflake version of a K_r-excluded metric embeds into ℓ_2 with distortion at most $C(r)/\sqrt{\varepsilon}$.*

Our argument is based on the following lemma, the proof of which is contained in [19].

Lemma 4. *For every $r \in \mathbb{N}$ there is a constant $\delta = \delta(r)$ such that for every $\rho > 0$ and every K_r-excluded metric (X, d) there exists a finitely supported probability distribution μ on partitions of X with the following properties:*

1. *For every $P \in \mathrm{supp}(\mu)$, and for every $C \in P$, $\mathrm{diam}(C) \leq \rho$.*
2. *For every $x \in X$, $\mathbb{E}_\mu \sum_{C \in P} d(x, X \setminus C) \geq \delta\rho$.*

Observe that the sum under the expectation in (2) above actually consists of only one summand.

Proof (Proof of Theorem 4). Let X be a K_r-excluded metric. For each $n \in \mathbb{Z}$, we define a map ϕ_n as follows. Let μ_n be the probability distribution on partitions of X from Lemma 4 with $\rho = 2^{n/(1-\varepsilon)}$. Fix a partition $P \in \mathrm{supp}(\mu_n)$. For any $\sigma \in \{-1, +1\}^{|P|}$, consider σ to be indexed by $C \in P$ so that σ_C has the obvious meaning. Following Rao [19], define

$$\phi_P(x) = \bigoplus_{\sigma \in \{-1,+1\}^{|P|}} \sqrt{\frac{1}{2^{|P|}}} \sum_{C \in P} \sigma_C \cdot d(x, X \setminus C),$$

and write $\phi_n = \bigoplus_{P \in \mathrm{supp}(\mu_n)} \sqrt{\mu_n(P)}\, \phi_P$ (here the symbol $\oplus$ refers to the concatenation operator).

Now, following Assouad [18], let $\{e_i\}_{i \in \mathbb{Z}}$ be an orthonormal basis of ℓ_2, and set

$$\Phi(x) = \sum_{n \in \mathbb{Z}} 2^{-n\varepsilon/(1-\varepsilon)} \phi_n(x) \otimes e_n$$

Claim. For every $n \in \mathbb{Z}$, and $x, y \in X$, we have $\|\phi_n(x) - \phi_n(y)\|_2 \leq 2 \cdot \min\{d(x,y), 2^{n/(1-\varepsilon)}\}$. Additionally, if $d(x,y) > 2^{n/(1-\varepsilon)}$, then $\|\phi_n(x) - \phi_n(y)\|_2 \geq \delta\, 2^{n/(1-\varepsilon)}$.

Proof. For any partition $P \in \mathrm{supp}(\mu_n)$, let C_x, C_y be the clusters of P containing x and y, respectively. Note that since for every $C \in P$, $\mathrm{diam}(C) \le 2^{n/(1-\varepsilon)}$, when $d(x,y) > 2^{n/(1-\varepsilon)}$, we have $C_x \ne C_y$. In this case,

$$\|\phi_P(x) - \phi_P(y)\|_2^2 = \mathbb{E}_{\sigma \in \{-1,+1\}^{|P|}} |\sigma_{C_x} d(x, X \setminus C_x) - \sigma_{C_y} d(y, X \setminus C_y)|^2$$
$$\ge \frac{d(x, X \setminus C_x)^2 + d(y, X \setminus C_y)^2}{2}.$$

It follows that

$$\|\phi_n(x) - \phi_n(y)\|_2^2 = \mathbb{E}_{\mu_n} \|\phi_P(x) - \phi_P(y)\|_2^2$$
$$\ge \frac{\mathbb{E}_{\mu_n} d(x, X \setminus C_x)^2 + \mathbb{E}_{\mu_n} d(y, X \setminus C_y)^2}{2} \ge \left(\delta \, 2^{n/(1-\varepsilon)}\right)^2.$$

On the other hand, for every $x,y \in X$, since $d(x, X \setminus C_x), d(y, X \setminus C_y) \le 2^{n/(1-\varepsilon)}$, we have that $\|\phi_P(x) - \phi_P(y)\|_2 \le 2 \cdot \min\left\{d(x,y), 2^{n/(1-\varepsilon)}\right\}$, hence $\|\phi_n(x) - \phi_n(y)\|_2 \le 2 \cdot \min\left\{d(x,y), 2^{n/(1-\varepsilon)}\right\}$. $\quad\square$

To finish the analysis, let us fix $x,y \in X$ and let m be such that $d(x,y)^{1-\varepsilon} \in \left(2^m, 2^{m+1}\right]$. In this case,

$$\|\Phi(x) - \Phi(y)\|_2^2 = \sum_{n \in \mathbb{Z}} 2^{-2n\varepsilon/(1-\varepsilon)} \|\phi_n(x) - \phi_n(y)\|_2^2$$
$$\le 4 \sum_{n<m} 2^{2n} + 4d(x,y)^2 \sum_{n \ge m} 2^{-2n\varepsilon/(1-\varepsilon)}$$
$$= 2^{2m+1} + 4d(x,y)^2 \frac{2^{-2m\varepsilon/(1-\varepsilon)}}{1 - 2^{-2\varepsilon/(1-\varepsilon)}}$$
$$= O\left(1/\varepsilon\right) \cdot d(x,y)^{2(1-\varepsilon)}.$$

On the other hand,

$$\|\Phi(x) - \Phi(y)\|_2 \ge 2^{-m\varepsilon/(1-\varepsilon)} \|\phi_m(x) - \phi_m(y)\|_2 \ge \delta 2^m \ge \frac{\delta}{2} d(x,y)^{1-\varepsilon}.$$

The proof is complete. $\quad\square$

Remark 2. The $O\left(1/\sqrt{\varepsilon}\right)$ upper bound in Theorem 4 is tight. In fact, for $i \approx 1/\varepsilon$, the $1 - \varepsilon$ snowflake version of the Laakso graph G_i (presented in Section 4) has Euclidean distortion $\Omega\left(1/\sqrt{\varepsilon}\right)$. To see this, let $f : G_i \to \ell_2$ be any non-contracting embedding of $(G_i, d_{G_i}^{1-\varepsilon})$ into ℓ_2. For $j \le i$ denote by K_j the Lipschitz constant of the restriction of f to $(G_j, d_{G_i}^{1-\varepsilon})$ (as before, we think of G_j as a subset of G_i). Clearly $K_0 = 1$, and the same reasoning as in the proof of Theorem 2 shows that for $j \ge 1$, $K_j^2 \ge \frac{K_{j-1}^2}{4^\varepsilon} + \frac{1}{4}$. This implies that $K_i^2 \ge \frac{1}{4} + \frac{1}{4^\varepsilon} + \ldots + \frac{1}{4^{i\varepsilon}} = \Omega(1/\varepsilon)$, as required.

References

1. Charikar, M., Sahai, A.: Dimension reduction in the ℓ_1 norm. In: Proceedings of the 43rd Annual IEEE Conference on Foundations of Computer Science, ACM (2002)
2. Charikar, M., Brinkman, B.: On the impossibility of dimension reduction in ℓ_1. In: to appear in Proceedings of the 44th Annual IEEE Conference on Foundations of Computer Science, ACM (2003)
3. Lee, J.R., Naor, A.: Embedding the diamond graph in L_p and dimension reduction in L_1. Preprint (2003)
4. Linial, N., London, E., Rabinovich, Y.: The geometry of graphs and some of its algorithmic applications. Combinatorica **15** (1995) 215–245
5. Aumann, Y., Rabani, Y.: An $O(\log k)$ approximate min-cut max-flow theorem and approximation algorithm. SIAM J. Comput. **27** (1998) 291–301
6. Johnson, W.B., Lindenstrauss, J.: Extensions of Lipschitz mappings into a Hilbert space. In: Conference in modern analysis and probability (New Haven, Conn., 1982). Volume 26 of Contemp. Math. Amer. Math. Soc., Providence, RI (1984) 189–206
7. Matoušek, J.: Lectures on discrete geometry. Volume 212 of Graduate Texts in Mathematics. Springer-Verlag, New York (2002)
8. Matoušek, J.: Open problems, workshop on discrete metric spaces and their algorithmic appl ications, Haifa (2002)
9. Linial, N.: Finite metric spaces - combinatorics, geometry and algorithms. In: Proceedings of the International Congress of Mathematicians III. (2002) 573 586
10. Indyk, P.: Algorithmic applications of low-distortion geometric embeddings. In: Proceedings of the 42nd Annual IEEE Symposium on Foundations of Computer Science. (2001) 10 33
11. Rabinovich, Y.: On average distorsion of embedding metrics into the line and into l_1. In: Proceedings of the 35th Annual ACM Symposium on Theory of Computing, ACM (2003)
12. Marcus, M.B., Pisier, G.: Characterizations of almost surely continuous p-stable random Fourier series and strongly stationary processes. Acta Math. **152** (1984) 245–301
13. Gutpa, A., Krauthgamer, R., Lee, J.R.: Bounded geometries, fractals, and low-distortion embeddings. In: Proceedings of the 44th Annual Symposium on Foundations of Computer Science. (2003)
14. Krauthgamer, R., Lee, J.R.: Navigating nets: Simple algorithms for proximity search. Submitted (2003)
15. Laakso, T.J.: Ahlfors Q-regular spaces with arbitrary $Q > 1$ admitting weak Poincaré inequality. Geom. Funct. Anal. **10** (2000) 111–123
16. Lang, U., Plaut, C.: Bilipschitz embeddings of metric spaces into space forms. Geom. Dedicata **87** (2001) 285–307
17. Figiel, T., Lindenstrauss, J., Milman, V.D.: The dimension of almost spherical sections of convex bodies. Acta Math. **139** (1977) 53–94
18. Assouad, P.: Plongements lipschitziens dans $\mathbf{R}^n$. Bull. Soc. Math. France **111** (1983) 429–448
19. Rao, S.: Small distortion and volume preserving embeddings for planar and Euclidean metrics. In: Proceedings of the 15th Annual Symposium on Computational Geometry, ACM (1999) 300–306

20. Widder, D.V.: The Laplace Transform. Princeton Mathematical Series, v. 6. Princeton University Press, Princeton, N. J. (1941)
21. Milman, V.D., Schechtman, G.: Asymptotic theory of finite-dimensional normed spaces. Springer-Verlag, Berlin (1986) With an appendix by M. Gromov.
22. Gupta, A., Newman, I., Rabinovich, Y., Sinclair, A.: Cuts, trees and ℓ_1 embeddings. In: Proceedings of the 40th Annual Symposium on Foundations of Computer Science. (1999)
23. Matoušek, J.: On embedding expanders into l_p spaces. Israel J. Math. **102** (1997) 189–197
24. Heinonen, J.: Lectures on analysis on metric spaces. Universitext. Springer-Verlag, New York (2001)

Nash Equilibria via Polynomial Equations

Richard J. Lipton[1,2] and Evangelos Markakis[1]

[1] Georgia Institute of Technology, Atlanta GA 30332, USA,
{rjl,vangelis}@cc.gatech.edu
[2] Telcordia Research, Morristown NJ 07960, USA

Abstract. We consider the problem of computing a Nash equilibrium in multiple-player games. It is known that there exist games, in which all the equilibria have irrational entries in their probability distributions [19]. This suggests that either we should look for symbolic representations of equilibria or we should focus on computing approximate equilibria. We show that every finite game has an equilibrium such that all the entries in the probability distributions are algebraic numbers and hence can be finitely represented. We also propose an algorithm which computes an approximate equilibrium in the following sense: the strategies output by the algorithm are close with respect to l_∞-norm to those of an exact Nash equilibrium and also the players have only a negligible incentive to deviate to another strategy. The running time of the algorithm is exponential in the number of strategies and polynomial in the digits of accuracy. We obtain similar results for approximating market equilibria in the neoclassical exchange model under certain assumptions.

1 Introduction

Noncooperative game theory has been extensively used for modeling and analyzing situations of strategic interactions. One of the dominant solution concepts in noncooperative games is that of a Nash equilibrium [19]. Briefly, a Nash equilibrium of a game is a situation in which no agent has an incentive to unilaterally deviate from her current strategy. A nice property of this concept is the well known fact that every game has at least one such equilibrium [19].

In this paper we consider the problem of computing a Nash equilibrium in finite games. The proof given by Nash for the existence of equilibria is based on Brouwer's fixed point theorem and is nonconstructive. A natural algorithmic question is whether a Nash equilibrium can be computed efficiently. Even for a 2-player game there is still no polynomial time algorithm. The running time of the known algorithms (see among others [12,13,14,15,16]) is either exponential or has not been determined yet (and is believed to be exponential). For m-person games, $m > 2$, the problem seems to be even more difficult [18]. Recently it has also been shown that finding equilibria with certain natural properties (e.g. maximizing payoff) is **NP**-hard [4,8]. The complexity of finding a single equilibrium has been addressed as one of the current challenges in computational complexity [20].

M. Farach-Colton (Ed.): LATIN 2004, LNCS 2976, pp. 413–422, 2004.
© Springer-Verlag Berlin Heidelberg 2004

An issue related to the complexity of the problem is that even for 3-player games, there exist examples [19] in which the payoff data are rational numbers but all the Nash equilibria have irrational entries. Hence it is still not clear whether an equilibrium can be finitely represented on a Turing machine.

The problems mentioned above suggest two potential directions for research. The first one (perhaps more interesting from a theoretical point of view) is to see whether there exist alternative symbolic representations of Nash equilibria. Symbolic representations of numbers have been used in many areas of mathematics such as algebra or algebraic geometry as well as in algorithmic problems involving symbolic computations. A second, more practical goal, is to look for *approximate* equilibria. An approximate equilibrium is usually defined in the literature as a set of strategies such that, either no player can increase her payoff by a nonnegligible amount if she deviates to another strategy, or the strategies, when seen as probability vectors, are close with respect to some norm to an exact Nash equilibrium.

We will address both objectives by using the observation that Nash equilibria are essentially the roots of a single polynomial equation. In particular we show that every game has at least one Nash equilibrium for which all the entries are algebraic numbers, hence it can be finitely represented. The current bounds for the size of the representation are exponential. We also use results from the existential theory of reals and propose an algorithm for computing an approximate equilibrium in time $poly(\log 1/\epsilon, L, m^n)$. Here m is the number of players, n is the total number of available strategies, ϵ is the degree of approximation and L is the maximum bit size of the payoff data. We show that for the case of two players we can compute an exact Nash equilibrium in time $2^{O(n)}$. This is yet another exponential algorithm for computing an equilibrium in 2-person games. We also note that similar algorithms can be obtained for computing market equilibria under certain assumptions.

1.1 Related Work

Recent algorithms for approximate equilibria but only for special classes of games have been obtained in [10,11]. The fact that Nash equilibria are fixed points of a certain map [19] gives rise to many algorithmic approaches that are based on Scarf's algorithm [22], which is a general algorithm for approximating fixed points. The worst case complexity of this algorithm is exponential in both the total number of strategies and the digits of accuracy [9]. A recent algorithm for approximate equilibria in m-player games with a provable upper bound on the running time is that of [17]. The running time is subexponential in the number of strategies and exponential in the accuracy parameter and the number of players. It is better than ours for games with a small number of players. Our algorithm is better in terms of the dependence on the digits of accuracy and for games with relatively small total number of strategies. Our result is also stronger in the sense that not only players have very small incentive to deviate from the approximate equilibrium, but also the set of strategies which are output are exponentially close to some exact Nash equilibrium. This is not ensured by the algorithms of [22] and [17]. More information on algorithmic approaches can be found in the surveys [18,25].

The algebraic characterization of Nash equilibria as the set of solutions to a system of polynomial inequalities has been used before. In [24], algebraic techniques are presented for counting the number of completely mixed equilibria. In [5] it is shown that every real algebraic variety is isomorphic to the set of completely mixed Nash equilibria of some three-person game. However representation and complexity issues are not addressed there.

2   Notation and Definitions

2.1   Nash Equilibria

Consider a game with m players. Suppose that the number of available (pure) strategies for player i is n_i. Let $n_0 = \max n_i$. An m-dimensional payoff matrix A^i is associated with each player. If players $1, \cdots, m$ play the pure strategies $j_1, \cdots, j_m$ respectively, player i receives a payoff equal to $A^i(j_1, \cdots, j_m)$. For simplicity we assume that the entries of the matrices are integers, at most L bits long and $H = 2^L$ is their maximum absolute value.

A *mixed strategy* for player i is a probability distribution over the set of her pure strategies and will be represented by a vector $x_i = (x_{i1}, x_{i2}, \cdots, x_{i,n_i})$, where $x_{ij} \geq 0$ and $\sum x_{ij} = 1$. Here x_{ij} is the probability that the player will choose her jth pure strategy. The *support* of x_i ($Supp(x_i)$) is the set $\{j : x_{ij} > 0\}$. We will denote by $\mathcal{S}_i$ the strategy space of player i, i.e., the $(n_i - 1)$-dimensional unit simplex. For an m-tuple of mixed strategies $x = (x_1, \cdots, x_m) \in \mathcal{S}_1 \times \cdots \times \mathcal{S}_m$, the expected payoff to the ith player is:

$$P^i(x) = \sum_{j_1=1}^{n_1} \cdots \sum_{j_m=1}^{n_m} A^i(j_1, \cdots, j_m) x_{1,j_1} \cdots x_{m,j_m} \tag{1}$$

Following standard notation, for a tuple of mixed strategies $x = (x_1, \cdots, x_m)$, we will denote by x^{-i} the set of strategies: $\{x_j : j \neq i\}$. We will also denote by (x^{-i}, x_i') the tuple $(x_1, \cdots, x_{i-1}, x_i', x_{i+1}, \cdots, x_m)$, i.e., the ith player switches to the strategy x_i' while all other players keep playing the same strategy as in x.

The notion of a Nash equilibrium [19] is formulated as follows:

Definition 1. *A tuple of strategies* $x = (x_1, \cdots, x_m) \in \mathcal{S}_1 \times \cdots \times \mathcal{S}_m$ *is a Nash equilibrium if for every player i and for every mixed strategy $x_i' \in \mathcal{S}_i$, $P^i(x^{-i}, x_i') \leq P^i(x)$.*

The definition states that x is a Nash equilibrium if no player has an incentive to unilaterally defect to another strategy. It is easily seen that it is enough to consider only deviations to pure strategies. For a player i, let s_i^j denote her jth pure strategy. Then an equivalent definition is the following: x is a Nash equilibrium if for any player i and any pure strategy of player i, s_i^j: $P^i(x^{-i}, s_i^j) \leq P^i(x)$.

Similarly we can formalize the notion of an ϵ-Nash equilibrium (or simply ϵ-equilibrium), in which players have only a small incentive to deviate:

Definition 2. *For $\epsilon \geq 0$, a tuple of strategies $x = (x_1, \cdots, x_m)$ is an ϵ-Nash equilibrium if for every player i and for every pure strategy s_i^j, $P^i(x^{-i}, s_i^j) \leq P^i(x) + \epsilon$.*

Another notion of approximation is that of ϵ-closeness:

Definition 3. *A point $x = (x_1, \cdots, x_m) \in \mathcal{S}_1 \times \cdots \times \mathcal{S}_m$ is ϵ-close to a point $y \in \mathcal{S}_1 \times \cdots \times \mathcal{S}_m$ if $\|x_i - y_i\|_\infty \leq \epsilon$ forall $i = 1, ..., m$*

Note that an ϵ-equilibrium is not necessarily close to a real Nash equilibrium.

2.2   Algebra

We give some definitions of basic algebraic concepts that we are going to use in the later sections. For a more detailed exposition we refer the reader to [3].

Definition 4. *A real number α is an algebraic number if there exists a univariate polynomial P with integer coefficients such that $P(\alpha) = 0$.*

Definition 5. *An ordered field R is a real closed field if*

1. *every positive element $x \in R$ is a square (i.e. $x = y^2$ for some $y \in R$).*
2. *every univariate polynomial of odd degree with coefficients in R has a root in R.*

Obviously the real numbers are an example of a real closed field.

3   Nash Equilibria Are Roots of a Polynomial

A Nash equilibrium is a solution of the following system of polynomial inequalities and equalities:

$$x_{ij} \geq 0 \qquad\qquad i = 1, ..., m, j = 1, ..., n_i$$

$$\sum_{j=1}^{n_i} x_{ij} = 1 \qquad\qquad i = 1, ..., m \qquad\qquad (2)$$

$$P^i(x^{-i}, s_i^j) \leq P^i(x) \quad i = 1, ..., m, j = 1, ..., n_i$$

Let $n = \sum n_i$. The system has n variables and $2n + m = O(n)$ multilinear constraints. By adding slack variables we can convert every constraint to an equation, where each polynomial is of degree at most m (the degree of a polynomial is the maximum total degree of its monomials). Note that the slack variables are squared so that we do not have to add any more constraints for their nonnegativity:

$$B_{ij} = x_{ij} - \beta_{ij}^2 = 0 \qquad\qquad i = 1, ..., m, j = 1, ..., n_i$$

$$\Gamma_i = \sum_{j=1}^{n_i} x_{ij} - 1 = 0 \qquad\qquad i = 1, ..., m \qquad\qquad (3)$$

$$\Delta_{ij} = P^i(x) - P^i(x^{-i}, s_i^j) - \delta_{ij}^2 = 0 \; i = 1, ..., m, j = 1, ..., n_i$$

We can now combine all the polynomial equations into one by taking the sum of squares ($P_1 = 0$ and $P_2 = 0$ is equivalent to $P_1^2 + P_2^2 = 0$). Therefore we have the following polynomial which we will refer to as the polynomial of the game $(A^1, ..., A^m)$:

$$\Phi(A^1, ..., A^m) = \sum_{i=1}^{m} \sum_{j=1}^{n_i} B_{ij}^2 + \sum_{i=1}^{m} \Gamma_i^2 + \sum_{i=1}^{m} \sum_{j=1}^{n_i} \Delta_{ij}^2 \qquad (4)$$

Claim. $\Phi(A^1, ..., A^m)$ has degree $2m$, $O(n)$ variables, $n_0^{O(m)}$ monomials and maximum absolute value of its coefficients $O(nH^2)$.

3.1   Finite Representation of Nash Equilibria

Irrationality is not necessarily an obstacle towards obtaining a finite representation for a Nash equilibrium. For example, a real algebraic number α can be uniquely specified by the irreducible polynomial with integer coefficients, P, for which $P(\alpha) = 0$ and an interval which isolates the root α from the other roots of P. In the next Theorem we show that every game has a Nash equilibrium that can be finitely represented. The proof is based on a deep result from the theory of real closed fields known as the transfer principle [3]. We also need to use the fact that equilibria always exist. We are not aware if there is an alternative way of proving Theorem 1. The original topological proof of existence by Nash via Brouwer's fixed point theorem, though powerful enough to guarantee an equilibrium, does not seem to give any further information on the algebraic properties of the equilibria.

Theorem 1. *For every finite game there exists a Nash equilibrium* $x = (x_1, ...,$ $x_m)$ *such that every entry in the probability distributions* $x_1, \cdots, x_m$ *is an algebraic number.*

Proof. Given a game $(A^1, ..., A^m)$, the set of its Nash equilibria is the set of roots of the corresponding polynomial Φ (excluding the slack variables). By Nash's proof [19] we know that the equation $\Phi(A^1, ..., A^m) = 0$ has a solution over the reals. Consider the field of the real algebraic numbers R_{alg}. It is known that R_{alg} is a real closed field [3]. The Tarski-Seidenberg theorem, also known as the transfer principle (see [3]), states that for two real closed fields R_1, R_2 such that $R_2 \subseteq R_1$, a polynomial with coefficients in R_2 has a root in R_2 if and only if it has a root in R_1. The real numbers form a real closed field which contains R_{alg}. Since the coefficients of Φ are integers, it follows immediately that there exists a Nash equilibrium in R_{alg}.

A natural question is whether there are reasonable upper bounds for the degree and the coefficient size of the polynomials that represent the entries of an equilibrium. The known upper bounds are exponential. In particular, it follows by [3][Chapter 13] and by the Claim in Section 3 that the degrees of the polynomials will be $m^{O(n)}$ and the coefficient size will be $O(L + \log n)m^{O(n)}$.

3.2   Algorithmic Implications

A more practical goal to pursue is to compute an approximate equilibrium. For this we will use as a subroutine a decision algorithm for the existential theory of reals.

A special case of the decision problem for the existential theory of reals is to decide whether the equation $P(x_1, ..., x_k) = 0$ has a solution over the reals. Here P is a polynomial in k variables of degree d and with integer coefficients. The best upper bound for the complexity of the problem is $d^{O(k)}$, as provided by the algorithms of Basu et al. [2] and Renegar [21].

Theorem 2. *For an m-person game, $m \geq 2$, and for $0 < \epsilon < 1$, there is an algorithm which runs in time $poly(\log 1/\epsilon, L, m^n)$ and computes an m-tuple of strategies $x \in S_1 \times \cdots \times S_m$ such that:*

1. *x is ϵ/d-close to some Nash equilibrium y, where $d = 2^{m+1}n_0^m H$.*
2. *$|P^i(x) - P^i(y)| < \epsilon/2$ for all $i = 1, ..., m$.*
3. *x is an ϵ-Nash equilibrium.*

To prove Theorem 2, we need the following Lemma:

Lemma 1. *Let $y = (y_1, ..., y_m)$ be a Nash equilibrium. Let $x = (x_1, ..., x_m)$ be Δ-close to y, where $\Delta < 1$. Then:*

1. *x is an ϵ-Nash equilibrium for $\epsilon = 2^{m+1}n_0^m H\Delta$.*
2. *$|P^i(x) - P^i(y)| < \epsilon/2$ for all $i = 1, ..., m$.*

Proof. We give a sketch of the proof. Since x is Δ-close to y, each x_i can be written in the form $x_i = y_i + e_i$, where $e_i = (e_{i1}, ..., e_{i,n_i})$ and $|e_{ij}| \leq \Delta$. For condition 1, we need to prove that for every player i, $P^i(x) \geq P^i(x^{-i}, s_i^j) - \epsilon$, for every pure strategy s_i^j. Fix a pure strategy s_i^j. Then:

$$P^i(x) = \sum_{j_1} \cdots \sum_{j_m} A^i(j_1, ..., j_m)(y_{1,j_1} + e_{1,j_1}) \cdots (y_{m,j_m} + e_{m,j_m})$$

$$= P^i(y) + E_1 + \cdots + E_{2^m - 1}$$

where each term E_i is an m-fold sum. Since y is a Nash equilibrium we have:

$$P^i(x) \geq P^i(y^{-i}, s_i^j) + \sum E_i = P^i(x^{-i}, s_i^j) + \sum F_i + \sum E_i$$

where each F_i is an $(m-1)$-fold sum similar to the E_i terms. By performing some simple calculations we can actually show that: $|\sum E_i + \sum F_i| \leq \epsilon$. Hence $\sum E_i + \sum F_i \geq -\epsilon$. Due to lack of space we omit the details for the final version. The second claim can also be verified along the same lines.

From now on, let $\mathcal{A}$ be an algorithm that decides whether $P(x_1, ..., x_k) = 0$ has a solution over the reals in time $d^{O(k)}$, for a degree d polynomial P (either the algorithm of [2] or [21] will do).

Proof of Theorem 2:   By Lemma 1, we only need to find an m-tuple x such that x is ϵ/d-close to some Nash equilibrium y. Let $\Phi(A^1, ..., A^m)$ be the corresponding polynomial of the game. By the Claim in Section 3, the time to compute the coefficients of all the monomials of Φ, given the payoff matrices, is $n_0^{O(m)}$ which is $poly(m^n)$. We can now use $\mathcal{A}$ combined with binary search to compute a rational approximation of some root. Suppose we start with the variable x_{11}. We can add two more constraints to Φ expressing the fact that $x_{11} \in [0, 1/2]$. We then run $\mathcal{A}$ for the new polynomial and if the answer is yes we know that there exists an equilibrium with $x_{11} \in [0, 1/2]$. We can replace the constraints that we added with the ones corresponding to $x_{11} \in [0, 1/4]$. If the answer is no then there exists an equilibrium with $x_{11} \in [1/4, 1/2]$, hence we can continue our binary search in that interval. Proceeding in this manner we will find an interval I_{11} with length at most $\epsilon/(n_1 d)$. For this we need to run $O(\log n_1 d/\epsilon) = O(\log 1/\epsilon + m + m \log n + L) = poly(\log 1/\epsilon, L, m, n)$ times the algorithm $\mathcal{A}$. We will then add to Φ the constraints corresponding to $x_{11} \in I_{11}$ and we will go on to the next variable. When we are done with the variable x_{1,n_i-1}, the interval I_{1,n_i} for x_{1,n_i} is also determined. This is because x_{1,n_i} should be equal to $1 - \sum_{j \neq n_i} x_{1j}$, so that x_1 is a probability distribution. Therefore the length of I_{1,n_i} will be at most ϵ/d. Hence by the end of this step we know that we can select a probability distribution x_1 for the first player such that $|x_1 - y_1|_\infty \leq \epsilon/d$ for some Nash equilibrium y. We continue the procedure to determine an interval for every variable x_{ij}. We can then output a rational number in I_{ij} for each variable so as to ensure that $x_1, ..., x_m$ are probability distributions. Note that by the end we have only added $O(n)$ additional slack variables and constraints. Therefore the total running time will be $poly(\log 1/\epsilon, L, m^n)$.

An exact algorithm for 2-person games.   We can show that for 2-person games we can compute an exact Nash equilibrium using algorithm $\mathcal{A}$ as a subroutine. The crucial observation is that for 2-person games, if we know the support of the Nash equilibrium strategies, the exact strategies can be computed by solving a linear program. This is true because an equilibrium strategy for player 2 equalizes the payoff that player 1 receives for every pure strategy in her support and vice versa. Hence we can write a linear program and compute the Nash equilibrium with the given support since all the constraints are now linear. By adding constraints of the form $x_{ij} = 0$ and by running $\mathcal{A}$ a linear number of times, we can identify the support of some Nash equilibrium.

Theorem 3. *There exists an algorithm that runs in time $2^{O(n)}$ and computes an exact Nash equilibrium.*

Due to lack of space we omit the proof. This is yet another exponential algorithm for computing an equilibrium in 2-person games. An upper bound on the compexity of the problem can be obtained by the naive algorithm that tries all possible pairs of supports for the two players, which is $O(2^n LP_n^n) = 2^{O(n)}$, where LP_n^n is the time to solve a linear program with $O(n)$ variables and $O(n)$ constraints. Our algorithm achieves the same asymptotic bound but is in fact worse since the constant in the exponent is bigger than two. However we would still like to bring it to the attention of the community firstly because it is a different approach that has not been addressed before to the best of our knowledge and secondly because a future improvement in decision algorithms for low degree polynomial equations would directly imply an improvement in our algorithm too.

4   Approximation of Economic Equilibria

Similar algorithms can be obtained for computing market equilibria in exchange economies as well as in other economic models. We will briefly mention a special case of the neoclassical exchange model. More information on the general model can be found in [23].

Consider a market of m agents and n commodities (or goods). Each agent has an initial endowment $e_i \in R_+^n$. A continuous, strictly concave and increasing utility function $u_i : R_+^n \to R_+$ is associated with each agent.

Given a price vector $p \in R_+^n$, there exists a unique allocation of goods x to each agent i that maximizes her happiness subject to her spending constraints ($px \le pe_i$). Given a price vector p and an agent i we denote by $S_i(p)$ the desired allocation:

$$S_i(p) = (S_{i1}(p), ..., S_{in}(p)) = arg \max_{x \in R_+^n} u_i(x) \text{ s. t. } px \le pe_i \tag{5}$$

For commodity j, let $D_j(p)$ be the total demand for j, i.e., $D_j(p) = \sum_i S_{ij}(p)$. Finally $D(p) = (D_1(p), ..., D_n(p))$ will be called the demand function. We will make the assumption that for each commodity j the demand $D_j(p_1, ..., p_n)$ is a polynomial of degree d.

A price vector at which the market *clears* (goods can be exchanged such that all the agents maximize their total happiness) is called a *market equilibrium*. It is easy to see that such a vector will satisfy the conditions: $p \ge 0$, $D(p) - \sum e_i \le 0$, $p(D(p) - \sum e_i) = 0$.

Without loss of generality we can assume that the price vector lies on a simplex, i.e., $\sum p_i = 1$. That such an equilibrium always exists follows from the celebrated Arrow-Debreu theorem [1], which in turn is based on Kakutani's fixed point theorem. By using the same argument as in Theorem 1 we can show that there is always an equilibrium in which all the prices are algebraic numbers. Concerning the complexity of the problem, since all the equations above involve polynomials of degree at most $d + 1$ we have that:

Theorem 4. *For any $\epsilon > 0$, there is an algorithm that runs in time poly($\log 1/\epsilon$, d^n), and computes a vector p such that p is ϵ-close to a market equilibrium.*

This improves the bound that can be obtained by Scarf's algorithm, which is exponential in both $\log 1/\epsilon$ and n. More efficient algorithms for market equilibria have been recently obtained (see among others [6,7]) but only for linear utilities.

5   An Application: Systems of Polynomial Inequalities

Much of the research on equilibria in economic models has focused on the algorithmic problem of computing an equilibrium. A common approach has been to reduce the question to an already known and studied problem (e.g. fixed point approximations, linear and nonlinear complementarity problems, systems of polynomial equations and many others). In this section we would like to propose an alternative viewpoint and take advantage of the fact that Nash or market equilibria always exist. In particular, if a problem can be reduced to the existence of an equilibrium in a game or market, then we are guaranteed that a solution exists. As an example, we give the following theorem:

Theorem 5. *Let A be a $n \times n$ matrix and a_i be the i-th row of A. Let $S \subseteq \{1, ..., n\}$. Then the following system of inequalities in n variables $x = (x_1, ..., x_n)$*

$$x^T A x - a_i x \geq 0, \quad i \in S$$

has a nonzero solution. In fact it has a probability distribution as a solution.

Proof. Consider the *symmetric* game (A, A^T). It is known that every symmetric game has an equilibrium in which both players play the same strategy. The inequalities of the system correspond to the constraints that if both players play strategy x, a deviation to a pure strategy i, for $i \in S$ does not make a player better off.

Deciding whether a set of polynomial equations and inequalities has a solution (or a non-trivial solution) has been an active research topic. Similar theorems can be obtained for any system that corresponds to partial constraints for the existence of Nash equilibria or market equilibria. We do not know if an algebraic proof of Theorem 5 is already known. We believe that the existence of equilibria in games and markets can yield a way of providing simple proofs for the existence of solutions in certain systems of polynomial inequalities.

Acknowledgements. We would like to thank Saugata Basu and Aranyak Mehta for many useful comments and discussions.

References

1. Arrow, K., J., Debreu, G.: Existence of an Equilibrium for a Competitive Economy. Econometrica **22** (1954) 265-290
2. Basu, S., Pollack, R., Roy, M., F.: On the Combinatorial and Algebraic Complexity of Quantifier Elimination. Journal of the ACM **43**(6) (1996) 1002-1045
3. Basu, S., Pollack, R., Roy, M., F.: Algorithms in Real Algebraic Geometry. Springer-Verlag (2003)
4. Conitzer, V., Sandholm, T.: Complexity Results about Nash Equilibria. Technical report CMU-CS-02-135 (2002)
5. Datta, R.: Universality of Nash Equilibria. Math. Op. Res. **28**(3) (2003) 424-432

6. Deng, X., Papadimitriou, C., H., Safra S.: On the Complexity of Equilibria. Annual ACM Symposium on the Theory of Computing (2002) 67-71
7. Devanur, N., Papadimitriou, C., H., Saberi A., Vazirani, V., V.: Market Equilibria via a Primal-Dual-Type Algorithm. Annual IEEE Symposium on Foundations of Computer Science (2002) 389-395
8. Gilboa, I., Zemel, E.: Nash and Correlated Equilibria: Some Complexity Considerations. Games and Economic Behavior (1989)
9. Hirsch, M. D., Papadimitriou, C. H., Vavasis, S.A.: Exponential Lower Bounds for Finding Brouwer Fixed Points. Journal of Complexity 5 (1989) 379-416
10. Kearns, M., J., Littman, M., L., Singh, S., P.: Graphical Models for Game Theory. UAI (2001) 253-260
11. Kearns, M., J., Mansour, Y.: Efficient Nash Computation in Large Population Games with Bounded Influence. UAI (2002) 259-266
12. Koller, D., Megiddo, N.: Finding Mixed Strategies with Small Support in Extensive Form Games. International Journal of Game Theory 25 (1996) 73-92
13. Koller, D., Megiddo, N., von Stengel, B.: Efficient Computation of Equilibria for Extensive Two-person Games. Games and Economic Behavior 14(2) (1996) 247-259
14. Kuhn, H. W.: An Algorithm for Equilibrium Points in Bimatrix Games. Proceedings of the National Academy of Sciences 47 (1961) 1657-1662
15. Lemke, C. E.: Bimatrix Equilibrium Points and Mathematical Programming. Management Science 11 (1965) 681-689
16. Lemke, C. E., Howson, J. T.: Equilibrium Points of Bimatrix Games. Journal of the Society of Industrial and Applied Mathematics 12 (1964) 413-423
17. Lipton, R., J., Markakis, E., Mehta, A.: Playing Large Games using Simple Strategies. ACM Conference on Electronic Commerce (2003) 36-41
18. McKelvey, R., McLennan, A.: Computation of Equilibria in Finite Games. In Amman, H., Kendrick, D., Rust, J. eds Handbook of Computational Economics 1 (1996)
19. Nash, J. F.: Non-Cooperative games. Annals of Mathematics 54 (1951) 286-295
20. Papadimitriou, C., H.: Algorithms, Games, and the Internet. Annual ACM Symposium on the Theory of Computing (2001) 749-753
21. Renegar, J.: On the Computational Complexity and Geometry of the First Order Theory of Reals. Journal of Symbolic Computation 13(3) (1992) 255-352
22. Scarf, H.: The Approximation of Fixed Points of a Continuous Mapping. SIAM Journal of Applied Mathematics 15 (1967) 1328-1343
23. Scarf, H.: The Computation of Economic Equilibria (in collaboration with T. Hansen). Cowles Foundation Monograph, Yale University Press (1973)
24. Sturmfels, B.: Solving Systems of Polynomial Equations. Regional Conference Series in Mathematics, 97 (2002)
25. von Stengel, B.: Computing Equilibria for two-person Games. In Aumann, R. and Hart, S. eds, Handbook of Game Theory 3 (2002) 1723-1759

Minimum Latency Tours and the k-Traveling Repairmen Problem[*]

Raja Jothi and Balaji Raghavachari

Department of Computer Science, University of Texas at Dallas
Richardson, TX 75083-0688
{raja, rbk}@utdallas.edu

Abstract. Given an undirected graph $G = (V, E)$ and a source vertex $s \in V$, the k-traveling repairman (KTR) problem, also known as the minimum latency problem, asks for k tours, each starting at s and covering all the vertices (customers) such that the sum of the latencies experienced by the customers is minimum. *Latency* of a customer p is defined to be the distance (time) traveled before visiting p for the first time. Previous literature on the KTR problem has considered the version of the problem in which the repairtime of a customer is assumed to be zero for latency calculations. We consider a generalization of the problem in which each customer has an associated repairtime. In this paper, we present constant factor approximation algorithms for this problem and its variants.

1 Introduction

Given a finite metric on a set of vertices V and a source vertex $s \in V$, the k-traveling repairman (KTR) problem, a generalization of the metric traveling repairman problem (also known as the *minimum latency problem*, the *delivery man problem*, and the *school bus-driver problem*), asks for k tours, each starting at s (depot) and covering all the vertices (customers) such that the sum of the latencies experienced by the customers is minimum. *Latency* of a customer p is defined to be the distance traveled before visiting p for the first time. The KTR problem is NP-hard [10], even for $k = 1$. The problem remains NP-hard even for weighted trees [11].

The KTR problem with $k = 1$ is known as the minimum latency problem (MLP) in the literature. The first constant factor approximation for MLP was given by Blum et al. [3]. Goemans and Kleinberg [8] improved the ratio for MLP to 3.59α. In the following discussion, let α be the best achievable approximation ratio for the i-MST problem. The current best approximation ratio for the i-MST problem is $(2 + \epsilon)$, due to Arora and Karakostas [2], an improvement over the previous best ratio of 3, due to Garg [7]. Archer, Levin and Williamson [1] presented faster algorithms for MLP with a slightly better approximation ratio

[*] Research supported in part by the National Science Foundation under grant CCR-9820902.

M. Farach-Colton (Ed.): LATIN 2004, LNCS 2976, pp. 423–433, 2004.
© Springer-Verlag Berlin Heidelberg 2004

of 7.18. Recently, Chaudhuri et al. [4] have reduced the ratio by a factor of 2, to 3.59. They build on Archer, Levin and Williamson's techniques with the key improvement being that they bound the cost of their i-trees by the cost of a minimum cost path visiting i nodes, rather than twice the cost of a minimum cost tree spanning i nodes.

For the KTR problem, Fakcharoenphol, Harrelson and Rao [6], presented a 8.497α-approximation algorithm. Their ratio was recently improved to $2(2 + \alpha)$ by Chekuri and Kumar [5]. For a multidepot variant of the KTR problem, in which k repairmen start from k different starting locations, Chekuri and Kumar [5] presented a 6α-approximation algorithm. Recently, Chaudhuri et al. [4] have reduced the ratio to 6 for both the KTR problem and its multidepot variant.

1.1 Problem Statement

The Generalized KTR Problem. Literature on the KTR problem shows that all the results thus far are based on the assumption that the repairtime of a customer is zero for latency calculations. In this paper, we consider a generalization of the KTR problem (GKTR), the problem definition of which may be formalized as follows.

GKTR: *Given a metric defined on a set of vertices, V, a source vertex $s \in V$ and a positive number k. Also given is a non-negative number for each vertex $v \in \{V - s\}$, denoting the repairtime at v. The objective is to find k tours, each starting at s, covering all the vertices such that the sum of the latencies of all the vertices is minimum.*

It is easy to see that the GKTR problem resembles most real-life situations, one of which is that the repairmen have to spend some time at each customer's location, say, for the repair or installation of equipment. This applies even for a deliveryman who spends some time delivering goods. Hence, it is natural to formulate the repairman problem with repairtimes.

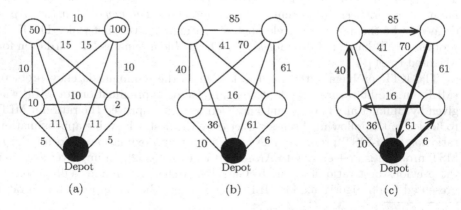

Fig. 1. (a) Original graph G. (b) Transformed graph G^*. (c) Optimal tour for G^*.

At first, even though it looks like that the GKTR problem can be reduced to the KTR problem in a straight forward manner, taking a deeper look into the problem reveals that such a reduction might not be possible without a compromise in the approximation ratio. One trivial idea would be to incorporate the repairtimes associated with vertices into edge weights (where the weight of an edge represents the time to traverse that edge), which can be done by boosting the edge weights as follows: for every edge e incident on vertices i and j in the given graph G, increase the weight (or distance) of e by the sum of $r_i/2$ and $r_j/2$, where r_i and r_j are the repairtimes of i and j respectively. Fig. 1 depicts such a transformation for a sample instance with $k = 1$. The resultant graph G^* after such a transformation will still obey triangle inequality, which allows us to use any of the KTR algorithms, say, with approximation guarantee β. The solution obtained would be a β-approximation for the modified graph G^*. However, the obtained solution will not be a β-approximation for the original problem G. This is due to the reason that the lower bounds for the problems defined as G and G^* are different, as can be seen from the fact that the latency of a customer v in an optimal solution to G^* comprises half of v's repairtime, while this is not the case with an optimal solution to G. At first, even though it looks like an optimal solution to G will be off by just a small constant when compared to an optimal solution to G^*, in reality, it could be arbitrarily large.

In this paper, we present a 3β-approximation algorithm[1] for the GKTR problem, where β is the best achievable approximation ratio (currently 6) for the KTR problem. When the repairtimes of all the customers are the same, we present an approximation algorithm with better ratio[2]. Our ratios hold for the respective multidepot variants of the GKTR problem as well.

The Bounded-Latency Problem. This problem is a complementary version of the KTR problem, in which we are given a latency bound L and are asked to find the minimum number of repairmen required to service all the customers such that the latency of no customer is more than L. More formally, we can define the bounded-latency problem (BLP) as follows:

BLP: *Given a metric defined on a set of vertices, V, a source vertex $s \in V$ and a positive number L. The objective is to find k tours, each starting at s, covering all the vertices such that the latency of no customer is more than L and k is minimum.*

The bounded-latency problem is very common in real-life as most service providers work only during the day, generally a 8-hour work day. Under these circumstances, the service provider naturally wants to provide service to all its outstanding customers within the work day, by using as small a number of repairmen as possible. For the BLP, we present a bicriteria approximation algorithm that finds a solution with at most $2/\rho$ times the number of repairmen required by an optimal solution, with the latency of no customer exceeding $(1 + \rho)L$, $\rho > 0$.

---

[1] Recently, Gubbala and Pursnani [9] have improved our analysis to obtain a ratio of $\frac{3}{2}\beta + \frac{3}{2}$.

[2] The ratio is 7.25, based on the current best β value, which is 6.

2   The GKTR Problem

Our algorithms for the GKTR problem uses the KTR algorithm as a black-box. The current best approximation ratio for the KTR problem is 6, due to Chaudhuri et al. [4]. Their ratio holds for the multidepot case as well.

Throughout this paper, the terms vertex and customer will be used inter-changeably. Let s denote the depot or the starting vertex. Let G be the complete graph induced by the vertex set V. Let r_i denote the repairtime of the customer i (repairtime of s is zero). Let $l(v)$ denote v's latency. Let $|ab|$ denote the weight of the edge connecting vertices a and b, which is the metric distance between a and b.

2.1   Non-uniform Repairtimes

Let $G = (V, E)$ be the given graph for which a solution is sought. Let $M \subset V$ be the set of vertices with k largest repairtimes. Let G' be the graph induced by $V \backslash M$. Construct a new graph G^* from G' such that for every edge e' incident on vertices i and j in G', introduce an edge e^* connecting i and j in G^* with weight $|ij| + \frac{r_i}{2} + \frac{r_j}{2}$. Make the repairtimes of all the vertices in G^* to be zero. It can be easily seen that the edges in G^* obey triangle inequality, and that G^* is a KTR instance. Let opt, opt' and opt^* denote the total latencies of all the customers in an optimum solution for G, G' and G^*, respectively. Let apx and apx' denote the total latencies of all the customers in our solution for G and G', respectively. Let APX' and APX^* denote the respective approximate solutions for G' and G^*. Before we proceed to the algorithm and its analysis, we present the following lemmas.

Lemma 1. $opt \geq opt'$.

Lemma 2. *Let $V = \{x_1, \ldots, x_n\}$ be the set of vertices in G. Let r_i denote the repairtime of x_i and let R_k denote the sum of the k largest repairtimes among the repairtimes of all vertices. Let opt be the sum of the latencies of all vertices in an optimal solution using k repairmen. Then,*

$$opt \geq \left[\sum_{i=1}^{n} \left(|sx_i| + r_i \right) \right] - R_k.$$

Proof. The fact that the latency of every vertex in an optimal solution is at least $|sx_i|$ and that such a solution has to include at least all, but the k largest, repairtimes proves the lemma.

Lemma 3. $opt' = opt^* - \sum_{i \in V \backslash M} \frac{r_i}{2}$.

Proof. We prove the lemma by showing that

$$opt' \geq opt^* - \sum_{i \in V \backslash M} \frac{r_i}{2} \quad \text{and} \quad opt' \leq opt^* - \sum_{i \in V \backslash M} \frac{r_i}{2}.$$

– $opt' \geq opt^* - \sum_{i \in V \setminus M} \frac{r_i}{2}$. Suppose $opt' < opt^* - \sum_{i \in V \setminus M} \frac{r_i}{2}$. Then, we can construct the same set of k tours in G^* as in G' such that the ith tour in G^* visits the same set of vertices as visited by the ith tour in G', and in the same order. The sum of the latencies of all the customers in such a solution for G^* will be $opt' + \sum_{i \in V \setminus M} \frac{r_i}{2}$, which contradicts the fact that opt^* is the optimal sum of latencies for G^*.

– $opt' \leq opt^* - \sum_{i \in V \setminus M} \frac{r_i}{2}$. Suppose $opt' > opt^* - \sum_{i \in V \setminus M} \frac{r_i}{2}$. Then, we can construct the same set of k tours in G' as in G^* such that the ith tour in G' visits the same set of vertices as visited by the ith tour in G^*, and in the same order. The sum of the latencies of all the customers in such a solution for G' will be $opt^* - \sum_{i \in V \setminus M} \frac{r_i}{2}$, which contradicts the fact that opt' is the optimal sum of latencies for G'.

We first obtain a β-approximate solution APX^* to G^*. Let $t_1, t_2, \dots t_k$ be the set of k tours in APX^*. Construct the same set of k tours in G' as in G^* such that the ith tour in G' visits the same set of vertices as visited by the ith tour in G^*, and in the same order. It can be seen that the sum of the latencies of all the customers in G' is

$$apx' = \beta opt^* - \sum_{j \in V \setminus M} \frac{r_j}{2}. \tag{1}$$

Let $M = \{v_1, v_2, \dots v_k\}$ be the set of vertices, with k largest repairtimes in G. Extending the tour t_i in G' to include v_i as its last vertex, for all i, gives a feasible set of k tours for G, with apx denoting the sum of the latencies of all the customers in G. The latency of vertex v_i, $l(v_i)$, added to the ith tour will be at most the sum of the latency of its predecessor vertex p_i (vertex visited by the ith tour just before visiting v_i), p_i's repairtime r_{p_i} and $|p_i v_i|$. Let p_i be the jth vertex visited in the ith tour and let $\{u_1, \dots, u_{j-1}\}$ be the other vertices visited by the ith tour before visiting p_i. Since $|sp_i| + |sv_i| \geq |p_i v_i|$, where s is the central depot, the latency of v_i can be written as follows.

$$l(v_i) \leq l(p_i) + r_{p_i} + |sp_i| + |sv_i|$$
$$\leq \left(\sum_{g=1}^{j-1} l(u_g) \right) + l(p_i) + r_{p_i} + |sp_i| + |sv_i|.$$

The sum of the latencies of all v_i's, where $i = 1 \dots k$, is given by

$$\sum_{i=1}^{k} l(v_i) \leq \sum_{i=1}^{k} \left[\left(\sum_{g=1}^{j-1} l(u_g) \right) + l(p_i) + r_{p_i} + |sp_i| + |sv_i| \right]$$
$$= \sum_{i=1}^{k} \left[\left(\sum_{g=1}^{j-1} l(u_g) \right) + l(p_i) \right] + \sum_{i=1}^{k} \left[r_{p_i} + |sp_i| + |sv_i| \right]$$
$$= apx' + opt \quad \text{(by Lemma 2)}$$
$$= \beta opt^* - \sum_{i \in V \setminus M} \frac{r_i}{2} + opt \quad \text{(by substituting (1))} \tag{2}$$

Substituting Lemma 3 in equations (1) and (2), we get

$$apx' = \beta\Big(opt' + \sum_{i\in V\setminus M}\frac{r_i}{2}\Big) - \sum_{i\in V\setminus M}\frac{r_i}{2}.$$

$$\sum_{i=1}^{k} l(v_i) \le \beta\Big(opt' + \sum_{i\in V\setminus M}\frac{r_i}{2}\Big) - \sum_{i\in V\setminus M}\frac{r_i}{2} + opt.$$

Recall that $\{v_1, v_2, \ldots v_k\}$ is the set of vertices with k largest repartimes in G. The sum of the latencies of all the customers in G is given by,

$$
\begin{aligned}
apx &= apx' + \sum_{i=1}^{k} l(v_i) \\
&\le 2\beta\Big(opt' + \sum_{i\in V\setminus M}\frac{r_i}{2}\Big) - \sum_{i\in V\setminus M} r_i + opt \\
&= 2\beta opt' + (\beta - 1)\sum_{i\in V\setminus M} r_i + opt \\
&\le 2\beta opt + (\beta - 1)\sum_{i\in V\setminus M} r_i + opt \quad \text{(by Lemma 1)} \\
&\le 2\beta opt + (\beta - 1)opt + opt \quad \text{(by Lemma 2)} \\
&= 3\beta opt.
\end{aligned}
$$

2.2   Uniform Repairtimes

At first, even though it looks like we can convert the original problem into one in which the repairtime is added to the length of all edges except those with s as an endpoint, and use the KTR algorithm as it is to get a ratio of β, it is not possible since doing such a transformation violates the triangle inequality property, which is a requirement for using the KTR algorithm. We do not know how to obtain a ratio same as that of the KTR problem.

When the repairtimes of all customers are the same, we show that the approximation guarantee can be improved considerably. To achieve a smaller approximation ratio, we present two algorithms that work at tandem. As before, let β be the best achievable approximation ratio for the KTR problem

Let $G = (V, E)$ be the given graph for which a solution is sought. Let r denote the repairtime of the customer i.e., $\forall_{i\neq s} r_i = r$ (repairtime of s is zero). Construct a new graph G^* from G such that for every edge e incident on vertices i and j in G, introduce an edge e^* connecting on i and j in G^* with weight $|ij| + \frac{r_i}{2} + \frac{r_j}{2} = |ij| + r$. Make the repairtimes of all the vertices in G^* to be zero. It can be easily seen that the edges in G^* obey triangle inequality and that G^* is a KTR instance. Let opt and opt^* denote the total latencies of all the customers in an optimum solution for G and G^*, respectively. Let apx denote the total latency of all the customers in our solution for G. Let n denote the number of vertices in the graph.

Proposition 1. *Let C be a positive constant. Let x and y be variables such that $x + y = C$. Then $x(x - 1) + y(y - 1)$ is minimum when $x = y$.*

Lemma 4. *In any optimal solution, the contribution to the sum of latencies due to repairtimes alone is at least $\frac{rn(\frac{n}{k}-1)}{2}$.*

Proof. Now suppose that for a given n and k, there exists an optimal solution for which the contribution due to repairtimes alone is opt_r. If, in such an optimal solution, there exists two repairmen who visit different number of customers, say y and z, then, we can always construct an alternate solution ALT from the optimal solution by making those two repairmen visit $\frac{y+z}{2}$ customers each. By Proposition 1, the contribution to the sum of latencies, in ALT, due to repairtimes alone will be less than opt_r. We can continue to find a feasible solution in this manner, until all repairmen visit the same number of customers, which is $\frac{n}{k}$. That brings us to the following equation, which proves the lemma.

$$opt_r \geq rk\frac{\frac{n}{k}(\frac{n}{k}-1)}{2} = \frac{rn(\frac{n}{k}-1)}{2}.$$

Algorithm 1. This algorithm proceeds on a case-by-case basis, based on the value of k with respect to n. Let the customers be sorted in non-decreasing order with respect to their distances to the depot. Let $A = \{c_1, \ldots, c_n\}$ be the sorted set of n customers, i.e., $|sc_1| \leq |sc_2| \leq \ldots \leq |sc_n|$.

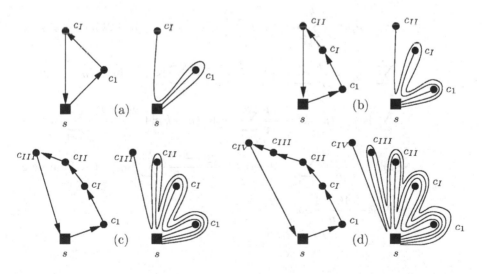

Fig. 2. (a) $\frac{n}{3} < k \leq \frac{n}{2}$ (b) $\frac{n}{3} < k \leq \frac{n}{2}$ (c) $\frac{n}{4} < k \leq \frac{n}{3}$ (d) $0.22598n < k \leq \frac{n}{4}$

1. *Case $k \geq \frac{n}{2}$.* The ith repairman visits customer c_i first, $\forall_{i \leq k}$. In addition, repairmen 1 to $n - k$ are assigned to visit one customer each, from the remaining pool of $n - k$ unassigned customers, as their second customer.

   Let t_1 be one of k such tours constructed in this manner. Let c_1, and maybe c_I, be the customers visited by tour t_1, in that order. The latency of c_1 would just be $|sc_1|$ and the latency of c_I would be at most $|sc_1| + r + |sc_1| + |sc_I|$. Fig. 2(a) depicts this pictorially. The sum of the latencies of customers c_1 and c_I is $|sc_1| + |sc_I| + r + 2|sc_1|$. The sum of the latencies of all the customers visited by k tours is then at most $\sum_{j=1}^{n} |sc_j| + (n - k)r + 2\sum_{j=1}^{n-k} |sc_j|$. By Lemma 2, the sum of the latencies is at most $opt + 2\sum_{j=1}^{n-k} |sc_j|$. Since A is sorted in non-decreasing order, the approximation ratio is less than or equal to $1 + 2(\frac{n-k}{n}) = 3 - 2(\frac{k}{n})$. As $k \geq \frac{n}{2}$, the ratio is at most 2.

2. *Case $\frac{n}{3} \leq k < \frac{n}{2}$.* The ith repairmen visits customer c_i first, $\forall_{i \leq k}$. Each repairmen picks one customer out of $\{c_{k+1}, \ldots, c_{2k}\}$ to be his next customer. In addition, repairmen 1 to $(n - 2k)$ are assigned to visit one customer each, from the remaining pool of $n - 2k$ unassigned customers, as their third customer.

   Let t_1 be one of k such tours constructed in this manner. Let c_1, c_I, and maybe c_{II}, be the customers visited by tour t_1, in that order. The latency of c_1 would just be $|sc_1|$. The latencies of c_I and c_{II} would be at most $|sc_1| + r + |sc_1| + |sc_I|$ and $|sc_1| + r + |sc_1| + |sc_I| + r + |sc_I| + |sc_{II}|$, respectively (see Fig. 2(b)). The sum of the latencies of customers c_1, c_I and c_{II} is $|sc_1| + |sc_I| + |sc_{II}| + 4|sc_1| + 2|sc_I| + r + 2r$. The sum of the latencies of all the customers visited by k tours is given by

$$
\begin{aligned}
apx &= \sum_{j=1}^{n} |sc_j| + 4\sum_{j=1}^{k} |sc_j| + 2\sum_{j=k+1}^{n-k} |sc_j| + kr + (n - 2k)2r \\
&\leq \sum_{j=1}^{n} |sc_j| + \frac{4k}{n}\sum_{j=1}^{n} |sc_j| + \frac{2(n - 2k)}{n - k}\sum_{j=1}^{n} |sc_j| + kr + (n - 2k)2r \\
&= \sum_{j=1}^{n} |sc_j| + (n-k)r + \frac{4k}{n}\sum_{j=1}^{n} |sc_j| + (n - k)r + \frac{2(n-2k)}{n-k}\sum_{j=1}^{n} |sc_j| - kr \\
&< opt + \frac{4k - 1}{n}\sum_{j=1}^{n} |sc_j| + opt + \frac{2(n - 2k)}{n - k}\sum_{j=1}^{n} |sc_j| \quad \text{(by Lemma 2)} \\
&\leq 2opt + \frac{4k - 1}{n}opt + \frac{2(n - 2k)}{n - k}opt \\
&= \left[1 + \frac{4k}{n} + \frac{2(n - 2k)}{n - k}\right]opt \\
&= \left[1 + 4x + \frac{2(1 - 2x)}{1 - x}\right]opt
\end{aligned}
$$

where $x = \frac{k}{n}$. Since $\frac{n}{3} \leq k < \frac{n}{2}$, it turns out to be that $apx \leq \frac{10}{3}opt$.

3. *Case* $\frac{n}{4} \leq k < \frac{n}{3}$. The algorithm (see Fig. 2(c)) and its analysis proceeds in the same manner as in case 2 and the sum of the latencies of all the customers visited by k tours is given by

$$apx < \left[1 + 6x + \frac{4x}{1-x} + \frac{2(1-3x)}{1-2x}\right]opt$$

where $x = \frac{k}{n}$. Since $\frac{n}{4} < k \leq \frac{n}{3}$, it turns out to be that $apx \leq 5.027opt$.

4. *Case* $\frac{n}{5} \leq k < \frac{n}{4}$. The algorithm (see Fig. 2(d)) and its analysis proceeds in the same manner as in case 2 and the sum of the latencies of all the customers visited by k tours is given by

$$apx \leq \left[1 + 8x + \frac{6x}{1-x} + \frac{4x}{1-2x} + \frac{2(1-4x)}{1-3x}\right]opt$$

where $x = \frac{k}{n}$. Since $\frac{n}{5} < k \leq \frac{n}{4}$, it turns out to be that $apx \leq 7opt$.

Algorithm 2. Just like in non-uniform repairtime case, we find a β-approximate solution APX^* to G^*. Let $t_1, t_2, \ldots t_k$ be the set of k tours in it. Construct the same set of k tours in G as in G^* such that the ith tour in G visits the same set of vertices as visited by the ith tour in G^*, and in the same order. It can be seen that the sum of the latencies of all the customers in G is

$$apx = \beta opt^* - \sum_{i=1}^{n} \frac{r_i}{2} = \beta opt^* - \frac{nr}{2}. \tag{3}$$

By Lemma 3,

$$opt = opt^* - \sum_{i=1}^{n} \frac{r_i}{2} = opt^* - \frac{nr}{2}.$$

Substituting for opt^* in equation (3), we get

$$apx = \beta opt + \left(\frac{\beta-1}{2}\right)nr.$$

Using Lemma 4, the approximation ratio of Algorithm 2 can be calculated from the above equation as follows.

$$\frac{apx}{opt} = \frac{\beta opt + \left(\frac{\beta-1}{2}\right)nr}{opt} \leq \beta + \frac{\left(\frac{\beta-1}{2}\right)nr}{\frac{rn(\frac{n}{k}-1)}{2}} \leq \beta + \left(\frac{\beta-1}{\frac{n}{k}-1}\right). \tag{4}$$

Substituting $\beta = 6$, it can be easily verified that the ratio is at most 7.25 for values of $k \leq \frac{n}{5}$. Since the ratio is bounded by 7 for values of $k \leq \frac{n}{5}$ (Algorithm 1) and 7.25 for values of $k \leq \frac{n}{5}$ (Algorithm 2), we get the following theorem.

3   The Bounded-Latency Problem

The bounded-latency problem (BLP) is a complementary version of the KTR problem, in which we are given a latency bound L and are asked to find the minimum number of repairmen required to service all the customers such that the latency of no customer is more than L. Unlike the GKTR problem where the sum of the latencies are minimized, the objective function of BLP is to minimize the number of repairmen with the constraint that the latency of no customer exceeds L. For this problem, we present a bicriteria algorithm that finds a solution with at most $2/\rho$ times the number of repairmen required by an optimal solution, with the latency of no customer exceeding $(1 + \rho)L$, $\rho > 0$.

Proposition 2. *Length of any optimal set of tours for the BLP is at least the length of an MST.*

Given below is an algorithm which groups the customers, so that a repairman can be assigned to each of the groups. Let $\rho > 0$.

1. Construct a tour for the given set of vertices (depot and customers) using the best available approximation algorithm for TSP.
2. Remove the depot from the tour.
3. Set `lengthTraveled = 0`.
4. Starting from some vertex, traverse the tour.
5. While not all edges in the tour are traversed, traverse the next edge e on the tour.
   a) If `lengthTraveled` + length(e) $\leq$ ρL, set `lengthTraveled +=` length(e).
   b) Else remove e from the tour, and set `lengthTraveled = 0`.

At the end of the above algorithm, we will be left with segments, each of length at most ρL. For each segment, introduce two edges to connect its endpoints (vertices) to the depot. Since our tour is of length at most twice than that of an MST, by Proposition 2, our solution will require at most $2/\rho$ repairmen to traverse the $2/\rho$ tours. Assuming that there exists a feasible solution for a given instance, the length of an edge connecting any vertex to the depot is at most L. Hence, regardless of the which direction each tour in our solution is traversed, each customer will have a latency of at most $(1 + \rho)L$.

References

1. A. Archer, A. Levin and D.P. Williamson, *Faster approximation algorithms for the minimum latency problem*, SODA 2003.
2. S. Arora and G. Karakostas, *A 2+epsilon approximation for the k-MST problem*, SODA 2000.
3. A. Blum, P. Chalasani, D. Coppersmith, B. Pulleyblank, P. Raghavan and M. Sudan, *The minimum latency problem*, SODA 1994.
4. K. Chaudhuri, B. Godfrey, S. Rao and K. Talwar, *Paths, tours, and minimum latency tours*, FOCS 2003.

5. C. Chekuri and A. Kumar, *A note on the k-traveling repairmen problem*, Manuscript, 2003.
6. J. Fakcharoenphol, C. Harrelson and S. Rao, *The k-traveling repairman problem*, SODA 2003.
7. N. Garg, *A 3-approximation for the minimum tree spanning k vertices*, FOCS 1996.
8. M. Goemans and J. Kleinberg, *An improved approximation ratio for the minimum latency problem*, SODA 1996.
9. P. Gubbala and H. Pursnani, Personal communication, November 2003.
10. S. Sahni and T. Gonzales, *P-complete approximation problems*, JACM, **23**(3), pp. 555-565, 1976.
11. R. Sitters, *The minimum latency problem is NP-hard for weighted trees*, IPCO 2002.

Server Scheduling in the Weighted ℓ_p Norm

Nikhil Bansal[1] and Kirk Pruhs[2*]

[1] Department of Computer Science, Carnegie Mellon University, USA
nikhil@cs.cmu.edu
http://www.cs.cmu.edu/~nikhil
[2] Department of Computer Science, University of Pittsburgh, USA
kirk@cs.pitt.edu
http://www.cs.pitt.edu/~kirk

Abstract. We explain how the apparent goals of the Unix CPU scheduling policy can be formalized using the weighted ℓ_p norm of flows. We then show that the online algorithm, Highest Density First (HDF), and the nonclairvoyant algorithm, Weighted Shortest Elapsed Time First (WSETF), are almost fully scalable. That is, they are $(1 + \epsilon)$-speed $O(1)$-competitive. Even for unit weights, it was known that there is no $O(1)$-competitive algorithm. We also give a generic way to transform an algorithm A in an algorithm B in such a way that if A is $O(1)$-speed $O(1)$-competitive with respect to some ℓ_p norm of flow then B is $O(1)$-competitive with respect to the ℓ_p norm of completion times. Further, if A is online (nonclairvoyant) then B is online (nonclairvoyant). Combining these results gives an $O(1)$-competitive nonclairvoyant algorithm for ℓ_p norms of completion times.

1 Introduction

1.1 Motivation

Tanenbaum [15, page 704] describes the generic Unix CPU scheduling policy as follows. Each process initially has a *nice* value in the range -20 to 20. Lower *nice* values correspond to processes that are more important. Users can set the *nice* value of a process to be in the range from 0 to 20 with a `nice` system call. Only the system administrator can give a process a negative *nice* value. Once a second the *priority* of a process is recalculated using the formula:

$$priority = CPUusage + nice + base$$

Here the *CPUusage* parameter is an exponential weighted moving average of past CPU usage, the *nice* parameter is the *nice* value for the process, and the *base* parameter is used to give higher priority to jobs that have just returned from some sort of interruption (say for I/O). Confusingly enough, the high priority

* Supported in part by NSF grant CCR-0098752, NSF grant ANI-0123705, and NSF grant ANI-0325353.

M. Farach-Colton (Ed.): LATIN 2004, LNCS 2976, pp. 434–443, 2004.
© Springer-Verlag Berlin Heidelberg 2004

jobs are those whose computed *priority* value is smallest. The jobs with highest priority are then scheduled using a Round Robin(RR) policy, typically with the quantum on order of 100 milliseconds. Round Robin shares the processor equally among all processes.

Round Robin represents an apparent effort to balance between optimizing the worst case Quality of Service (QoS) and optimizing the average case QoS. If the goal was to optimize worst case QoS then the best algorithm would be First Come First Served (FCFS). If the goal was to optimize average QoS then Shortest Elapsed Time First (*SETF*) is generally considered to be the best non-clairvoyant algorithm. Processes with lower *nice* values get more of the CPU, but the *CPUusage* parameter works to try to prevent starvation. That is, the *CPUusage* parameter will be high for processes that have been run a lot recently, and thus these processes will have a higher computed *priority*, and thus these processes will be given less CPU time in the near future. So it seems that the Unix system designers' goals for the process scheduling policy were:

Goal A: Amongst jobs of the same priority, there should be some balance between optimizing for average QoS and optimizing for worst case QoS.

Goal B: Higher priority jobs should get a greater share of the CPU resources, but lower priority jobs should not be starved.

In this paper we try to formalize these goals and then analyze algorithms with respect to this formalization.

In the literature, the most common QoS measure for a single process/job J_i is clearly flow/response/waiting time $f_i = c_i - r_i$, where c_i is the time that the job completes and r_i is the time that the job enters the system. The most common way to compromise between optimizing for the average and optimizing for the worst case is to optimize the ℓ_p norm, generally for something like $p = 2$ or $p = 3$. For example, the standard way to fit a line to collection of points is to pick the line with minimum least squares, equivalently ℓ_2, distance to the points, and Knuth's TeXtypesetting system uses the ℓ_3 metric to determine line breaks [12, page 97]. The ℓ_p, $1 < p < \infty$, metric still considers the average in the sense that it takes into account all values, but because x^p is strictly a convex function of x, the ℓ_p norm more severely penalizes outliers than the standard ℓ_1 norm. Analyses of algorithms for optimizing $(\sum F_i^p)^{1/p}$, the ℓ_p norms of flow, can be found in [3].

The most common way that priorities of jobs is formalized is to assume that each job J_i has a positive weight w_i and then to have the objective function be maximizing the weighted QoS. By far the most commonly studied QoS measure for a collection of equal priority jobs is average flow time, and logically enough, the most commonly studied QoS measure for jobs with variable priorities is weighted flow time $\sum w_i \cdot F_i$, e.g. [2,5,6,7]. It is easy to see that even an optimal algorithm for optimizing weighted flow time does not in general accomplish **Goal B** as it can starve low weight jobs if there are always higher weight jobs to be run.

If one wishes wishes to achieve both **Goal A** and **Goal B**, then the appropriate objective function to optimize would be something like the weighted ℓ_p

norms of flow, that is, $(\sum w_i F_i^p)^{1/p}$, where $p > 1$ is some small constant. Note that in any competitive schedule for the weighted ℓ_p norm of flow, a low weight job J_i would eventually be scheduled even in the face of a constant stream of high weight jobs.

In [3] it was shown that there is no $O(1)$-competitive online scheduling algorithm for any unweighted ℓ_p norm of flow. This motivated the authors of [3], and us, to fall back to resource augmentation analysis [9]. In the context of a scheduling minimization problem with an objective function F, an algorithm A is s-speed c-competitive if

$$\max_{\mathcal{I}} \frac{F(A_s(\mathcal{I}))}{F(Opt_1(\mathcal{I}))} \leq c$$

where $A_s(\mathcal{I})$ denotes the the schedule that algorithm A with a speed s produces on input $\mathcal{I}$, and similarly $Opt_1(\mathcal{I})$ denotes the adversarial schedule for $\mathcal{I}$ with a unit speed processor. A $(1 + \epsilon)$-speed $O(1)$-competitive algorithm is said to be *almost fully scalable* [13]. The intuition is that such an algorithm should perform well up to load close to the capacity of the system since increasing speed corresponds to lowering the load. This intuition is borne out in the lower bound instances, such as those in [3], that show no algorithm can be $O(1)$-competitive. In the lower bound instances, the system is fully loaded, so that there are no spare resources to recover from even small mistakes in scheduling decisions. For a more in depth discussion of this motivation see [9,3,13]. In [3] it is shown that several standard algorithms — SETF, Shortest Remaining Processing Time(SRPT), and Shortest Job First(SJF) — are almost fully scalable for any ℓ_p norm of flow. Surprisingly, RR is not almost fully scalable for any ℓ_p norm of flow. Note that this result would argue against the use of RR by Unix.

1.2   Our Results

We first show in section 3 that the results in [3] can be extended to the case where the objective function is the weighted ℓ_p norm of flow. In particular, we show that the algorithm Highest Density First(HDF) is almost fully scalable. HDF always runs the job that has the largest weight to work ratio. HDF is the natural generalization of SJF. Note however that HDF is clairvoyant, that is, it needs to know the work of a job at its release time. While this might be reasonable in a web server serving static documents, this is not reasonable in the context of an operating system.

We then show in section 4 that the obvious nonclairvoyant generalization of the nonclairvoyant algorithm SETF, Weighted Shortest Elapsed Time First (WSETF), is almost fully scalable. For a job J_i, let $x_i(t)$ denote the amount of work done on that job by time t. We define the measure of a job J_i as $\|J_i\|_t = \frac{x_i(t)}{w_i}$. Amongst the jobs with the smallest measure, WSETF splits the processor proportionally to weights of the jobs. So, if $J_1, \ldots, J_k$ are the jobs that have the smallest measure, then the job J_j will receive a $w_j/(\sum_{i=1}^k w_i)$ fraction of the processor. Thus this result suggests the adoption of the algorithm WSETF by Unix.

An interesting aspect of our analysis of *HDF* and *WSETF* is that we first transform the problem on the weighted instance to a related problem on the unweighted instance. This makes the problem simpler and also allows us to use previous results on unweighted scheduling.

There is a lot of literature on scheduling to minimize total/average completion time (a nice survey can be found in [11]), and average weighted completion time [8,1]. While this does not appear to be an interesting objective function from a computer systems point of view, it seems to be of general academic interest. So one natural academic question to ask is whether there are good online algorithms when the objective is the ℓ_p norm of completion time, or the weighted ℓ_p norm of completion time. In section 5 we give a rather generic way to transform an algorithm for a flow time problem, which possibly uses resource augmentation, to obtain an algorithm for the corresponding completion time problem, which *does not* use resource augmentation. A nice property of our transformation is that online algorithms are transformed to online algorithms, and non-clairvoyant algorithms are transformed to non-clairvoyant algorithms. As a corollary of this result, we will obtain $O(1)$ competitive online and non-clairvoyant algorithms for minimizing the ℓ_p norms of weighted completion time.

1.3   Other Related Results

The following results are known about online algorithms when the objective function is average flow time. The competitive ratio of every deterministic nonclairvoyant algorithm is $\Omega(n^{1/3})$, the competitive ratio of every randomized nonclairvoyant algorithm against an oblivious adversary is $\Omega(\log n)$ [14]. The randomized nonclairvoyant algorithm *RMLF*, proposed in [10], is $O(\log n)$-competitive against an oblivious adversary [4]. The online clairvoyant algorithm *SRPT* is optimal. The online clairvoyant algorithm *SJF* is almost fully scalable [5]. The nonclairvoyant algorithm *SETF* is almost fully scalable [9].

For online weighted flow time, the best known competitive ratio is $O(\log W)$ [2]. It is an outstanding open question whether an $O(1)$-competitive algorithm exists.

2   Definitions

We assume a collection of jobs $\mathcal{J} = J_1, \ldots, J_n$. For J_i, the release time is denoted by r_i, the work/size by p_i, and weight by w_i. Without loss of generality we assume that all job sizes and job weights are integers. The completion time c_i^S of a job J_i in a schedule S is the first time after r_i where J_i has been processed for p_i time units. The flow time of J_i in S is $f_i = c_i^S - r_i$. A clairvoyant algorithm learns p_i at time r_i. A nonclairvoyant algorithm only knows a lower bound on p_i equal to the length of time that it has run J_i. For an algorithm A on an input instance $\mathcal{I}$ with an s speed processor, let $F^p(A, \mathcal{I}, s)$ denote the sum of the pth powers of the flow time of all jobs. Similarly, $WF^p(A, \mathcal{I}, s)$ will denote the sum of weighted p^{th} powers of the flow time (i.e. $\sum_i w_i f_i^p$) of all jobs. Finally, for the measure F^p, let $Opt(F^p, \mathcal{I}, s)$ denote the value of the optimum schedule for the F^p measure on $\mathcal{I}$ with a speed s processor. Similarly, let $Opt(WF^p, \mathcal{I}, s)$ denote the optimum value for the WF^p measure.

3   Analysis of *HDF*

In this section we show that *HDF*, a natural generalization of *SJF* is a $(1 + \epsilon)$-speed $O(1/\epsilon^2)$-competitive online algorithm for minimizing the weighted ℓ_p norms of flow time.

The algorithm *HDF* at any time works on the job which has the largest weight to processing time ratio. The ties are broken in favor of the partially executed job. We will show that

Theorem 1. *HDF is $(1 + \epsilon)$-speed, $O(1/\epsilon^2)$-competitive for minimizing the weighted ℓ_p norms of flow time.*

The main idea of the proof will be to reduce the weighted problem to an unweighted problem and then invoke the result for ℓ_p norms of unweighted flow time. We first define the relevant notation.

Given an instance $\mathcal{I}$, we define an instance $\mathcal{I}'$ obtained by applying the following transformation to each job in $\mathcal{I}$: Consider a job $J_i \in \mathcal{I}$. The instance $\mathcal{I}'$ is obtained by replacing J_i by w_i identical jobs each of size p_i/w_i and weight 1, and release time r_i. We denote these w_i jobs by $J'_{i1}, \ldots, J'_{iw_i}$. Let $X_i = \{J'_{i1}, \ldots, J'_{iw_i}\}$ denote this collection of jobs obtained from J_i. Note that all jobs in $\mathcal{I}'$ have the same weight.

Lemma 1. *For $\mathcal{I}$ and $\mathcal{I}'$ as defined above,*

$$Opt(F^p, \mathcal{I}', 1) \le Opt(WF^p, \mathcal{I}, 1) \tag{1}$$

Proof. Let S be the schedule which minimizes the weighted ℓ_p norm of flow time for $\mathcal{I}$. Given S, we create a schedule for $\mathcal{I}'$ as follows. At any time t, work on a job in X_i if and only if J_i is executed at time t under S. Clearly, all jobs in X_i finish when J_i finishes execution, thus no job in X_i has a flow time higher than that of J_i. By definition, the contribution of J_i to WF^p is $w_i f_i^p$. Also, the contribution to the measure F^p of each of the w_i jobs in X_i will be at most f_i^p, and hence the total contribution of jobs in X_i to F^p is at most $w_i f_i^p$. Since the optimum schedule for $\mathcal{I}'$ can be no worse than the schedule constructed above, the result follows.

From Theorem 3 in [3] we know that *SJF* is $(1+\epsilon)$-speed, $O(1/\epsilon)$ competitive for the (unweighted) ℓ_p norms of flow time , or equivalently *SJF* is $(1 + \epsilon)$-speed $O(1/\epsilon^p)$ competitive for the F^p measure. This implies that,

$$F^p(SJF, \mathcal{I}', 1 + \epsilon) = O(\frac{1}{\epsilon^p})Opt(F^p, \mathcal{I}', 1) \tag{2}$$

We now relate the performance of *HDF* on $\mathcal{I}$ with a $(1 + \epsilon)$ times faster processor to that of *SJF* on $\mathcal{I}'$.

Lemma 2.

$$WF^p(HDF, \mathcal{I}, 1 + \epsilon) \leq (1 + \frac{1}{\epsilon})^p F^p(SJF, \mathcal{I}', 1) \tag{3}$$

Proof. We claim that for every job $J_i \in \mathcal{I}$ and every time t, if J_i is alive at time t under HDF with a $1 + \epsilon$ speed processor, then at least $\frac{\epsilon}{1+\epsilon} w_i$ jobs in $X_i \in \mathcal{I}'$ are alive at time t under SJF with a 1 speed processor.

The claim above immediately implies the result for the following reason. Consider the time $t^- = (f_i + r_i)^-$ just before J_i finishes execution under HDF. Then J_i contributes exactly $w_i f_i^p$ to $WF^p(HDF, \mathcal{I}, 1+\epsilon)$, while the $\geq \epsilon w_i/(1+\epsilon)$ jobs in X_i that are unfinished by time t contribute at least $\epsilon w_i/(1 + \epsilon) f_i^p$ to $F^p(SJF, \mathcal{I}', 1)$. Taking the contribution over each job, the result follows.

We now prove the claim. Suppose for the sake of contradiction that t is the earliest time when J_i is alive under HDF and there are fewer than $\epsilon/(1 + \epsilon) w_i$ jobs from X_i left under SJF. Since J_i is alive under HDF and HDF has a $1 + \epsilon$ faster processor, it has spent less than $p_i/(1 + \epsilon)$ time on J_i, whereas SJF has spent strictly more than $p_i/(1 + \epsilon)$ time on X_i. Thus there was a some time t', such that $r_i \leq t' < t$ during which HDF was running $J_j \neq J_i$ while SJF was working on some job from X_i. Since $t' \geq r_i$, it follows from the property of HDF that J_j has higher density than that of J_i. This implies that jobs in X_j have smaller size than X_i. Since SJF works on X_i at time t', it must have already finished all the jobs in X_j by t'. Since J_j is alive at time t', this contradicts our assumption of the minimality of t.

Proof. (of Theorem 1) By Equations 2 and 3 we have that

$$WF^p(HDF, \mathcal{I}, (1 + \epsilon)^2) = O(1/\epsilon)^{2p} Opt(F^p, \mathcal{I}', 1)$$

Combining this with Equation 1 gives us the result.

4   Analysis of *WSETF*

4.1   Algorithm Description

For a job J_i with weight w_i, let $p_i(t)$ denote the amount of work done on J_i by time t. We define the norm of a job J_i as $\|J_i\|_t = \frac{p_i(t)}{w_i}$.

Algorithm *WSETF*: At all times, *WSETF* splits the processor, proportional to weights of the jobs, among the jobs J_i that have the smallest norm $\|J_i\|_t$. So, if $J_1, \ldots, J_k$ are the jobs that have the smallest norm. Then J_j, for $i = 1, \ldots, k$, will receive $w_j/(\sum_{i=1}^{k} w_i)$ fraction of the processor.

Note that for all jobs J_i that *WSETF* executes, the norm increases at the same rate and thus stays the same.

4.2   Analysis

As in the analysis of *HDF* the main step of our analysis will be to relate the behavior of *WSETF* on an instance $\mathcal{I}$ with weighted jobs to that of *SETF* on another instance $\mathcal{I}'$ which consists of unweighted jobs. We then use the results about (unweighted) ℓ_p norms of flow time under *SETF* to obtain results for *WSETF*.

Given an instance $\mathcal{I}$ consisting of weighted jobs, let $\mathcal{I}'$ denote the instance defined as in Section 3 which consists of unweighted jobs. Suppose we run *WSETF* on $\mathcal{I}$ and *SETF* on $\mathcal{I}'$ with the same speed processor. Then the schedules produced by *WSETF* and *SETF* are related by the following simple observation.

Lemma 3. *At any time t, a job $J_i \in \mathcal{I}$ is alive and has received $p_i(t)$ units of service if and only if each job in $X_i \in \mathcal{I}'$ is alive and has received exactly $p_i(t)/w_i$ amount of service. In particular, this implies that if J_i has flow time f_i then each $J'_{ik} \in X_i$ for $k = 1, \ldots, w_i$ has flow time f_i.*

Proof. We view the execution of *WSETF* on $\mathcal{I}$ as follows: If at any time *WSETF* allocates x units of processing to a job of weight w_i, then we think of it as allocating x/w_i units of processing to each of the w_i jobs in the collection X_i. Thus the norm of job J_i under *WSETF* is exactly equal to the amount of service received by a job in X_i. Since *WSETF* at any time shares the processor among jobs with the smallest norm in the ratio of their weights, this is identical to the behavior of *SETF* on $\mathcal{I}'$ which works equally on the jobs which have received the smallest amount of service.

Theorem 2. *WSETF is a $1 + \epsilon$-speed, $O(1/\epsilon^{2+2/p})$-competitive non-clairvoyant algorithm for minimizing the weighted ℓ_p norms of flow time.*

Proof. By Lemma 3 we know that if $J_i \in \mathcal{I}$ has flow time f_i, then the w_i jobs in X_i have flow time f_i. Thus the ℓ_p norm of unweighted flow time for $\mathcal{I}'$ is $(\sum_i w_i f_i^p)^{1/p}$ which is identical to the weighted flow time for $\mathcal{I}$ under *WSETF*, which implies that

$$WF^p(WSETF, \mathcal{I}, 1) = F^p(SETF, \mathcal{I}', 1) \qquad (4)$$

By Equation 1 we know that $Opt(F^p, \mathcal{I}', 1) \le Opt(WF^p, \mathcal{I}, 1)$. By the main result of Section 7 in [3] about the competitiveness of *SETF* for unweighted ℓ_p norms of flow time we know that

$$F^p(SETF, \mathcal{I}', (1 + \epsilon)) = O(1/\epsilon^{2p+2})Opt(F^p, \mathcal{I}', 1) \qquad (5)$$

Now, by Equations 4, 5 and 1 we get that

$$WF^p(WSETF, \mathcal{I}, 1 + \epsilon) = O(1/\epsilon^{2p+2})Opt(WF^p, \mathcal{I}, 1)$$

Thus the result follows.

5   Completion Time Scheduling

In this section, we give a rather generic way to transform an algorithm for a flow time problem that possibly uses resource augmentation to obtain an algorithm for the corresponding completion time problem that does not use resource augmentation. Our transformation carries online algorithms to online algorithms and also preserves non-clairvoyance. As a corollary of this result we will obtain $O(1)$-competitive online and non-clairvoyant algorithms for minimizing the weighted ℓ_p norms of completion time.

We first make precise the notion of a completion time measure corresponding to a flow time measure. Given a schedule S for n jobs, this determines the flow times $f_1, \ldots, f_n$ and the completion times $c_1, \ldots, c_n$. Let $\mathcal{G}$ be some function that takes as input n real numbers and outputs another real number. Given a schedule S, we define the functions $\mathcal{F}$ and $\mathcal{C}$ as follows:

$$\mathcal{F}(S) = \mathcal{G}(f_1, f_2, \ldots, f_n)$$

$$\mathcal{C}(S) = \mathcal{G}(c_1, c_2, \ldots, c_n)$$

For example, if $\mathcal{G}(x_1, \ldots, x_n) = (\sum_i w_i x_i^p)^{1/p}$, then $\mathcal{F}$ and $\mathcal{C}$ are simply the weighted ℓ_p norms of flow time and completion time respectively.

Our technique for converting a flow time result to a completion time result will require two properties from the function $\mathcal{G}$.

Scalability: For any positive real number k, $\mathcal{G}(kx_1, \ldots, kx_n) = k\mathcal{G}(x_1, \ldots, x_n)$. In particular, if we scale all the flow times in a schedule by k times then $\mathcal{F}(S)$ increases by k times.

We now motivate the next property that we require from the function $\mathcal{G}$. We first point out a somewhat surprising property of the ℓ_p norms of the completion time measure. While it is easy to see that minimizing the total weighted flow time (i.e. ℓ_p norm with $p - 1$) is equivalent to minimizing the total weighted completion time, this is not the case for $p > 1$. In particular, it could be the case that a schedule which is optimum for the $\sum_i f_i^2$ measure is suboptimal for $\sum_i c_i^2$ measure and vice versa.

Consider the following instance with just two jobs. The first job has size 10 and arrives at $t = 0$, the second job has size 1 and arrives at $t = 8$. A simple calculation shows that in order to minimize the total flow time squared, it is better to first finish the longer job and then the smaller job. This incurs a total flow time squared of $10^2 + 3^2 = 109$, where as the other possibility which is to finish the small job as soon as it arrives an then finish the big job incurs a total flow time squared of $11^2 + 1^2 = 122$. On the other hand, if we consider completion time squared, finishing the larger job first incurs a cost of $10^2 + 11^2$. If instead if finish the smaller job first, this incurs a cost of $9^2 + 11^2$. Thus the optimal schedule for ℓ_p norms of flow time need not be optimal for ℓ_p norms of completion time and vice versa.

We say that a function $\mathcal{G}$ is $\rho - good$ if is satisfies the following condition:

Given a problem instance $\mathcal{I}$ and any two arbitrary schedules S and S' for $\mathcal{I}$. If $\mathcal{F}(S) \leq c\mathcal{F}(S')$, then $\mathcal{C}(S) \leq \rho c \mathcal{C}(S')$.

Lemma 4. $\mathcal{G}(x_1, \ldots, x_n) = (\sum_i w_i x_i^p)^{1/p}$ *is* $2 - good$ *for all* $p \geq 1$.

Our main result is the following:

Theorem 3. *Let* $\mathcal{G}$ *be a* $\rho - good$ *function. If there is an* s-*speed,* c-*competitive online algorithm with respect to the measure* $\mathcal{F}$ *(derived from* $\mathcal{G}$*), then this algorithm can be transformed into another online algorithm which is* 1-*speed,* ρcs-*competitive with respect to the corresponding completion time measure* $\mathcal{C}$. *Moreover, non-clairvoyant algorithms are transformed into non-clairvoyant algorithms.*

We now describe the transformation:

Let A be a s-speed, c-competitive algorithm for a flow time problem. Let $\mathcal{I}$ be the original instance where job J_i has release date r_i and size p_i. The online algorithm (which we call B) is the defined as follows:

1. When a job arrives at time r_i, pretend that it has not arrived till time sr_i.
2. At any time t, run A on the jobs for which $t \geq sr_i$

Proof. (of Theorem 3) Let I' be the instance obtained from $\mathcal{I}$ by replacing job $J_i \in \mathcal{I}$ by a job J_i' that has release date sr_i and size sp_i. Also, let $\mathcal{I}''$ be the instance from $\mathcal{I}$ by replacing the job $J_i \in \mathcal{I}$ with a job J_i'' that has release date sr_i and size p_i.

Let $Opt(\mathcal{F}, \mathcal{I}, x)$ (resp $Opt(\mathcal{C}, \mathcal{I}, x)$) denote the flow time cost (resp completion time cost) of the optimum schedule on $\mathcal{I}$ run using an x speed processor. We first relate the values of the optimum schedules for $\mathcal{I}$ and $\mathcal{I}'$.

Fact 4 $Opt(\mathcal{C}, \mathcal{I}', 1) = sOpt(\mathcal{C}, \mathcal{I}, 1)$

By our resource augmentation guarantee for the algorithm A, we know that

$$\mathcal{F}(A, \mathcal{I}', s) \leq cOpt(\mathcal{F}, \mathcal{I}', 1)$$

By the $\rho - goodness$ of $\mathcal{G}$ the above guarantee on flow time implies that

$$\mathcal{C}(A, \mathcal{I}', s) \leq c\rho Opt(\mathcal{C}, \mathcal{I}', 1) \tag{6}$$

We now relate $\mathcal{I}'$ to $\mathcal{I}''$.

Fact 5 $\mathcal{C}(A, \mathcal{I}', s) = \mathcal{C}(A, \mathcal{I}'', 1)$

Now, by definition of the algorithm B, executing the algorithm A on $\mathcal{I}''$ with a speed 1 processor is exactly the schedule produced by B on $\mathcal{I}$ using a 1 speed processor. So the completion times are identical. This implies that

$$\mathcal{C}(B, \mathcal{I}, 1) = \mathcal{C}(A, \mathcal{I}'', 1) \tag{7}$$

Now using Facts 4 and 5 and Equations 6 and 7 it follows that

$$\mathcal{C}(B, \mathcal{I}, 1) \leq c\rho s Opt(\mathcal{C}, \mathcal{I}, 1)$$

Thus we are done.

For $\mathcal{G}(x_1,\ldots,x_n) = (\sum_i w_i x_i^p)^{1/p}$, it is easily seen that the scalability property is satisfied, and Lemma 4 implies that it is $2 - good$. Thus by Theorems 1, 2 and 3 we get that

Corollary 1. *There exist $O(1)$-competitive clairvoyant and non-clairvoyant algorithms for minimizing the weighted ℓ_p norms of completion time.*

References

1. F. Afrati, E. Bampis, C. Chekuri, D. Karger, C. Kenyon, S. Khanna, I. Milis, M. Queyranne, M. Skutella, C. Stein, M. Sviridenko, "Approximation Schemes for Minimizing Average Weighted Completion Time with Release Dates", Foundations of Computer Science (FOCS), 32-44, 1999.
2. N. Bansal, K. Dhamdhere, "Minimizing Weighted Flow Time", ACM/SIAM Symposium on Discrete Algorithms (SODA), 508–516, 2003.
3. N. Bansal, K. Pruhs, "Server scheduling in the L_p norm: a rising tide lifts all boats", ACM Symposium on Theory of Computing (STOC), 242–250, 2003.
4. L. Becchetti, and S. Leonardi, "Non-Clairvoyant Scheduling to Minimize the Average Flow Time on Single and Parallel Machines" ACM Symposium on Theorcy of Computing (STOC), 2001.
5. L. Becchetti, S. Leonardi, A. Marchetti–Spaccamela, K. Pruhs, "Online weighted flow time and deadline scheduling", Workshop on Approximation Algorithms for Combinatorial Optimization Problems (APPROX), 2001.
6. C. Chekuri and S. Khanna, "Approximation schemes for preemptive weighted flow time", ACM Symposium on Theory of Computing (STOC), 2002.
7. C. Chekuri, S. Khanna and A. Zhu, "Algorithms for weighted flow time", ACM Symposium on Theory of Computing (STOC), 2001.
8. L.A. Hall, A. Schulz, D.B. Shmoys and J. Wein, "Scheduling to minimize average completion time: off-line and on-line approximation algorithms", Mathematics of Operations Research 22, 513–549, 1997.
9. B. Kalyanasundaram, and K. Pruhs, "Speed is as powerful as clairvoyance", *Journal of the ACM*, **47**(4), 617 – 643, 2000.
10. B. Kalyanasundaram, and K. Pruhs, "Minimizing flow time nonclairvoyantly", *Journal of the ACM*, July 2003.
11. D. Karger, C. Stein and J. Wein, "Scheduling algorithms", CRC handbook of theoretical computer science, 1999.
12. D. Knuth, *The TeXbook*, Addison Wesley, 1986.
13. K. Pruhs, J. Sgall, E. Torng, "Online Scheduling", to appear in *Handbook on Scheduling: Algorithms, Models and Performance Analysis*, CRC press.
14. R. Motwani, S. Phillips, and E. Torng, "Non-clairvoyant scheduling", *Theoretical Computer Science*, **130**, 17–47, 1994.
15. A. Tanenbaum, "Operating systems: design and implementation", Prentice-Hall, 2001.

An Improved Communication-Randomness Tradeoff

Martin Fürer*

Department of Computer Science and Engineering
Pennsylvania State University
University Park, PA 16802, USA
furer@cse.psu.edu,
http://www.cse.psu.edu/~furer

Abstract. Two processors receive inputs X and Y respectively. The communication complexity of the function f is the number of bits (as a function of the input size) that the processors have to exchange to compute $f(X, Y)$ for worst case inputs X and Y. The List-Non-Disjointness problem ($X = (x^1, \ldots, x^n)$, $Y = (y^1, \ldots, y^n)$, $x^j, y^j \in \mathbf{Z_2^n}$, to decide whether $\exists j \; x^j = y^j$) exhibits maximal discrepancy between deterministic n^2 and Las Vegas ($\Theta(n)$) communication complexity. Fleischer, Jung, Mehlhorn (1995) have shown that if a Las Vegas algorithm expects to communicate $\Omega(n \log n)$ bits, then this can be done with a small number of coin tosses.

Even with an improved randomness efficiency, this result is extended to the (much more interesting) case of efficient algorithms (i.e. with linear communication complexity). For any $R \in \mathbb{N}$, R coin tosses are sufficient for $O(n + n^2/2^R)$ transmitted bits.

1 Introduction

For many computations, communication is the decisive bottleneck. For example, in order to multiply two integers of length n each on a VLSI chip, it is necessary and sufficient to have an area time product $AT^2 = \Theta(n^2)$, for a the whole range of meaningful times ($c \log n \leq T \leq c'n$). Such a result is obtained by partitioning the chip into two parts, viewing each part as a separate computing agent and considering the required communication between the two agents. For a given time, this communication requires a certain bandwidth implying a bound on the width and area of the chip.

This is just one example illustrating the fact that to analyze the complexity of certain algorithms, it is useful to study the communication complexity in the two agent model introduced by Yao [11]. Two agents want to compute a function f. They receive inputs X and Y respectively. The communication complexity C assigns to every input size n the number of bits that have to be transmitted between the two agent, before one of them knows the value $f(X, Y)$ for worst case inputs X and Y of size n each (see Figure 1).

---

* Research supported in part by NSF Grant CCR-0209099

M. Farach-Colton (Ed.): LATIN 2004, LNCS 2976, pp. 444–454, 2004.

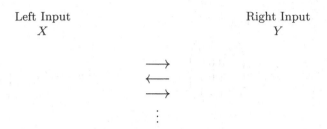

Left Input
X

Right Input
Y

back and forth messages until
one agent knows $f(X, Y)$

Fig. 1. Two agent communication complexity

The purpose of this paper is to study the effect of limited randomness on the communication complexity for reliable (i.e., Las Vegas) computations. Rather than using communication complexity as a tool to analyze the speed of algorithms, we want to design clever communication algorithms (protocols) to speed up the communication time, in order to study the tradeoff between communication and randomness.

We focus on one particular example that exhibits extremal behavior. The inputs X and Y are $n \times n$ matrices with entries from $\{0, 1\}$. Note that in this case, the size n of the input is not the number of bits in its representation. The List-Non-Disjointness problem (LND) asks whether there is an j such that matrices X and Y agree in their jth columns x^j and y^j respectively.

$$\mathrm{LND}(X, Y) = \begin{cases} 0 \text{ if } \forall j \; x^j \neq y^j \\ 1 \text{ if } \exists j \; x^j = y^j) \end{cases}$$

Figure 2 presents the List-Non-Disjointness problem with an example for which $\mathrm{LND}(X, Y)$ is 1 (or true).

Mehlhorn and Schmidt [8] have introduced the LND problem. They have determined its deterministic communication complexity

$$C_{\mathrm{Det}}(\mathrm{LND}) = n^2$$

and provided a good upper bound on its Las Vegas communication complexity

$$\bar{C}_{\mathrm{Las\ Vegas}}(\mathrm{LND}) = O(n \log^2 n)$$

The communication complexity $\bar{C}$ measures the expected number of bits transferred for a worst case input.

Aho, Ullman and Yannakakis [1] have shown that this result is almost optimal. They have proved that for any function f, even the nondeterministic communication complexity (where both computing agents can guess, but are not allowed to output a wrong value for $f(X, Y)$) is bounded by

$$C_{\mathrm{NDet}}(f) = \Omega(\sqrt{C_{\mathrm{Det}}(f)})$$

Left Input   Right Input

$$X = (x^1 \cdots x^n) \underset{\text{e.g.}}{=} \begin{pmatrix} 0\,1\,1\,0 \\ 1\,0\,0\,0 \\ 1\,1\,0\,1 \\ 0\,1\,1\,1 \end{pmatrix} \qquad Y = (y^1 \cdots y^n) \underset{\text{e.g.}}{=} \begin{pmatrix} 0\,0\,1\,1 \\ 1\,0\,0\,1 \\ 0\,1\,0\,1 \\ 1\,0\,1\,1 \end{pmatrix}$$

$$\longrightarrow$$
$$\longleftarrow$$
$$\longrightarrow$$
$$\vdots$$

One agent knows $\mathrm{LND}(X, Y)$

Fig. 2. The List-Non-Disjointness Problem (LND) with an example

trivially implying

$$\bar{C}_{\text{Las Vegas}}(f) = \Omega(\sqrt{C_{\text{Det}}(f)})$$

Indeed, using a more efficient communication algorithm, it has been shown that this lower bound is tight [6], because

$$\bar{C}_{\text{Las Vegas}}(\text{LND}) = \Theta(\sqrt{C_{\text{Det}}(\text{LND})})$$

i.e., LND exhibits the maximal possible discrepancy between deterministic and Las Vegas communication complexity. The communication algorithm exhibiting this discrepancy has not been optimized for its use of randomness.

In a more recent paper, Fleischer, Jung and Mehlhorn [5] have shown that instead of jumping directly from deterministic to Las Vegas protocols, one could interpolate between these two extremes by considering coin tosses as a limited resource. The paper shows that by increasing the number of random bits from R to $R + \log(R + 1) + O(1)$[1] one can provably decrease the communication complexity of LND for every R with

$$0 \le R \le \log n - \log \log n$$

This range of randomness corresponds to a range of communication between $\Theta(n^2)$ (deterministic case, $R = 0$) and $\Theta(n \log n)$ (using almost $\log n$ random bits). The interesting case of communication efficient computations, i.e., communication $O(n)$, is left out, because it cannot be handled by the communication protocol based on prime numbers [6,5].

Solving two of the three open problems in Fleischer, Jung and Mehlhorn [5], we improve their results in two respects:

- The range of the communication complexity is extended all the way down to the interesting region of $O(n)$. Before, only the number of random bits required for inefficient algorithms has been determined.

---

[1] log denotes the logarithm to the base 2

- The randomness is used more efficiently such that even increasing the number of random bits by 3 provably decreases the communication complexity.

Our improved communication algorithm is based on a simplified protocol [7] for the LND problem using ideas from universal hashing [2]. This protocol had been developed to exhibit a function with

$$AT^2_{\text{Det}} = \Theta(n^2)$$

but

$$AT^2_{\text{Las Vegas}} = \Theta(n \operatorname{polylog} n)$$

It is conjectured that for all functions f

$$AT^2_{\text{Las Vegas}} = \Omega\left(\sqrt{AT_{\text{Det}}}\right)$$

2 Definitions

There are two precise definitions of communication complexity in use. In the traditional definition, the two agents alternate between sending and receiving one bit [11]. In some literature focusing on the number of communication rounds [10], obviously longer messages are allowed. Both definitions are fine for most of our purposes, as we do not focus on communication rounds. But some of our results are so precise that constant factors matter. For these results, we don't want to use the first definition, as it might waste up to a factor of 2, and we don't want to use the second definition, as it implicitly includes an end-of-message signal that is sent for free.

Definition 1. *A two agent Communication Algorithm A (also called protocol) for a decision problem (i.e., a function $f : \{0, 1\}^* \times \{0, 1\}^* \to \{accept, reject\}$) is given by 3 functions $p_{\mathcal{L}}$, $p_{\mathcal{R}} : \{0, 1\}^* \times \{0, 1\}^* \to \{0, 1, accept, reject\}$ and $t : \{0, 1\}^* \to \{\mathcal{L}, \mathcal{R}\}$.*
When the input is (X, Y), and so far the string $w \in \{0, 1\}^*$ has been communicated (initially, the empty string has been communicated), then

- *if $t(w) = \mathcal{L}$ and $p_{\mathcal{L}}(X, w) \in \{0, 1\}$ then agent $\mathcal{L}$ sends bit $p_{\mathcal{L}}(X, w)$ to $\mathcal{R}$ (and concatenates this bit to w),*
- *if $t(w) = \mathcal{R}$ and $p_{\mathcal{R}}(Y, w) \in \{0, 1\}$ then agent $\mathcal{R}$ sends bit $p_{\mathcal{R}}(Y, w)$ to $\mathcal{L}$ (and concatenates this bit to w),*
- *if $p_{t(w)} \notin \{0, 1\}$ then the communication ends with output $f(X, Y) = t(w)$.*

The function t tells whose turn it is (to send a message or to stop). Both agents know the communicated string w and thus $t(w)$. The agent whose turn it is, may stop accepting or rejecting or send a bit to the other agent. Maximal consecutive bits sent by the same agent are called a message. Intuitively, the algorithm stops when one agent knows the output while the other agent waits for a message.

The *length of a communication* is the length $|w|$ of the communicated string w when Algorithm A stops. We write $w = A(X, Y)$, as w is the function of (X, Y) computed by Algorithm A.

Note that as we are using a non-uniform computation model, there really is a separate algorithm for every input size n.

This definition truly counts the number of bits sent, because the information about the end of a message is contained in the message itself. Note that both agents can compute the function t telling whose turn it is to send the next bit.

Definition 2. *For a given communication algorithm A, the* deterministic communication complexity *is defined by*

$$C_{Det}[A](n) = \max\{|w| : w = A(X, Y) \wedge size(X, Y) = n\}$$

For a function f, the deterministic communication complexity *is defined by*

$$C_{Det}(n) = \min_A C_{Det}[A](n)$$

where A ranges over all communication algorithms computing f.

Communication algorithms defined so far are also called "deterministic communication algorithms."

In a Las Vegas communication algorithm for f, the agents are allowed to toss coins. Nevertheless, the computed function has to be f for every outcome of the coin tosses. In this case, the communication complexity is the expected number of bits transmitted (expectation over the sequence of coin tosses). Here, we only consider ideal coins. Furthermore, we restrict ourselves to randomized algorithms where the number of coin tosses only depends on the input size, and both agents want to know the values of all random bits. Thus w.l.o.g., all coin tosses are done at the beginning by agent $\mathcal{L}$ who sends the outcome to $\mathcal{R}$.

Our assumption that all coin tosses are done at the beginning, implies that a Las Vegas algorithm A (for some input size n) really consist of a collection of deterministic algorithms A_r (one for each sequence of coin tosses $r \in \{0, 1\}^R$).

Definition 3. *A standard Las Vegas communication algorithm using $R = R(n)$ random bits is defined by $2^{R(n)}$ deterministic communication algorithms A_r ($r \in \{0, 1\}^{R(n)}$) for every input size n. It starts with agent $\mathcal{L}$ tossing $R(n)$ coins to obtain the string r and sending r to agent $\mathcal{R}$. Then the agents simulate Algorithm A_r without any further coin tosses. Furthermore, the complete algorithm is required to be Las Vegas, i.e., to produce the same output for every random sequence r.*

Our Las Vegas communication algorithms are standard, while for lower bounds, also algorithms compete where both agents may toss coins at any time. Thus for a given standard Las Vegas communication algorithm A, using $R(n)$

random bits, the Las Vegas communication complexity $\bar{C}^R_{\text{Las Vegas}}[A]$ is defined by

$$\bar{C}^R_{\text{Las Vegas}}[A](n) = R(n) + \frac{1}{2^{R(n)}} \sum_{r \in \{0,1\}^{R(n)}} C_{\text{Det}}[A_r](n)$$

For given functions f and R, the Las Vegas communication complexity is defined by

$$\bar{C}^R_{\text{Las Vegas}}(n) = \min_A \bar{C}^R_{\text{Las Vegas}}[A](n)$$

where A ranges over all Las Vegas communication algorithms computing f using $R(n)$ random bits. Finally, we define

$$\bar{C}_{\text{Las Vegas}}(n) = \min_R \bar{C}^R_{\text{Las Vegas}}(n)$$

3   Results

Various rank functions produce lower bounds for the communication complexity [8,4,5]. The rank function $2^{C_{\text{Det}}(f)}$ has been used to get the following lower bound for the Las Vegas communication complexity of LND with limited randomness.

Theorem 1. *(Lower Bound [5])*
For $0 \leq R \leq \log n$ $\bar{C}^R_{Las\ Vegas}(LND) \geq n^2/2^R$.

Actually, this lower bound holds for all non-negative integers R, but the result is trivial for $R \geq \log n$, because every Las Vegas algorithm has to transfer at least n bits before accepting an input of LND.

For $R < \log n$, this lower bound has been partially matched by the following upper bound [5].

$$\bar{C}^R_{\text{Las Vegas}}(\text{LND}) = O(\frac{Rn^2}{2^R})$$

The best result implied by this theorem without a limit on the coin tosses is

$$\bar{C}_{\text{Las Vegas}}(\text{LND}) = O(n \log n)$$

We want to present a communication algorithm A providing a better match for the lower bound. Not only is the communication algorithm A itself a randomized algorithm, but also the selection of A is done with a stochastic process. We just want to show the existence of a good algorithm. This is guaranteed as soon as the probability for the stochastic process to deliver a good algorithm is positive.

If the stochastic process did not guarantee anything better, one might rightly be hesitant to use such an algorithm A. Fortunately, things are much better. First of all, the stochastically produced algorithm A is unconditionally correct, only its running time is in question. Furthermore, with high probability the expected

Parameter Selection:
for $(r, i) \in \{0, \ldots, 2^R - 1\} \times \{1, \ldots, n\}$
      choose $z_i^{(r)} \in \mathbb{Z}_2^n$ uniformly at random
      if $z_i^{(r)}$ is linearly dependent of $z_1^{(r)}, \ldots, z_{i-1}^{(r)}$
            then $z_i^{(r)}$ is replaced by any vector of $\mathbb{Z}_2^n$ which is
            linearly independent of $z_1^{(r)}, \ldots, z_{i-1}^{(r)}$

Fig. 3. The Randomized Parameter Selection Algorithm

time complexity of the stochastically produced algorithm A does not exceed the expected time complexity of an optimal algorithm by more than a tiny fraction.

We interpret the randomized selection of an algorithm A, as the randomized selection of certain parameters $Z^{(r)}$ of a single algorithm A. For each $r \in \{0, \ldots, 2^R - 1\}$, $Z^{(r)}$ is an $n \times n$ $\{0, 1\}$-matrix with $z_i^{(r)}$ being its ith row. The selection of these parameters is described in the Parameter Selection Algorithm of Figure 3.

The basic idea of Algorithm A is quite simple. Instead of transmitting a column x^j of X to compare it with column y^j of Y, we could just as well transmit the matrix product $Z^{(r)} x^j$ for a regular matrix $Z^{(r)}$ and compare it with $Z^{(r)} y^j$, but obviously, there would be no advantage in doing so. But if we replace the whole matrix $Z^{(r)}$ by just some rows of it, then we have a nice short hash value to transmit instead of the whole x^j. If y^j hashes to a different value (for the same hash function), then we have discovered a difference between x^j and y^j without transferring all of x^j. To reduce the chance of a collision of hash values for distinct x^j and y^j, we employ a randomized selection of the hash functions defined by the matrices $Z^{(r)}$.

The Randomized Parameter Selection Algorithm (Figure 3) preselects for each $r \in \{0, \ldots, 2^R - 1\}$ a random sequence

$$z_i^{(r)} \quad (j = 1, \ldots, n)$$

of vectors to hash the columns x^j of X and Y by scalar products. This initial random selection can be slightly improved by replacing useless vectors (linearly dependent on previous ones) $z_i^{(r)}$ by good ones.

With these parameters selected, the randomized algorithm A (Figure 4) starts by first tossing R coins to select $r \in \{0, \ldots, 2^R - 1\}$. It then proceeds deterministically, transmitting for each column x^j a sequence of scalar products until a difference to column y^j is discovered. If instead, one pair of columns is found equal, the algorithm stops immediately.

To obtain the precise constant factors as claimed in the theorem, Algorithm A is slightly optimized. $\mathcal{L}$ transmits several bits as one message, but receives a one bit answer indicating whether a difference between x^j and y^j has just been detected. Giving an answer after every bit is a good strategy for a random input, but giving an answer after several bits is a slightly better strategy for worst case inputs with respect to the previously selected parameters $Z^{(r)}$.

Algorithm A:
```
L tosses R coins to form the binary number
r ∈ {0,...,2^R − 1}
L sends r (to R)
m ← 0
for j ← 1 to n do
      for i ← 1 to n do
            m ← m + 1
            L sends wm ← z_i^(r) x^j   (inner product of vectors)
            if z_i^(r) x^j = z_i^(r) y^j then equal = 1 else equal = 0
            if i = n ∧ equal = 1 then R halts accepting
            if j = n ∧ equal = 0 then R halts rejecting
            m ← m + 1
            L receives the answer wm ← equal (from R)
```

Fig. 4. The Algorithm

We use a Chernoff bound to obtain our upper bound. We use the version on page 70 in the book of Motwani and Raghavan [9] specialized to simple case of a sequence of Bernoulli trials.

Theorem 2. *(Chernoff Bound [3,9])*
Let $X_1, X_2, \ldots, X_m$ be independent Bernoulli trials, such that, for $1 \le i \le m$, $Pr[X_i = 1] = p$, where $0 < p < 1$. Then, for $X = \sum_{i=1}^{m} X_i$, $\mu = E[X] = mp$, and $0 < \delta \le 1$,

$$Pr[X < (1 - \delta)\mu] < e^{-\mu\delta^2/2)}$$

We only use the case $p = \frac{1}{2}$ implying $\mu = \frac{1}{2}m$. Furthermore, we let $m = \lfloor \alpha n \rfloor$ for some constant $\alpha > 1$ and $(1 - \delta)\mu = n$.

Theorem 3. *For $m = \lfloor \alpha n \rfloor$, let $X_1, X_2, \ldots, X_m$ be independent Bernoulli trials, such that, for $1 \le i \le m$, $Pr[X_i = 1] = \frac{1}{2}$. Let $X = \sum_{i=1}^{m} X_i$. Then*

$$Pr[X < 2^R] < 2^{-n}$$

holds for $\alpha \ge \beta = 2(1 + \ln 2 + \sqrt{2\ln 2 + \ln^2 2})$ (which is < 6.11888) and $2^R \le n$.

$$Pr[X < 2^R] < 2^{-n}$$

also holds for $\alpha > \gamma = 4\ln 2$ (which is < 2.77259), $R = O(1)$ and n sufficiently large.

Proof. We apply Theorem 2 with $p = \frac{1}{2}$ (implying $\mu = \frac{1}{2}m$), $m = \lceil \alpha n \rceil$, and $(1 - \delta)\mu = 2^R$. Then

$$\delta = 1 - \frac{2^R}{\mu} = 1 - \frac{2^{R+1}}{m} = 1 - \frac{2^{R+1}}{\lceil \alpha n \rceil}$$

implying

$$\Pr[X < 2^R] = \Pr[X < (1+\delta)\mu]$$
$$< \exp(-\mu\delta^2/2)$$
$$= \exp(-\lceil\alpha n\rceil\delta^2/4)$$
$$\leq \exp(-\alpha n\delta^2/4)$$

Thus

$$\Pr[X < 2^R] < 2^{-n}$$

if

$$\frac{1}{4}(\log e)\alpha\delta^2 \geq 1$$

or equivalently

$$\frac{1}{4}(\log e)\alpha\left(1 - \frac{2^{R+1}}{\lceil\alpha n\rceil}\right)^2 \geq 1$$

In other words, the quadratic inequality in α

$$\frac{1}{4}(\log e)\alpha\left(1 - \frac{2^{R+1}}{\alpha n}\right)^2 \geq 1$$

is sufficient to obtain

$$\Pr[X < 2^R] < 2^{-n}$$

For this quadratic inequality in α, we are only interested in the solutions α with $\delta > 0$, i.e.,

$$\frac{2^R}{\mu} = \frac{2^{R+1}}{\lceil\alpha n\rceil} < 1$$

These solutions are

$$\alpha \geq 2\left(\frac{2^R}{n} + \ln 2 + \sqrt{\frac{2^{R+1}}{n}\ln 2 + \ln^2 2}\right)$$

Hence, for $2^R \leq n$ and $\alpha \geq 2(1 + \ln 2 + \sqrt{2\ln 2 + \ln^2 2}) < 6.11888$ (e.g., $\alpha = 7$) is sufficient for all n. For constant R, and n sufficiently large, any $\alpha > 4\ln 2 < 2.77259$ (e.g., $\alpha = 2.8$) is sufficient. □

Theorem 4. *With R bits of randomness, the LND problem can be decided within communication complexity*

$$\bar{C}^R_{Las\ Vegas}(n) \leq R + \alpha\frac{n^2}{2^R} + 2\sqrt{\frac{\alpha n^3}{2^R}}$$

for any $\alpha \geq \beta$ of Theorem 3.

Proof. The first term is from sending the random bits, while the second term is a bound on the expected number of hash bits for a worst case input (X, Y) with disjoint lists. The third term represents the expected number of answering bits after selecting a message length of $\sqrt{an/2^R}$. An extra term for the case of non-disjoint lists can be avoided with a simple trick. With probability $1/2$ (just reuse the first random bit) the left and right halves of the matrix X are swapped. The same is done with matrix Y. As a consequence, it is always expected to discover equal columns sufficiently early. □

4 Open Problems

The main open problem in this area is the question whether there is a more intelligent combinatorial construction of the Las Vegas algorithm (parameter selection) than by random choice. Upper and lower bounds are already very close together. Nevertheless such a construction might decrease the constant factor gap between upper and lower bounds. More important, a better explicit construction seems like an interesting and challenging problem in combinatorics.

Clearly for very small R some improvement is possible. For $R = 1$ an optimal choice of parameters is the following: $z_1^{(0)}, \dots, z_n^{(0)}$ is any sequence of basis vectors of $\mathbb{Z}_2^n$, and $z_1^{(1)}, \dots, z_n^{(1)}$ is the reverse sequence $z_n^{(0)}, \dots, z_1^{(0)}$.

Another remaining open problem in this area is the extension of limited randomness results to Las Vegas bounded round communication complexity.

References

1. Alfred V. Aho, Jeffrey D. Ullman, and Mihalis Yannakakis. On notions of information transfer in VLSI circuits. In *Proceedings of the fifteenth annual ACM symposium on Theory of computing*, pages 133–139, 1983.
2. J. Lawrence Carter and Mark N. Wegman. Universal classes of hash functions. *Journal of Computer and System Sciences*, 18(2):143–154, April 1979.
3. H. Chernoff. A measure of asymptotic efficiency for tests of a hypothesis based on the sum of observations. *Annals of Math. Stat.*, 23:493–509, 1952.
4. Rudolf Fleischer. Communication complexity of multi-processor systems. *Information Processing Letters*, 30(2):57–65, January 1989.
5. Rudolf Fleischer, Hermann Jung, and Kurt Mehlhorn. A communication-randomness tradeoff for two-processor systems. *Information and Computation*, 116(2):155–161, 1 February 1995.
6. M. Fürer. The power of randomness for communication complexity. In *Proceedings of the nineteenth annual ACM conference on Theory of computing*, pages 178–181. ACM Press, 1987.
7. M. Fürer. Universal hashing in VLSI. In John H. Reif, editor, *VLSI algorithms and architectures: Proceedings of the 3rd Aegean Workshop on Computing, AWOC 1988, Corfu, Greece*, pages 312–318. Springer-Verlag, LNCS 319, 1988.
8. Kurt Mehlhorn and Erik M. Schmidt. Las Vegas is better than determinism in VLSI and distributed computing (extended abstract). In *Proceedings of the Fourteenth Annual ACM Symposium on Theory of Computing*, pages 330–337, 1982.

9. Rajeev Motwani and Prabhakar Raghavan. *Randomized algorithms*. Cambridge University Press, 1995.
10. Christos H. Papadimitriou and Michael Sipser. Communication complexity. In *Proceedings of the fourteenth annual ACM symposium on Theory of computing*, pages 196–200, 1982.
11. A. C. Yao. Some complexity questions related to distributive computing (preliminary report). In *Proceedings of the eleventh annual ACM symposium on Theory of computing*, pages 209–213, 1979.

Distributed Games and Distributed Control for Asynchronous Systems

Paul Gastin*, Benjamin Lerman*, and Marc Zeitoun*

LIAFA, Université Paris 7 & CNRS
2, pl. Jussieu, case 7014
F-75251 Paris cedex 05, France
{Paul.Gastin, Benjamin.Lerman, Marc.Zeitoun}@liafa.jussieu.fr

Abstract. We introduce distributed games over *asynchronous* transition systems to model a distributed controller synthesis problem. A game involves two teams and is not turn-based: several players of both teams may simultaneously be enabled. We define distributed strategies based on the *causal* view that players have of the system. We reduce the problem of finding a winning distributed strategy with a given memory to finding a memoryless winning distributed strategy in a larger distributed game. We reduce the latter problem to finding a strategy in a classical 2-players game. This allows to transfer results from the sequential case to this distributed setting.

Keywords. Distributed game, distributed control, distributed strategy.

1 Introduction

The controller synthesis problem has been widely investigated by many authors for different system types (sequential, concurrent, timed, probabilistic) and different specification languages (linear or branching temporal logics for instance). The variant addressed here, called distributed controller synthesis problem, is the following. We are given a *distributed reactive system* executing a program in some environment, modeled by an *asynchronous* transition system [13] made up of several processes. It can perform local actions and synchronization actions involving several processes. Such an action first reads states of the participating processes and, depending on what is read, chooses a transition changing the states of the involved processes. Interpreting processes as memory locations, this suggests communication via shared memory, and explains the terminology adopted in the paper. However, one can as well simulate with these asynchronous systems other communication paradigms, such as point-to-point channels. Another advantage of this model is to handle actions of the environment just as actions of the reactive system. We are also given a *specification*, i.e., a property expressing behaviors one wants to ensure for the system. The distributed

* Work partly supported by the European research project HPRN-CT-2002-00283 GAMES and by the ACI Sécurité Informatique 2003-22 (VERSYDIS).

M. Farach-Colton (Ed.): LATIN 2004, LNCS 2976, pp. 455–465, 2004.

controller synthesis problem is then to compute a distributed controller on the same process set (actions of the controller observe local states of some processes). With this information, the controller has to enable or disable controllable actions of the system, so that the *overall* system behaves correctly according to the specification.

The problem is known to be decidable and to have an optimal solution in the sequential case [10], *i.e.*, when there is a single process. For distributed systems however, the situation is more involved. Several models have been considered so far (formulated in control or in game theory terminology). For synchronous processes communicating via buffers, the problem is undecidable [9] for LTL specifications except for very few communication architectures. Recent works [5, 4] extend this result, *e.g.*, for local specifications (talking only of actions of single processes). The approach of [7] unifies [9,5,4]. In all settings, a major reason for undecidability is when the specification language makes it possible to express properties of an *observed linearization* of process actions, ignoring their possible concurrency. Another distributed model is studied in [6], and it is shown that the existence of specific controllers is decidable for specification languages making no distinction between linearizations of the same concurrent execution.

Systems used in this paper subsume those of [9,5,4,7] and [6]. In [6], global transitions of the system are obtained by synchronizing local transitions of processes, and transitions of the environment are local. Here in contrast, a transition of a synchronizing action also depends on the states of other involved processes, so that transition functions of actions are not necessarily a cartesian product of local transition functions. Further, environment moves can be defined globally. Systems of [7], in which the environment is global and transitions of the system are purely local (process communication flows through the environment) can also be modeled naturally in our framework. Another difference is that [6,7] use *local* memory controllers, based on the history of process local states (cf. Sec. 5). We use *causal* memory: a controller can remember information collected from other processes along the computation. The existence of a distributed controller in the settings of [6,7] implies the existence of a controller in our setting. As the converse does not hold, one cannot transfer immediately the undecidability results of [6,7] to our case.

Our primary goal is to model the distributed controller synthesis problem by *games*. Sequential 2-players games already provide a natural and widely used context to model sequential reactive systems [8,11,14,12]. Player 0 represents the system and player 1 represents the environment. The rules of the game describe the possible interactions between them, and the winning condition for player 0 expresses the specification that the system should meet. Thus, deciding whether player 0 has a winning strategy corresponds to deciding whether the system can be controlled to meet the specification, and computing a winning strategy for player 0 corresponds to solving the controller synthesis problem.

Distributed games proposed in this paper fit suitably to the model of asynchronous systems and supply a natural framework for studying the distributed controller problem. Two teams play one against the other. Players of team 0 may be viewed as controllable actions of a distributed system which cooperate in order to meet the specification, no matter how the environment (team 1) behaves. All players use a pool of shared variables to transmit information. The game is not turn-based: in each position of the game, several players of team 0 or team 1 may be simultaneously enabled. Thus, the game is asynchronous, contrary to the setting of [7] where at each stage players act synchronously. Assuming that dependencies between actions are fixed (*i.e.*, do not depend on the context), a play is then a Mazurkiewicz trace.

We next define the notion of distributed strategy for a team in such a game. Roughly speaking, a strategy is distributed if any move it predicts for a player only depends on the causal view of that player. In this context, there exist games in which neither team 0 nor team 1 have a winning distributed strategy.

We first show that, as in the sequential framework, one can transform a game G for which team 0 has a distributed strategy with memory μ into a game G^μ for which team 0 has a memoryless strategy. If G and the memory are finite, then so is G^μ. We further transform a distributed game G into a classical 2-players game $\widetilde{G}$, such that team 0 has a memoryless distributed strategy in G if and only if player 0 has a memoryless winning strategy in $\widetilde{G}$. This result is effective: $\widetilde{G}$ can be effectively constructed and from a winning strategy of $\widetilde{G}$ we can effectively construct a winning distributed strategy for G and vice versa. We then show that if the winning condition is a recognizable trace language, then one can decide whether team 0 has a memoryless distributed winning strategy, and compute it. The restriction to recognizable specifications is not artificial and makes it possible to express a relationship between the architecture of the game and the specification. Moreover, in practice, recognizable languages cover most interesting properties. As in [6,7], specifications depending on the order of independent actions lead to undecidability.

Finally, we explain how to simulate distributed games of [7] in our context. The goals of the two kinds of games are quite different. The aim of [7] is to unify different approaches and to find generic transformations on distributed games (*e.g.*, the reduction of the number of players) to get decidability results, while we first focused on a natural model which we then reduced to sequential games. Due to space constraints, proofs are omitted.

2   Preliminaries

In this section, we briefly recall definitions of pomsets and Mazurkiewicz traces. The reader is referred to [3,2] for details.

If $(V, \leqslant)$ is a poset and $S \subseteq V$, we let $\downarrow S = \{e \in V \mid \exists s \in S, e \leqslant s\}$. When $e \in V$ then we simply write $\downarrow e$ for $\downarrow\{e\}$ and we let $\Downarrow e = \downarrow e \setminus \{e\}$. The successor relation associated with the partial order $<$ is $\lessdot\, =\, <\, \setminus <^2$.

A *pomset* over an alphabet Σ is a tuple $(V, \leqslant, \ell)$ where $(V, \leqslant)$ is a poset, and $\ell : V \to \Sigma$ is a mapping called the *labeling*. Elements of V are called *events* or *vertices*. Two pomsets $t = (V, \leqslant, \ell)$, and $t' = (V', \leqslant', \ell')$ are *isomorphic*, written $t \sim t'$ if there exists a bijection $\varphi : V \to V'$ such that $\ell' \circ \varphi = \ell$ and for all $e, f \in V$ $e \leqslant f$ iff $\varphi(e) \leqslant' \varphi(f)$. If $\Sigma = \Sigma_1 \times \Sigma_2$ and $\ell(e) = (\ell_1(e), \ell_2(e))$, we write (ℓ_1, ℓ_2)

(or even ℓ_1, ℓ_2) instead of ℓ. If $t = (V, \leqslant, \ell)$ is a pomset, we denote by $\max(t)$ (resp. by $\min(t)$) the set of maximal (resp. minimal) elements of t. The *alphabet* of t is $\mathrm{alph}(t) = \ell(V)$. We let $\mathrm{alphinf}(t) = \{a \in \mathrm{alph}(t) \mid \ell^{-1}(a) \text{ is infinite}\}$.

A *dependence alphabet* is a pair (Σ, D) where Σ is a finite alphabet and D is a reflexive, symmetric binary relation over Σ, called the *dependence relation*. We let $\mathrm{I} = \Sigma \times \Sigma \setminus \mathrm{D}$ be the independence relation. A *(Mazurkiewicz) trace* over (Σ, D) is an isomorphism class of a pomset $(V, \leqslant, \ell)$ such that, for all $e, f \in V$: (1) $\ell(e)\,\mathrm{D}\,\ell(f) \Rightarrow e \leqslant f$ or $f \leqslant e$, (2) $e \lessdot f \Rightarrow \ell(e)\,\mathrm{D}\,\ell(f)$ and (3) $\downarrow e$ is finite. Two traces t, t' are *independent* if $(\mathrm{alph}(t) \times \mathrm{alph}(t')) \cap \mathrm{D} = \emptyset$. We denote by $\mathbb{R}(\Sigma, \mathrm{D})$ (resp. by $\mathbb{M}(\Sigma, \mathrm{D})$) the set of traces (resp. of finite traces) over (Σ, D). It is well-known that $\mathbb{M}(\Sigma, \mathrm{D})$ is a monoid. The free monoid (resp. the free semigroup) over Σ is denoted by Σ^* (resp. by Σ^+).

A *prefix* of $t = (V, \leqslant, \ell)$ is a trace $(U, \leqslant, \ell)$, where $U \subseteq V$ satisfies $\downarrow U = U$. We write $s \leqslant t$ is s is a prefix of t. A *linearization* of t is a labeled total order $(V, \preccurlyeq, \ell)$ such that $e \leqslant f$ implies $e \preccurlyeq f$. For any $w \in \Sigma^*$, there exists a unique trace $[w]$ of which w is a linearization.

3   Distributed Games

A distributed system made up of asynchronous processes interacting together and with the environment may be viewed as a single asynchronous model having controllable actions (the system's ones) and uncontrollable actions (the environment's ones). In the game setting, one views actions as players, which are split in two teams Σ_0 (actions of the system) and Σ_1 (actions of the environment). An execution of the system inside the environment corresponds then to a play, a property of the executions to a winning condition, and a distributed controller to a winning distributed strategy for team 0.

If X and I are sets and $J \subseteq I$, then for $x = (x_i)_{i \in I} \in X^I$ we let $x_J = (x_i)_{i \in J} \in X^J$. Given sets $(X_i)_{i \in I}$, and $J \subseteq I$, we let $X_J = \prod_{i \in J} X_i$.

An *architecture* is a tuple $(\Sigma, \mathcal{P}, R, W)$ such that Σ is a finite set of *actions* or *players*, $\mathcal{P}$ is a finite set of *processes*, $R : \Sigma \to 2^{\mathcal{P}}$ assigns to each $a \in \Sigma$ its read domain $R(a)$, $W : \Sigma \to 2^{\mathcal{P}}$ assigns to each $a \in \Sigma$ its write domain $W(a)$. We only consider architectures satisfying the following natural restriction, already considered in [13], and sufficient to get a dependence alphabet on actions.

$$\forall a \in \Sigma, \qquad \emptyset \neq W(a) \subseteq R(a)$$
$$\forall a, b \in \Sigma, \qquad R(a) \cap W(b) = \emptyset \iff R(b) \cap W(a) = \emptyset$$

We define the dependence relation over Σ as $\mathrm{D} = \{(a, b) \mid R(a) \cap W(b) \neq \emptyset\}$.

Let $(\Sigma, \mathcal{P}, R, W)$ be an architecture. A *distributed game* over $(\Sigma, \mathcal{P}, R, W)$ is given by a tuple $G = (\Sigma_0, \Sigma_1, (Q_i)_{i \in \mathcal{P}}, (T_a)_{a \in \Sigma}, q^0, \mathcal{W})$ where Σ_0 and Σ_1 are the players of teams 0 and 1 respectively and we have $\Sigma = \Sigma_0 \uplus \Sigma_1$, $\forall i \in \mathcal{P}$, Q_i is the set of local states for process i, $\forall a \in \Sigma$, $T_a \subseteq Q_{R(a)} \times Q_{W(a)}$ gives the local moves of player a, $q^0 \in Q = \prod_{i \in \mathcal{P}} Q_i$ is the *starting position* of G, and $\mathcal{W}$ defines the winning condition of G.

The easiest way to define the semantics of the distributed game is via its sequential game graph whose set of positions is Q, the initial position is q^0 and there is an a-move from $p \in Q$ to $q \in Q$ (denoted $p \xrightarrow{a} q$) if $(p_{R(a)}, q_{W(a)}) \in T_a$ and $q_{\mathcal{P} \setminus W(a)} = p_{\mathcal{P} \setminus W(a)}$. A sequential play is a sequence $q^0 \xrightarrow{a_1} q^1 \xrightarrow{a_2} q^2 \cdots$. Note that in a position $p \in Q$, several players of team 0 and of team 1 may be simultaneously enabled, hence this sequential game graph does not correspond to a conventional (sequential) game in which each position is either a position of player (team) 0 or a position of player (team) 1.

We consider a new symbol $\perp \notin \Sigma$ with $R(\perp) = W(\perp) = \mathcal{P}$ and the alphabet $\Sigma' = \{(a,p) \mid a \in \Sigma$ and $p \in Q_{W(a)}\} \cup \{(\perp, q^0)\}$ with the dependence relation $D' = \{((a,p),(b,q)) \mid R(a) \cap W(b) \neq \emptyset\}$. The winning condition of the game is a set of words $\mathcal{W} \subseteq \Sigma'\infty$ which is closed under the usual trace equivalence (see [2,3]). With the sequential play $\pi = q^0 \xrightarrow{a_1} q^1 \xrightarrow{a_2} q^2 \cdots$ of G we associate the word $w = (\perp, q^0)(a_1, q^1_{W(a_1)})(a_2, q^2_{W(a_2)}) \cdots$ over Σ'. Note that the word w faithfully encodes the sequential play π and team 0 wins the play π if $w \in \mathcal{W}$.

A better semantics of these distributed games is to view a play directly as a *rooted* trace over the alphabet Σ'. A finite or infinite trace $t = (V, \leqslant, (\ell, \sigma)) \in \mathbb{R}(\Sigma', D')$ is *rooted* if $\ell^{-1}(\perp) = \{x_\perp\}$ is a singleton and $x_\perp \leqslant y$ for all $y \in V$. If $s = (U, \leqslant, (\ell, \sigma))$ is a nonempty prefix of t then $\bar{\sigma}(s) = (\bar{\sigma}(s)_i)_{i \in \mathcal{P}} \in Q$ is defined by $\bar{\sigma}(s)_i = \sigma(y)_i$ where y is the maximal vertex in $\{x \in U \mid i \in W(\ell(x))\}$.

A distributed play of G is a finite or infinite rooted trace $t = (V, \leqslant, (\ell, \sigma)) \in \mathbb{R}(\Sigma', D')$ such that for each $a \in \Sigma$ and $x \in \ell^{-1}(a)$, we have $(\bar{\sigma}(\Downarrow x)_{R(a)}, \sigma(x)) \in T_a$. The winning condition $\mathcal{W}$ is now a subset of $\mathbb{R}(\Sigma', D')$ and team 0 wins the distributed play t if $t \in \mathcal{W}$. The two definitions above are indeed equivalent but the second one is better suited to distributed games and allows a natural definition of a distributed strategy. In sequential games, one often considers infinite plays only. We also consider finite plays because it can be more convenient.

Let $t = (V, \leqslant, \ell, \sigma) \in \mathbb{R}(\Sigma, D)$ be a rooted trace and let $J \subseteq \mathcal{P}$. The trace $\partial_J t$ is the prefix of t defined by the set of vertices $U = \Downarrow\{x \in V \mid W(\ell(x)) \cap J \neq \emptyset\}$.

An asynchronous mapping [1] is a function $\mu : \mathbb{M}(\Sigma, D) \to M$ such that $\mu(\partial_{A \cup B} t)$ only depends on $\mu(\partial_A t)$ and $\mu(\partial_B t)$, and $\mu(\partial_{R(a)} t.a)$ only depends on $\mu(\partial_{R(a)} t)$ and a. Asynchronous mappings can be computed by deterministic asynchronous transition systems. A *distributed memory* on a game G is an asynchronous mapping $\mu : \mathbb{M}(\Sigma', D') \to M$. It will be used by players of team 0 as an abstraction (computed in M) of their causal view of the play. A *distributed strategy with memory* μ for team 0 (μ-DS) is a pair (f, μ) where f is a partial function $f : \bigcup_{a \in \Sigma_0} Q_{R(a)} \times M^{R(a)} \times \{a\} \to Q_{W(a)}$ such that if $f(p, m, a) = q$, then $(p, q) \in T_a$. Intuitively, if $f(p, m, a) = q$, then the strategy f dictates an a-move to $q \in Q_{W(a)}$ when the memory of the play that a can observe using μ is $m \in M^{R(a)}$. If $f(p, m, a)$ is undefined, the a-move is disabled by the strategy. Note that several players of team 0 may be simultaneously enabled by f during a play. In the sequel we write f instead of (f, μ), μ being understood.

Let f be a μ-DS. Let $\bar{\mu}(t) = (\mu(\partial_i(t)))_{i \in \mathcal{P}}$. A distributed play $t = (V, \leqslant, \ell, \sigma) \in \mathbb{R}(\Sigma', \mathrm{D}')$ is an f-*play* if

$$\forall x \in V, \ \sigma(x) = f(\bar{\sigma}(\Downarrow x)_{R(a)}, \bar{\mu}(\Downarrow x)_{R(a)}, a)$$

The play t is f-*maximal* if $f(\bar{\sigma}(\partial_{R(a)}t)_{R(a)}, \bar{\mu}(\partial_{R(a)}t)_{R(a)}, a)$ is undefined for all $a \in \Sigma_0$ such that $\partial_{R(a)}t$ is finite. The *maximality* condition is natural: if the DS of team 0 dictates some a-moves at some f-play t, then the f-play t is not over and we do not have to decide whether it is winning or not for team 0. Note that this applies also if t is infinite and corresponds to some fairness condition: along an infinite f-play, a move of team 0 cannot be ultimately enabled by f. Observe that any f-play t is the prefix of some f-maximal f-play. If each f-maximal f-play is in $\mathcal{W}$ then f is a *winning* distributed strategy (WDS) for team 0.

A distributed game is not necessarily determined in the sense that it is possible that neither team 0 nor team 1 have a WDS, even with perfect memory. For instance, consider $G = (\Sigma_0, \Sigma_1, (Q_i)_{i \in \mathcal{P}}, (T_a)_{a \in \Sigma}, q^0, \mathcal{W})$ with $\Sigma_0 = \{a\}$, $\Sigma_1 = \{b\}$, $\mathcal{P} = \{1, 2\}$, $R(a) = W(a) = \{1\}$, $R(b) = W(b) = \{2\}$, $Q_1 = Q_2 = \{1\}$, $T_a = Q_1^2$, $T_b = Q_2^2$, $q^0 = (1, 1)$, and $\mathcal{W} = \mathbb{M}(\Sigma', \mathrm{D}') \cup \{(\perp, q^0)(a, 1)^\omega(b, 1)^\omega\}$. Assume that team 0 has a DS. If $f((\perp, q^0)(a, 1)^n, a) \neq \emptyset$ for all $n \geqslant 0$, then team 0 loses if team 1 does not play at all, yielding the play $(\perp, q^0)(a, 1)^\omega$. Conversely, if there exists $n \geqslant 0$ such that $f((\perp, q^0)(a, 1)^n, a) = \emptyset$, then team 0 loses if team 1 makes infinitely many moves. Symmetrically, team 1 does not have a WDS.

Actually, this non-determinacy is not a problem. For the distributed control problem, we are looking for a WDS allowing controllable events (team 0) to enforce good behaviors but we are not interested in a winning *distributed* strategy for the uncontrollable events: uncontrollable events are played by an environment, and there is no reason to consider only distributed environments.

A *memoryless distributed strategy (MDS)* is a μ-DS with $|\mu(\mathbb{M}(\Sigma', \mathrm{D}'))| = 1$, that is, the memory does not record any information. In this case, we write $f(p, a)$ instead of $f(p, m, a)$. A *perfect-memory distributed strategy* is a μ-DS with $\mu(t) = t$. It provides for a move to x with $\ell(x) = a$ the full causal view $\bar{\mu}(\Downarrow x)_{R(a)} = (\partial_i \Downarrow x)_{i \in R(a)}$. Since $\bar{\sigma}(\Downarrow x)_{R(a)}$ can be computed from $(\partial_i \Downarrow x)_{i \in R(a)}$, one can drop the state component in f and write $f(m, a)$ instead of $f(p, m, a)$. As in the sequential case, one can embed a given memory into the game.

Proposition 1. *Let G be a distributed game and let μ be a distributed memory on G. One can construct a distributed game G^μ such that there exists a μ-WDS for G iff there exists a WMDS in G^μ. Moreover, if G is finite and μ is realized by a finite asynchronous automaton, then G^μ is finite.*

4   Global Game

In order to use known results of game theory, we want to define a classical two-players global game $\widetilde{G} = (Z, T)$ such that team 0 has a WMDS in the distributed game G iff player 0 has a winning memoryless strategy in the global game $\widetilde{G}$.

The positions of the global game are $Z = Z_0 \cup Z_1$ where $Z_0 = Q \times \Sigma_0$ are the positions of player 0 and $Z_1 = Q \times (\Sigma_1 \cup \{0, 1, 2\})$ are the positions of player 1. The initial position is $(q^0, 0) \in Z_1$. In a position (q, a), the first component describes the current global state of the play and the second component is used both to determine whose turn it is and which action should be executed. The set $T \subseteq (Z_0 \times Z_1) \cup (Z_1 \times Z)$ of moves is defined as follows:

- $(p, b) \to (p, a)$ with $b \in \{0, 1, 2\}$ and $a \in \Sigma$. Player 1 decides that the next move should be an a-move. In this global game, player 1 is in charge of deciding which actions are used and in which order. This allows him to investigate all possible linearizations of distributed plays.
- $(p, a) \to (q, 1)$ with $a \in \Sigma$, $(p_{R(a)}, q_{W(a)}) \in T_a$ and $q_{\mathcal{P} \setminus W(a)} = p_{\mathcal{P} \setminus W(a)}$. This a-move is executed by player 0 or player 1 depending on whether $a \in \Sigma_0$ or $a \in \Sigma_1$.
- $(p, a) \to (p, 2)$ with $a \in \Sigma_0$. Player 0 *refuses* to make an a-move.
- $(q, b) \to (q^0, 0)$ with $b \in \{0, 1, 2\}$. These *reset-moves* are used by player 1 to show that player 0 is not following a *distributed* strategy.

Note that player 1 may perform several consecutive moves.

A global play is a finite or infinite sequence $z = z_0 z_1 z_2 \cdots \in Z^\infty$ starting from the initial position $z_0 = (q^0, 0)$ and such that $z_n \to z_{n+1}$ is a move for all $n \geqslant 0$. Let $z = z_0 z_1 z_2 \cdots \in Z^\infty$ be a global play and let $z_n = (q^n, a_n) \in Z$ for $n \geqslant 0$. We define by induction the sequence $(t_n)_{n \geqslant 0} \in \mathbb{M}(\Sigma', D')^{\mathbb{N}}$ associated with z. If $z_n = (q^0, 0)$ then $t_n = (\perp, q^0)$. If $a_{n+1} \in \Sigma \cup \{2\}$ then $t_{n+1} = t_n$. Finally, if $a_n = a \in \Sigma$ and $a_{n+1} = 1$ then $t_{n+1} = t_n \cdot (a, q_{W(a)}^{n+1})$. We prove by induction that t_n is a distributed play and $\bar{\sigma}(t_n) = q^n$ for all $n \geqslant 0$. The only non trivial case is when $a_n = a \in \Sigma$ and $a_{n+1} = 1$. By induction, t_n is a distributed play and $\bar{\sigma}(t_n) = q^n$. We have $(q_{R(a)}^n, q_{W(a)}^{n+1}) \in T_a$ and $q_{R(a)}^n = \bar{\sigma}(\partial_{R(a)} t_n)_{R(a)}$. Therefore, t_{n+1} is a distributed play and using $q_{\mathcal{P} \setminus W(a)}^{n+1} = q_{\mathcal{P} \setminus W(a)}^n$ we get $\bar{\sigma}(t_{n+1}) = q^{n+1}$.

The global play z is *consistent* if for all $j, k \geqslant 0$ with $a_j = a_k = a \in \Sigma_0$ and $q_{R(a)}^j = q_{R(a)}^k$ we have $a_{j+1} = a_{k+1}$ and $q_{W(a)}^{j+1} = q_{W(a)}^{k+1}$. The global play z is *fair* if $\{n \geqslant 0 \mid a_n = 0\}$ is finite and for all $a \in \Sigma_0$, $\{n \geqslant 0 \mid a_n = a\}$ is infinite. If z is both consistent and fair then we let $N(z) = \max\{n \mid a_n = 0\}$. The sequence $(t_n)_{n \geqslant N(z)}$ is increasing and admits a least upper bound $t(z)$ which is a distributed play of G.

The winning condition $\widetilde{\mathcal{W}}$ of $\widetilde{G}$ only involves infinite plays $z \in Z^\omega$. If z is not consistent then player 0 loses the game since this reveals that he does not mimic a memoryless *distributed* strategy. If z is not fair then player 1 loses the game. Finally, if z is both consistent and fair then player 0 wins the game iff $t(z) \in \mathcal{W}$.

A (global) *strategy* (S) for player 0 in $\widetilde{G}$ is a mapping $g : Z^* Z_0 \to Z_1$ such that $g(z(p, a)) = (q, b)$ implies $(p, a) \to (q, b)$. A global play $z = z_0 z_1 z_2 \cdots \in Z^\infty$ is *played according to* g (g-play) if each move of player 0 is done according to g: $z_k \in Z_0$ implies $z_{k+1} = g(z_0 \cdots z_k)$. If player 0 wins all infinite g-plays then g is a winning strategy (WS) for player 0. A strategy g is *memoryless* if for all $x, x' \in Z^*$ and $y \in Z_0$, we have $g(xy) = g(x'y)$. We write MS and WMS for *memoryless strategy* and *winning memoryless strategy*.

We can now state the main result of this section.

Theorem 1. *The following conditions are equivalent for a distributed game G:*

1. *There exists a WMDS for team 0 in the distributed game G.*
2. *There exists a WMS for player 0 in the global game $\widetilde{G}$.*
3. *There exists a WS for player 0 in the global game $\widetilde{G}$.*

The following proposition gives the construction used for the implication $(1 \Rightarrow 2)$.

Proposition 2. *Let f be a deterministic WMDS for team 0 in G. For $(p, a) \in Z_0$, we define $g((p, a)) = (p, 2)$ if $f(p_{R(a)}, a) = \emptyset$ and $g((p, a)) = (q, 1)$ with $q_{\mathcal{P} \setminus W(a)} = p_{\mathcal{P} \setminus W(a)}$ and $f(p_{R(a)}, a) = \{q_{W(a)}\}$ otherwise. Then, g is a WMS for player 0 in the global game $\widetilde{G}$.*

To prove the implication $(2 \Rightarrow 1)$ of Theorem 1, we exploit reset-moves.

Lemma 1. *Let g be a WMS of player 0 in the global game $\widetilde{G}$. Let $(p^1, a) \in Z_0$ and $(p^2, a) \in Z_0$ be accessible in g-plays and such that $p^1_{R(a)} = p^2_{R(a)}$. Then, $g(p^1, a) = (p^1, 2)$ iff $g(p^2, a) = (p^2, 2)$ and if $g(p^1, a) = (q^1, 1)$ and $g(p^2, a) = (q^2, 1)$ then $q^1_{W(a)} = q^2_{W(a)}$.*

Using this lemma, we can now transform a WMS in $\widetilde{G}$ into a WMDS in G.

Proposition 3. *Let g be a WMS of player 0 in the global game $\widetilde{G}$. For $(p, a) \in Z_0$ accessible by a g-play, we define $f(p_{R(a)}, a) = \emptyset$ if $g(p, a) = (p, 2)$ and $f(p, a) = \{q_{W(a)}\}$ if $g(p, a) = (q, 1)$. Then, f is a WMDS of team 0 in the distributed game G.*

Even if $\mathcal{W}$ is rational ($\mathcal{W} = [L]$, where $L \in \mathrm{Rat}(\Sigma^*)$), determining if team 0 has a W(M)DS is undecidable. Indeed, on $\mathbb{M}(\Sigma, D) = A^* \times B^*$, determining if a rational trace language $\mathcal{L}$ is $[\Sigma^*]$ is undecidable [2].

From such a language $\mathcal{L}$, we construct a 2-processes game in which team 0 has a WMDS iff $\mathcal{L} = [\Sigma^*]$: $\Sigma_0 = \emptyset$, $\Sigma_1 = A \uplus B$, $R(a) = W(a) = \{1\}$ for $a \in A$ and $R(b) = W(b) = \{2\}$ for $b \in B$. Finally, $|Q| = 1$ (so that we identify Σ' and Σ), and $\mathcal{W} = \mathcal{L} \cup (\mathbb{R}(\Sigma, D) \setminus \mathbb{M}(\Sigma, D))$. Players of team 1 nondeterministically choose a move in some finite local game, so that any possible trace is a play. Now, team 0 has a WMDS iff he has a WDS iff team 1 cannot generate a finite trace outside $\mathcal{L}$, that is, iff $\mathcal{L} = [\Sigma^*]$.

We now explain how to use Theorem 1 to decide if team 0 has a WDS. Denote by $\mathrm{Lin}(t)$ the set of all linearizations of $t \in \mathbb{R}(\Sigma, D)$. Properties considered in practice are recognizable (*i.e.*, $\mathrm{Lin}(\mathcal{W})$ is rational), and as noted in [6], there are many temporal logics expressing only recognizable specifications. To determine whether team 0 has a WMDS in G with $\mathcal{W}$ recognizable, we could enumerate all memoryless distributed strategies for player 0, and check whether one is winning. This amounts to testing an inclusion between recognizable trace languages. Theorem 1 provides a better algorithm. The principle is to build the global game $\widetilde{G}$, to transform it into a parity game and to apply known algorithms.

Let $\mathcal{A}$ be a parity automaton accepting $\mathrm{Lin}(\mathcal{W})$. The winning condition $\widetilde{\mathcal{W}}$ for player 0 on the global game $\widetilde{G}$ can be defined by $\widetilde{\mathcal{W}} = \widetilde{\mathcal{W}}_c \cap (\mathrm{Lin}(W) \cup \widetilde{\mathcal{W}}_{\mathrm{nf}})$, where $\widetilde{\mathcal{W}}_c = \{z \in Z^\omega \mid z \text{ is consistent}\}$, and $\widetilde{\mathcal{W}}_{\mathrm{nf}} = \{z \in Z^\omega \mid z \text{ is not fair}\}$. We describe informally how to construct a parity automaton accepting $\widetilde{\mathcal{W}}$. One can build a Büchi automaton accepting consistent plays in Z^ω: it records all transitions $(p_{R(a)}, q_{W(a)})$ performed by player 0, as well as the refused transitions. It falls into the unique rejecting state as soon as an inconsistent move is detected. One can also build a parity automaton checking that a play, supposed consistent, is not fair, by checking that there is an infinite number of reset moves (a Büchi condition) or that, for some $a \in \Sigma_0$, there is only a finite number of states of the form (p, a) (co-Büchi conditions). From the parity automaton $\mathcal{A}$ and from these automata, it should now be clear how to build a parity automaton for $\widetilde{\mathcal{W}}$.

Observe that using Proposition 1, one can also determine whether team 0 has a μ-WDS for a given finite distributed memory μ.

5   Related Approaches

A distributed game $G = \langle P, E, \mathrm{Tr}, \mathrm{Acc}, q^0 \rangle$ as in [7] is built from n local games $G_1, \ldots, G_n$, where $G_i = \langle P_i, E_i, \mathrm{Tr}_i, q_i^0 \rangle$ with $\mathrm{Tr}_i \subseteq (P_i \times E_i)$. Positions of the environment are $E = \prod_i E_i$. The position set of the players is $P = \prod_i (P_i \cup E_i) \setminus E$. Transitions of the players are defined with a cartesian product: $\mathrm{Tr}_p = (\prod_i (\mathrm{Tr}_i \cup \Delta_i)) \cap (P \times E)$ where $\Delta_i = \{(x_i, x_i) \mid x_i \in E_i\}$ is the diagonal. Transitions of the environment are simply given by a subset Tr_e of $E \times P$, and $\mathrm{Tr} = \mathrm{Tr}_e \uplus \mathrm{Tr}_p$. A play of G starts in position $q^0 \in E$, and moves from the environment and from the players alternate. Hence, any infinite play is in $(E \cdot P)^\omega$, and the winning condition is a subset $\mathrm{Acc} \subseteq (E \cdot P)^\omega$.

There is a natural translation from these games to our setting. With G, we associate the distributed game $\bar{G} = (\Sigma_0, \Sigma_1, (T_a)_{a \in \Sigma}, q^0, \mathcal{W})$ as follows. The set of processes is $\mathcal{P} = \{1, \ldots, n\}$ and the local states are $Q_i = P_i \cup E_i$ for $i \in \mathcal{P}$. Team 0 is defined by $\Sigma_0 = \{1, \ldots, n\}$ with $R(i) = W(i) = \{i\}$ for all $i \in \Sigma_0$. The transitions for player i are simply $T_i = \mathrm{Tr}_i$. Team 1 consists of a single player e (the environment) with $R(e) = W(e) = \mathcal{P}$. Its transitions are $T_e = \mathrm{Tr} \cap (E \times P)$.

To define the winning condition $\mathcal{W}$, we associate with each infinite play $w = e^0 x^1 e^1 \cdots \in (E \cdot P)^\omega$ of G with $e^0 = q^0$ a distributed play $\mathrm{trace}(w) \in \mathbb{R}(\Sigma', D')$ as the least upper bound of $(t_i)_{i \geqslant 0}$ where the increasing sequence $(t_n)_{n \geqslant 0}$ is defined inductively by $t_0 = (\bot, q^0)$ and $t_{n+1} = t_n \cdot (e, x^{n+1}) \cdot \prod_{i \mid x_i^{n+1} \in P_i} (i, e_i^{n+1})$. Finally, the winning condition is $\mathcal{W} = \mathrm{trace}(\mathrm{Acc}) \cup \{t \in \mathbb{M}(\Sigma', D') \mid \bar{\sigma}(t) \in E\}$.

The distributed games of [7] are thus a special case of our games in which all players of team 0 are completely local and the environment consists of a single global player. Note that, in the game G, information between players can only flow through environment moves and the environment can decide which information is exchanged between players.

However, the crucial difference between games of [7] and the ones presented here concerns the definition of strategies. In [7], a strategy is a tuple of mappings

$f_i : (E_i P_i)^+ \rightarrow E_i$. Hence a move of player i only depends on its *local* view consisting of the history on its process only. In our setting, the strategy of player i which is the mapping $f(-, i)$ is based on its causal view $\partial_{R(i)} t$. Since actions of the environment are global, the causal view is almost the complete global view of the game. It is therefore clear that if there exists a WS for the players in G, then there exists a WDS for team 0 in $\bar{G}$. The converse is false: it is easy to find a game G not determined while there is a winning strategy for team 0 in $\bar{G}$. Yet, one can prove that if a game G of [7] is determined, then there is a WDS for team 0 in $\bar{G}$ iff players have a WS in G.

If we want to get an equivalence, we have to restrict the memory used by our strategies to the local view, that is, to change the notion of memory used by the strategies. The i-projection of $t \in \mathbb{M}(\Sigma, \mathrm{D}')$ is $\Pi_i(t)$ where Π_i is the morphism from $\mathbb{M}(\Sigma', \mathrm{D}')$ to Q_i^* defined by $\Pi_i(x, q) = q_i$ if $x = e$ or $x = i$ and $\Pi_i(x, q) = \varepsilon$ otherwise. Since player i is only aware of move he takes part in, we have to abstract away from unobservable stuttering of the environment. For this we use the congruence on Q_i^* generated by $p_i^3 = p_i$ for $p_i \in E_i$ and we write $w_{\equiv}$ for the equivalence class of $w \in Q_i^*$. We say that a distributed strategy f is local if for all $i \in \Sigma_0$, $f(t, i)$ depends only on i and $\Pi_i(t)_{\equiv}$.

Proposition 4. *The players have a WS in G iff team 0 has a local WDS in $\bar{G}$.*

In the distributed control problem presented in [6], the environment performs local moves and the transitions of a controllable action is defined by a cartesian product of local transition functions. It is therefore straightforward to translate as above these games to our framework. Since local strategies are used in [6], Propositions 4 also holds.

Acknowledgment. We thank Madhavan Mukund for fruitful discussions.

References

1. R. Cori, Y. Métivier, and W. Zielonka. Asynchronous mappings and asynchronous cellular automata. *Inform. and Comput.*, 106:159–202, 1993.
2. V. Diekert and Y. Métivier. *Handbook of Formal languages*, volume 3, chapter Partial Commutations and Traces, pages 457–533. Springer, 1997.
3. V. Diekert and G. Rozenberg, editors. *The Book of Traces.* World Scientific, 1995.
4. O. Kupferman and M. Y. Vardi. Synthesizing distributed systems. In *Proc. of the LICS '01.* Computer Society Press, 2001.
5. P. Madhusudan and P. S. Thiagarajan. Distributed controller synthesis for local specifications. In *Proc. of the ICALP '01*, volume 2076 of *Lect. Notes Comp. Sci.*, pages 396–407. Springer, 2001.
6. P. Madhusudan and P. S. Thiagarajan. A decidable class of asynchronous distributed controllers. In *Proc. of the CONCUR '02*, volume 2421 of *Lect. Notes Comp. Sci.*, pages 145–160. Springer, 2002.
7. S. Mohalik and I. Walukiewicz. Distributed games. In *FSTTCS '03*, Lect. Notes Comp. Sci. Springer, 2003.

8. J. Pin and D. Perrin. *Infinite words. Automata, Semigroups, Logic and Games.* Elsevier, To appear.
9. A. Pnueli and R. Rosner. Distributed reactive systems are hard to synthetize. In *Proc. of the 31th IEEE Symp. FOCS*, pages 746–757, 1990.
10. P. Ramadge and W. Wonham. The control of discrete event systems. In *Proceedings of the IEEE*, volume 77, pages 81–98, 1989.
11. W. Thomas. Infinite games and verification. In *Proc. of CAV'02*, volume 2404 of *Lect. Notes Comp. Sci.*, pages 58–64. Springer, 2002.
12. J. Vöge and M. Jurdziński. A discrete strategy improvement algorithm for solving parity games. In *Proc. of CAV '00*, volume 1855 of *Lect. Notes Comp. Sci.*, pages 202–215. Springer, 2000.
13. W. Zielonka. Asynchronous automata. In G. Rozenberg and V. Diekert, editors, *Book of Traces*, pages 175–217. World Scientific, Singapore, 1995.
14. W. Zielonka. Infinite games on finitely coloured graphs with applications to automata on infinite trees. *Theoretical Computer Science*, 200(1–2):135–183, 1998.

A Simplified and Dynamic Unified Structure

Mihai Bădoiu and Erik D. Demaine

MIT Computer Science and Artificial Intelligence Laboratory,
200 Technology Square, Cambridge, MA 02139, USA, {mihai,edemaine}@mit.edu

Abstract. The unified property specifies that a comparison-based search structure can quickly find an element nearby a recently accessed element. Iacono [Iac01] introduced this property and developed a static search structure that achieves the bound. We present a dynamic search structure that achieves the unified property and that is simpler than Iacono's structure. Among all comparison-based dynamic search structures, our structure has the best proved bound on running time.

1 Introduction

The classic *splay conjecture* says that the amortized performance of splay trees [ST85] is within a constant factor of the optimal dynamic binary search tree for any given request sequence. This conjecture has motivated the study of sublogarithmic time bounds that capture the performance of splay trees and other comparison-based data structures. For example, it is known that the performance of splay trees satisfies the following two upper bounds. The *working-set bound* [ST85] says roughly that recently accessed elements are cheap to access again. The *dynamic-finger bound* [CMSS00,Col00] says roughly that it is cheap to access an element that is nearby the previously accessed element. These bounds are incomparable: one does not imply the other. For example, the access sequence $1, n, 1, n, 1, n, \ldots$ has a small working-set bound (constant amortized time per access) because each accessed element was accessed just two time units ago. In contrast, for this sequence the dynamic-finger bound is large (logarithmic time per access) because each accessed element has rank distance $n - 1$ from the previously accessed element. On the other hand, the access sequence $1, 2, \ldots, n, 1, 2, \ldots, n, \ldots$ has a small dynamic-finger bound because most accessed elements have rank distance 1 to the previously accessed element, whereas it has a large working-set bound because each accessed element was accessed n time units ago.

In SODA 2001, Iacono [Iac01] proposed a *unified bound* (defined below) that is strictly stronger than all other proved bounds about comparison-based structures. Roughly, the unified bound says that it is cheap to access an element that is nearby a recently accessed element. For example, the access sequence $1, \frac{n}{2} + 1, 2, \frac{n}{2} + 2, 3, \frac{n}{2} + 3, \ldots$ has a small unified bound because most accessed elements have rank distance 1 to the element accessed two time units ago, whereas it has large working-set and dynamic-finger bounds. It remains open whether

M. Farach-Colton (Ed.): LATIN 2004, LNCS 2976, pp. 466–473, 2004.
© Springer-Verlag Berlin Heidelberg 2004

splay trees satisfy the unified bound. However, Iacono [Iac01] developed the *unified structure* which attains the unified bound. Among all comparison-based data structures, this structure has the best proved bound on running time.

The only shortcomings of the unified data structure are that it is static (keys cannot be inserted or deleted), and that both the algorithms and the analysis are complicated. We improve on all of these shortcomings with a simple dynamic unified structure. Among all comparison-based dynamic data structures, our structure has the best proved bound on running time.

2 Unified Property

Our goal is to maintain a dynamic set of elements from a totally ordered universe in the (unit-cost) comparison model on a pointer machine. Consider a sequence of m operations—insertions, deletions, and searches—where the ith operation involves element x_i. Let S_i denote the set of elements in the structure just before operation i (at time i). Define the *working-set number* $t_i(z)$ of an element z at time i to be the number of distinct elements accessed since the last access to z and prior to time i, including z. Define the *rank distance* $d_i(x,y)$ between elements x and y at time i to be the number of distinct elements in S_i that fall between x and y in rank order. A data structure has the *unified property* if the amortized cost of operation i is $O(\lg \min_{y \in S_i}[t_i(y) + d_i(x_i, y) + 2])$, the *unified bound*. Intuitively, the unified bound for accessing an element x_i is small if any element y is nearby x in both time and space.

3 New Unified Structure

In this section, we develop our dynamic unified structure which establishes the following theorem:

Theorem 1. *There is a dynamic data structure in the comparison model on a pointer machine that supports insertions and searches within the unified bound and supports deletions within the unified bound plus $O(\lg \lg |S_i|)$ time (amortized).*

An interesting open problem is to attain the unified bound for all three operations simultaneously.

3.1 Data Structure

The bulk of our unified structure consists of $\Theta(\lg \lg |S_i|)$ balanced search trees and linked lists whose sizes increase doubly exponentially; see Fig. 1. Each tree T_k, $k \geq 0$, stores between 2^{2^k} and $2^{2^{k+1}} - 1$ elements, ordered by their rank, except that the last tree may have fewer elements. We can store each tree T_k using any balanced search tree structure supporting insertions, deletions, and searches in $O(\lg |T_k|)$ time, e.g., B-trees [BM72]. List L_k stores exactly the same

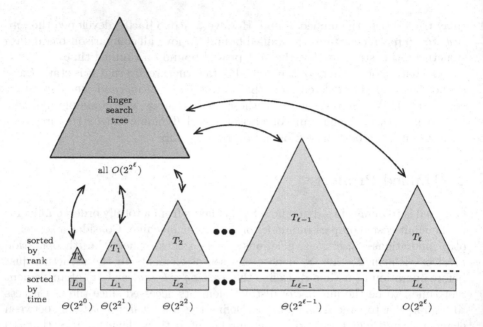

Fig. 1. Overview of our dynamic unified structure. In addition to a single finger search tree storing all elements in the dynamic set S_i, there are $\ell + 1 = \Theta(\lg \lg |S_i|)$ balanced search trees and lists whose sizes grow doubly exponentially. (As drawn, the heights accurately double from left to right.)

elements stored in T_k, but ordered by the time of access. We store pointers between corresponding nodes in T_k and L_k.

Each element x may be stored by several nodes in various trees T_k, or possibly none at all, but x appears at most once in each tree T_k. Each tree node storing element x represents an access to x at a particular time. At most one tree node represents each access to x, and some accesses to x have no corresponding tree node. We maintain the invariant that the access times of nodes in tree T_k are all more recent than access times of nodes in tree T_{k+1}. Thus, the concatenation of corresponding nodes in lists $L_0, L_1, L_2, \ldots$ is also ordered by access time.

Our unified structure also stores a single finger search tree containing all n elements. We can use any finger search tree structure supporting insertions, deletions, and searches within rank distance r of a previously located element in $O(\lg(r + 2))$ amortized time, e.g., level-linked B-trees [BT80]. Each node in tree T_k stores a pointer to the unique node in this finger search tree corresponding to the stored element.[1]

---

[1] In fact, because nodes in the finger search tree may move (e.g., from a B-tree split), each node in tree T_k stores a pointer to an indirect node, and each indirect node is connected by pointers to the corresponding node in the finger search tree. The former pointers never change, and the latter pointers can be easily maintained when nodes move in the finger search tree.

3.2 Search

Up to constant factors, the unified property requires us to find an element $x = x_i$ in $O(2^k)$ time if it is within rank distance 2^{2^k} of an element y with working-set number $t_i(y) \leq 2^{2^k}$. We maintain the invariant that all such elements x are within rank distance $3 \cdot 2^{2^k}$ of some element y' in $T_0 \cup T_1 \cup \cdots \cup T_k$. (This invariant is proved below in Lemma 1.)

At a high level, then, our search algorithm will investigate the elements in $T_0, T_1, \ldots, T_k$ and, for each such element, search among the elements within rank distance $3 \cdot 2^{2^k}$ for the query element x. The algorithm cannot perform this procedure exactly, because it does not know k. Thus we perform the procedure for each $k = 0, 1, 2, \ldots$ until success. To avoid repeated searching around the elements in T_j, $j \leq k$, we maintain the two elements so far encountered among these T_j's that are closest to the target x, and just search around those two elements. If any of the searches from any of the elements would be successful, one of these two searches will be successful.

More precisely, our algorithm to search for an element x proceeds as shown in Algorithm 1. The variables L and U store pointers to elements in the finger search tree such that $L \leq x \leq U$. These variables represent the tightest known bounds on x among elements that we have located in the finger search tree as predecessors and successors of x in $T_0, T_1, \ldots, T_k$. In each round, we search for x in the next tree T_k, and update L and/or U if we find elements closer to x. Then we search for x in the finger search tree within rank distance $3 \cdot 2^{2^k}$ of L and U.

Thus, if x is within rank distance $3 \cdot 2^{2^k}$ of an element in $T_0 \cup T_1 \cup \cdots \cup T_k$, then the search algorithm will complete in round k. The total running time of k rounds is $\sum_{i=0}^{k} O(\lg |T_i|) = O(2^k)$. Thus, the search algorithm attains the unified bound, provided we have the invariant in Lemma 1 below.

When the search algorithm finds x, it records this most recent access by inserting a node storing x into the smallest tree T_0. This insertion may cause T_0 to grow too large, triggering the overflow algorithm described next.

3.3 Overflow

It remains to describe what we do when a tree T_k becomes too full; see Algorithm 2. The main idea is to promote all but the most recent 2^{2^k} elements from T_k to T_{k+1}, by repeated insertion into T_{k+1} and deletion from T_k. In addition, we discard elements that would be promoted but are within $2^{2^{k+1}}$ of other promoted elements. Such discards are necessary to prevent excessive overflows in the future. The intuition of why discards do not substantially slow future searches is that, for the purposes of searching for an element x within rank distance $2^{2^{k+1}}$ of elements in T_{k+1}, it is redundant up to a factor of 2 to have more than one element in T_{k+1} within a rank range of $2^{2^{k+1}}$. This intuition is formalized by the following lemma:

Algorithm 1. Searching for an element x.

- Initialize $L \leftarrow -\infty$ and $U \leftarrow \infty$.
- For $k = 0, 1, 2, \ldots :$[a]
  1. Search for x in T_k to obtain two elements L_k and U_k in T_k such that $L_k \leq x \leq U_k$.
  2. Update $L \leftarrow \max\{L, L_k\}$ and $U \leftarrow \min\{U, U_k\}$.
  3. Finger search for x within the rank ranges $[L, L+3 \cdot 2^{2^k}]$ and $[U - 3 \cdot 2^{2^k}, U]$.
  4. If we find x in the finger search tree:
     a) Insert x into tree T_0 and at the front of list L_0, unless x is already in T_0.
     b) If T_0 is too full (storing 2^{2^1} elements), *overflow* T_0 as described in Algorithm 2.
     c) Return a pointer to x in the finger search tree.

___
[a] If we reach a k for which T_k does not exist, then $k = \Theta(\lg \lg n)$ and we can afford to search in the global finger tree.

Algorithm 2. Overflowing a tree T_k.

1. Remove the 2^{2^k} most recently accessed elements from list L_k and tree T_k.
2. Build a balanced search tree T_k' and a list L_k' on these 2^{2^k} elements.
3. For each remaining element z in T_k, in rank order, if the predecessor of z in T_k (among the elements not yet deleted) is within rank distance $2^{2^{k+1}}$ of z, then delete z from T_k and L_k.
4. For each remaining element z in L_k, in access-time order:
   a) Search for z in T_{k+1}.
   b) If found, remove z from L_{k+1}.
   c) Otherwise, insert z into T_{k+1}.
5. Concatenate L_k and L_{k+1} to form a new list L_{k+1}.
6. Replace $L_k \leftarrow L_k'$; $T_k \leftarrow T_k'$.
7. If T_{k+1} is now too full (stores at least $2^{2^{k+2}}$ elements), recursively overflow T_{k+1}.

Lemma 1. *All elements within rank distance 2^{2^k} of an element y with working-set number $t_i(y) \leq 2^{2^k}$ are within rank distance $3 \cdot 2^{2^k}$ of some element y' in $T_0 \cup T_1 \cup \cdots \cup T_k$.*

Proof. We track the evolution of y or a nearby element from when it was last accessed and inserted into T_0, to when it moved to T_1, T_2, and so on, until access i. If the tracked element y' is ever discarded from some tree T_j, we continue by tracking the promoted element within rank distance $2^{2^{j+1}}$ of y'. The tracked element y' makes monotone progress through $T_0, T_1, T_2, \ldots$ because, even if y' is accessed and inserted into T_0, the tracked node storing y' is not deleted. The tracked node also cannot expire from T_k (and get promoted or discarded), because at most 2^{2^k} distinct elements have been accessed in the time window under consideration, so y' must be among the first 2^{2^k} elements in the list L_k when it reaches L_k. Therefore, y' remains within rank distance $2^{2^k} + 2^{2^{k-1}} + \cdots + 2^{2^1} + 2^{2^0}$ of y, so we obtain the stronger bound that all elements within rank distance 2^{2^k} of y are within rank distance $2 \cdot 2^{2^k} + 2^{2^{k-1}} + \cdots + 2^{2^1} + 2^{2^0}$ of an element y' in $T_0 \cup T_1 \cup \cdots \cup T_k$.

3.4 Overflow Analysis

To analyze the amortized cost of the overflow algorithm, we consider the cost of overflowing T_k into T_{k+1} for each k separately. To be sure that we do not charge to the same target for multiple k, we introduce the notion of a *coin* c_k which can be used to pay for one node to overflow from T_k to T_{k+1} as well as for the node to later be discarded from T_{k+1}. A coin c_k cannot pay for overflows or discards at different levels. We assign coin c_k an intrinsic value of $\Theta(2^k)$ time units, but twice what is required to pay for a node to overflow and be discarded, so that whenever paying with a coin c_k we are also left with a fractional coin $\frac{1}{2}c_k$.

For each k, we consider the time interval after the previous overflow from T_k into T_{k+1} up to the next overflow from T_k into T_{k+1}. At the beginning of this time interval, we have just completed an overflow involving $\Theta(2^{2^{k+1}})$ elements each with a c_k coin. From the use of these coins we obtain fractional coins of half the value, which we can combine to give two whole c_k coins to every node of $T_0, T_1, \ldots, T_k$, because there are only $\sum_{j=0}^{k} 2^{2^j} = O(2^{2^k}) = o(2^{2^{k+1}})$ such nodes.

Consider a search for an element x during the time interval between the previous overflow and the next overflow of T_k. Suppose x was found at round ℓ of the search algorithm. The cost of searching for x is $O(2^\ell)$ time. We can therefore afford to give x two c_m coins for each $m \le \ell$. We also award x with fractional coins $(6/2^{2^m})c_m$ for each $m > \ell$, which have total worth $o(1)$. We know that x was within rank distance $3 \cdot 2^{2^\ell}$ of an element y in $T_0 \cup T_1 \cup \cdots \cup T_\ell$. If $\ell < k$, we assign y as the *parent* of x. (This assignment may later change if we search for x again.)

Now consider each element x that gets promoted from T_k to T_{k+1} when T_k next overflows. If x has not been searched for since the previous overflow of T_k, then it was in $T_0 \cup T_1 \cup \cdots \cup T_k$ right after the previous overview, so x has two coins c_k. If the last search for x terminated in round ℓ with $\ell \ge k$, then x also has two coins c_k. In either of these cases, x uses one of its own c_k coins to pay for the cost of its overflow (and wastes the other c_k coin).

If x is within rank distance $2^{2^{k+1}}$ of a such an element y in $T_0 \cup T_1 \cup \cdots \cup T_k$ with two c_k coins, then y must not be promoted during this overflow. For if y expires from T_k during this overflow, then at most one of x and y can be promoted (whichever is larger), and we assumed it is x. Thus, x can steal one of y's c_k coins and use it for promotion. Furthermore, y can have a c_k coin stolen at most twice, once by an element $z < y$ and once by an element $z > y$, so we cannot over-steal. If y remains in $T_0 \cup T_1 \cup \cdots \cup T_k$, its c_k coins will be replenished after this overflow, so we also need not worry about y.

If x has no nearby element y with two c_k coins, we consider the chain connecting x to x's parent to x's grandparent, etc. Because every element without a c_k coin has a parent, and because we already considered the case in which an element with a c_k coin is within rank distance $2^{2^{k+1}}$ of x, the chain must extend so far as to reach an element with rank distance more than $2^{2^{k+1}}$ from x. Because every edge in the chain connects elements within rank distance $3 \cdot 2^{2^k}$, the chain must consist of at least $2^{2^{k+1}}/(3 \cdot 2^{2^k}) = 2^{2^k}/3$ elements within rank distance $2^{2^{k+1}}$ of x. Because each of these elements has a parent, they must have been searched for since the last overflow of T_k, and were therefore assigned fractional coins of $(6/2^{2^k})c_k$. As before, none of these elements could be promoted from

T_k during this overflow because they are too close to the promoted element x. Thus, x can steal a fractional coin of $(3/2^{2^k})c_k$ from each of these $2^{2^k}/3$ elements' fractional $(6/2^{2^k})c_k$ coins. Again, this stealing can happen at most twice for each fractional $(6/2^{2^k})c_k$ coin, so we do not over-steal.

Therefore, a promoted element x from the overflow of T_k can find a full coin c_k to pay for its promotion. The $O(2^k)$ cost of discarding an element x from T_k can be charged to the coin c_{k-1} that brought x there, or if $k = 0$, to the search that brought x there. This concludes the amortized analysis of overflow.

3.5   Insert

To insert an element x, we first call a slight variation of the search algorithm from Section 3.2 to find where x fits in the finger search tree. Specifically, we modify Step 4 to realize when the search has gone beyond x, at which point we can find the predecessor and successor of x in the finger search tree. Then, as part of Step 4, we insert x at that position in the finger search tree in $O(1)$ amortized time. We execute Steps 4(a–c) as before, inserting x into tree T_0 and list L_0.

Because this algorithm is almost identical to the search algorithm, we can use essentially the same analysis as in Section 3.4. More precisely, when we insert an element x, suppose we find where x fits during round ℓ of the algorithm Then we assign x a parent as before, and award x two c_m coins for each $m \leq \ell$ and fractional coins $(6/2^{2^m})c_m$ for each $m > \ell$. The only new concern in the amortized analysis is that the rank order changes by an insertion. Specifically, the rank distance between an element z and its parent y can increase by 1 because of an element x inserted between z and y. In this case, we set z's parent to x immediately after the insertion, and the proof goes through. Thus, the amortized cost of an insertion is proportional to the amortized cost of the initial search.

3.6   Delete

Finally we describe how to delete an element within the unified bound plus $O(\lg \lg |S_i|)$ time. Once we have found the element x to be deleted within the unified bound via the search algorithm, we remove x from the finger tree in $O(1)$ time and replace all instances of x in the T_k's with the successor or predecessor of x in the finger tree. To support each replacement in $O(1)$ time, and obtain a total bound of $O(\lg \lg |S_i|)$,[2] we maintain a list of back pointers from each element in the finger tree to the instances of that element as tree nodes in the T_k's. If more than one node in the same tree T_k ever points to the same element, we remove all but one of them.

The amortized analysis is again similar to Section 3.4, requiring only the following changes. Whenever we delete an element x and replace all its instances

---

[2] We maintain the invariant that the number of trees T_k is at most $1 + \lg \lg |S_i|$ simply by removing a tree T_k if k becomes larger than $\lg \lg |S_i|$. Such trees are not necessary for achieving the unified bound during searches.

by its rank predecessor or successor y, element y inherits all of x's coins and takes over all of x's responsibilities in the analysis. We can even imagine x and y as both existing with equal rank, and handling their own responsibilities, with the additional advantage that if either one gets promoted the other one will be discarded (having the same rank) and hence need not be accounted for. An edge of a chain can only get shorter by this contraction in rank space, so the endpoints remain within rank distance $3 \cdot 2^{2^k}$ as required in the analysis. The unified bound to access an element z may also go down because it is closer in rank space to some elements, but this property is captured by the removal of x in the finger tree, and hence finger searches are correspondingly faster. Each tree T_k might get smaller (if both x and y were in the same tree), requiring us to break the invariant that T_k stores at least 2^{2^k} elements. However, we use this invariant only in proving Lemma 1, which remains true because the working-set numbers $t_i(z)$ count accesses to deleted elements and hence do not change.

Acknowledgments. We thank John Iacono for helpful discussions.

References

[BM72]   Rudolf Bayer and Edward M. McCreight. Organization and maintenance of large ordered indexes. *Acta Informatica*, 1(3):173–189, February 1972.

[BT80]   Mark R. Brown and Robert Endre Tarjan. Design and analysis of a data structure for representing sorted lists. *SIAM Journal on Computing*, 9(3):594–614, 1980.

[CMSS00] Richard Cole, Bud Mishra, Jeanette Schmidt, and Alan Siegel. On the dynamic finger conjecture for splay trees. Part I: Splay sorting $\log n$-block sequences. *SIAM Journal on Computing*, 30(1):1–43, 2000.

[Col00]   Richard Cole. On the dynamic finger conjecture for splay trees. Part II: The proof. *SIAM Journal on Computing*, 30(1):44–85, 2000.

[Iac01]   John Iacono. Alternatives to splay trees with $O(\log n)$ worst-case access times. In *Proceedings of the 12th Annual ACM-SIAM Symposium on Discrete Algorithms*, pages 516–522, Washington, D.C., January 2001.

[ST85]   Daniel Dominic Sleator and Robert Endre Tarjan. Self-adjusting binary search trees. *Journal of the ACM*, 32(3):652–686, July 1985.

Another View of the Gaussian Algorithm

Ali Akhavi and Céline Moreira Dos Santos

GREYC – Université de Caen, F-14032 Caen Cedex, France
{ali.akhavi,moreira}@info.unicaen.fr

Abstract. We introduce here a rewrite system in the group of unimodular matrices, *i.e.*, matrices with integer entries and with determinant equal to ±1. We use this rewrite system to precisely characterize the mechanism of the Gaussian algorithm, that finds shortest vectors in a two–dimensional lattice given by any basis. Putting together the algorithmic of lattice reduction and the rewrite system theory, we propose a new worst–case analysis of the Gaussian algorithm. There is already an optimal worst–case bound for some variant of the Gaussian algorithm due to Vallée [16]. She used essentially geometric considerations. Our analysis generalizes her result to the case of the usual Gaussian algorithm. An interesting point in our work is its possible (but not easy) generalization to the same problem in higher dimensions, in order to exhibit a tight upper-bound for the number of iterations of LLL–like reduction algorithms in the worst case. Moreover, our method seems to work for analyzing other families of algorithms. As an illustration, the analysis of sorting algorithms are briefly developed in the last section of the paper.

1 Introduction

This paper deals with extracting worst-cases of some algorithms. Our method is originally proposed by the first author [1,2] as a possible approach for solving the difficult and still open problem of exhibiting worst-cases of lattice reduction algorithms (LLL and its variants). Here the method is applied first to the Gaussian algorithm that solves the two-dimensional lattice problem and that is also intensively used by LLL-like algorithms when reducing higher-dimensional lattices. As another illustration of the method, three sorting algorithms (bubble, insertion and selection sorts) are also considered. In the sequel, we first briefly recall the problem of lattice reduction and our motivation to exhibit worst-cases of LLL-like algorithms.

A Euclidean lattice is the set of all integer linear combinations of a set of linearly independent vectors in $\mathbb{R}^p$. The independent vectors are called *a basis* of the lattice. Any lattice can be generated by many bases. All of them have the same cardinality, that is called the *dimension* of the lattice. If B and B' represent matrices of two bases of the same lattice in the canonical basis of $\mathbb{R}^p$, then there is a unimodular matrix U such that $B' = UB$. A unimodular matrix is a matrix with integer entries and with determinant equal to ±1.

The lattice basis reduction problem is to find bases with good Euclidean properties, that is, with sufficiently short and almost orthogonal vectors.

M. Farach-Colton (Ed.): LATIN 2004, LNCS 2976, pp. 474–487, 2004.

In two dimensions, the problem is solved by the Gaussian algorithm, that finds in any two-dimensional lattice, a basis formed with the shortest possible vectors. The worst-case complexity of Gauss' algorithm (explained originally in the vocabulary of quadratic forms) was first studied by Lagarias [7], who showed that the algorithm is polynomial with respect to its input. The worst-case complexity of Gauss' algorithm was also studied later more precisely by Vallée[16].

In 1982, Lenstra, Lenstra and Lovász [10] gave a powerful approximation reduction algorithm for lattices of arbitrary dimension. Their famous algorithm, called LLL, was an important breakthrough to numerous theoretical and practical problems in computational number theory and cryptography [13,6,8]. The LLL algorithm seems difficult to analyze precisely, both in the worst-case [2,9,10] and in average-case [1,3,4]. In particular when the dimension is higher than two, the problem of the real worst-case of the algorithm is completely open. However, LLL-like reduction algorithms are so widely used in practice that the analyzes are a real challenge, both from a theoretical and practical point of view. To finish this brief presentation, we recall that the LLL algorithm is a possible generalization of its 2-dimensional version, which is the Gaussian algorithm. Moreover the Gaussian algorithm is intensively used (as a black box) by the LLL algorithm.

In this paper, we propose a new approach to the worst-case analyze of LLL-like lattice reduction algorithms. For the moment this approach is presented only in two dimensions. We have to observe here that the worst case of some variant of the Gaussian algorithm is already known: In [16], Vallée studied a variant of this algorithm whose elementary transforms are some integer matrices of determinant equal to 1. In the case of the usual Gaussian algorithm, elementary transforms are integer matrices of determinant either 1 or −1. Even if our paper generalizes [16] to the case of the usual Gaussian algorithm, we do not consider this as its most important point. Our aim here is to present our new approach.

An LLL-like lattice reduction algorithm or a sorting algorithm uses some atomic transforms. In both cases, the monoid of finite sequences of atomic transforms is a group. A trace of execution of the algorithm is always a sequence of atomic transforms. But each such sequence is not necessarily a trace of the algorithm. We exhibit a family of rewriting rules over the group generated by the atomic transforms corresponding to the mechanism of the algorithm: The rewriting rules make some sequences forbidden in the sense that possible executions will be exactly normal forms of the rewrite system. Thus the length of a valid word (or a normal form or a reduced word) over the set of generators, *i.e.*, the number of atomic transforms that compose the word, becomes very close to the number of steps of the algorithm. In this paper, we present some rewrite systems over $GL_2(\mathbb{Z})$ and over the permutations group, that make us *predict* how the Gaussian algorithm and some sorting algorithms are running on an arbitrary input.

Then we consider the variation of the length of the input with respect to the length of the reduced word issued by the trace of the algorithm running on that input. In the case of the reduction algorithm, an input is a basis of a lattice. The length of an input is naturally related to the number of bits needed to store it. For an input basis it is for instance the sum of the square of lengths of the basis' vectors. We make appear inputs whose length is minimal among all inputs demanding a given number of steps to the algorithm. We deduce from this the worst-case configuration of the usual Gaussian algorithm and give an "optimal" bound for the number of steps.

Let us explain this last point more precisely. Usually when counting the number of steps of an algorithm, one considers all inputs of length less than a fixed bound, say M. Then one estimates the maximum number of steps taken over all these inputs by:

$$f(M) := \max_{\text{all inputs of length at most } M} \text{number of steps of the algorithm}[1].$$

$$(1)$$

Here to exhibit the precise real worst-case, we first proceed in "the opposite way". Consider k a fixed number of steps. We will estimate the minimum length of those inputs demanding at least k steps to be processed by the algorithm:

$$g(k) := \min_{\text{all inputs demanding at least } k \text{ steps}} \text{length of the input.} \quad (2)$$

Clearly $f(g(k)) = k$. Otherwise there would be an input of length less than $g(k)$ demanding more than k steps. But $g(k)$ is by definition the minimal length of such inputs. So by inverting the function g, we can compute f.

Plan of the paper. Section 2 introduces the Gaussian algorithm and outlines our method. Section 3 is the crucial point of our method: We identify all the executions of the Gaussian algorithm with normal forms of 4 rewrite systems. Section 4 exhibits some particular inputs whose length is minimal among the lengths of all inputs requiring at least k steps. Then we recall a result of [16] that estimates the length of the particular basis exhibited before and deduce an upper-bound for the maximal number of steps of the Gaussian algorithm with respect to the length of the input. Finally in Section 5 our method is briefly applied to three sorting algorithms: For each sorting algorithm, all possible executions are identified with normal forms of a rewrite system.

2   Gaussian Algorithm and the New Approach to Its Worst-Case Analysis

Let $\mathbb{R}^2$ be endowed with the usual scalar product $(\,,\,)$ and Euclidean length $|\mathbf{u}| = (\mathbf{u}, \mathbf{u})^{1/2}$. A two-dimensional lattice is a discrete additive subgroup of $\mathbb{R}^2$. Equivalently, it is the set of all integer linear combinations of two linearly independent vectors. Generally it is given by one of its bases $(\mathbf{b}_1, \mathbf{b}_2)$. Let $(\mathbf{e}_1, \mathbf{e}_2)$ be the canonical basis of $\mathbb{R}^2$. We often associate to a lattice basis $(\mathbf{b}_1, \mathbf{b}_2)$ a matrix B, such that *the vectors of the basis are the rows of the matrix:*

$$B = \begin{matrix} & \mathbf{e}_1 & \mathbf{e}_2 \\ \mathbf{b}_1 \\ \mathbf{b}_2 \end{matrix} \begin{pmatrix} b_{1,1} & b_{1,2} \\ b_{2,1} & b_{2,2} \end{pmatrix}. \quad (3)$$

---

[1] When dealing with a non-trivial algorithm f is always an increasing function.

The *length ℓ of the previous basis* (or the *length of the matrix B*) is defined here by $\ell(B) := |\mathbf{b}_1|^2 + |\mathbf{b}_2|^2$.

The usual Gram-Schmidt orthogonalization process builds, in polynomial-time, from a basis $b = (\mathbf{b}_1, \mathbf{b}_2)$ an orthogonal basis $b^* = (\mathbf{b}_1^*, \mathbf{b}_2^*)$ and a lower-triangular matrix M that expresses the system b into the system b^{*2}. Let m be equal to $\frac{(\mathbf{b}_2, \mathbf{b}_1)}{(\mathbf{b}_1, \mathbf{b}_1)}$. By construction, the following equalities hold:

$$
\begin{cases} \mathbf{b}_1^* = \mathbf{b}_1 \\ \mathbf{b}_2^* = \mathbf{b}_2 - m\,\mathbf{b}_1 \end{cases} \text{and } M = \begin{matrix} \\ \mathbf{b}_1 \\ \mathbf{b}_2 \end{matrix} \overset{\begin{matrix} \mathbf{b}_1^* & \mathbf{b}_2^* \end{matrix}}{\begin{pmatrix} 1 & 0 \\ m & 1 \end{pmatrix}}. \tag{4}
$$

The ordered basis $B = (\mathbf{b}_1, \mathbf{b}_2)$ is called *proper* if the quantity m satisfies

$$
-1/2 \le m < 1/2. \tag{5}
$$

There is a natural representative of all the bases of a given two-dimensional lattice. This basis is composed of two shortest vectors generating the whole lattice. It is called the Gauss-reduced basis and the Gaussian algorithm outputs this reduced basis running on any basis of the lattice. Any lattice basis in two dimensions can always be expressed as

$$
B = U\,R, \tag{6}
$$

where R is the so-called Gaussian reduced basis of the same lattice and U is a unimodular matrix, *i.e.*, an element of $GL_2(\mathbb{Z})$. The goal of a reduction algorithm, the Gaussian algorithm in two dimensions, is to find R given B. The Gaussian algorithm is using two kinds of elementary transforms, explained in the sequel of this paper. Let $(\mathbf{b}_1, \mathbf{b}_2)$ be an input basis of a lattice and the matrix B expressing $(\mathbf{b}_1, \mathbf{b}_2)$ in the canonical basis of $\mathbb{R}^2$ as specified by (3).

The algorithm first makes an integer translation of $\mathbf{b}_2$ in the direction of $\mathbf{b}_1$ in order to make $\mathbf{b}_2$ as short as possible. This is done just by computing the integer x nearest to $m = (\mathbf{b}_2, \mathbf{b}_1)/(\mathbf{b}_1, \mathbf{b}_1)$ and replacing $\mathbf{b}_2$ by $\mathbf{b}_2 - x\mathbf{b}_1$. Notice that, after this integer translation, the basis $(\mathbf{b}_1, \mathbf{b}_2)$ is proper.

The second elementary transform is just the swap of the vectors $\mathbf{b}_1$ and $\mathbf{b}_2$ in case when after the integer translation we have $|\mathbf{b}_1| > |\mathbf{b}_2|$. The algorithm iterates these transforms, until after the translation, $\mathbf{b}_1$ remains still smaller than $\mathbf{b}_2$, *i.e.*, $|\mathbf{b}_1| \le |\mathbf{b}_2|$.

The Gaussian algorithm can also be regarded (especially for the analysis purposes) as an algorithm that gives a decomposition of the unimodular matrix U of relation (6) by means of some basic transforms:

$$
\begin{aligned} \text{Input:} \quad & B = U\,R. \\ \text{Output:} \quad & R = T^{x_{k+1}} S T^{x_k} S T^{x_{k-1}} \ldots S T^{x_2} S T^{x_1} B; \end{aligned} \tag{7}
$$

where the matrix T corresponds to an integer translation of $\mathbf{b}_2$ in the direction of $\mathbf{b}_1$ by one and the matrix S represents a swap:

$$
S = \begin{pmatrix} 0 & 1 \\ 1 & 0 \end{pmatrix} \quad \text{and} \quad T = \begin{pmatrix} 1 & 0 \\ 1 & 1 \end{pmatrix}. \tag{8}
$$

---

[2] Of course, b^* is generally not a basis for the lattice generated by b.

Each step of the algorithm is indeed an integer translation followed by a swap, represented by[3] ST^x, $x \in \mathbb{Z}^*$.

Writing the output as in (7) shows not only the output but how precisely the algorithm is working since T and S represent the only elementary transforms made during the execution of the Gaussian algorithm.

So when studying the mechanism of a reduction algorithm in two dimensions and for a fixed reduced basis R, the algorithm can be regarded as a decomposition algorithm over $GL_2(\mathbb{Z})$. The integer $k+1$ in (7) denotes the number of steps. Indeed the algorithm terminates [2,7,16]. The unimodular group in two dimensions $GL_2(\mathbb{Z})$ has been already studied [11,12,14,15] and it is well-known that $\{S,T\}$ is a possible family of generators for $GL_2(\mathbb{Z})$. Of course there are relators associated to these generators and there is no uniqueness of the decomposition of an element of $GL_2(\mathbb{Z})$ in terms of S and T. But the Gaussian algorithm gives one precise of these possible decompositions. When the algorithm is running on an input UR, the decomposition of U could a priori depend on the reduced basis R. We will show that the decomposition of U does not depend strongly on the reduced basis R: The next fact divides all reduced bases of $\mathbb{R}^2$ into 4 classes. Inside one fixed class of reduced bases the decomposition of U output by the Gaussian algorithm does not depend at all on the reduced basis R.

Fact. Let $R = (\mathbf{b_1}, \mathbf{b_2})$ be any reduced basis of $\mathbb{R}^2$. Then one of the following cases occurs:

$$|\mathbf{b_1}| < |\mathbf{b_2}| \quad \text{and} \quad m \neq -1/2; \tag{9}$$

$$|\mathbf{b_1}| = |\mathbf{b_2}| \quad \text{and} \quad m \neq -1/2; \tag{10}$$

$$|\mathbf{b_1}| < |\mathbf{b_2}| \quad \text{and} \quad m = -1/2; \tag{11}$$

$$|\mathbf{b_1}| = |\mathbf{b_2}| \quad \text{and} \quad m = -1/2. \tag{12}$$

In the sequel we completely characterize the decomposition of unimodular matrix output by the Gaussian algorithm and we will call it *the Gaussian decomposition of a unimodular matrix.* Roughly speaking, we exhibit forbidden sequences of values for the x_i-s.

More precisely, we exhibit in Section 3 a set of rewriting rules that leads to the formulation output by the Gaussian algorithm, from any product of matrices involving S and T. The precise characterization of the Gaussian decomposition that we give makes appear the slowest manner the length of a unimodular matrix can grow with respect to its Gaussian decomposition: We consider unimodular matrices whose length of Gaussian decomposition is fixed, say k:

$$U := T^{x_{k+1}} ST^{x_k} ST^{x_{k-1}} \ldots ST^{x_2} ST^{x_1}.$$

We exhibit in Section 4 the Gaussian word of length k with minimal length. We naturally deduce the minimum length $g(k)$ of all inputs demanding at least k steps. Finally by "inverting" the function g we find the maximum number of steps of the Gaussian algorithm.

---

[3] A priori x_1 and x_{k+1} in (7) may be zero so the algorithm may start by a swap or end with a translation.

3   The Gaussian Decomposition of a Unimodular Matrix

Let Σ be a (finite or infinite) set. A word ω on Σ is a finite sequence $\alpha_1\alpha_2\ldots\alpha_n$ where n is a positive integer, and $\alpha_i \in \Sigma$, for all $i \in \{1,\ldots,n\}$. Let Σ^* be the set of finite words on Σ. We introduce for convenience the *empty word* and we denote it by 1.

Consider the alphabet $\Sigma = \{S, T, T^{-1}\}$. We recall that the Gaussian decomposition of a unimodular matrix U is the decomposition of U corresponding to the trace of the algorithm when running on an input basis $B = UR$ where R is a reduced basis. In the sequel we show that there are at most 4 Gaussian decompositions for a unimodular matrix U. In the following subsections, 4 sets of rewriting rules depending on the form of R are given. Any word in which none of these rewriting rules can be applied is proven to be Gaussian. Since the results of these subsections are very similar, we only give sketches[7] of proofs for Subsection 3.1.

3.1   The Basis R Is Such That $|b_1| < |b_2|$ and $m \neq -1/2$

We say that a word ω is a *normal form* or *reduced word* or a *reduced decomposition* of the unimodular matrix U, if ω is a decomposition of U in which none of the rewriting rules of Theorem 1 can be applied.

Theorem 1 shows that the Gaussian decomposition and a reduced decomposition of a unimodular matrix are the same. By recalling that the Gaussian algorithm is deterministic, *i.e.*, for an input basis B, there is a couple (U, R) such that U and R are output by the Gaussian algorithm, the next theorem shows also that the reduced decomposition of a unimodular matrix is unique.

Theorem 1. *Let ω_1 be any decomposition of U in terms of the family of generators $\{S, T\}$. The Gaussian decomposition of U is obtained from ω_1 by applying repeatedly the following set of rules:*

$$S^2 \longrightarrow 1; \tag{13}$$

$$T^x T^y \longrightarrow T^{x+y}; \tag{14}$$

$$\forall x \in \mathbb{Z}_-^*, \qquad ST^2 ST^x \longrightarrow TST^{-2}ST^{x+1}; \tag{15}$$

$$\forall x \in \mathbb{Z}_+^*, \qquad ST^{-2}ST^x \longrightarrow T^{-1}ST^2ST^{x-1}; \tag{16}$$

$$\forall x \in \mathbb{Z}^*, \forall k \in \mathbb{Z}_+, \; STST^x \prod_{i=k}^{1} ST^{y_i} \longrightarrow TST^{-x-1}\prod_{i=k}^{1} ST^{-y_i}; \tag{17}$$

$$\forall x \in \mathbb{Z}^*, \forall k \in \mathbb{Z}_+, \; ST^{-1}ST^x \prod_{i=k}^{1} ST^{y_i} \longrightarrow T^{-1}ST^{-x+1}\prod_{i=k}^{1} ST^{-y_i}. \tag{18}$$

Let us consider the Gaussian algorithm running on inputs UR where U is any unimodular matrix and R a reduced basis (b_1, b_2) satisfying (9). As explained in the last section, an execution of the Gaussian algorithm is always expressed as a (finite) word on the alphabet $\Sigma = \{S, T, T^{-1}\}$. As a direct consequence of the previous theorem, a word on this alphabet is associated to a possible execution if and only if the word is a normal form of the previous rewrite system.

The trivial rules (13) and (14) have to be applied whenever possible. So any word ω_1 on the alphabet Σ can trivially be written as

$$T^{x_{k+1}} \prod_{i=k}^{1} ST^{x_i}, \tag{19}$$

with $x_i \in \mathbb{Z}^*$ for $2 \leq i \leq k$ and $(x_1, x_{k+1}) \in \mathbb{Z}^2$. The integer k is called *the length*[4] *of* ω_1. Notice that usually the length of a word is the number of its letters, which would be here equal to $2k + 1$. Here the length is k, which corresponds to the number of iterations of the algorithm (eventually minus 1).

The proof of Theorem 1 is based on the next Lemmata.

Lemma 1. *Let ω_1 be a word as in (19). Then the rewriting process of Theorem 1 always terminates*[5].

Proof (Sketch of the proof). Let k be a nonnegative integer, and let $x_1, \ldots, x_{k+1}$ be integers such that $x_2, \ldots, x_k$ are nonzero. Let ω_1 be a word on $\{S, T\}$ expressed by $\omega_1 = T^{x_{k+1}} \prod_{i=k}^{1} ST^{x_i}$. We consider the index sets $S_1 := \{i : 2 \leq i \leq k$ and $|x_i| = 1\}$ and $S_2 := \{i : 2 \leq i \leq k, \ x_i x_{i-1} < 0$ and $|x_i| = 2\}$. Finally for a word ω_1, the quantity $d(\omega_1)$ is $\sum_{i \in S_1 \cup S_2} i$. The Lemma is shown[7] by induction on the length of ω_1, and on the integer quantity $d(\omega_1)$.

The next lemma is crucial in the proof of Theorem 1. Indeed, as a direct Corollary of the next Lemma, normal forms of the rewrite system proposed in Theorem 1 are possible traces of the Gaussian algorithm[7]. Let us observe that the proof of Lemma 2 is closely related to the mechanism of the algorithm. Even slightly modifying the Gaussian algorithm may make the lemma fail.

Lemma 2. *Let B be the matrix of a proper basis $(\mathbf{b_1}, \mathbf{b_2})$ (see (3), (4) and (5). Let $x \in \mathbb{Z}^*$ be a non zero integer and $\tilde{B}$ defined by $\tilde{B} = (\tilde{\mathbf{b}_1}, \tilde{\mathbf{b}_2}) := ST^x B$.*

1. *If $|x| \geq 3$, then $\tilde{B}$ is still proper. Moreover,*
   - *if $(\mathbf{b_1}, \mathbf{b_2})$ and x are both positive or both negative, then $\tilde{B}$ is proper whenever $|x| \geq 2$.*
   - *if B is reduced, $|\mathbf{b_1}| < |\mathbf{b_2}|$ and $m \neq -1/2$, then $\tilde{B}$ is proper for all $x \in \mathbb{Z}$.*
2. *If $|x| \geq 2$, then $|\tilde{\mathbf{b}_2}| < |\tilde{\mathbf{b}_1}|$. Moreover, if $(\mathbf{b_1}, \mathbf{b_2})$ and x are both positive or both negative, it is true provided that $|x| \geq 1$.*

---

[4] The length of a word which is a decomposition of a unimodular matrix has of course to be distinguished from what we call the length of a unimodular matrix, that is closely related to the number of bits to store the matrix.

[5] Of course saying that the rewriting process presented by the previous Theorem always terminates has a priori nothing to do with the well-known fact that the Gaussian algorithm always terminates.

[7] A detailed proof is available in the full version of the paper.

3. If $|x| \geq 2$, then $\max(|\tilde{\mathbf{b}_1}|, |\tilde{\mathbf{b}_2}|) \geq \max(|\mathbf{b}_1|, |\mathbf{b}_2|)$.

4. If $|x| \geq 1$, then $(\tilde{\mathbf{b}_1}, \tilde{\mathbf{b}_2})$ and x are both positive or both negative.

Let us explain how the previous lemma shows that a normal form is a possible trace of the algorithm. First consider a proper (see (5) and non reduced basis P. Given to the Gaussian algorithm, the first operation done will be the swap. Now for any non proper basis B there is a unique integer x and a unique proper basis P such that B is expressed as $T^x P$. So if B is given to the algorithm the first operation done will be T^{-x}. Now consider for instance $B' = ST^5 P$, where P is a proper basis. The previous lemma asserts that B' is also proper and non reduced. So if B' is given to the algorithm, the first operation is the swap. More generally if

$$U := T^{x_{k+1}} ST^{x_k} ST^{x_{k-1}} \ldots ST^{x_2} ST^{x_1}.$$

is a normal form of the rewrite system and R a reduced basis satisfying (9), then thanks to the previous lemma all the intermediate bases $ST^{x_1} R$, $ST^{x_2} ST^{x_1} R$, $\ldots$, $ST^{x_k} ST^{x_{k-1}} \ldots ST^{x_2} ST^{x_1} R$ are proper. So the algorithm will perform the exact following sequence of operations:

$$T^{-x_{k+1}} ST^{-x_k} ST^{-x_{k-1}} \ldots ST^{-x_2} ST^{-x_1}.$$

Corollary 1. *Any normal form of the rewrite system defined in Theorem 1 is Gaussian.*

Proof (Sketch of the proof of Theorem 1). Consider an input $B = UR$ where U is an unimodular matrix and R is a reduced basis of a lattice L satisfying (9).

Lemma 1 asserts that for any decomposition ω of U in terms of S and T, there is a normal form ω' (and a unimodular matrix $U' = \omega'$). Notice that the Lemma does not show the uniqueness of the normal form.

Corollary 1 of Lemma 2 asserts that normal forms of the rewrite system are Gaussian words (traces of the Gaussian algorithm).

Now observe that the use of a nontrivial rewriting rule changes a base of the lattice into another base of the same lattice and the way the basis is changed is totally explicit. So for an input UR there is a couple (U', R') such that

(i) $UR = U'R'$,

(ii) the matrix U' is unimodular and its decomposition is a normal form of the rewrite system,

(iii) and R' is also a reduced basis of the same lattice L.

Finally by recalling that the Gaussian algorithm is deterministic, the decomposition ω' of U' is the trace of the Gaussian algorithm when running on $B = UR$. (Of course the output is R'.)

Proofs of Theorems 2, 3 and 4, which are presented in the following subsections, are very similar.

3.2   The Basis R Is Such That $|b_1| = |b_2|$ and $m \neq -1/2$

Theorem 2. *Let ω_1 be any decomposition of U in terms of the family of generators $\{S, T\}$. The Gaussian decomposition of U is obtained from ω_1 by applying repeatedly the set of rules (13) to (18) of Theorem 1, together with the following rules:*

$$\omega S \longrightarrow \omega; \tag{20}$$

$$\omega ST \longrightarrow \omega TST^{-1}. \tag{21}$$

3.3   The Basis R Is Such That $|b_1| < |b_2|$ and $m = -1/2$

Theorem 3. *Let R be a reduced basis and let U be a unimodular matrix, i.e., an element of $GL_2(\mathbb{Z})$. Let ω_1 be any decomposition of U in terms of the family of generators $\{S, T\}$. The Gaussian decomposition of U is obtained from ω_1 by applying repeatedly the rules (13) to (16) of Theorem 1 together with the rules (22) and (23) defined here, until no one of these rules applies. Then if we have $\omega_1 = \omega ST^2 S$, the ending rule (24) applies once and the rewriting process is over.*

$$x \in \mathbb{Z}^*, k \in \mathbb{Z}_+, \; STST^x \left(\prod_{i=k}^{1} ST^{y_i} \right) \longrightarrow TST^{-x-1} \left(\prod_{i=k}^{1} ST^{-y_i} \right) T; \tag{22}$$

$$x \in \mathbb{Z}^*, k \in \mathbb{Z}_+ ST^{-1} ST^x \left(\prod_{i=k}^{1} ST^{y_i} \right) \longrightarrow T^{-1} ST^{-x+1} \left(\prod_{i=k}^{1} ST^{-y_i} \right) T; \tag{23}$$

$$\omega ST^2 S \longrightarrow \omega T ST^{-2} ST. \tag{24}$$

3.4   The Basis R Is Such That $|b_1| = |b_2|$ and $m = -1/2$

Theorem 4. *Let R be a reduced basis and let U be a unimodular matrix, i.e., an element of $GL_2(\mathbb{Z})$. Let ω_1 be any decomposition of U in terms of the family of generators $\{S, T\}$. The Gaussian decomposition of U is obtained from ω_1 by applying repeatedly Rules (13) to (16) of Theorem 1, together with Rules (22), (23) and (20) and the following set of rules:*

$$\omega ST \longrightarrow \omega T; \tag{25}$$

$$\omega ST^2 \longrightarrow \omega TST^{-1}. \tag{26}$$

4   The Length of a Unimodular Matrix with Respect to Its Gaussian Decomposition and the Maximum Number of Steps of the Algorithm

Let $B = (\mathbf{b_1}, \mathbf{b_2})$ be a basis. The *length* of B, denoted by $\ell(B)$, is the sum of the squares of the norms of its vectors, that is, $\ell(B) = |\mathbf{b_1}|^2 + |\mathbf{b_2}|^2$.

The easy but tedious proof of the following theorem is given[7] in the full version of the paper.

Theorem 5. *Let $R = (\mathbf{b_1}, \mathbf{b_2})$ be a reduced basis, let k be a positive integer, and let $x_1, \ldots, x_{k+1}$ be integers such that the word $\omega = \prod_{i=k}^{1} S T^{x_i}$ is Gaussian. Then the following properties hold:*

1. *if $|\mathbf{b_1}| < |\mathbf{b_2}|$ and $m \geq 0$ then $\ell(\omega R) \geq \ell((ST^{-2})^{k-1} S R)$;*
2. *if $|\mathbf{b_1}| < |\mathbf{b_2}|$ and $-1/2 < m < 0$ then $\ell(\omega R) \geq \ell((ST^2)^{k-1} S R)$;*
3. *if $|\mathbf{b_1}| < |\mathbf{b_2}|$ and $m = -1/2$ then $\ell(\omega R) \geq \ell((ST^{-2})^{k-1} ST R)$;*
4. *if $|\mathbf{b_1}| = |\mathbf{b_2}|$ then $\ell(\omega R) \geq \ell((ST^{-2})^{k-1} ST^{-1} R)$.*

The previous theorem provides bases whose length are minimal among all bases requiring at least k iterations to the Gaussian algorithm. These are essentially $(ST^2)^{k-1} S R$ where R is a reduced basis. We have then to lower bound the length $\ell((ST^2)^{k-1} S R)$. In the next section, we just recall how to evaluate such a length.

The next lemma is exactly Lemma 4 of [16]. A sketch of the proof is recalled in the full version of the paper.

Lemma 3. *Let $k > 2$ be a fixed integer. There exists an absolute constant A such that any input basis demanding more than k steps to the Gaussian algorithm has a length greater than $A(1 + \sqrt{2})^{2k-1}$.*

It follows that any input with length less than $A(1 + \sqrt{2})^{2k-2}$ is demanding less than k steps. We deduce the following corollary.

Corollary 2. *There is an absolute constant A such that the number of steps of the Gaussian algorithm on inputs of length less than M is bounded from above by*

$$\frac{1}{2}\left(\log_{(1+\sqrt{2})}\left(\frac{M}{A}\right) + 1\right).$$

5   Sorting Algorithms

In the previous sections, we proposed a method for worst–case analyzing the Gaussian algorithm. We hope to generalize the approach to the LLL algorithm in higher dimensions (a still open problem even in three dimensions). On the other hand, our method can be applied to other families of algorithms. In this Section we consider some the bubble sort algorithm (in the full vesion of the paper we consider also the insertion sort and the selection sort algorithms). Of course worst-cases of these sorting algorithms are very well-known. Here the aim is to use our method to recover these well-known worst cases.

A sorting algorithm (as the Gaussian algorithm) uses some atomic transforms. Once more the monoid of finite sequences of atomic transforms is a group: The permutation group plays here the role played by $GL_2(\mathbb{Z})$ in Section 3.

For each considered sorting algorithm we propose a set of rewriting rules over the group of permutations represented by a family of generators that is precisely the set of atomic transforms of the algorithm. Clearly an execution of the algorithm is a finite word on the alphabet of these atomic transforms. But any such word is not necessarily an execution of the algorithm. We prove that a word on the alphabet of atomic transforms is associated to an execution of the algorithm if and only if the word is a normal form of the rewrite system we propose.

In a second step, as for the analysis of the Gaussian algorithm, we have to consider the variation of the length of the input with respect to the length of the reduced word issued by the trace of the algorithm running on that input. So in our method we have to deal with two notions of length: the length of normal forms, that counters the number of iterations of the algorithm and the length of the inputs that is classically associated to the number of bits to store the input.

Let us observe that here this step is somehow trivial. Indeed when running on n items to be sorted, the length of the input of a sorting algorithm is always n (at least in usual analyzes), no matter how many steps are needed to sort the n input items. In other words, the length of the input is in the case of sorting algorithm constant and does not depend on the length of the normal form associated to this input, as it did when considering the Gaussian algorithm.

So clearly the longest length of the normal forms is here exactly the maximal number of iterations of the sorting running on inputs of length n.

The sketch of the proof is the same than the one of the Gaussian algorithm: We first prove that the rewriting process always terminates. Then, we show that the reduced words are also the normal forms given by the corresponding algorithm.

In the sequel, we first recall some useful definitions and notations. Then we analyze the bubble sort algorithm. The rewrite systems for the insertion and selection sort algorithms, which are close to the rewrite system associated to the bubble sort are also given in the full vesion of this paper.

Let n be a positive integer, and let $[1, \ldots, n]$ be the sorted list of the n first positive integers. Let $\mathcal{S}_n$ be the set of all permutations on $[1, \ldots, n]$, and let $\mathcal{S}$ be the set of all permutations on a list of distinct integers of variable size. Let us denote by t_i the transposition which swaps the elements in positions i and $i+1$ in the list , for all $i \in \{1, \ldots, n\}$. Any permutation can be written in terms of the t_i-s. Let Σ_n be defined by $\Sigma_n = \{t_1, \ldots, t_n\}$ and Σ denote $\Sigma = \{t_i : i \in \mathbb{N}^*\}$. Thus Σ_n (resp. Σ) is a generating set of $\mathcal{S}_n$ (resp. $\mathcal{S}$).

As in previous sections, any word ω on Σ will be denoted as following:

$$\omega = t_{i_1} t_{i_2} \ldots t_{i_k} = \prod_{j=1}^{k} t_{i_j},$$

where k and $i_1, \ldots, i_k$ are positive integers.

Definition 1. *Let $\omega_1 = t_{i_1} t_{i_2} \ldots t_{i_k}$, $\omega_2 = t_{j_1} t_{j_2} \ldots t_{j_l}$ and $\omega_3 = t_{r_1} t_{r_2} \ldots t_{r_m}$ be words on Σ.*

1. *The length of ω, denoted by $|\omega|$, is k;*
2. *the distance between ω_1 and ω_2, denoted by $Dist(\omega_1, \omega_2)$, is given by $\min_{t_i \in \omega_1, t_j \in \omega_2} |i - j|$;*

3. *the* maximum *(resp.* minimum*) of* ω_1, *denoted by* $\max(\omega_1)$ *(resp.* $\min(\omega_1)$)*,
   is given by* $\max_{t_i \in \omega_1}(i)$ *(resp.* $\min_{t_i \in \omega_1}(i)$*)*;
4. ω_1 *is an* increasing *(resp.* decreasing*) word if* $i_p < i_{p+1}$ *(resp.* $i_p > i_{p+1}$*),
   for all* $p \in \{1, \ldots, k-1\}$;
5. ω_2 *is a maximally increasing factor of* $\omega_1\omega_2\omega_3$ *if* ω_2 *is increasing and both*
   $t_{i_k}\omega_2$ *and* $\omega_2 t_{r_1}$ *are not increasing.*

Any word ω on Σ is uniquely expressed on the form

$$\omega = \omega_1\omega_2 \ldots \omega_m, \tag{27}$$

where each ω_i is a maximally increasing factor of ω. We will call (27) the *increasing decomposition* of ω. We define $s \colon \Sigma^* \to \mathbb{N}$ as the map given by the rule $s(\omega) = m$.

The basic idea of the bubble sort algorithm is the following: pairs of adjacent values in the list to be sorted are compared and interchanged if they are out of order, the process starting from the beginning of the list. Thus, list entries 'bubble upward' in the list until they bump into one with a higher sort value. The algorithm first compares the two first elements of the list and swaps them if they are in the wrong order. Then, the algorithm compares the second and the third elements of the list and swaps them if necessary. The algorithms continues to compare adjacent elements from the beginning to the end of the list. This whole process is iterated until no changes are done.

Let σ be a permutation on $[1, \ldots, n]$. There exists a unique decomposition ω of σ on the alphabet Σ corresponding to the sequence of elementary transforms performed by the bubble sort algorithm on $\sigma[1, \ldots, n]$. We will call it the *bubblian decomposition* of σ. Notice that $(\omega)^{-1}\sigma = 1$. The bubble sort algorithm can be regarded as an algorithm giving the bubblian decomposition of a permutation.

Definition 2. *A word* ω *on* Σ *is a bubblian word if it corresponds to a possible execution of the bubble sort algorithm.*

Let us define some rewriting rules on Σ^*. In the following equations, i, j and k are arbitrary positive integers and ω is a word on Σ:

$$t_i\, t_i \longrightarrow 1; \tag{28}$$

$$\text{if } \mathrm{Dist}(t_{i+1}, \omega) > 1, \quad t_{i+1}\, \omega\, t_i\, t_{i+1} \longrightarrow \omega t_i\, t_{i+1}\, t_i; \tag{29}$$

$$\mathrm{Dist}(t_i, \omega) > 1 \text{ and } \omega \text{ maximally increasing factor}, \quad \omega\, t_i \longrightarrow t_i\, \omega; \tag{30}$$

$$\mathrm{Dist}(t_j, t_k\omega) > 1, \quad i \le j \le k \text{ or } k < i \le j, \quad t_i\, t_k\, \omega\, t_j \longrightarrow t_i\, t_j\, t_k\, \omega. \tag{31}$$

Theorem 6. *Let* σ *be a permutation and let* $\omega \in \Sigma^*$ *be a decomposition of* σ *on* Σ. *The bubblian decomposition of* σ *is obtained from* ω *by applying repeatedly the rules (28) to (31).*

Remark 1. Let ω and ω' be words on Σ. It is well known that a presentation of S on Σ is the following:

- $t_i t_i = 1$;
- $t_i t_j = t_j t_i$;
- $t_i t_{i+1} t_i = t_{i+1} t_i t_{i+1}$;

for all positive integers i, j such that $|i - j| = 1$. Thus, it is easy to prove that if ω' is obtained from ω, then $\omega = \omega'$ in $\mathcal{S}$.

The proof[7] of Theorem 6 is very similar to the proof of Theorem 1. It is based on the following lemmata.

The next Lemma shows first that the rewrite process terminates: Given a permutation and any decomposition of this permutation in terms of t_i's , one obtains a normal form of the rewrite system represented by rules (28), (29), (30) and (31) by finitely many times applying the rules (in an arbitrary order). For any word $\omega = t_{\alpha_1} \ldots t_{\alpha_{|\omega|}} \in \Sigma^*$, we define the two quantities $l(\omega)$ and $h(\omega)$ by

$$l(\omega) = \sum_{i=1}^{|\omega|} \alpha_i \quad \text{and} \quad h(\omega) = \sum_{i=1}^{s(\omega)} (s(\omega) - i)(\max(\omega) - |\omega_i|),$$

where $\omega_1 \omega_2 \ldots \omega_m$ is the increasing decomposition of ω. The proof of the next lemma is a double induction on the positive integer quantities $l(\omega)$ and $h(\omega)$.

Lemma 4. *The rewriting process defined by rules (28), (29), (30) and (31) always terminates.*

The next Lemma shows that any normal form of the rewrite system is a bubblian word (a possible execution of the bubble sort algorithm). The proof[7] of this lemma is of course related to the bubble sort algorithm.

Lemma 5. *Let ω be a reduced word. Then ω is a bubblian word.*

Since the bubblian word associated to a given permutation is unique and the rewriting process terminates, the bubblian words are exactly normal forms of the rewrite system. Notice that we can easily deduce from Theorem 6 the worst-case for the bubble sort algorithm.

6   Conclusion

In this paper we studied the Gaussian algorithm by considering a rewriting system over $GL_2(\mathbb{Z})$. We first believe that our method should be applied to other variants of the Gaussian algorithm (for example, Gaussian algorithm with other norms [5]): For each variant there is an adequate rewriting system over $GL_2(\mathbb{Z})$.

The most important and interesting continuation to this work is to generalize the approach in higher dimensions. Even in three dimensions, the worst-case configuration of all possible generalization of the Gaussian algorithm is completely unknown for the moment. Although the problem is really difficult, we have already achieved a step, since the LLL algorithm uses the Gaussian algorithm as an elementary transform.

The group of n-dimensional lattice transformations has been studied first by Nielsen [14] ($n = 3$) and for an arbitrary n by Magnus [11,12], based on the work of Nielsen[15]. Their work should certainly help to exhibit such rewrite systems on $GL_n(\mathbb{Z})$ if there exists.

This approach may also be an insight to the still open problem of the complexity of the optimal LLL algorithm [2,9].

Acknowledgments. The authors are indebted to Brigitte Vallée for drawing their attention to algorithmic problems in lattice theory and for regular helpful discussions.

References

1. A. AKHAVI. *Étude comparative d'algorithmes de réduction sur des réseaux aléatoires*. PhD thesis, Université de Caen, 1999.
2. A. AKHAVI. Worst-case complexity of the optimal LLL algorithm. In *Proceedings of LATIN'2000 - Punta del Este*. LNCS 1776, pp 476–490.
3. H. DAUDÉ, PH. FLAJOLET, AND B. VALLÉE. An average-case analysis of the Gaussian algorithm for lattice reduction. *Comb., Prob. & Comp.*, 123:397–433, 1997.
4. H. DAUDÉ, AND B. VALLÉE. An upper bound on the average number of iterations of the LLL algorithm. *Theoretical Computer Science 123(1)* (1994), pp. 95–115.
5. M. KAIB AND C. P. SCHNORR. The generalized Gauss reduction algorithm. *J. of Algorithms*, 21:565–578, 1996.
6. R. KANNAN. Improved algorithm for integer programming and related lattice problems. In *15th Ann. ACM Symp. on Theory of Computing*, pages 193–206, 1983.
7. J. C. LAGARIAS. Worst-case complexity bounds for algorithms in the theory of integral quadratic forms. *J. Algorithms*, 1:142–186, 1980.
8. H.W. LENSTRA. Integer programming with a fixed number of variables. *Math. Oper. Res.*, 8:538–548, 1983.
9. H.W. LENSTRA. Flags and lattice basis reduction. In *Proceedings of the 3rd European Congress of Mathematics - Barcelona July 2000* I: 37–51, Birkhäuser Verlag, Basel
10. A. K. LENSTRA, H. W. LENSTRA, AND L. LOVÁSZ. Factoring polynomials with rational coefficients. *Math. Ann.*, 261:513-534, 1982.
11. W. MAGNUS. Über n-dimensionale Gittertransformationen. *Acta Math.*, 64:353–357, 1934.
12. W. MAGNUS, A. KARRASS, AND D. SOLITAR. Combinatorial group theory. Dover, New York, 1976 (second revised edition).
13. P. NGUYEN. AND J. STERN. The two faces of lattices in cryptology. In *Proceedings of CALC'01*. LNCS 2146.
14. J. NIELSEN. Die Gruppe der dreidimensionale Gittertransformationen. *Kgl Danske Videnskabernes Selskab., Math. Fys. Meddelelser*, V 12: 1–29, 1924.
15. J. NIELSEN. Die Isomorphismengruppe der freien Gruppen. *Math. Ann.*, 91:169–209, 1924. translated in english by J. Stillwell in *J. Nielsen collected papers*, Vol 1.
16. B. VALLÉE. Gauss' algorithm revisited. *J. of Algorithms*, 12:556–572, 1991.

Generating Maximal Independent Sets for Hypergraphs with Bounded Edge-Intersections*

Endre Boros[1], Khaled Elbassioni[1], Vladimir Gurvich[1], and Leonid Khachiyan[2]

[1] RUTCOR, Rutgers University, 640 Bartholomew Road, Piscataway NJ 08854-8003;
{boros,elbassio,gurvich}@rutcor.rutgers.edu
[2] Department of Computer Science, Rutgers University, 110 Frelinghuysen Road, Piscataway NJ 08854-8003; leonid@cs.rutgers.edu

Abstract. Given a finite set V, and integers $k \geq 1$ and $r \geq 0$, denote by $\mathbb{A}(k,r)$ the class of hypergraphs $\mathcal{A} \subseteq 2^V$ with (k,r)-bounded intersections, i.e. in which the intersection of any k distinct hyperedges has size at most r. We consider the problem $MIS(\mathcal{A},\mathcal{I})$: given a hypergraph $\mathcal{A}$ and a subfamily $\mathcal{I} \subseteq \mathcal{I}(\mathcal{A})$, of its maximal independent sets (MIS) $\mathcal{I}(\mathcal{A})$, either extend this subfamily by constructing a new MIS $I \in \mathcal{I}(\mathcal{A}) \setminus \mathcal{I}$ or prove that there are no more MIS, that is $\mathcal{I} = \mathcal{I}(\mathcal{A})$. We show that for hypergraphs $\mathcal{A} \in \mathbb{A}(k,r)$ with $k+r \leq const$, problem $MIS(\mathcal{A},\mathcal{I})$ is NC-reducible to problem $MIS(\mathcal{A}',\emptyset)$ of generating a single MIS for a partial subhypergraph $\mathcal{A}'$ of $\mathcal{A}$. In particular, for this class of hypergraphs, we get an incremental polynomial algorithm for generating all MIS. Furthermore, combining this result with the currently known algorithms for finding a single maximal independent set of a hypergraph, we obtain efficient parallel algorithms for incrementally generating all MIS for hypergraphs in the classes $\mathbb{A}(1,c)$, $\mathbb{A}(c,0)$, and $\mathbb{A}(2,1)$, where c is a constant. We also show that, for $\mathcal{A} \in \mathbb{A}(k,r)$, where $k+r \leq const$, the problem of generating all MIS of $\mathcal{A}$ can be solved in incremental polynomial-time with space polynomial only in the size of $\mathcal{A}$.

1 Introduction

Let $\mathcal{A} \subseteq 2^V$ be a hypergraph (set family) on a finite vertex set V. A vertex set $I \subseteq V$ is called *independent* if I contains no hyperedge of $\mathcal{A}$. Let $\mathcal{I}(\mathcal{A}) \subseteq 2^V$ denote the family of all maximal independent sets (MIS) of $\mathcal{A}$. We assume that $\mathcal{A}$ is given by the list of its hyperedges and consider problem $MIS(\mathcal{A})$ of incrementally generating all sets in $\mathcal{I}(\mathcal{A})$. Clearly, this problem can be solved by performing $|\mathcal{I}(\mathcal{A})| + 1$ calls to the following problem:

$MIS(\mathcal{A},\mathcal{I})$: *Given a hypergraph $\mathcal{A}$ and a collection $\mathcal{I} \subseteq \mathcal{I}(\mathcal{A})$ of its maximal independent sets, either find a new maximal independent set $I \in \mathcal{I}(\mathcal{A}) \setminus \mathcal{I}$, or prove that the given collection is complete, $\mathcal{I} = \mathcal{I}(\mathcal{A})$.*

* This research was supported in part by the National Science Foundation, grant IIS-0118635. The research of the first and third authors was also supported in part by the Office of Naval Research, grant N00014-92-J-1375. The second and third authors are also grateful for the partial support by DIMACS, the National Science Foundation's Center for Discrete Mathematics and Theoretical Computer Science.

M. Farach-Colton (Ed.): LATIN 2004, LNCS 2976, pp. 488–498, 2004.
© Springer-Verlag Berlin Heidelberg 2004

Note that if $I \in \mathcal{I}(\mathcal{A})$ is an independent set, the complement $B = V \setminus I$ is a *transversal* to $\mathcal{A}$, that is $B \cap A \neq \emptyset$ for all $A \in \mathcal{A}$, and vice versa. Hence $\{B \mid B = V \setminus I, \ I \in \mathcal{I}(\mathcal{A})\} = \mathcal{A}^d$, where $\mathcal{A}^d \overset{\text{def}}{=} \{B \mid B \text{ is a minimal transversal to } \mathcal{A}\}$ is the *transversal* or *dual* hypergraph of $\mathcal{A}$. For this reason, MIS$(\mathcal{A}, \mathcal{I})$ can be equivalently stated as the *hypergraph dualization problem*:

$DUAL(\mathcal{A}, \mathcal{B})$: *Given a hypergraph $\mathcal{A}$ and a collection $\mathcal{B} \subseteq \mathcal{A}^d$ of minimal transversals to $\mathcal{A}$, either find a new minimal transversal $B \in \mathcal{A} \setminus \mathcal{B}$ or show that $\mathcal{B} = \mathcal{A}$.*

This problem has applications in combinatorics, graph theory, artificial intelligence, reliability theory, database theory, integer programming, and learning theory (see, e.g. [5,9]). It is an open question whether problem DUAL$(\mathcal{A}, \mathcal{B})$, or equivalently MIS$(\mathcal{A}, \mathcal{I})$, can be solved in polynomial time for arbitrary hypergraphs. The fastest currently known algorithm [11] for DUAL$(\mathcal{A}, \mathcal{B})$ is quasi-polynomial and runs in time $O(nm) + m^{o(\log m)}$, where $n = |V|$ and $m = |\mathcal{A}| + |\mathcal{B}|$.

It was shown in [6,9] that in the case of hypergraphs of bounded dimension, $\dim(\mathcal{A}) \overset{\text{def}}{=} \max_{A \in \mathcal{A}} |A| \leq const$, problem MIS$(\mathcal{A}, \mathcal{I})$ can be solved in polynomial time. Moreover, [4] shows that the problem can be efficiently solved in parallel, MIS$(\mathcal{A}, \mathcal{I}) \in NC$ for $\dim(\mathcal{A}) \leq 3$ and MIS$(\mathcal{A}, \mathcal{I}) \in RNC$ for $\dim(\mathcal{A}) = 4, 5 ...$ Let us also mention that for graphs, $\dim(\mathcal{A}) \leq 2$, all MIS can be generated with polynomial delay, see [13] and [19].

In [8], a total polynomial time generation algorithm was obtained for the hypergraphs of bounded degree, $\deg(\mathcal{A}) \overset{\text{def}}{=} \max_{v \in V} |\{A : v \in A \in \mathcal{A}\}| \leq const$. This result was recently strengthened in [10], where a polynomial delay algorithm was obtained for a wider class of hypergraphs.

In this paper we consider the class $\mathbb{A}(k, r)$ of hypergraphs with (k, r)-*bounded intersections*: $\mathcal{A} \in \mathbb{A}(k, r)$ if the intersection of each (at least) k distinct hyper edges of $\mathcal{A}$ is of cardinality at most r. We will always assume that $k > 1$ and $r \geq 0$ are fixed integers whose sum is bounded, $k + r \leq c = const$. Note that

$$\dim(\mathcal{A}) \leq r \quad \text{iff} \quad \mathcal{A} \in \mathbb{A}(1, r) \quad \text{and} \quad \deg(\mathcal{A}) < k \quad \text{iff} \quad \mathcal{A} \in \mathbb{A}(k, 0),$$

and hence, the class $\mathbb{A}(k, r)$ contains both the bounded-dimension and bounded-degree hypergraphs as subclasses. It will be shown that problem MIS$(\mathcal{A}, \mathcal{I})$ can be solved in polynomial time for hypergraphs with (k, r)-bounded intersections. It is not difficult to see that for any hypergraph $\mathcal{A} \in \mathbb{A}(k, r)$ the following property holds for every vertex-set $X \subseteq V$: X is contained in a hyperedge of $\mathcal{A}$ whenever each subset of X of cardinality at most $c = k + r$ is contained in a hyperedge of $\mathcal{A}$. [Indeed, suppose that X is a minimal subset of V not contained in any hyperedge of $\mathcal{A}$, and that every subset of X of cardinality at most $k + r$ is contained in a hyperedge of $\mathcal{A}$. Note that $|X| \geq k + r + 1$. Let $e_1, \ldots, e_k$ be distinct elements of X. Then the exist distinct hyperedges $A_1, \ldots, A_k \in \mathcal{A}$ such that $X \setminus \{e_i\} \subseteq A_i$, for $i = 1, \ldots, k$. Now we get a contradiction to the fact that $\mathcal{A} \in \mathbb{A}(k, r)$ since $|A_1 \cap \ldots \cap A_k| \geq r + 1$.] Hypergraphs $\mathcal{A} \subseteq 2^V$ with this property were introduced by Berge [3] under the name of *c-conformal hypergraphs*, and clearly define a wider class of hypergraphs than $\mathbb{A}(k, r)$ with

$k + r = c$. In fact, we will prove our result for this wider class of c-*conformal* hypergraphs.

Theorem 1. *For the c-conformal hypergraphs, $c \leq const$, and in particular for $\mathcal{A} \in \mathbb{A}(k, r)$, $k + r \leq c = const$, problem $MIS(\mathcal{A}, \mathcal{I})$ is polynomial and hence $\mathcal{I}(\mathcal{A})$, the set all MIS of $\mathcal{A}$, can be generated in incremental polynomial time.*

Theorem 1 is a corollary of the following stronger theorem which will be proved in Section 2.

Theorem 2. *For any c-conformal hypergraph $\mathcal{A}$, where c is a constant, problem $MIS(\mathcal{A}, \mathcal{I})$ is NC-reducible to $MIS(\mathcal{A}', \emptyset)$, where $\mathcal{A}'$ is a partial sub-hypergraph of $\mathcal{A}$.*

In Section 2, we also derive some further consequences of Theorem 2, related to the parallel complexity of problem $MIS(\mathcal{A}, \mathcal{I})$ for certain classes of hypergraphs.

Let us note that our algorithm of generating $I(\mathcal{A})$ based on Theorem 1 is incremental, since it requires solving problem $MIS(\mathcal{A}, \mathcal{I})$ iteratively $|I(\mathcal{A})| + 1$ times. Thus, this algorithm may require space exponential in the size of the input hypergraph $N = N(\mathcal{A}) = \sum_{A \in \mathcal{A}} |A|$. A generation algorithm for $\mathcal{I}(\mathcal{A})$ is said to work in *polynomial space* if the total space required by the algorithm to output all the elements of $\mathcal{I}(\mathcal{A})$ is polynomial in N. In Section 3, we prove the following.

Theorem 3. *For the hypergraphs of bounded intersections, $\mathcal{A} \in \mathbb{A}(k, r)$, where $k + r \leq const$, all MIS of $\mathcal{A}$ can be enumerated in incremental polynomial time and with polynomial space.*

Finally, we conclude in Section 4, with a third algorithm for generating all maximal independent sets of a hypergraph $\mathcal{A} \in \mathbb{A}(k, r)$, $k + r \leq const$.

2   NC-Reduction for c-Conformal Hypergraphs

The results of [4] show that, for hypergraphs of bounded dimension $\mathcal{A}(1, c)$, there is an NC-reduction from $MIS(\mathcal{A}, \mathcal{I})$ to $MIS(\mathcal{A}', \emptyset)$, where $\mathcal{A}'$ is a partial sub-hypergraph of $\mathcal{A}$. In other words, the problem of extending in parallel a given list of MIS of $\mathcal{A}$ can be reduced to the problem of generating in parallel a single MIS for a partial sub-hypergraph of $\mathcal{A}$. In this section we extend this reduction to the class of c-conformal hypergraphs, when c is a constant.

2.1 c-Conformal Hypergraphs

Given a hypergraph $\mathcal{A} \subseteq 2^V$, we say that $\mathcal{A}$ is *Sperner* if no hyperedge of $\mathcal{A}$ contains another hyperedge. By definition, for every hypergraph $\mathcal{A}$, its MIS hypergraph $\mathcal{I}(\mathcal{A})$ is Sperner. Let us inverse the operator $\mathcal{I}$. Given a Sperner hypergraph $\mathcal{B} \subseteq 2^V$, introduce the hypergraph $\mathcal{A} = \mathcal{I}^{-1}(\mathcal{B}) \subseteq 2^V$ whose hyper-edges are all minimal subsets $A \subseteq V$ which are not contained in any hyperedge of $\mathcal{B}$, that is $A \subseteq B$ for no $A \in \mathcal{A}, B \in \mathcal{B}$ and $A' \subseteq B$ for some $B \in \mathcal{B}$ for each proper subset $A' \subset A \in \mathcal{A}$. The hypergraph $\mathcal{A} = \mathcal{I}^{-1}(\mathcal{B})$ is Sperner by definition, too. It is also easy to see that $\mathcal{B}$ is the MIS hypergraph of $\mathcal{A}$. In other words, for Sperner hypergraphs $\mathcal{B} = \mathcal{I}(\mathcal{A})$ if and only if $\mathcal{A} = \mathcal{I}^{-1}(\mathcal{B})$. In [3], Berge introduced the class of c-conformal hypergraphs and characterized them in several equivalent ways as follows.

Proposition 1 ([3]). *For each hypergraph $\mathcal{A} \subseteq 2^V$ the following statements are equivalent: (i) $\mathcal{A}$ is c-conformal; (ii) The transposed hypergraph $\mathcal{A}^T$ (whose incidence matrix is the transposed incidence matrix of $\mathcal{A}$) satisfies the $(c-1)$-dimensional Helly property: a subset of hyperedges from $\mathcal{A}^T$ has a common vertex whenever every at most c hyperedges of this subset have one; (iii) For each partial hypergraph $\mathcal{A}' \subseteq \mathcal{A}$ having $c+1$ edges, the set $\{x \in V \mid d_{\mathcal{A}'}(x) \geq c\}$ of vertices of degree at least c in $\mathcal{A}'$, is contained in an edge of $\mathcal{A}$.*

It is not difficult to see that we can add to the above list the following equivalent characterization:
(iv) $\dim(\mathcal{I}^{-1}(\mathcal{A})) \leq c$.

Note also that (iii) gives a polynomial-time membership test for c-conformal hypergraphs, for a fixed constant c. Thus even though, given a hypergraph $\mathcal{A}$, the precise computation of $\dim(\mathcal{I}^{-1}(\mathcal{A}))$ is an NP-complete problem (it can be reduced from *stability number for graphs*), verifying condition (iv) is polynomial for every fixed c by Proposition 1.

Given a hypergraph $\mathcal{A} \subseteq 2^V$, let us introduce the *complementary* hypergraph $\mathcal{A}^c = \{V \setminus A \mid A \in \mathcal{A}\}$ whose hyperedges are complementary to the hyperedges of $\mathcal{A}$. It is easy to see that $\mathcal{A}^{dd} = \mathcal{A}$, $\mathcal{A}^{cc} = \mathcal{A}$ for each Sperner hypergraph $\mathcal{A}$. In other words, both operations, duality and complementation, are involutions. It is also clear that $\mathcal{A}^{dc} = \mathcal{I}(\mathcal{A})$ and $\mathcal{A}^{cd} = \mathcal{I}^{-1}(\mathcal{A})$.

A vertex set S is called a *sub-transversal* of $\mathcal{A}$ if $S \subseteq B$ for some minimal transversal $B \in \mathcal{A}^d$. Our proof of Theorem 2 makes use of a characterization of sub-transversals suggested in [6].

2.2   Characterization of Sub-transversals to a Hypergraph

Given a hypergraph $\mathcal{A} \subseteq 2^V$, a subset $S \subseteq V$, and a vertex $v \in S$, let $\mathcal{A}_v(S) = \{A \in \mathcal{A} \mid A \cap S = \{v\}\}$ denote the family of all hyperedges of $\mathcal{A}$ whose intersection with S is exactly v. Let further $\mathcal{A}_0(S) = \{A \in \mathcal{A} \mid A \cap S = \emptyset\}$ denote the partial hypergraph consisting of the hyperedges of $\mathcal{A}$ disjoint from S. A selection of $|S|$ hyperedges $\{A_v \in \mathcal{A}_v(S) \mid v \in S\}$ is called *covering* if there exists a hyperedge $A \in \mathcal{A}_0(S)$ such that $A \subseteq \bigcup_{v \in S} A_v$.

Proposition 2 (cf. [6]). *Let $S \subseteq V$ be a non-empty vertex set in a hypergraph $\mathcal{A} \in 2^V$.*

i) *If S is a sub-transversal for $\mathcal{A}$ then there exists a non-covering selection $\{A_v \in \mathcal{A}_v(S) \mid v \in S\}$ for S.*

ii) *Given a non-covering selection* $\{A_v \in \mathcal{A}_v(S) \mid v \in S\}$ *for S, we can extend S to a minimal transversal of $\mathcal{A}$ by solving problem* $MIS(\mathcal{A}', \emptyset)$ *for the induced partial hypergraph* $\mathcal{A}' = \{A \cap U \mid A \in \mathcal{A}_0(S)\} \subseteq 2^U$, *where* $U = V \setminus \bigcup_{v \in S} A_v$.

Unfortunately, finding a non-covering selection for S (or equivalently, testing if S is a sub-transversal) is NP-hard if the cardinality of S is not bounded (see [4]). However, if the size of S is bounded by a constant then there are only polynomially many selections $\{A_v \in \mathcal{A}_v(S) \mid v \in S\}$ for S. All of these selections, including the non-covering ones, can be easily enumerated in polynomial time (moreover, it can be done in parallel).

Corollary 1. *For any fixed c there is an NC algorithm which, given a hypergraph $\mathcal{A} \subseteq 2^V$ and a set S of at most c vertices, determines whether S is a sub-transversal to $\mathcal{A}$ and if so finds a non-covering selection* $\{A_v \in \mathcal{A}_v(S) \mid v \in S\}$.

Note that this Corollary holds for hypergraphs of arbitrary dimension.

2.3 Proof of Theorem 2

We prove the theorem for the equivalent problem $DUAL(\mathcal{A}, \mathcal{B})$. We may assume without loss of generality that $\mathcal{A}$ is Sperner. Our reduction consists of the following steps:

Step 1. By definition, each set $B \in \mathcal{B}$ is a minimal transversal to $\mathcal{A}$. This implies that each set $A \in \mathcal{A}$ is transversal to $\mathcal{B}$. Check whether each $A \in \mathcal{A}$ is a *minimal* transversal to $\mathcal{B}$. If not, a new element in $\mathcal{A}^d \setminus \mathcal{B}$ can be found be calling problem $MIS(\mathcal{A}', \emptyset)$, for some induced partial hypergraph $\mathcal{A}'$ of $\mathcal{A}$. We may assume therefore that each set in $\mathcal{A}$ is a minimal transversal to $\mathcal{B}$, i.e. $\mathcal{A} \subseteq \mathcal{B}^d$. Recall that $\mathcal{A}^{dd} = \mathcal{A}$ for each Sperner hypergraph $\mathcal{A}$. Therefore, if $\mathcal{B} \neq \mathcal{A}^d$ then $\mathcal{A} \neq \mathcal{B}^d$, and thus $\mathcal{B}^d \setminus \mathcal{A} \neq \emptyset$. Hence we arrive at the following duality criterion: $\mathcal{A}^d \setminus \mathcal{B} \neq \emptyset$ iff there is a sub-transversal S to $\mathcal{B}$ such that

$$S \subseteq A \quad \text{for} \quad \text{no} \quad A \in \mathcal{A}. \tag{1}$$

Hence we can apply the sub-transversal test only to S such that

$$|S| \leq \dim(\mathcal{I}^{-1}(\mathcal{A})). \tag{2}$$

Step 2 (Duality test.) For each set S satisfying (1), (2) and the condition that

$$A \not\subseteq S \quad \text{for all} \quad A \in \mathcal{A}, \tag{3}$$

check whether or not

$$S \text{ is a sub-transversal to } \mathcal{B}. \tag{4}$$

We need the assumption that $\dim(\mathcal{I}^{-1}(\mathcal{A}))$ is bounded to guarantee that this step is polynomial (and moreover, is in NC). Recall that by Proposition 2, S satisfies (4) iff there is a selection

$$\{B_v \in \mathcal{B}_v(S) \mid v \in S\} \tag{5}$$

which covers no set $B \in \mathcal{B}_0(S)$. Here as before, $\mathcal{B}_0(S) = \{B \in \mathcal{B} \mid B \cap S = \emptyset\}$ and $\mathcal{B}_v(S) = \{B \in \mathcal{B} \mid B \cap S = \{v\}\}$ for $v \in S$.

If conditions (1)-(4) cannot be met, we conclude that $\mathcal{B} = \mathcal{A}^d$ and halt.

Step 3. Suppose we have found a non-covering selection (5) for some set S satisfying (1)-(4). Then it is easy to see that the set $Z = S \bigcup \left[V \setminus \bigcup_{v \in S} B_v\right]$ is independent in $\mathcal{A}$. Furthermore, Z is transversal to $\mathcal{B}$, because selection (5) is non-covering. Let $\mathcal{A}' = \{A \cap U \mid A \in \mathcal{A}\}$, where $U = V \setminus Z$, and let T be a minimal transversal to $\mathcal{A}'$. (As before, we can let $T = U \setminus output(MIS(\mathcal{A}', \emptyset))$.) Since Z is an independent set of $\mathcal{A}$, we have $T \cap A \neq \emptyset$ for all $A \in \mathcal{A}$, that is T is transversal to $\mathcal{A}$. Clearly, T is minimal, that is $T \in \mathcal{A}^d$. It remains to argue that T is a *new* minimal transversal to $\mathcal{A}$, that is $T \notin \mathcal{B}$. This follows from the fact that Z is transversal to $\mathcal{B}$ and disjoint from T. $\qquad\square$

Note that Theorem 2 does not imply that $MIS(\mathcal{A}, \mathcal{I}) \in NC$ because the parallel complexity of the resulting problem $MIS(\mathcal{A}', \emptyset)$ is not known. The question whether it is in NC in general (for arbitrary hypergraphs) was raised in [14]. The affirmative answers were obtained in [1,7,15,17] for the following special cases: For hypergraphs of bounded dimension, $\mathcal{A} \in \mathbb{A}(1, c)$, it is known that $MIS(\mathcal{A}', \emptyset) \in NC$ for $c \leq 3$, and $MIS(\mathcal{A}', \emptyset) \in RNC$ for $c = 4, 5, \ldots$, see [2,15]. Furthermore, it was shown in [17,18] that $MIS(\mathcal{A}', \emptyset) \in NC$ for the so-called *linear* hyperedges, in which each two hyperedges intersect in at most one vertex, that is for $\mathcal{A}' \in \mathbb{A}(2, 1)$. Finally, it follows from [12] that $MIS(\mathcal{A}', \emptyset) \in NC$ for hypergraphs of bounded degree, that is for $\mathcal{A}' \in \mathbb{A}(c, 0)$. Combining the above results with Theorem 2, we obtain the following corollary.

Corollary 2. *Problem $MIS(\mathcal{A}, \mathcal{I})$ is in RNC for $\mathcal{A} \in \mathbb{A}(1, c)$, where c is a constant (hypergraphs of bounded dimension). Furthermore, $MIS(\mathcal{A}, \mathcal{I})$ is in NC for $\mathcal{A} \in \mathbb{A}(1, c)$, $c \leq 3$ (hypergraphs of $\dim \leq 3$), for $\mathcal{A} \in \mathbb{A}(c, 0)$, where c is a constant (hypergraphs of bounded degree), and for $\mathcal{A} \in \mathbb{A}(2, 1)$ (linear hypergraphs).*

Yet, for a hypergraph $\mathcal{A}$ satisfying $\dim(\mathcal{I}^{-1}(\mathcal{A})) \leq const$, or even more specifically for $\mathcal{A} \in \mathbb{A}(k, r)$, $k + r \leq const$, we only have an NC-reduction of $MIS(\mathcal{A}, \mathcal{I})$ to $MIS(\mathcal{A}', \emptyset)$, where the parallel complexity of the latter problem is not known.

3   Polynomial Space Algorithm for Generating $\mathcal{A}^d$

For $i = 1, \ldots, n$ denote by $[i : n]$ the set $\{i, i + 1, \ldots, n\}$, where $[n + 1 : n]$ is assumed to be the empty set. Given a hypergraph $\mathcal{A} \subseteq 2^{[n]}$, we shall say that $X \subseteq [n]$ is an *i-minimal transversal* for $\mathcal{A}$ if $X \supseteq [i : n]$, X is a transversal of $\mathcal{A}$, and $X \setminus \{j\}$ is not a transversal for all $j \in X \cap [1 : i - 1]$. Thus, $n + 1$-minimal transversals are just the minimal transversals of $\mathcal{A}$. For $i = 1, \ldots, n$, let $\mathcal{A}^{d_i}$ be the family of *i-minimal transversals* for $\mathcal{A}$.

Given $i \in [n]$ and $X \in \mathcal{A}^{d_i}$, let $\mathcal{A}_i(X)$ be the hypergraph

$$\mathcal{A}_i(X) = \{A \setminus \{i\} \ : \ A \in \mathcal{A}, \ A \cap X = \{i\}\}.$$

Proposition 3 (see [10,16]).
(i) $|\mathcal{A}^{d_i}| \leq |\mathcal{A}^d|$, *for* $i = 1, \ldots, n+1$.
(ii) $|\mathcal{A}_i(X)^d| \leq |\mathcal{A}^{d_{i+1}}|$, *for* $i \in [n]$ *and* $X \in \mathcal{A}^{d_i}$.

Now consider the following generalization of an algorithm in [19] for generating maximal independent sets in graphs (see also [13] and [16]). Given $i \in [n]$, and $X \in \mathcal{A}^{d_i}$, we assume in the algorithm that the minimal transversals $\mathcal{A}_i(X)^d$ are computed by calling a process $P(i, X)$ that invokes the same algorithm recursively on the partial hypergraph $\mathcal{A}_i(X)$. We further assume that, once $P(i, X)$ finds an element $Y \in \mathcal{A}_i(X)^d$, it returns control to the calling process $\text{GEN}(\mathcal{A}, i, X)$. When called for the next time, $P(i, X)$ returns the next element of $\mathcal{A}_i(X)^d$ that has not been generated yet, if such an element exists.

Algorithm GEN($\mathcal{A}, i, X$):
Input: A hypergraph $\mathcal{A}$, an index $i \in [n]$, and an i-minimal transversal $X \in \mathcal{A}^{d_i}$.
Output: All minimal transversals of $\mathcal{A}$.

1. if $i = n + 1$ then
2.   output X;
3. else
4.   if $X \setminus \{i\}$ is a transversal of $\mathcal{A}$ then
5.   GEN($\mathcal{A}, i + 1, X \setminus \{i\}$);
6.   else
7.   GEN($\mathcal{A}, i + 1, X$);
8.   for each minimal transversal $Y \in \mathcal{A}_i(X)^d$ (found recursively) do
9.   if $X \cup Y \setminus \{i\} \in \mathcal{A}^{d_{i+1}}$ then
10.   Compute the *lexico. largest* set $Z \subseteq X \cup Y$ such that $Z \in \mathcal{A}^{d_i}$;
11.   if $Z = X$ then
12.   GEN($\mathcal{A}, i + 1, X \cup Y \setminus \{i\}$);

Lemma 1. *When called with $i = 1$ and $X = [n]$, Algorithm GEN($\mathcal{A}, i, X$) outputs all minimal transversals of $\mathcal{A}$ with no repetitions.*

Proof. Consider the recursion tree **T** traversed by the algorithm. Label each node of tree by the pair (i, X) which represents the input to the algorithm at this node. Clearly i represents the level of node (i, X) in the tree (where the root of **T** is at level 1). By induction on $i = 1, \ldots, n+1$, we can verify the following statement:

$$\mathcal{A}^{d_i} = \{X \subseteq [n] \ : \ (i, X) \in \mathbf{T}\}. \tag{6}$$

Indeed, this trivially holds at $i = 1$. Assume now that (6) holds for a specific $i \in [n-1]$. It is easy to see that any node $(i+1, X) \in \mathbf{T}$ generated at level $i+1$ of the tree must have $X \in \mathcal{A}^{d_{i+1}}$. Thus it remains to verify that $\mathcal{A}^{d_{i+1}} \subseteq \{X \ : \ (i+1, X) \in \mathbf{T}\}$. To see this, let X' be an arbitrary element of $\mathcal{A}^{d_{i+1}}$. Note first that if $X' \ni i$ then $X' \setminus \{i\}$ is not a transversal of $\mathcal{A}$ and $X' \in \mathcal{A}^{d_i}$, and therefore by induction we have a node $(i, X') \in \mathbf{T}$. Consequently, we get a node $(i+1, X') \in \mathbf{T}$ as a child of $(i, X') \in \mathbf{T}$, by Step 7 of the algorithm. Let us therefore assume that $X' \not\ni i$. Note that X' must contain a subset $X \setminus \{i\}$, for some $X \in \mathcal{A}^{d_i}$. This is because $X' \cup \{i\}$ is a transversal and therefore it contains an i-minimal transversal X of $\mathcal{A}$. Among all the sets X satisfying this property, let Z be the lexicographically largest. Now, if $Z \setminus \{i\}$ is a transversal of $\mathcal{A}$, then $Z \setminus \{i\} = X'$ and Step 5 will create a node $(i+1, X') \in \mathbf{T}$ as the only child of $(i, Z) \in \mathbf{T}$. On the other hand, if $Z \setminus \{i\}$ is not a transversal, then X' can be written as $X' = Z \cup Y \setminus \{i\}$, for some $Y \in \mathcal{A}_i(Z)^d$. But then node $(i+1, X')$ will be generated as a child of $(i, Z) \in \mathbf{T}$ by Step 12 of the Algorithm. This completes the proof of (6). Finally, it follows from Step 10 that each node in the tree is generated as the child of exactly one other node. Consequently each leaf is visited, and hence each set $X \in \mathcal{A}^d$ is output, only once and the lemma follows. □

The next lemma states that, for hypergraphs $\mathcal{A}$ of (k, r)-bounded intersections, Algorithm GEN is a polynomial-space, *output*-polynomial time algorithm for generating all minimal transversals of $\mathcal{A}$.

Lemma 2. *The time taken by Algorithm GEN until it outputs the last minimal transversal of a hypergraph $\mathcal{A} \in \mathbb{A}(k, r)$ is $O(n^{k+r-1} |\mathcal{A}^d|^{r+1})$, and the total space required is $O(N^{r+1})$.*

Proof. For a hypergraph $\mathcal{A} \in \mathbb{A}(k, r)$, let $T(\mathcal{A})$ and $M(\mathcal{A})$ be respectively the time and space required by Algorithm GEN to output the last minimal transversal of $\mathcal{A}$. Note that the algorithm basically performs depth-first search on the tree $\mathbf{T}$ (whose leaves are the elements of $\mathcal{A}^d$), and only generates nodes of $\mathbf{T}$ as needed during the search. Since each node of the tree $\mathbf{T}$, which is not a leaf, has at least one child, the time between two successive outputs generated by the algorithm does not exceed the time required to generate the children of nodes along a complete path of the tree $\mathbf{T}$ from the root to a leaf. But, as can be seen from the algorithm, for a given node $v = (i, X)$ in $\mathbf{T}$, where $i \in [n]$ and $X \in \mathcal{A}^{d_i}$, the time required to generate all the children of v, is bounded by the time to output all the elements of $\mathcal{A}_i(X)^d$. Since the depth of the tree is $n+1$, we get the recurrence

$$T(\mathcal{A}) \leq n |\mathcal{A}^d| \max\{T(\mathcal{A}_i(X)) \ : \ i \in [n], \ \ X \in \mathcal{A}^{d_i}\}. \tag{7}$$

Note that $\mathcal{A}_i(X) \in \mathbb{A}(k, r-1)$. Furthermore, by Proposition 3, we have $|\mathcal{A}_i(X)^d| \leq |\mathcal{A}^d|$, and thus (7) gives $T(\mathcal{A}) \leq (n|\mathcal{A}^d|)^r T(\mathcal{A}')$, for some subhypergraph $\mathcal{A}' \in \mathbb{A}(k, 0)$ of $\mathcal{A}$ which satisfies $|(\mathcal{A}')^d| \leq |\mathcal{A}^d|$. Now, we observe that for any $i \in [n]$ and $X \in (\mathcal{A}')^{d_i}$, we have $|\mathcal{A}'_i(X)| \leq k-1$, and hence it follows that $T(\mathcal{A}') = O(n^{k-1} |(\mathcal{A}')^d|)$. The bound on the running time follows.

Now let us consider the total memory required by the algorithm. Since, for each recursion tree (corresponding to a (sub-)hypergraph that is to be dualized), the algorithm maintains only the path from the root to a leaf of the tree, we get the recurrence $M(\mathcal{A}) \leq N \max\{M(\mathcal{A}_i(X)) \ : \ i \in [n], \ \ X \in \mathcal{A}^{d_i}\}$. This recurrence again gives $M(\mathcal{A}) \leq N^r M(\mathcal{A}')$, for some sub-hypergraph $\mathcal{A}' \in \mathbb{A}(k, 0)$ of $\mathcal{A}$. But $M(\mathcal{A}') = O(N)$ and the bound on the space follows. □

Now Theorem 3 follows by combining Lemma 2 with the following reduction.

Proposition 4. *Let $\mathcal{A} \subseteq 2^{[n]}$ be a hypergraph. Suppose that there is an algorithm P that generates all minimal transversals of $\mathcal{A}$ in time $p(n, |\mathcal{A}^d|)$ and space $q(N(\mathcal{A}))$, for some polynomials $p(\cdot, \cdot)$ and $q(\cdot)$. Then for any integer k, we can generate at least k minimal transversals of $\mathcal{A}$ in time $2n(p(n, k) + 1)$ and space $q(N(\mathcal{A}))$.*

Note that it is implicit in the proof of Lemma 2 that, for both graphs $\mathcal{A} \in \mathbb{A}(1, 2)$ and hypergraphs of bounded degree $\mathcal{A} \in \mathbb{A}(c, 0)$, Algorithm GEN is in fact a polynomial delay and polynomial space algorithm for generating $\mathcal{A}^d$. In particular, Theorem 3 implies the following previously known results [10,13,19].

Corollary 3. *For graphs, $\mathcal{A} \in \mathbb{A}(1, 2)$, and also for the hypergraphs of bounded degree, $\mathcal{A} \in \mathbb{A}(c, 0)$, all minimal transversals of $\mathcal{A}$ can be enumerated with polynomial delay and polynomial space.*

4   Generating $\mathcal{A}^d$ Using the Supergraph Approach

Let $\mathcal{A} \subseteq 2^V$ be a hypergraph. In this section, we sketch another algorithm to list all minimal transversals of $\mathcal{A}$. The algorithm works by building a *strongly connected directed supergrah* $\mathcal{G} = (\mathcal{A}^d, \mathcal{E})$ on the set of minimal transversals, in which a pair of vertices (X, X') forms an edge in $\mathcal{E}$ if and only if X' can be obtained from X by deleting an element from $X \setminus X'$, adding a minimal subset of elements from $X' \setminus X$ to obtain a transversal, and finally reducing the resulting set to a minimal feasible solution in a specified way (say in reverse-lexicographic order). In other words, $(X, X') \in \mathcal{E}$ if and only if $X' \subseteq X \cup Z \setminus \{e\}$, for some $e \in X \setminus X'$ and $Z \subseteq X' \setminus X$, such that Z is minimal with the property that $X \cup Z \setminus \{e\}$ is a transversal.

The strong connectivity of $\mathcal{G}$ can be proved as follows. Given two vertices $X_0, X_l \in \mathcal{A}^d$ of $\mathcal{G}$, there exists a set $\{X_1, \ldots, X_{l-1}\}$ of elements of $\mathcal{F}$, where for all $i = 1, \ldots, l$, X_i is obtained from X_{i-1} by deleting an element $e_i \in X_{i-1} \setminus X_l$ (thus making $X_{i-1} \setminus \{e_i\}$ non-transversal), adding a minimal subset of elements $Z_i \subseteq X_l \setminus X_{i-1}$ to obtain a transversal $X_{i-1} \setminus \{e_i\} \cup Z_i$, and finally, reducing the resulting set to a minimal transversal $X_i \subseteq X_{i-1} \cup Z_i \setminus \{e_i\}$. Note that, for $i = 1, \ldots, l$, $|X_i \setminus X_l| < |X_{i-1} \setminus X_l|$ and therefore $l \leq |X_0 \setminus X_l|$. In other words, $\mathcal{G}$ has *diameter* at most n.

The minimal transversals of $\mathcal{A}$ can thus be generated by performing breadth-first search on the vertices of $\mathcal{G}$, starting from an arbitrary vertex. Such a procedure can be executed in incremental polynomial time if the neighbourhood of every vertex in $\mathcal{G}$ can also be generated in (incremental) polynomial time. Given a hypergraph $\mathcal{A} \in \mathcal{A}(k, r)$, and a minimal transversal $X \in \mathcal{A}^d$, all neighbours of X in $\mathcal{G}$ can be generated in time $O(n^{k+r}|\mathcal{A}^d|^{r+1})$. Indeed, for any $e \in X$, all minimal subsets of vertices Z, such that $X \setminus \{e\} \cup Z$ is a transversal of $\mathcal{A}$, can be obtained by finding all minimal transversals for the hypergraph $\mathcal{A}_e(X) = \{A \setminus \{e\} : A \in \mathcal{A}, A \cap X = \{e\}\}$. But as noted before,

$\mathcal{A}_e(X) \in \mathcal{A}(k, r-1)$ and $|\mathcal{A}_e(X)^d| \leq |\mathcal{A}^d|$. We conclude therefore, as in the proof of Lemma 2, that the time required to produce all the neighbours of X by applying the algorithm recursively on each of the hypergraphs $\mathcal{A}_e$, for $e \in X$, is $O(n^{k+r}|\mathcal{A}^d|^{r+1})$.

References

1. N. Alon, L. Babai, A. Itai, A fast randomized parallel algorithm for the maximal independent set problem, *J. Algorithms* 7 (1986), pp. 567–583
2. P. Beame, M. Luby, Parallel search for maximal independence given minimal dependence, *Proc. 1st SODA Conference* (1990), pp. 212–218.
3. C. Berge, *Hypergraphs*, North Holland Mathematical Library, Vol. 445, Elsevier-North Holland, Amestrdam, 1989.
4. E. Boros, K. Elbassioni, V. Gurvich, and L. Khachiyan, An efficient incremental algorithm for generating all maximal independent sets in hypergraphs of bounded dimension, *Parallel Processing Letters*, 10 (2000), pp. 253–266.
5. E. Boros, K. Elbassioni, V. Gurvich, L. Khachiyan and K.Makino, Dual-bounded generating problems: All minimal integer solutions for a monotone system of linear inequalities, *SIAM Journal on Computing*, 31 (5) (2002) pp. 1624–1643.
6. E. Boros, V. Gurvich, and P.L. Hammer, Dual subimplicants of positive Boolean functions, *Optimization Methods and Software*, 10 (1998), pp. 147–156.
7. E. Dahlhaus, M. Karpinski, P. Kelsen, An efficient parallel algorithm for computing a maximal independent set in a hypergraph of dimension 3, *Information Processing Letters* 42(6) (1992), pp. 309–313.
8. C. Domingo, N. Mishra and L. Pitt, Efficient read-restricted monotone CNF/DNF dualization by learning with membership queries, *Machine learning* 37 (1999), pp. 89–110.
9. T. Eiter and G. Gottlob, Identifying the minimal transversals of a hypergraph and related problems, *SIAM Journal on Computing*, 24 (1995), pp. 1278–1304.
10. T. Eiter, G. Gottlob and K. Makino, New results on monotone dualization and generating hypergraph transversals, in *Proc. 34th Annual ACM STOC Conference*, (2002), pp. 14–22.
11. M. L. Fredman and L. Khachiyan, On the complexity of dualization of monotone disjunctive normal forms. *J. Algorithms*, 21 (1996), pp. 618–628.
12. O. Garrido, P. Kelsen, A. Lingas, A simple NC-algorithm for a maximal independent set in a hypergraph of polylog arboricity, *Information Processing Letters* 58(2) (1996), pp. 55–58.
13. D. S. Johnson, M. Yannakakis and C. H. Papadimitriou, On generating all maximal independent sets, *Information Processing Letters*, 27 (1988), pp. 119–123.
14. R. Karp and A. Wigderson, A fast parallel algorithm for the maximal independent set problem, *JACM* 32 (1985), pp. 762–773.
15. P. Kelsen, On the parallel complexity of computing a maximal independent set in a hypergraph, *Proc. the 24-th Anual ACM STOC Conference*, (1992), pp. 339–350.
16. E. Lawler, J. K. Lenstra and A. H. G. Rinnooy Kan, Generating all maximal independent sets: NP-hardness and polynomial-time algorithms, *SIAM Journal on Computing*, 9 (1980), pp. 558–565.
17. T. Luczak , E. Szymanska, A parallel randomized algorithm for finding a maximal independent set in a linear hypergraph, *J. Algorithms*, v.25 (1997) n.2, pp. 311–320.

18. E. Szymanska, Derandomization of a Parallel MIS Algorithm in a Linear Hypergraph, in *Proc. 4th International Workshop on Randomization and Approximation Techniques in Computer Science*, (2000), pp. 39–52.

19. S. Tsukiyama, M. Ide, H. Ariyoshi and I. Shirakawa, A new algorithm for generating all maximal independent sets, *SIAM Journal on Computing*, 6 (1977), pp. 505–517.

Rooted Maximum Agreement Supertrees

Jesper Jansson[1], Joseph H.-K. Ng[1], Kunihiko Sadakane[2], and Wing-Kin Sung[1]

[1] School of Computing, National University of Singapore, 3 Science Drive 2,
Singapore 117543. {jansson,nghonkeo,ksung}@comp.nus.edu.sg
[2] Department of Computer Science and Communication Engineering,
Kyushu University, Japan. sada@csce.kyushu-u.ac.jp

Abstract. Given a set $\mathcal{T}$ of rooted, unordered trees, where each $T_i \in \mathcal{T}$ is distinctly leaf-labeled by a set $\Lambda(T_i)$ and where the sets $\Lambda(T_i)$ may overlap, the *maximum agreement supertree problem* (MASP) is to construct a distinctly leaf-labeled tree Q with leaf set $\Lambda(Q) \subseteq \bigcup_{T_i \in \mathcal{T}} \Lambda(T_i)$ such that $|\Lambda(Q)|$ is maximized and for each $T_i \in \mathcal{T}$, the topological restriction of T_i to $\Lambda(Q)$ is isomorphic to the topological restriction of Q to $\Lambda(T_i)$. Let $n = \left| \bigcup_{T_i \in \mathcal{T}} \Lambda(T_i) \right|$, $k = |\mathcal{T}|$, and $D = \max_{T_i \in \mathcal{T}} \{\deg(T_i)\}$. We first show that MASP with $k = 2$ can be solved in $O(\sqrt{D} n \log(2n/D))$ time, which is $O(n \log n)$ when $D = O(1)$ and $O(n^{1.5})$ when D is unrestricted. We then present an algorithm for MASP with $D = 2$ whose running time is polynomial if $k = O(1)$. On the other hand, we prove that MASP is NP-hard for any fixed $k \geq 3$ when D is unrestricted, and also NP-hard for any fixed $D \geq 2$ when k is unrestricted even if each input tree is required to contain at most three leaves. Finally, we describe a polynomial-time $(n/\log n)$-approximation algorithm for MASP.

1 Introduction

An important objective in phylogenetics is to develop methods for merging a collection of phylogenetic trees on overlapping sets of taxa into a single supertree so that no (or as little as possible) branching information is lost. Ideally, the resulting supertree can then be used to deduce evolutionary relationships between taxa which do not occur together in any one of the input trees. Supertree methods are useful because most individual studies investigate relatively few taxa [22] and because sample bias leads to certain taxa being studied much more frequently than others [4]. Also, supertree methods can combine trees constructed for different types of data or under different models of evolution. Furthermore, although computationally expensive methods for constructing reliable phylogenetic trees are infeasible for large sets of taxa, they can be applied to obtain highly accurate trees for smaller, overlapping subsets of the taxa that may then be merged using less computationally intense, supertree-based techniques (see, e.g., [7,16,20]).

Since the set of trees which is to be combined may in practice contain contradictory branching structure (for example, if the trees have been constructed from data originating from different genes or if the experimental data contains errors), a supertree method needs to specify how to resolve conflicts. In this paper, we consider *maximum agreement supertrees*. The intuitive idea is to identify

M. Farach-Colton (Ed.): LATIN 2004, LNCS 2976, pp. 499–508, 2004.

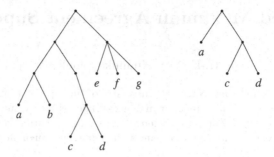

Fig. 1. Let T be the tree on the left. Then $T \mid \{a, c, d, h\}$ is the tree shown on the right.

and remove a smallest possible subset of the taxa so that the remaining taxa can be combined without conflicts. In this way, one would get an indication of which ancestral relationships can be regarded as resolved and which taxa need to be subjected to further experiments. We formalize the above as a computational problem called *the maximum agreement supertree problem* (MASP).

Further motivation for studying maximum agreement supertrees comes from the relation to a well-studied problem known as *the maximum agreement subtree problem* (MAST) in which the input is a set of leaf-labeled trees and the goal is to compute a tree contained in all of the input trees with as many labeled leaves as possible. Our results in this paper complement those previously known for MAST. The computational complexity of MAST has been closely investigated (see Section 1.2), motivated by the practical usefulness of maximum agreement subtrees. For example, maximum agreement subtrees can be used not only to identify small problematic subsets of taxa during phylogenetic reconstruction, but also to measure the similarity of a given set of trees [9,11,19] or to estimate a classification's stability to small changes in the data [11]. Moreover, MAST-based algorithms have been used to prepare and improve bilingual context-using dictionaries for automated language translation systems [8,21].

1.1   Problem Definitions

Let T be a tree whose leaves are labeled by a set S. We say that T is *distinctly leaf-labeled by S* if no two leaves in T have the same label. Below, each leaf in such a tree is identified with its corresponding label in S. Given a rooted, unordered, distinctly leaf-labeled tree T and a set S', *the topological restriction of T to S'* (denoted by $T \mid S'$) is the tree obtained by deleting from T all nodes which are not on any path from the root to a leaf in S' along with their incident edges, and then contracting every edge between a node having just one child and its child (see Fig. 1). For any tree T, denote its set of leaves by $\Lambda(T)$.

Let $\mathcal{T} = \{T_1, T_2, ..., T_k\}$ be a set of rooted, unordered trees, where each T_i is distinctly leaf-labeled and where the sets $\Lambda(T_i)$ may overlap. A *total agreement supertree of $\mathcal{T}$* is a tree Q such that Q is distinctly leaf-labeled by $\bigcup_{T_i \in \mathcal{T}} \Lambda(T_i)$ and $Q \mid \Lambda(T_i)$ is isomorphic to T_i for every $T_i \in \mathcal{T}$. Note that two or more trees in $\mathcal{T}$ may contain conflicting branching information, in which case a total agreement supertree of $\mathcal{T}$ does not exist. *The total agreement supertree problem* (TASP) is: Given a set $\mathcal{T}$ of distinctly leaf-labeled, rooted, unordered trees, output a total agreement supertree of $\mathcal{T}$ if one exists, otherwise output null.

When $\mathcal{T} = \{T_1, T_2, ..., T_k\}$ is specified, we write $S = \bigcup_{T_i \in \mathcal{T}} \Lambda(T_i)$ and call S *the leaf set of* $\mathcal{T}$. For any $S' \subseteq S$, we let $\mathcal{T} \mid S'$ denote the set $\{T_1 \mid S', T_2 \mid S', ...,$ $T_k \mid S'\}$. If there exists a total agreement supertree Q of $\mathcal{T} \mid S'$ then we say that S' is *consistent with* $\mathcal{T}$ and call Q an *agreement supertree of* $\mathcal{T}$. A *maximum agreement supertree of* $\mathcal{T}$ is an agreement supertree of $\mathcal{T}$ with as many leaves as possible. *The maximum agreement supertree problem* (MASP) is: Given a set $\mathcal{T}$ of distinctly leaf-labeled, rooted, unordered trees, output a maximum agreement supertree of $\mathcal{T}$. An *agreement subtree of* $\mathcal{T}$ is a tree U such that for some $S' \subseteq S$ it holds that U is distinctly leaf-labeled by S' and $T_i \mid S'$ is isomorphic to U for every $T_i \in \mathcal{T}$. A *maximum agreement subtree of* $\mathcal{T}$ is an agreement subtree of $\mathcal{T}$ with the maximum possible number of leaves. *The maximum agreement subtree problem* (MAST), also referred to in the literature as *the maximum homeomorphic subtree problem* (MHT), is to find a maximum agreement subtree of $\mathcal{T}$.

Throughout this paper, we let n denote the cardinality of the leaf set and k the number of input trees, i.e., $n = |\bigcup_{T_i \in \mathcal{T}} \Lambda(T_i)|$ and $k = |\mathcal{T}|$ in the problem definitions above. We let $D = \max_{T_i \in \mathcal{T}} \{\deg(T_i)\}$, where $\deg(T_i)$ is the degree[1] of T_i. We assume that none of the trees in $\mathcal{T}$ have a node with degree 1, so that each tree contains $O(n)$ nodes. Note that if we are given a subset S' of S which is consistent with $\mathcal{T}$, then we can efficiently construct a total agreement supertree of $\mathcal{T} \mid S'$ using the algorithm for TASP by Henzinger *et al.* [16] (see also Lemma 7 in Section 5). Hence, we focus on the subproblem of MASP of computing a maximum cardinality subset S' of S such that S' is consistent with $\mathcal{T}$.

A *rooted triplet* is a distinctly leaf-labeled, binary, rooted, unordered tree with three leaves. The unique rooted triplet on $\{a, b, c\}$ in which the lowest common ancestor of a and b is a proper descendant of the lowest common ancestor of a and c (or equivalently, where the lowest common ancestor of a and b is a proper descendant of the lowest common ancestor of b and c) is denoted by $(\{a, b\}, c)$.

1.2   Previous Results

Comprehensive surveys of existing methods for constructing supertrees can be found in [4,20,22]. Below, we mention some known results related to MASP.

Aho, Sagiv, Szymanski, and Ullman [1] presented an algorithm which can be used to solve TASP in $O(kn)$ time when all trees in $\mathcal{T}$ are rooted triplets. Several years later, Henzinger, King, and Warnow [16] showed how to modify the algorithm to solve TASP for any $\mathcal{T}$ in $\min\{O(Nn^{0.5}), O(N + n^2 \log n)\}$ time, where $N = \sum_{T_i \in \mathcal{T}} |T_i|$ is the total number of nodes in $\mathcal{T}$. In contrast, the analog of TASP for *unrooted* trees is NP-hard, even if all of the input trees are *quartets* (distinctly leaf-labeled, unrooted trees each having four leaves and no nodes with precisely two neighbors) [23]. A polynomial-time algorithm for

---

[1] The *degree of a node* u in a rooted tree is the number of children of u. The *degree of a rooted tree* T is the maximum degree of all nodes in T.

computing an unrooted total agreement supertree if one exists when all k input trees are binary and $k = O(1)$ was given by Bryant in [6].

The computational complexity of MAST has been studied extensively (e.g., [3,5,8,9,10,11,14,15,19,24]). Today, the fastest known algorithm for MAST for two trees, invented by Kao, Lam, Sung, and Ting [19], runs in $O(\sqrt{D}\, n \log(2n/D))$ time, which is $O(n \log n)$ when $D = O(1)$ and $O(n^{1.5})$ when D is unrestricted.

Amir and Keselman [3] considered the case of $k \geq 3$ input trees. They proved that MAST is NP-hard for three trees with unrestricted degrees, but solvable in polynomial time for three or more trees if the degree of at least one of the trees is bounded by a constant. For the latter case, Farach, Przytycka, and Thorup [9] gave an algorithm with improved efficiency running in $O(kn^3 + n^d)$ time, where d is an upper bound on at least one of the input trees' degrees; Bryant [5] proposed a conceptually different algorithm with the same running time.

Hein, Jiang, Wang, and Zhang [15] proved the following inapproximability result: MAST for three trees with unrestricted degrees cannot be approximated within a factor of $2^{\log^\delta n}$ in polynomial time for any constant $\delta < 1$, unless NP $\subseteq$ DTIME$[2^{\text{polylog} n}]$. Gąsieniec, Jansson, Lingas, and Östlin [14] proved that MAST cannot be approximated within a factor of n^ε for any constant ε where $0 \leq \varepsilon < \frac{1}{9}$ in polynomial time unless P $=$ NP, even for instances containing only trees of height 2, and showed that if the number of trees is bounded by a constant and all the input trees' heights are bounded by a constant then MAST can be approximated within a constant factor in $O(n \log n)$ time.

A problem related to MASP and MAST is *the maximum refinement subtree problem* (MRST). Its goal is to construct a tree W with $\Lambda(W) \subseteq S$ which maximizes $|\Lambda(W)|$ such that for each $T_i \in \mathcal{T}$, $T_i\,|\,\Lambda(W)$ can be obtained from W by applying a series of edge contractions. MRST is NP-hard for $k = 2$ if D is unrestricted [15] but solvable in polynomial time if $k = O(1)$ and $D = O(1)$ [12]. Another related problem is *the maximum compatible subset of rooted triplets problem* (MCSR) in which the input is a set $\mathcal{T}$ of rooted triplets and the objective is to find a $\mathcal{T}' \subseteq \mathcal{T}$ of maximum cardinality such that there exists a total agreement supertree of $\mathcal{T}'$. MCSR is NP-hard [5,18]; two polynomial-time approximation algorithms for MCSR were given in [14].

1.3 Our Results and Organization of Paper

In Section 2, we make use of known positive and negative results for MAST to obtain an efficient algorithm for MASP restricted to $k = 2$ and an NP-hardness proof for MASP restricted to any fixed $k \geq 3$, respectively. The algorithm for $k = 2$ runs in $O(\sqrt{D}\, n \log(2n/D))$ time, which is $O(n \log n)$ when $D = O(1)$ and $O(n^{1.5})$ when D is unrestricted. Then, in Section 3, we present a more complex MAST-based algorithm for solving MASP with $D = 2$. It runs in $O(k(2n)^{3k^2})$ time, which is polynomial when $k = O(1)$. In Section 4, we prove that MASP is NP-hard even if all of the input trees are required to be rooted triplets (i.e., $D = 2$ and k is unrestricted). Finally, in Section 5, we describe a simple polynomial-time approximation algorithm for MASP which is guaranteed to find an approximate solution with at least $\frac{\log n}{n}$ times the number of leaves in an optimal solution.

2   Preliminaries

We first investigate the close relationship between MASP and MAST.

Lemma 1. *For any set $\mathcal{T} = \{T_1, T_2, ..., T_k\}$ of distinctly leaf-labeled, rooted, unordered trees such that $\Lambda(T_1) = \Lambda(T_2) = ... = \Lambda(T_k)$, an optimal solution to MASP for $\mathcal{T}$ is an optimal solution to MAST for $\mathcal{T}$ and vice versa.*

Proof. Write $S = \Lambda(T_1) = \Lambda(T_2) = ... = \Lambda(T_k)$, let Q be any agreement supertree of $\mathcal{T}$, and let $S' = \Lambda(Q)$. Then, by definition, $Q \mid \Lambda(T_i \mid S') = T_i \mid S'$ for every $T_i \in \mathcal{T}$. Now, $\Lambda(T_i \mid S') = S \cap S' = S'$, so $T_i \mid S' = Q \mid S' = Q$ for every $T_i \in \mathcal{T}$, which means that Q is an agreement subtree of $\mathcal{T}$. Conversely, let U be an agreement subtree of $\mathcal{T}$ whose leaves are distinctly labeled by some set S'. For every $T_i \in \mathcal{T}$, we have $T_i \mid S' = U$. Then $U \mid \Lambda(T_i \mid S') = (T_i \mid S') \mid \Lambda(T_i \mid S') = T_i \mid S'$ for every $T_i \in \mathcal{T}$, i.e., U is an agreement supertree of $\mathcal{T}$.   □

Theorem 1. *MASP with $k = 2$ can be solved in $O(\sqrt{D} \, n \log(2n/D))$ time.*

Proof. Given an instance $\mathcal{T} = \{T_1, T_2\}$ of MASP with $k = 2$, let $L = \Lambda(T_1) \cap \Lambda(T_2)$ and run the algorithm of Kao, Lam, Sung, and Ting [19] on the instance $\mathcal{T} \mid L$ to obtain a maximum agreement subtree U of $\mathcal{T} \mid L$. This takes $O(\sqrt{D} \, n \log(2n/D))$ time. By Lemma 1, U is also a maximum agreement supertree of $\mathcal{T} \mid L$. Next, for every leaf which appears in exactly one of T_1 and T_2, insert it into U according to its position in T_1 or T_2. More precisely, let $X = L \setminus \Lambda(U)$ and first compute $T_1' = T_1 \mid (\Lambda(T_1) \setminus X)$ and $T_2' = T_2 \mid (\Lambda(T_2) \setminus X)$ in $O(n)$ time. For any node $u \in U$, let $T_1'(u)$ and $T_2'(u)$ be the node in T_1' and T_2' respectively corresponding to u. Construct a tree Q as follows: initially, set $Q = T_1'$, then for each edge (u, v) of U, where we assume u is the parent of v, replace the edge in Q between $T_1'(v)$ and its parent with the path in T_2' between $T_2'(v)$ and $T_2'(u)$. Q can be constructed using a total of $O(n)$ time. It is straightforward to show that Q is a maximum agreement supertree of $\mathcal{T}$.   □

The running time given in Theorem 1 is $O(n \log n)$ for two trees whose degrees are bounded by a constant and $O(n^{1.5})$ for two trees with unrestricted degrees.

The NP-hardness of MAST for any fixed $k \geq 3$ when D is unrestricted [3] together with Lemma 1 yield the following theorem (in fact, Lemma 1 can be used to show that the inapproximability results of [14] and [15] for MAST mentioned in Section 1.2 hold for MASP as well).

Theorem 2. *For any fixed $k \geq 3$, MASP with unrestricted D is NP-hard.*

3   A Polynomial-Time Algorithm for $D = 2$, $k = O(1)$

In this section, we show how MASP restricted to $D = 2$ can be reduced to MAST for a set of k distinctly leaf-labeled binary trees having $O((2n)^{k^2})$ leaves.[2] Hence, we can solve MASP with $D = 2$ in polynomial time if $k = O(1)$.

---

[2] The proofs and figures in this section have been omitted due to space constraints. They can be found in the full-length version of our paper.

Without loss of generality, assume that every $a \in S$ appears in at least two trees in $\mathcal{T}$. (If a appears in exactly one tree in $\mathcal{T}$, we can obtain a maximum agreement supertree of $\mathcal{T}$ as follows: (1) Remove a from $\mathcal{T}$; (2) compute a maximum agreement supertree T' for the modified $\mathcal{T}$; and (3) insert a into T' according to its position in the original $\mathcal{T}$, as described in the proof of Theorem 1 above.)

MASP is first transformed to MAST for non-distinctly leaf-labeled trees; then, the latter problem is transformed to MAST. Here, by an agreement subtree of a set $\mathcal{R} = \{R_1, R_2, \ldots, R_k\}$ of *non-distinctly* leaf-labeled trees, we mean a distinctly leaf-labeled tree which is a homeomorphic subtree of every $R_i \in \mathcal{R}$.

We now describe our transformation from MASP to MAST for a set $\mathcal{R} = \{R_1, R_2, \ldots, R_k\}$ of non-distinctly leaf-labeled binary trees. To obtain each R_i:

1. Set $R_{i,0} = T_i$.
2. For $j = 1$ to k, do
   a) Let $L = \Lambda(T_j) \setminus \bigcup_{j' \in \{1, \ldots, j-1\} \cup \{i\}} \Lambda(T_{j'})$ and let $U = T_j | L$.
   b) Initially, set $R_{i,j} = R_{i,j-1}$. Generate $|R_{i,j-1}| - 1$ copies of U and attach one to every edge of $R_{i,j}$. Let r be a new node having the current $R_{i,j}$ and another copy of U as its two subtrees, and make r the root of $R_{i,j}$.
3. Set $R_i = R_{i,k}$.

Based on the above construction, for every i, any label in $\Lambda(T_i)$ appears exactly once in R_i, and T_i is a homeomorphic subtree of R_i. Also, $\mathcal{R}$ satisfies:

Lemma 2. *For every $R_i \in \mathcal{R}$, the number of leaves in R_i is at most $(2n)^k$ and the height of R_i is at most $2^k n$.*

Lemma 3. *For any tree X which is distinctly leaf-labeled by some $S' \subseteq S$, X is an agreement supertree of $\mathcal{T}$ if and only if X is an agreement subtree of $\mathcal{R}$.*

Next, we transform MAST for the set $\mathcal{R}$ of non-distinctly leaf-labeled binary trees to MAST for a set $\mathcal{P} = \{P_1, P_2, \ldots, P_k\}$ of binary trees which are distinctly leaf-labeled by $\{a^1_{b_1, b_2, \ldots, b_k}, a^2_{b_1, b_2, \ldots, b_k} \mid a \in S, 1 \leq i \leq k, 1 \leq b_i \leq \gamma_a[i]\}$, where $\gamma_a[i]$ is the number of occurrences of leaf label a in R_i.

To describe the transformation, we need some additional notation. For every $a \in S$, define $a([b_1..d_1], [b_2..d_2], \ldots, [b_k..d_k])$, where $b_i \leq d_i$ for all $1 \leq i \leq k$, to be a rooted caterpillar with $\prod_{i=1}^{k} (2(d_i - b_i + 1))$ leaves labeled (in order of non-decreasing distance from the root) by $a^1_{b_1, b_2, \ldots, b_k}$, $a^2_{b_1, b_2, \ldots, b_k}$, $a^1_{b_1, b_2, \ldots, b_k+1}$, $a^2_{b_1, b_2, \ldots, b_k+1}$, $\ldots$, $a^1_{d_1, d_2, \ldots, d_k}$, $a^2_{d_1, d_2, \ldots, d_k}$. Define $\bar{a}([b_1..d_1], [b_2..d_2], \ldots, [b_k..d_k])$ as the reversed caterpillar of $a([b_1..d_1], [b_2..d_2], \ldots, [b_k..d_k])$. For every leaf in R_i labeled by a, such a leaf is called *the jth occurrence of a in R_i* if, according to pre-order traversal of R_i, it is the jth visited leaf which is labeled by a.

For $i = 1, 2, \ldots, k$, the tree P_i is constructed from R_i by replacing, for every $a \in S$, the leaf labeled by a with a caterpillar tree $a()$ or $\bar{a}()$ as follows.

1. Set $P_i = R_i$.
2. For every $a \in S$,
   - if T_i is the first tree containing a among $T_1, T_2, \ldots, T_i$, then (in this case, P_i contains exactly one a, that is, $\gamma_a[i] = 1$) replace a in P_i by the caterpillar $\bar{a}([1..\gamma_a[1]], \ldots, [1..\gamma_a[i-1]], [1..1], [1..\gamma_a[i+1]], \ldots, [1..\gamma_a[k]])$.

– else for $j = 1, 2, \ldots, \gamma_a[i]$, replace the jth occurrence of a in P_i by the caterpillar $a([1..\gamma_a[1]], \ldots, [1..\gamma_a[i-1]], [j..j], [1..\gamma_a[i+1]], \ldots, [1..\gamma_a[k]])$.

It is easy to check that each P_i is distinctly labeled by $\{a^1_{b_1,b_2,\ldots,b_k}, a^2_{b_1,b_2,\ldots,b_k} \mid a \in S, 1 \leq i \leq k, 1 \leq b_i \leq \gamma_a[i]\}$. In addition, for every label $a \in S$, there exists exactly one tree P_i which contains the caterpillar $\bar{a}()$ while the rest of the trees in $\mathcal{P}$ contain caterpillars of the form $a()$. Below, more properties of $\mathcal{P}$ are described.

Lemma 4. *For every P_i, $|\Lambda(P_i)| = O((2n)^{k^2})$.*

Lemma 5. *For any $a \in S$, a MAST of $\mathcal{P}$ has ≤ 2 leaves of the form $a^\ell_{b_1,b_2,\ldots,b_k}$.*

Lemma 6. *For any integer x, the size of the MAST of $\mathcal{R}$ is $\geq x$ if and only if the size of the MAST of $\mathcal{P}$ is $\geq 2x$.*

A MASP of $\mathcal{T}$ can now be computed by applying the algorithm of Bryant [5] or Farach *et al.* [9] (see Section 1.2) to $\mathcal{P}$. Since the number of leaves in $\mathcal{P}$ is less than $(2n)^{k^2}$ and all trees are binary, we obtain the main theorem of this section.

Theorem 3. *Given a set of k binary trees $\mathcal{T}$ which are labeled by n distinct labels, their maximum agreement supertree can be computed in $O(k(2n)^{3k^2})$ time.*

4   MASP with $D = 2$ Is NP-Hard

Theorem 2 states that MASP is an NP-hard problem for any fixed $k \geq 3$ when D is unrestricted. We now show that MASP remains NP-hard if restricted to instances with $D = 2$ but where k is left unrestricted. In fact, we prove that MASP is NP-hard even if all of the input trees are required to be rooted triplets. Our NP-hardness proof consists of a polynomial-time reduction from the independent set problem which is known to be NP-hard (see, e.g., [13]).

The independent set problem

Instance: An undirected graph $G = (V, E)$ and a positive integer I.
Question: Is there a subset V' of V with $|V'| = I$ such that V' is an independent set, i.e., such that no two vertices in V' are joined by an edge in E?

The maximum agreement supertree problem restricted to rooted triplets, decision problem version (MASPR-d)

Instance: A set $\mathcal{T}$ of rooted triplets with leaf set S and a positive integer K.
Question: Is there a subset S' of S with $|S'| = K$ which is consistent with $\mathcal{T}$?

Theorem 4. *MASP is NP-hard even if restricted to rooted triplets.*

Proof. Given an arbitrary instance (G, I) of the independent set problem, construct an instance of MASPR-d as follows. Let $S = V \cup \{z_e \mid e \in E\}$ and set $K = I + |E|$. For each edge e in E, include the two rooted triplets $(\{a, z_e\}, b)$ and $(\{b, z_e\}, a)$ in $\mathcal{T}$, where $e = \{a, b\}$. Claim: G has an independent set of size I if and only if there exists a subset S' of S of size K which is consistent with $\mathcal{T}$.

Proof of claim: Suppose there exists an independent set W in G of size I. Then $S' = W \cup \{z_e \mid e \in E\}$ with $|S'| = I + |E|$ is consistent with $\mathcal{T}$ since $\mathcal{T} \mid S'$ contains no rooted triplets (if $\mathcal{T} \mid S'$ had a rooted triplet $(\{x, z_{\{x,y\}}\}, y)$ then x and y would be joined by an edge in E and thus could not both belong to W).

Conversely, suppose there exists a consistent subset S' of S of size K. For each $\{x, y\} \in E$, if $z_{\{x,y\}} \notin S'$ but at least one of x and y belongs to S' then replace x or y in S' by $z_{\{x,y\}}$, and if none of x, y, and $z_{\{x,y\}}$ are contained in S' then replace any element in S' belonging to V by $z_{\{x,y\}}$ (such an element always exists because $K > |E|$). The resulting set S'' will have the form $W \cup \{z_e \mid e \in E\}$ with $W \subseteq V$ and $|S''| = K$, and will still be consistent with $\mathcal{T}$. Next, observe that by the construction of $\mathcal{T}$, for each $\{x, y\} \in E$ at most two of x, y, and $z_{\{x,y\}}$ can be included in any subset of S which is consistent with $\mathcal{T}$. Therefore, for each $\{x, y\} \in E$, since $z_{\{x,y\}} \in S''$ it holds that S'' cannot contain both x and y. Thus, W is an independent set and $|W| = K - |E| = I$.

Hence, MASPR-d is NP-hard and the theorem follows. □

5 A Polynomial-Time $(n/\log n)$-Approximation Algorithm

By the comments preceding Theorem 2, it is highly unlikely that MASP in its general form can be solved exactly or even approximated efficiently (say, within a constant factor) in polynomial time. However, we can adapt one of Akutsu and Halldórsson's [2] algorithms for the largest common subtree problem to obtain the following polynomial-time $(n/\log n)$-approximation algorithm for MASP:

> Arbitrarily partition S into $\lfloor n/\log n \rfloor$ sets $S_1, S_2, ..., S_{\lfloor n/\log n \rfloor}$, each of size at most $\lceil \log n \rceil + 1$. Then, check every subset S'_i of every set S_i to see if S'_i is consistent with $\mathcal{T}$, and let Z be one such subset of maximum cardinality. Return Z.

To see that this algorithm always returns a solution with at least $\frac{\log n}{n}$ times the number of leaves in an optimal solution, let S^* be a maximum consistent leaf subset. Because of the pigeonhole principle, at least one of $S_1, S_2, ..., S_{\lfloor n/\log n \rfloor}$ contains $\geq \frac{1}{\lceil n/\log n \rceil}$ of the elements in S^*; thus, $|Z| \geq \frac{|S^*|}{\lceil n/\log n \rceil} \geq \frac{|S^*|}{n/\log n}$.

To implement the algorithm efficiently, we first note that the deterministic algorithm for dynamic graph connectivity employed in the algorithm for TASP of Henzinger *et al.* [16] can be replaced with a more recent one due to Holm *et al.* [17] to yield the following improvement. We then obtain Theorem 5 below.

Lemma 7. *TASP is solvable in* $\min\{O(N \log^2 n), O(N + n^2 \log n)\}$ *time, where* $N = \sum_{T_i \in \mathcal{T}} |T_i|$ *is the total number of nodes in* $\mathcal{T}$.

Theorem 5. *MASP can be approximated within a factor of* $\frac{n}{\log n}$ *in* $O(n^2 \cdot \log \log n) \cdot \min\{O(k \log \log n), O(k + \log n)\}$ *time. MASP restricted to rooted triplets can be approximated within a factor of* $\frac{n}{\log n}$ *in* $O(k + n^2 \log^2 n)$ *time.*

Finally, we remark that MAST can be approximated within a factor of $\frac{n}{\log n}$ in $O(kn^2)$ time using the same technique.

6 Concluding Remarks

Below, we summarize our results on how restricting the parameters D and k affects the computational complexity of MASP. Arrows indicate when a result follows directly from another by generalization (for example, MASP with $D = 2$ and unrestricted k is NP-hard, so the more general case $D = O(1)$ and unrestricted k cannot be any easier) or by specialization (e.g., the algorithm for $D = O(1)$ and $k = 2$ still works for the more restricted case $D = 2$ and $k = 2$).

MASP	$k = 2$	$k = O(1)$	k unrestricted
$D = 2$	$O(n \log n)$ ($\downarrow$)	$O(k(2n)^{3k^2})$ (Theorem 3)	NP-hard (Theorem 4)
$D = O(1)$	$O(n \log n)$ (Theorem 1)	Open	NP-hard ($\uparrow$)
D unrestricted	$O(n^{1.5})$ (Theorem 1)	NP-hard (Theorem 2)	NP-hard ($\leftarrow$ or $\uparrow$)

We have also described a polynomial-time $(n/\log n)$-approximation algorithm for MASP (Theorem 5).

It is interesting to note that MASP with $D = 2$ and unrestricted k is NP-hard while on the other hand, MAST with $D = 2$ and unrestricted k can be solved in $O(kn^3)$ time, i.e., in polynomial time, using the algorithm of Bryant [5] or Farach et al. [9] (see Section 1.2). This means that for certain restrictions on the parameters D and k, MASP and MAST cannot have the same computational complexity unless P = NP. Furthermore, although our results indicate that MASP is computationally harder than MAST, the maximum refinement subtree problem (see Section 1.2) does not seem any easier than MASP since it is NP-hard already for $k = 2$ when D is unrestricted [15].

An open problem is to determine the computational complexity of MASP with $D = O(1)$ and $k = O(1)$. We believe that this case is solvable in polynomial time. We would also like to know if the running time of our algorithm for the case $D = 2$ and $k = O(1)$ can be improved.

References

1. A. V. Aho, Y. Sagiv, T. G. Szymanski, and J. D. Ullman. Inferring a tree from lowest common ancestors with an application to the optimization of relational expressions. *SIAM Journal on Computing*, 10(3):405–421, 1981.
2. T. Akutsu and M. M. Halldórsson. On the approximation of largest common subtrees and largest common point sets. *Theoretical Computer Science*, 233(1–2):33–50, 2000.
3. A. Amir and D. Keselman. Maximum agreement subtree in a set of evolutionary trees: Metrics and efficient algorithms. *SIAM Journal on Computing*, 26(6):1656–1669, 1997.
4. O. Bininda-Emonds, J. Gittleman, and M. Steel. The (super)tree of life: Procedures, problems, and prospects. *Annual Review of Ecology and Systematics*, 33:265–289, 2002.

5. D. Bryant. *Building Trees, Hunting for Trees, and Comparing Trees: Theory and Methods in Phylogenetic Analysis.* PhD thesis, Univ. of Canterbury, N.Z., 1997.

6. D. Bryant. Optimal agreement supertrees. In *Proc. of the 1st International Conference on Biology, Informatics, and Mathematics* (JOBIM 2000), volume 2066 of *LNCS*, pages 24–31. Springer, 2001.

7. B. Chor, M. Hendy, and D. Penny. Analytic solutions for three-taxon ML_{MC} trees with variable rates across sites. In *Proc. of the 1st Workshop on Algorithms in Bioinformatics* (WABI 2001), volume 2149 of *LNCS*, pages 204–213. Springer, 2001.

8. R. Cole, M. Farach-Colton, R. Hariharan, T. Przytycka, and M. Thorup. An $O(n \log n)$ algorithm for the maximum agreement subtree problem for binary trees. *SIAM Journal on Computing*, 30(5):1385–1404, 2000.

9. M. Farach, T. Przytycka, and M. Thorup. On the agreement of many trees. *Information Processing Letters*, 55:297–301, 1995.

10. M. Farach and M. Thorup. Sparse dynamic programming for evolutionary-tree comparison. *SIAM Journal on Computing*, 26(1):210–230, 1997.

11. C. R. Finden and A. D. Gordon. Obtaining common pruned trees. *Journal of Classification*, 2:255–276, 1985.

12. G. Ganapathysaravanabavan and T. Warnow. Finding a maximum compatible tree for a bounded number of trees with bounded degree is solvable in polynomial time. In *Proc. of the 1st Workshop on Algorithms in Bioinformatics* (WABI 2001), volume 2149 of *LNCS*, pages 156–163. Springer, 2001.

13. M. Garey and D. Johnson. *Computers and Intractability – A Guide to the Theory of NP-Completeness.* W. H. Freeman and Company, New York, 1979.

14. L. Gąsieniec, J. Jansson, A. Lingas, and A. Östlin. On the complexity of constructing evolutionary trees. *Journal of Combinatorial Optimization*, 3:183–197, 1999.

15. J. Hein, T. Jiang, L. Wang, and K. Zhang. On the complexity of comparing evolutionary trees. *Discrete Applied Mathematics*, 71:153–169, 1996.

16. M. R. Henzinger, V. King, and T. Warnow. Constructing a tree from homeomorphic subtrees, with applications to computational evolutionary biology. *Algorithmica*, 24(1):1–13, 1999.

17. J. Holm, K. de Lichtenberg, and M. Thorup. Poly-logarithmic deterministic fully-dynamic algorithms for connectivity, minimum spanning tree, 2-edge, and biconnectivity. *Journal of the ACM*, 48(4):723–760, 2001.

18. J. Jansson. On the complexity of inferring rooted evolutionary trees. In *Proc. of the Brazilian Symp. on Graphs, Algorithms, and Combinatorics* (GRACO'01), volume 7 of *Electronic Notes in Discrete Mathematics*, pages 121–125. Elsevier, 2001.

19. M.-Y. Kao, T.-W. Lam, W.-K. Sung, and H.-F. Ting. An even faster and more unifying algorithm for comparing trees via unbalanced bipartite matchings. *Journal of Algorithms*, 40(2):212–233, 2001.

20. P. Kearney. Phylogenetics and the quartet method. In T. Jiang, Y. Xu, and M. Q. Zhang, editors, *Current Topics in Computational Molecular Biology*, pages 111–133. The MIT Press, Massachusetts, 2002.

21. A. Meyers, R. Yangarber, and R. Grishman. Alignment of shared forests for bilingual corpora. In *Proc. of the 16th International Conference on Computational Linguistics* (COLING-96), pages 460–465, 1996.

22. M. J. Sanderson, A. Purvis, and C. Henze. Phylogenetic supertrees: assembling the trees of life. *TRENDS in Ecology & Evolution*, 13(3):105–109, 1998.

23. M. Steel. The complexity of reconstructing trees from qualitative characters and subtrees. *Journal of Classification*, 9(1):91–116, 1992.

24. M. Steel and T. Warnow. Kaikoura tree theorems: Computing the maximum agreement subtree. *Information Processing Letters*, 48:77–82, 1993.

Complexity of Cycle Length Modularity Problems in Graphs*

Edith Hemaspaandra[1], Holger Spakowski[2**], and Mayur Thakur[3]

[1] Department of Computer Science, Rochester Institute of Technology,
Rochester, NY 14623, USA.
eh@cs.rit.edu
[2] Institut für Informatik, Heinrich-Heine-Universität Düsseldorf,
40225 Düsseldorf, Germany.
spakowsk@cs.uni-duesseldorf.de,
[3] Department of Computer Science, University of Rochester,
Rochester, NY 14627, USA.
thakur@cs.rochester.edu

Abstract. The even cycle problem for both undirected [Tho88] and directed [RST99] graphs has been the topic of intense research in the last decade. In this paper, we study the computational complexity of *cycle length modularity problems*. Roughly speaking, in a cycle length modularity problem, given an input (undirected or directed) graph, one has to determine whether the graph has a cycle C of a specific length (or one of several different lengths), modulo a fixed integer. We denote the two families (one for undirected graphs and one for directed graphs) of problems by (S, m)-UC and (S, m)-DC, where $m \in \mathbb{N}$ and $S \subseteq \{0, 1, \ldots, m-1\}$. (S, m)-UC (respectively, (S, m)-DC) is defined as follows: Given an undirected (respectively, directed) graph G, is there a cycle in G whose length, modulo m, is a member of S? In this paper, we fully classify (i.e., as either polynomial-time solvable or as NP-complete) each problem (S, m)-UC such that $0 \in S$ and each problem (S, m)-DC such that $0 \notin S$. We also give a sufficient condition on S and m for the following problem to be polynomial-time computable: (S, m)-UC such that $0 \notin S$.

1 Introduction

In this paper we study the complexity of problems related to lengths of cycles, modulo a fixed integer, in undirected and directed graphs. Given $m \in \mathbb{N}$, and $S \subseteq \{0, 1, \ldots, m-1\}$, we define the following two cycle length modularity problems.

(S, m)-UC $= \{G \mid G$ is an undirected graph such that there exists an $\ell \in \mathbb{N}$ such that $\ell \bmod m \in S$, and there exists a cycle of length ℓ in $G\}$.

(S, m)-DC $= \{G \mid G$ is a directed graph such that there exists an $\ell \in \mathbb{N}$ such that $\ell \bmod m \in S$, and there exists a directed cycle of length ℓ in $G\}$.

The most basic cases of cycle length modularity problems are the following problems for undirected (respectively, directed) graphs: deciding whether a given undirected

---

* Supported in part by grants NSF-INT-9815095/DAAD-315-PPP-gü-ab, NSF-CCR-0311021, and a DAAD grant.
** Work done while visiting the University of Rochester.

M. Farach-Colton (Ed.): LATIN 2004, LNCS 2976, pp. 509–518, 2004.

(respectively, directed) graph has a cycle of odd length, and deciding whether a given undirected (respectively, directed) graph has a cycle of even length. We will refer to these problems as the odd cycle problem for undirected (respectively, directed) graphs, and the even cycle problem for undirected (respectively, directed) graphs, respectively. In our notation, these problems are denoted by $(\{1\}, 2)$-UC (respectively, $(\{1\}, 2)$-DC) and $(\{0\}, 2)$-UC (respectively, $(\{0\}, 2)$-DC). All these four problems are now known to be in P.

The odd cycle problem and the even cycle problem are quite different in nature. The reason is that if a closed walk of odd length is decomposed into cycles, then there is at least one odd cycle in the decomposition. The corresponding statement for even walks and even cycles is not true. Since odd closed walks can easily be found in polynomial time, it is easy to detect odd cycles.

It is well known that an undirected graph has an odd cycle if and only if the graph is not bipartite. No such simple characterization is known for the case of even cycles. However, Thomassen [Tho88] showed that the family of cycles of length divisible by m has the Erdös-Pósa property [EP65], and then used results from Robertson and Seymour [RS86] to prove that the even cycle problem for undirected graphs is in P. In fact, Thomassen proved that, for each $m \in \mathbb{N}$, $(\{0\}, m)$-UC is in P. Even though Thomassen's graph minor and tree-width approach to solving the even cycle problem is elegant, it means that the algorithm has the drawback of having huge constants in its running time. Arkin, Papadimitriou, and Yannakakis [APY91] used a simpler approach to give an efficient algorithm for the even cycle problem for undirected graphs. Their algorithm is based on their characterization of undirected graphs that do not contain even cycles with certain efficiently checkable properties of the biconnected components of these graphs. However, unlike Thomassen's approach, their approach does not seem to generalize beyond the $(\{0\}, 2)$-UC case to, say, $(\{0\}, 3)$-UC. A related result from Yuster and Zwick [YZ97] shows that, for each k, the problem of deciding if a given undirectd graph has a cycle of length $2k$ is in P.

It is interesting to note that even though the algorithms for the two odd cycle problems (i.e., undirected and directed) are similar, neither of the two algorithms mentioned above (namely, Thomassen [Tho88] and Arkin, Papadimitriou, and Yannakakis [APY91]) seems to be able to handle the even cycles case for directed graphs. However, Robertson, Seymour, and Thomas [RST99] (the conference version of the paper is by McCuaig et al. [MRST97]) prove, via giving a polynomial time algorithm for the problem of deciding whether a bipartite graph has a Pfaffian orientation, that the even cycle problem for directed graphs is also in P. We note that Vazirani and Yannakakis [VY89] proved the polynomial-time equivalence of the even cycle problem for directed graphs with the following problems:

1. The problem of checking whether a bipartite graph has a Pfaffian orientation [Kas67],
2. Polya's problem [Pol13], i.e., given a square $(0, 1)$ matrix A, is there a $(-1, 0, 1)$ matrix B that can be obtained from A by changing some of the 1's to -1's such that the determinant of B is equal to the permanent of A, and
3. Given a square matrix A of nonnegative integers, determine if the determinant of A equals the permanent of A.

Until the even cycle problem was shown to be in P by Thomassen [Tho88] (for undirected graphs) and Robertson, Seymour, and Thomas [RST99] (for directed graphs) and even since then, a lot of interesting research on the even cycle problem and related problems led to the study of other cycle length modularity problems. Some problems have been shown to be in P, while others have been shown to be NP-complete. We mention some that have a close relationship with the problems studied in this paper. As stated earlier, Thomassen [Tho88] proved that, for each $m \in \mathbb{N}$, the problem of deciding whether an undirected graph has a cycle of length $\equiv 0 \pmod{m}$ is in P. That is, for each $m \in \mathbb{N}$, $(\{0\}, m)$-UC $\in$ P. Arkin, Papadimitriou, and Yannakakis [APY91] study, among other problems, the problem of deciding whether a directed graph contains cycles of length $p \pmod{m}$. They prove, via reduction from the directed subgraph homeomorphism problem [FHW80], that for all $m > 2$ and for all p such that $0 < p < m$, $(\{p\}, m)$-DC is NP-complete. They also give a polynomial-time algorithm for the problem of finding the greatest common divisor of all cycles in graphs, a problem motivated by the problem of finding the period of a Markov chain. Furthermore, they prove that, for all $m > 2$ and for all $0 < p < m$, the problem of deciding whether all cycles in an undirected graph are of length $p \pmod{m}$ is in P. Galluccio and Loebl [GL96] study the complexity of the corresponding problem in directed graphs. They prove that for the case of planar directed graphs, checking whether all cycles in the input graph are of length $p \pmod{m}$ can be done in polynomial time.

In this paper, we resolve the complexity (as either in P or NP-complete) of the following problems:

1. (S, m)-DC, where $0 \notin S$, and
2. (S, m)-UC, where $0 \in S$.

We prove that each problem in 2 is in P. For 1, we classify each problem as either in P or NP-complete, depending only on the properties of S and m: If there exist $0 \leq d_1, d_2 < m$ such that $d_1, d_2 \notin S$ and $(d_1 + d_2) \bmod m \in S$, then (S, m)-DC is NP-complete, otherwise (S, m)-DC is in P. We also prove a sufficient condition for (S, m)-UC (with $0 \notin S$) to be polynomial-time computable: for each $p \in S$, and for each d_1, d_2 such that $0 \leq d_1, d_2 < m$ and $d_1 + d_2 \equiv p \pmod{m}$, it holds that either $d_1 \in S$ or $d_2 \in S$. Note that this condition is exactly the same as that for the "in P" case of 1 given above.

The paper is organized as follows. In Section 2, we introduce the definitions and notations that will be used in the rest of the paper. In Section 3, we present results for cycle length modularity problems in directed graphs and in Section 4 we present results for cycle length modularity problems in undirected graphs. Finally, in Section 5 we present some open problems and future research directions.

2   Definitions and Notations

In this section we describe the notations used in the rest of the paper. For each finite set S, let $\|S\|$ denote the cardinality of S.

An *undirected graph* G is a pair (V, E), where V is a finite set (the set of vertices or nodes) and $E \subseteq V \times V$ (the set of edges) with the following properties.

1. For each $u, v \in V$, if $(u, v) \in E$, then $(v, u) \in E$.
2. For each $v \in V$, $(v, v) \notin E$, that is, self-loops are not allowed.

A *directed graph* G is a pair (V, E), where V is a finite set and $E \subseteq V \times V$. For each graph G, let $V(G)$ denote the set of vertices of G, and let $E(G)$ denote the set of edges of G. A *walk* of length k in a graph G is a sequence of vertices $(u_0, u_1, \ldots, u_k)$ with $k \geq 1$ in G such that, for each $0 \leq i < k$, $(u_i, u_{i+1}) \in E(G)$. A *path* is a walk where all vertices are distinct. A *closed walk* in a graph G is a walk $(u_0, u_1, \ldots, u_k)$ in G such that $u_0 = u_k$. A cycle in an undirected graph is a closed walk $(u_0, u_1, \ldots, u_{k-1}, u_0)$ of length ≥ 3 such that $u_0, u_1, \ldots, u_{k-1}$ are k distinct vertices. A cycle in a directed graph is a closed walk $(u_0, u_1, \ldots, u_{k-1}, u_0)$ such that $u_0, u_1, \ldots, u_{k-1}$ are k distinct vertices. It should be noted that this definition of a cycle is sometimes called a *simple cycle*.

3   Cycle Length Modularity Problems in Directed Graphs

In this section, we study the complexity of cycle length modularity problems in directed graphs. Arkin, Papadimitriou, and Yannakakis [APY91] proved that, for each $m \in \mathbb{N}$ and each $0 < r < m$, $(\{r\}, m)$-DC is NP-complete. In Theorem 1, we generalize their result. For each m and S such that $0 \notin S$, we give a condition on S and m for (S, m)-DC to be NP-complete. Furthermore, we prove that if the stated conditions on S and m are not satisfied, then (S, m)-DC is in P.

Theorem 1. *For all $m \geq 1$ and $S \subseteq \{1, \ldots, m - 1\}$, the following is true:*

(i) *If there is a $p \in S$, and $d_1 \notin S, d_2 \notin S$ such that $0 \leq d_1, d_2 < m$ and $d_1 + d_2 \equiv p$ (mod m), then (S, m)-DC is NP-complete.*
(ii) *Otherwise, (S, m)-DC is in P.*

Proof. To prove (i), let $p \in S$ and $d_1, d_2 \notin S$ be such that $0 \leq d_1, d_2 < m$ and $d_1 + d_2 \equiv p \pmod{m}$. We closely follow the proof of Theorem 1 in Arkin, Papadimitriou, and Yannakakis [APY91]. Fortune, Hopcroft, and Wyllie [FHW80] showed that the directed subgraph homeomorphism problem is NP-complete for any fixed directed graph that is not a tree of depth 1. In particular, the following problem is NP-complete:

> Given a directed graph G and vertices s and t in $V(G)$, does G contain a cycle through both s and t?

We now specify a polynomial-time function σ that reduces this problem to (S, m)-DC. Given a directed graph G and vertices $s, t \in V(G)$, $\sigma(\langle G, s, t \rangle)$ outputs the graph G' where $G' = (V', E')$ is defined as follows. (Note that in the steps below, we can assume that $d_1 \neq 0$ and $d_2 \neq 0$, because if either d_1 or d_2 is equal to 0, then the preconditions of (i) cannot be satisfied.)

1. Set $V' := V$. Set $E' := \emptyset$.
2. For every edge $(v, s) \in E(G)$, do the following.
   a) Set $V' := V' \cup \{w_j \mid 1 \leq j \leq d_1 - 1\}$, where the w_j's are new vertices.
   b) Set $E' := E' \cup \{(v, w_1), (w_1, w_2), \ldots, (w_{d_1-1}, s)\}$.
3. For every edge $(v, t) \in E(G)$, do the following.
   a) Set $V' := V' \cup \{w_j \mid 1 \leq j \leq d_2 - 1\}$, where the w_j's are new vertices.
   b) Set $E' := E' \cup \{(v, w_1), (w_1, w_2), \ldots, (w_{d_2-1}, t)\}$.

4. For every edge $(v, w) \in E(G)$ such that $v, w \notin \{s, t\}$, do the following.
   a) Set $V' := V' \cup \{w_j \mid 1 \leq j \leq m - 1\}$, where the w_j's are new vertices.
   b) Set $E' := E' \cup \{(v, w_1), (w_1, w_2), \ldots, (w_{m-1}, w)\}$.

It is easy to see that the cycles in G' have the following properties.

1. All cycles in G' going through neither s nor t have length $\equiv 0 \pmod{m}$.
2. All cycles in G' going through s but not through t have length $\equiv d_1 \pmod{m}$.
3. All cycles in G' going through t but not through s have length $\equiv d_2 \pmod{m}$.
4. All cycles in G' going through s and t have length $\equiv (d_1 + d_2) \pmod{m} \equiv p \pmod{m}$

Roughly speaking, we replace each edge $e \in E(G)$ that ends in s, by a series of d_1 edges in G' such that the series of edges ends in s. Similarly, we replace each edge $e \in E(G)$ that ends in t, by a series of d_2 edges in G' that ends in t. It is clear from the construction of G' that there is a cycle through s and t in G if and only if there is a cycle through s and t in G'. Since $\{0, d_1, d_2\} \cap S = \emptyset$ and $p \in S$, it follows from the properties stated above that there is a cycle through s and t in G if and only if there is a cycle of length $\equiv p \pmod{m}$ in G'. Also, it is clear that G' can be computed from G in polynomial time. It follows that (S, m)-DC is NP-hard. Note that, for each S and m, (S, m)-DC is clearly in NP. Thus, (S, m)-DC is NP-complete.

We will now prove (ii). Let $m \geq 1$ and $S \subseteq \{1, \ldots, m - 1\}$ be such that for all $p \in S$ and all d_1, d_2, if $0 \leq d_1, d_2 < m$ and $d_1 + d_2 \equiv p \pmod{m}$, then $d_1 \in S$ or $d_2 \in S$.

We claim that the following algorithm solves (S, m)-DC in polynomial time:

Input: A directed graph G.

1. **for** each $p \in S$ **do**
2.   **if** G has a closed walk of length $\equiv p \pmod{m}$ **then** accept.
3. reject.

Clearly, step 2 can be done in polynomial time. If the algorithm rejects, then obviously G is not in (S, m)-DC. To complete the proof of (ii), we will prove the following claim.

Claim. If G has a closed walk W of length $\equiv p \pmod{m}$ for some $p \in S$, then G has a cycle of length $\equiv p' \pmod{m}$ for some $p' \in S$.

Proof. The proof is by induction on the length of W. The claim is certainly true for all closed walks W of length 1. Assume that the claim is true for all closed walks W whose length is less than k. Suppose G has a closed walk W of length k with $k \bmod m = p$ and $p \in S$. Distinguish the following two cases.

Case 1: W is a cycle.
    Then we are done.
Case 2: W is not a cycle.
    Then there exist $\ell_1 > 0$, $\ell_2 > 0$, $d_1 < m$, and $d_2 < m$ such that W can be decomposed into a simple cycle C of length ℓ_1 and a closed walk W' of length ℓ_2 such that $\ell_1 \equiv d_1 \pmod{m}$, $\ell_2 \equiv d_2 \pmod{m}$. Since $\ell_1 + \ell_2 = k$, it follows that $\ell_1 + \ell_2 \equiv d_1 + d_2 \equiv p \pmod{m}$. We know that $d_1 \in S$ or $d_2 \in S$. If $d_1 \in S$ then we are done. If $d_2 \in S$ we are done by the induction hypothesis.

Thus, this claim holds, and so Theorem 1 holds.   □

As an immediate corollary, we get that the problem of deciding whether all cycles in a directed graph have length $\equiv 0 \pmod{m}$ is in P.

Corollary 2. *For each* $m \in \mathbb{N}$, $(\{1, 2, \dots, m-1\}, m)$-*DC* $\in$ P.

We note that Corollary 2 also follows from the fact that finding the period (greatest common divisor of all cycle lengths) of a graph is in P [BV73] (see also [Knu73, Tar72]).

Yuster and Zwick [YZ97] proved that for directed graphs, a shortest odd length cycle can be found in time $O(\|V\| \cdot \|E\|)$. We show that for all (S, m)-DC-problems satisfying the condition (ii) of Theorem 1, a shortest cycle with length, modulo m, in S can be found in time $O(M(\|V\|) \cdot \log \|V\|)$, where $M(n) = n^{2.376}$ is the complexity of boolean matrix multiplication. For the special case $m = 2$, $S = \{1\}$, the algorithm is for dense graphs an improvement over the one given in [YZ97].

Theorem 3. *For all* $m \geq 2$ *and* $S \subseteq \{0, \dots, m-1\}$ *with* $0 \notin S$ *the following is true: If for all* $p \in S$, *and all* d_1, d_2, *such that* $0 \leq d_1, d_2 < m$ *and* $d_1 + d_2 \equiv p \pmod{m}$, *it holds that* $d_1 \in S$ *or* $d_2 \in S$, *then there is an* $O(M(\|V\|) \cdot \log \|V\|)$ *time algorithm that computes a shortest cycle* C *such that the length of* C, *modulo* m, *is in* S.

Proof. If the precondition of Theorem 3 holds, every closed walk whose length, modulo m, belongs to S, is a cycle or decomposes into cycles such that the length of at least one of these cycles, modulo m, belongs to S. Hence the problem reduces to finding a shortest closed walk whose length, modulo m, belongs to S.

Let $G = (V, E)$, where $V = \{v_1, \dots, v_n\}$. For every $r \in \{0, \dots, m-1\}$ and $0 < k \leq n$, we define the boolean matrix $A_{k,r}$ by $A_{k,r}(i, j) \stackrel{df}{=} 1$ iff there is a walk of length ℓ from v_i to v_j in G with $0 < \ell \leq k$ and $\ell \equiv r \pmod{m}$. With $O(\log n)$ boolean matrix multiplications we can determine k_{min}, the length of the desired closed walk. The value of k_{min} equals the smallest k with $A_{k,r'}(i, i) = 1$ for some $i \in \{1, \dots, n\}$ and $r' \in S$. First, compute the matrices $A_{k,r}$ where k is a power of 2, using the identity

$$A_{2k,r} = \bigvee_{i=0}^{m-1} (A_{k,i} \wedge A_{k,r-i}) \vee A_{k,r},$$

where $\wedge$ and $\vee$ stand for boolean matrix multiplication and componentwise 'or', respectively. Note that $A_{1,1}$ is the adjacency matrix of G, and $A_{1,r}$, $r \neq 1$, is a zero matrix. After that, apply binary search to determine k_{min}, and a representation of $A_{k_{min},r'}$ as product of matrices $A_{k,r}$ with k being a power of 2. A specific closed walk with length k_{min} (which we know, is a cycle) can now easily be found in additional $O(\|V\|^2)$ time.   □

4   Cycle Length Modularity Problems in Undirected Graphs

In this section, we study the complexity of problems (S, m)-UC, for different S and m. The case when $S = \{0\}$ has been shown to be in P by Thomassen [Tho88]. We extend Thomassen's result and prove that for all S such that $0 \in S$, (S, m)-UC is in P.

Theorem 4. *For each m, and each $S \subseteq \{0, \dots, m-1\}$ such that $0 \in S$, (S, m)-UC $\in$ P.*

The proof of Theorem 4 is an extension of the proof of Thomassen's result for $(\{0\}, m)$-UC, which in turn is based on the result from Robertson and Seymour [RS86] for the k-disjoint paths problem. We will need the following results related to tree-widths for the proof of Theorem 4. Tree-width is an invariant of graphs that has been a central concept in the development of algorithms for fundamental problems in graph theory. See [RS86] for a definition of tree-width, and [RS85] for a survey on graph minor results. We will not define tree-widths because the definition is rather involved and for the proof of Theorem 4 we need to know only the following fact about tree-widths of graphs.

Theorem 5 ([RS86]). *For each $t \in \mathbb{N}$, there is a polynomial-time algorithm for deciding whether an undirected graph has tree-width at least t.*

The following theorem shows that, for fixed m, all graphs of sufficiently large tree-width have a cycle whose length is a multiple of m.

Theorem 6 ([Tho88]). *For each m, there exists a $t_m \in \mathbb{N}$ such that, for each undirected graph G with tree-width at least t_m, G contains a cycle of length $\equiv 0 \pmod m$.*

Roughly speaking, Theorem 6 allows us to handle those graphs that have large tree-widths. Theorem 8 allows us to handle small tree-widths.

Definition 7. *For each $t, m \in \mathbb{N}$, $d_1, d_2, \dots d_k$ such that, for each $1 \le i \le k$, $d_i < m$, DISJ-PATH$_{\langle t, m, d_1, d_2, \dots, d_k \rangle}$ is defined as follows: DISJ-PATH$_{\langle t, m, d_1, d_2, \dots, d_k \rangle} = \{\langle G, x_1, y_1, \dots, x_k, y_k \mid G$ is an undirected graph such that (a) G has tree width at most t, (b) for each i, x_i and y_i are vertices in G, and (c) there exist k node-disjoint paths $P_1, P_2, \dots, P_k$ in G such that, for each $1 \le i \le k$, P_i is a path connecting x_i and y_i such that P_i has length $d_i \pmod m)\}$.*

Theorem 8 ([Tho88]). *Let $t, m, d_1, d_2, \dots, d_k \in \mathbb{N}$ be such that, for each $1 \le i \le k$, $d_i < m$. Then, DISJ-PATH$_{\langle t, m, d_1, d_2, \dots, d_k \rangle}$ is in P.*

Proof of Theorem 4. Let m and S be such that $m \in \mathbb{N}$, $S \subseteq \{0, 1, \dots, m - 1\}$, and $0 \in S$. We will now describe a polynomial-time algorithm that decides (S, m)-UC. Let G be the input graph. Check, using algorithm in Theorem 5, if G has tree-width at least t_m, where t_m is as in Theorem 6. If so, then, by Theorem 6, G has a cycle of length $0 \pmod m$. Otherwise, G has tree-width at most t_m. So, we use Theorem 8 to check if G has a cycle of length ℓ such that $\ell \in S$. For all distinct vertices v_1, v_2, v_3, v_4 in G such that $\{(v_1, v_2), (v_3, v_4)\} \subseteq E(G)$, for each $\ell \in S$, and for each $0 \le d_1, d_2 \le m - 1$ such that $d_1 + d_2 + 2 = \ell \pmod m$, we do the following. Check, using Theorem 8, whether there are 2 disjoint paths P_1 and P_2, P_1 between v_1 and v_3 of length $d_1 \pmod m$ and P_2 between v_2 and v_4 of length $d_2 \pmod m$. If there are such disjoint paths, then there is a cycle of length $d_1 + d_2 + 2 = \ell \pmod m$, namely the cycle consisting of the edges in P_1, the edges in P_2, and the edges (v_1, v_2) and (v_3, v_4). Note that we may be missing cycles consisting of 3 nodes or less, but that can be easily handled by checking brute-force for all cycles of 3 nodes or less. $\qquad\square$

Let us consider the complements of the following cycle length modularity problems (for fixed $m \geq 2$ and fixed $0 \leq r < m$): $(\{0, 1, \ldots, m-1\} - \{r\}, m)$-UC. For any m and r such that $0 \leq r < m$, these problems asks whether all cycles in the given graph are of length $\equiv r \pmod{m}$. For $m = 2$ and $r = 0$, this problem is the odd cycle problem in undirected graphs, which as noted in the introduction is easily seen to be in P based on the simple observation that any closed walk of odd length in a graph must contain a simple cycle of odd length. For $m = 2$ and $r = 1$, this problem is the even cycle problem, which is also in P [APY91]. Arkin, Papadimitriou, and Yannakakis in fact prove, via using the properties of triconnected components of graphs, that for each m, and each $0 \leq r < m$, finding whether all cycles in a graph are of length $\equiv r \pmod{m}$ can be done in polynomial time.

Theorem 9 ([APY91]). *For each $m \in \mathbb{N}$, and each r such that $0 \leq r < m$, $(\{0, 1, \ldots, m-1\} - \{r\}, m)$-UC $\in$ P.*

Corollary 10. $(\{1, 2, \ldots, m-1\}, m)$-UC $\in$ P.

The following theorem is an analog of Theorem 1(ii) for undirected graphs.

Theorem 11. *For all $m > 2$ and $S \subseteq \{1, \ldots, m-1\}$, the following is true: If for all $p \in S$, and all d_1, d_2, such that $0 \leq d_1, d_2 < m$ and $d_1 + d_2 \equiv p \pmod{m}$, it holds that $d_1 \in S$ or $d_2 \in S$, then (S, m)-UC $\in$ P.*

The proof given for the corresponding statement regarding directed graphs does not work here. The reason is that closed walks in undirected graphs need not decompose properly into cycles. To see why this is true, consider a closed walk C of length 5 in an undirected graph: $v_1 v_2 v_3 v_4 v_2 v_1$. Note that even though C is a closed walk of length 5, it is neither a cycle nor does it decompose properly into cycles, basically because $v_1 v_2 v_1$ is not a valid cycle.

In order to prove Theorem 11, we reduce the problem to the problem of determining the period of an undirected graphs, which is solvable in polynomial time by the algorithm from Arkin, Papadimitriou, and Yannakakis [APY91]. We need the following lemma.

Lemma 12. *For all $m \geq 1$ and $S = \{a_1, \ldots, a_n\} \subseteq \{0, \ldots, m-1\}$, $0 \in S$, the following is true:*
If for all $d_1 \in S, d_2 \in S$ it holds that $(d_1 + d_2) \bmod m \in S$, then $S = \{\ell \mid 0 \leq \ell < m$ and $g \mid \ell\}$ for some g with $g \mid m$.

Proof. Let $S - \{0\} = \{a_1, a_2, \ldots, a_n\}$ Let $g = \gcd(a_1, \ldots, a_n, m)$. From number theory (see [Apo76]) we know that there exist $k_1, \ldots, k_n, k_{n+1} \in \mathbb{Z}$, such that

$$k_1 a_1 + k_2 a_2 + \cdots + k_n a_n + k_{n+1} m = g.$$

For all i, $1 \leq i \leq n + 1$, let

$$k_i' \stackrel{df}{=} k_i + m|k_i|.$$

Then

$$(k_1' a_1 + \cdots + k_n' a_n) \bmod m = g, \tag{1}$$

where $k_1', \ldots, k_n' \geq 0$.

For all $d_1 \in S$, $d_2 \in S$ it holds that $(d_1 + d_2) \bmod m \in S$. Hence Eq. (1) implies that $g \in S$. Furthermore, $sg \in S$ for all $s \in \mathbb{Z}$ such that $0 \le sg < m$. Since for each $1 \le i \le n$, $g|a_i$, it follows that

$$S = \{\ell \mid 0 \le \ell < m \text{ and } g|\ell\}.$$

This concludes the proof of Lemma 12   □

Proof of Theorem 11. Let $\overline{S} \stackrel{df}{=} \{0, \dots, m-1\} - S$. Lemma 12 implies that

$$\overline{S} = \{\ell \mid 0 \le \ell < m \text{ and } g|\ell\}$$

for some g with $g|m$. Hence

$$S = \{\ell \mid 0 \le \ell < m \text{ and } g \nmid \ell\}.$$

Define

$$S' = \{1, \dots, g-1\}.$$

Since $g|m$ holds

$$x \bmod m \in S \iff x \bmod g \in S'$$

for all $x \in \mathbb{N}$. Hence (S, m)-UC is equivalent to $(\{1, \dots, g-1\}, g)$-UC. However, $(\{1, \dots, g-1\}, g)$-UC is the set of graphs containing a cycle not divisible by g, which is in P since the period of a graph (the gcd of all cycle lengths) can be determined in polynomial time [APY91]. This concludes the proof of Theorem 11.   □

5 Conclusion and Open Problems

In this paper, we studied the complexity of cycle length modularity problems. We completely characterized (i.e., as either polynomial-time computable or as NP-complete) each problem (S, m)-DC, where $0 \notin S$. We also proved that, for each S such that $0 \in S$, (S, m)-UC is in P, and we proved a sufficient condition on S and m for the problem (S, m)-UC ($0 \notin S$) to be in P. We mention several open problems.

1. Theorem 1 completely characterizes all modularity problems in directed graphs when $0 \notin S$. Robertson, Seymour, and Thomas [RST99] prove that $(\{0\}, 2)$-DC is in P. In light of these results, it is natural to ask if $(\{0\}, m)$-DC $\in$ P, for some or all $m > 2$. Also, the complexity of (S, m)-DC such that $0 \in S$ and $m > 2$ is still open, except for trivial ($S = \{0, 1, \dots, m-1\}$) cases.
2. Theorem 4 shows that all cycle length modularity problems in undirected graphs (S, m)-UC such that $0 \in S$ are solvable in polynomial time. What about the complexity of the (S, m)-UC problems with $0 \notin S$ which are not covered by Theorem 9 or 11?
3. Theorem 9 shows that, for undirected graphs, the problem of finding whether all cycles have length $\equiv r \pmod m$ is in P. What is the complexity of the corresponding problem for directed graphs?

References

[Apo76]   T. Apostol. *Introduction to Analytic Number Theory*. Undergraduate Texts in Mathematics. Springer-Verlag, 1976.

[APY91]   E. Arkin, C. Papadimitriou, and M. Yannakakis. Modularity of cycles and paths in graphs. *Journal of the ACM*, 38(2):255–274, April 1991.

[BV73]   Y. Balcer and A. Veinott. Computing a graph's period quadratically by node condensation. *Discrete Mathematics*, 4:295–303, 1973.

[EP65]   P. Erdös and L. Pósa. On independent circuits contained in a graph. *Canadian Journal on Mathematics*, 17:347–352, 1965.

[FHW80]   S. Fortune, J. Hopcroft, and J. Wyllie. The directed subgraph homeomorphism problem. *Theoretical Computer Science*, 10:111–121, 1980.

[GL96]   A. Galluccio and M. Loebl. Cycles of prescribed modularity in planar digraphs. *Journal of Algorithms*, 21:51–70, 1996.

[Kas67]   P. Kasteleyn. Graph theory and crystal physics. In F. Harary, editor, *Graph Theory and Theoretical Physics*, pages 43–110. Academic Press, New York, 1967.

[Knu73]   D. Knuth. Strong components. Technical Report 004639, Computer Science Department, Stanford University, Stanford, California, 1973.

[MRST97]   W. McCuaig, N. Robertson, P. Seymour, and R. Thomas. Permanents, pfaffian orientations, and even directed circuits. In *Proceedings of the 29th ACM Symposium on Theory of Computing*, pages 402–405, 1997.

[Pol13]   G. Polya. Aufgabe 424. *Arch. Math. Phys.*, 20(3):271, 1913.

[RS85]   N. Robertson and P. Seymour. Graph minors—a survey. In I. Anderson, editor, *Surveys in Combinatorics 1985: Invited Papers for the Tenth British Combinatorial Conference*, pages 153–171. Cambridge University Press, 1985.

[RS86]   N. Robertson and P. Seymour. Graph minors II. Algorithmic aspects of tree-width. *Journal of Algorithms*, 7:309–322, 1986.

[RST99]   N. Robertson, P. Seymour, and R. Thomas. Permanents, pfaffian orientations, and even directed circuits. *Annals of Mathematics*, 150:929–975, 1999.

[Tar72]   R. Tarjan. Depth first search and linear graph algorithms. *SIAM Journal on Computing*, 2:146–160, 1972.

[Tho88]   C. Thomassen. On the presence of disjoint subgraphs of a specified type. *Journal of Graph Theory*, 12(1):101–111, 1988.

[VY89]   V. Vazirani and M. Yannakakis. Pfaffian orientations, 0-1 permanents, and even cycles in directed graphs. *Discrete Applied Mathematics*, 25:179–190, 1989.

[YZ97]   R. Yuster and U. Zwick. Finding even cycles even faster. *SIAM Journal on Discrete Mathematics*, 10(2):209–222, May 1997.

Procedural Semantics for Fuzzy Disjunctive Programs on Residuated Lattices

Dušan Guller

Institute of Informatics, Comenius University, **
Mlynská dolina, 842 15 Bratislava, Slovakia
guller@fmph.uniba.sk

Abstract. In the paper, we present a procedural semantics for fuzzy disjunctive programs - sets of graded implications of the form:

$$(h_1 \vee \cdots \vee h_n \longleftarrow b_1 \& \cdots \& b_m, c) \qquad (n > 0, m \geq 0)$$

where h_i, b_j are atoms and c a truth degree from a complete residuated lattice

$$\boldsymbol{L} = (L, \leq, \vee, \wedge, *, \Rightarrow, 0, 1).$$

A graded implication can be understood as a means of the representation of incomplete and uncertain information; the incompleteness is formalised by the consequent disjunction of the implication, while the uncertainty by its truth degree. We generalise the results for Boolean lattices in [3] to the case of residuated ones. We take into consideration the non-idempotent triangular norm $*$, instead of the idempotent $\wedge$, as a truth function for the strong conjunction $\&$. In the end, the coincidence of the proposed procedural semantics and the generalised declarative, fixpoint semantics from [4] will be reached.

Keywords: Disjunctive logic programming, multivalued logic programming, fuzzy logic, knowledge representation and reasoning

1   Introduction

The complexity of the real world causes that our knowledge about it is unfortunately ambiguous. The complete and certain description of even a 'small' part of the reality requires much more detailed information than humans or computer systems are capable to recognise simultaneously. From the philosophical point of view, such the exhaustive description of the world is an unreachable platonic ideal. Nevertheless, humans can understand complex real systems thanks to their ability of maintaining only a generic comprehension of them and of approximative reasoning about them. Generally speaking, the ambiguity in our knowledge is of various natures; we could roughly distinguish the incompleteness and uncertainty of information. Consequently, in the knowledge representation and commonsense reasoning, we should be able to handle such incomplete and uncertain

** Partially supported by Slovak project VEGA 1/1055/04

M. Farach-Colton (Ed.): LATIN 2004, LNCS 2976, pp. 519–529, 2004.
© Springer-Verlag Berlin Heidelberg 2004

information. Attempts to solve many real-world problems bring us to storing and retrieving incomplete and uncertain knowledge in deductive databases in order to represent an incomplete and uncertain model of the world, and carry out a reasonable inference of new facts from this model. Expert systems, cognitive robots, mechatronic systems, sensor data, sound and image databases, temporal indeterminacy are only a few of the fields dealing with incomplete and uncertain information. In recent years, disjunctive [1,2,11, etc.] and multivalued [9,10,7,8, etc.], annotated [12,5,6, etc.] logic programming have been recognised as powerful tools for maintenance of such knowledge bases.

In the paper, we shall aim to combine both the disjunctive and multivalued approaches and provide some formalisation of reasoning with incomplete and uncertain information represented by graded implications (disjunctions if $m = 0$) of the form:

$$(h_1 \vee \cdots \vee h_n \longleftarrow b_1 \& \cdots \& b_m, c) \qquad (n > 0, m \geq 0)$$

where h_i, b_j are atoms and c a truth degree from a complete residuated lattice

$$L = (L, \leq, \vee, \wedge, *, \Rightarrow, 0, 1)$$

with a (triangular) t-norm $*$ and its residuum $\Rightarrow$. Fuzzy disjunctive programs, from which we shall infer incomplete and uncertain information, will be viewed as sets of graded implications.

Similar approaches can be found in [9,10,7,8]. The papers [9] and [10] describe minimal model and stable semantics based on t-norms for multivalued disjunctive logic programs; while, [7] and [8] provide probabilistic semantics of minimal, stable, perfect models, and least model states. In addition, this paper introduces a procedural counterpart to the mentioned minimal model semantics for positive programs (without negation).

2   Basic Notions and Notation

2.1   Predicate Fuzzy Logic

Throughout the paper, we shall use common notions of predicate fuzzy logic. Let $\mathcal{L}$ denote a predicate language.

We shall assume that truth values (degrees) of our fuzzy logic constitute a lattice $L = (L, \leq, \vee, \wedge, *, \Rightarrow, 0, 1)$ where

- $L = (L, \leq, \vee, \wedge, 0, 1)$ is a complete lattice;
- the supremum operator $\vee$ and the infimum operator $\wedge$ are infinitely distributive, i.e. for all $K \subseteq L$, $a \in L$,

$$a \vee \bigwedge K = \bigwedge_{k \in K}(a \vee k), \qquad a \wedge \bigvee K = \bigvee_{k \in K}(a \wedge k);$$

- the binary operation $*$ over L is commutative, associative, and non-decreasing in both the arguments; its neutral element is 1, i.e. $1 * a = a$;

Table 1. Truth functions.

connective	truth function
∨	∨
∧	∧
&	*
$x \longleftarrow y$	$y \Rightarrow x$

- the binary operation $\Rightarrow$ over L is non-increasing in the first argument and non-decreasing in the second one; and
- for all $\alpha, \beta, \gamma \in L$, $\alpha * \beta \leq \gamma$ iff $\alpha \leq \beta \Rightarrow \gamma$.

In other words, L is a residuated lattice with the t-norm $*$ and the residuum $\Rightarrow$ which is furthermore complete and infinitely distributive.

The language $\mathcal{L}$ contains the following connectives: $\vee$ (disjunction), $\wedge$ (conjunction), $\&$ (strong conjunction), and $\longleftarrow$ (implication). The truth functions of the connectives are defined in Tab. 1 in the usual way.

An L-interpretation, say $\mathfrak{A}$, for $\mathcal{L}$ is a structure

$$(\mathcal{U}_{\mathfrak{A}}, \{f^{\mathfrak{A}} \mid f \in Func_{\mathcal{L}}\}, \{p^{\mathfrak{A}} \mid p \in Pred_{\mathcal{L}}\})$$

defined as follows:

- $\mathcal{U}_{\mathfrak{A}} \neq \emptyset$ is the universum (domain) of the interpretation $\mathfrak{A}$;
- a function symbol $f \in Func_{\mathcal{L}}$ is interpreted as a function $f^{\mathfrak{A}} : \mathcal{U}_{\mathfrak{A}}^{ar(f)} \longrightarrow \mathcal{U}_{\mathfrak{A}}$;
- a predicate symbol $p \in Pred_{\mathcal{L}}$ is interpreted as a function, an L-fuzzy relation, $p^{\mathfrak{A}} : \mathcal{U}_{\mathfrak{A}}^{ar(p)} \longrightarrow L$.

A variable assignment in $\mathfrak{A}$ is a mapping $Var_{\mathcal{L}} \longrightarrow \mathcal{U}_{\mathfrak{A}}$ assigning each variable an element of the universum $\mathcal{U}_{\mathfrak{A}}$. Let t be a term and ϕ a formula of $\mathcal{L}$. In the standard manner [4], we assign t an element of $\mathcal{U}_{\mathfrak{A}}$ and ϕ a truth value of L in $\mathfrak{A}$ with respect to e, denoted by $\|t\|_e^{\mathfrak{A}}$ and $\|\phi\|_e^{\mathfrak{A}}$, respectively.

By a graded formula of $\mathcal{L}$ we mean a pair (ϕ, c) consisting of a formula ϕ of $\mathcal{L}$ and of a truth degree $c \in L$.

We say that a graded formula (ϕ, c) of $\mathcal{L}$ is true in an L-interpretation $\mathfrak{A}$ for $\mathcal{L}$ with respect to a variable assignment e in $\mathfrak{A}$, written as $\mathfrak{A} \models_e (\phi, c)$, iff $\|\phi\|_e^{\mathfrak{A}} \geq c$. We define the truth value of ϕ, denoted by $\|\phi\|^{\mathfrak{A}}$, as follows:

$$\|\phi\|^{\mathfrak{A}} = \bigwedge \{\|\phi\|_e^{\mathfrak{A}} \mid e \text{ is a variable assignment in } \mathfrak{A}\}.$$

We say that (ϕ, c) is true in $\mathfrak{A}$, written as $\mathfrak{A} \models (\phi, c)$, iff $\|\phi\|^{\mathfrak{A}} \geq c$.

A graded theory of $\mathcal{L}$ is a set of graded formulae of $\mathcal{L}$. An L-interpretation $\mathfrak{A}$ for $\mathcal{L}$ is an L-model of a graded theory T, in symbols $\mathfrak{A} \models T$, iff $\mathfrak{A} \models (\phi, c)$ for all $(\phi, c) \in T$.

We say that a graded formula (ϕ, c) of $\mathcal{L}$ is a fuzzy logical consequence of a graded theory T of $\mathcal{L}$, written as $T \models (\phi, c)$, iff for every L-interpretation $\mathfrak{A}$ for $\mathcal{L}$, $\mathfrak{A} \models T$ implies $\mathfrak{A} \models (\phi, c)$.

2.2   Graded Disjunctions and Implications

Let $D = d_1 \vee \cdots \vee d_n$ be a disjunction of atoms of $\mathcal{L}$. If $n = 0$, the empty disjunction is denoted by $\square$. We put: $\|\square\|_e^{\mathfrak{A}} = \mathbf{0}$, $|D| = n$, and $Atom(D) = \{d_i \mid 1 \leq i \leq n\}$. We say that the disjunction D is a disjunctive factor iff there do not exist indices $i \neq j$, $1 \leq i, j \leq n$, such that $d_i = d_j$. A disjunction of atoms D is called a subdisjunction of another disjunction of atoms D', written as $D \sqsubseteq D'$, iff $Atom(D) \subseteq Atom(D')$. We often say that D subsumes D' or D' is subsumed by D.

Let $C = c_1 \& \cdots \& c_n$ be a conjunction of atoms of $\mathcal{L}$. If $n = 0$, the empty conjunction is denoted by $\top$. We put: $\|\top\|_e^{\mathfrak{A}} = \mathbf{1}$, $|C| = n$, and $Atom(C) = \{c_i \mid 1 \leq i \leq n\}$.

An implication of atoms of $\mathcal{L}$ is an implication of the form $D \longleftarrow C$ where D is a non-empty disjunction of atoms and C a conjunction of atoms of $\mathcal{L}$. An implication of atoms $D \longleftarrow C$ is said to be a tautology iff $Atom(D) \cap Atom(C) \neq \emptyset$.

Let D, D' and C, C' be disjunctions and conjunctions of atoms, respectively. Let A be a set of atoms. We say that

- D is a disjunctive factor of A iff D is a disjunctive factor and $Atom(D) = A$.
- D is a disjunctive factor of D' iff D is a disjunctive factor and $Atom(D) = Atom(D')$;
- C is a conjunctive factor of C' iff $C = c_1 \& \cdots \& c_n$, $n \geq 0$, there exists a permutation π on the set $\{1, \ldots, n\}$, and $C' = c_{\pi(1)} \& \cdots \& c_{\pi(n)}$;
- $D \longleftarrow C$ is an implication factor of $D' \longleftarrow C'$ iff D is a disjunctive factor of D' and C is a conjunctive factor of C'.

For the sake of simplicity, we shall abbreviate the expression *a graded implication (disjunction) of atoms* by *a graded implication (disjunction)*.

A fuzzy disjunctive program of $\mathcal{L}$ is an arbitrary set of graded implications of $\mathcal{L}$.

2.3   Substitutions

We shall use the standard concepts of substitutions [4]:

A substitution ϑ of $\mathcal{L}$ on a finite set $X \subseteq Var_{\mathcal{L}}$ is a mapping $\vartheta : X \longrightarrow Term_{\mathcal{L}}$. The domain of ϑ $dom(\vartheta) = X$ and $range(\vartheta)$ is the set of all the variables of $Var_{\mathcal{L}}$ occurring in the terms $\vartheta(x)$, $x \in X$. The set of all substitutions of $\mathcal{L}$ is denoted as $Subst_{\mathcal{L}}$.

Let ϑ and ϑ' be substitutions of $\mathcal{L}$. ϑ' is a regular extension of ϑ iff

- $dom(\vartheta) \subseteq dom(\vartheta')$, $\vartheta'|_{dom(\vartheta)} = \vartheta$,
- $\vartheta'|_{dom(\vartheta') - dom(\vartheta)}$ is   a   variable   renaming,   and $range(\vartheta) \cap range(\vartheta'|_{dom(\vartheta') - dom(\vartheta)}) = \emptyset$.

Let ϕ be an open formula. An open formula ϕ' is a variant of ϕ iff there exists a variable renaming ρ such that $\phi' = \phi\rho$.

Let S be a finite non-empty set of terms or open formulae of $\mathcal{L}$, or tuples of them. A substitution θ of $\mathcal{L}$ is called a unifier for S iff $S\theta$ is a singleton. A unifier

θ for S is said to be a most general unifier, mgu, for S iff for every unifier ϑ of $\mathcal{L}$ for S, there exists a substitution γ of $\mathcal{L}$ such that $\vartheta|_{vars(S)} = \theta|_{vars(S)} \circ \gamma$.

3   Declarative and Fixpoint Semantics

In this section, we generalise the declarative and fixpoint semantics for fuzzy disjunctive programs proposed in [4]. We proceed from the declarative one:

Definition 1 (Declarative semantics). *Let P be a fuzzy disjunctive program of $\mathcal{L}$.*

$$\mathcal{DS}(P) = \{(D,c) \mid (D,c) \text{ is a graded disjunction of } \mathcal{L} \text{ and } P \models (D,c)\}.$$

Consider a complete lattice $\mathbf{L} = (L, \leq, \vee, \wedge, \mathbf{0}, \mathbf{1})$. A mapping $T : L \longrightarrow L$ is ω-continuous iff for all ω-chains $X \subseteq L$,[1] $T(\bigvee X) = \bigvee \{T(x) \mid x \in X\}$. We denote the α-th iteration power of T on $\mathbf{0}$ as T^{α}. We say that $x \in L$ is a fixpoint of T iff $T(x) = x$. By the fixpoint theorem (Knaster, Tarski), if T is ω-continuous, the least fixpoint of T $lfp(T) = T^{\omega}$.

Denote the set of all graded disjunctions of $\mathcal{L}$ as $Dis_{\mathcal{L}}$. Let P be a fuzzy disjunctive program of $\mathcal{L}$. We now generalise the hyperresolution operator $\mathcal{C}_P$, proposed in [4], which computes over $Dis_{\mathcal{L}}$. To sketch how the operator works, let us consider Fig. 1. An input of $\mathcal{C}_P$ comprises:

- an implication factor $(a_1 \vee \cdots \vee a_n \longleftarrow b_1 \& \cdots \& b_m, cv_0)$ of a variant of a graded implication in the program P and
- disjunctive factors $(c_1^i \vee \cdots \vee c_{u^i}^i \vee D^i, cv_i)$, $i = 1 \ldots m$, of some variants of graded disjunctions in $I \subseteq Dis_{\mathcal{L}}$.

Each disjunctive factor is divided into two parts. The first parts enter into the unification with the body of the input implication; and the rest together with the head of the implication create the output disjunction, which is instantiated by a regular extension θ' of the most general unifier θ of the framed columns. The output is formed of a disjunctive factor of the output disjunction and of the resulting truth value $*_{i=0}^m cv_i$. We next give a more formal treatment:

Definition 2. *Let P be a fuzzy disjunctive program of $\mathcal{L}$ and $I \subseteq Dis_{\mathcal{L}}$.*
   A graded disjunction (D,c) of $\mathcal{L}$ is said to be an immediate consequence of I and P iff there exist

- *$(H \longleftarrow b_1 \& \cdots \& b_m, cv_0)$, an implication factor of a variant of a graded implication in P not being a tautology;*
- *$(C_i \vee D_i, cv_i)$, $i = 1, \ldots, m$, disjunctive factors of some variants of graded disjunctions in I;*
- *the implication and disjunctions do not share variables in common;*
- *$\theta = mgu((\underbrace{b_1 \vee \cdots \vee b_1}_{|C_1|}, \ldots, \underbrace{b_m \vee \cdots \vee b_m}_{|C_m|}), (C_1, \ldots, C_m)),$*

   $dom(\theta) = vars(b_1, \ldots, b_m, C_1, \ldots, C_m);$

---

[1] $X = \{x_i \mid i \in \omega\}$ is simply ordered by $\leq$ that is $i \leq i' \implies x_i \leq x_{i'}$.

Output

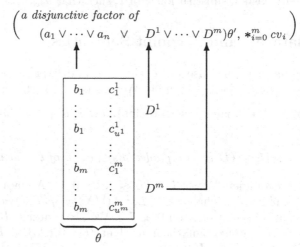

$$\left(\begin{array}{c}\text{a disjunctive factor of}\\ (a_1 \vee \cdots \vee a_n \quad \vee \quad D^1 \vee \cdots \vee D^m)\theta', \, *_{i=0}^m \, cv_i\end{array}\right)$$

Input

$$(a_1 \vee \cdots \vee a_n \longleftarrow b_1 \, \& \cdots \& \, b_m, cv_0)$$
$$(c_1^i \vee \cdots \vee c_{u^i}^i \vee D^i, cv_i)$$

Fig. 1. $\mathcal{C}_P$-operator.

- θ', its regular extension to $vars(H \longleftarrow b_1 \, \& \cdots \& \, b_m) \cup \bigcup_{i=1}^m vars(C_i \vee D_i)$;

such that

- D is a disjunctive factor of $Atom(H\theta') \cup \bigcup_{i=1}^m Atom(D_i\theta')$; and
- $c = *_{i=0}^m \, cv_i$.

The hyperresolution operator $\mathcal{C}_P$ is defined as follows: $\mathcal{C}_P : 2^{Dis\mathcal{L}} \longrightarrow 2^{Dis\mathcal{L}}$, $\mathcal{C}_P(I) = \{(D, c) \,|\, (D, c)$ is an immediate consequence of I and $P\}$.

From Definition 2, we can easily see that $\mathcal{C}_P$ is monotonic and ω-continuous. So, the fixpoint semantics is based on the least fixpoint $\mathcal{C}_P^\omega$ of the hyperresolution operator:

Definition 3 (Fixpoint semantics). *Let P be a fuzzy disjunctive program of $\mathcal{L}$.*

$$\mathcal{F}S(P) = \{(D, c) \,|\, (D, c) \text{ is a graded disjunction of } \mathcal{L},$$
$$c \leq \bigvee \{c' \,|\, (D', c') \in \mathcal{C}_P^\omega, \vartheta \in Subst_\mathcal{L}, D'\vartheta \sqsubseteq D\}\}.$$

We close the section with an equivalence theorem:

Theorem 1 (Equivalence Theorem). *Let P be a fuzzy disjunctive program of $\mathcal{L}$. For any graded disjunction (D, c) of $\mathcal{L}$,*

$$P \models (D, c) \text{ if and only if } c \leq \bigvee \{c' \,|\, (D', c') \in \mathcal{C}_P^\omega, \vartheta \in Subst_\mathcal{L}, D'\vartheta \sqsubseteq D\}.$$

Proof. See `http://www.ii.fmph.uniba.sk/~guller/res04c.ps`.   □

As a consequence, we obtain the coincidence of the proposed semantics:

Corollary 1. *Let P be a fuzzy disjunctive program of $\mathcal{L}$.*

$$\mathcal{DS}(P) = \mathcal{FS}(P).$$

4   Procedural Semantics

A most important aim of a resolution procedure is to provide a computed answer satisfying some suitable logical conditions for a query and program. In case of Horn programs and *SLD*-resolution, a computed answer for a query $\leftarrow G$ (G is a conjunction of atoms) and from a program P, is a substitution ϑ such that $G\vartheta$ is a common logical consequence of P and G. Concerning disjunctive programs, our intention is to compute all disjunctions being common logical consequences of a program and query disjunction. For this reason, we should consider not only the instantions of the query disjunction, but also some 'suitable' disjunctions being subsumed by these instantions. For example, let $P = \{p(f(x)) \vee q(f(x)) \vee s(x)\}$ and a query be of the form $\leftarrow p(y) \vee r(y)$. A computed answer will consist of the substitution $\{y/f(x), x/x\} = mgu(p(y), p(f(x)))$ and of the 'remainder' disjunction $q(f(x)) \vee s(x)$. Indeed, the resulting disjunction $q(f(x)) \vee s(x) \vee (p(y) \vee r(y))\{y/f(x), x/x\}$, formed of the instantiated query disjunction and of the 'remainder' disjunction, is a common logical consequence of P and the query disjunction.

In order to compute such compound answers, *ULSLD*-resolution has been developed for fuzzy disjunctive programs on Boolean lattices in [3]. The abbreviation *ULSLD* stands for **Un**Limited **Sele**ction rule driven **L**inear resolution for **D**isjunctions. To outline how the resolution works, let us consider Fig. 2, where an unlimited fuzzy derivation step is drawn. Let D_q be the selected disjunction by some selection rule R from the old goal. We choose some sequence of atoms $d_1, \ldots, d_{|H|}$ from D_q and unify the disjunction $d_1 \vee \cdots \vee d_{|H|}$ with the subdisjunction H from the head of the input implication. Let θ be some most general unifier. Then the resulting new goal is formed of the new subgoals of the form $D_q \vee b_i$ and the remaining ones D_i, $i \neq q$. All the subgoals are instantiated by a regular extension θ' of θ to the variables appearing in the old goal and the input implication.[2] A new part of an unlimited derivation step is a step remainder disjunction. In our example, the step remainder disjunction $Z\theta'$ consists of the remaining atoms from the head of the input implication instantiated by θ'.

Let P be a fuzzy disjunctive program and $\leftarrow D_1, \ldots, D_k$ a goal (D_i are disjunctions of atoms). A refutation (computed) answer will consist of a refutation answer substitution θ, a refutation remainder $(rd_1, \ldots, rd_k)$ (rd_i are disjunctions of atoms), and moreover of a refutation truth vector $(c_1, \ldots, c_k)$ where $c_i \in \boldsymbol{L}$. Each remainder disjunction rd_i is associated with the corresponding subgoal D_i. By the remainder disjunctions, we supplement the subgoals instantiated with

---

[2] They do not share variables in common.

Old goal

$$\longleftarrow D_1, \ldots, D_{q-1}, \qquad D_q, \qquad D_{q+1}, \ldots, D_k$$

$$\boxed{d_1 \vee \cdots \vee d_{|H|}}$$

$$\theta$$

Implication

$$Z \qquad \vee \qquad \boxed{H} \longleftarrow b_1 \& \cdots \& b_m$$

New goal

$$\longleftarrow (D_1, \ldots, D_{q-1},$$

New subgoals $\begin{cases} D_q \vee b_1, \\ \vdots \quad \vdots \\ D_q \vee b_m, \end{cases}$

$$D_{q+1}, \ldots, D_k)\theta'$$

Step remainder disjunction

$$Z\theta'$$

Fig. 2. Unlimited fuzzy derivation step.

the substitution θ. Also, each truth value c_i is assigned to the corresponding subgoal D_i so that

$$(rd_i \vee D_i\theta, c_i)$$

is a common fuzzy logical consequence of the program P and the graded subgoal (disjunction) (D_i, c_i).

We now generalise $ULSLD$-resolution to the case of residuated lattices:

A finite sequence $D_1, \ldots, D_k$, $k \geq 0$, of disjunctions of atoms of $\mathcal{L}$ is called a goal of $\mathcal{L}$. We denote the goal by $\longleftarrow D_1, \ldots, D_k$. The empty goal is denoted as $\square$.

By a selection rule R we mean a function which returns an index q, $1 \leq q \leq k$, for a non-empty goal $\longleftarrow D_1, \ldots, D_k$, $k \geq 1$. For the empty goal $\square$, we put $R(\square) = 0$.

Definition 4. *For a given graded implication (Im, c) of $\mathcal{L}$, a selection rule R, a goal $\longleftarrow D_1, \ldots, D_k$, $k \geq 1$, of $\mathcal{L}$ where Im and $\longleftarrow D_1, \ldots, D_k$ do not share variables in common; an unlimited fuzzy derivation step is defined as follows:*

Let

- *Im be of the form $Z \vee H \longleftarrow b_1 \& \cdots \& b_m$, $m \geq 0$;*
- *$R(\longleftarrow D_1, \ldots, D_k) = q$, $1 \leq q \leq k$;*
- *there exist $\theta = mgu(H, d_1 \vee \cdots \vee d_{|H|})$ where $d_i \in Atoms(D_q)$, $i = 1, \ldots, |H|$, and $dom(\theta) = vars(H, d_1 \vee \cdots \vee d_{|H|})$;*
- *θ' be a regular extension of θ to $vars(Im, \longleftarrow D_1, \ldots, D_k)$.*

The resulting new goal is of the form

$$\leftarrow (D_1, \ldots, D_{q-1}, D_q \vee b_1, \ldots, D_q \vee b_m, D_{q+1}, \ldots, D_k)\theta';$$

the step remainder disjunction is defined as $Z\theta'$.
The entire derivation step is denoted as:

$$\leftarrow D_1, \ldots, D_k \vdash \xrightarrow{\quad Z\theta' \mid (Im, c), \theta', R \quad}$$
$$\leftarrow (D_1, \ldots, D_{q-1}, D_q \vee b_1, \ldots, D_q \vee b_m, D_{q+1}, \ldots, D_k)\theta'.$$

See Fig. 2.

Definition 5. *Let P be a fuzzy disjunctive program of $\mathcal{L}$, $G_0 = \leftarrow D_1, \ldots, D_k$, $k \geq 0$, be a goal of $\mathcal{L}$, and R a selection rule.*
   A ULSLD-derivation for G_0 of length n, $n \geq 0$, is a finite sequence of goals $G_0, \ldots, G_n$ satisfying

$$for\ 0 \leq i < n:\ G_i \vdash \xrightarrow{\quad rm_i \mid (Im_i, c_i), \theta'_i, R \quad} G_{i+1}$$

where (Im_i, c_i) is an implication factor of a variant of a graded implication in P.

Definition 6. *Let P be a fuzzy disjunctive program of $\mathcal{L}$, $G_0 = \leftarrow D_1, \ldots, D_k$, $k \geq 0$, a goal of $\mathcal{L}$, $G_0, \ldots, G_n$ be a derivation for G_0, and R a selection rule.*
   If $G_n = \square$, the derivation is called a refutation for G_0. In this case, we define

- *the refutation answer substitution ϑ, $dom(\vartheta) = vars(G_0)$;*
- *the refutation remainder, a tuple of disjunctions $(rd_1, \ldots, rd_k)$ where rd_i are of $\mathcal{L}$; and*
- *the refutation truth vector $(cv_1, \ldots, cv_k)$, $cv_i \in \mathbf{L}$,*

by recursion on the length n of the refutation:

- *If $n = 0$, then $G_0 = \square$ and $k = 0$. The refutation answer substitution $\vartheta = \emptyset$, the refutation remainder and truth vector are $()$.*
- *If $n \geq 1$, then $G_0 \neq \square$ and $k \geq 1$.[3] Let $R(\leftarrow D_1, \ldots, D_k) = q$, $1 \leq q \leq k$, and Im_0 be of the form $H \leftarrow b_1 \& \cdots \& b_m$. Then the first derivation step is of the form:*

$$G_0 \vdash \xrightarrow{\quad rm_0 \mid (Im_0, c_0), \theta'_0, R \quad} G_1,$$

$$G_1 = \leftarrow (D_1, \ldots, D_{q-1}, D_q \vee b_1, \ldots, D_q \vee b_m, D_{q+1}, \ldots, D_k)\theta'_0.$$

Denote the rest of the refutation of the length $n - 1$ as follows:

$$G_1 \vdash \xrightarrow{\quad \substack{(rd'_1, \ldots, rd'_{q-1}, rd'_{q_1}, \ldots, rd'_{q_m}, rd'_{q+1}, \ldots, rd'_k) \\ (cv'_1, \ldots, cv'_{q-1}, cv'_{q_1}, \ldots, cv'_{q_m}, cv'_{q+1}, \ldots, cv'_k)} \mid P, \vartheta', R \quad} \square$$

where, by recursion,

---
[3] We may execute derivation steps only for non-empty goals.

- ϑ', $dom(\vartheta') = vars(G_1)$, is the refutation answer substitution;
- $(rd'_1, \ldots, rd'_{q-1}, rd'_{q_1}, \ldots, rd'_{q_m}, rd'_{q+1}, \ldots, rd'_k)$ is the refutation remainder;
- $(cv'_1, \ldots, cv'_{q-1}, cv'_{q_1}, \ldots, cv'_{q_m}, cv'_{q+1}, \ldots, cv'_k)$ the refutation truth vector for G_1.

Denote $V_r = vars(rd'_1, \ldots, rd'_{q-1}, rd'_{q_1}, \ldots, rd'_{q_m}, rd'_{q+1}, \ldots, rd'_k)$. Let ϑ^* be a regular extension of ϑ' to $range(\theta'_0) \supseteq vars(G_1)$ such that

$$range(\vartheta^*|_{range(\theta'_0) - vars(G_1)}) \cap V_r = \emptyset.$$

Then
- $\vartheta = \theta'_0|_{vars(G_0)} \circ \vartheta^*$;
- $(rd_1, \ldots, rd_k) = (rd'_1, \ldots, rd'_{q-1}, rd'_q, rd'_{q+1}, \ldots, rd'_k)$ where rd'_q is a disjunctive factor of $rm_0\vartheta^* \vee rd'_{q_1} \vee \cdots \vee rd'_{q_m}$; and
- $(cv_1, \ldots, cv_k) = (cv'_1, \ldots, cv'_{q-1}, c_0 * *_{i=1}^m cv'_{q_i}, cv'_{q+1}, \ldots, cv'_k)$.

The entire refutation is denoted as

$$G_0 \vdash \frac{\genfrac{}{}{0pt}{}{(rd_1, \ldots, rd_k)}{(cv_1, \ldots, cv_k)} \Big| P, \vartheta, R}{} \longrightarrow \square.$$

In the end, we state that $ULSLD$-resolution is sound and complete to the least fixpoint $\mathcal{C}_P^\omega$.

Theorem 2 (Soundness and Completeness of $ULSLD$). *Let P be a fuzzy disjunctive program of $\mathcal{L}$, R a selection rule, and $G = \longleftarrow D_1, \ldots, D_k$, $k \geq 0$, be a goal of $\mathcal{L}$.*

There exist
- $(C_i \vee Z_i, cv_i)$, $i = 1, \ldots, k$, *disjunctive factors of some variants of graded disjunctions in $\mathcal{C}_P^\omega$, the goal and disjunctions do not share variables in common;*
- $\theta = mgu((d_1^1 \vee \cdots \vee d_{|C_1|}^1, \ldots, d_1^k \vee \cdots \vee d_{|C_k|}^k), (C_1, \ldots, C_k))$, *where $d_j^i \in Atoms(D_i)$ and $dom(\theta) = vars(d_1^1, \ldots, d_{|C_k|}^k, C_1, \ldots, C_k)$;*
- θ', *its regular extension to $vars(G) \cup \bigcup_{i=1}^k vars(C_i \vee Z_i)$;*

if and only if there exists a $ULSLD$-refutation

$$G \vdash \frac{\genfrac{}{}{0pt}{}{(rd_1, \ldots, rd_k)}{(dv_1, \ldots, dv_k)} \Big| P, \vartheta, R}{} \longrightarrow \square;$$

so that $\vartheta = \theta'|_{vars(G)}$, $rd_i = Z_i\theta'$, and $dv_i = cv_i$ for $i = 1, \ldots, k$.

Proof. See http://www.ii.fmph.uniba.sk/~guller/res04c.ps. $\square$

The procedural semantics can be proposed by means of $ULSLD$-refutation:

Definition 7 (Procedural semantics). *Let P be a fuzzy disjunctive program of $\mathcal{L}$.*

$$\mathcal{PS}(P) = \{(D, c) \,|\, (D, c) \text{ is a graded disjunction of } \mathcal{L},$$

$$c \leq \bigvee\{c' \,|\, \longleftarrow D \vdash \frac{\genfrac{}{}{0pt}{}{(\square)}{(c')} \Big| P, id|_{vars(D)}, R}{} \longrightarrow \square \text{ for some } R\}.$$

Using Theorem 2, we conclude:

Corollary 2. *Let P be a fuzzy disjunctive program of $\mathcal{L}$.*

$$\mathcal{D}S(P) = \mathcal{F}S(P) = \mathcal{P}S(P).$$

Thereby we have reached the coincidence of the presented semantics.

5 Conclusions

In the paper, we have proposed a procedural semantics for fuzzy disjunctive programs. We considered the non-idempotent t-norm $*$ as a truth function for the strong conjunction &. The coincidence of the proposed procedural semantics and the generalised declarative, fixpoint semantics from [4] has been reached.

References

1. Brass, S., Dix, J. and Przymusinski, T. Computation of the Semantics for Autoepistemic Beliefs. *Artificial Intelligence*, 112(1-2):104-123, 1999.
2. Brewka, G. and Dix, J. Knowledge Representation with Logic Programs. *Handbook of Phil. Logic, 2nd ed., vol. 6, chap. 6, Oxford University Press*, 2001.
3. Guller, D. Procedural semantics for fuzzy disjunctive programs. *Proc. of the 9th International Conference LPAR, LNAI vol. 2514, Springer-Verlag*, 247-261, Tbilisi, 2002.
4. Guller, D. Model and fixpoint semantics for fuzzy disjunctive programs with weak similarity. In *Innovations in Intelligent Systems*, Abraham, A., Jain, L. C. and Zwaag, B. J. v.d., eds., *Studies in Fuzziness and Soft Computing vol. 140, Springer-Verlag*, to appear, 2004.
5. Kifer, M. and Lozinskii, E. L. A logic for reasoning with inconsistency. *Journal of Automated Reasoning, Kluwer Ac. Publ.*, 9(2):179-215, 1992.
6. Kifer, M. and Subrahmanian, V. S. Theory of the generalized annotated logic programming and its applications. *Journal of Logic Programming*, 12, 335-367, 1992.
7. Lukasiewicz, T. Many-valued disjunctive logic programs with probabilistic semantics. *Proc. of the 5th International Conference LPNMR, LNAI vol. 1730, Springer-Verlag*, 277-289, El Paso, USA, 1999.
8. Lukasiewicz, T. Fixpoint characterizations for many-valued disjunctive logic programs with probabilistic semantics. *Proc. of the 7th International Conference LPNMR, LNAI vol. 2173, Springer-Verlag*, 336-350, 2001.
9. Mateis, C. Extending disjunctive logic programming by t-norms. *Proc. of the 5th International Conference LPNMR, LNAI vol. 1730, Springer-Verlag*, 290-304, El Paso, USA, 1999.
10. Mateis, C. Quantitative disjunctive logic programming: semantics and computation. *AI Communications*, 13(4):225-248, 2000.
11. Minker, J. and Seipel, D. Disjunctive Logic Programming: A Survey and Assessment. In *Computational Logic: Logic Programming and Beyond, Essays in Honour of Robert A. Kowalski, Part I.*, Kakas, A. C. and Sadri, F., eds., *LNCS vol. 2407, Springer-Verlag*, 472-511, 2002.
12. Subrahmanian, V. S. On the semantics of quantitative logic programs. *Proc. of the 4th IEEE Symposium on Logic Programming, Computer Society Press*, 173-182, Washington DC, 1987.

A Proof System and a Decision Procedure for Equality Logic

Olga Tveretina and Hans Zantema

Department of Computer Science, TU Eindhoven, P.O. Box 513,
5600 MB Eindhoven, The Netherlands
{o.tveretina,h.zantema}@tue.nl

Abstract. Equality Logic with uninterpreted functions is used for proving the equivalense or refinement between systems (hardware verification, compiler translation, etc). Current approaches for deciding this type of formulas use a transformation of an equality formula to the propositional one of larger size, and then any standard SAT checker can be applied. We give an approach for deciding satisfiability of equality logic formulas (E-SAT) in conjunctive normal form. Central in our approach is a single proof rule called *ER*. For this single rule we prove soundness and completeness. Based on this rule we propose a complete procedure for E-SAT and prove its correctness. Applying our procedure on a variation of the pigeon hole formula yields a polynomial complexity contrary to earlier approaches to E-SAT.

Keywords: Equality logic, satisfiability, resolution.

1 Introduction

The logic of equality with uninterpreted functions (UIFs) has been proposed for verifying hardware [5]. This type of logic is mainly used for proving equivalence between systems. When verifying equivalence between two formulas it is often possible to abstract away functions replacing them with UIFs. In [1] Ackermann showed that the problem of deciding the validity of the formula in equality logic with UIFs can be reduced to checking satisfiability of formulas without function symbols. These formulas are called equality logic formulas. Bryant et al. [3] presented an alternative approach.

In the past several years various procedures for checking satisfiability of equality logic formulas have been suggested. Barrett at al. [2] proposed a decision procedure based on computing congruence closure in combination with case splitting. Goel et al. [6] and Bryant et al. [4] use transformation of equality logic to propositional logic by adding transitivity constraints and analyzing which transitivity properties may be relevant. In approach called range allocation [8,11] a formula structure is analyzed to define a small domain for each variable. Then a standard BDD based tool is used to check satisfiability of the formula under the domain. Another approach is given in [7]. This approach is based on BDD computation, with some extra rules for dealing with transitivity.

M. Farach-Colton (Ed.): LATIN 2004, LNCS 2976, pp. 530–539, 2004.

The problem of deciding whether a given equality formula is satisfiable or not we call *E-SAT*, similar to the way propositional satisfiability is called SAT. Analogously to propositional logic, every equality logic formula can be transformed to an equality formula in conjunctive normal form (E-CNF) such that the original formula is satisfiable if and only if the E-CNF is satisfiable. Hence we may, and shall concentrate on satisfiability of E-CNFs.

We present a single-rule inference system for equality logic. Our rule, called ER, incorporates some ideas similar to paramodulation and resolution. But it is different from them and from other proof systems for first order logic with equality such as hyperresolution, etc. Special axioms for equality, i.e. reflexivity, symmetry and transitivity axioms, are not required to be added to the original set of clauses. Since the equality substitution mechanism is not applied, ER does not generate new literals. The rule is sound and complete.

A decision procedure is an essential component of formal verification systems. We propose a procedure based on the ER rule. Since checking satisfiability of equality formula is NP-complete it is not expected that a general efficient algorithm exists. As an example we apply this procedure to a formula parameterized by n that is a variation of the well-known pigeon hole formula. It turns out that our procedure can prove unsatisfiability of this formula very efficiently, even quadratic in n, while standard approaches fail to efficiently prove unsatisfiability of this formula.

Our paper is organized as follows. In section 2 we give basic definitions. In Section 3 we present a general theorem globalizing a local commutation criterion for different proof systems. In section 4 we present the ER rule, and we prove its soundness and completeness in section 5. The E-SAT procedure is described in section 6. In section 7 we prove soundness and completeness of the procedure, in section 8 we give an example, and some concluding remarks are in section 9. In this version of the paper some details in proofs are omitted; all full proofs can be found in [13].

2   Basic Definitions and Preliminaries

Any formula in equality logic, as in propositional logic, can be straightforwardly converted to an equivalent E-CNF . In the worst case the size of the result is exponential in the size of the original formula. This can be avoided by adding extra variables. The well-known Tseitin transformation [12] transforms an arbitrary propositional formula to a CNF in such a way that the original formula is satisfiable if and only if the CNF is satisfiable. Both the size of the resulting CNF and the complexity of the transformation procedure are linear in the size of the original formula. In this transformation new propositional symbols are introduced, so applying it directly to equality formulas will yield a CNF in which the atoms are both equalities and propositional variables. However, if we have n propositional variables $p_1, \ldots, p_n$ we can introduce $n + 1$ fresh domain variables $a, x_1, \ldots x_n$ and replace every propositional variable p_i by the equality $x_i \approx a$. In this way satisfiability is easily seen to be maintained. Hence we may and shall restrict ourselfs to satisfiability of E-CNFs.

An E-CNF F is a conjunction of clauses. A *clause* C is a disjunction of literals. The *empty clause* is denoted by $\perp$. A *literal* l is a an *atom* $x \approx y$ or a *negated atom* $x \not\approx y$, where x and y belong to a set of *variables* V. We consider $x \approx y$ and $y \approx x$ as the same atom. Since conjunction and disjunction are associative and commutative, an E-CNF can be viewed as a set of literals sets. We denote by V_F the set of all variables which occur in F and by L_F the set of all literals which occur in F.

A *domain* D is defined to be a non-empty set. For every domain we define an *assignment* as a function $A : V \to D$. For an assignment A we define the corresponding *interpretation* I_A on literals by:

$I_A(x \approx y) = $ true if $A(x) = A(y)$
$I_A(x \approx y) = $ false if $A(x) \neq A(y)$
$I_A(x \not\approx y) = \neg I_A(x \approx y)$

We define $I_A(C) = $ true if $I_A(l) = $ true for some $l \in C$, otherwise $I_A(C) = $ false, and $I_A(F) = $ true if $I_A(C) = $ true for any $C \in F$, otherwise $I_A(F) = $ false.

An E-CNF F is called *satisfiable* if $I_A(F) = $ true for some assignment A, otherwise it is called *unsatisfiable*.

Since $x \approx x$ can be replaced by true, and $x \not\approx x$ can be replaced by false we will consider E-CNFs not containing the literals of the shape $x \approx$ and $x \not\approx x$.

3   Commutation of Proof Systems

In this section we present the desired commutation result for arbitrary proof systems. It will be used in following sections for proving completeness of the ER rule and the decision procedure based on the rule.

Here a proof system may be anything by which new statements, e.g. clauses, may be deduced from existing statements. For such a proof system s we use the notation $F \to_s G$ for $G = F \cup \{C\}$, where C is a statement deduced from F by the proof system s.

For every relation $\to$ we write $\to^*$ for its reflexive transitive closure, i.e., we write $F \to^* G$ if $F_0, \dots, F_n$ exist for $n \geq 0$ satisfying

$$F = F_0 \to F_1 \to \cdots \to F_n = G.$$

We write $F \sqsubseteq G$ if for each $C \in G$ there is $D \in F$ such that $D \subseteq C$.

Definition 1. *Suppose*

- *s is a proof system,*
- *$F \to_s^* F'$ for some sets of statements F and F',*
- *G is an arbitrary set of statements such that $G \sqsubseteq F$.*

We say that s is $\sqsubseteq$-monotonic proof system if there is a set of statements G' such that

- *$G \to_s^* G'$, and*
- *$G' \sqsubseteq F'$.*

Now we will give the formal definition of local commutation.

Definition 2. (*local commutation*) *Let s_1 and s_2 be proof systems and*

$$F \to_{s_1} F' \to_{s_2} F''.$$

We say that a proof system s_1 commutes over a proof system s_2 if for some finite n there exist $G_1, \ldots, G_n, G$ such that

- $F \to_{s_2} G_i$ *for each $i \in \{1, \ldots, n\}$, and*
- $\bigcup_{i=1}^{n} G_i \to_{s_1}^* G$, *where $G \sqsubseteq F''$.*

Local commutation means that for two proof systems s_1 and s_2, doing one step of s_1 followed by one step of s_2 can be simulated by first doing s_2 and then s_1.

Theorem 1. (*global commutation*) *Let s be a union of $\sqsubseteq$-monotonic proof systems $s_1, \ldots, s_n$ such that s_i commute over s_j for each $i > j$. Suppose $F \to_s^* G$ for some F and G; then there are $F_1, \ldots, F_n$ such that*

$$F \to_{s_1}^* F_1 \to_{s_2}^* \cdots \to_{s_n}^* F_n,$$

where $F_n \sqsubseteq G$.

Proof. The proof is given in [13].

4 Resolution for Equality Logic

An important notion in this paper is a *contradictory cycle*.

Definition 3. *A contradictory cycle θ is defined to be a set of literals*

$$x_1 \approx x_2, \ldots, x_{n-1} \approx x_n, x_1 \not\approx x_n,$$

where $x_1, \ldots, x_n$ are distinct variables, and $n > 1$.

When drawing a graph consisting of the variables from an E-CNF F as nodes, equalities contained in F as solid edges, and disequalities contained in F as dashed edges, then a contradictory cycle of F corresponds exactly to a cycle in this graph in which one edge is dashed and all other edges are solid. For a given E-CNF such a graph is easily made, and such cycles are easily established by looking for solid paths from one end of a dashed edge to the other end.

The principle of a contradictory cycle enables us to establish the following resolution-based inference rule for equality logic:

ER: $$\frac{\{x_1 \approx x_2\} \cup C_1, \ldots, \{x_{n-1} \approx x_n\} \cup C_{n-1}, \{x_1 \not\approx x_n\} \cup C_n}{C_1 \cup \cdots \cup C_n}$$

where $x_1 \approx x_2, \ldots, x_{n-1} \approx x_n, x_1 \not\approx x_n$ is a contradictory cycle. The newly obtained clause $C_1 \cup \cdots \cup C_n$ is called an *ER-resolvent*.

Clearly for every contradictory cycle $\theta = \{x_1 \approx x_2, \ldots, x_{n-1} \approx x_n, x_1 \not\approx x_n\}$ we have a corresponding instance of ER.

We write $F \to_{er} F_e$ if $F_e = F \cup \{C\}$, where C is an ER-resolvent. We write $F \to_\theta F_e$ if ER is applied for a fixed contradictory cycle θ, and the transition from F to F_e in this case is called a θ-step.

5   Soundness and Completeness of the ER Rule

The following theorems show that the ER rule is sound and complete. It is well-known that resolution together with paramodulation is complete for FOL with equality [9]. Using this fact and Theorem 1 for global commutation property of proof systems, we will prove completeness of the ER rule. Proving of soundness is straightforward.

Theorem 2. (*soundness*) *Let* $F \to_{er} F_e$. *Then* F *is satisfiable iff* F_e *is satisfiable.*

Proof. ($\Rightarrow$) Suppose F is a set of clauses that is satisfiable by some assignment A. Let the clauses

$$\{l_1\} \cup C_1, \ldots, \{l_n\} \cup C_n$$

be the members of F, where the set $\theta = \{l_1, \ldots, l_n\}$ is a contradictory cycle. Obviously, the set θ is unsatisfiable but any its subset is satisfiable. Now, A does not satisfy at least one literal from θ. Let us say that $I_A(l_1) = false$. Then $I_A(C_1) = true$, as A satisfies $\{l_1\} \cup C_1$. Then A also satisfies $C_1 \cup \cdots \cup C_n$.

($\Leftarrow$) Let F_e be satisfiable. Then F is satisfiable as a subset of a satisfiable set of clauses. $\qquad\qquad\qquad\qquad\qquad\qquad\qquad\qquad\qquad\qquad\qquad\qquad\qquad\square$

As we mentioned above, the combination of resolution and paramodulation is complete for FOL with equality.

Paramodulation:
$$\frac{\{l\} \cup C, \{y \approx z\} \cup D}{\{l[y := z]\} \cup C \cup D}$$

where $l[y := z]$ is a literal l such that y is substituted by z.

For equality logic resolution can be presented as follows.

Resolution:
$$\frac{\{x \approx y\} \cup C, \{x \not\approx y\} \cup D}{C \cup D}$$

Let $F' = F \cup \{C\}$. We wil use the notation $F \to_p F'$ if C was derived from F using paramodulation and $F \to_r F'$ if C was obtained using resolution.

It is easily observed that paramodulation and ER rule are $\sqsubseteq$-monotonic. In order to use completeness of resolution and paramodulation for FOL with equality we have to prove that paramodulation and the ER rule satisfy local commutation property, i.e. Definition 2.

Lemma 1. *Paramodulation commutes over ER.*

Proof. The prove is given in [13].

Theorem 3. *(Completeness) An E-CNF* F *is unsatisfiable iff there is a derivation of the empty clause from* F *using ER.*

Proof. ($\Rightarrow$) Suppose F is unsatisfiable set of clauses. Then there is a derivation of the empty clause from F using both paramodulation and resolution. Since resolution is a particular case of the ER rule for $n = 2$ than by Lemma 1 there is a derivation of the empty clause from F at first applying ER rule and then paramodulation. Since there is no derivation of the empty clause using just paramodulation there is a derivation of the empty clause from F using ER.

($\Leftarrow$) Assume that the empty clause can be derived from the E-CNF by ER. Then by Theorem 2 the original set of clauses is unsatisfiable.   □

6   The E-SAT Procedure

In this section we shall describe the E-SAT procedure and prove its correctness.

Given a nonempty E-CNF containing nonempty clauses the E-SAT procedure forms the set of all contradictory cycles Θ and then repeats the following steps.

- Choose a contradictory cycle $\theta \in \Theta$ and remove θ from Θ.
- Add all possible clauses derived from F by the ER rule over θ.

We give a precise version of the procedure.

```
Procedure E-SAT(F);
    begin
        Θ := ContrCycle(F);
        while (Θ ≠ ∅) do
        begin
            choose θ ∈ Θ;
            Θ := Θ\{θ};
            F := F∪ ER(F,θ);
            if ⊥ ∈ F return(unsatisfiable);
        end
        return(satisfiable);
    end
```

Fig. 1. The E-SAT procedure

In this procedure the function `ContrCycle(F)` forms the set of all possible contradictory cycles. The function $\mathrm{ER}(F, \theta)$ forms the set of clauses derived from F by all possible θ-steps.

The procedure ends when either the empty clause is derived or no contradictory cycle is left. If the empty clause is derived the output the procedure "unsatisfiable". If the empty clause is not derived during the procedure the output is "satisfiable".

The search space of the saturation-based procedures can grow very rapidly. The procedure becomes more efficient when we have criteria to remove redundant clauses from the search space. One can use *subsumption* introduced by Robinson [10] for general resolution.

Example 1. As an example we have taken the formula from [8] raised during the process of translation validation. After abstracting functions and performing the Ackermann reduction the following E-CNF is obtained:

$$F = (x_1 \not\approx x_2 \lor x_3 \not\approx x_4 \lor y_1 \approx y_2) \land (y_1 \not\approx y_3 \lor y_2 \not\approx y_4 \lor z_1 \approx z_2) \land$$
$$y_1 \approx y_3 \land y_2 \approx y_4 \land z_1 \approx z_3 \land z_2 \not\approx z_3.$$

A current approaches for proving unsatisfiability of the formula require to transform it to propositional formula first and then to apply any standard SAT-checker. We will show how unsatisfiability of F can be proven by the E-SAT procedure.

(1) $x_1 \not\approx x_2 \lor x_3 \not\approx x_4 \lor y_1 \approx y_2$
(2) $y_1 \not\approx y_3 \lor y_2 \not\approx y_4 \lor z_1 \approx z_2$
(3) $y_1 \approx y_3$
(4) $y_2 \approx y_4$
(5) $z_1 \approx z_3$
(6) $z_2 \not\approx z_3$

(7) $y_2 \not\approx y_4 \lor z_1 \approx z_2$ (2,3)
(8) $z_1 \approx z_2$ (4,7)
(9) $\perp$ (5,6,8)

One can see that after three ER-steps the empty clause was derived.

7   Soundness and Completeness of the Procedure

We will prove the completeness of the E-SAT procedure.

Let $\theta_1, \ldots, \theta_n$ be all contradictory cycles of an unsatisfiable E-CNF F_0. Based on the completeness of the ER rule we will show that there is a finite sequence $F_1, \ldots, F_n$ such that for each $i \in \{1, \ldots, n\}$ F_i consists of all clauses contained in F_{i-1} and clauses derived from F_{i-1} in one θ_i-step, and F_n contains the empty clause.

At first we will prove the local commutation property.

Lemma 2. *Let s_i be a proof system consisting of θ_i-step for $i \in \{1, 2\}$. Then s_1 commutes over s_2.*

Proof. The proof is given in [13]. □

Theorem 4. *Let $\{\theta_1, \ldots, \theta_n\}$ be the set of all contradictory cycles in F. Let $F \rightarrow_{er}^* G$. Then for some $m \leq n$ there exist $F_1, \ldots, F_m$ such that*

$$F \rightarrow_{\theta_1}^* F_1 \rightarrow_{\theta_2}^* \cdots \rightarrow_{\theta_m}^* F_m,$$

where $F_m \sqsubseteq G$.

Proof. The prove follows immediately from Theorem 1 and Lemma 2.   □

Theorem 5. *Let F and G be E-CNFs, θ be a contradictory cycle, and $G = F \cup ER(F, \theta)$. If $G \to_\theta G'$ then $G \sqsubseteq G'$.*

Proof. The proof is given in [13].

Theorem 6. (*Soundness and completeness of the E-SAT procedure*) *Let F be an E-CNF. Then F is unsatisfiable iff the output of the basic procedure is the empty clause.*

Proof. ($\Rightarrow$) If F is unsatisfiable then by Theorem 3 there is a derivation of the empty clause from F by the ER rule, i.e. $F \to_{er}^* G$, where $\bot \in G$.

Let $\{\theta_1, \ldots, \theta_n\}$ be the set of all contradictory cycles in F. Then by Theorem 4 for some $m \leq n$ there are $F_1, \ldots, F_m$ such that $F \to_{\theta_1}^* F_1 \to_{\theta_2}^* \cdots \to_{\theta_m}^* F_m$, where $F_m \sqsubseteq G$.

Since $\bot \in G$, we obtain that $\bot \in F_m$. By Theorem 5, $F_i = F_{i-1} \cup ER(F_{i-1}, \theta_i)$ for each $i \in \{1, \ldots, m\}$. It implies that the empty clause can be derived by the E-SAT procedure.

($\Leftarrow$) If there is a derivation of the empty clause by the ER rule then F is unsatisfiable by Theorem 2.   □

8   Example

As an example we consider a formula that is related to the pigeon hole formula in proposition calculus. This formula has been studied in [14] too. Just like the pigeon hole formula our formula is parameterized by a number n, it is easily seen to be contradictory by a meta argument, and its shape is the conjunction of two subformulas. In our formula there are $n + 1$ variables $x_1, \ldots, x_n, y$. The first subformula states that all values of $x_1, \ldots, x_n$ are different. The second subformula states that the value of y occurs in every subset of size $n - 1$ of $\{x_1, \ldots, x_n\}$, hence it will occur at least twice in $\{x_1, \ldots, x_n\}$, contradicting the property of the first subformula. Hence the total formula

$$\Phi_n \equiv \bigwedge_{1 \leq i < j \leq n} x_i \not\approx x_j \wedge \bigwedge_{j=1}^{n} (\bigvee_{i \in \{1, \ldots, n\}, i \neq j} x_i \approx y)$$

is unsatisfiable as an E-CNF . It is easy to see that Φ_n is minimally unsatisfiable, hence in any proof of unsatisfiability all $\frac{n(n+1)}{2}$ clauses have to be used. The goal now is to prove unsatisfiability of Φ_n automatically.

We applied the bit vector encoding to this formula, i.e., in this formula every $z \approx w$ is replaced by $\bigwedge_i (z_i \leftrightarrow w_i)$ for i running from 1 to $\lceil \log(n + 1) \rceil$ and then a standard SAT approach is applied for the resulting propositional formula. It turned out that both for a BDD-based approach and a resolution based approach this is a hard job. For $n = 50$ or even lower a combinatory explosion comes up.

However, by applying the approach introduced in this paper proving unsatisfiability of Φ_n can be done polynomial in n. It turns out that all contradictory cycles in Φ_n are of length 3 and are of the shape $\theta_{ij} = \{x_i \approx y, x_j \approx y, x_i \not\approx x_j\}$ for $1 \leq i < j \leq n$; the total number of these contradictory cycles is $\frac{n(n-1)}{2}$. Now we will study the behavior of our procedure consecutively proceeding all these contradictory cycles. Write C_j for the clause $\bigvee_{i \in \{x_1,\dots,x_n\}, i \neq j} x_i \approx y$ for $j = 1, \dots, n$, and write C_{jn} for the clause obtained from C_j by removing $x_n \approx y$, for $j = 1, \dots, n-1$. As a first contradictory cycle choose $\theta_{1,n}$. Then by applying a $\theta_{1,n}$-step on C_1, C_n and $x_1 \not\approx x_n$ we obtain the new clause C_{1n}. Another number of $\theta_{1,n}$-steps is possible, but each of them yields a clause in which C_{1n} is contained, hence will be removed. Also C_1 and C_n are supersets of C_{1n} and will be removed. So after treating this first contradictory cycle apart from the inequalities only the following $n-1$ clauses remain: $C_2, \dots, C_{n-1}, C_{1n}$. As a second contradictory cycle choose $\theta_{2,n}$. Applying a corresponding step on C_2, C_{1n} and $x_2 \not\approx x_n$ yields the new clause C_{2n}. Since this is a subclause of all other clauses generated by $\theta_{2,n}$-steps, and also of C_2, after treating this second contradictory cycle apart from the inequalities only the following $n-1$ clauses remain: $C_3, \dots, C_{n-1}, C_{1n}, C_{2n}$.

This pattern continues after choosing the $n-1$-th contradictory cycle $\theta_{n-1,n}$ apart from the inequalities only the following $n-1$ clauses remain: $C_{1n}, C_{2n}, \dots, C_{n-1,n}$. Since now no equality occurs any more involving the variable x_n, there is no contradictory cycle any more containing the inequalities $x_i \not\approx x_n$ for $i = 1, \dots, n-1$. It turns out that the remaining E-CNF is exactly Φ_{n-1}. Continuing with consecutively choosing $\theta_{1,n-1}$, $\theta_{2,n-1}, \dots$, after $n-2$ steps the remaining E-CNF is exactly Φ_{n-2}. This goes on until the remaining E-CNF is exactly Φ_2 consisting of the three unit clauses $x_1 \approx y$, $x_2 \approx y$, $x_1 \not\approx x_2$ from which the empty clause is derived in one single θ_{12}-step.

We conclude that all $\frac{n(n-1)}{2}$ contradictory cycles were proceeded before the empty clause was derived. Surprisingly, after removing redundant clauses, in intermediate steps the total number of clauses was never greater than the original number of clauses.

9   Concluding Remarks and Further Research

We developed a new rule for reasoning with E-CNFs. We proved its soundness and completeness. We proposed an algorithm based on this rule for satisfiability of E-CNFs, and also proved soundness and completeness of this procedure. So far we have this procedure only in a high-level pseudo-code. Many implementation details have not yet been considered. However, on a theoretical level we analyzed the complexity of our procedure when applied to a particular formula, yielding a polynomial complexity, while standard approaches applied to this formula show up an exponential behavior. This is quite hopeful for our new approach, and as a next step we will implement our procedure and will do experiments with real benchmarks. Our procedure can also be modified, i.e., if removing redundant clauses is applied in a repeated manner.

References

1. ACKERMANN, W. *Solvable cases of the decision problem.* Studies in Logic and the Foundations of Mathematics. North-Holland, Amsterdam, 1954.
2. BARRETT, C. W., DILL, D., AND LEVITT, J. Validity checking for combinations of theories with equality. In *Formal Methods in Computer-Aided Design (FMCAD'96)* (November 1996), M. Srivas and A. Camilleri, Eds., vol. 1166 of *LNCS*, Springer-Verlag, pp. 187–201.
3. BRYANT, R., GERMAN, S., , AND VELEV, M. Processor verification using efficient reductions of the logic of uninterpreted functions to propositional log. *ACM Transactions on Computational Logic 2*, 1 (January 2001), 93–134.
4. BRYANT, R., AND VELEV, M. Boolean satisfiability with transitivity constraints. *ACM Transactions on Computational Logic 3*, 4 (October 2002), 604–627.
5. BURCH, J., AND DILL, D. Automated verification of pipelined microprocesoor control. In *Computer-Aided Verification (CAV'94)* (June 1994), D. Dill, Ed., vol. 818 of *LNCS*, Springer-Verlag, pp. 68–80.
6. GOEL, A., SAJID, K., ZHOU, H., AZIZ, A., AND SINGHAL, V. BDD based procedures for a theory of equality with uninterpreted functions. In *Computer-Aided Verification (CAV'98)* (1998), A. J. Hu and M. Y. Vardi, Eds., vol. 1427 of *LNCS*, Springer-Verlag, pp. 244–255.
7. GROOTE, J., AND VAN DE POL, J. Equational binary decision diagrams. In *Logic for Programming and Reasoning (LPAR'2000)* (2000), M. Parigot and A. Voronkov, Eds., vol. 1955 of *LNAI*, pp. 161–178.
8. PNUELI, A., RODEH, Y., SHTRICHMAN, O., AND SIEGEL, M. Deciding equality formulas by small domains instantiations. In *Computer Aided Verification (CAV'99)* (1999), vol. 1633 of *LNCS*, Springer-Verlag, pp. 455–469.
9. ROBINSON, G., AND WOS, L. Paramodulation and theorem-proving in first-order theories with equality. *Machine inteligence 4* (1969), 135–150.
10. ROBINSON, J. A machine-oriented logic based on the resolution principle. *Journal of the ACM 12(1)* (1965), 23–41.
11. RODEH, Y., AND SHTRICHMAN, O. Finite instantiations in equivalence logic with uninterpreted functions. In *Computer Aided Verification (CAV'01)* (July 2001), vol. 2102 of *LNCS*, Springer-Verlag, pp. 144–154.
12. TSEITIN, G. On the complexity of derivation in propositional calculus. In *Studies in Constructive Mathematics and Mathematical Logic, Part 2.* Consultant Bureau, New York-London, 1968, pp. 115–125.
13. TVERETINA, O., AND ZANTEMA, H. A proof system and a decision procedure for equality logic. Tech. rep., 2003. http://www.tue.nl/bib/indexen.html.
14. ZANTEMA, H., AND GROOTE, J. F. Transforming equality logic to propositional logic. In *Proceedings of 4th International Workshop on First-Order Theorem Proving (FTP'03)* (2003), vol. 86 of *Electronic Notes in Theoretical Computer Science.*

Approximating the Expressive Power of Logics in Finite Models

Argimiro Arratia[1]* and Carlos E. Ortiz[2]

[1] Departamento de Matemáticas, Universidad Simón Bolívar, Venezuela, and Depto. de Matemáticas Aplicadas y Computación, Universidad de Valladolid, España
arratia@mac.cie.uva.es,
[2] Department of Mathematics and Computer Science, Arcadia University, 450 S. Easton Road, Glenside, PA 19038-3295, U.S.A. ortiz@arcadia.edu

Abstract. We present a *probability logic* (essentially a first order language extended with quantifiers that count the fraction of elements in a model that satisfy a first order formula) which, on the one hand, captures uniform circuit classes such as AC^0 and TC^0 over *arithmetic models*, namely, finite structures with linear order and arithmetic relations, and, on the other hand, their semantics, with respect to our arithmetic models, can be closely approximated by giving interpretations of their formulas on finite structures where all relations (including the order) are restricted to be "modular" (i.e. to act subject to an integer modulo). In order to give a precise measure of the proximity between satisfaction of a formula in an arithmetic model and satisfaction of the same formula in the "approximate" model, we define the *approximate formulas* and work on a notion of approximate truth. We also indicate how to enhance the expressive power of our probability logic in order to capture polynomial time decidable queries,
There are various motivations for this work. As of today, there is not known logical description of any computational complexity class below **NP** which does not requires a built-in linear order. Also, it is widely recognized that many model theoretic techniques for showing definability in logics on finite structures become almost useless when order is present. Hence, if we want to obtain significant lower bound results in computational complexity via the logical description we ought to find ways of by-passing the ordering restriction. With this work we take steps towards understanding how well can we approximate, without a true order, the expressive power of logics that capture complexity classes on ordered structures.

1 Introduction

The logical description of many computational complexity classes is based on the fact that the possible domains of interpretations must be at least partially ordered. This is certainly the case for logics meant for describing complexity classes

---

* Supported by grant *Ramón y Cajal 2003* from Ministerio Ciencia y Tecnología, España

M. Farach-Colton (Ed.): LATIN 2004, LNCS 2976, pp. 540–556, 2004.

below **NP**, for it is still unknown whether such classes can be described without any order, and it is further believed that is not the case (further comments in [5] and see also [4]). However, a negative aspect of describing low complexity classes by logics with built-in order is that model theoretic techniques for showing inexpressibility, such as Ehrenfeucht–Fraïssé games and its variations, becomes almost useless; thus, in turn, hopeless for leading into significant complexity lower bounds. (For an illustration of how difficult is to play Ehrenfeucht–Fraïssé games on ordered structures see Section 6.6 of [5].)

This dichotomy with the order had led researchers into exploring ways of keeping some order in the models for various forms of extensions of first order logic, and yet obtain some significant lower bound results (for example, see [3] and [7]). The results presented in this paper are inscribed in that line of research. We introduce a probability logic $\mathcal{LP}$, which is, essentially, first order logic extended with quantifiers that count the fraction of elements in a model that satisfy a first order formula. Our definition of the logic $\mathcal{LP}$ is inspired on the probability logic of Keisler (see [6]), who conceived it as a logic appropriate for his investigations on *probability hyperfinite spaces*, or infinite structures suitable for approximating large finite phenomena of applied mathematics. In order to suit our need of this logic for describing computability problems, we restrict our use of relation symbols to a finite set and mainly of the arithmetic type: addition, multiplication and order. With this ability to approximately count and in the presence of built-in order, addition, and multiplication, fragments of this $\mathcal{LP}$ logic are capable of fully describing circuit classes such as AC^0 and TC^0, since they coincide with known logics that capture these computational complexity classes, for example, first order logic extended with threshold quantifiers. Following our programme of studying possible ways of reducing the scope of the order and other arithmetic relations within our models, we group in the same set of witnesses of a formula all those elements that are congruent modulo the value of a sublinear function F, and define the concept of an F-modular approximation of a finite structure $\mathcal{A}$. The F-modular approximation of $\mathcal{A}$ thus obtained do not have the order built-in but just approximations of it, and subject to these interpretations we do get separation results among fragments of the corresponding logic $\mathcal{LP}_F$, for a particular family of (sublinear) functions F.

Having satisfied our goal of obtaining inexpressibility results within our probability logic under a weaker interpretation of the atomic symbols, we wonder how to translate that result to an inexpressibility of the same query (or similar query) in the logic with the unrestricted interpretation of symbols (e.g. full linear order). As a partial answer to this question we introduce the notion of approximate formulas and through them we establish a bridge between satisfaction in the structures with natural interpretations of the symbols and their corresponding F-modular approximation. In the last section of the paper we show how to extend this probability logic and approximations to capture **P**.

2   Logic of Probability Quantifiers

We work with finite vocabularies and finite models. A vocabulary or signature τ is a set of relation symbols and constant symbols. The models for τ will be denoted by $\mathcal{A}_m$, $\mathcal{B}_n$, $\mathcal{C}_k$, etc. where the subscripts refer to the cardinality of the model. A logic over the vocabulary τ will be denoted $\mathcal{L}(\tau)$. In particular $FO(\tau)$ is the set of first order formulas over τ (or τ-formulas). The logic we are mainly concerned in this paper is the logic of probability quantifiers which we define below. Given a natural number m and a set $C \subseteq \{0, \ldots, m-1\}$ we can define the natural probability $\mu_m(C)$ as just the cardinality of C divided by m. Likewise, for $s > 0$, we can define, for every set $C \subseteq \{0, \ldots, m-1\}^s$ the natural probability $\mu_m^s(C)$ as the cardinality of C divided by m^s.

Definition 1. *For a vocabulary τ, we define the logic of probability quantifiers (or probability logic) over τ, as the set of formulas $\mathcal{LP}(\tau)$ formed as follows:*

Atomic formulas. *Formulas of the form $R(\overline{x}, \overline{c})$, where R is a relation symbol in τ, $\overline{x}$ is a vector of variables, $\overline{c}$ is a vector of constants from τ, are in $\mathcal{LP}(\tau)$.*
Conjunction. *If $\phi_1(\overline{x}), \phi_2(\overline{x}) \in \mathcal{LP}(\tau)$ then $\phi_1(\overline{x}) \wedge \phi_2(\overline{x})$ is in $\mathcal{LP}(\tau)$.*
Negation. *If $\phi(\overline{x}) \in \mathcal{LP}(\tau)$ then $\neg\phi(\overline{x}) \in \mathcal{LP}(\tau)$.*
Existential quantification. *If $\phi(\overline{x}, z) \in \mathcal{LP}(\tau)$ and z is a variable not appearing in $\overline{x}$, then $\exists z\phi(\overline{x}, z) \in \mathcal{LP}(\tau)$.*
Probability quantification. *Fix a rational number r, $0 \le r < 1$. If $\phi(\overline{x}, z) \in \mathcal{LP}(\tau)$ and z is a variable not appearing in $\overline{x}$, then*

$$(P(z) > r)\phi(\overline{x}, z) \, and \, (P(z) \ge r)\phi(\overline{x}, z) \, are \, in \, \mathcal{LP}(\tau).$$

We define the following abbreviations: $(P(z) < r)\phi(\overline{x}, z)$ stands for $\neg(P(z) \ge r)\phi(\overline{x}, z)$, and $(P(z) \le r)\phi(\overline{x}, z)$ stands for $\neg(P(z) > r)\phi(\overline{x}, z)$. Likewise $\forall z\phi(\overline{x}, z)$ stands for $\neg\exists z\neg\phi(\overline{x}, z)$ and $\phi \vee \psi$ stands for $\neg(\neg\phi \wedge \neg\psi)$.

We define the interpretation of the formulas in $\mathcal{LP}(\tau)$ in a finite structure $\mathcal{B}_m$ ($m \in \mathbb{N}$) by induction in formulas, with the usual interpretations for conjunction, negation and existential quantifier. The interpretation for a formula $(P(z) > r)\phi(\overline{x}, z)$ in $\mathcal{B}_m$ is as follows:

$$\mathcal{B}_m \models (P(z) > r)\phi(\overline{a}, z) \text{ iff } \mu_m(\{z < m : \mathcal{B}_m \models \phi(\overline{a}, z)\}) > r$$

Likewise, the interpretation of the formula $(P(z) \ge r)\phi(\overline{x}, z)$ is as follows:

$$\mathcal{B}_m \models (P(z) \ge r)\phi(\overline{a}, z) \text{ iff } \mu_m(\{z < m : \mathcal{B}_m \models \phi(\overline{a}, z)\}) \ge r$$

Observe that under this interpretation, $\neg(P(z) \ge r)\phi(\overline{x}, z)$ is equivalent to $(P(z) > 1 - r)\neg\phi(\overline{x}, z)$, and $\neg(P(z) > r)\phi(\overline{x}, z)$ is equivalent to $(P(z) \ge 1 - r)\neg\phi(\overline{x}, z)$.

By $\mathcal{LP}$ we denote the union of all probability logics $\mathcal{LP}(\tau)$ taken over all finite vocabularies. We shall also deal with the following fragments of $\mathcal{LP}$:

Definition 2. *Let τ be a finite vocabulary. Let $r_1, r_2, \ldots r_k$ be distinct natural numbers. By $\mathcal{LP}(\tau)[r_1, r_2, \ldots, r_k]$ we understand the smallest subset of $\mathcal{LP}(\tau)$ containing the atomic formulas and closed under conjunction, negation, existential quantification and the probability quantifiers $P(z) > q_{ij}/r_i$, $P(z) \geq q_{ij}/r_i$ where $i \leq k$ and q_{ij} are natural numbers such that $0 \leq q_{ij} < r_i$.*

We had in mind using this type of logic for describing computational properties and for that matter we restrict semantics to finite models and also the kind of relation symbols for building our formulas. In general we restrict our symbols to be numerical (in a sense as explained in [5]), and in particular we fix throughout this paper the vocabulary $\Gamma = \{\oplus, \otimes, \lhd, 0, 1\}$, where $\oplus, \otimes$ are ternary relation symbols and $\lhd$ is a binary relation symbol and 0 and 1 are constant symbols. Furthermore, we fix a generic vocabulary Γ^+ that contains Γ and a set $\{R_s\}_{s=1}^k$ of other numerical relation symbols and a set $\{c_w\}_{w=1}^u$ of other constant symbols. We define the **arithmetic** structures over Γ^+ as the finite structures $\mathcal{A}_m$ of the form: $\mathcal{A}_m = \langle \{0, 1, \ldots m-1\}, \oplus, \otimes, \lhd, \{R_s\}_{s=1}^k, \{c_w\}_{w=1}^u, 0, 1 \rangle$, where the relation symbols $\oplus, \otimes, \lhd$ are interpreted as the usual addition, multiplication and order in the set $\{0, 1, \ldots m-1\}$.

We will refer to the probability logic restricted to finite structures that are arithmetic as $\mathcal{LP}_A$. The following examples show that the logic $\mathcal{LP}_A$ contains fragments that are relevant to Descriptive Complexity Theory.

Example 1. Let $FO(\Gamma)$ be the first order logic over Γ and consider the interpretation of the symbols in Γ as natural addition, multiplication and linear order. It is shown in [1] (see also [5]) that this logic captures the complexity class DLOGTIME–uniform AC^0, where AC^0 is defined as the class of problems accepted by polynomial size, constant depth circuits with unbounded fan-in. This logic corresponds to the smallest subset of $\mathcal{LP}_A$ that contains the atomic Γ-formulas and is closed under $\wedge, \neg$ and $\exists z$.

Example 2. Let $FO(\Gamma) + M$ be the first order logic over Γ with the interpretations of the symbols in Γ fixed as in the previous example, extended with the majority quantifier M which is defined as follows: If $\phi(\overline{x}, z)$ is a formula with one free variable z, then $(Mz)\phi(\overline{a}, z)$ is a well defined sentence, which is true if and only if $\phi(\overline{a}, z)$ is true for more than half of the possible values for z. It is shown in [1] (see also [5]) that this logic captures the complexity class DLOGTIME-uniform TC^0, where TC^0 is the class of problems accepted by circuits of polynomial size, constant depth and unbounded fan-in threshold gates (gates which counts its Boolean inputs of value 1 and compares the total with some prefixed number to determine its output). Note that this logic is the fragment of $\mathcal{LP}_A$ that contains the atomic Γ-formulas and is closed under $\wedge, \neg, \exists z$ and the quantifier $P(z) > \frac{1}{2}$; that is $\mathcal{LP}(\Gamma)[2]$.

Our purpose is to approximate the expressive power of arithmetic relations occurring naturally in finite model theory by arithmetic relations that are "weaker" yet perform better under definability tools such as Ehrenfeucht–Fraissé games. Our choice of candidates for relations to approximate the natural arithmetic relations are those that are "modular" in a number theoretic sense.

By $a \equiv_q b$ we mean that the number a is congruent to the number b modulo q. Furthermore, given $\bar{a} = (a_1, \ldots, a_k)$ and $\bar{b} = (b_1, \ldots, b_k)$ two vectors of natural numbers of equal length, we write $\bar{a} \equiv_q \bar{b}$ as an abbreviation of $a_1 \equiv_q b_1$, $a_2 \equiv_q b_2$, $\ldots$, $a_k \equiv_q b_k$. Also, whenever we write $\bar{a} < m$ for some number m, we mean that $a_1 < m, \ldots, a_k < m$.

We understand that a function $F : \mathbb{N} \to \mathbb{N}$ is **sublinear**, if for every natural $m > 0$, $0 < F(m) \leq m$

Definition 3. *Fix a sublinear function F, a formula $\theta(\bar{x}) \in \mathcal{LP}(\Gamma^+)$ and a Γ^+-model $\mathcal{B}_m$. The formula $\theta(\bar{x})$ is F-**modular** in $\mathcal{B}_m$ iff the following condition holds:*

– *For every $\bar{a}, \bar{b} < m$, if $\bar{a} \equiv_{F(m)} \bar{b}$ then $(\mathcal{B}_m \models \theta(\bar{a})$ iff $\mathcal{B}_m \models \theta(\bar{b}))$.*

We will say that a collection of formulas $\{\theta_i(x)\}_{i=1}^r \subseteq \mathcal{LP}(\Gamma^+)$ is F-modular in $\mathcal{B}_m$ iff every formula θ_i is F-modular in $\mathcal{B}_m$.

The next lemma states that modularity is preserved by the logical operations and quantification of $\mathcal{LP}(\Gamma^+)$. The proof is an easy induction on formulas.

Lemma 1. *If the collection of atomic Γ^+-formulas is F-modular for a structure $\mathcal{B}_m$ then every formula in $\mathcal{LP}(\Gamma^+)$ is F-modular for $\mathcal{B}_m$.* $\square$

The direct consequence of the above lemma is that the F-modularity of the formulas in $\mathcal{LP}(\Gamma^+)$ in a model $\mathcal{B}_m$ depends only on the modularity of the interpretation of the relation symbols in $\mathcal{B}_m$. Because of this fact, every model where all the interpretations of the relation symbols are F-modular will be called an F-**modular structure**.

Remark 1. For every natural numbers e and $f > 0$ we understand $[e]_f$ to be the reminder of dividing e by f. For any vector of natural numbers $\bar{a} = (a_1, a_2, \ldots, a_d)$, we understand by $[a]_f$ the vector $([a_1]_f, [a_2]_f, \ldots, [a_d]_f)$.

Definition 4. *Fix a sublinear function F and an arithmetic structure $\mathcal{A}_m$. The F-modular approximation of $\mathcal{A}_m$ is a structure*

$$\mathcal{A}_m^F = \langle \{0, 1, \ldots, m-1\}, \oplus, \otimes, \lhd, \{R_s\}_{s=1}^k, \{c_w\}_{w=1}^u, 0, 1 \rangle$$

such that for every $a, b, c, a_1, \ldots, a_r < m$,

– $\mathcal{A}_m^F \models \oplus(a, b, c)$ *iff* $\mathcal{A}_m \models \oplus([a]_{F(m)}, [b]_{F(m)}, [c]_{F(m)})$.
– $\mathcal{A}_m^F \models \otimes(a, b, c)$ *iff* $\mathcal{A}_m \models \otimes([a]_{F(m)}, [b]_{F(m)}, [c]_{F(m)})$.
– $\mathcal{A}_m^F \models \lhd(a, b)$ *iff* $\mathcal{A}_m \models \lhd([a]_{F(m)}, [b]_{F(m)})$.
– $\mathcal{A}_m^F \models R_s(a_1, \ldots, a_r)$ *iff* $\mathcal{A}_m \models R_s([a_1]_{F(m)}, \ldots, [a_r]_{F(m)})$.

It is easy to see that for every arithmetic structure $\mathcal{A}_m$, the structure $\mathcal{A}_m^F$ is F-modular. We also remark that for every s, for every relation symbol R_s, the set $\{(a_1, \ldots, a_r) < m : \mathcal{A}_m^F \models R_s^F(a_1, \ldots, a_r)\}$ and the set $\{(a_1, \ldots, a_r) < m : \mathcal{A}_m \models R_s(a_1, \ldots, a_r)\}$ coincide in the set $\{(a_1, \ldots, a_r) : a_1, \ldots, a_r < F(m)\}$. These two remarks justify the name of F-modular approximation of $\mathcal{A}_m$.

3   Modular Logics

Here is an example of a class of sublinear functions with some nice properties. These functions will play an important role in the rest of this paper.

Example 3. Fix a natural $n > 0$. For every natural number m, let t, r be the unique natural numbers such that $m = tn + r$ and $0 \leq r < n$. Define the function $g_n : \mathbb{N} \mapsto \mathbb{N}$ by
$$g_n(m) = \begin{cases} tn & \text{if } m \geq n \\ 1 & \text{otherwise} \end{cases}.$$

For every n, g_n is sublinear. Furthermore, $\lim_{m \to \infty} \frac{g_n(m)}{m} = 1$. Also, for every n, g_n is first order definable in the following sense: there exists a formula $\theta_n(x) \in FO$ with built-in order, addition and multiplication such that for every arithmetic model $\mathcal{A}_m$, for any $a < m$, $\mathcal{A}_m \models \theta_n(a)$ iff $a+1 = g_n(m)$. Here is why: Note first that in every $\mathcal{A}_m$ it is possible to capture the property that x is the maximal element with a formula $Max(x) \in FO(\{\oplus, \otimes, \triangleleft, 0, 1\})$ that says that $\neg \exists z \oplus (x, 1, z)$. Likewise, we can say that "the size of the model $= tn+r$", with $r < n < $ (size of the model), by a formula $DIVSIZE(t, n, r) \in FO(\{\oplus, \otimes, \triangleleft, 0, 1\})$ that says that there exists x such that $Max(x)$ and

$$0 < r < n \text{ and } x = tn + (r - 1) \text{ or}$$
$$r = 0 \text{ and } x = (t - 1)n + (n - 1).$$

It follows then that the statement $g_n(\text{size of model}) = h + 1$, for $n < m$, is definable in the models $\mathcal{A}_m$ by a formula in $FO(\Gamma)$ that says that:

$$G(h, n) := \exists t, r, z (DIVSIZE(t, n, r) \wedge$$
$$[(\oplus(h, 1, z) \wedge \neg \oplus (0, 0, r) \wedge \otimes(t, n, z)) \vee (\oplus(0, 0, r) \wedge Max(h))]$$

For the case when $n = m$, we know that $g_n(m) = m$ in which case we can define h as $Max(h)$. Finally, if $m < n$ we know that $g_n(m) = 1$ and we can define $h = 0$.

Recall that we refer to the probability logic restricted to finite structures that are arithmetic as $\mathcal{LP}_A$. The related logic restricted to modular approximations of arithmetic structures is formalise below.

Definition 5. *We denote by $\mathcal{LP}_F$ the probability logic restricted to structures that are F-modular approximations of arithmetic structures, for F a sublinear function. Likewise, by FO_F we understand the smallest fragment of $\mathcal{LP}_F$ that contains the atomic formulas and is closed under $\exists, \neg$ and $\wedge$. Similarly, we define $\mathcal{LP}_F[r_1, \ldots, r_k]$ as the smallest fragment of $\mathcal{LP}_F$ that is closed under $\exists, \neg, \wedge$ and $(P(z) \geq q_{ij}/r_i)$ and $(P(z) > q_{ij}/r_i)$ for $i \leq k$ and natural numbers $0 \leq q_{ij} < r_i$. In particular, we define the **modular** probability logic*

$$\mathcal{LP}_{MOD} = \bigcup_{n \in \mathbb{N}} \mathcal{LP}_{g_n}.$$

Likewise, we define

$$FO_{MOD} = \bigcup_{n \in \mathbb{N}} FO_{g_n} \quad \text{and} \quad \mathcal{LP}_{MOD}[r_1, \ldots, r_k] = \bigcup_{n \in \mathbb{N}} \mathcal{LP}_{g_n}[r_1, \ldots, r_k].$$

Note that the logics $FO_{MOD}, \mathcal{LP}_{MOD}[r_1, \ldots, r_k], \mathcal{LP}_{MOD}$ do not have built-in order nor built-in addition nor built-in multiplication. Instead, for each n, $FO_{g_n}, \mathcal{LP}_{g_n}[r_1, \ldots, r_k], \mathcal{LP}_{g_n}$ have built-in g_n-modular approximations of the order, addition and multiplication.

We now show that the expressive power of $\mathcal{LP}_{MOD}$ (respectively $\mathcal{LP}_{MOD}[r_1, \ldots, r_k], FO_{MOD}$) is contained in the expressive power of $\mathcal{LP}_A$ (respectively $\mathcal{LP}_A[r_1, \ldots, r_k], FO$). Before proceeding, however, we need to clarify the meaning of a boolean query in the context of modular logics.

Definition 6. *Fix a vocabulary* $\Gamma^+ = \Gamma \cup \{R_s\}_{s=1}^k \cup \{c_w\}_{w=1}^u$. *A* **boolean query** *for the modular logic* $\mathcal{LP}_{MOD}(\Gamma^+)$ *is a map* $I : \{\mathcal{A}_m^{g_n} : m, n \in \mathbb{N}\} \to \{0,1\}$, *with the additional property that for every* $1 < n_1 < n_2$, *for every* $m > n_2$, $I(\mathcal{A}_m^{g_{n_1}}) = I(\mathcal{A}_m^{g_{n_2}})$. *We say that a boolean query is expressible in* $\mathcal{LP}_{MOD}(\Gamma^+)$ *(respectively* $FO_{MOD}(\Gamma^+)$*) iff there exists a sentence* $\theta \in \mathcal{LP}(\Gamma^+)$ *(respectively* FO*) such that for* $n \in \mathbb{N}$, *for every arithmetic structure* $\mathcal{A}_m$ *with* $m > n$, $I(\mathcal{A}_m^{g_n}) = 1$ *iff* $\mathcal{A}_m^{g_n} \models \theta$.

The idea behind the above definition of a boolean query for $\mathcal{LP}_{MOD}$ is to capture the notion that a query does not depend on the built-in order or arithmetic predicates, instead it depends on notions that remain constant for all the approximations $\mathcal{A}_m^{g_n}$. For the rest of this section we fix again a vocabulary of the form $\Gamma^+ = \Gamma \cup \{R_s\}_{s=1}^k \cup \{c_w\}_{w=1}^u$, where R_s and c_w are numeric relations and constants.

Lemma 2. *There exist formulas* $ADD(x_1, x_2, x_3, y), PRODUCT(x_1, x_2, x_3, y),$ $ORDER(x_1, x_2, y)$ *and for every* s, *formulas* $PRED_s(\bar{x}, y)$ *in* $FO(\Gamma^+)$, *such that for natural* n, *for every arithmetic structure* $\mathcal{A}_m$ *with* $m > n$,

- *for every* $a, b, c < m$, $\mathcal{A}_m^{g_n} \models \oplus(a, b, c)$ *iff* $\mathcal{A}_m \models ADD(a, b, c, n)$.
- *For every* $a, b, c < m$, $\mathcal{A}_m^{g_n} \models \otimes(a, b, c)$ *iff* $\mathcal{A}_m \models PRODUCT(a, b, c, n)$.
- *For every* $a, b, c < m$, $\mathcal{A}_m^{g_n} \models \triangleleft(a, b)$ *iff* $\mathcal{A}_m \models ORDER(a, b, n)$.
- *For every index* s *and every* $\bar{a} < m$, $\mathcal{A}_m^{g_n} \models R_s(\bar{a})$ *iff* $\mathcal{A}_m \models PRED_s(\bar{a}, n)$.
  □

The previous lemma allow us to translate modular interpretations to natural interpretations.

Corollary 1. *Let* B *be a boolean query expressible in* $\mathcal{LP}_{MOD}$. *Then this query is also expressible in* $\mathcal{LP}_A$. *Likewise, any boolean query expressible in* FO_{MOD} *(respectively* $\mathcal{LP}_{MOD}[r_1, \ldots, r_k]$*) is also expressible in* FO_A *(respectively* $\mathcal{LP}_A[r_1, \ldots, r_k]$*).* □

The logic $\mathcal{LP}_{MOD}$ is capable of expressing queries as the evenness of the cardinality of a set, as we show in the next example.

Example 4. We claim that there exists a sentence θ_2 in $\mathcal{LP}(\{\oplus, \otimes, \lhd, 0, 1\})$ such that for all n, for every arithmetic structure $\mathcal{A}_m$, with $m > n$,

$$\mathcal{A}_m^{g_n} \models \theta_2 \text{ iff } m \text{ is even}$$

To prove this, note first that for every naturals $m > n > 1$ and every c such that $g_n(m) > c > m - g_n(m)$,

$$\{y < m : \mathcal{A}_m^{g_n} \models c \lhd y \vee \oplus(0, y, c)\} = \{y < m : c \le y \le g_n(m) - 1\}$$

and this implies that for every c such that $g_n(m) > c > m - g_n(m)$,

$$\mu_m(\{y < m : \mathcal{A}_m^{g_n} \models c \lhd y \vee \oplus(0, y, c)\}) = \frac{g_n(m) - c}{m}. \tag{1}$$

Fix now a natural n. Then there exists a natural k such that for every $m > k$, $g_n(m) > (3/4)m$ (since $\lim_{m \to \infty} \frac{g_n(m)}{m} = 1$). Let $m > k$ and consider the formula

$$\theta_2 := \exists x[(P(y) \ge 1/2)(x \lhd y \vee \oplus(0, x, y)) \wedge (P(y) \le 1/2)(x \lhd y \vee \oplus(0, x, y))]$$

We claim that for $m > n$, $\mathcal{A}_m^{g_n} \models \theta_2$ iff m is even. One direction goes as follows: If $m = 2s$ and $g_n(m) > (3/4)m$ then $m - g_n(m) < \frac{1}{2}s$. Taking c as $s - m + g_n(m)$, we have by equation (1) that

$$\mu_m(\{y < m : \mathcal{A}_m^{g_n} \models (c \lhd y \vee \oplus(0, y, c)\}) = \frac{m - s}{m} = \frac{1}{2}$$

For the other direction, suppose that there exists a $d < m$ such that

$$\mu_m(\{y < m : \mathcal{A}_m^{g_n} \models d \lhd y \vee \oplus(0, y, d)\}) = \frac{1}{2}$$

From the fact that $\mathcal{A}_m^{g_n}$ is g_n-modular we obtain that there exists an $a < g_n(m)$ such that

$$\mu_m(\{y < m : \mathcal{A}_m^{g_n} \models a \lhd y \vee \oplus(0, y, a)\}) = \frac{1}{2}$$

which implies that

$$\mu_m(\{y < m : \mathcal{A}_m^{g_n} \models \neg(a \lhd y) \wedge \neg(\oplus(0, y, a))\})$$
$$= \mu_m(\{y < m : \mathcal{A}_m^{g_n} \models y \lhd a\}) = \frac{1}{2} \tag{2}$$

Note now that a cannot be $\le m - g_n(m)$ because, if this was the case then from g_n-modularity we have that

$$\mu_m(\{y < m : \mathcal{A}_m^{g_n} \models y \lhd a\}) \le \mu_m(\{y < m : \mathcal{A}_m^{g_n} \models y \lhd m - g_n(m)\})$$
$$\le \frac{2(m - g_n(m))}{m} = 2(1 - \frac{g_n(m)}{m}) < 2(1 - \frac{3}{4}) = \frac{1}{2}$$

since for sufficiently large m, $\frac{g_n(m)}{m} > \frac{3}{4}$, but this contradicts (2).

Thus $m - g_n(m) < a < g_n(m)$. We can apply now equation (1) to obtain that

$$\frac{1}{2} = \mu_m(\{y < m : \mathcal{A}_m^{g_n} \models a \lhd y \lor y = a\}) = \frac{g_n(m) - a}{m}$$

Hence $\frac{1}{2} = (g_n(m) - a)/m$, that is, $g_n(m) - a = m/2$, so m must be even.

In a similar way, one can prove that for every natural $d > 2$, there exists a formula θ_d in $FO + \{P(z) \geq 1/d, P(z) > (d-1)/d\}(\{\oplus, \otimes, \lhd, 0, 1\})$ such that for every natural n, for every arithmetic structure $\mathcal{A}_m$ with $m > n$, $\mathcal{A}_m^{g_n} \models \theta_d$ iff m is a multiple of d.

A consequence of the above example is that the boolean query "the size of the model is divisible by d", for $d > 1$, is expressible in $(FO + \{P(z) \geq 1/d, P(z) > (d-1)/d\})_{MOD}$.

4   Separation Results for Modular Logics

In this section we prove separation results between fragments of $\mathcal{LP}_{MOD}$ defined in Definition 2. Since a formula such as $\neg((P(z) > \epsilon)\varphi)$ is equivalent to $(P(z) \geq 1 - \epsilon)\neg\varphi$ (and $\neg((P(z) \geq \epsilon)\varphi)$ is equivalent to $(P(z) > 1 - \epsilon)\neg\varphi$) we can push all negation symbols inside and together with all well known ways of manipulating quantifiers in a formula, we get the following prenex normal form for formulas in $\mathcal{LP}$.

Theorem 1. *For every formula* $\phi(\overline{x}) \in \mathcal{LP}(\Gamma^+)[r_1, r_2, \ldots, r_k]$ *there exists a quantifier free formula* $\theta(y_1, \ldots, y_w, \overline{x}) \in \mathcal{LP}(\Gamma)[r_1, r_2, \ldots, r_k]$ *such that for every structure* $\mathcal{B}_m$, *for every vector of naturals* $\overline{a} < m$,

$$\mathcal{B}_m \models \phi(\overline{a}) \leftrightarrow Q_1 y_1 Q_2 y_2 \ldots Q_w y_w \theta(\overline{y}, \overline{a}),$$

where each quantifier Q_s *is either* $\exists$ *or* $\forall$ *or* $(P(z) > q_{ij}/r_i)$ *or* $(P(z) \geq q_{ij}/r_i)$, *for some* $i \in \{1, \ldots, k\}$ *and some* $0 \leq q_{ij} < r_i$.   □

We proceed now to define the notion of an F-chain of models and the stronger notion of a chain.

Definition 7. *Fix a sublinear function* F. *An* **F-chain of models C** *is a collection of finite structures for* $\Gamma^+ = \Gamma \cup \{R_s\}_{s=1}^n \cup \{c_r\}_{r=1}^t$ *with the following property:*

- *For every relation symbol* $R(\overline{x})$ *of* Γ^+, *for every two models* $\mathcal{B}_m, \mathcal{B}_n$ *in* C *with* $m \leq n$ *and* $F(m) = F(n)$, *and for every* $\overline{a} < F(m)$, $\mathcal{B}_m \models R(\overline{a})$ *iff* $\mathcal{B}_n \models R(\overline{a})$.

A **chain of models C** *is a collection of finite structures for* Γ^+ *with the following property:*

- *For every relation symbol* $R(\overline{x})$ *of* Γ^+, *for every two models* $\mathcal{B}_m, \mathcal{B}_n$ *in* C *with* $m \leq n$ *and for every* $\overline{a} < m$, $\mathcal{B}_m \models R(\overline{a})$ *iff* $\mathcal{B}_n \models R(\overline{a})$.

In other words, chains are collections of models with inter-compatibility for its predicates.

Remark 2. If $\mathbf{C}$ is a chain of arithmetic models then, for every sublinear function F, $\mathbf{C}^F = \{\mathcal{A}_m^F : \mathcal{A}_m \in \mathbf{C}\}$ is an F-chain.

Example 5. Let $\{\mathcal{A}_m\}_{m=1}^{\infty}$ be the collection of arithmetic models for $\Gamma = \{\oplus, \otimes, \lhd, 0, 1\}$. It is easy to check that this collection is a chain.

We are ready to obtain separation results for the expressive power of the different modular logics. Our main tool is the following lemma which establish conditions for elementary equivalence. It states that for every sentence ϕ in $\mathcal{LP}(\Gamma^+)$, models that are in the same chain and have almost the same size can not distinguish ϕ.

Lemma 3. *Let F be a sublinear function and $\mathbf{C}$ an F-chain of models. Let r_1, $r_2, \ldots, r_k$ be distinct non zero natural numbers. Let $\phi(x_1, \ldots, x_s)$ be any formula in $\mathcal{LP}(\Gamma^+)[r_1, r_2, \ldots, r_k]$. Then one of the following two possibilities hold:*

1. *For every two F-modular models $\mathcal{B}_m$ and $\mathcal{B}_{m+1}$ in $\mathbf{C}$ such that $m + 1 > r_i$ and $m \equiv_{r_i} -1$, for every $i \leq k$ and $F(m) = F(m+1)$, we have that, for every $a_1, \ldots, a_s < m$, $\mathcal{B}_m \models \phi(a_1, \ldots, a_s)$ implies $\mathcal{B}_{m+1} \models \phi(a_1, \ldots, a_s)$,   or*
2. *For every two F-modular models $\mathcal{B}_m$ and $\mathcal{B}_{m+1}$ in $\mathbf{C}$ such that $m + 1 > r_i$ and $m \equiv_{r_i} -1$, for every $i \leq k$ and $F(m) = F(m + 1)$, we have that, for every $a_1, \ldots, a_s < m$, $\mathcal{B}_{m+1} \models \phi(a_1, \ldots, a_s)$ implies $\mathcal{B}_m \models \phi(a_1, \ldots, a_s)$.*

Proof. We proceed by induction on the quantifier rank of ϕ.

Quantifier Free Formulas: By definition of F-chain, if $\phi(x_1, \ldots, x_s)$ is quantifier free and $a_1, \ldots, a_s < F(m)$, we have that

$$\mathcal{B}_m \models \phi(a_1, \ldots, a_s) \text{ if and only if } \mathcal{B}_{m+1} \models \phi(a_1, \ldots, a_s),$$

We prove that this equivalence holds for $a_1, \ldots, a_s < m$. For each coordinate a_i such that $F(m) \leq a_i < m$, pick $b_i < F(m)$ such that $b_i \equiv_{F(m)} a_i$, and otherwise take $b_i = a_i$. Since $\mathcal{B}_m$ and $\mathcal{B}_{m+1}$ are F-modular, $\mathcal{B}_k \models \phi(a_1, \ldots, a_s) \Longleftrightarrow \mathcal{B}_k \models \phi(b_1, \ldots, b_s)$ for $k = m, m+1$. From this it follows the desired equivalence for $a_1, \ldots, a_s < m$ and $F(m) = F(m + 1)$.

Existentially or Universally Quantified Formulas: These two cases are not difficult to prove and we omit the proofs for lack of space. (Hint: the direction from $\mathcal{B}_{m+1}$ to $\mathcal{B}_m$ use that $\mathcal{B}_{m+1}$ is F-modular.)

Probability Quantifiers: We assume that case 1. holds, that is, for F, m, r_1, $\ldots$, r_k as in the hypothesis and for every $a_1, \ldots, a_s < m$ and every $b < m$:

$$\mathcal{B}_m \models \phi(a_1, \ldots, a_s, b) \text{ implies } \mathcal{B}_{m+1} \models \phi(a_1, \ldots, a_s, b).$$

We have two cases to consider under these hypothesis.

We consider first the formula, $(P(z) \geq q_{ij}/r_i)\phi(\bar{a}, z)$. Fix an arbitrary m satisfying that $F(m) = F(m+1)$, $m+1 > r_i$ and $m \equiv_{r_i} -1$ for every $i \leq k$, fix $a_1, \ldots, a_s < m$. Let t be a natural number such that $m = tr_i + r_i - 1$. Now, if $\mathcal{B}_m \models (P(z) \geq q_{ij}/r_i)\phi(\bar{a}, z)$ and since $gcd(r_i, m) = 1$, then

$$|\{z < m : \mathcal{B}_m \models \phi(\bar{a}, z)\}| > \frac{q_{ij}m}{r_i} = \frac{q_{ij}(tr_i + r_i - 1)}{r_i} = q_{ij}(t+1) - \frac{q_{ij}}{r_i}$$

and since $q_{ij} < r_i$, we obtain that $|\{z < m : \mathcal{B}_m \models \phi(\bar{a}, z)\}| \geq q_{ij}(t+1)$. By induction hypothesis we get that

$$|\{z < m+1 : \mathcal{B}_{m+1} \models \phi(\bar{a}, z)\}| \geq q_{ij}(t+1) = \frac{q_{ij}}{r_i}(t+1)(r_i) = \frac{q_{ij}}{r_i}(m+1),$$

which implies that $\mu(\{z < m+1 : \mathcal{B}_{m+1} \models \phi(\bar{a}, z)\}) \geq q_{ij}/r_i$, that is $\mathcal{B}_{m+1} \models (P(z) \geq q_{ij}/r_i)\phi(\bar{a}, z)$, which is the desired result.

Next we consider the formula $(P(z) > q_{ij}/r_i)\phi(\bar{a}, z)$ and we shall prove that case 2. holds for this formula. Fix an arbitrary m satisfying that $F(m) = F(m+1)$, $m+1 > r_i$ and $m \equiv_{r_i} -1$ for every $i \leq k$, fix $a_1, \ldots, a_s < m$. Let t be a natural number such that $m = tr_i + r_i - 1$. If $\mathcal{B}_m \models (P(z) \leq q_{ij}/r_i)\phi(\bar{a}, z)$ and since $gcd(r_i, m) = 1$, then

$$|\{z < m : \mathcal{B}_m \models \phi(\bar{a}, z)\}| < \frac{q_{ij}m}{r_i} = \frac{q_{ij}(tr_i + r_i - 1)}{r_i} = q_{ij}(t+1) - \frac{q_{ij}}{r_i}$$

and since $q_{ij} < r_i$, we obtain that

$$|\{z < m : \mathcal{B}_m \models \phi(\bar{a}, z)\}| \leq q_{ij}(t+1).$$

By induction hypothesis we get that

$$|\{z < m+1 : \mathcal{B}_{m+1} \models \phi(\bar{a}, z)\}| \leq q_{ij}(t+1) = \frac{q_{ij}}{r_i}(t+1)(r_i) = \frac{q_{ij}}{r_i}(m+1).$$

which implies that $\mu(\{z < m+1 : \mathcal{B}_{m+1} \models \phi(\bar{a}, z)\}) \leq q_{ij}/r_i$, that is, $\mathcal{B}_{m+1} \models (P(z) \leq q_{ij}/r_i)\phi(\bar{a}, z)$, which give us case 2. for this formula. The proofs for both type of probability quantifiers under the assumption that case 2. holds for ϕ are just the counterpositive versions of the two cases just proved. $\square$

The above lemma can be used to prove separation of different fragments of $\mathcal{LP}_{MOD}$.

Theorem 2. *Let $r, r_1, r_2, \ldots, r_k$ be distinct non zero natural numbers, and such that r is relatively prime with each $r_1, \ldots, r_k$. Then $\mathcal{LP}_{MOD}[r_1, \ldots, r_k]$ is properly contained in $\mathcal{LP}_{MOD}[r_1 \ldots r_k, r]$.*

Proof. It is obvious that $\mathcal{LP}_{MOD}[r_1, \ldots, r_k]$ is contained in $\mathcal{LP}_{MOD}[r_1, \ldots, r_k, r]$. Furthermore, we saw (Example 4) that the query: "the size of the model is a multiple of r" is expressible in $\mathcal{LP}_{MOD}(\Gamma)[r]$. We will show that this query is not expressible in $\mathcal{LP}[r_1 \ldots, r_k]_{MOD}(\Gamma)$. More specifically, we will show that there is no sentence ϕ in $\mathcal{LP}_{g_n}[r_1 \ldots, r_k](\Gamma)$ that defines the above query, where g_n is the sublinear function defined in Example 3, for all $n > (\prod_{i=1}^{k} r_i)r$.

Recall that the collection of all arithmetic models $\mathbf{C} = \{\mathcal{A}_m\}_{m=1}^{\infty}$ forms a chain. It follows that for every n, the collection $\mathbf{C}^{g_n} = \{\mathcal{A}_m^{g_n}\}_{m=1}^{\infty}$ forms a g_n-chain. Suppose now that there exists a sentence ϕ in $\mathcal{LP}_{g_n}[r_1 \ldots, r_k](\Gamma)$ that captures the query "the size of the model is a multiple of r" for all (except finitely many) structures $\mathcal{A}_m^{g_n}$. Then we can apply Lemma 3 and get the following:

> For every two models $\mathcal{A}_m^{g_n}$ and $\mathcal{A}_{m+1}^{g_n}$ in $\mathbf{C}^{g_n}$ such that $m + 1 > r_i$, $m \equiv_{r_i} -1$ for every i, and $g_n(m) = g_n(m + 1)$, we have that at least one of the following two cases hold
> (1) $\mathcal{A}_m^{g_n} \models \phi$ implies $\mathcal{A}_{m+1}^{g_n} \models \phi$, or (2) $\mathcal{A}_{m+1}^{g_n} \models \phi$ implies $\mathcal{A}_m^{g_n} \models \phi$.

Suppose it is case 1. that is true. Using that r is relatively prime with the r_i's together with the Generalized Chinese Remainder Theorem we can get a natural number $b \leq (\prod_{i=1}^{k} r_i)r$ such that $b \equiv_{r_i} -1$ for every i and $b \equiv_r 0$. Let D be the collection of naturals m such that $m = r(\prod_{i=1}^{k} r_i)tn + b$ for some natural $t > 0$. Clearly $m + 1 > r_i$, $m \equiv_{r_i} -1$ for every i, and $g_n(m) = g_n(m+1)$. Furthermore, D is infinite and for every $m \in D$, $m \equiv_r 0$. It follows that for almost all the $m \in D$, $\mathcal{A}_m^{g_n} \models \phi$ and, in consequence, for almost all the $m \in D$, $\mathcal{A}_{m+1}^{g_n} \models \phi$, i.e. for almost all elements m of D, $m + 1$ is a multiple of r, which is impossible.

Suppose it is case 2. that is true. Then by a similar argument as above we prove the existence of $b \leq (\prod_{i=1}^{k} r_i)r$ such that $b \equiv_{r_i} -1$ for every i and $b \equiv_r -1$. Let D be the same as above. Then D is infinite and for every $m \in D$, $m \equiv_r -1$. It follows that for almost all the $m \in D$, $\mathcal{A}_{m+1}^{g_n} \models \phi$ and, in consequence, for almost all the $m \in D$, $\mathcal{A}_m^{g_n} \models \phi$, i.e. for almost all elements m of D, m is a multiple of r, which is impossible.

We conclude that such sentence ϕ can not exists in $\mathcal{LP}_{g_n}[r_1 \ldots, r_k](\Gamma)$. $\square$

Corollary 2. *The expressive power of* FO_{MOD} *is strictly weaker than the expressive power of* $\mathcal{LP}_{MOD}[2]$. $\square$

This last result, for modular logics, corresponds to the separation of FO and FO + M in the context of arithmetic models, which in turn is equivalent to the separation of AC^0 from TC^0 shown by Ajtai and independently by Furst, Saxe and Sipser (see [5] for a nice exposition of this result and references).

5   Approximating $\mathcal{LP}_A$ with $\mathcal{LP}_{MOD}$

We introduce the notion of approximate formulas. This concept will provide a link between satisfaction in arithmetic structures and satisfaction in modular approximations of these arithmetic structures.

Definition 8 (Approximate Formulas). *For every formula in prenex normal form* $\theta(\overline{x}) \in \mathcal{LP}(\Gamma^+)$, *for every* $0 \leq \epsilon < 1$, *we define the ϵ-approximation of $\theta(\overline{x})$ as follows:*

Atomic formulas. *If* $\theta(\overline{x}) := R_s(\overline{x}, \overline{c})$ *then* $\theta_\epsilon(\overline{x}) := R_s(\overline{x}, \overline{c})$.
Negation of atomic formulas. *If* $\theta(\overline{x}) := \neg R_s(\overline{x}, \overline{c})$ *then* $\theta_\epsilon(\overline{x}) := \neg R_s(\overline{x}, \overline{c})$.
Conjunction. *If* $\theta(\overline{x}) := \phi(\overline{x}) \wedge \psi(\overline{x})$ *then* $\theta_\epsilon(\overline{x}) := \phi_\epsilon(\overline{x}) \wedge \psi_\epsilon(\overline{x})$.
Disjunction. *If* $\theta(\overline{x}) := \phi(\overline{x}) \vee \psi(\overline{x})$ *then* $\theta_\epsilon(\overline{x}) := \phi_\epsilon(\overline{x}) \vee \psi_\epsilon(\overline{x})$.
Existential quantification. *If* $\theta(\overline{x}) := \exists z \phi(\overline{x}, z)$ *then* $\theta_\epsilon(\overline{x}) := \exists z \phi_\epsilon(\overline{x}, z)$.
Universal quantification. *If* $\theta(\overline{x}) := \forall z \phi(\overline{x}, z)$ *then*
$$\theta_\epsilon(\overline{x}) := (P(z) > 1 - \epsilon)\phi_\epsilon(\overline{x}, z).$$
Probability quantifiers. *If* $\theta(\overline{x}) := (P(z) > r)\phi(\overline{x}, z)$ *then*
$$\theta_\epsilon(\overline{x}) := (P(z) > r - \min(\epsilon, r))\phi_\epsilon(\overline{x}, z).$$
If $\theta(\overline{x}) := (P(z) \geq r)\phi(\overline{x}, z)$ *then* $\theta_\epsilon(\overline{x}) := (P(z) \geq r - \min(\epsilon, r))\phi_\epsilon(\overline{x}, z)$.

The next lemma provides the basic operational properties of the approximate formulas.

Lemma 4. *For every formula (in prenex normal form)* $\theta(\overline{x}) \in \mathcal{LP}(\Gamma^+)$, *for every* $0 < \epsilon < 1$, *for every finite structure* $\mathcal{B}_m$ *and every vector* $\overline{a} < m$ *the following holds:*

- *If* $0 < \epsilon < \delta < 1$ *then* $\mathcal{B}_m \models \theta(\overline{a}) \to \theta_\epsilon(\overline{a}) \to \theta_\delta(\overline{a})$.
- *If* $\{\epsilon_i\}_{i=1}^{\infty}$ *is a sequence of real numbers less than 1 and converging to 0, then*

$$If\ (\forall i \in \mathbb{N},\ \mathcal{B}_m \models \theta_{\epsilon_i}(\overline{a}))\ then\ \mathcal{B}_m \models \theta(\overline{a}).$$

The purpose of the next theorem is to establish an "approximation" relationship between satisfaction in the modular logic $\mathcal{LP}_{MOD}$ and satisfaction in $\mathcal{LP}_A$ via the approximate formulas.

Theorem 3. (Bridge Theorem). *Fix a natural n. For every formula in prenex normal form* $\theta(\overline{x}) \in \mathcal{LP}(\Gamma^+)$, *for every arithmetic model* $\mathcal{A}_m$ *with* $m > n^2$, *for every* $\overline{a} < g_n(m)$, *the following holds:* $\mathcal{A}_m^{g_n} \models \theta(\overline{a})$ *implies* $\mathcal{A}_m \models \theta_{1/n}(\overline{a})$.

Proof. By induction in the complexity of the formula.

Atomic formulas and negation of atomic formulas. (Hint: for atomic
   formulas and their negation $\theta_{1/n}$ is the same as θ.)
Conjunction, disjunction. Direct.
Existential quantifier. (Hint: Suppose $\mathcal{A}_m^{g_n} \models \exists z \theta(\overline{a}, z)$. Then use Lemma 1
   and that $\mathcal{A}_m^{g_n}$ is g_n-modular to conclude $\theta(\overline{x}, z)$ is g_n-modular for $\mathcal{A}_m^{g_n}$ and,
   hence, $\mathcal{A}_m^{g_n} \models \theta(\overline{a}, [c]_{g_n(m)})$ for some $c < m$.)
Universal quantifier. Suppose that $\mathcal{A}_m^{g_n}$ satisfies the formula $\forall z \theta(\overline{a}, z)$. Then
   for every $c < g_n(m)$ we have that $\mathcal{A}_m^{g_n} \models \theta(\overline{a}, c)$. We can apply now the
   induction hypothesis to obtain that for every $c < g_n(m)$ we have that $\mathcal{A}_m \models$
   $\theta_{1/n}(\overline{a}, c)$. Since $\frac{m - g_n(m)}{m} \leq \frac{n}{m}$ and $m > n^2$ we get that $\frac{g_n(m)}{m} > 1 - \frac{1}{n}$, which
   implies $\mathcal{A}_m \models (P(z) > 1 - \frac{1}{n})\theta_{1/n}(\overline{a}, c)$.

Probability quantification. Suppose that $\mathcal{A}_m^{g_n}$ satisfies the formula $(P(z) > r)\theta(\bar{a}, z)$ for $0 < r < 1$. It follows that $|\{c < m : \mathcal{A}_m^{g_n} \models \theta(\bar{a}, c)\}| > rm$. Then we get that

$$|\{c < g_n(m) : \mathcal{A}_m^{g_n} \models \theta(\bar{a}, c)\}| > rm - (m - g_n(m)).$$

Applying the induction hypothesis we obtain that

$$|\{c < m : \mathcal{A}_m \models \theta_{1/n}(\bar{a}, c)\}| > rm - (m - g_n(m)).$$

It follows that

$$\mu_m\left(\{c < m : \mathcal{A}_m \models \theta_{1/n}(\bar{a}, c)\}\right) > \frac{rm - (m - g_n(m))}{m} =$$

$$r - \frac{(m - g_n(m))}{m} = r - \frac{1}{n} \text{ since } m > n^2.$$

But this last statement is just $\mathcal{A}_m \models (P(z) > r - \frac{1}{n})\theta_{1/n}(\bar{a}, z)$. $\square$

The gist of the above result is to give a quantifiable relationship between satisfaction of a formula in the structures $\mathcal{A}_m^{g_n}$ and satisfaction of its approximation in $\mathcal{A}_m$. It implies the following relationship between boolean queries captured by $\mathcal{LP}_A$ and the boolean queries captured in $\mathcal{LP}_{MOD}$. (We will abbreviate by $(\neg\theta)_\epsilon$, for $\theta \in \mathcal{LP}(\Gamma^+)$, the ϵ-approximation of the formula equivalent to $\neg\theta$.)

Corollary 3. *Assume there is a boolean query B, a natural n and a formula $\theta \in \mathcal{LP}(\Gamma^+)$ such that for every arithmetic model $\mathcal{A}_m$, with $m > n^2$, if $\mathcal{A}_m \models \theta_{1/n}$ then $\mathcal{A}_m \in B$, and if $\mathcal{A}_m \models (\neg\theta)_{1/n}$ then $\mathcal{A}_m \notin B$. Then for every $m > n^2$, $\mathcal{A}_m \in B$ iff $\mathcal{A}_m^{g_n} \models \theta$.* $\square$

6   P and the Logic LP Extended

The first problem shown to be complete for the class **P**, deterministic polynomial time, was *Path System Accessibility* due to Cook [2]. An instance of the Path System Accessibility problem, which we abbreviate from now on as PS, is a finite structure $\mathcal{A} = \langle A, R, T, s \rangle$, or a *path system*, where the universe A consists of, say, n vertices, a relation $R \subseteq A \times A \times A$ (the *rules* of the system), a *source* $s \in A$, and a set of *targets* $T \subseteq A$ such that $s \notin T$. A positive instance of PS is a path system $\mathcal{A}$ where some target in T is *accessible* from the source s, where a vertex v is accessible if it is the source s or if $R(x, y, v)$ holds for some accessible vertices x and y, possibly equal. In [8], Stewart shows that PS is complete for **P** via quantifier free first order reductions; in fact, via *projections* (see [8] for definitions and also [5] Section 11.2), and we will use that result to show that an approximation version of PS which we present in Example 6 below is also complete for **P** via reductions that are projections, and that would help us to show that a certain extension of our $\mathcal{LP}$ logic captures **P** on finite ordered structures. (We remark that Stewart considers the path systems in [8] as having only one target, and not a set of targets as we do here. However one can see that his results on completeness of PS via first order reductions holds also for our version of this problem.)

Definition 9. *Let X be a second order variable of arity 1, and $\alpha(\overline{x}, X)$ a first order formula over some (finite) vocabulary τ with first order variables $\overline{x} = (x_1, \ldots, x_m)$ and second order variable X. Let $r \in [0,1]$. Then*

$$(P(X) > r)\alpha(\overline{x}, X) \quad and \quad (P(X) \geq r)\alpha(\overline{x}, X)$$

are new formulas with the following semantic. For an appropriate finite τ-model $\mathcal{A}_n$, and elements $\overline{a} = (a_1, \ldots, a_m)$ from $\{0, \ldots, n-1\}$, the universe of $\mathcal{A}_n$,

$$\mathcal{A}_n \models (P(X) > r)\alpha(\overline{a}, X)$$
$$\Longleftrightarrow \text{ the least subset } A \subseteq \{0, \ldots, n-1\} \text{ such that}$$
$$\mathcal{A}_n \models \alpha(\overline{a}, A) \text{ has } |A|/n > r$$

Similarly for $(P(X) \geq r)\alpha(\overline{a}, X)$.

Example 6. Let $\tau = \{R, T, s\}$ where R is a ternary relation symbol, T is a unary relation symbol and s is a constant symbol. We think of τ-structures as path systems with source s, a target set T and set of rules R. Let r be a rational with $0 < r < 1$. We define

NPS$_{\geq r} := \{\mathcal{A} = \langle A, R, T, s \rangle : \mathcal{A}$ is a path system and **at least** a fraction r of the elements accessible from s are **not** in $T\}$

Let $\alpha_{nps}(X)$ be the following formula (the constant symbol $\bot$ stands for false),

$$\alpha_{nps}(X) := \forall x(x = s \longrightarrow X(x))$$
$$\wedge \ \forall x \forall y \forall z(X(x) \wedge X(y) \wedge R(x,y,z) \longrightarrow X(z))$$
$$\wedge \ \forall x(X(x) \wedge T(x) \longrightarrow \bot)$$

Then

$$\mathcal{A}_n \in \text{NPS}_{\geq r} \iff \mathcal{A}_n \models (P(X) \geq r)\alpha_{nps}(X)$$

NPS$_{\geq r}$ is an approximation version of the problem PS, definable by our probability quantifiers over unary second order variables acting on formulas with a particular form to which we give a name below.

Definition 10. *Let $\tau = \{R_1, \ldots, R_m, C_1, \ldots, C_k\}$ be some vocabulary with relation symbols R_1, ..., R_m, and constant symbols C_1, ..., C_k, and let X be a unary second order variable. A first order formula α over $\tau \cup \{X\}$, and extra symbols as $=$ (equality) and the constant $\bot$ (standing for false), is a universal Horn formula, if α is the conjunction of universally quantified formulas over $\tau \cup \{X\}$ of the form*

$$\psi_1 \wedge \psi_2 \wedge \ldots \wedge \psi_s \longrightarrow \varphi$$

where φ is either $X(\overline{u})$ or $\bot$, and ψ_1, ..., ψ_s are atomic $(\tau \cup \{X\})$-formulas with any occurrence of the variable X being positive (there are no restrictions on the predicates in τ or $=$).

The logic $\mathcal{LP}_{Horn}$ is the set of formulas

$$FO + \{(P(X) > r)\alpha_1(\overline{x}, X), (P(X) \geq r)\alpha_2(\overline{x}, X) : \alpha_i(\overline{x}, X) \text{ is universal}$$
$$\text{Horn (first order) formula with second order variable } X\}$$

Example 6 shows that the problem $\text{NPS}_{\geq r}$ is definable in $\mathcal{LP}_{Horn}$. We shall see that this is true of all problems in **P**

Lemma 5. *The set of finite structures that satisfy a sentence θ in $\mathcal{LP}_{Horn}$ is in* **P**.

Proof. Let $\theta \in \mathcal{LP}_{Horn}$ be of the form

$$(P(X) > r)[\bigwedge_{i=1}^{m} \forall \overline{x}_i (\psi_{i1} \wedge \ldots \wedge \psi_{is} \longrightarrow \varphi_i)],$$

and let $\mathcal{A}_n$ be a model of the appropriate vocabulary of size n. Then it's not difficult to describe a polynomial time procedure that decides whether $\mathcal{A}_n$ satisfies the above sentence. $\square$

Thus, according to this lemma, our problem $\text{NPS}_{\geq r}$ is in **P**. We show next that it is hard for **P**.

Lemma 6. *The problem* $\text{NPS}_{\geq r}$ *is complete for* **P** *via projections.*

Proof. We exhibit a (successor free) projection from the complement of the problem PS to $\text{NPS}_{\geq r}$. Let $\mathcal{A} = \langle A, R, T, s \rangle$ be an instance of PS. Define $\mathcal{A}' = \langle A', R', T', s' \rangle$ as follows: its universe $A' = A^2$, and

$$T' = T \times s = \{(x, s) : x \in T\}$$
$$R' = \{((x, s), (y, s), (z, s)) : (x, y, z) \in R\} \cup$$
$$\{((x, s), (y, s), (z, s)) : x \in T \wedge x \neq s \wedge y \in T \wedge y \neq s \wedge z \neq s\}$$
$$s' = (s, s)$$

Then, $\mathcal{A} \in \text{PS} \iff \mathcal{A}' \notin \text{NPS}_{\geq r}$. $\square$

Corollary 4. *Every problem in* **P** *is a set of finite ordered structures that satisfy a sentence in* $\mathcal{LP}_{Horn}$

Proof. Every problem in **P** is reducible to $\text{NPS}_{\geq r}$ via projections; $\text{NPS}_{\geq r}$ is definable in $\mathcal{LP}_{Horn}$ and this logic is closed via projections. $\square$

Corollary 5. *Over finite ordered structures, the logic* $\mathcal{LP}_{Horn}$ *captures* **P**. $\square$

The logic $\mathcal{LP}_{Horn}$ verifies Lemma 1; namely, for a sublinear function F, F-modularity is preserved. Indeed, we need only to check for formulas of the form $(P(X) > r)\alpha(\overline{z}, X)$: Suppose $\overline{a}, \overline{b} < m$, $\overline{a} \equiv_{F(m)} \overline{b}$ and $\mathcal{B}_m \models (P(X) > r)\alpha(\overline{a}, X)$. Then there exists a $B \subseteq \{0, 1, \ldots, m-1\}$, such that $\mathcal{B}_m \models \alpha(\overline{a}, B)$ and $|B| > rm$. The parameters in $\overline{a}$ do not occur in B; hence, by inductive hypothesis $\mathcal{B}_m \models \alpha(\overline{b}, B)$. Thus, $\mathcal{B}_m \models (P(X) > r)\alpha(\overline{b}, X)$. $\square$

References

1. Barrington, D., Immerman, N., Straubing, H.: On uniformity within NC^1. J. Computer and Syst. Sci. **41** (1990) 274–306.
2. Cook, S. A.: An observation on time-storage trade off, J. Comput. System Sci. **9** (1974) 308–316.
3. Etessami, K., Immerman, N.: Reachability and the power of local ordering, Theo. Comp. Sci. **148**, 2 (1995) 261–279.
4. Gurevich, Y.: Logic and the challenge of computer science. In: "Current trends in theoretical computer science" (E. Börger, Ed.) Computer Science Press. (1988) 1-57.
5. Immerman, N.: Descriptive Complexity. Springer (1998).
6. Keisler, H. J.: Hyperfinite model theory. In: "Logic Colloquium 76" R.C. Gandy and J.M. E. Hyland, Eds.), North-Holland (1977) .
7. Libkin, L., Wong, L.: Lower bounds for invariant queries in logics with counting. Theoretical Comp. Sci. **288** (2002), 153-180.
8. Stewart, I.: Logical description of monotone NP problems, J. Logic Computat. **4**, 4 (1994) 337-357.

Arithmetic Circuits for Discrete Logarithms

Joachim von zur Gathen

University of Paderborn, Germany
gathen@upb.de
http://www-math.uni-paderborn.de/~aggathen/

Abstract. We introduce a new model of "generic discrete log algorithms" based on arithmetic circuits. It is conceptually simpler than previous ones, is actually applicable to the natural representations of the popular groups, and we can derive upper and lower bounds that differ only by a constant factor, namely 10.

Keywords. Discrete logarithm, generic algorithm, arithmetic circuit, cyclic group

1 Introduction

Discrete logarithm computations and their presumed difficulty are a central topic in cryptography. Let G be a finite cyclic group of order d, p the largest prime divisor of d, and n the bit length of d (that is, n is the "private key length"). There are three types of results:

- "Generic" algorithms such as baby-step giant-step, Pollard rho, and Pohlig-Hellman. Together they provide a solution with $O(n\sqrt{p} + n^2)$ group operations.
- Algorithms for special groups, such as the index calculus for the group of units in a finite field, and Weil descent for special elliptic curves.
- Lower bounds $\Omega(\sqrt{p})$ on "generic" algorithms.

This paper proposes a new solution to the last point.

Babai & Szemerédi (1984) first proposed a model in which even a lower bound $\Omega(p)$ holds. Then Nechaev (1994) suggested a deterministic model with an $\Omega(\sqrt{p})$ bound, and Boneh & Lipton (1996) considered finite fields. The most popular model was invented by Shoup (1997). It is probabilistic, has an $\Omega(\sqrt{p})$ lower bound, and also works for the Diffie-Hellman problem. Maurer & Wolf (1998, 1999) continued to work on this, in particular by relating the two questions of discrete logarithms and the Diffie-Hellman task. See also Schnorr & Jakobsson (2000) and Schnorr (2001).

An essential ingredient of Shoup's method is a bit representation of the group elements, and his lower bound holds for a random description of this form. The standard "generic" algorithms consist of two phases: first some group calculations are performed, and in a second phase the resulting lists of group elements

M. Farach-Colton (Ed.): LATIN 2004, LNCS 2976, pp. 557–566, 2004.

are sorted, with the goal of finding a collision. Of course, when one wants to implement such an algorithm, one has to use some bit representation of the group elements in computer memory. But the algorithms will use one "natural" representation, not random ones. Strictly speaking, Shoup's result does not apply to this situation, and thus does not provide a lower bound in the natural setting.

This paper repairs this state of affairs by presenting a new model for "generic" discrete log computations which is both technically simpler and more powerful. It has the following properties:

- the known "generic" algorithms fit in,
- a lower bound of $\Omega(\sqrt{p})$ holds,
- it does not make assumptions about the representation of groups,
- there is a matching upper bound, larger only by a constant factor.

This is basically achieved by ignoring the second phase, where sorting occurs. Then one can do away with the group representation, and describe the first phase in a simple arithmetic model.

It is important to note that the goal here is **not** a way of describing useful discrete log computations. In fact, our computations do not calculate discrete logs, but any "generic" discrete log computation yields one of our type. The asymptotically matching upper and lower bounds are an indication that this may be the "right" level of abstraction.

The most natural way of saying that we "only want to use group operations" is by using arithmetic circuits (a.k.a. straight-line programs) with group operations. This model was introduced in great generality by Strassen (1972). However, a circuit computes only group elements and not discrete logs, which live in the "exponent group". Success in the usual algorithms is signalled by a collision, where the same group element is calculated in two different ways. The basic idea is to declare a circuit as successful if it produces such a collision. One has to be a bit careful: it is easy to produce trivial collisions, say by calculating the group element 1 in two different ways. This leads to our notion of a collision "respecting" a divisor q of the group order: it is not trivial in the "exponent group modulo q".

In Section 2, we set up the required notions. Section 3 starts with the usual "nonzero preservation" result modulo a prime power; it is somewhat simplified in comparison with other generic models by considering only linear polynomials. Technically, this Lemma 7 is the main overlap with Shoup's method. Then we prove the main result, a lower bound of $\Omega(\sqrt{p})$ in Theorem 8. The model is sufficiently powerful (or weak, as you have it) that essentially matching upper and lower bounds hold; they differ only by a constant factor, namely 10 (Corollary 10).

The model so far is deterministic; Section 4 extends it to probabilistic computations. The same lower bound holds. This is no surprise, since randomized algorithms such as Pollard's rho method do not reduce the computing time. This method is important because it reduces the required memory to a constant number of group elements, but we do not consider this resource.

2   Arithmetic Circuits for Discrete Logarithms

We fix the following notation:

$$G = \langle g \rangle \text{ is a finite cyclic group}, d = \#G, \tag{1}$$
$$p \text{ is the largest prime divisor of } d, \text{ and } n \text{ is the binary length of } d.$$

We consider algorithms that use only the group operations, starting with three special group elements: 1, the generator g, and x. From these we build further group elements by multiplication and inversion.

Example 2. Here is a formulation of the baby-step giant-step algorithm for $d = 20$:

instruction	trace	trace exponent
$y_{-2} \longleftarrow 1$	1	0
$y_{-1} \longleftarrow g$	g	1
$y_0 \longleftarrow x$	x	t
$y_1 \longleftarrow y_0 \cdot y_{-1}$	xg	$t+1$
$y_2 \longleftarrow y_1 \cdot y_{-1}$	xg^2	$t+2$
$y_3 \longleftarrow y_2 \cdot y_{-1}$	xg^3	$t+3$
$y_4 \longleftarrow y_3 \cdot y_{-1}$	xg^4	$t+4$
$y_5 \longleftarrow y_4 \cdot y_{-1}$	xg^5	$t+5$
$y_6 \longleftarrow y_5 \cdot y_0^{-1}$	g^5	5
$y_7 \longleftarrow y_6 \cdot y_6$	g^{10}	10
$y_8 \longleftarrow y_7 \cdot y_6$	g^{15}	15
$y_9 \longleftarrow y_8 \cdot y_6$	g^{20}	20
$y_{10} \longleftarrow y_9 \cdot y_6$	g^{25}	25

The "trace" gives the group element computed in each step. The "trace exponent" is explained below. The algorithm is in its simplest form, ignoring shortcuts like $g^{20} = 1$.

If $\log_g x = 5b + c$, with $0 \leq b, c < 5$, then $x = g^{5b+c}$, hence $xg^{5-c} = g^{5(b+1)}$, and both elements appear in the computation. If we take $G = \mathbb{Z}_{25}^{\times} = \langle 2 \rangle$, a group of order 20, and $x = 19 = 2^{18}$, then we have $18 = 5 \cdot 3 + 3$ and $y_2 = xg^2 = g^{20} = y_9$.

How do we express that the algorithm successfully computes $\log_z x$? We are very generous: we say that the algorithm is **successful** if a "collision" $u = v$ occurs for two previously computed results u and v for which "$u = v$ is not trivial". If we computed $y_1 = y_{-1} \cdot y_{-1}^{-1}$, $y_2 = y_0 \cdot y_0^{-1}$, then $y_1 = y_2$ would be trivial. We will make this precise in a minute.

The type of computation shown in the table above could be called an "arithmetic group circuit with inputs 1, g, and x". We abbreviate the assignment $y_k \longleftarrow y_i \cdot y_j^{\pm 1}$ as $(i, j, \pm 1)$, and also trace the exponents of g and x in the circuit. Then we arrive at the following notion.

Definition 3. *(i) An **arithmetic circuit** is a finite sequence $C = (I_1, \ldots, I_\ell)$ of instructions $I_k = (i, j, \varepsilon)$, with $-2 \leq i, j < k$ and $\varepsilon \in \{1, -1\}$. The **size** of C is ℓ. Note that C is not connected to any particular group.*

*(ii) If $C = (I_1, \ldots, I_\ell)$ is an arithmetic circuit, G a group and $g, x \in G$, then the **trace** of C on input (g, x) is the following sequence $z_{-2}, z_{-1}, \ldots, z_\ell$ of elements z_k of G:*

$$z_{-2} = 1, z_{-1} = g, z_0 = x, z_k = z_i \cdot z_j^\varepsilon \text{ for } k \geq 1 \text{ and } I_k = (i, j, \varepsilon).$$

*(iii) For an arithmetic circuit $C = (I_1, \ldots, I_\ell)$, the **trace exponents** consist of the following sequence $\tau_{-2}, \tau_{-1}, \ldots, \tau_\ell$ of linear polynomials τ_k in $\mathbb{Z}[t]$:*

$$\tau_{-2} = 0, \tau_{-1} = 1, \tau_0 = t, \tau_k = \tau_i + \varepsilon \cdot \tau_j \text{ for } k \geq 1 \text{ and } I_k = (i, j, \varepsilon).$$

We think of g as fixed, and also write $z_k(x)$ for the trace elements z_k in (ii).

The connection between the trace and the trace exponents is clear: if $x = g^a$ and $\tau_k = c \cdot t + b$, then

$$z_k(x) = g^b x^c = g^b \cdot g^{ac} = g^{\tau_k(a)}.$$

Recall that in the exponents, we may calculate modulo the group order d, once we consider a fixed group.

Example 4. Here are two more examples of trivial collisions.

(i) We take $g, x = g^a$ in a group of order d, and an arithmetic circuit which computes $y_m = g^d$ with an addition chain of some length m, and also $y_{2m} = x^d$. Then $\tau_m = d$ and $\tau_{2m} = dt$, $y_m = g^d = 1 = x^d = y_{2m}$, and we take the congruence $\tau_m - \tau_{2m} \equiv 0 \bmod d$ as an indicator for the triviality of this collision.

(ii) Now let q be an arbitrary prime divisor of d, maybe a small one, and assume that $d \neq q$. Again we calculate some $y_m = g^{d/q}$ and $y_{2m} = x^{d/q}$. Now both results lie in the subgroup $H = \langle g^{d/q} \rangle$ of order q, and we can find a collision with a further q (or even $O(\sqrt{q})$) steps. But we have only calculated a discrete logarithm in H, not in G. If, say, $q = 2$, then $y_m = g^{d/2} \neq 1$ and y_{2m} is either y_m or 1. Thus we have a collision, either $y_{-2} = y_{2m}$ or $y_m = y_{2m}$. $\diamond$

How do we express that "$u = v$ is trivial"? We certainly want to say that "the collision $y_i = y_j$ is trivial" if $\tau_i = \tau_j$, or even if $\tau_i \equiv \tau_j \bmod d$, but this is not quite enough. We have to rule out unpleasant cases like the one at the end of Example 4, where a collision occurs but the discrete logarithm is not really computed.

Definition 5. *Let C be an arithmetic circuit of size ℓ, $G = \langle g \rangle$, q an arbitrary divisor of the group order $d = \#G$, and $i, j \leq \ell$.*

*(i) Then (i, j) is said to **respect** q if and only if $\tau_i - \tau_j \not\equiv 0 \bmod q$.*

(ii) If on input some $g, x \in G$, a collision $y_i = y_j$ occurs, then this collision respects q if and only if (i, j) respects q.

Thus we have the linear polynomial $\tau_i - \tau_j \in \mathbb{Z}[t]$ which is nonzero modulo q, hence modulo d, and if a collision occurs for $x = g^a$, then $g^{\tau_i(a)} = z_i(x) = z_j(x) = g^{\tau_j(a)}$, so that $(\tau_i - \tau_j)(a) \equiv 0 \bmod d$.

If $q_1 \mid q_2 \mid d$, and (i, j) respects q_1, then it also respects q_2.

Example 4 continued. (ii) For $q = 2$, we have $\tau_m = d/2$, $\tau_{2m} = dt/2$, and $\tau_m - \tau_{2m} \equiv d/2 \cdot (1 - t) \bmod d$. We assume that d is not a power of 2, and take a prime divisor $q \neq 2$ of d. Then q divides $d/2$, and $\tau_m - \tau_{2m} \equiv 0 \bmod q$. Thus $(m, 2m)$ does not respect q, and if on some input x from some group G, the collision $g^{d/2} = z_m(x) = z_{2m}(x) = x^{d/2}$ occurs, then this does not respect q, either.   ◇

Definition 6. *Let $G = \langle g \rangle$ be a finite cyclic group, $\mathcal{C}$ an arithmethic circuit, and q an arbitrary divisor of the group order $d = \#G$. Then the success rate $\sigma_{\mathcal{C}, q}$ of $\mathcal{C}$ over G respecting q is the fraction of group elements for which a collision respecting q occurs:*

$$\sigma_{\mathcal{C}, q} = d^{-1} \cdot \#\{x \in G : \text{on input } x, \text{ a collision respecting } q \text{ occurs in } \mathcal{C}\}.$$

Thus $0 \leq \sigma_{\mathcal{C}, q} \leq 1$, and a circuit, for which a collision respecting q occurs for every input x, has $\sigma_{\mathcal{C}, q} = 1$. If $q_1 \mid q_2 \mid d$, then $\sigma_{\mathcal{C}, q_1} \leq \sigma_{\mathcal{C}, q_2}$. Example 2 indicates that the baby-step giant-step algorithm gives a circuit of size $O(\sqrt{d})$, where $d = \#G$ and $\sigma_{\mathcal{C}, d} = 1$. For simplicity, our notation does not reflect the dependence of the success rate on the group.

Also, the Pohlig–Hellman algorithm is a generic algorithm. But index calculus in $G = \mathbb{F}_p^\times$ is not generic; it makes essential use of the representation of the elements of G as integers less than p, and the ability to compute with these integers, say to check whether they factor over the factor base.

3   The Deterministic Lower Bound

"Nonzero preservation" is a generally useful tool. It says that the value of a nonzero polynomial at a random point is likely to be nonzero. It is well-known over integral domains; we need a slight generalization here. See Shoup (1997) for a more general version.

Lemma 7. *Let $d \geq 2$ be an integer, p^e a prime power divisor of d, where p is a prime, and $\tau = c_1 t + c_0 \in \mathbb{Z}[t]$ a linear polynomial with $\tau \not\equiv 0 \bmod p^e$. Then*

$$\#\{a \in \mathbb{Z}_d : \tau(a) \equiv 0 \bmod p^e\} \leq d/p.$$

Proof. Let $i \geq 0$ be the largest exponent with $\tau \equiv 0 \bmod p^i$. Thus $i < e$, and we can write $\tau = p^i \cdot (c_1' t + c_0')$, with $c_0', c_1' \in \mathbb{Z}_{d/p^i}$ and at least one of them nonzero modulo p^{e-i}. If $c_1' \equiv 0 \bmod p$, then there is no $a \in \mathbb{Z}_d$ with

$\tau(a) \equiv 0 \bmod p^{i+1}$, let alone modulo p^e. Otherwise there is exactly one $a_0 \in \mathbb{Z}_p$ with $c_1' a_0 + c_0' \equiv 0 \bmod p$, namely $a_0 \equiv -c_0' \cdot c_1^{-1} \bmod p$. The residue class mapping $\mathbb{Z}_d \longrightarrow \mathbb{Z}_p$ maps any $a \in \mathbb{Z}_d$ to $a \bmod p$. Exactly d/p elements of $\mathbb{Z}_d$ are mapped to the same element of $\mathbb{Z}_p$. Now if $p^i(c_1' a + c_0') = \tau(a) \equiv 0 \bmod p^e$, then $c_1' a + c_0' \equiv 0 \bmod p$, and hence $a \bmod p = a_0$. There are exactly d/p such a, and the claim follows.   □

Theorem 8. *Let $G = \langle g \rangle$ be a finite cyclic group, $q = p^e$ a prime power divisor of the group order $d = \#G$, $\mathcal{C}$ an arithmetic circuit over G of size ℓ, and $\sigma_{\mathcal{C},q}$ its success rate respecting q. Then*

$$\ell \geq \sqrt{2\sigma_{\mathcal{C},q}p} - 3.$$

When $\sigma_{\mathcal{C},q}$ is a positive constant, then $\ell \in \Omega(\sqrt{p})$.

Proof. On some input x, a collision in $\mathcal{C}$ is of the form $y_i(x) = y_j(x)$ with $-2 \leq i < j \leq \ell$. There are $(\ell+2)(\ell+3)/2$ such (i,j). Any (i,j) which respects q leads to a collision for at most d/p values of x, by Lemma 7, since the exponents $a \in \mathbb{Z}_d$ correspond bijectively to the group elements $x = g^a$. Thus the total number of possible collisions respecting q is at most $(\ell+2)(\ell+3)/2 \cdot d/p$, and hence

$$\sigma_{\mathcal{C},q} \leq (\ell+2)(\ell+3)/2p,$$
$$(\ell+3)^2 \geq (\ell+2)(\ell+3) \geq 2\sigma_{\mathcal{C},q}p.$$ □

The various well-known algorithms yield an $O(n\sqrt{p} + n^2)$ upper bound for discrete logarithm computations, and we now have a lower bound $\Omega(\sqrt{p})$ where p is the largest prime divisor of d. In what follows, we derive upper and lower bounds that differ only by a constant factor. We start with a lower bound different from Theorem 8, namely $\Omega(n)$. This is not of direct cryptographic interest, since $n \approx \log_2 d$ is roughly the "key length" or "input length"—in contrast to $\sqrt{p}$ which will usually be chosen so that it is exponentially large in n. The interest is a desire to understand the complexity of discrete logarithms as well as possible.

Theorem 9. *Let $\mathcal{C}$ be an arithmetic circuit of size ℓ, $G = \langle g \rangle$ a cyclic group of order $d \geq 3$, with $\sigma_{\mathcal{C},d} = 1$, and let $n = \lfloor \log_2 d \rfloor + 1$ be the binary length of d. Then*

$$\ell \geq \frac{n}{2} - 2,$$

and hence $\ell \in \Omega(n)$.

Proof. Any element a of $\mathbb{Z}_d$ has exactly one *balanced representative* $b \in \mathbb{Z}$ with

$$a = (b \bmod d), \quad -d/2 < b \leq d/2.$$

For $-2 \leq k \leq \ell$, we write the trace exponent $\tau_k \in \mathbb{Z}_d[t]$ as $\tau_k = (c_k \bmod d) \cdot t + (b_k \bmod d)$, where $c_k, b_k \in \mathbb{Z}$ are balanced representatives. By induction on k it follows that $|b_k|, |c_k| \leq 2^k$ for $0 \leq k \leq \ell$ (and $|b_k|, |c_k| \leq 1$ for $k = -2, -1$). Now let $a_0 = \lfloor \sqrt{d} \rfloor$, $a = (a_0 \bmod d) \in \mathbb{Z}_d$ and $x = g^a \in G$. The assumption $\sigma_{C,d} = 1$ implies that there are $i, j \leq \ell$ with $\tau_i - \tau_j \not\equiv 0 \bmod d$ and $(\tau_i - \tau_j)(a) \equiv 0 \bmod d$. We let

$$u = (c_i - c_j) \cdot a_0 + (b_i - b_j) \in \mathbb{Z}.$$

The above implies that $u \equiv 0 \bmod d$.

If $c_i = c_j$, then $b_i \equiv b_j \bmod d$ and $\tau_i - \tau_j \equiv 0 \bmod d$, which is ruled out. Thus $c_i \neq c_j$. If $u = 0$, then

$$\sqrt{d} - 1 \leq |a_0| = \frac{|b_i - b_j|}{|c_i - c_j|} \leq |b_i - b_j| \leq |b_i| + |b_j| \leq 2^{\ell+1}.$$

If $u \neq 0$, then $|u| \geq d$, and

$$2^{\ell+1}(\sqrt{d} + 1) = 2^{\ell+1}\sqrt{d} + 2^{\ell+1} \geq |c_i - c_j|a_0 + |b_i - b_j|$$
$$\geq |(c_i - c_j)a_0 + (b_i - b_j)| = |u| \geq d,$$
$$2^{\ell+1} \geq \frac{d}{\sqrt{d} + 1} \geq \sqrt{d} - 1.$$

Thus $\ell \geq \log(\sqrt{d} - 1) - 1$ in both cases. The claim now follows from

$$\log(\sqrt{d} - 1) > \frac{1}{2}\log d - \frac{1}{2} \geq \frac{1}{2}\lfloor \log d \rfloor - \frac{1}{2} = \frac{n}{2} - 1$$

for $d \geq 12$. (One checks the cases $3 \leq d \leq 11$ separately.) □

For an upper bound in our model, we just compute $g^{d/p}$ and $x^{d/p}$, and then perform a baby-step giant-step search in the subgroup $\langle g^{d/p} \rangle$ of p elements. The total cost is $2(n + \sqrt{p})$, and we have the lower bounds of $n/2$ and $\sqrt{2p}$, approximately. Thus the gap is a factor of about 4 or $\sqrt{2}$, depending on whether n or $\sqrt{p}$ is larger. We can obtain a specific estimate as follows.

Corollary 10. *Let G be a cyclic group with d elements, $n - \lfloor \log_2 d \rfloor + 1$ the binary length of d, p the largest prime divisor of d, e the multiplicity of p in d,*

$$m = \max\{\sqrt{2p} - 3, n/2 - 2\},$$

and assume that $m \geq 37$. Then there exists an arithmetic circuit C with success rate $\sigma_{C,p^e} = 1$ over G and size at most $10m$. Any circuit C with $\sigma_{C,p^e} = 1$ has size at least m.

Proof. The last claim follows from Theorems 8 and 9. For C we take the circuit described above. Then $\sigma_{C,p^e} = 1$, and its size ℓ is at most $2 \cdot 2\log(d/p) + 2\sqrt{p}$. Thus

$$\ell \leq 4\log d + 2\sqrt{p} \leq 8(n/2 - 2) + \sqrt{2} \cdot (\sqrt{2p} - 3) + 17 + 3\sqrt{2}$$
$$\leq (8 + \sqrt{2})m + 17 + 3\sqrt{2} \leq 10m. \qquad \square$$

In usual models of computation, upper bounds come from algorithms—the real thing—and lower bounds impose barriers on improving these. But here, the lower bound is the real thing, and the upper bound a barrier on deriving better ones. As stated before, the above circuit cannot claim to actually compute discrete logarithms in G.

4   Probabilistic Arithmetic Circuits

We now have a model for discrete log computations with essentially matching upper and lower bounds. However, Pollard's rho method works in any group, but does not fit into our model because it makes random choices. The method does not make progress over the baby-step giant-step method in terms of time (= size of circuit), but cuts the space dramatically down to a constant. Space is not accounted for in our model, but we now adapt it to allow probabilistic choice. Once the model is appropriately set up within our framework, it is easy to obtain the same lower bound as before. Thus random choices do not help, in this specific sense.

We allow two types of random choices in our algorithms: random group elements, and random exponents. For the first, we might allow a new instruction

$$y_k \longleftarrow \text{rand}(G)$$

which assigns an element of G to y_k. On executing the circuit, this element is chosen uniformly at random, independent of other executions of the circuit. Actually, this feature is not used in any discrete log algorithm that we are aware of. For the corresponding trace exponent, we take new variables $t_1, \ldots, t_s$ if s instructions $\text{rand}(G)$ occur. Thus $\tau_k = t_i$ if $t_1, \ldots, t_{i-1}$ have been used so far. But actually this feature is not required, because the next one subsumes it.

We also want to allow random exponents, that is, an element y^e with random e and previously computed y. When $y = g$, this may be thought of as a random element of G with known discrete logarithm. So, as a new feature we allow our circuits to use a string

$$b = (b_1, \ldots, b_r) \in \{0, 1\}^r$$

of random bits, via assignments

$$y_k \longleftarrow y_i^{b_j}$$

with $i < k$ and $1 \leq j \leq r$. The corresponding trace exponent is

$$\tau_k = b_j \cdot \tau_i.$$

Example 11. In a probabilistic version of Pollard's rho method, the next element y_{k+1} is calculated as one of $y_k \cdot g$, y_k^2, or $y_k \cdot x$, each with probability $1/3$. This is easy to simulate, using two random bits b and c. If we set

$$y_{k+1} = g^{(1-b)(1-c)} y_k^{(1-b)c+1} x^{b(1-c)}, \tag{12}$$

then y_{k+1} will take one of the three required values for $(b, c) = (0, 0), (0, 1), (1, 0)$, respectively. We set the probability of $(b, c) = (1, 1)$ to 0. The formula can be implemented with an arithmetic circuit of size 11. In another version of Pollard's rho method, one divides the group into three parts and makes the three-fold choice according to where y_k has landed. This does not fit into our model.   ◇

Definition 13. *(i) A **probabilistic arithmetic circuit** is a pair $C = (C_r, u)$ consisting of a probability distribution u on $\{0,1\}^r$ for some nonnegative integer r and an arithmetic circuit C_r as in Definition 3 except that in addition the following type of assignment is allowed:*

$$y_k \longleftarrow y_i^{b_j} \tag{14}$$

with $-2 \leq i < k$ and $1 \leq j \leq r$.

(ii) The size ℓ of C is the number of group operations performed. Operations of the type (14) are not counted. (Formally, we might give appropriate rational indices k to their results.)

(iii) If $b \in \{0,1\}^r$ is provided, then we obtain a circuit $C(b)$ of size ℓ as follows. If y_k is given by (14), then we replace all references to y_k by a reference to $y_{-2} (= 1)$ if $b_j = 0$, and by a reference to y_i if $b_j = 1$. This replacement is performed recursively starting at the beginning of the instruction list until no more references to an assignment of type (14) exist. The new instructions are denoted as $y_k(b) \longleftarrow y_i(b) \cdot y_j(b)^\varepsilon$.

(iv) For a divisor q of d, the success rate of C with respect to q is

$$\sigma_{C,q} = \sum_{b \in \{0,1\}^r} u(b) \cdot \sigma_{C(b),q}.$$

Thus $\sigma_{C,q}$ is the average success rate of the $C(b)$ for random b. Recall that

$$\sigma_{C(b),q} = d^{-1} \cdot \#\{x \in G : \text{there is a collision } y_i(b)(x) = y_j(b)(x) \text{ respecting } q\},$$

and such a collision respects q if and only if $\tau_i(b) - \tau_j(b) \not\equiv 0 \bmod q$, with the usual trace exponent $\tau_i(b), \tau_j(b) \in \mathbb{Z}[t]$. These are defined only when some $b \in \{0,1\}^r$ is fixed, not for C itself.

Theorem 15. *Let $G = \langle g \rangle$ be a finite cyclic group, p a prime divisor of the group order $d = \#G$, C a probabilistic arithmetic circuit of size ℓ, and $\sigma_{C,p}$ it success rate respecting p. Then*

$$\ell \geq \sqrt{2\sigma_{C,p}p} - 3.$$

When $\sigma_{C,p}$ is a constant, then $\ell \in \Omega(\sqrt{p})$.

Proof. The probabilistic circuit $C = (C_r, u)$ and each (deterministic) circuit $C(b)$ have size ℓ. From the proof of Theorem 8, we have

$$\frac{(\ell + 3)^2}{2p} \geq \sigma_{C(b),p}$$

for each $b \in \{0,1\}^r$. Hence

$$\frac{(\ell + 3)^2}{2p} = \frac{(\ell + 3)^2}{2p} \sum_{b \in \{0,1\}^r} u(b) \geq \sum_{b \in \{0,1\}^r} u(b)\sigma_{C(b),p} = \sigma_{C,p}. \qquad \square$$

References

1. LÁSZLÓ BABAI & ENDRE SZEMERÉDI (1984). On the complexity of matrix group problems I. In *Proceedings of the 25th Annual IEEE Symposium on Foundations of Computer Science*, Singer Island FL, 229–240. IEEE Computer Society Press. ISBN 0-8186-0591-X. ISSN 0272-5428.
2. DAN BONEH & RICHARD J. LIPTON (1996). Algorithms for Black-Box Fields and their Application to Cryptography. In *Advances in Cryptology: Proceedings of CRYPTO '96*, Santa Barbara CA, NEAL KOBLITZ, editor, number 1109 in Lecture Notes in Computer Science, 283–297. Springer-Verlag. ISSN 0302-9743.
3. UELI MAURER & STEFAN WOLF (1998). Lower Bounds on Generic Algorithms in Groups. In *Advances in Cryptology: Proceedings of EUROCRYPT 1998*, Santa Barbara, CA, KAISA NYBERG, editor, number 1403 in Lecture Notes in Computer Science, 72–84. Springer-Verlag. ISSN 0302-9743.
http://link.springer.de/link/service/series/0558/bibs/1403/14030072.htm.
4. UELI M. MAURER & STEFAN WOLF (1999). The relationship between breaking the Diffie-Hellman protocol and computing discrete logarithms. *SIAM Journal on Computing* **28**(5), 1689–1721.
5. V. I. NECHAEV (1994). К вопросу о сложности детерминированного алгоритма для дискретного логарифма. Российская Академия Наук. Математические Заметки **55**(2), 91–101, 189. ISSN 0025-567X. Complexity of a determinate algorithm for the discrete logarithm, *Mathematical Notes* **55**(2) (1994), 165-172.
5. CLAUS PETER SCHNORR (2001). Security of DL-encryption and signatures against generic attacks-a survey. In *Public-Key Cryptography and Computational Number Theory Conference 2000*, 257–282.
http://www.mi.informatik.uni-frankfurt.de/research/papers.htm\%1.
6. CLAUS PETER SCHNORR & MARKUS JAKOBSSON (2000). Security Of Discrete Log Cryptosystems in the Random Oracle and the Generic Model. Technical report, Universität Frankfurt/Main and Bell Laboratories, Murray Hill, New Jersey. http://www.mi.informatik.uni-frankfurt.de/research/papers.htm\%1. The Mathematics of Public-Key Cryptography, The Fields Institute, Toronto.
7. VICTOR SHOUP (1997). Lower Bounds for Discrete Logarithms and Related Problems. In *Advances in Cryptology: Proceedings of EUROCRYPT 1997*, Konstanz, Germany, SPRINGER-VERLAG, editor, number 1233 in Lecture Notes in Computer Science, 256–266. ISSN 0302-9743. http://www.shoup.net/papers/.
8. V. STRASSEN (1972). Berechnung und Programm. I. *Acta Informatica* **1**, 320–335.

On the Competitiveness of AIMD-TCP within a General Network

Jeff Edmonds*

Department of Computer Science, York University, Toronto, Canada
jeff@cs.yorku.ca
http://www.cs.yorku.ca/~jeff

Abstract. This paper presents a new mathematical model of AIMD (Additive Increase Multiplicative Decrease) TCP for general networks that we believe is better than those previously used when it is driven by bottleneck capacities. Extending the paper by Edmonds, Datta, and Dymond that solves the single bottleneck case, we view AIMD as a distributed scheduling algorithm and prove that with extra resources, it is competitive against the optimal global algorithm in minimizing the average flow time of the jobs.

Keywords: AIMD, TCP, online competitive ratio, flow time, fairness, multi-bottleneck.

1 Introduction

AIMD (Additive Increase Multiplicative Decrease) is the core algorithmic component of TCP (Transport Control Protocol) for allocating bandwidth or transmission rate to the different jobs. In this algorithm, each job J_i increases his bandwidth linearly at a rate of $\delta b_{i,t}/\delta t = \alpha$ (typically $\alpha = 1$) until he detects that one of the bottlenecks that his transmission passes through has reached capacity, at which point he cuts his bandwidth by a multiplicative factor of β (typically $\beta = \frac{1}{2}$).

This simple algorithm is understood quite well when the network is restricted to a single bottleneck. [3] proves that even though each sender has no global knowledge of the state of the network, the allocation converges quite quickly to EQUI, which partitions the bandwidth equally between the active jobs. Though this is fair to all users, it does not perform well at minimizing the average flow/response/waiting time of the jobs, which is the standard measure both in the systems and the scheduling communities. In fact, [14] proves the *competitive ratio* of this online, non-clairvoyant scheduler can be as bad as $\mathcal{O}(\frac{n}{\log n})$ when measured against the optimal all-powerful, all knowing, off-line scheduler, which in this case is Shortest Remaining Work First. When there is such a negative result, a typical way to prove that the scheduler does perform well is to give it some extra resources before comparing it to the optimal scheduler, [8]. (See

* Supported in part by NSERC grant.

M. Farach-Colton (Ed.): LATIN 2004, LNCS 2976, pp. 567–576, 2004.
© Springer-Verlag Berlin Heidelberg 2004

Section 4 for additional motivation.) [4] does this proving that EQUI is $(2 + \epsilon)$-speed $\mathcal{O}(1+\frac{1}{\epsilon})$-competitive, meaning that when EQUI is given $2+\epsilon$ times as much bandwidth, it performs within a constant as well as the optimal. AIMD, however, is different from EQUI. Its allocations continually increase and decrease and it takes some time for it to reconverge after jobs arrive or depart. [6] proves that if AIMD is given a constant number of adjustment periods per job to converge than it is also $\mathcal{O}(1)$-speed $\mathcal{O}(1)$-competitive.

The main purpose of this paper is to extend these results to the multi-bottleneck case. There is little work done in this area. It is much harder, because it is not at all clear what either the steady state of AIMD, "EQUI", or the optimal are. We help to answer each of these three questions.

Surprisingly there has not previously been a model of how AIMD changes or to what it converges. Kelly in [12,10] does a good job, but the algorithm they consider is different. In their AIMD, the frequency at which a bottleneck drop packets, instructing its jobs to decrease their bandwidth changes as fixed function that depends only on the current total traffic through the bottleneck in question. In contrast, in the standard AIMD algorithm for TCP, a bottleneck instructs its jobs to back off only when it reaches its capacity. The frequency at which this occurs is a much more complex function of what the other bottlenecks are doing. In Section 2, we define a new continuous model of how AIMD evolves on a general network within this setting and also define the scheduler, AIMDEQUI, to be that to which it converges.

Because different jobs pass through different bottlenecks, the notion of the *fairness* of bandwidth allocation is not well define. Section 3 considers three notions of fairness. According to a *socialist* view of fairness, [7] prove that AIMD can be unfair by a factor of m, where m is a bound on the number of bottlenecks that a job goes through. We show that according to a *local* view of fairness, it is never more than a factor of m unfair and that according to a *free market* view, it is perfectly fair.

Finally, Section 4 proves that AIMDEQUI is $\mathcal{O}(m^3)$-speed $\mathcal{O}(m)$-competitive, meaning that with $\mathcal{O}(m^3)$ times the bandwidth, the flow time under AIMDEQUI is within a factor of $\mathcal{O}(m)$ of that of the optimal all knowing scheduler. We believe that it is not unreasonable to assume that m is a constant because within the actual internet no transmission hops more than a half dozen times. We are also able to prove that AIMDEQUI is $\mathcal{O}(1)$-speed $\mathcal{O}(1)$-competitive independent of m. However, this result requires the assumption that the adjustment frequencies of the bottlenecks do not change much within the life of an individual job. This we believe is a reasonable assumption because the adjustment frequencies are a global property that should not be greatly effected by the arrival and the completion of individual jobs. We believe that the result is true without this assumption or minimally when given speed $s = \mathcal{O}(m)$, however, as of yet this has been unattainable.

2   The Continuous AIMD Model for General Networks

In this section, we propose two new models of AIMD through a general network. The first model is a set of differential equations similar to those given by Kelly in [12,10]. We argue, however, that ours is a better model of AIMD when it is driven by bottleneck capacities. Unlike Kelly, however, we are unable to prove that the system converges, though we have strong arguments that it does. To avoid this problem, we will simply define another model, denoted AIMDEQUI, which is the previous model at its steady state. It is this second model that we prove is competitive against the optimal bandwidth scheduling algorithm. We use the following notation:

- $\mathcal{B}$ is set of routers that act as bottlenecks, the kth of which has maximum bandwidth B_k. When the scheduler has "speed" s, this maximum bandwidth is increased to $s \cdot B_k$.
- $\mathcal{J} = \{J_i\}$ is the set of jobs (or sessions). Each job J_i is defined by its arrival time a_i, its file length l_i, and as done in [12,10], the subset of the bottlenecks $\mathcal{B}(i)$ that it passes through. Conversely let $\mathcal{J}_t(k)$ denote the set of jobs J_i that pass through the kth bottleneck and are active at time t. Note that as a simplifying assumption, we are ignoring the path that a job takes through these bottlenecks and any delays caused by transmission times. In particular, we are ignoring the fact that different jobs may have different transmission times.
- We denote by $b_{i,t}$ the bandwidth or transmission rate used by job J_i at time t. The restriction for the kth bottleneck is that $\sum_{i \in \mathcal{J}_t(k)} b_{i,t} \le sB_k$.
- We denote by c_i the time that the transmission of job J_i is completed. To accomplish this, the algorithm must allocate enough bandwidth so that $\int_{t \in [a_i, c_i]} b_{i,t} = l_i$.
- We measure the quality of a scheduling algorithm using the average flow/response/waiting time of the jobs, i.e. $\mathrm{Avg}_{i \in \mathcal{J}}[c_i - a_i]$.
- α is the *additive increase* and β the *multiplicative decrease* parameter set by the AIMD algorithm. Namely, each user increases his transmission rate linearly at a constant rate of $\delta b_{i,t}/\delta t = \alpha$ (typically $\alpha = 1$) until he detects that one of the bottlenecks that his transmission passes through has reached capacity. At this point, the sender cuts his own rate $b_{i,t}$ by a multiplicative factor of β (typically $\beta = \frac{1}{2}$).
- $f_{k,t}$, the *adjustment frequency*, will denote the instantaneous frequency at time t at which the event occurs in which the kth bottleneck reaches capacity and instructs its users to back off.

The equations relating these values are as follows.

$$\forall k \ \left(f_{k,t} \ge 0 \text{ and } \sum_{i \in \mathcal{J}_t(k)} b_{i,t} = sB_k \right) \text{ or } \left(f_{k,t} = 0 \text{ and } \sum_{i \in \mathcal{J}_t(k)} b_{i,t} < sB_k \right) \quad (1)$$

$$\forall i \ \frac{\delta b_{i,t}}{\delta t} = \alpha - (1 - \beta)b_{i,t} \sum_{k \in \mathcal{B}(i)} f_{k,t} \quad (2)$$

Equation 1 states that the total bandwidth $\sum_{i \in \mathcal{J}_t(k)} b_{i,t}$ through the kth bottleneck is bounded by its capacity sB_k. More over, this bottleneck instructs its users to back off if and only if it is at capacity. Equation 2 states that each job J_i continually increases his bandwidth linearly at a rate of $\delta b_{i,t}/\delta t = \alpha$ and approximates the effect of the multiplicative deceases. When any one of the bottlenecks that J_i passes through reaches capacity, its bandwidth $b_{i,t}$ decreases by a multiplicative factor of β, i.e. from $b_{i,t}$ to $\beta b_{i,t}$, which is a decrease of $(1 - \beta)b_{i,t}$. The number of times that this occurs during a time period of length δt is $\left[\sum_{k \in \mathcal{B}(i)} f_{k,t}\right] \delta t$ for a total decrease of $\left[(1 - \beta)b_{i,t} \sum_{k \in \mathcal{B}(i)} f_{k,t}\right] \delta t$. Clearly, Equation 2 is only a differential approximation of the decreases that occur at discrete points in time. This same approximation was made in [12,10].

The main difference between this model and Kelly's in [12,10] is that Kelly has a single equation $(f_{k,t}) = \mu_k = p_k(\sum_{i \in \mathcal{J}_t(k)} b_{i,t})$ defining a bottleneck's adjustment frequency $f_{k,t}$ as a function of the total flow $\sum_{i \in \mathcal{J}_t(k)} b_{i,t}$ through the bottleneck. Though Kelly defines μ_k instead to be "the proportion of marked packets", it is used in the same way in Equation 2 as we do and we assume that this quantity reflects the proportion of the jobs passing through the bottleneck that will adjust and hence is related to our frequency $f_{k,t}$. Moreover, Kelly does not speak of the bottlenecks having a capacity, but presumably this fixed non-negative, continuous, strictly increasing function p_k can be such that as this total flow increases towards the bottleneck's "capacity", a sufficiently strong message is given to the jobs to back off that this capacity is never exceeded.

In contrast, our model does not have a single equation defining a bottleneck's adjustment frequency $f_{k,t}$. We feel that this is a better model for AIMD when it is driven by bottleneck capacities, because when an individual bottleneck adjusts in practice does depend in an intricate way on when the other bottlenecks adjust. For example, having a job pass through a long line of m bottlenecks with the same capacities, should be equivalent to passing through only one. In Kelly's model, each of these bottlenecks will send the same message as if it were the only bottleneck and hence the job will back off m times more often. On the other hand, in our model, it is irrelevant and undefined which one of bottlenecks will adjust. We can only make claims about $\sum_k f_{k,t}$.

Not knowing which bottlenecks are at capacity adds extra complications. One way to ensures that each bottleneck is at capacity is to assume that each bottleneck k, has a local job $i(k)$ that goes only through the kth bottleneck. This job will be free to increase its bandwidth filling any remaining space in the bottleneck. This change allows us to ignore the second half of Equation 1.

Given the current bandwidth allocations $b_{i,t}$, the next values are determined by first solving a system of equation for the adjusting frequencies $f_{k,t}$ and then using these to compute $\delta b_{i,t}/\delta t$. The following matrix notation is useful. Let M denote the 0/1 matrix such that $M_{k,i} = 1$ iff the ith job is in the kth bottleneck. Similarly, define the vectors $B = \langle B_k \rangle$, $f = \langle f_{k,t} \rangle$, $b' = \langle \delta b_{i,t}/\delta t \rangle$, $0_K = \langle 0, \ldots, 0 \rangle$, and $1_n = \langle 1, \ldots, 1 \rangle$. In contrast, represent the bandwidths $b_{i,t}$ as an $n \times n$ matrix with diagonals $b_{i,t}$ and the rest zero. Equations 1 and 2 translate into $Mb1_n = sB$ and $b' = \alpha 1_n - (1 - \beta)bM^T f$. Note there is one equation and one unknown b'_i and $f_{k,t}$ for each job and for each bottleneck. We can solve these as follows. Differentiating the first gives $Mb' = 0_K$. Substituting the second into this gives $M(\alpha 1_n - (1-\beta)bM^T f) = 0_K$ or $\alpha M 1_n = (1-\beta)MbM^T f$. Solving this gives the required values $f = \frac{\alpha}{(1-\beta)}(MbM^T)^{-1}M1_n$. These values are used in $b' = \alpha 1_n - (1-\beta)bM^T f = \alpha 1_n - \alpha bM^T(MbM^T)^{-1}M1_n$ to compute b'. These in turn gives us the next values for $b_{i,t}$, namely $b_{i,t+\delta t} = b_{i,t} + b'_{i,t}\delta t$. (This can't easily be represented as a matrix because b is square and b' is a vector.)[1]

The steady state of this system occurs when $\delta b_{i,t}/\delta t = 0$. Equation 2, then gives $b_{i,t} = \frac{\alpha}{(1-\beta)} / \left(\sum_{k \in \mathcal{B}(i)} f_{k,t} \right)$. It is our strong belief, that this system quickly converges to this state. If the dynamic system allocates job J_i an amount that is different from this then Equation 2 automatically moves it closer. Assume, for example that job J_i just arrived and hence, $b_{i,t_0} = 0$. If we assume that the total frequency $f_{i,t} = \sum_{k \in \mathcal{B}(i)} f_{k,t}$ remains relatively constant for a few adjustment periods, then the single differential equation $\delta b_{i,t}/\delta t = \alpha - (1-\beta)b_{i,t}f_{i,t}$ can be solved in isolation from the others, giving $b_{i,(t_0+d)} = \frac{\alpha}{1-\beta}\frac{1}{f_{i,t}}(1 - e^{-(1-\beta)f_{i,t}d})$. The time until the AIMD allocation to the job is within a factor $1 - e^{-(1-\beta)q} \approx 1 - \beta^{-q}$ of the steady state allocation is $d_i = \frac{q}{f_{i,t}}$. In the single bottleneck case, this equals q adjustment periods, which corresponds exactly to the results given in [6].

To avoid the problem of whether the system quickly converges, we will simply define another model, denoted AIMDEQUI, which is the previous model at its steady state. Replacing Equation 2 with Equation 4 gives the equations defining AIMDEQUI to be:

$$\forall k \left(f_{k,t} \geq 0 \text{ and } \sum_{i \in \mathcal{J}_t(k)} b_{i,t} = sB_k \right) \text{ or } \left(f_{k,t} = 0 \text{ and } \sum_{i \in \mathcal{J}_t(k)} b_{i,t} < sB_k \right) \quad (3)$$

$$\forall i \ b_{i,t} = \frac{\alpha}{(1-\beta)} / \left(\sum_{k \in \mathcal{B}(i)} f_{k,t} \right) \quad (4)$$

In the matrix notation, these translate into $Mb1_n = B$ and $bM^T f = \frac{\alpha}{(1-\beta)}1_n$.

3   Socialistic, Local, and Free Market Views of Fairness

It is clear what a fair distribution is of a single resource like the bandwidth of a bottleneck. However, when different jobs are restricted by different bottle-

---

[1] If M were square and invertible then $b' = \alpha 1_n - \alpha bM^T[MbM^T]^{-1}M1_n = \alpha 1_n - \alpha bM^T[(M^T)^{-1}b^{-1}M^{-1}]M1_n = \alpha 1_n - \alpha 1_n = 0$. But we already know this from $Mb' = 0_K$. However, I confess, I do not fully understand what happens when M is not square or invertible? Also we need to be able to invert MbM^T. For what it is worth, MbM^T is positive semi-definite. In fact, $\forall z, z^T[MbM^T]z > 0$. Does this mean that MbM^T is invertible? I have no proof that $f_{k,t}$ does not go negative, which would go against the intuition.

necks with different capacities, it is not clear what is "fair". This section defines three views of fairness: *socialistic*, *local*, and a *Free Market*, with corresponding "Equal Partition" schedulers: SEQUI_s, LEQUI_s, and FEQUI_S. AIMDEQUI will be evaluated with respect to each.

The *socialistic* view attempts to give each job the same bandwidth. A job is only given more bandwidth than another if they are not in competition for this extra bandwidth. SEQUI_s achieves such a distribution of bandwidth as follows. Starting with zero bandwidth to each job, increase the bandwidth of each job equally, except fixing that to any job passing through a bottleneck that is at capacity. According to this view, [7] proves that AIMD can be unfair by a factor of m to jobs that pass through m bottlenecks. An open problem is to prove that this is the worst case.

In the *local* view, a bottleneck never gives a job more bandwidth than is fair from its local information. In the scheduler LEQUI_s, the kth bottleneck tries to allocate a fair share $\frac{sB_k}{n_{k,t}}$ of its bandwidth to each of the $n_{k,t} = |\mathcal{J}_t(k)|$ jobs that pass through it. A job, however, may not be able to receive this high of a bandwidth because of the constraints of its other bottlenecks. Therefore, LEQUI_s allocates to job J_i the minimum allocated by each of the bottlenecks though which it passes, i.e. $b_{i,t} = \min_{k\in\mathcal{B}(i)} \frac{sB_k}{n_{k,t}}$. This locality of the fairness is used to reduce a schedule on the general network G to one separate single bottleneck network for each of G's bottlenecks. Using this, Theorem 3 proves that though LEQUI sometimes allocates less bandwidth than it could, it is $\mathcal{O}(m^2)$-speed $\mathcal{O}(m)$-competitive. The same result automatically applies for SEQUI because it never allocates less bandwidth to any job. Lemma 3 proves that AIMDEQUI allocates at least $\frac{1}{m}$ as much. Theorem 2, stating that AIMDEQUI is $\mathcal{O}(m^3)$-speed $\mathcal{O}(m)$-competitive, follows.

The *free market* view of fair argues that it is not fair to allocate the same bandwidth to every job when the jobs pass through different numbers of bottlenecks with different demands on their bandwidth. Instead, in this view each job is charged by each bottleneck it passes through for the bandwidth that it uses at a cost which decreases proportional to the *supply*, namely its capacity sB_k, and increases proportional to the *demand*, namely the number of jobs $n_{k,t} = |\mathcal{J}_t(k)|$ passing through it or perhaps on the number $n_{k,t}^{max}$ that are most constrained by it. Then each job is allocated the same cost of bandwidth. AIMDEQUI itself is a scheduler that once the costs are rigged slightly is completely fair in this sense. The adjustment frequency $f_{k,t}$ of a bottleneck is a reasonable cost for its bandwidth because Lemma 2 proves that it is bounded within $\frac{\alpha}{(1-\beta)}[\frac{1}{m}\frac{n_{k,t}^{max}}{sB_k}, \frac{n_{k,t}}{sB_k}]$ and Lemma 1 proves equality of this relationship on average, i.e. $n_t = \frac{s(1-\beta)}{\alpha}\sum_k f_{k,t}B_k$. Being charged for its bandwidth by each bottleneck it passes through, Job J_i is charged a total of $(\sum_{k\in\mathcal{B}(i)} f_{k,t})b_{i,t}$. Equation 4 then enforces that the allocations of bandwidth are such that this charge is the same for all jobs. The global aspect of this view of fairness is used to reduce AIMDEQUI on the entire network G to a single network with a single bottleneck. This is used to prove Theorem 4, which states that AIMDEQUI is $\mathcal{O}(1)$-speed $\mathcal{O}(1)$-competitive when these adjustment frequencies $f_{k,t}$ do not change much within the life of an individual job.

4 The Competitiveness of AIMDEQUI

To understand the worst-case analysis results in the literature, we need to introduce and motivate resource augmentation analysis [8]. A scheduling algorithm A is said to be s-speed c-competitive if $\max_J \frac{A_s(J)}{\mathrm{OPT}_1(J)} \le c$ where $A_s(J)$ denotes the average flow time for the schedule given by A with a speed s on input J, and similarly $\mathrm{OPT}_1(J)$ denotes the flow time of the adversarial schedule for J with a unit speed.

Though most scheduling papers consider the allocation of a fixed number of processors between the active jobs, the results hold for our setting of allocating the fixed bandwidth of a single bottleneck network. It is shown in [4] that the algorithm, EQUI, which devotes an equal amount of processing power to each job, is a $(2+\epsilon)$-speed $O(1+1/\epsilon)$-competitive algorithm for scheduling of jobs with "natural" speed-up curves. The result in the original paper [4] stated $\frac{\mathrm{EQUI}_s(\mathcal{J})}{\mathrm{OPT}_1(\mathcal{J})} \le \frac{2s}{(s-2)}$. This was improved in [5] for the purpose of proving Theorem 2 to $1 + \mathcal{O}(\frac{\sqrt{s}}{s-2})$, which does not change the result $\mathcal{O}(\frac{1}{\epsilon})$ when the speed s is $2 + \epsilon$, but when the speed s is large, the improvement is from $2 + \mathcal{O}(\frac{1}{s})$ to $1 + \mathcal{O}(\frac{1}{\sqrt{s}})$. It is likely that the competitive ratio should be $1 + \mathcal{O}(\frac{1}{s})$, but as of yet that is unattainable.

To be more complete, all the results allows arbitrary "natural" speed-up curves. For example, fully parallelizable work is the usual in which the rate at which the work gets completed is in proportion $\angle$ to the amount of bandwidth/processors allocated. In contrast, sequential work $\llcorner$ gets completed at a fixed rate independent of the amount of resources allocated. Given that any non-clairvoyant algorithm (limited knowledge about jobs) is bound to waist lots of resources on sequential jobs, it is surprising that the algorithm is competitive against an all knowing adversary when given only a little extra resources. In fact, more general speedup curves are also allowed. One might not think that such a result would have any direct application to the problem of transmitting files. However, it does. Each sender may have a different upper bound on the rate at which it can transmit data. This can be modeled by representing the transmission with a job whose speedup function is fully parallelizable up to the senders capacity and then becomes sequential for any additional bandwidth allocated to it, namely $\angle$. It is not needed, but to make the proofs simpler, Lemma 1 in [4] proves that the worst cast set of jobs is such that every phase of every job is either fully parallelizable or is sequential. Hence, we will restrict our attention to these. Another improvement needed to prove Theorem 2 that [5] provides over [4] is that it allows the optimal scheduler to complete the fully parallelizable work and the sequential work independently. The formal statement needed is as follows.

Theorem 1 ([5]) *Let $\mathcal{J}$ be any set of jobs in a single bottleneck network in which each phase of each job can have an arbitrary sublinear-nondecreasing speedup function.* $\frac{\text{EQUI}_s(\mathcal{J})}{\text{OPT}_1(\mathcal{J}_{par})+\text{OPT}_1(\mathcal{J}_{seq})} \leq 1 + \mathcal{O}(\frac{\sqrt{s}}{s-2})$, *where $\mathcal{J}_{par}$ and $\mathcal{J}_{seq}$ contain respectively only the non-sequential and the sequential phases of the jobs $\mathcal{J}$.*

The main result of this paper is that AIMDEQUI despite being online, non-clairvoyant, and distributed is $\mathcal{O}(m^3)$-speed $\mathcal{O}(m)$-competitive.

Theorem 2 *Let G be any general network. Let $\mathcal{J}$ be any set of jobs in which each phase of each job can have an arbitrary sublinear-nondecreasing speedup function. Let m denote the maximum number of bottlenecks that a job passes through. It follows that* $\frac{\text{AIMDEQUI}_{\mathcal{O}(m^3)}(\mathcal{J})}{\text{OPT}_1(\mathcal{J})} \leq \mathcal{O}(m)$.

Proof of Theorem 2: The result follows from Theorem 3 stating that LEQUI is $\mathcal{O}(m^2)$-speed $\mathcal{O}(m)$-competitive and from Lemma 3 stating that AIMDEQUI allocates at least $\frac{1}{m}$ as much bandwidth as LEQUI to each job. ∎

Theorem 3 $\frac{\text{LEQUI}_{\mathcal{O}(m^2)}(\mathcal{J})}{\text{OPT}_1(\mathcal{J})} \leq \mathcal{O}(m)$.

Proof of Theorem 3: This proof uses the fact that LEQUI is locally fair at each bottleneck. It is a reduction to many instances of Theorem 1. For each bottleneck within the general network G, the proof reduces what occurs in that bottleneck to a separate single bottleneck network with capacity B_k on a job set denoted $\mathcal{J}^k$. The proof is two pages and is quite involved. It is omitted because of the page restriction. ∎

The previous result can be completely tightened giving that AIMDEQUI is $(2 + \epsilon)$-speed $\mathcal{O}(1)$-competitive if we assume that the adjustment frequencies of the bottlenecks do not to change much within the life of an individual job. This we believe is a reasonable assumption because the adjustment frequencies are a global property that should not be greatly effected by the arrival and the completion of individual jobs.

Theorem 4 *Let G be any general network. Let $\mathcal{J}$ be any set of jobs in which each phase of each job can have an arbitrary sublinear-nondecreasing speedup function. Suppose for each job J_i, the ratio $(\sum_{k\in\mathcal{B}(i)} f_{k,t})/(\sum_k f_{k,t}B_k)$ between adjustment frequencies does not change by more than a factor of r through out the life of a job, where $r \geq 1$ is some constant. It follows that* $\frac{\text{AIMDEQUI}_{r(2+\epsilon)}(\mathcal{J})}{\text{OPT}_1(\mathcal{J})} \leq \mathcal{O}(1+\frac{1}{\epsilon})$.

Proof of Theorem 4: This proof uses the fact that AIMDEQUI is Free Market Fair. It is a simple reduction to Theorem 1 by reducing everything that is occurring within the general network G to a single network with a single bottleneck with capacity $B = 1$, namely $\frac{\text{AIMDEQUI}_{r(2+\epsilon)}(G,\mathcal{J})}{\text{OPT}_1(G,\mathcal{J})} \leq \frac{\text{EQUI}_{(2+\epsilon)}(1,\mathcal{J}^f)}{\text{OPT}_1(1,\mathcal{J}^f)} = \mathcal{O}\left(1+\frac{1}{\epsilon}\right)$. The last step is a direct application of Theorem 1.

Define $F_{i,t} = (\sum_{k\in\mathcal{B}(i)} f_{k,t})/(\sum_k f_{k,t}B_k)$ to be a needed comparison between the adjusting frequency of job J_i at time t and that of the overall network. By the statement of the theorem, this does not change by more than r through out the life of the job and hence $F_i \leq F_{i,t} \leq rF_i$ for some F_i. We construct another set of jobs $\mathcal{J}^f$ by scaling the fully parallel work in job $J_i \in \mathcal{J}$ by this constant F_i.

The first step is to prove that $\text{AIMDEQUI}_{r(2+\epsilon)}(G, \mathcal{J}) \leq \text{EQUI}_{(2+\epsilon)}(1, \mathcal{J}^f))$. By induction on t, assume that at time t $\text{AIMDEQUI}_{r(2+\epsilon)}(G, \mathcal{J})$ has completed at least as much work on each job as $\text{EQUI}_{(2+\epsilon)}(1, \mathcal{J}^f)$. We prove as follows that the first algorithm allocates at least $\frac{1}{F_i}$ times more bandwidth to job J_i at this time than the second does, i.e. $b_{i,t}^A \geq \frac{1}{F_i} \cdot b_{i,t}^E$. By the bound on $F_{i,t}$ given by the theorem and that on $b_{i,t}^A$ given in Equation 4,

$F_i \cdot b_{i,t}^A \geq \frac{1}{r} F_{i,t} \cdot b_{i,t}^A = \frac{1}{r} \left[(\sum_{k \in \mathcal{B}(i)} f_{k,t})/(\sum_k f_{k,t} B_k) \right] \cdot \left[\frac{\alpha}{(1-\beta)}/(\sum_{k \in \mathcal{B}(i)} f_{k,t}) \right] =$

$(2 + \epsilon)/ \left(r(2+\epsilon) \frac{(1-\beta)}{\alpha} (\sum_k f_{k,t} B_k) \right)$. By Lemma 1, this is $(2 + \epsilon)/n_t^A$. By the induction hypothesis $n_t^A \leq n_t^E$ and hence this is at least $(2 + \epsilon)/n_t^E$, which is the bandwidth $b_{i,t}^E$, allocated by $\text{EQUI}_{(2+\epsilon)}(1, \mathcal{J}^f)$. Because $b_{i,t}^A \geq \frac{1}{F_i} \cdot b_{i,t}^E$, $\text{AIMDEQUI}_{r(2+\epsilon)}(G, \mathcal{J})$ continues to keep up. The second algorithm has F_i times as much fully parallelizable work and by definition sequential work completes at a fixed rate independent of the number of processors allocated. This completes the proof by induction.

The final step in the proof is to compare the optimal algorithms $\text{OPT}_1(G, \mathcal{J}) \geq \text{OPT}_1(1, \mathcal{J}^f)$. This is done by constructing another algorithm OPT_1^f for which $\text{OPT}_1(G, \mathcal{J}) = \text{OPT}_1^f(1, \mathcal{J}^f)$. Because $\text{OPT}_1(1, \mathcal{J}^f)$ is the optimal algorithm, $\text{OPT}_1^f(1, \mathcal{J}^f) \geq \text{OPT}_1(1, \mathcal{J}^f)$. $\text{OPT}_1^f(1, \mathcal{J}^f))$ is defined to be the same as $\text{OPT}_1(G, \mathcal{J})$ except that the bandwidth allocated to job J_i is scaled by F_i, i.e. $b_{i,t}^f = F_i \cdot b_{i,t}^G$. $\text{OPT}_1(G, \mathcal{J}) = \text{OPT}_1^f(1, \mathcal{J}^f)$, because both the amount of parallel work and the number of processors have been scaled by F_i. What remains is to show that $\text{OPT}_1^f(1, \mathcal{J}^f)$ does not allocate more than a total of one bandwidth at any given time, namely $\sum_i b_{i,t}^f = \sum_i F_i \cdot b_{i,t}^G \leq \sum_i F_{i,t} \cdot b_{i,t}^G = \sum_i (\sum_{k \subset \mathcal{B}(i)} f_{k,t})/(\sum_k f_{k,t} B_k) \cdot b_{i,t}^G = (\sum_k f_{k,t}(\sum_{i \in \mathcal{J}_t(k)} b_{i,t}^G))/(\sum_k f_{k,t} B_k)$. Because $\text{OPT}_1(G, \mathcal{J})$ cant exceed the capacity of the kth bottleneck, this is at most $(\sum_k f_{k,t}(B_k))/(\sum_k f_{k,t} B_k) = 1$. This completes all the required steps of the proof. ∎

Lemma 1 *The number n_t of jobs active at time t under AIMDEQUI_s is $n_t = \frac{s(1-\beta)}{\alpha} \sum_k f_{k,t} B_k$.*

Proof of Lemma 1: $n_t = \sum_{\text{active } i} 1$, which by both the left and right sides of Equation 4 equals $\sum_i \sum_{k \in \mathcal{B}(i)} f_{k,t} b_{i,t} \frac{(1-\beta)}{\alpha} = \frac{(1-\beta)}{\alpha} \sum_k f_{k,t}(\sum_{i \in \mathcal{J}_t(k)} b_{i,t})$, which by Equation 3 equals $\frac{(1-\beta)}{\alpha} \sum_k f_{k,t} s B_k$. ∎

Lemma 2 *The adjustment frequency for the kth bottleneck is bounded by $f_{k,t} \in \frac{\alpha}{(1-\beta)} \left[\frac{1}{m} \frac{n_{k,t}^{max}}{s B_k}, \frac{n_{k,t}}{s B_k} \right]$.*

The proof is a simple quarter page algebric proof. It is omitted because of the page restriction.

Lemma 3 *The bandwidth allocated by $\text{AIMDEQUI}_s(\mathcal{J})$ to job J_i at time t is at least $\frac{1}{m}$ that allocated by $\text{LEQUI}_s(\mathcal{J})$, i.e. $b_{i,t} \geq \frac{1}{m} \min_{k \in \mathcal{B}(i)} \frac{s B_k}{n_{k,t}}$. It follows that $\frac{\text{AIMDEQUI}_{ms}(\mathcal{J})}{\text{LEQUI}_s(\mathcal{J})} \leq 1$.*

Proof of Lemma 3:   Consider a job J_i. Equation 4 gives $(m \max_{k \in \mathcal{B}(i)} f_{k,t}) b_{i,t}$ $\geq (\sum_{k \in \mathcal{B}(i)} f_{k,t}) b_{i,t} = \frac{\alpha}{(1-\beta)}$ and hence $b_{i,t} \geq \frac{\alpha}{1-\beta} \frac{1}{m} \min_{k \in \mathcal{B}(i)} \frac{1}{f_{k,t}}$. Applying the bound in Lemma 2 gives $b_{i,t} \geq \frac{1}{m} \min_{k \in \mathcal{B}(i)} \frac{sB_k}{n_{k,t}}$. ∎

References

1. F. Baccelli and D. Hong. AIMD, Fairness and Fractal Scaling of TCP Traffic RR-4155 INRIA Rocquencourt and Infocom, June 2002
2. A. Borodin and R. El-Yaniv. *Online Computation and Competitive Analysis.* Cambridge University Press, 1998.
3. D.M. Chiu and R. Jain. Analysis of the increase and decrease algorithms for congestion avoidance in computer networks. *Computer networks and ISDN systems*, 17(1):1–14, 1989.
4. Jeff Edmonds. Scheduling in the dark. In *Journal of Theoretic Computer Science*, 1999 and *ACM Symposium on Theory of Computing*, pages 179–188, 1999.
5. Jeff Edmonds.   Scheduling in the dark – improved results: manuscript, http://www.cs.yorku.ca/~jeff, 2001.
6. J. Edmonds, S. Datta, and P. Dymond, TCP is Competitive Agains a Limited Adversary, *Proc. 15th Ann. ACM Symp. of Parallelism in Algorithms and Achitectures*, pp. 174-183, 2003.
7. S. Floyd. Connections with multiple congested gateways in packet-switched networks, part I: One-way traffic. *Computer communications review*, 21(5):30–47, October 1991.
8. Bala Kalyanasundaram and Kirk Pruhs. Speed is as powerful as Clairvoyance. *Journal of the ACM*, 47(4):617–643, 2000.
9. R. Karp, E. Koutsoupias, C. Papadimitriou, and S. Shenker. Optimization problems in congestion control. In *IEEE Symposium on Foundations of Computer Science*, pages 66–74, 2000.
10. F. Kelly. Mathematical modelling of the internet. In *Bjorn Engquist and Wilfried Schmid (Eds.), Mathematics Unlimited – 2001 and Beyond.* Springer, 2001.
11. F. Kelly. Fairness and stability of end-to-end congestion control European Control Conference, Cambridge, 2003
12. F. Kelly, A. Maulloo, and D. Tan. Rate control in communication networks: shadow prices, proportional fairness and stability. In *Journal of the Operational Research Society*, volume 49, 1998.
13. J. Kurose, and K. Ross, "Computer networking: A top-down approach featuring the Internet", Addison-Wesley, 2002.
14. R. Motwani, S. Phillips, and E. Torng. Non-clairvoyant scheduling. *Theoretical computer science (Special issue on dynamic and on-line algorithms)*, 130:17–47, 1994.

Gathering Non-oblivious Mobile Robots

Mark Cieliebak

Institute of Theoretical Computer Science
ETH Zurich
cieliebak@inf.ethz.ch

Abstract. We study the GATHERING PROBLEM, where we want to gather a set of n autonomous mobile robots at a point in the plane. This point is not fixed in advance. The robots are very weak, in the sense that they have no common coordinate system, no identities, no central coordination, no means of direct communication, and no synchronization. Each robot can only sense the positions of the other robots, perform a deterministic algorithm, and then move towards a destination point. It is known that these simple robots cannot gather if they have no additional capabilities. In this paper, we show that the GATHERING PROBLEM can be solved if the robots are non-oblivious, i.e., if they are equipped with memory.

1 Introduction

We consider a distributed system whose entities are autonomous mobile robots, where the robots can freely move in the two-dimensional plane. The coordination mechanism for these robots is totally *decentralized*, i.e., the robots are completely *autonomous* and no central control is used. The research interest is to establish a minimal set of capabilities the robots need to have to be able to perform a certain task, like forming a pattern. In this paper, we study the problem of gathering the robots at a point. This problem is known as GATHERING PROBLEM (or rendezvous, or point-formation problem) and is obviously one of the most primitive tasks that a set of robots might perform. The GATHERING PROBLEM has been studied intensively in the literature, in particular in the realm of distributed computing [2,4,5,7,8], but also in robotics [3] and artificial intelligence [6].

We study the GATHERING PROBLEM for a set of weak robots: the robots are anonymous (i.e., identical), they have no common coordinate system, and they have no means of direct communication. All robots operate individually, according to the following cycle: Initially, they are in a waiting state. They wake up independently and asynchronously, observe the other robots' positions, and compute a point in the plane. They start moving towards this points, but may not reach it (e.g. because of limits to the robot's motion energy). Then they become waiting again. Details of the model are given in Section 2. For these robots, the GATHERING PROBLEM is defined as follows:

Definition 1. *Given n robots $r_1, \ldots, r_n$, arbitrarily placed in the plane, with no two robots at the same position, make them gather at one point.*

M. Farach-Colton (Ed.): LATIN 2004, LNCS 2976, pp. 577–588, 2004.

If the robots are asked only to move "very close" to each other, this task is easily solved: each robot computes the center of gravity[1] of all robots, and moves towards it. However, in the GATHERING PROBLEM we ask the robots to meet at *exactly* one point.

If the robots are oblivious, i.e., if they do not remember previous observations and calculations, then the GATHERING PROBLEM is *unsolvable* [7,8]. On the other hand, the problem can be solved if we change the nature of the robots: If we assume a common coordinate system, gathering is possible even with limited visibility [5]; if the robots are synchronous and movements are instantaneous, then the GATHERING PROBLEM has a simple solution [8] and can be achieved even with limited visibility [2]; finally, the problem can be solved for more than two robots if the robots can detect how many robots are at a certain point (*multiplicity detection*) [4]. Recently, the GATHERING PROBLEM was studied in the presence of faulty robots; assuming a strong model of synchronizity, then the non–faulty robots can gather if at most one third of the robots are faulty [1].

In this paper, we show that the GATHERING PROBLEM is solvable for $n \geq 2$ non–oblivious robots. First, we present in Section 4 an algorithm that gathers $n = 2$ robots. At the beginning, two robots move on a line ℓ, which connects their initial positions, away from each other. As soon as both robots have observed the configuration at least once (hence, they know ℓ), they start moving on lines perpendicular to ℓ until, again, both have seen both perpendicular lines. Finally, they meet on ℓ in the center between the two perpendicular lines.

For more than two robots, we distinguish in Section 5 how many robots are on the smallest enclosing circle SEC of the positions of all robots in the initial configuration. If there are more than two robots on SEC, then each robot moves on a circle around the center of SEC until all robots have seen SEC. Hereby, we use the fact that the smallest enclosing circle of the robots positions does not change. Then all robots gather at the center of SEC. On the other hand, if there are only two robots on SEC, then the robots that are not on SEC move perpendicular to the line ℓ connecting the two robots on SEC, while the robots on SEC move on line ℓ away from each other. The smallest enclosing circle increases, but ℓ remains invariant. As soon as all robots have seen line ℓ and the configuration, they gather at the intersection between ℓ and a line k, which is the median perpendicular line of the robots, if n is odd, or the center between the two median perpendicular lines, if n is even.

2   Autonomous Mobile Robots

A robot is a mobile computational unit provided with sensors, and it is viewed as a point in the plane. Once activated, the sensors return the set of all points in the plane occupied by at least one robot. This forms the current *local view* of the robot. The local view of each robot also includes a unit of length, an origin (which we assume w.l.o.g. to be the position of the robot in its current observation),

---

[1] For n points $p_1, \ldots, p_n$ in the plane, the center of gravity is $c = \frac{1}{n} \sum_{i=1}^{n} p_i$.

and a coordinate system (e.g. Cartesian). There is no a priori agreement among the robots on the unit of length, the origin, or the coordinate systems.

A robot is initially in a *waiting* state (*Wait*). Asynchronously and independently from the other robots, it *observes* the environment (*Look*) by activating its sensors. The sensors return a snapshot of the world, i.e., the set of all points that are occupied by at least one other robot, with respect to the local coordinate system. The robot then *calculates* its destination point (*Compute*) according to its deterministic algorithm (the same for all robots), based only on its local view of the world. It then *moves* towards the destination point (*Move*); if the destination point is the current location, the robot stays still. A move may stop before the robot reaches its destination. The robot then returns to the waiting state. The sequence *Wait - Look - Compute - Move* forms a *cycle* of a robot.

The robots are *fully asynchronous*, i.e., the amount of time spent in each state of a cycle is finite but otherwise unpredictable. In particular, the robots do not have a common notion of time. As a result, robots can be seen by other robots while moving, and thus computations can be made based on obsolete observations. The robots are *anonymous*, meaning that they are a priori indistinguishable by their appearance, and they do not have any kind of identifiers that can be used during the computation. Finally, the robots have *no means of direct communication*: any communication occurs in a totally implicit manner, by observing the other robots' positions.

There are two limiting assumptions concerning *infinity*: The amount of time required by a robot to complete a cycle is not infinite, nor infinitesimally small; and the distance traveled by a robot in a cycle is not infinite, nor infinitesimally small (unless it brings the robot to the destination point). As no other assumptions on space exist, the distance traveled by a robot in a cycle is unpredictable. All times and distances are under control of the adversary. We assume in our algorithms that the adversary is *fair*, in the sense that he respects the previous assumptions, and that no robot sleeps forever, since otherwise no algorithm can guarantee to gather the robots.

For the remainder of this paper, we assume that the robots are *non–oblivious*, meaning that each robot is equipped with infinite memory, and its computation in each cycle can be based on its observations and computation results from previous cycles.

3   Notation

In general, r indicates any robot in the system; when no ambiguity arises, r is used also to represent the point in the plane occupied by that robot. A *configuration* of the robots at a given time instant t is the set of positions in the plane occupied by the robots at time t.

We say that a point p is *on a circle* if it is on the circumference of the circle, and that p is *inside the circle* if it is strictly inside the circle. Given three distinct points p, q and c, we denote by $\lessdot(p, c, q)$ the *convex angle* (i.e., the angle that is

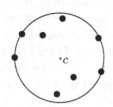

Fig. 1. Smallest enclosing circle SEC for 8 points.   **Fig. 2.** Proof of Lemma 2. Center of SEC cannot be at q.

at most 180°) between p and q, centered in c. The Euclidean distant between p and q is denoted by $dist(p, q)$.

Given a set of n distinct points P in the plane, the *smallest enclosing circle* of the points is the circle with minimum radius such that all points from P are inside or on the circle (see Figure 1). We denote it by $SEC(P)$, or SEC if set P is unambiguous from the context. The smallest enclosing circle of a set of n points is unique and can be computed in polynomial time [9].

The smallest enclosing circle of P remains invariant if we move some of the points from P that are inside SEC such that they remain inside SEC; moreover, the maximum angle between any two adjacent points on SEC w.r.t. the center of SEC is 180°, since otherwise there would be a smaller circle enclosing all points. The following lemma shows that the smallest enclosing circle remains invariant even if we move the points along the rim of SEC, as long as no angle of more than 180° between adjacent points occurs.

Lemma 1. *Let $P = \{p_1, \ldots, p_k\}$ be k points on a circle C with center c. If the maximum angle between any two adjacent points w.r.t c is at most 180°, then C is the smallest enclosing circle of the points.*

Proof (sketch). The idea of the proof is as follows (cf. Figure 2): Assume that the center of $SEC(P)$ would be at some point $q \neq c$. Then there are two adjacent points $x, y \in P$ such that their angle w.r.t. c is minimum (and at most 180°), and such that q is within the sector of C that is beyond c and delimited by the lines ℓ_x and ℓ_y from x and y, respectively, through c (bottom sector in Figure 2). Let ℓ be the perpendicular line that bisects the angle between x and y (dashed line ℓ in Figure 2). If x and q are not on the same side of ℓ, then $dist(x, c) \leq dist(x, q)$; otherwise, y and q are not on the same side of ℓ, and $dist(y, c) \leq dist(y, q)$. In both cases, the radius of C is at most the radius of $SEC(P)$. Thus, since the smallest enclosing circle is unique, we have $C = SEC(P)$.   □

4   Gathering Two Robots

In this section, we present an algorithm that solves the GATHERING PROBLEM for two robots. The idea of our algorithm, which is similar to the algorithm

---

Algorithm 1 Gathering two robots

If first observation **Then**
$\quad$ x_0 := my position; $y_0 \leftarrow$ other robot's position;
$\quad$ $\ell \leftarrow$ line through x and y; d_0 := distance between x and y;
$\quad$ $state \leftarrow 1$; move on ℓ by distance $\frac{d_0}{100}$ away from y;
5: **If** $state = 1$ **Then**
$\quad$ **If** other robot is at y_0 **Then** do nothing;
$\quad$ **Else**
$\quad\quad$ $x_{perp} \leftarrow$ my position; y_1 := other robot's position;
$\quad\quad$ d_1 := distance between x_{perp} and y_1;
10: $\quad\quad$ **If** other robot is on ℓ **Then** $state \leftarrow 2$; move perpendicular to ℓ by $\frac{d_1}{100}$;
$\quad\quad$ **Else**
$\quad\quad\quad$ $y_{perp} \leftarrow$ intersection between ℓ and line through other robots position
$\quad\quad\quad$ perpendicular to ℓ; $d_{perp} \leftarrow$ distance between x_{perp} and y_{perp};
$\quad\quad\quad$ $state \leftarrow 3$; move perpendicular to ℓ by distance $\frac{d_{perp}}{100}$;
If $state = 2$ **Then**
15: $\quad$ **If** other robot is on ℓ **Then** do nothing
$\quad$ **Else**
$\quad\quad$ $y_{perp} \leftarrow$ intersection between ℓ and line through other robots position per-
$\quad\quad$ pendicular to ℓ; $d_{perp} \leftarrow$ distance between x_{perp} and y_{perp};
$\quad\quad$ $state \leftarrow 3$; do nothing;
If $state = 3$ **Then**
20: $\quad$ **If** other robot is on the line perpendicular to ℓ through y_{perp} and less than d_{perp}
$\quad$ away from ℓ **Then** move perpendicular to ℓ to distance d_{perp};
$\quad$ **Else**
$\quad\quad$ $g \leftarrow$ center point between x_{perp} and y_{perp};
$\quad\quad$ $state \leftarrow 4$; do nothing;
If $state = 4$ **Then**
25: $\quad$ **If** I am not at g **Then** move to g **Else** $state \leftarrow STOP$; do nothing;
End.

---

presented in [8], is as follows: The two robots move away from each other until both have seen the configuration at least once. Then they know the connecting line ℓ through their initial positions. In a next phase, they both move on lines that are perpendicuar to ℓ, again until both have seen the other robot at least once on its perpendicular line. Then they both know ℓ and its intersection with the two perpendicular lines, hence, they can gather on ℓ in the center between the perpendicular lines.

Lemma 2. *Two robots can gather at a point.*

Proof. Both robots perform Algorithm 1. Here, we use $\leftarrow$ to assign a value to a variable that is stored in the permanent memory of the robot (and is available in subsequent cycles), while we use := to assign values to variables that are only used in the current cycle.

We now prove that this algorithms gathers the two robots. Let r and s be the two robots. The following proofs are presented from the point of view of one robot r; analogous proofs yield the same propositions for the other robot. We denote

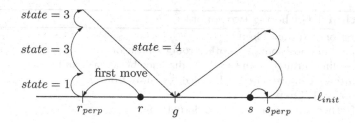

Fig. 3. Illustration of the algorithm for two robots. Distances are not drawn to scale.

the variables of robot r and s with superscript r and s, respectively. Let ℓ_{init} be the line through the initial positions of the robots, before any of the robots made its first movement. A schematic illustration of the robots' movements can be found in Figure 3.

1. If robot r is the first robot that leaves ℓ_{init}, then both robots agree on ℓ, i.e., $\ell^r = \ell^s = \ell_{init}$.

   Proof. If r leaves ℓ_{init} while s is still on the line, then r is in state 1 and $\ell^r = \ell_{init}$. Moreover, r has seen s in two different positions on ℓ_{init}, thus, s has moved on ℓ_{init} *before* r leaves ℓ_{init}. Hence, s has seen ℓ_{init} already, and we have $\ell^s = \ell_{init}$.

2. Both robots eventually leave ℓ_{init}.

   Proof. Assume that robot r wakes up first. Then it moves by $\frac{d_0^r}{100}$ and enters state 1. As soon as s has moved at least once, r moves away from ℓ_{init} by either $\frac{d_1^r}{100}$, if s is still on ℓ_{init}, or by $\frac{d_{perp}^r}{100}$, if s has left ℓ_{init}. Hence, as soon as r has observed the first movement of s, it leaves ℓ_{init}. If robot s has left ℓ_{init} at that time already, we are done. Otherwise, we know from Item 1 that s is in state 1, since it knows already ℓ_{init}, but it is still on ℓ_{init}. Hence, when s wakes up the next time, it observes that r has left ℓ_{init}, and s moves away from ℓ_{init} by $\frac{d_{perp}^s}{100}$.

3. Every subsequent movement of robot r after it left line ℓ_{init} is perpendicular away from ℓ_{init}, until it reaches state 4.

   Proof. When r moves away from ℓ_{init} for the first time, it is in state 1. By construction, this movement is perpendicular away from ℓ_{init}, starting in x_{perp}, by either distance $\frac{d_1^r}{100}$ or $\frac{d_{perp}^r}{100}$. Afterwards, robot r moves only if it is in state 3, and there the movements are by definition perpendicular to ℓ_{init}. It remains to show that r always moves *away* from ℓ_{init}. Since d_{perp}^r never changes, it is sufficient to show that $\frac{d_1^r}{100} < d_{perp}^r$. To see this, let d_{init} be the distance between the initial positions of the robots. When the robots wake up first, each of them makes one movement by at most $\frac{d_0^r}{100}$ and $\frac{d_0^s}{100}$, respectively, on ℓ_{init}, away from the other robot. Afterwards, all movements are perpendicular to ℓ_{init}. Hence, we have $d_{init} \leq d_1^r \leq d_{init} + \frac{d_0^r}{100} + \frac{d_0^s}{100}$.

With $d_0^r \leq d_{init} + \frac{d_0^s}{100}$ and $d_0^s \leq d_{init} + \frac{d_0^r}{100}$, straight–forward analysis shows that $d_1^r \leq \frac{10}{9} d_{init}$. This yields the claim, since d_{perp}^r is obviously greater than d_{init}.

4. Both robots eventually agree on point g, and gather there.

Proof. Due to Item 2, both robots eventually leave line ℓ_{init}, say at positions r_{perp} and s_{perp}. Let r be the robot that leaves ℓ_{init} first. Then r stores value r_{perp} in x_{perp}^r, moves by $\frac{d_1^r}{100}$ away from ℓ_{init}, and enters state 2, where it remains until s will have left ℓ_{init}. When s wakes up the next time, it observes that r has left ℓ_{init}, and moves perpendicular away from ℓ_{init}, too. Moreover, it stores s_{perp} in x_{perp}^s, and r_{perp} in y_{perp}^s, since robot r has moved only perpendicular to ℓ_{init} due to Item 3. The next time robot r wakes up, it observes that s has left ℓ_{init}, too, and stores $y_{perp}^r = s_{perp}$ (again, since s moved only perpendicular to ℓ_{init}). Hence, both robots agree on the points r_{perp} and s_{perp} where they left ℓ_{init}, on distance d_{perp} between these points, and on the center point g. Moreover, both robots move on their perpendicular line until at least one of them, say s, has reached distance d_{perp} from ℓ_{init} (state 3). When this is observe by the other robot r, it enters state 4 and moves straight towards g, hence, r leaves its perpendicular line. When s wakes up the next time, it observes that r has left its perpendicular line, and s starts moving towards g, too. Eventually, both robots reach g and gather there.

$\square$

5 Gathering $n > 2$ Robots

We now show how to gather more than two robots. We split the algorithm up into two separate cases, depending on the number of robots on the smallest enclosing circle SEC in the initial configuration: if there are at least three robots on SEC, we make all robots move on circles around the center of SEC until all robots know SEC (which does not change during the movements); then we gather the robots at the center of SEC. This is shown in the following Lemma 3. On the other hand, if there are exactly two robots on SEC, then we adapt the algorithm for two robots from Section 4 to gather all robots at the line connecting the two robots on SEC. This is shown in Lemma 4.

Lemma 3. *If there are more than 2 robots on the smallest enclosing circle in the initial configuration, then the robots can gather at a point.*

Proof. Given a configuration of the robots, we define a *movement angle* γ and a *movement direction moveDir* for each robot r as follows (cf. Figure 4): Let c be the center of the smallest enclosing circle of all robots. Let C be the circle with center c such that r is on C. If there is no other robot on C, then let $\gamma = \frac{1}{360n}$ and *moveDir* be an arbitrary direction on C, say clockwise. If there are exactly two robots on C, then let s be the other robot. [By assumption, C is not the

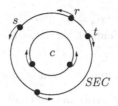

Fig. 4. Idea of Algorithm 2. Arrows indicate movement directions of the robots.

smallest enclosing circle of the robots.] Let α and β be the two angles between r and s w.r.t. c. Assume w.l.o.g. $\alpha \leq \beta$. Let $\gamma = \frac{\alpha}{360n}$, and let $moveDir$ be in the direction of angle β. If there are more than two robots on C, then let s and t be the two robots on C that are adjacent to r. Let α be the angle between r and s w.r.t. c, and β be the angle between r and t w.r.t. c. Assume w.l.o.g. that $\alpha \leq \beta$. Then $\alpha < 180°$. If $\alpha \leq 178°$, then let $\gamma = \frac{\alpha}{360n}$ and $moveDir = t$. If $178° < \alpha$, then $\gamma = \frac{180° - \alpha}{360n}$ and $moveDir = t$.

If robot r observes the configation of all robots, it can order the other robots in a unique way, for instance by using the coordinates of the robots positions in the local coordinate system of robot r. We assume w.l.o.g. that robot r has index 1 in this ordering. Recall that different robots may have different coordinate systems, hence, the robots do not agree on this ordering. We will ensure in our algorithm that the basic configuration remains invariant; in particular, robots will stay on the same circle with center c, and no two robots on the same circle will interchange their position. Each robot stores the positions of all robots that it observes in its first cycle in an array $posns$, where $posns_j$ denotes the position of robot r_j. Hence, in later cycles robot r can compare the current position of a robot r_j with the position of r_j observed in its first cycle. This allows r to determine whether r_j has made at least one movement at some time. In addition, robot r maintains a vector $hasMoved$, such that $hasMoved_j$ is set to $true$ if r has observed at least once that robot r_j has moved.

The algorithm that the robots perform is shown in Algorithm 2, and an illustration can be found in Figure 4. We prove that the robots gather at c, the center of the smallest enclosing circle of the robots initial positions, by showing the following items:

1. Every robot makes at most n moves by angle γ in its direction $moveDir$.

   Proof. A robot only moves in direction $moveDir$ in states 2 and 3. If it is in state 2, then it moves once in direction $moveDir$, sets $hasMoved_1 = true$, and changes into state 3. In state 3, it moves in direction $moveDir$ if a value $hasMoved_j$ has changed from $false$ to $true$ (i.e., if another robot has moved). This can happen at most $n - 1$ times, once for each other robot.

2. The angles between two adjacent robots on the same circle changes at most by $1°$.

   Proof. We have $\gamma \leq \frac{180°}{360n}$ by definition, and each robot moves at most n times by its angle γ. Hence, the movement of a single robot changes the

Algorithm 2 Gathering with more than 2 robots on SEC

If this is my first observation **Then**
$n \leftarrow$ number of robots;
$SEC \leftarrow$ smallest enclosing circle of all robots; $c \leftarrow$ center of SEC;
If I am at c **Then**
5:       $d :=$ minimum distance of any other robot to c;
$state \leftarrow 2$; move away from c by distance $\frac{d}{2}$;
Else
If some robot is at c **Then** $state \leftarrow 1$; do nothing;
Else $state \leftarrow 2$; do nothing;
10: **If** $state = 1$ **Then**
If a robot is at c **Then** do nothing;
Else $state \leftarrow 2$; do nothing;
If $state = 2$ **Then**
$posns \leftarrow$ all robots positions, with $posns_1$ my own position;
15:    $\forall j : hasMoved_j \leftarrow false$; $hasMoved_1 \leftarrow true$;
$\gamma \leftarrow$ my movement angle; $moveDir \leftarrow$ my movement direction;
$state \leftarrow 3$; move by angle γ in direction $moveDir$;
If $state = 3$ **Then**
If a robot decreased its distance from c **Then** $state \leftarrow 4$; do nothing;
20:   **Else**
$\forall j$ such that robot r_j changed its position w.r.t $posns_j$: $hasMoved_j \leftarrow true$;
If $\forall j : hasMoved_j = true$ **Then** $state \leftarrow 4$; do nothing;
If at least one value $hasMoved_j$ changed to $true$ in this step **Then**
move by angle γ in direction $moveDir$;
25:    **Else** do nothing;
If $state = 4$ **Then**
If I am not at c **Then** move to c **Else** $state \leftarrow STOP$; do nothing;
End.

angle between itself and its neighbors by at most $n\gamma \leq \frac{1}{2}°$. Thus, even if two adjacent robots move in opposite directions, the angle between them changes by at most $1°$.

3. No two robots on the same circle interchange their position.

Proof. Let v, w, x and y be adjacent robots (in this ordering) on the same circle with center c. Assume by contradiction that w and x interchange their positions. We show that this cannot happen even if w and x move towards each other. The other cases, where either x and w move in the same direction, or they move away from each other, can be shown analogous. If w and x move towards each other, then $\sphericalangle(w, c, x) > \sphericalangle(v, c, w)$ and $\sphericalangle(w, c, x) > \sphericalangle(x, c, y)$. By construction, we have $\gamma_w \leq \frac{\sphericalangle(w,c,x)}{360n}$: if $\sphericalangle(v, c, w) \leq 178°$, then $\gamma_w = \frac{\sphericalangle(v,c,w)}{360n}$; on the other hand, if $\sphericalangle(v, c, w) > 178°$, then $\gamma_w = \frac{180° - \sphericalangle(w,c,x)}{360n} \leq \frac{\sphericalangle(w,c,x)}{360n}$. Analogously, $\gamma_x \leq \frac{\sphericalangle(w,c,x)}{360n}$. Robot w moves at most by angle $n \cdot \gamma_w$ towards x, and robot x moves at most by angle $n \cdot \gamma_x$ towards w (due to Item 1). Hence, the new angle between w and x is at least

$\sphericalangle(w,c,x) - n\gamma_w - n\gamma_x \geq \sphericalangle(w,c,x)(1 - \frac{1}{180}) > \frac{\sphericalangle(w,c,x)}{2} > 0°$, i.e., the two robots do not interchange their position.

4. *SEC* remains invariant until at least one robot has reached its state 4.

Proof. Until some robots reach their state 4, all robots move on circles with center c. Hence, the smallest enclosing circle can only change if the maximum angle between the robots on *SEC* becomes larger than 180° (cf. Lemma 1). By previous Item 2, the angle between adjacent robots changes by at most 1°; thus, if all adjacent robots on *SEC* in the initial configuration have angle at most 178°, the smallest enclosing circle cannot change. If in the initial configuration there is exactly one angle between adjacent robots on *SEC* that is greater than 178°, say between robots x and y, then this is for both x and y the maximum adjacent angle. Hence, the moving direction of x is towards y by definition, and the moving direction of y is towards x. Thus, the angle between x and y decreases, and no angle of more that 180° can occur.

For the case that there are 2 angles of more than 178°, first assume that there is no robot on *SEC* that has an angle of more than 178° to both neighbors (see Figure 5). Then there are two disjoint pairs of robots x, y and u, v such that the angle between x and y is greater than 178°, and the angle between u and v is greater than 178°. By construction, x moves towards y and y towards x, decreasing the angle between them. Likewise, u and v move towards each other. Hence, no angle greater than 180° occurs.

Now assume that there is one robot r on *SEC* such that both angles α and β to its two neighbors s and t, respectively, are greater than 178° (see Figure 6). Assume that $\alpha \leq \beta$. Both s and t move towards r. By definition, the movement angle for robot r is $\gamma = \frac{180° - \alpha}{360n}$, and r moves towards t. Hence, the angle between r and t decreases. On the other hand, even if s does not move at all, and even if r moves by maximum angle $n\gamma$ towards t, then the new angle between s and r is at most $\alpha + n\gamma \leq 180°$. Hence, *SEC* does not change.

5. If a robot reaches its state 4, then all robots agree on *SEC* and c.

Proof. A robot r reaches its state 4 only if $hasMoved_j^r = true$ for all $1 \leq j \leq n$. This yields the claim, since $hasMoved_j^r$ is set to $true$ only if robot r_j has made a move, i.e., if it was awake and had observed the configuration, including *SEC* and c.

6. At least one robot eventually reaches its state 4.

Proof. Let r be the first robot to wake up. Then r observes the initial configuration of the robots. If there is a robot at c in the initial configuration, then this robot moves away from c in its first cycle. Afterwards, every robot that wakes up moves on its circle by its movement angle γ. Assuming a fair schedule where no robot sleeps for an infinite time, after some finite time every robot has woken up at least once. If some other robot but r reaches its state 4, then the claim is true. Otherwise, as soon as robot r wakes up

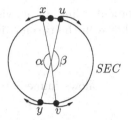 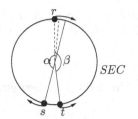 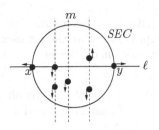

Fig. 5. Proof of Lemma 3, Item 4, for angles $\alpha, \beta > 178°$. Angles are not drawn to scale.

Fig. 6. Proof of Lemma 3, Item 4, for angles $\alpha, \beta > 178°$. Angle between the dashed lines is $n\gamma$. Angles are not drawn to scale.

Fig. 7. Idea of algorithm for two robots on SEC. Line m is the median perpendicular line.

the next time, it observes that all other robots have moved since its first observation (i.e., $hasMoved^r_j = true$ for all $1 \leq j \leq n$), and r enters state 4.

7. All robots eventually reach their state 4.

Proof. Due to Item 6, at least one robot r reaches its state 4. In its next cycle, this robot moves towards c, i.e., it decreases its distance from c. Hence, all other robots that wake up afterwards observe this decrease of the distance, and enter their state 4. Assuming a fair schedule where no robot sleeps forever yields the claim.

8. All robots gather at c and stop there.

Proof. This is obvious, since all robots agree on c due to Item 5, all robots reach their state 4 due to Item 7, and each robot that is in state 4 moves towards c.

□

We now show how to solve the GATHERING PROBLEM if only two robots are on the smallest enclosing circle in the initial configuration.

Lemma 4. *If $n > 2$, and there are exactly 2 robots on the smallest enclosing circle in the initial configuration, then the robots can gather at a point.*

Proof (sketch). Let x and y be the two robots on smallest enclosing circle, and let ℓ be the line through x and y. Our algorithm works as follows (see Figure 7). First, all robots move "a little bit" until each robot has moved at least once. Here, both x and y move on ℓ away from each other. Every other robots r moves on a line perpendicular to ℓ, without reaching the next robot (if any) on the same line. The movement of x and y changes the smallest enclosing circle (in fact, it increases the radius of the circle), but x and y remain the only robots on this circle. Hence, each of the other robots moves always on the *same* line perpendicular to ℓ. As soon as all robots have made one move, they all know ℓ and all perpendicular lines. If the number of robots n is odd, then all robots gather at the intersection of ℓ and the median perpendicular line. Otherwise, they gather at the intersection of ℓ and the center line between the two median perpendicular lines.

□

We summarize our result in the following theorem, which follows immediately from Lemmas 2, 3 and 4.

Theorem 1. *The* GATHERING PROBLEM *can be solved for $n \geq 2$ non–oblivious robots.*

6 Conclusion

We have presented an algorithm that gathers a set of n non–oblivious mobile robots. Thus, it is sufficient to equip the robots with memory to make the GATHERING PROBLEM become solvable. Moreover, our results indicates that memory is a more powerful capability than multiplicity detection, since we have shown that two robots with memory can gather, while two robots with multiplicity detection cannot [8].

Our algorithm makes generous use of memory, as it stores, among others, the exact positions of all robots. It would be interesting to see whether this could be significantly reduced. What is the minimum amount of memory necessary to solve the GATHERING PROBLEM?

References

1. N. Agmon and D. Peleg. Fault–tolerant gathering algorithms for autonomous mobile robots. In *ACM–SIAM Symposium on Discrete Algorithms (SODA 2004)*, to appear.
2. H. Ando, Y. Oasa, I. Suzuki, and M. Yamashita. A distributed memoryless point convergence algorithm for mobile robots with limited visibility. *IEEE Transaction on Robotics and Automation*, 15(5):818–828, 1999.
3. T. Balch and R. C. Arkin. Behavior-based formation control for multi-robot teams. *IEEE Transaction on Robotics and Automation*, 14(6):926–939, 1998.
4. M. Cieliebak, P. Flocchini, G. Prencipe, and N. Santoro. Solving the robots gathering problem. In *Proc. of the 30th Intern. Colloquium on Automata, Languages and Programming (ICALP 2003)*, pages 1181–1196, 2003.
5. P. Flocchini, G. Prencipe, N. Santoro, and P. Widmayer. Gathering of autonomous mobile robots with limited visibility. In *Proc. of the 18th International Symposium on Theoretical Aspects of Computer Science (STACS 2001)*, pages 247–258, 2001.
6. M. J. Matarić. Designing emergent behaviors: From local interactions to collective intelligence. In *From Animals to Animats 2: Int. Conf. on Simulation of Adaptive Behavior*, pages 423–441, 1993.
7. G. Prencipe. *Distributed Coordination of a Set of Autonomous Mobile Robots*. PhD thesis, Università di Pisa, 2002.
8. I. Suzuki and M. Yamashita. Distributed anonymous mobile robots: Formation of geometric patterns. *Siam Journal of Computing*, 28(4):1347–1363, 1999.
9. E. Welzl. Smallest enclosing disks (balls and ellipsoids). In H. Maurer, editor, *New Results and New Trends in Computer Science*, pages 359–370. Springer, 1991.

Bisecting and Gossiping in Circulant Graphs

Bernard Mans and Igor Shparlinski

Department of Computing, Macquarie University,
Sydney, NSW 2109, Australia
{bmans,igor}@ics.mq.edu.au

Abstract. Circulant graphs are popular network topologies that arise in
distributed computing. In this paper, we show that, for circulant graphs,
a simple condition for isomorphism, combined with lattices reduction
algorithms, can be used to develop efficient distributed algorithms. We
improve the known upper bounds on the vertex-bisection (respectively
the edge-bisection) width of circulant graphs. Our method is novel and
provides a polynomial-time algorithm to partition the set of vertices
(respectively the set of edges) to obtain these bounds and the respective
sets. By exploiting the knowledge of the bisection width of this topology,
we introduce generic distributed algorithms to solve the gossip problem
in these networks. We present lower and upper bounds of the number of
rounds in the vertex-disjoint and the edge-disjoint paths communication
models when the number of nodes is prime.

1 Introduction

Circulant graphs are popular network topologies that arise in distributed computing (e.g., supercomputer architectures [14]) and in quantum walk analysis [3].
Unlike other highly regular network topologies, computing the shortest paths in
circulant graphs can be challenging (e.g., NP-hard [11]).

The bisection width of a network topology is an important factor for determining the complexity of distributed algorithms in which information has to be
exchanged between two subsets of the networks. In this paper, we give new upper
bounds on the vertex-bisection (respectively the edge-bisection) width of circulant graphs. Our upper bound on the vertex-bisection width of circulant graphs
with n-nodes provides an improvement of a factor $O(\ln n)$ compared to the best
known results when n is prime. Moreover, our method provides a polynomial-
time algorithm to partition the set of vertices (respectively the set of edges) to
obtain these bounds and the respective sets. By exploiting this knowledge, we
introduce generic distributed algorithms to solve the gossip problem. We give
lower and upper bounds of the number of rounds required by these algorithms
in the vertex-disjoint and the edge-disjoint paths communication models.

Circulant graphs are regular graphs based on Cayley graphs defined on the
Abelian group $\mathbb{Z}_n$. We recall that an n-vertex *circulant graph* G is a graph whose
adjacency matrix $A = (a_{ij})_{i,j=1}^{n}$ is a circulant. That is, the ith row of A is the
cyclic shift of the first row by $i - 1$, $a_{ij} = a_{1,j-i+1}$, with $i, j = 1, \ldots, n$. In this

M. Farach-Colton (Ed.): LATIN 2004, LNCS 2976, pp. 589–598, 2004.
© Springer-Verlag Berlin Heidelberg 2004

section, the subscripts are taken modulo n, that is $a_{i,j} = a_{i+n,j} = a_{i,j+n}$ for all integers i and j (the interval $[1, n]$ is more convenient here). We also assume that $a_{ii} = 0$, $i = 1, \ldots, n$. Therefore, with every circulant graph one can associate a set $S \subseteq \mathbb{Z}_n$ of the positions of non-zero entries of the first row of the adjacency matrix of the graph. Respectively we denote by $\langle S \rangle_n$ the corresponding graph. The elements of the generating set S are called *chords*.

We recall that two graphs G_1, G_2 are *isomorphic*, and write $G_1 \simeq G_2$, if their adjacency matrices differ by a permutation of their rows and columns. For general graphs the isomorphism problem is known to be in **NP**, not known to be in **P**, and probably is not **NP**-complete (e.g., see [6, Section 6]). We say that sets $S, T \subseteq \mathbb{Z}_n$ are *proportional*, and write $S \sim T$, if for some integer l with $\gcd(l, n) = 1$, $S = lT$ where the multiplication is taken over $\mathbb{Z}_n$. Obviously, $S \sim T$ implies $\langle S \rangle_n \simeq \langle T \rangle_n$. For example $(S_1 = \{\pm 2, \pm 10\}$, $S_2 = \{\pm 3, \pm 8\}$, and $n = 23)$, $\langle S_1 \rangle_n \simeq \langle S_2 \rangle_n$ since $S_1 \sim S_2$ (with $l = 16$). Although it was conjectured by Ádám [1], the inverse statement is not true as counterexamples exist for any values of n except some n of the form $n = 2^\alpha 3^\beta m$, where $\alpha \in \{0, 1, 2, 3\}$, $\beta \in \{0, 1, 2\}$, $\gcd(m, 6) = 1$ and m is squarefree. For example in $\mathbb{Z}_{16}$, the 6-element sets $S_1 = \{\pm 1, \pm 2, \pm 7\}$ and $S_2 = \{\pm 1, \pm 6, \pm 7\}$, verify the isomorphism $\langle S_1 \rangle_{16} \simeq \langle S_2 \rangle_{16}$ but $S_1 \not\sim S_2$. However the simple isomorphism rule holds for important special cases (for example, circulant graphs with prime number of vertices [12] or with 4-element sets S). Under some additional restrictions, the isomorphism property of graphs can be replaced by the property of their isospectrality [22]. The relative independence of link length from delay time opens up the possibility of distinguishing among isomorphic networks on the basis of their algorithmic performance. A network that provides labelled edges should be able to exploit the same properties as one with different labelling if the graphs are isomorphic.

2   Bisection of Circulant Graphs

For any graph $G = (V, E)$, a *vertex bisector* of G is a set of vertices $V' \subseteq V$ such that the removal of the edges incident to the vertices of V' splits G into two components G^1 and G^2 of the same size (that is, $||V(G^1)| - |V(G^2)|| \leq 1$). G^1 and G^2 are called the two *halves* of the bisection. The *vertex-bisection width* $\mathrm{vw}(G)$ of G is defined as: $\mathrm{vw}(G) = \min\{|V'| \text{ such that } V' \text{ is a vertex bisector of } G\}$. The *edge-bisection width* $\mathrm{ew}(G)$ of G is the minimum number of edges whose deletion yields two components G^1 and G^2 such that $|V(G^1)| = \lfloor \frac{n}{2} \rfloor$ and $|V(G^2)| = \lceil \frac{n}{2} \rceil$ where $n = |V(G)|$. The problems are not equivalent: the complete graph has no vertex bisector, whilst it has an edge-bisection set of size $\lfloor \frac{n}{2} \rfloor \lceil \frac{n}{2} \rceil$. Both problems are **NP**-complete, but lower and upper bounds are known for most of the regular topologies of networks (for example, see [20]). Upper bounds on the vertex-bisection width of Cayley graphs (with a generating set of cardinality r), given in [4] and improved in [9] in the relaxed case where $|V(G^1)| \geq \frac{1}{3}n$ and $|V(G^2)| \geq \frac{1}{3}n$ to: $\mathrm{vw}(G) \leq c(r)n^{1-1/r}$ where $c(r)$ is a constant depending only on r. In [15], it has been proved that an Abelian Cayley graph G can be separated into two equal parts by deleting less than $\frac{8e}{r}n^{1-\frac{1}{r}} \ln \frac{n}{2}$ vertices.

As circulant graphs are vertex-transitive, an upper bound of the edge-bisection width of any circulant graph of n vertices can be given from the partitioning of vertex set (cyclically labelled $\{0, \ldots, n-1\}$) into two halves: $V(G^1) = [0, \ldots, \lfloor n/2 \rfloor - 1]$ and $V(G^2) = [\lfloor n/2 \rfloor, \ldots, (n-1)]$ or any rotation of such a cut, as shown in Figure 1, say $V(G^1) = [a, \ldots, b]$ and $V(G^2) = [b+1, \ldots, a-1]$ where all operations are taken modulo n and $b = a + \lfloor n/2 \rfloor - 1 \pmod{n}$. Without loss of generality, let us label the nodes on the ring cyclically and clockwise, and let us assume $1 \le s_1 < s_2 < \ldots < s_r < n/2$.

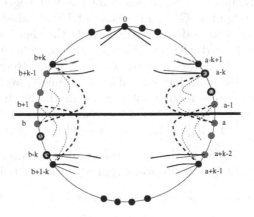

Fig. 1. Bisection of a Circulant Graph.

Lemma 1. *The edge-bisection width of any circulant graph of degree $2r$ with n vertices and chord set $S = \{\pm s_1, \ldots, +s_r\}$ is at most $2(|s_1| + \ldots + |s_r|)$.*

Proof. Let us partition the vertex set $\{0, \ldots, n-1\}$ into two disjoint sets with the same order (within one), as described above (see Figure 1). We count the number of chords of type s_i which are "*cut*" at a and b. All the positive chords (clockwise on the figure) outgoing from node b are cut. In particular, the largest positive chord s_r of length k outgoing from node b is cut. In fact, the same type of chord s_r outgoing from nodes $[b+1-k, \ldots, b]$ is cut. Similarly, in the neighbouring of node a, all the nodes $[a-k, \ldots, a-1]$ have their outgoing chord s_r cut. No other node have their outgoing chord s_r cut (otherwise, it will require their s_r chords to be larger than k). Hence, the number of chords of type s_r, of length $k = s_r$, which need to be deleted to bisect the graph is: $|[b+1-k, \ldots, b]| + |[a-k, \ldots, a-1]| = k + k = 2k = 2s_r$. Similarly, the edge bisecting set includes $2s_i$ of each type of chord s_i (of length $k_i = s_i$), and the lemma follows. Note that "negative" chords (i.e., $\{-s_1, -s_2, \ldots, -s_r\}$) have been already counted as they correspond to incoming edges while labelling clockwise.

Similarly, we give an upper bound of the vertex-bisection width.

Lemma 2. *The vertex-bisection width* $\mathrm{vw}(G)$ *of any circulant graph of degree* $2r$ *with* n *vertices and chord set* $S = \{\pm s_1, \ldots, \pm s_r\}$ *is at most* $2 \max_{1 \le i \le r} |s_i|$.

Proof. Let us partition the vertex set $\{0, \ldots, n-1\}$ into three disjoint subsets of V: V', $V(G^1)$ and $V(G^2)$ such as V' is a vertex bisector. Our proof is constructive. Initially, set $V' = \emptyset$ and let $V(G^1)$ and $V(G^2)$ be the two sets of vertices obtained by an edge-bisection of G (as described in Lemma 1). We remove the nodes $b+1, \ldots, b+k-1$ from V^2 and add them to V' (and delete all incident edges accordingly). Similarly, remove the nodes $a, \ldots, a+k-2$ from $V(G^1)$ and add them to V' (and delete all incident edges accordingly). Any path between a node of $V(G^1)$ and a node of $V(G^2)$ must either include the chord s_r from node b to node $b+k$, or include the chord s_r from node $a-1$ to node $a+k-1$. Indeed, the path can neither use a chord larger than k, nor use an intermediate node (as they are all in V' now). By adding nodes $a-1$ and b to V', and deleting all incident edges accordingly, we bisect G as desired. As we removed the same number of vertices in the original sets $V(G^1)$ and $V(G^2)$, it is easy to verify that they are of the same size (within one). Clearly, $|V'| = |[b, \ldots, b+k-1]| + |[a-1, \ldots, a+k-2]| = k+k = 2k = 2|s_r|$.

Using Lemmas 1 and 2, the isomorphic rule and lattices reduction algorithms [21], we can find an appropriate representation in polynomial time.

Corollary 1. *For any circulant graph of degree* $2r$ *with* n *vertices and chord set* $S = \{\pm s_1, \ldots, \pm s_r\}$:

$$\mathrm{ew}(G) \le 2 \min_{T \sim S} \sum_{i=1}^{r} |t_i| \quad \text{and} \quad \mathrm{vw}(G) \le 2 \min_{T \sim S} \max_{1 \le i \le r} |t_i|$$

where the minima are taken over all sets $T = \{t_1, \ldots, t_r\}$ *with* $T \sim S$.

Theorem 1. *Let* $n = p$ *be prime and let* $r = o(\log p)$. *For any circulant graph of degree* $2r$ *with* p *vertices and chord set* $S = \{\pm s_1, \ldots, \pm s_r\}$:

$$\mathrm{ew}(G) \le (r^{\log r + O(\log \log r)} r!)^{1/r} p^{1-1/r}.$$

Proof. Let us consider the family of p points $(ls_1, \ldots, ls_r), l \in \mathbb{Z}_p$. They all belong (after reduction modulo p) to the r-dimensional cube $[0, p-1]^r$ with side length p. Let us consider the r dimensional octahedron $\mathcal{O}$ of diameter $2L$ centred at the origin which is defined as the set of points $(x_1, \ldots, x_r) \in \mathbb{R}^r$ with $|x_1| + \ldots + |x_r| \le L$. The volume of $\mathcal{O}$ is $\mathrm{vol}\,\mathcal{O} = \frac{2^r L^r}{r!}$. Because $\mathcal{O}$ is convex, from Theorem 5.8 of [23] we derive that the cube $[0, p-1]^r$ is covered by at most

$$M = r^{\log r + O(\log \log r)} \frac{p^r}{\mathrm{vol}\,\mathcal{O}} < p$$

parallel translates of $\mathcal{O}$ for some $L = 0.5(r^{\log r + O(\log \log r)} r!)^{1/r} p^{1-1/r}$. Therefore there is at least one translate of $\mathcal{O}$ containing at least two points corresponding to some $0 \le l_1 < l_2 \le p-1$. Therefore, putting $l = l_2 - l_1$, we see that $\sum_{i=1}^{r} |ls_i| \le L$ and from Corollary 1 we obtain the desired result.

Remarks: (i) Using the known inequality $r! < r^{r+1}e^{-r}$ which holds for all integers $r \geq 7$ we obtain that, provided that r is large enough: $\mathrm{ew}(G) \leq 0.4rp^{1-1/r}$. (ii) If $r \to \infty$ such that $r = o(\log p)$ then the bound of Theorem 1 becomes of the form $\mathrm{ew}(G) \leq (r/e + o(1))p^{1-1/r}$.

Theorem 2. *Let $n = p$ be prime. For any circulant graph of degree $2r$ with p vertices and chord set $S = \{\pm s_1, \ldots, \pm s_r\}$: $\mathrm{vw}(G) \leq 4p^{1-\frac{1}{r}}$.*

Proof. Let $N = \lceil p^{1/r} \rceil - 1$. Separating this cube into $N^r < p$ equal subcubes with the side length $h = p/N$, we see that there is at least one subcube that contains at least two points corresponding to some $0 \leq l_1 < l_2 \leq p-1$. Therefore, as in proof of Theorem 1, putting $l = l_2 - l_1$, we obtain $|ls_i| \leq h, i = 1, \ldots, r$. Because p is prime and $1 \leq l \leq p-1$, $\gcd(l, p) = 1$ and, thus, Corollary 1 can be applied to the set lS. If $p^{1/r} \leq 2$, the bound is trivial. If $p^{1/r} \geq 2$ then we have $N = \lceil p^{1/r} \rceil - 1 \geq 0.5p^{1/r}$. Thus $h \leq 2p^{1-1/r}$ and the result follows.

Remark. This is a improvement of a factor $\frac{2e \ln(n/2)}{r}$ compared to the best known bound, and if $r = o(\log p)$ then $\mathrm{vw}(G) \leq (2 + o(1))p^{1-\frac{1}{r}}$.

Composite values of n. It would be natural to try to extend our method to composite values of n. This does not seem to be possible, as if $n = 2m$ is even, and $S_0 = \{\pm 1, \pm(m+1)\}$ then $\min_{T \sim S_0} \max_{i=1,2} |t_i| \geq n/4$. Indeed, let $T = lS$. If $|l| = |t_1| < m/2 = n/4$, then, because the condition $\gcd(l, n) = 1$ implies that l is odd, we have $|t_2| = m + l > m/2 = n/4$. Thus any methods using our Lemmas 1 and 2 will lead to very weak results.

The difficulty of finding a precise bisection width when n is composite is not really surprising. There is nothing new in the fact that the arithmetic structure of n (e.g., primality) plays an important role. Let us recall that the simple isomorphism rule only holds for special cases, including circulant graphs with prime number of vertices [12], but is not true for most of values of n (see previous section). More than 35 years after being conjectured [1], there is still some cases for which it is unknown if it holds. In composite cases, the intricate relationship between chords requires a particular method for each case.

3   Gossiping in Circulant Graphs

Information dissemination is the most important communication problem in interconnection networks. Three basic communication problems are: *Broadcast* (one-to-all): one node has a piece of information and has to communicate this information to all the other nodes; *Accumulation* (all-to-one): all the nodes have a different piece of information and want to communicate this information to the same particular node; *Gossip* (all-to-all): each node has a piece of information and wants to communicate this information to all the other nodes (such that all nodes learn the cumulative message).

Gossiping is the communication problem where each node of a network has a piece of information and wants to communicate this information to all the other nodes (such that all nodes learn the cumulative message). A communication algorithm consists of a number of communication *rounds* during which nodes are involved in communications. Let $g(G)$ denote the number of rounds of the optimal gossip algorithm for G. The communication algorithm necessary to solve this problem depends on the communication model.

Several communication modes exist. The *vertex-disjoint paths mode* (VDP) assumes: (i) a communication involves exactly two nodes which can be can be at distance more than 1, (ii) any two paths corresponding to simultaneous communications must be vertex-disjoint. Similarly, the *line mode* or *edge-disjoint paths mode* (EDP) assumes: (i) a communication involves exactly two nodes which can be can be at distance more than 1, (ii) any two paths corresponding to simultaneous communications must be edge-disjoint. The mode of communication also depends on the type of communication links available: (a) *half-duplex* (or *1-way*) or (b) *full-duplex mode* (or *2-way*). In the 2-VDP mode (resp., 2-EDP), two nodes involved in a 2-way VDP communication (resp., EDP communication) can exchange their information. In the 1-VDP mode (resp., 1-EDP), the information will flow in the 1-way direction from one node to the other.

3.1   Bisection Lower Bounds

In this paper, we give lower bounds of the gossip complexity for the circulant graphs of $|V| = n = p$ vertices with p prime. A direct relationship exists between the bisection width and the gossip complexity. Let $\vartheta = (\log(1 + 5^{1/2}) - 1)^{-1} = 1.440\ldots$. The following statement gives a summary of several known bounds from [18] (for the EDP mode) and from [19] (for the VDP mode):

Lemma 3. *Let G be a network of edge-bisection $\mathrm{ew}(G)$ and of vertex-bisection $\mathrm{vw}(G)$.*

- *In the 2-EDP mode, $g(G) \geq 2\log n - \log \mathrm{ew}(G) - \log\log n - 4$.*
- *In the 2-VDP mode, $g(G) \geq 2\log n - \log \mathrm{vw}(G) - \log\log \mathrm{vw}(G) - 6$.*
- *In the 1-VDP mode, for any "well-structured" gossip algorithm,*
  $g_w(G) \geq 2\log n - (2 - \vartheta)(\log \mathrm{vw}(G) + \log\log \mathrm{vw}(G)) - 15$.

By using our Theorems 1 and 2, we obtain the following lower bounds.

Theorem 3. *Let $n = p$ be prime. For any gossip algorithm running on a circulant graph of degree $2r$ with p vertices and chord set $S = \{\pm s_1, \ldots, \pm s_r\}$, the number of rounds is at least:*

- *In the 2-EDP mode, provided that $r = o(\log p)$, $g(G) \geq (1 + \frac{1}{r})\log p + o(\log p)$.*
- *In the 2-VDP mode, $g(G) \geq (1 + \frac{1}{r})\log p - \log\log p - 8$.*
- *In the 1-VDP mode, provided that $r = o(\log p)$,*
  $g_w(G) \geq (\vartheta + \frac{2 - \vartheta}{r})\log p - (2 - \vartheta)\log\log p - 17$.

Of course, if $r = o(\log p)$ then, in the 2-VDP mode, the slightly sharper bound $g(G) \geq (1 + \frac{1}{r})\log p - (\log\log p + \log r + 5)$ holds.

3.2   Generic 2-VDP Gossiping in Circulant Graphs

It is clear that "specialised" algorithms must be used to obtain tight complexity bounds for specific chord sets. (For example, it is easy to see that a circulant graph of 2^r nodes with chord set $S = \{\pm 1, \pm 2, \pm 2^2, \ldots, \pm 2^{r-1}\}$ perfectly embeds a hypercube, and thus, can gossip with the minimum number of rounds.)

In this paper, we only focus on "generic" algorithms that works correctly for any circulant graph given as input. Although, we can only bound the number of rounds required by the algorithm when n is prime, let us emphasize that the algorithms described below completes the gossip correctly even if n is composite.

A popular strategy, introduced in [16], to solve the gossip problem is to use a 3-phase algorithm described as follows. Let $G(V, E)$ be the graph corresponding to the topology of the network. Let $a(G)$ be any subset of nodes of G (called the *accumulation set*). Divide G into $m = |a(G)|$ connected components (called *accumulation components*) of size $\lceil n/m \rceil$, such that each connected component contains exactly one accumulation node of $a(G)$. The 3-phase gossip algorithm for G with respect to $a(G)$ follows the phases, (where $1 \leq i \leq m$): (1) Accumulation: each accumulation node a_i accumulates the information from the nodes lying in its accumulation component A_i; (2) Gossip performs a gossip algorithm among the nodes a_i of $a(G)$; (3) Broadcast: each node a_i broadcasts the cumulative message in its component A_i.

For the 2-VDP mode, the accumulation problem can be considered as the reverse of broadcast problem, and hence, a similar strategy can be used. In [16], the authors proved that, with the 2-VDP mode, broadcasting (resp. accumulating) in an Hamiltonian path of k nodes can be done in $\lceil \log k \rceil$ rounds. By analogy to notations in the 1-VDP mode [17], we call a 3-phase algorithm *well-structured* if the gossip phase is an implementation of an optimal gossip algorithm (for example, in the complete graph K_m, or a hypercube $Q_{\lceil \log m \rceil}$), and takes $\lceil \log m \rceil$.

With the 2-VDP mode a well-structured gossip algorithm g_w only requires $\lceil \log n/m \rceil + \lceil \log m \rceil + \lceil \log n/m \rceil \leq 2\lceil \log n \rceil - \lceil \log m \rceil$ rounds, when the accumulating components are Hamiltonian paths. An upper bound on $g(G)$ can be obtained by constructing a well-structured gossip algorithm g_w by taking the m nodes of the vertex bisector as accumulation nodes. It is easy to see that, when $m = n^{1-1/r}$, the number of rounds is nearly optimal: $g_w(G) \leq (1 + \frac{1}{r}) \log n$.

In the following, we will exploit the knowledge of the existence of such m nodes when n is prime by computing (in polynomial time) a bisector set of size m (as described in Theorem 2). We first describe a well-structured gossip algorithm for a specific infinite family of circulant graphs of degree four with n prime. We then show how to extend this strategy to other circulant graphs. Let us consider the circulant graph G of degree 4 with n vertices and chord set $S = \{\pm 1, \pm s_2\}$ (that is, $r = 2$) with $s_2 = 2^d$ and $s_2(s_2 - 1) \leq n \leq s_2^2$ (that is, $s_2 \sim n^{1/2}$). Without loss of generality we can label the nodes $[0, \ldots, (n-1)]$ cyclically along the chord $+1$. We partition G into $m = s_2$ connected components of, up to, $\lceil n/s_2 \rceil \leq s_2 = 2^d$ nodes:

$$
\begin{aligned}
A_i &= [i, s_2 + i, 2s_2 + i, \ldots, (s_2 - 1)s_2 + i], \quad 0 \leq i < n - (s_2 - 1)s_2, \\
A_i &= [i, s_2 + i, 2s_2 + i, \ldots, (s_2 - 2)s_2 + i], \quad n - (s_2 - 1)s_2 \leq i \leq s_2 - 1.
\end{aligned}
\tag{1}
$$

where each component has an accumulating node $a_i = i$, $0 \le i \le (s_2 - 1)$. This also defines s_2 consecutive segments S_k, $0 \le k \le (s_2 - 1)$ along the $+1$ chords. The first segment, S_0, corresponds to the accumulation nodes, and, all segments, but possibly the last one S_{s_2-1}, are of size s_2. Clearly, each accumulating component is connected by an Hamiltonian path along the chord $+s_2$. With the 2-VDP mode, the accumulating phase and the broadcasting phase take at most $\log s_2 = d \le \log n^{1/2} + 1 = \frac{1}{2} \log n + 1$ rounds respectively.

We say that the nodes i and j exchange information through the segment S_k, if the information is passed first through k chords ($+s_2$), then through the chord (± 1) between the nodes $ks_2 + i$ and the node $ks_2 + j$ (in the segment S_k of nodes $[ks_2, \ldots, ks_2 + (s_2 - 1)]$), and finally back through k chords ($-s_2$). For the gossip phase of the accumulating nodes, we present the recursive Algorithm (see also Figure 2) similar to the algorithm presented in [17] for square grids Gr_n^2 of $n = 2^{2d}$ nodes and side-length $n^{1/2} = 2^d$. Initially, the algorithm is started by running $\mathtt{Gossip}(0, s_2 - 1)$.

Procedure $\mathtt{Gossip}(a, b)$
  if $(b - a) > 1$ do in parallel
      $\mathtt{Gossip}(a, a + \lfloor \frac{b-a}{2} \rfloor)$ and $\mathtt{Gossip}(a + \lceil \frac{b-a}{2} \rceil, b)$
  endo in parallel
  for $a \le i \le a + \lfloor \frac{b-a}{2} \rfloor$ do in parallel
      exchange information between i and $j = b - i$ through segment S_k, $k = \lfloor \frac{j-i}{2} \rfloor$
  endo in parallel

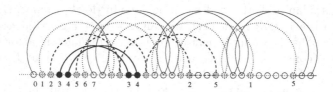

Fig. 2. The communication paths in the 2-VDP mode.

By induction, it is easy to see that each node learns the cumulative message and the communication at each round is vertex-disjoint. Any two pairs of nodes (i, j) and (i', j') exchanging information through a segment k define two non-intersecting sections in the segment S_k, as $k = \lfloor \frac{j-i}{2} \rfloor = \lfloor \frac{j'-i'}{2} \rfloor$ we have $i < j < i' < j'$. In the case $i < i' < j' < j$, we have $\lfloor \frac{j-i}{2} \rfloor = k \ne k' = \lfloor \frac{j'-i'}{2} \rfloor$.

Obviously, $\log s_2 \le \frac{1}{2} \log n + 1$ rounds are sufficient to gossip among the accumulating nodes, and thus, the total number of rounds of this well-structured gossip algorithm is $g_w(G) \le (1 + \frac{1}{2}) \log n + 3 = \frac{3}{2} \log n + 3$.

This is near optimal as it almost match the lower bound introduced in Theorem 3 and can be generalised to other cases. Note that the last section S_{s_2-1} is not used and does not require to be full. In fact, only the first k segments, $k = \lfloor \frac{n-1}{2} \rfloor$, are used by the algorithm. Hence the algorithm can be adapted to run correctly for $\frac{1}{2} s_2^2 \le n \le s_2^2$, and takes either $\frac{3}{2} \log n$ or $\frac{3}{2} \log n + 3$ rounds.

In the case that s_2 is not a power of 2 (i.e., $2^d < s_2 < 2^{d+1}$ for some d), a similar algorithm can be used with two extra rounds. After the accumulation phase, each accumulation node j, $2^d \leq j \leq s_2 - 1$, gossips its information to the accumulation node i, $2^d - 1 \geq i \geq 2^d - 1 - (s_2 - 1 - 2^d)$, through the segment S_k, $k = (j - i)/2$. After the gossip phase $\texttt{Gossip}(0, 2^d - 1)$, and before the broadcast phase, each node i sends the cumulative message to the respective node j.

When considering circulant graphs with n prime and chord set $S = \{\pm s_1, \pm s_2\}$, $s_1 < s_2$, (instead of the specific chord set $S = \{\pm 1, \pm s_2\}$), Theorem 2 proves that for circulant graphs G of degree 4 with n vertices, where $n = p$ is prime, it is possible to generate a representation of G with a chord set $T = \{\pm t_1, \pm t_2\}$ such that $\langle S \rangle_n \simeq \langle T \rangle_n$ and $t_1 < t_2 \leq 4n^{1-\frac{1}{2}} = 4n^{1/2}$. Using T as the chord set, and if $\gcd(t_1, t_2) = 1$, we can partition G into $m - \lceil n/t_2 \rceil$ connected components (along the chord $+t_1$) of, up to, t_2 nodes. Each component has an accumulating node $a_i = it_2$, $0 \leq i \leq (m-1)$. Clearly, each accumulating component is connected by an Hamiltonian path along the chord $+t_1$. With the 2-VDP mode, the accumulating phase and the broadcasting phase take at most $\lceil \log t_2 \rceil$. Representing G with chord set $W = \{\pm 1, \pm w_2\}$ where $w_2 \equiv t_1 t_2^{-1}$ (mod n), the accumulating nodes a_i, $0 \leq i \leq m-1$, are now consecutive along the chord $+1$. It is easy to see that if $2n/w_2 < m < w_2$, the generic $\texttt{Gossip}$ algorithm will run correctly within a number of rounds close to the minimum.

With chord set $S = \{\pm 1, \pm n^{1/r}, \pm n^{2/r}, \dots, \pm n^{(r-1)/r}\}$, $r \geq 3$, we can give tight upper bounds on the gossiping complexity in the 2-VDP mode using an algorithm similar to [17] for r-dimensional grids Gr_n^r of n nodes and side-length $n^{1/r}$, $r \geq 3$, when n is an rth power, and extend from these algorithms.

4   Concluding Remarks

We introduced a novel approach to compute (in polynomial time) a convenient isomorphic representation of a circulant graph to exploit the knowledge of its bisector. This method is general and may be of use in other applications. As an example, we introduced a "generic" algorithm that gossip efficiently in circulant graphs of prime order. Although "specialised" algorithms should be used for specific chord sets to obtain tight bounds, our algorithms work in all cases and are nearly optimal in important cases. We showed that in composite cases, the intricate relationship between chords requires a particular method for each case.

Let us note that the problem of finding l for which $T = lS$ minimizes one of the expressions of Corollary 1 is an instance of the famous *shortest vector problem* for metrics $\mathcal{L}_\infty$ and $\mathcal{L}_1$. Thus one can use the variety of the algorithms available for this problem, e.g. [2,5,21] in order to find an optimal value of l or a value which gives almost optimal results. Typically these algorithms target the metric $\mathcal{L}_2$ but some can apparently be adjusted to the metrics $\mathcal{L}_\infty$ and $\mathcal{L}_1$ and in any case they can be used directly for approximate solutions if one uses that:
$$\sum_{i=1}^{r} |t_i| \leq r^{1/2} \left(\sum_{i=1}^{r} |t_i|^2\right)^{1/2} \quad \text{and} \quad \max_{1 \leq i \leq r} |t_i| \leq \left(\sum_{i=1}^{r} |t_i|^2\right)^{1/2}.$$
Unfortunately, when r is growing, all known algorithms either have exponential time or output (in polynomial time) a vector which is an exponential factor

longer than the shortest vector. However for fixed r or slowly growing with p (as about $\log \log p$) these algorithms are polynomial in $\log p$.

References

1. A. Ádám, 'Research problem 2-10', *J. Combinatorial Theory*, **3** (1967), 393.
2. M. Ajtai, R. Kumar and D. Sivakumar, 'A sieve algorithm for the shortest lattice vector problem' *Proc. 33rd ACM STOC*, Crete, Greece, July 6-8, 2001, 601–610.
3. A. Ahmadi, R. Belk, C. Tamon and C. Wendler, 'Mixing in Continuous Quantum Walks on Graphs', *xxx.arxiv.cornell.edu/pdf/quant-ph/0209106*, Apr 2003.
4. F. Annexstein and M. Baumslag, 'On the diameter and bisector size of Cayley graphs', *Mathematical Systems Theory*, **26** (1993), 271–291.
5. L. Babai, 'On Lovász' lattice reduction and the nearest lattice point problem', *Combinatorica*, **6** (1986), 11–13.
6. L. Babai, 'Automorphism groups, isomorphism, reconstruction' *Handbook of Combinatorics*, Elsevier, Amsterdam, 1995, 1749–1783.
7. L. Barrière, J. Cohen and M. Mitjana, 'Gossiping in chordal rings under the line model', *Theoretical Computer Science*, **264(1)** (2001), 53–64.
8. L. Barrière and J. Fàbrega, 'Edge-bisection of chordal rings', *Proc. of the Math. Found. of Comp. Science (MFCS'2000)*, Bratislava, LNCS-1893, 2000, 162–171.
9. S. R. Blackburn, 'Node Bisectors of Cayley Graphs', *Mathematical Systems Theory*, **29** (1996), 589–598.
10. J.-C. Bermond, F. Comellas and D. F. Hsu, 'Distributed loop computer networks: A survey', *Journal of Parallel and Distributed Computing*, **24** (1995), 2–10.
11. J.-Y. Cai, G. Havas, B. Mans, A. Nerurkar, J.-P. Seifert and I. Shparlinski, 'On routing in circulant graphs', *Proc. of the Computing and Combinatorics Conference (Cocoon'99)*, Tokyo, LNCS-1627, 1999, 370-378.
12. B. Elspas and J. Turner, 'Graphs with circulant adjacency matrices', *J. Comb. Theory*, **9** (1970), 229–240.
13. A. Farley, 'Minimum-time line broadcast networks', *Networks*, **10** (1980), 57–70.
14. R. Gruber and A. Gunzinger, 'The Swiss-Tx Supercomputer Project', *Speedup*, **11(2)** (1997), 20–26.
15. Y. O. Hamidoune and O. Serra, 'On small cuts separating an Abelian Cayley graph into two equal parts', *Mathe. Systems Theory*, **29(4)** (1996), 407–409.
16. J. Hromkovič, R. Klasing and E. A. Stöhr, 'Dissemination of information in generalized communication modes', *Comp. Art. Intell.*, **15(4)** (1996), 295–318.
17. J. Hromkovič, R. Klasing, E. A. Stöhr and H. Wagener, 'Gossiping in vertex-disjoint paths mode in d-dimensional grids and planar graphs', *Information and Computation*, **123** (1995), 17–28.
18. J. Hromkovič, R. Klasing, W. Unger and H. Wagener, 'Optimal algorithms for broadcast and gossip in the edge-disjoint path modes', *Information and Computation*, **133** (1997), 1–33.
19. R. Klasing, 'The relationship between the gossip complexity in the vertex-disjoint paths mode and the vertex bisection width', *D.A.M.*, **83** (1998), 229–246.
20. F. T. Leighton, *Introduction to parallel algorithms and architectures: Arrays, trees, hypercubes*, M. Kaufmann, 1992.
21. A. K. Lenstra, H. W. Lenstra and L. Lovász, 'Factoring polynomials with rational coefficients', *Mathematische Annalen*, **261** (1982), 515–534.
22. B. Mans, F. Pappalardi and I. Shparlinski, 'On the spectral Ádám property for circulant graphs', *Discrete Math.*, **254(1-3)** (2002), pp. 309-329.
23. C. A. Rogers, *Packing and covering*, Cambridge Univ. Press, NY, 1964.

Multiple Mobile Agent Rendezvous in a Ring

Paola Flocchini[1], Evangelos Kranakis[2], Danny Krizanc[3], Nicola Santoro[2], and
Cindy Sawchuk[2]

[1] SITE, University of Ottawa, Ottawa, ON, Canada.
flocchin@site.uottawa.ca
[2] School of Computer Science, Carleton University, Ottawa, ON, Canada.
{kranakis,santoro,sawchuk}@scs.carleton.ca
[3] Department of Mathematics and Computer Science, Wesleyan University,
Middletown, Connecticut 06459, USA.
dkrizanc@wesleyan.edu

Abstract. We study the rendezvous search problem for $k \geq 2$ mobile
agents in an n node ring. Rather than using randomized algorithms or different deterministic algorithms to break the symmetry that often arises
in this problem, we investigate how the mobile agents can use identical
stationary tokens to break symmetry and solve the rendezvous problem.
After deriving the conditions under which identical stationary tokens
can be used to break symmetry, we present several solutions to the rendezvous search problem. We derive the lower bounds of the memory required for mobile agent rendezvous and discuss the relationship between
rendezvous and leader election for mobile agents.

1 Introduction

In the mobile agent rendezvous search problem, k mobile agents located on an
n node network are required to meet or rendezvous. When the mobile agents
or the network nodes are uniquely numbered, solving the rendezvous search
problem is trivial. When the mobile agents are identical and the network nodes
are anonymous, however, the resulting symmetry can make the problem difficult
to solve.

Symmetry in the rendezvous search problem is typically broken by using randomized algorithms or different deterministic algorithms [2]. While Baston and
Gal [5] mark the starting points of the searchers, they still rely on randomization
or different deterministic algorithms. Kranakis et al [7], however, studied how
two searchers, i.e., mobile agents, on an n node ring can use identical tokens to
break the symmetry.

Most of the literature on the rendezvous search problem deals with the case
of $k = 2$ searchers. The few exceptions include Lim, Beck, and Alpern [8],
Alpern [1], Pikounis and Thomas [9], and Gal [6], but that research focuses
almost exclusively on the line. In this paper, we investigate the mobile agent
rendezvous search problem for $k \geq 2$ mobile agents on an n node ring, where
the mobile agents use tokens to break symmetry.

M. Farach-Colton (Ed.): LATIN 2004, LNCS 2976, pp. 599–608, 2004.
© Springer-Verlag Berlin Heidelberg 2004

1.1 The Network Model

The model consists of $k \geq 2$ identical mobile agents that are located on separate nodes of an anonymous, synchronous n node ring. The mobile agents may or may not share a common orientation, i.e., agree on the direction that is clockwise. A given node requires only enough memory to host a token and at most k mobile agents. Each mobile agent, MA, owns a single identical stationary token, i.e., the tokens are indistinguishable and once they are placed on a node, they must remain in place. A token or MA at a given node is visible to all MAs on the same node, but is not visible to any other MAs. When a MA is visible, its state is also visible. The MAs follow the **same** deterministic algorithm and begin execution of that algorithm at the same time.

Memory permitting, a MA can count such things as the number of nodes visited, the number of tokens discovered, the number of MAs discovered, the number of nodes between tokens, or the total number of nodes in the network. In addition, a MA might already know the number of nodes in the network or some other network parameter and requires sufficient memory to store this information. Since the MAs are identical, they face the same limitations on their knowledge of the network. Rendezvous occurs when all the MAs meet on a network node.

An instance of the mobile agent rendezvous problem is **solvable** when the mobile agents

1. can correctly determine whether or not rendezvous is possible and then
2. rendezvous or stop as appropriate.

Solving the mobile agent rendezvous problem involves making the correct choice, i.e., stopping if rendezvous is impossible and achieving rendezvous if possible.

In this paper, we assume that the MAs always place their tokens on their respective starting nodes in the first step of any algorithm they execute. The tokens are identical and stationary so that intertoken distances that exist at the beginning of any algorithm persist throughout the algorithm.

1.2 Our Contribution

In this paper, we continue the study of the mobile agent rendezvous search problem in the ring [7]. Our model consists of $k \geq 2$ identical MAs in an anonymous, synchronous, and possibly oriented n node ring. Rather than using randomized algorithms or different deterministic algorithms, we use identical stationary tokens to break symmetry so that the MAs can run the same deterministic algorithm.

First, we prove that if neither k, the number of MAs, nor n, the number of nodes in the ring, are known then the rendezvous problem is unsolvable. For the remainder of the paper, we assume that k is known. Next, we prove that when k is known, rendezvous is possible if and only if $\mathcal{S}$, the sequence of intertoken distances, is aperiodic.

We present three algorithms that solve the rendezvous problem when k is known. Either k or n must be known so when n is known, k can be determined in one traversal of the ring. When $\mathcal{S}$, the sequence of intertoken distances between the MAs, is aperiodic, the algorithms guarantee that rendezvous occurs. However, if $\mathcal{S}$ is periodic and thus rendezvous is impossible, the algorithms guarantee that the MAs stop. The memory and time complexities for these algorithms are presented in table 1. We also present three algorithms that solve the rendezvous problem for values of k and n that satisfy various conditions such as primality. The memory and time complexities for these algorithms, numbered 4 through 6, are also presented in table 1.

Table 1. The Rendezvous Search Problem with $k \geq 2$ Mobile Agents

Algorithm	Memory	Time
1	$O(k \lg n)$	$O(n)$
2	$O(\lg n)$	$O(kn)$
3	$O(k \lg \lg n)$	$O(\frac{n \lg n}{\lg \lg n})$
4	$O(\lg n)$	$O(n)$
5	$O(\lg k)$	$O(n \lg k)$
6	$O(\lg k)$	$O(n)$

Kranakis et al [7] proved that solving the rendezvous problem with $k = 2$ mobile agents requires at least $O(\lg \lg n)$ memory. In this paper, we prove that solving the rendezvous problem with $k \geq 2$ mobile agents requires at least $\Omega(\lg \lg n + \lg k)$ memory.

Finally, we prove that if the MAs share a common orientation, then the rendezvous problem and the leader election problem for MAs are equivalent. A solution to the first problem can be used to derive a solution for the second problem and vice versa. If the MAs do not share a common orientation, however, then the leader election problem for MAs is strictly more complex than the rendezvous problem since a solution to the latter problem does not always imply a solution to the former problem.

1.3   Outline of the Paper

In section 2, we present the impossibility results for the MA rendezvous search problem with $k \geq 2$ MAs. In section 3, we prove that rendezvous is possible if and only if the sequence of intertoken distances is aperiodic. We present unconditional solutions for rendezvous in section 4 and derive the lower bounds on memory necessary for rendezvous in section 5. In section 6, we present solutions to the rendezvous problem for specific circumstances and then, in section 7, we discuss the relationship between leader election and rendezvous for mobile agents.

2   Impossibility Results

To solve the rendezvous problem, mobile agents must recognize when rendezvous is possible. As the next theorem shows, knowledge of k or n is a necessary condition for the rendezvous of identical MAs in an anonymous, synchronous ring.

Theorem 1. *When each MA in the ring knows neither n, the number of nodes, nor k, the number of mobile agents, the mobile agent rendezvous search problem is unsolvable.*

Theorem 1 indicates that knowing either k or n is a necessary condition for solving the rendezvous problem. For the remainder of the paper, we shall assume that k is known.

3   Solving the Rendezvous Problem

As stated in section 1.1, we assume that the MAs always place their tokens on their respective starting nodes in the first step of any algorithm they execute. The tokens are identical and stationary so that the intertoken distances that exist at the beginning of an algorithm persist throughout the algorithm. Since the MAs are identical and run the same deterministic algorithm in an anonymous ring, rendezvous can only occur if the intertoken distances can be used to break the resulting symmetry.

Theorem 2. *Rendezvous is guaranteed if and only if $\mathcal{S}$, the sequence of intertoken distances, is aperiodic.*

4   Unconditional Solutions

Given the results of section 3, the mobile agent rendezvous problem can be solved when the MAs can determine that $\mathcal{S}$, the sequence of intertoken distances, is aperiodic.

First, assume that each MA has $O(k \log n)$ memory. The MAs know k but do not necessarily know n or share a common orientation. Consider the following algorithm.

Algorithm 1

1. Release the token at the starting node.
2. Choose a direction and begin to walk around the ring.
3. Compute the k intertoken distances $d_1, \ldots, d_k$.
4. If $\mathcal{S} = d_1, \ldots, d_k$ is periodic, then stop. (Rendezvous is not guaranteed.)
5. Set *forward* $= d_1, \ldots, d_k$ and *reverse* $= d_k, \ldots, d_1$.
6. Let *lexi(someSequence)* denote the lexicographically maximum rotation of the sequence *someSequence*.
7. Set *forward* $=$ *lexi(forward)* and *reverse* $=$ *lexi(reverse)*.

8. If *forward* and *reverse* differ, then
 i) determine which of these sequences is the lexicographic maximum
 and rendezvous at the node where this sequence starts.
 ii) else let MA_i and MA_j denote the MAs at the beginning
 of *forward* and *reverse* respectively.
9. If MA_i and MA_j are the same MA, then rendezvous at the node
 where MA_i resides.
10. If MA_i and MA_j are distinct MAs, then look at the two paths
 between MA_i and MA_j in the ring.
 i) If only one of the paths had an odd number of nodes,
 then rendezvous at the node in the midpoint of that path.
 ii) If both paths have an odd number of nodes, then
  a) if the paths differ in length, rendezvous at the
  midpoint of the shorter path,
  b) else compare the sequences of intertoken distances for
  the two paths and rendezvous at the node in the midpoint
  of the path that is the lexicographic maximum.
 iii) If both paths have an even number of nodes, then
 rendezvous at the node in the midpoint of the path that
 contains an odd number of MAs.

Theorem 3. *If each MA has memory $O(k \lg n)$, then the mobile agent rendezvous problem can be solved in time $O(n)$.*

If the MAs are restricted to memory $O(\lg n)$, the mobile agent rendezvous problem is still solvable.

Consider the following algorithm. In each round, a MA may become inactive and thus spend the rest of the algorithm at its starting node. If the MAs share a common orientation, then the MAs travel in the same direction, so an active MA can identify an inactive MA because the former will find the latter stopped. If the MAs do not share a common orientation, however, an active MA that meets another MA can not tell if the latter MA is inactive or merely travelling in the opposite direction. In this case, each MA needs to set a bit that indicates whether it is active or inactive.

With only $O(\lg n)$ memory, a MA needs more than one traversal of the ring to determine if S is aperiodic. Consider the following algorithm.

Algorithm 2

1. Release the token at the starting node.
2. Set $c = 1$.(The number of the current round.)
3. Set $active = 1$. (A bit to indicate whether the MA is active.)
4. Set $inactive = 0$. (Count the number of inactive MA.)
5. Choose a direction and begin to walk around the ring.
6. Increment $inactive$ each time an inactive MA is met.
7. Compute the distance to the cth token, d_c, i.e., if $c = 1$, count the distance to the first token and if $c = 2$, count the distance to the second token, etc.

8. Continue to walk around the ring and compare d_c to each intertoken distance between c tokens.

9. If MA sees an intertoken distance d_i such that $d_i > d_c$, then the MA continues in the same direction and becomes inactive, i.e., sets $active = 0$, when it reaches its starting node.

10. If MA did not see an intertoken distance d_i such that $d_i > d_c$, then the MA remains active when it returns to its starting node.

11. If only one MA remains active, i.e., $inactive = k - 1$, then walk around the ring and arrange the rendezvous. (A MA that reaches this point is the sole active MA.)

12. If $c == 2$ and $inactive == 0$, then stop. (All the intertoken distances are equal and thus rendezvous is impossible.)

13. Else set $c = c + 1$ and $inactive = 0$.

14. Repeat from step 5.

Theorem 4. *If each MA has memory $O(\lg n)$, then the mobile agent rendezvous problem is solvable in time $O(kn)$.*

Algorithms 1 and 2 solve the mobile agent rendezvous problem when each MA has memory $O(k \lg n)$ and $O(\lg n)$ respectively. It is also possible, however, to solve the mobile agent rendezvous problem when each MA has memory $O(k \lg \lg n)$.

Let $p_1, ..., p_r$ denote the first r prime numbers such that $\prod_{i=1}^{r} p_i > n$. An active MA needs to recognize if another MA is active. Without a common orientation, each MA needs to set a bit to indicate whether it is active.

Algorithm 3

1. Release token at the starting node.

2. Set $active = 1$.

3. Set p_r to the first prime such that $\prod_{i=1}^{r} p_i > n$.

4. Set $p_i = p_1 = 2$.

5. Set $\alpha = k$, the number of active MAs.

6. Walk around the ring and compute the intertoken distances mod p_i between the α active MAs, i.e., $d_1, ..., d_\alpha$ mod p_i.

7. Set $forward = d_1, ..., d_\alpha$ mod p_i.

8. Set $reverse = d_\alpha, ..., d_1$ mod p_i.

9. If $forward$ is periodic, i.e., $forward = (d_1, ..., d_{\alpha/a})^a$ mod p_i, then
   i) if at start of a block $(d_1, ..., d_{\alpha/a})$, remain active.
   ii) else set $active = 0$.
   iii) if $p_i < p_r$, then
       a) set $p_i = p_{i+1}$, $\alpha = a$, and repeat from step 6.
       b) else stop, since rendezvous is impossible.

10. If $forward$ is aperiodic, then let $lexi(someSequence)$ denote the lexicographically maximum rotation of the sequence $someSequence$.

11. Follow steps 7 through 10 of Algorithm 1.

The following lemma is necessary for the proof of Theorem 5 below.

Lemma 1. *Consider a prime p such that $2 < p \leq r$. Assume that for all primes $p_i < p$, the sequence of distances mod p_i between the α_i active MAs is periodic such that $d_1, \ldots, d_{\alpha_i} \equiv (d_1, \ldots, d_{\alpha_i/a})^a \bmod p_i$ and $a \mid \alpha_i$. Let the first MA in each occurence of the block $d_1, \ldots, d_{\alpha_i/a}$ remain active, while the remaining MAs become inactive. If the sequence of distances mod p between the a active MAs is periodic, then the original intertoken distances can be partitioned into $t \mid k$ equal length blocks with sums $\sigma_1, \sigma_2, \ldots, \sigma_t$ that are congruent modulo all the primes $p_1, p_2, \ldots, p$.*

Theorem 5. *If each MA has memory $O(k \lg \lg n)$, then the rendezvous problem can be solved in time $O(\frac{n \lg n}{\lg \lg n})$.*

Proof of Theorem 5.
Algorithm 3 solves the rendezvous if it stops the MAs when rendezvous is impossible and otherwise ensures a rendezvous. Suppose that for all $p_i \leq p_r$, the sequence of distances mod p_i between the active MAs is periodic. Algorithm 3 will stop the MAs in step 8 and indicate that rendezvous is impossible. Lemma 1 implies that in the last round of Algorithm 3, where $p = p_r$, the original intertoken distances can be partitioned into $a_r \mid k$ equal length blocks with sums σ_i^r, $i = 1, \ldots, a_r$, that are congruent mod p for all $p \leq p_r$. The Chinese Remainder Theorem then implies that the sums σ_i^r, $i = 1, \ldots, a_r$, are congruent mod $\prod_{i=1}^{r} p_i$. Since $\prod_{i=1}^{r} p_i > n$, then the original intertoken distances can be partitioned $a_r \mid k$ into equal length blocks with sums σ_i^r that are equal for all i, $i = 1, \ldots, a_r$. This implies, however, that $n = a_r \sigma_i^r$ for any i and thus $a_r \mid n$. Since $\gcd(k, n) = g > 1$, S is periodic, and the algorithm 3 correctly stops the MAs in step 8.

Algorithm 3 must also guarantee that rendezvous occurs when possible, i.e., when S is aperiodic. Suppose not, i.e., S is aperiodic but rendezvous does not occur. This implies that for all $p_i \leq p_r$, the sequences calculated in step 7 of Algorithm 3 are periodic and thus all rounds of the algorithm will be executed. In the final round, where $p_i == p_r$, algorithm 3 will stop the MAs and indicate that rendezvous is impossible. By the Chinese Remainder Theorem, however, this implies that S is periodic and thus contradicts the fact that S is aperiodic. Thus algorithm 3 solves the mobile agent rendezvous problem.

If each mobile agent has memory $O(k \lg \lg n)$, then Algorithm 3 correctly determines whether rendezvous is possible and instructs the MAs to stop or rendezvous as appropriate. In the worst case, rendezvous is impossible and the MAs must complete all r rounds of Algorithm 3, where r is the smallest number of prime numbers such that $\prod_{i=1}^{r} p_i > n$. Each of the r rounds takes n steps so the time complexity is $O(rn)$. Kranakis et al [7] proves that $r \in O(\frac{\lg n}{\lg \lg n})$, so the time complexity of Algorithm 3 is $O(\frac{n \lg n}{\lg \lg n})$. This completes the proof of Theorem 5. ∎

5 Lower Bound on Memory

In Theorem 1 of section 2, we prove that the mobile agent rendezvous problem can only be solved if either k or n is known. When $k = 2$, Kranakis et al [7] prove that $\Omega(\lg \lg n)$ memory is required to solve the rendezvous problem when k is known.

Theorem 6. *Solving the mobile agent rendezvous problem for $k > 2$ MAs requires $\Omega(\lg \lg n + \lg k)$ memory.*

6 Conditional Solutions

In the preceding two sections, we proved that the mobile agent rendezvous problem is unconditionally solvable if each MA has memory $\Omega(\lg \lg n + \lg k)$. In this section, however, we explore conditional solutions, i.e., solutions for cases where k and n satisfy various conditions such as primality. A network designer or administrator may be able to choose the values of k and n so as to meet these conditions.

The mobile agent rendezvous problem can be solved correctly whenever the sequences S_i are aperiodic for all i, e.g., n is prime or n is the product of two or more primes larger than k. A network designer or administrator, however, probably cannot directly dictate that S_i is aperiodic for all i but they are able to choose k and n. If $\gcd(k', n) = 1$, $\forall k' \le k$, e.g., n is prime or is the product of two primes greater than k, then S_i is aperiodic for all i. The following algorithm assumes an oriented ring. An *active* token in unoccupied while an *inactive* token has a MA residing on it.

Algorithm 4

1. Release the token at the starting node.
2. Set *active* = 1.
3. Set *count* = 0.
4. Begin to walk around the ring in the clockwise direction.
5. Compute the intertoken distances to the next three active tokens, i.e., d_1, d_2, d_3, and increment *count* for each inactive token passed.
6. If *count* == $k - 1$, arrange rendezvous. (Only active MA remaining.)
7. If $d_2 > d_1$ and $d_2 \ge d_3$, then remain active.
8. Else become inactive, i.e., set *active* = 0, continue in current direction to starting node, and wait for further instructions.
9. Repeat from step 3.

Theorem 7. *When the MAs share a common orientation and $\gcd(k', n) = 1$, $\forall k' \le k$, then the mobile agent rendezvous problem can be solved with $O(\lg n)$ memory and $O(n)$ time in an oriented ring.*

When k is prime, the ring is oriented, and $\gcd(k', n) = 1$, $\forall k' \le k$, then a variation of algorithm 4 solves the mobile agent rendezvous problem with $O(\lg k)$ memory and $O(n)$ time.

Algorithm 5

1. Release the token at the starting node.
2. Set active = 1.
3. Begin to walk around the ring in a clockwise direction.
4. Execute round 1 of algorithm 4 but calculate the intertoken distances mod k. (All MAs will return to their starting nodes and those that became inactive have set active = 0.)
5. (Now execute algorithm 4 as if on a ring of size k). The distances of interest are now the number of inhabited tokens between pairs of empty tokens.)
6. Compute the number of inhabited tokens, i.e., tokens hosting inactive MAs, met on the path to the next three uninhabited tokens, i.e., m_1, m_2, m_3.
7. If $m_1 == k - 1$, arrange the rendezvous. (Only one active MA left.)
8. If $m_2 > m_1$ and $m_2 \geq m_3$, then remain active.
9. Else become inactive, i.e., set $active = 0$, continue in current direction to starting node, and wait for further instructions.
10. Repeat from step 5.

Theorem 8. *When the MAs share a common orientation, k is prime, and $\gcd(k',n) = 1$, $\forall k' \leq k$, the mobile agent rendezvous problem can be solved with $O(\log k)$ memory and $O(n)$ time.*

The following algorithm solves the rendezvous problem when $\gcd(k',n) = 1$, $\forall k' \leq k$, but k is not prime.

Algorithm 6

1. Release the token at the starting node.
2. Set active = 1 and count = 0.
3. Begin to walk around the ring in the clockwise direction.
4. Compute the intertoken distances mod k to the next three active tokens, i.e., d_1, d_2, d_3 mod k, and increment *count* for each inactive token passed.
5. If $count == k - 1$, arrange rendezvous. (Only active MA remaining.)
6. If $d_2 > d_1$ mod k and $d_2 \geq d_3$ mod k, then remain active.
7. Else become inactive, i.e., set $active = 0$, and wait for further instructions.
8. Repeat from step 4.

Theorem 9. *When the MAs share a common orientation and $\gcd(k',n) = 1$, $\forall k' \leq k$, then the mobile agent rendezvous problem can be solved with $O(\log k)$ memory and $O(n \log k)$ time.*

7 Leader Election and Rendezvous

The relationship between the rendezvous problem and the leader election problem among the k MAs depends on whether the MAs share a common orientation.

Theorem 10. *If the MAs share a common orientation, then the leader election problem among k MAs is equivalent to the rendezvous problem for those MAs. If the MAs do not share a common orientation, however, then the leader election problem is strictly more complex than the rendezvous problem.*

8   Conclusion

After proving that the mobile agent rendezvous search problem is unsolvable when both k and n are unknown, we prove that rendezvous is possible if and only if the sequence of intertoken distances is aperiodic. We then present unconditional and conditional solutions for the rendezvous problem. We derive the lower bounds on the memory required for mobile agent rendezvous and then discuss the relationship between rendezvous and leader election for mobile agents.

In future research, it would be interesting to study how changes in the model affect the complexity of the mobile agent rendezvous search problem. For example, it would be interesting to study a network topology that differs from the ring or the case where each mobile agent has more than one token.

References

1. S. Alpern, Rendezvous Search: A Personal Perspective, Operations Research, 50, No. 5, pp. 772-795, 2002.
2. S. Alpern and S. Gal, The Theory of Search Games and Rendezvous, Kluwer Academic Publishers, London, 2003.
3. L. Barriere, P. Flocchini, P. Fraigniaud, and N. Santoro, Election and Rendezvous of Anonymous Mobile Agents in Anonymous Networks with Sense of Direction, Proceedings of the 9th International Colloquium on Structural Information and Communication Complexity (SIROCCO), pp. 17-32, 2003.
4. L. Barriere, P. Flocchini, P. Fraigniaud, and N. Santoro, Can We Elect If We Cannot Compare?, Proceedings of the 15th ACM Symposium on Parallel Algorithms and Architectures (SPAA), pp. 324-332, 2003.
5. V. Baston and S. Gal, Rendezvous Search When Marks are Left at the Starting Points, Naval Research Logistics, 47, No. 6, pp. 722-731, 2001.
6. S. Gal, Rendezvous Search on the Line, Operations Research, 47, No. 6, pp. 849-861, 1999.
7. E. Kranakis, D. Krizanc, N. Santoro, and C. Sawchuk, Mobile Agent Rendezvous in the Ring, International Conference on Distributed Computing Systems (ICDCS 2003), pp. 592 - 599, Providence, RI, May 19 - 22, 2003.
8. Rendezvous Search on the Line with More than Two Players, Operations Research, 45, pp. 357-364, 1997.
9. M. Pikounis and L.C. Thomas, Many Player Rendezvous Search: Stick Together or Split and Meet? University of Edinburgh, Management School, Preprint, 1998.
10. X. Yu and M. Yung, Agent Rendezvous: A Dynamic Symmtery-Breaking Problem, Proceedings of ICALP'96, LNCS 1099, pp. 610-621, 1996.

Global Synchronization in Sensornets

Jeremy Elson, Richard M. Karp, Christos H. Papadimitriou, and Scott Shenker

UCLA and Information Sciences Institute (Elson), UC Berkeley and International Computer Science Institute (Karp, Papadimitriou, Shenker)

Abstract. Time synchronization is necessary in many distributed systems, but achieving synchronization in sensornets, which combine stringent precision requirements with severe resource constraints, is particularly challenging. This challenge has been met by the recent Reference-Broadcast Synchronization (RBS) proposal, which provides on-demand pairwise synchronization with low overhead and high precision. In this paper we introduce a model of the basic RBS synchronization paradigm. Within the context of this model we characterize the optimally precise clock synchronization algorithm and establish its global consistency. In the course of this analysis we point out unexpected connections between optimal clock synchronization, random walks, and resistive networks, and present a polynomial-time approximation scheme for the problem of calculating the effective resistance in a network based on min-cost flow. We also sketch a polynomial-time algorithm for finding a schedule of data acquisition giving the optimal trade-off between energy consumption and precision of clock synchronization. We also discuss synchronization in the presence of clock skews. In ongoing work we are adapting our synchronization algorithm for execution in a network of seismic sensors that requires global clock consistency.

1 Introduction

Many traditional distributed systems employ time synchronization to improve the consistency of data and the correctness of algorithms. Time synchronization plays an even more central role in sensornets, whose deeply distributed nature necessitates fine-grained coordination among nodes. Precise time synchronization is needed for a variety of sensornet tasks such as sensor data fusion, TDMA scheduling, localization, coordinated actuation, and power-saving duty cycling. Some of these tasks require synchronization precision measured in μsecs, which is far more stringent than the precision required in traditional distributed systems. Moreover, the severe power limitations endemic in sensornets constrain the resources they can devote to synchronization. Thus, sensornet time synchronization must be both more precise, and more energy-frugal, than traditional time synchronization methods.

The recent *Reference-Broadcast Synchronization* (RBS) design meets these two exacting objectives by producing on-demand pairwise synchronization with low overhead and high precision [7]. RBS is specifically designed for sensornet contexts in which (1) communications are locally broadcast, (2) the maximum speed-of-light delay between sender and receiver is small compared to the desired synchronization precision, and (3) the delays between time-stamping and sending a packet are significantly more

M. Farach-Colton (Ed.): LATIN 2004, LNCS 2976, pp. 609–624, 2004.
© Springer-Verlag Berlin Heidelberg 2004

variable than the delays between receipt and time-stamping a packet (so estimates of when a packet is sent are far noisier than estimates of when it is received). See [7] for a much fuller discussion of this last point, but measurements described therein suggest that the receiving delays can be reasonably modeled as a Gaussian centered around some mean, with the mean being the same for all nodes (assuming they share the same hardware/software).

There is a vast literature on clock synchronization in the theory and distributed systems literature [1,4,10,18,22]; see [6] and references therein for a comprehensive review. We note, however, that most traditional methods synchronize a receiver with a sender by transmitting current clock values, and are thus sensitive to transmission delay variability and asymmetry. In contrast, RBS avoids these vulnerabilities by synchronizing receivers with each other leveraging the special properties of sensornet communications. Reference broadcast signals are periodically sent in each region, and sensornet nodes record the times-of-arrival of these packets. Nodes within range of the same reference broadcast can synchronize their clocks by comparing their respective recent time-of-arrival histories. Nodes at distant locations (not in range of the same reference broadcast) can synchronize their clocks by following a chain of pairwise synchronizations. RBS is therefore completely insensitive to transmission delays and asymmetries. In fact, errors in RBS arise only from differences in time-of-flight to different receivers and delays in recording packet arrivals. In the contexts for which RBS is intended, both of these errors are quite small and the latter dominates the former. Therefore, most of the errors in synchronization are due to essentially random delays in recording times-of-arrival (which, as observed earlier, are reasonably modeled as Gaussian).

To penetrate this noise, RBS uses pairwise linear regressions of the time-of-arrival data from a shared broadcast source. While this seems like a very promising approach, and has been verified on real hardware, there are two aspects of RBS, and in fact of any similar synchronization algorithm, that we wish to improve upon. First, the resulting synchronization is purely pairwise, in that for any pair of nodes i, j RBS can compute coefficients $a_{i,j}, b_{i,j}$ that translate readings on i's clock into readings on j's clock via $t_j \approx t_i a_{ij} + b_{ij}$, but these pairwise translations are not necessarily globally consistent. Converting times from i to j, and then j to k can be different than directly converting from i to k; *i.e.* the transitive properties $a_{ij}a_{jk} = a_{ik}$ and $b_{ij}a_{jk} + b_{jk} = b_{ik}$ need not hold.[1] Second, the pairwise synchronizations are not optimally precise in that they do not have minimal variance from the truth. The RBS synchronization of two sensornet nodes is based only their time-of-arrival information from a single broadcast source. No information from other broadcast sources is used, nor is time-of-arrival information from other receivers. Thus, much relevant data is being ignored in the synchronization process, resulting in suboptimal precision.[2]

---

[1] Note that requiring the pairwise synchronizations to be globally consistent is equivalent to saying that there is some universal time standard to which all nodes are synchronized (e.g. the time of one particular node could serve as this universal time, though we choose to adopt a more distributed approach).

[2] Some of this is inherent in the RBS approach and some is an artifact of the particular design described in [7]. Using only a single synchronization source is an artifact; not incorporat-

We address these limitations in a simplified model of synchronization in which changes in clock skew (differences in the *rates* of clocks) occur at much longer time scales than changes in clock offset (differences in the current clock *values*);[3] that is, we assume that over short time scales the clock skews are known and synchronization is used only to adjust for clock offsets; estimates of clock skew are taken on much longer time scales. Thus, in what follows we will assume that all clocks advance at the same rate (because any differences in rate are explicitly compensated for); later, in Section 5, we will relax this assumption.

Our focus in this paper is primarily theoretical and we do not evaluate the feasibility (in terms of energy consumption) of this approach. However, we are planning to adopt this approach in a seismographic sensornet array. The requirements of optimally precise and globally consistent time synchronization are particularly acute in this context. We expect that there are ways to increase the energy efficiency of the approach described here without sacrificing significant precision or consistency.

While our discussion is focused entirely on RBS, our methods and results could be extended to any pairwise synchronization procedure whose errors were independent. In addition, our focus here is primarily theoretical and we do not evaluate the feasibility of this protocol. However, we are planning to implement this protocol, or some variant, in a seismographic sensornet array. The requirements of optimally precise and globally consistent time synchronization are particularly acute in this context. We expect that there are ways to increase the energy efficiency of the approach described here without sacrificing significant precision or consistency.

RBS is, of course, not the only approach to sensornet clock synchronization. In some contexts, Global Positioning System (GPS) can provide a universal clock signal, but GPS requires a clear sky view, and thus does not work inside buildings, underwater, or beneath dense foliage. Moreover, many current sensornet nodes (*e.g.*, the Berkeley Motes [12] are not equipped with GPS. There are several proposals for synchronizing clocks within a single broadcast domain [27,26,20], but they do not generalize to global synchronization, which is what we address here.

Two global synchronization protocols of note are [17] and [25]. The microsecond precision achieved in [17] is similar to our goals here, but the approach assumes a fixed topology and guarantees on latency and determinism in packet delivery. A very energy-efficient time diffusion algorithm is presented in [25], but the precision analysis assumes deterministic transmission times. Our interest here is in synchronization algorithms that do not require specific underlying networks to function.

Some synchronization designs, such as [11,8], integrate the medium-access control protocol (MAC) MAC with the time synchronization procedure. While our discussion does not make assumptions about the underlying hardware and MAC, the results would benefit from these MAC-specific features to the extent that they reduce the magnitude of the receive-time errors. We note that, while the discussion of our approach builds on RBS,

---

ing time-of-arrival data from other receivers is inherent in the general pairwise-comparison approach adopted by RBS.

[3] In our previous notation where $t_j \approx t_i a_{ij} + b_{ij}$, a_{ij} represents the relative clock skew and b_{ij} represents the relative clock offset.

our methods and results could be extended to any pairwise synchronization procedure whose errors were independent.

Another quite different approach is that taken in [21], which doesn't directly synchronize clocks but instead refers to events in terms of their age, not time. The problem of calibration [28] is related to that of synchronization, though ti differs in some essential details. The discussion in[3] is especially relevant to our discussion here, as it considers how to use nonlocal information across multiple calibration paths in a consistent manner.

2   Summary of Results

The core problem of the paper is the following. We are given a set of *receivers*. Each receiver r_i has a clock that is offset from a (fictitious) universal time standard by a constant amount T_i. We are also given a set of synchronization signals. Each synchronization signal s_k is transmitted at an unknown time U_k and is received by a set of receivers. the time-of-flight of s_k is negligible. If r_i is a receiver of s_k then r_i measures the arrival time of s_k on its local clock. Let this measured time be y_{ik}. We assume that $y_{ik} = U_k + T_i + e_{ik}$ where the error e_{ik} is a random variable with zero mean and variance V_{ik}. We also assume that the errors e_{ik} are independent.

The main results are as follows:

1. We define a resistive network with a node for each receiver and each signal, such that the minimum-variance estimator of $T_i - T_j$ is derivable from the distribution of current when one unit of current is inserted at node T_i and extracted at node T_j. The variance of the estimator is the effective resistance between T_i and T_j. The variances V_{ik} appear as resistances in this network.

2. The minimum-variance estimator is globally consistent, in the sense that for any triple (i, j, m) the estimates of $T_i - T_j$, $T_j - T_m$ and $T_m - T_i$ sum to zero.

3. The effective resistance between two nodes of a network can be approximated with relative error ϵ by performing $\sqrt{\frac{V}{\epsilon R}}$ flow augmentations on the network, where V is the sum of the resistances and R is the effective resistance.

4. The effective resistance of the regular infinite d-dimensional grid is given explicitly, and indicates the advantage of the proposed synchronization scheme over its predecessor RBS in this case. We believe that similar advantages will typically be realized in large-scale sensornets in which the sensors have a homogeneous spatial distribution.

5. Under the additional assumption that the errors e_{jk} are Gaussian, the maximum-likelihood joint choice of the T_i and U_k, subject to the convention that $T_1 = 0$, is the unique solution to a linear system of least-squares equations. This sparse system of equations can be solved iteratively by a distributed sensornet algorithm in which each receiver or generator of a signal is responsible for updating the corresponding variable T_i or U_k.

6. The maximum-likelihood joint choice of the T_i and U_k agrees exactly with the minimum-variance pairwise estimates of $T_i - T_j$. Therefore the minimum-variance estimator is what is known in statistics as an efficient estimator.

7. The maximum-likelihood estimate of T_i is exactly the hitting time of a random walk from r_i to r_0 on a weighted directed graph with 'delay' y_{ik} on edge $[r_i, s_k]$ and $-y_{ik}$ on edge $[s_k, r_i]$, where r_i is a receiver of signal s_k, and the transition probabilities out of each vertex are inversely proportional to the variances V_{ik}.

8. A polynomial-time algorithm is presented which, given a set of receivers and a set of potential signals, determines the optimal repetition rate of each signal to minimize energy consumption while keeping the variance of the estimate of each offset $T_i - T_j$ below a specified value.

9. A method is given for estimating clock skews under the assumption that the clock of each receiver r_i advances at a fixed rate α_i per unit time. The method is based on measurements of the time elapsed on the clock of each signal transmitter and the time elapsed on the clock of each receiver between two transmissions of the same signal widely separated in time. The method is based on an isomorphism between this version of the clock skew problem and the clock offset problem described above.

3   Optimal and Global Synchronization

In this section we consider a simple model where clocks all progress at the same rate (*i.e.* no skew), but have arbitrary offsets; we later, in Section 5, extend our results to the case of general clock skew. After describing the model and notation, we consider the question of optimal pairwise synchronization and then that of the most likely globally consistent synchronization. We then show their equivalence and end this section by describing a simple iterative computation of the solution and its variance.

3.1   Model and Notation

We consider the case where there are n sensornet nodes, and let r_i denote the i'th such node. These nodes use synchronization signals to align their clocks; let s_k denote the k'th synchronization signal. Our treatment does not care from whence these signals come, only which nodes hear them, so we don't identify the source of these signals. We let E be the set of pairs (r_i, s_k) such that node r_i receives signal s_k; in what follows, we will use the terms "node" and "receiver" interchangeably. In order to explain our theory, we make reference to a perfect universal time standard or clock; of course, no such clock exists and our results do not depend on such a clock, but it is a useful pedagogical fiction. In fact, the approximation of such a universal time standard is one of the goals of our approach.

We assume, in this section, that all clocks progress at the same rate and that propagation times are insignificant (or have been explicitly compensated for). We represent the offset of a node, or receiver, by the variable T_i. This offset is the difference between the local time on r_i's clock and the universal absolute time standard. Of course, there is a degree of freedom in choosing these T_i, as they could all be increased by the same constant without changing any of the pairwise conversions; the addition of such a constant term would reflect changing the setting of the global clock. We represent by U_k the time when synchronization signal s_k is sent (or, equivalently, received) according to the absolute time standard. The U_k's are not known, but are estimated as part of the synchronization process; thus, they are outputs, not inputs, of our theory.

Each node records the times-of-arrival of all synchronization messages it receives (*i.e.* all those that they are in range of). We let y_{ik} denote the measured time on r_i's clock when it receives signal s_k. The quantity y_{ik} is defined if and only if $(r_i, s_k) \in E$. The basic assumption we make about measurement errors is that:

$$y_{ik} = U_k + T_i + e_{ik} \qquad (1)$$

where e_{ik} is a random variable with mean zero and variance V_{ik}. We further assume that all these random variables are independent.

To find the optimal (*e.g.* the minimum-variance) pairwise synchronization between nodes i and j, we must produce the minimum-variance estimate of the difference $T_j - T_i$. In contrast, to produce a globally consistent synchronization, we must estimate all the T_i independently and seek a maximum-likelihood joint choice of all the offsets T_i. When we assume the measurement errors e_{ik} are Gaussian we are able to reduce this maximum-likelihood problem to a linear system of least-squares equations. Surprisingly, the solution to this system of equations also solves the flow problem used to produce minimum-variance estimators.

3.2   Minimum-Variance Pairwise Synchronization

Given two nodes r_1 and r_2 an unbiased estimator of $T_1 - T_2$ can be obtained from any appropriate path between r_1 and r_2. In general such a path is of the alternating form $r_{i_1}, s_{k_1}, r_{i_2}, s_{k_2}, \cdots, s_{k_t}, r_{i_{t+1}}$ where $r_{i_1} = r_1$ and $r_{i_{t+1}} = r_2$ and each adjacent pair is in E. The corresponding estimator is

$y_{i_1,k_1} - y_{i_2,k_1} + y_{i_2,k_2} - \cdots - y_{i_{t+1},k_t}$. which, in view of the equation $y_{ik} = U_k + T_i + e_{ik}$, is equal to $T_1 - T_2 + e_{i_1,k_1} - e_{i_2,k_1} + e_{i_2,k_2} - \cdots - e_{i_{t+1},k_t}$. This estimator is unbiased because each e_{ik} has zero mean.

By considering appropriate weighted combinations of alternating paths we can obtain an estimator of much lower variance than any single path can provide, thus providing a more accurate synchronization of the two nodes. Such a weighted combination of paths is a flow from r_1 and r_2, satisfying the *flow conservation requirement* that the net flow into any node except r_1 and r_2 is zero. In this subsection we characterize the minimum-variance estimator of $T_1 - T_2$ in terms of flows.

Consider an undirected flow network with edge set E. We will use the following convention regarding summations: $\sum_{ik}$ will denote a summation over all pairs (i, k) such that $\{r_i, s_k\} \in E$; when k is understood from context, $\sum_i$ will denote a summation over all i such that $\{r_i, s_k\} \in E$; and when i is understood from context, $\sum_k$ will denote a summation over all k such that $\{r_i, s_k\} \in E$.

We first state, without proof, a basic but straightforward fact about unbiased estimators:

Theorem 1. *The unbiased estimators of $T_1 - T_2$ are precisely the linear expressions $\sum_{ik} f_{ik} y_{ik}$ such that $\{f_{ik}\}$ is a flow of value 1 from r_1 to r_2. Here f_{ik} is positive if the flow on edge $\{r_i, s_k\}$ is directed from r_i to s_k, and negative if the flow is directed from s_k to r_i. The variance of the unbiased estimator $\{f_{ik}\}$ is $\sum f_{ik}^2 V_{ik}$. A similar statement holds for the unbiased estimators of $T_j - T_i$, for any i and j.*

The problem of finding a minimum-variance unbiased estimator of $T_1 - T_2$ is related to the problem of determining the effective resistance between two nodes of a resistor network. In order to sketch this connection we review some basic facts about resistive electric networks.

Let G be a connected undirected graph with vertex set V and edge set A, such that there is a resistance $R(u, v)$ associated with each edge $\{u, v\}$. An *applied current vector* is a vector e with a component $e(u)$ for each vertex, such that $\sum_{u \in V} e(u) = 0$; $e(u)$ represents the (steady-state) current (positive, negative or zero) injected into the network at vertex u. Associated with every applied current vector e is an assignment to each ordered pair $[u, v]$ of adjacent vertices of a current $c(u, v)$ and to each vertex u a potential $p(u)$ satisfying Kirchhoff's law (net current into a vertex = 0) and Ohm's law $p(v) - p(u) = c(u, v)R(u, v)$. The current is unique and the potential is unique up to an additive constant. When we want to identify the particular applied current vector e we write $c_e(u, v)$ and $p_e(v)$. A key property is the *superposition principle*:

$$c_{e_1 + e_2}(u, v) = c_{e_1}(u, v) + c_{e_2}(u, v)$$

and

$$p_{e_1 + e_2}(v) - p_{e_1 + e_2}(u) = (p_{e_1}(v) - p_{e_1}(u)) + (p_{e_2}(v) - p_{e_2}(u))$$

The *effective resistance* between u and v is the potential difference $p(v) - p(u)$ when the applied current vector is as follows: $e(u) = 1$, $e(v) = -1$ and all other components of e are zero; *i.e.* when one unit of current is injected at u and extracted at v.

The effective resistance between u and v can be characterized in terms of a minimum-cost flow problem with quadratic costs. It is the minimum, over all currents $c(u, v)$ satisfying Kirchhoff's law (with external current 1 at u and -1 at v) of $\sum_{(u,v) \in E} c(u, v)^2 R(u, v)$. This quadratic objective function represents the power dissipation in the network.

Now consider the undirected bipartite graph of signals $\{s_k\}$ and receivers $\{r_i\}$ as a resistor network, with the variance V_{ik} as the resistance of the edge $\{s_k, r_i\}$. Combining Theorem 1 with the minimum-cost-flow characterization of effective resistance we obtain the following theorem.

Theorem 2. *The minimum variance of an unbiased estimator of $T_1 - T_2$ is the effective resistance between r_1 and r_2, and the corresponding estimator is $\sum_{ik} f_{ik} y_{ik}$ where f_{ik} is the current along the edge from r_i to s_k when one unit of current is injected at r_1 and extracted at r_2.*

The following theorem establishes the mutual consistency of the minimum-variance estimators of the differences between offsets. Its proof is a simple application of the superposition principle. Let $A(i, j)$ be the minimum-variance estimator of $T_i - T_j$.

Theorem 3. *For any three indices i, m and j, we have $A(i, m) + A(m, j) = A(i, j)$.*

It follows from Theorem 3 that we can compute $A(i, j)$ for all i and j by computing $A(i, m)$ for all i and a fixed m and using the identity $A(i, j) = A(i, m) - A(j, m)$. This shows that the set of minimum-variance pairwise synchronizations are globally consistent. The question remains whether they are the maximally likely set of offset assignments.

3.3   Maximum-Likelihood Offset Assignments

We now seek the maximally likely set of offset assignments T_i. This approach is guaranteed to produce a globally consistent set of pairwise synchronizations, but it is not clear *a priori* that they are minimum-variance pairwise synchronizations. In this formulation we assume that the y_{ik} are independent Gaussian random variables such that y_{ik} has mean $U_k + T_i$ and variance V_{ik}. Then the joint probability density $\mathcal{P}$ of the y_{ik} given values T_i for the offsets of the receivers and U_k for the absolute transmission times of the signals is given by:

$$\mathcal{P} = \prod_{ik} \frac{1}{\sqrt{2\pi V_{ik}}} e^{-\frac{(y_{ik}-U_k-T_i)^2}{2V_{ik}}}$$

We shall derive a system of linear equations for the T_i and U_k that maximize this joint probability density.

Let C_{ik} denote the reciprocal of V_{ik}. We refer to C_{ik} as the *conductivity* between s_k and r_i.

Differentiating the logarithm of the joint probability density with respect to each of the U_k and T_i we find that the choice of $\{U_k\}$ and $\{T_i\}$ that maximizes the joint probability density is a solution to the following system of equations:

For each k,

$$\sum_i C_{ik}(U_k + T_i) = \sum_i C_{ik} y_{ik} \tag{2}$$

For each i,

$$\sum_k C_{ik}(U_k + T_i) = \sum_k C_{ik} y_{ik} \tag{3}$$

From these equations we can derive an interpretation of each T_i as the hitting time of a random walk from r_i to r_0 on a directed graph with a positive or negative 'delay' on each edge. Assume that the set of indices i associated with the receivers is disjoint from the set of indices k associated with the signals. Under this assumption there is no ambiguity in defining, for each signal index k, a new variable T_k equal to $-U_k$. The system of equations becomes:

For each k,

$$T_k = \frac{\sum_i C_{ik}(-y_{ik} + T_i)}{\sum_i C_{ik}} \tag{4}$$

For each $i \neq 0$,

$$T_i = \frac{\sum_k C_{ik}(y_{ik} + T_k)}{\sum_k C_{ik}} \tag{5}$$

Fixing T_0 at 0, it is clear by inspection that these equations support the following interpretation: for each receiver i, T_i is the expected total delay of a random walk starting at r_i and ending at the first visit to r_0, where the transition probability from r_i to s_k is $\frac{C_{ik}}{\sum_k C_{ik}}$, the transition probability from s_k to r_i is $\frac{C_{ik}}{\sum_i C_{ik}}$, the delay on a transition from r_i to s_k is y_{ik} and the (negative) delay on a transition from s_k to r_i is $-y_{ik}$.

3.4   Equivalence of the Two Formulations

The following theorem shows that, even though the minimum-variance pairwise synchronization and the maximum-likelihood offset assignment appear to be based on different principles, they determine the same values of $T_j - T_i$, for all i and j. Our proof is based on the superposition principle, but the theorem can also be seen as a consequence of the Cramer-Rao inequality [16], a general tool for eastablishing that the variance of an estimator is best possible.

Theorem 4. *For any fixed index m we obtain a solution to the system of equations 5 by setting $T(i) = A(i, m)$ for each i.*

3.5   Solving the Equations

The solution to the system of equations 2 and 3 can be found through a simple two-step iterative process. In the first step, the y_{ik} and T_i are used to estimate the U_k:

For each k,
$$U_k \leftarrow \frac{\sum_i C_{ik}(y_{ik} - T_i)}{\sum_i C_{ik}}$$

In the second step, the y_{ik} and U_k are used to estimate the T_i:

For each i,
$$T_i \leftarrow \frac{\sum_k C_{ik}(y_{ik} - U_k)}{\sum_k C_{ik}}$$

Each iteration reduces $\sum_i \sum_k C_{ik}(y_{ik} - U_k - T_i)^2$. It follows that the iterative process converges to a solution of the system. Convergence can be accelerated by over-relaxation techniques which are standard in numerical analysis [2].

While this theory produces optimal (in two senses of optimality, maximum-likelihood and minimum-variance) estimators, it does not directly reveal the quality of the estimated values. The variance of each the estimators can be obtained by computing the effective resistance between the corresponding receivers. This can be done exactly by solving a system of linear equations or approximately by a new approximation algorithm based on minimum-cost flow. these two approaches are described in the following subsections.

3.6   Computing Optimal and Near-Optimal Estimators and Their Variances

As we have seen, finding an optimal unbiased estimator of $T_2 - T_1$ and determining its variance is an equivalent problem to computing the distribution of currents, and the corresponding effective resistance, when a unit of current is injected at node s and extracted at node t of a resistive network. A standard approach is to set up and solve a system of linear equations for the currents and potentials using Ohm's Law and Kirchhoff's Law.

In certain special cases the effective resistance can be determined analytically. For example, in the infinite d-dimensional grid with unit resistors and unit distance between neighboring nodes, the effective resistance between two nodes at Manhattan distance L is $O(\log L)$. Thus, in the clock synchronization problem corresponding to this network, our approach would yield an estimator with variance $O(\log L)$ whereas RBS, which bases its estimator on a single path, would yield an estimator with variance L.

3.7   A PTAS for Effective Resistance

An alternate approach is to use the formulation of effective resistance as a flow problem in which each edge has unbounded capacity and cost quadratic in the flow. The quadratic edge costs can then be approximated by piecewise-linear functions, yielding a flow problem with finite capacities but linear costs [13]. Pursuing this idea, we have shown that the effective resistance can be approximated within relative error ϵ by performing $\sqrt{\frac{V}{\epsilon R}}$ flow augmentations in the linear-cost network, where V is the sum of the resistances and R is the effective resistance. Moreover, this bound can be achieved without knowing R in advance.

Given a resistive network $G = ([n], S, V)$, we denote by R the sought effective resistance between s and t, and by V the sum of the resistances V_{ij} of all edges of G.

Theorem 5. *R can be approximated within $\epsilon > 0$ by $\sqrt{\frac{V}{\epsilon R}}$ flow augmentations.*

Proof: For any positive real F, let $Q(F)$ be the problem of finding a flow of value F that minimizes the quadratic objective function $\sum_{i,j} f_{ik}^2 V_{ij}$ over all flows $\{f_{ij}\}$ of value F from s to t, and let $C^*(F)$ be the cost of a minimum-cost solution to $Q(F)$. Notice that the effective resistance is $C^*(1)$, and $C^*(F) = F^2 C^*(1)$.

For each F we shall define a linear-cost network flow problem $L(F)$ in which the quadratic objective function of $Q(F)$ is replaced by a piecewise-linear approximation which becomes very tight when F is sufficiently large.

The piecewise-linear function $G(x)$ is defined as follows: $G(0) = 0$; for any odd positive integer $2t + 1$, $G(2t + 1) = 4t^2$; over the interval $[0,1]$ and each interval $[2t + 1, 2t + 3]$ G is linear. Then, for all nonnegative x, $G(x) \leq x^2 \leq G(x) + 1$.

For any positive real F let $L(F)$ be the problem of minimizing $\sum_{ij} G(f_{ij}) V_{ij}$ over all flows from r_1 to r_2 of value F. Let $D(F)$ denote the cost of an optimal solution of $L(F)$. Then $D(F) \leq C^*(F) \leq C^*(F) + V$, since a minimum-cost flow in $L(F)$ will have cost less than or equal to $D(F) + V$ with respect to the quadratic cost function of $Q(F)$.

Our goal is to compute a solution to $Q(F)$ of cost less than or equal to $(1+\epsilon)C^*(F)$. By the above inequalities, it suffices to take an optimal flow for $L(F)$, for any F greater than or equal to $\sqrt{\frac{V}{\epsilon R}}$. Since R is initially unknown, we will solve the sequence of problems $L(1), L(2), L(3), \cdots$ until a solution is found that can be verified to solve some $Q(F)$ within the approximation ratio $1 + \epsilon$. This solution, scaled down by the factor F, provides the required approximate solution to the original Problem $Q(1)$.

To solve this sequence of linear-cost flow problems we construct a network in which, between any pair (r_i, r_j) of adjacent vertices there are (in principle) infinitely many parallel edges. The first of these has capacity 1, and each each subsequent edge has capacity 2. The cost coefficient of the first edge is 0, and the cost coefficient of the tth subsequent edge is $4t V_{ij}$. The cost of a flow of f along an edge is f times the cost coefficient of the edge. In an optimal solution to this linear network flow problem for a specified flow value F, the flow through this set of parallel edges will exhaust the capacities of these edges in increasing order of their cost coefficients. It is easy to check that this linear network flow problem is equivalent to Problem $L(F)$.

If we start with the zero flow and repeatedly augment the flow by sending one unit of flow along a minimum-cost flow-augmenting path from r_1 to r_2 then, after F augmentations, we will have a minimum-cost flow for $L(F)$. The computational cost of each augmentation is $O(m)$, where m is the number of edges in the resistor network for the original problem $Q(1)$.

For each successive F, an optimal flow for $L(F)$ is computed, and its cost is computed both in $Q(F)$ and in $L(F)$. When for some F, the ratio of these costs is less than or equal to $1 + \epsilon$, the current flow (scaled down by the factor F) achieves the desired approximation ratio, and the algorithm halts. This will happen after at most $\sqrt{\frac{V}{\epsilon R}}$ flow augmentations. In the case where all V_{ij} are equal to 1, $V = m$ and $R \geq 1/m$, so the number of flow augmentations is at most $m\epsilon^{-1/2}$, and the execution time of the polynomial-time approximation scheme is at most $O(m^2 \epsilon^{-1/2})$.

4   Optimal Synchronization Design

An interesting problem raised by the results above is to select a subset of the set of available signals, and their rate of synchronization messages, so as to minimize the energy consumption required to achieve a specified precision in the estimates of all offsets $T_i - T_j$. We formulate this problem as a continuous nonlinear optimization problem, and present a polynomial-time algorithm, based on the ellipsoid method, for approximating the solution to any desired accuracy.

We associate with each signal s_k a real variable x_k giving the frequency with which the signal is repeated. We assume that successive repetitions are independent, so that the composite signal obtained by averaging x_k repetitions of s_k reduces the variance of each measured value y_{ik} by the factor x_k, yielding a variance of $\frac{V_{ik}}{x_k}$. We also assume that the rate of power consumption for the network is proportional to the sum of the x_k.

We wish to minimize $\sum_k x_k$ subject to the requirement that, for the corresponding set of variances $\frac{V_{ik}}{x_k}$, the effective resistance between each pair of receivers is at most a specified value α. The joint choice of all the variables x_k yields a point in a euclidean space of dimension equal to the number of signals s_k. For a given pair r_i, r_j of receivers, let K_{ij} be the set of points in this space for which the effective resistance between r_i and r_j is less than or equal to α. Let K be the intersection of all the sets K_{ij}. Thus our *synchronization design problem* is:

$$\min \sum_k x_k$$

subject to $x \in K$.

Note that, because dividing all the resistances in a network by a factor t reduces the effective resistances by that same factor, the optimal choice of x for a bound β on the effective resistances is obtained from the optimal choice for α simply by multiplying each x_k by $\frac{\alpha}{\beta}$.

The following can be shown: each set K_{ij} is convex and possesses a polynomial-time *separation oracle*; *i.e.*, a polynomial-time algorithm which, given any point p not in K_{ij}, returns a hyperplane separating p from K_{ij}. It follows at once that K is convex and possesses a polynomial-time separation oracle. Given these facts, the following theorem is a consequence of general results on separation vs. optimization due to [9].

Theorem 6. *The optimum solution to the synchronization design problem can be approximated to any desired accuracy in polynomial time.*

5   From Theory to Protocol

We have described abstractly how one could optimally compute the appropriate clock offsets T_i from the measurement data y_{ik}. In this section we briefly discuss how one might transform this theory into a practical protocol. This discussion is by no means complete or definitive, and is completely untested; instead, we offer it only as providing some glimmer that the ideas of presented here could be successfully applied to real systems with their skewed clocks and energy constraints. The two issues we address are: (1) generalizing the theory to compensate for clock skew and (2) turning the abstract calculation into a series of practical message exchanges.

5.1   Clock Skew

The theoretical treatment assumed that all clocks progressed at the same rate. We now relax this assumption and describe how one can estimate the relative rates of clocks. In particular, we wish to estimate parameters α_i that describe the rate of the local clocks relative to the standard clock: if a time δ has elapsed on the universal standard clock then each local clock shows that time $\alpha_i \delta$ has elapsed (so large α_i reflect fast clocks). As with the offsets T_i, there is a degree of freedom in choosing these α_i; each could be multiplied by the same constant (which would only change the speed of the absolute clock).

Given the pair (α_i, T_i) for some node i, we can translate local times t_i into *standard* times τ: $\tau = \frac{t_i}{\alpha_i} - T_i$. Moreover, if one had the constants α_i, then one can estimate the T_i's as in the previous section by first dividing all local clocks by α_i. Thus, we must now describe how to obtain estimates of these skew values α_i, and do so without knowledge of the offsets T_i (since the computation of the T_i requires knowledge of the α_i).

To estimate clock rates, we use the same set of synchronization signals, but now select pairs of them originating from the same source spaced at sizable intervals (*i.e.*, large compared to the variances V_{ik} of the individual measurements). We label the k'th signal pair by p_k. We let W_k and w_{ik} represent the time elapsed between their transmission as measured by, respectively, the standard clock and i's local clock. In the notation of Section 3, W_k is the difference between the pair of signals of the U values; w_{ik} is the corresponding difference in the y values. We assume that the measurement errors, as expressed by the e_{ik}, are negligible compared to the magnitude of the W_k. If all clocks progressed at perfectly constant rates, then $w_{ik} = W_k \alpha_i$ for each i, k and we could estimate the variables α_i based on a single measurement for each i.

However, clock rates drift and wander over time in random and unpredictable ways. The skew variable α_i represent the long-time averages of the skew, and instantaneous estimates of the skew are affected by drifts in the clock rate. More specifically, we assume that clock rates vary in such a way that $w_{ik} = \alpha_i e^{\delta_{ik}} W_k$ where δ_{ik} is a random variable with mean zero and variance X_{ik}.

Note that, when taking the logs, the equation becomes:

$$\log w_{ik} = \log W_k + \log \alpha_i + \delta_{ik}$$

Note that this is exactly the form of Equation 1 with the following substitutions:

- $y_{ik} \to \log w_{ik}$
- $U_k \to \log W_k$
- $T_i \to \log \alpha_i$
- $e_{ik} \to \delta_{ik}$
- $V_{ik} \to X_{ik}$

Thus, we can apply all of the previous theory to the estimate of clock skew. The difference is that the basic measurements now are the locally measured *intervals* between two synchronization signals (and thus are unaffected by the offsets), and the magnitude of these intervals is much larger than the measurement errors (*i.e.*, $W_k \gg V_{ik}$) so the only significant errors arise from clock frequency drift. The same set of equations, and the same iterative procedure, will produce the optimal and globally consistent estimates of skews through the set of parameters α_i.

We can treat skew and offsets on different time scales. That is, we can adjust the parameters α_i roughly every τ_s time units, whereas we adjust the parameters T_i roughly every τ_o time units, with $\tau_s \gg \tau_o$; the absolute values of these quantities will depend on the nature of the clocks and the setting. When computing the offsets we treat the skew as constant (and known), so we can apply the theory we presented earlier. On longer time scales, we adjust the skew using the same iterative procedure (with different variables).

The result is that we can treat general clocks with both offsets and skews. Experiments with real clocks will be needed before we can fine tune the time constants and verify that this two-time-scale approach is valid.

5.2   Outline of a Synchronization Protocol

The calculations in Section 3 seem, at first glance, far too complex for implementation in actual sensornets. This may well be true, but here we sketch out how one might achieve the desired results in an actual sensornet protocol. None of the various parameters are specified; we only sketch out the structure of what a protocol might look like.

The synchronization process can use any message as a synchronizing signal. We will assume that all messages have unique identifiers, so different nodes can know that they are referring to the receipt of the same message. Also, in what follows pairs of nodes are considered to be in range of another node if and only if they can exchange messages; pairs where one node can hear another, but not vice versa, are not considered to be in range. We first describe the approach for estimating clock offsets, and then later describe how to use this for estimating clock skew.

Each node broadcasts a synchronization status message every τ_o (with some randomness), which contains data for the last τ_w seconds; τ_w represents a time window after which data is discarded. Each status message contains:

- Their current estimate of T_i.

- Their current estimates of U_k for all previous status messages sent within the last τ_w seconds.
- Their time-of-arrival data y_{ik} for all status messages received in the last τ_w seconds.

Upon receipt of a status message, node i uses the data to update their estimate of T_i and U_k as described in the iterative equations 2 and 3. Thus, each round of synchronization messages invokes another round in the iterative computation.

At longer intervals, τ_s, nodes send skew status messages that additionally contain the data on α_i, W_k, and w_{ik}. This data can be used to update the skew variables in the same way as for the offset variables.

The main open question is what rate of message passing is needed to achieve reasonable degrees of convergence and whether this entails too much energy consumption. The answer will depend greatly on the nature of clock drifts and measurement errors in real systems. If the rates of change are slow, then once the system is reasonably well synchronized only a slow rate of iterations will be required to stay converged. If the rates of change are high, then a much faster rate of iterations will be required to stay within the desired precision bounds. Because we don't know what the relevant rates of change will be, we don't offer any conjectures about the feasibility of this approach. Instead, we hope to investigate the issue empirically by deploying this approach in an experimental setting.

Our first planned real-world deployment is for an ad-hoc deployable distributed array for detecting seismic activity. Seismologists often perform source localization through coherent beam-forming, requiring time consistency within the array of order 10 microseconds. Traditionally, all nodes in a seismic array are time synchronized using satellites in the Global Positioning System, which provides the international UTC timescale to sub-microsecond precision. Interest in network time synchronization has grown because it allows instrumentation of areas that are seismically interesting but inaccessible to GPS (*e.g.,* within structures, canyons, or tunnels). As an array grows in network diameter, existing RBS implementations may prove insufficient because RBS does not optimize for global coherence that, unlike some other sensor network applications, is required for seismic arrays. This makes it an ideal test application for our scheme.

Acknowledgments. We would like to thank David Karger for suggesting the maximum likelihood formulation and for many other useful comments during the early stages of this work. We would also like to thank Deborah Estrin for stimulating conversations on this topic.

References

1. H. Attiya, A. Herzberg and S. Rajsbaum, "Clock Synchronization Under Different Delay Assumptions," SIAM Journal on Computing, Vol. 25, No. 2 (April 1996), pp. 369-389.
2. D. Bertsekas and J.Tsitsiklis, *Parallel and Distributed Computation – Numerical Methods.* Prentice Hall, 1989. ISBN 0-13-648759-9.
3. V. Bychkovskiy, S. Megerian, D. Estrin and M. Potkonjak. Colibration: A Collaborative Approach to In-Place Sensor Calibration. In *2nd International Workshop on Information Processing in Sensor Networks (IPSN'03).* Palo Alto, CA, USA, April, 2003.

4. F. Cristian, Probabilistic Clock Synchronization. In *Distributed Computing 3* (1989), 146–158.
5. P. G. Doyle and J. L. Snell. *Random Walks and Electric Networks,* Mathematical Association of America, Washington, D. C., 1984.
6. J.Elson. *Time Synchronization in Wireless Sensor Networks*, Ph.D. thesis, University of californaia, Los Angeles, 2003.
7. J. Elson, L. Girod and D. Estrin. Fine–grained Network Time Synchronization Using Reference Broadcasts. In *Proceedings of the Fifth Symposium on Operating Systems Design and Implementation (OSDI)* (Boston, MA, December 2002), pp. 147–163.
8. S. Ganeriwal, R. Kumar, S. Adlakha and M. Srivastava. Network-wide Time Synchronization in Sensor Networks. Technical Report, University of California, Dept. of Electrical Engineering, 2002.
9. M. Grotschel, L. Lovasz, and A. Schrijver, *Geometric Algorithms and Combinatorial Optimization*, Springer-Verlag, 1993
10. J. Halpern, N. Megiddo and A.Munshi. Optimal Precision in the Presence of Uncertainty. *J. Complexity 1* (1985), 170-196.
11. J. Hill, and D. Culler. A Wireless Embedded Sensor Architecture for System-level Optimization. Tech. rep., U.C. Berkeley, 2001.
12. J. Hill, R. Szewczyk, A. Woo, S. Hollar, D. Culler, and K. Pister. System Architecture Directions for Networked Sensors. In *Proceedings of the Ninth International Conference on Architectural Support for Programming Languages and Operating Systems (ASPLOS IX)*, 93-104, (November, 2000). ACM.
13. T.C. Hu, "Minimum Cost Flows in Convex Cost Networks," Naval Research Logistics Quarterlu, Vol. 13, No. 1, pp. 1-9 (1966).
14. David R. Karger, Matthew S. Levine "Random Sampling in Residual Graphs," *STOC 2002*, 63-66
15. R. Karp, J. Elson, D. Estrin, and S. Shenker, "Optimal and Global Clock Synchronization in Sensornets," preprint, 2003.
16. E. L. Lehmann, *Theory of Point Estimation*, Chapman and Hall, New York (1991).
17. C. Liao, M. Martonosi, and D. Clark. Experience with an Adaptive Globally-Synchronizing Clock Algorithm. In *Eleventh Annual Symposium on Parallel Algorithms and Architectures (SPAA '99)*, 106-114.
18. J. Lundelius and N. Lynch. An Upper and Lower Bound for Clock Synchronization. Information and Control, 62 (1984) 190-204.
19. D. Mills. Internet Time Synchronization: The Network Time Protocol. In *Zhonghua Yang and T. Anthony Marsland (Eds.), Global States and Time in Distributed Systems*. IEEE Computer Society Press (1994).
20. M. Mock, R. Frings, E. Nett, and S. Trikaliotis. Continuous Clock Synchronization in Wireless Real-time Applications. In *The 19th IEEE Symposium on Reliable Distributed Systems (SRDS'00)*, 125-133 (October, 2000).
21. Kay Römer. Time Synchronization in Ad Hoc Networks. In *ACM Symposium on Mobile Ad Hoc Networking and Computing (MobiHoc 01)*. (October, 2001).
22. B. Patt-Shamir and S. Rajsbaum. A Theory of Clock Synchronization. STOC 1994 810-819
23. T.K. Srikanth and S. Toueg. Optimal Clock Synchronization. *J-ACM 34*, 3 (July 1987), 626–645.
24. Barbara Simons, Jennifer L. Welch, Nancy A. Lynch: An Overview of Clock Synchronization. Fault-Tolerant Distributed Computing 1986: 84–96.
25. W. Su and I. Akylidis. Time-Diffusion Sensor Protocol for Sensor Networks. Technical report, Georgia Institute of Technology, 2002.

26. P. Verissimo and L. Rodrigues. A Posteriori Agreement for Fault-Tolerant Clock Synchronization on Broadcast Networks. In D. K. Pradhan, Editor, *Proceedings of the 22nd Annual International Symposium on Fault-Tolerant Computing (FCTS'92)* 85-85, IEEE Computer Society Press.
27. P. Verissimo, L. Rodrigues, and A. Casimiro. Cesiumspray: A Precise and Accurate Global Time Servive for Large-Scale Systems. Technical Report NAV-TR-97-0001, Universidade de Lisboa (1997).
28. Calibration as Parameter Estimation in Sensor Networks. In *Proceedings of the First ACM International Workshop on Sensor Networks and Applications (WSNA2002)*. (September, 2002).

Author Index

Lecture Notes in Computer Science

For information about Vols. 1–2854

please contact your bookseller or Springer-Verlag

Vol. 2908: K. Chae, M. Yung (Eds.), Information Security Applications. XII, 506 pages. 2004.

Vol. 2907: I. Lirkov, S. Margenov, J. Wasniewski, P. Yalamov (Eds.), Large-Scale Scientific Computing. XI, 490 pages. 2004.

Vol. 2906: T. Ibaraki, N. Katoh, H. Ono (Eds.), Algorithms and Computation. XVII, 748 pages. 2003.

Vol. 2905: A. Sanfeliu, J. Ruiz-Shulcloper (Eds.), Progress in Pattern Recognition, Speech and Image Analysis. XVII, 693 pages. 2003.

Vol. 2904: T. Johansson, S. Maitra (Eds.), Progress in Cryptology - INDOCRYPT 2003. XI, 431 pages. 2003.

Vol. 2903: T.D. Gedeon, L.C.C. Fung (Eds.), AI 2003: Advances in Artificial Intelligence. XVI, 1075 pages. 2003. (Subseries LNAI).

Vol. 2902: F.M. Pires, S.P. Abreu (Eds.), Progress in Artificial Intelligence. XV, 504 pages. 2003. (Subseries LNAI).

Vol. 2901: F. Bry, N. Henze, J. Ma luszyński (Eds.), Principles and Practice of Semantic Web Reasoning. X, 209 pages. 2003.

Vol. 2900: M. Bidoit, P.D. Mosses (Eds.), Casl User Manual. XIII, 240 pages. 2004.

Vol. 2899: G. Ventre, R. Canonico (Eds.), Interactive Multimedia on Next Generation Networks. XIV, 420 pages. 2003.

Vol. 2898: K.G. Paterson (Ed.), Cryptography and Coding. IX, 385 pages. 2003.

Vol. 2897: O. Balet, G. Subsol, P. Torguet (Eds.), Virtual Storytelling. XI, 240 pages. 2003.

Vol. 2896: V.A. Saraswat (Ed.), Advances in Computing Science – ASIAN 2003. VIII, 305 pages. 2003.

Vol. 2895: A. Ohori (Ed.), Programming Languages and Systems. XIII, 427 pages. 2003.

Vol. 2894: C.S. Laih (Ed.), Advances in Cryptology - ASIACRYPT 2003. XIII, 543 pages. 2003.

Vol. 2893: J.-B. Stefani, I. Demeure, D. Hagimont (Eds.), Distributed Applications and Interoperable Systems. XIII, 311 pages. 2003.

Vol. 2892: F. Dau, The Logic System of Concept Graphs with Negation. XI, 213 pages. 2003. (Subseries LNAI).

Vol. 2891: J. Lee, M. Barley (Eds.), Intelligent Agents and Multi-Agent Systems. X, 215 pages. 2003. (Subseries LNAI).

Vol. 2890: M. Broy, A.V. Zamulin (Eds.), Perspectives of System Informatics. XV, 572 pages. 2003.

Vol. 2889: R. Meersman, Z. Tari (Eds.), On The Move to Meaningful Internet Systems 2003: OTM 2003 Workshops. XIX, 1071 pages. 2003.

Vol. 2888: R. Meersman, Z. Tari, D.C. Schmidt (Eds.), On The Move to Meaningful Internet Systems 2003: CoopIS, DOA, and ODBASE. XXI, 1546 pages. 2003.

Vol. 2887: T. Johansson (Ed.), Fast Software Encryption. IX, 397 pages. 2003.

Vol. 2886: I. Nyström, G. Sanniti di Baja, S. Svensson (Eds.), Discrete Geometry for Computer Imagery. XII, 556 pages. 2003.

Vol. 2885: J.S. Dong, J. Woodcock (Eds.), Formal Methods and Software Engineering. XI, 683 pages. 2003.

Vol. 2884: E. Najm, U. Nestmann, P. Stevens (Eds.), Formal Methods for Open Object-Based Distributed Systems. X, 293 pages. 2003.

Vol. 2883: J. Schaeffer, M. Müller, Y. Björnsson (Eds.), Computers and Games. XI, 431 pages. 2003.

Vol. 2882: D. Veit, Matchmaking in Electronic Markets. XV, 180 pages. 2003. (Subseries LNAI).

Vol. 2881: E. Horlait, T. Magedanz, R.H. Glitho (Eds.), Mobile Agents for Telecommunication Applications. IX, 297 pages. 2003.

Vol. 2880: H.L. Bodlaender (Ed.), Graph-Theoretic Concepts in Computer Science. XI, 386 pages. 2003.

Vol. 2879: R.E. Ellis, T.M. Peters (Eds.), Medical Image Computing and Computer-Assisted Intervention - MICCAI 2003. XXXIV, 1003 pages. 2003.

Vol. 2878: R.E. Ellis, T.M. Peters (Eds.), Medical Image Computing and Computer-Assisted Intervention - MICCAI 2003. XXXIII, 819 pages. 2003.

Vol. 2877: T. Böhme, G. Heyer, H. Unger (Eds.), Innovative Internet Community Systems. VIII, 263 pages. 2003.

Vol. 2876: M. Schroeder, G. Wagner (Eds.), Rules and Rule Markup Languages for the Semantic Web. VII, 173 pages. 2003.

Vol. 2875: E. Aarts, R. Collier, E.v. Loenen, B.d. Ruyter (Eds.), Ambient Intelligence. XI, 432 pages. 2003.

Vol. 2874: C. Priami (Ed.), Global Computing. XIX, 255 pages. 2003.

Vol. 2871: N. Zhong, Z.W. Raś, S. Tsumoto, E. Suzuki (Eds.), Foundations of Intelligent Systems. XV, 697 pages. 2003. (Subseries LNAI).

Vol. 2870: D. Fensel, K.P. Sycara, J. Mylopoulos (Eds.), The Semantic Web - ISWC 2003. XV, 931 pages. 2003.

Vol. 2869: A. Yazici, C. Şener (Eds.), Computer and Information Sciences - ISCIS 2003. XIX, 1110 pages. 2003.

Vol. 2868: P. Perner, R. Brause, H.-G. Holzhütter (Eds.), Medical Data Analysis. VIII, 127 pages. 2003.

Vol. 2866: J. Akiyama, M. Kano (Eds.), Discrete and Computational Geometry. VIII, 285 pages. 2003.

Vol. 2865: S. Pierre, M. Barbeau, E. Kranakis (Eds.), Ad-Hoc, Mobile, and Wireless Networks. X, 293 pages. 2003.

Vol. 2864: A.K. Dey, A. Schmidt, J.F. McCarthy (Eds.), UbiComp 2003: Ubiquitous Computing. XVII, 368 pages. 2003.

Vol. 2863: P. Stevens, J. Whittle, G. Booch (Eds.), "UML" 2003 - The Unified Modeling Language. XIV, 415 pages. 2003.

Vol. 2860: D. Geist, E. Tronci (Eds.), Correct Hardware Design and Verification Methods. XII, 426 pages. 2003.

Vol. 2859: B. Apolloni, M. Marinaro, R. Tagliaferri (Eds.), Neural Nets. X, 376 pages. 2003.

Vol. 2857: M.A. Nascimento, E.S. de Moura, A.L. Oliveira (Eds.), String Processing and Information Retrieval. XI, 379 pages. 2003.

Vol. 2856: M. Smirnov (Ed.), Quality of Future Internet Services. IX, 293 pages. 2003.

Vol. 2855: R. Alur, I. Lee (Eds.), Embedded Software. X, 373 pages. 2003.